Designing Control Loops for Linear and Switching Power Supplies

A Tutorial Guide

Related Artech House Titles in Power Engineering

The Advanced Smart Grid: Edge Power Driving Sustainability,
Andres Carvallo, and John Cooper

Battery Management Systems for Large Lithium Ion Battery Packs,
Davide Andrea

Designing Control Loops for Linear and Switching Power Supplies: A Tutorial Guide,
Christophe Basso

Electric Systems Operations: Evolving to the Modern Grid,
Subramanian Vadari

Energy Harvesting for Autonomous Systems,
Stephen Beeby, and Neil White

Power Line Communications in Practice,
Xavier Carcelle

Designing Control Loops for Linear and Switching Power Supplies

A Tutorial Guide

Christophe Basso

ARTECH HOUSE
BOSTON | LONDON
artechhouse.com

Library of Congress Cataloging-in-Publication Data
A catalog record for this book is available from the U.S. Library of Congress.

British Library Cataloguing in Publication Data
A catalog record for this book is available from the British Library.

ISBN-13: 978-1-60807-557-7

Cover design by Vicki Kane

10 9 8 7 6 5 4 3 2

Contents

CHAPTER 4

CHAPTER 5

Foreword

In his previous book, *Switch-Mode Power Supplies: Spice Simulations and Practical Designs*, Christophe Basso gave valuable, numerous, and detailed explanations of converter topologies and PFC circuits by way of practical circuit designs. Although modeling and feedback control were well explained and utilized in that book, the emphasis was on SPICE simulations.

In this book, Christophe has delved deep into understanding and analyzing specific control circuits used with power converters. He has examined in detail the performance of converters ranging from dynamic load response and stability to line rejection for various controllers, including some unconventional ones for educational purposes to emphasize some of the subtle aspects of feedback control methods. He has derived numerous analytical results to design a compensator that achieves a particular type of performance goal or stability criterion.

In addition to the rigors of control theoretic aspects of this book, I was pleased to see the analysis of converter circuits being carried out using fast analytic circuit techniques with answers in low-entropy form very much in the spirit of Dr. R. D. Middlebrook. If design is the reverse of analysis, Middlebrook would say, then the only kind of analysis worth doing is design-oriented analysis, which yields low-entropy expressions. Christophe has taken this to heart and done a very fine job with it.

Such a book is necessary because of the continued isolation and demise of the field of analog electronics of which power converter design is part. Control theory books abound, whereas analog electronic textbooks are increasingly becoming less adequate and cookbook like. For the new graduate in electrical engineering who wants to work in the field of power electronics, the necessary knowledge to get started is diffused and rather difficult to consolidate. Good sources exist to teach one-week courses to practicing engineers, but these assume some working knowledge or experience with power converters. This book in my opinion fills the gap between control theory and converter design *rigorously*. Also, that such a book is necessary in my opinion is based on my experience, albeit limited, with a few control problems I have had to work on in addition to designing power converters full time. I have seen control systems or rather control schemes that are far too elaborate and unnecessary for stabilizing a platform or regulating the temperature of a laser. Their performance, often marginal, could have been tremendously improved by designing a far simpler feedback control circuit if only the designer understood the equivalent circuit model of the transducer in the first place. Christophe has done an excellent job in exposing the concept of modeling and rigorously compensating converter circuits. It is a book that I recommend to all power electronics engineers who, every once in a while, may have to tackle a control problem for a device that is not a power converter.

Vatché Vorpérian
Jet Propulsion Laboratory

Preface

When I started this book in January 2009, I had the intention to write a quick booklet exclusively covering compensator structures. The idea germinated as I realized that most of the available documents covered compensator examples implementing operational amplifiers. This is the way I learned how to stabilize a loop at university in the eighties. Later, as an engineer, I wanted to put my knowledge at work with a TL431 or a transconductance amplifier to which an optocoupler was hooked. As you can imagine, the connections were missing between my school books and the practical circuit I was working on.

Literature does not abound on the subject of compensator structures, so I had the choice to either plunge into analytical analysis or start tweaking the circuit through trial and error. Obviously, the second approach was wrong, but when time pressure becomes an unbearable situation, I understand that engineers have no other alternative for their ongoing design. I felt there was a gap to fill in the technical literature to show how compensation theory could apply to electronic circuits different than op amps. On the run, I wrote Chapters 5, 6, 7, and 8. Then, I decided to write a little about loop control theory, something engineers could use to refresh their memory on the topic. As I wrote Chapter 1, I discovered that most of the theory I knew on the subject was not really what I had been taught at school. In fact, when I graduated from Montpellier University (in France, not Maine!) and tried to apply that fresh knowledge to a project, I was stuck: I could not bridge the stuff I knew to what I was asked to do. My teachers talked about PID coefficients, and I had to place poles and zeros.

I want this book to be the companion you look at when you need to stabilize a power converter. For that purpose, I have tried to balance the useful theory—you do not need to know everything in the field of loop control to be a good engineer—and the necessity to make it work on real projects. In the nine chapters I wrote, Chapter 1 starts with generalities on the subject. If you are a beginner, you must read it. Chapter 2 introduces transfer functions and the formalism to write them correctly. Fast analytical techniques are used throughout this chapter, and I encourage you to dig further into the subject. Chapter 3 is an important part of the book, as it details the stability criteria to build rugged control systems. Back to my university time, I had been told to maintain the phase margin to at least 45° and that was it, with no further explanations or origins of this number. There is nothing new here, but I have derived the equations so you can link phase margin numbers and the expected closed-loop transient performance. The same applies to the crossover frequency that you will no longer arbitrarily select. Chapter 4 explains compensation basics, starting with the PID blocks and expanding to what you will deal with: poles and zeros placement. Then I introduce several compensation methods, including output impedance shaping for high-speed dc-dc converters.

The next chapters, 5, 6, 7, and 8, teach you how to compensate your converter with an op amp, a TL431, a transimpedance amplifier, or a shunt regulator. This is the strength of this book: I strived to exhaustively cover all the possible configurations, with and without optocoupler, regardless of the active compensation element. You will even find TL431 internals secrets that are not often disclosed in data-sheets or application notes. Finally, Chapter 9 closes the subject with measurement methods and design examples.

In the text, I will bring you to the important matter being examined but will suddenly digress to a mathematical tool needed for understanding. This is my writing style. A lot of books simply assume you know the concerned technique and continue the explanation, leaving the reader with a fragmented knowledge. I have tried to avoid this situation and it is the raison d'être of the appendixes at the end of some of the chapters.

In this book, I have derived more than 1,550 equations in three years. Despite all the reviewers' care, it is impossible to have trapped all the typos, missing signs, or wrong numerical results that could have escaped my attention. I sincerely apologize in advance: I know the frustration, as a reader, when you discover errors in a book you want to trust. To help improve the content, I would like you to kindly report the mistake you spot while reading and I will maintain an errata list in my webpage with credit given to the discoverer. It worked very well for the previous book and helped to maintain analytical integrity. Thank you in advance for your kind help. Please send comments to cbasso@wanadoo.fr.

As a conclusion, I have spent three great years writing this book. I learned a lot when tackling some of the subjects. A few of them were tough, and I confess there was highs and lows. But the final content confirms that I was on the right path. I hope your comments will acknowledge that point. Above all, I hope you often come back to this companion book while fulfilling your engineering tasks. Happy reading to you all!

Acknowledgments

There is no way I could have written a book like this without the help and involvement of many people. My warmest thanks and love go to my dear family: Anne, my wife, who let me write this book despite the numerous long nights during which I struggled with equations or lack of inspiration! My two children, Lucile and Paul, whom I scared by the amount of equations I derived: no, don't worry, kids—this does not happen in the real life!

I was also fortunate to exchange and debate ideas at work with my ON Semiconductor colleagues and friends, Thierry Sutto, Stéphanie Cannenterre, Yann Vaquette, and Dr. José Capilla. Special thanks go to my friend Joël Turchi, who spent long hours reviewing my work throughout the years, kindly pointing out mistakes or flaws in some of my unorthodox approaches! A lot of email exchanges as well as trainings with experts in this field like Dr. Ray Ridley (Ridley Engineering), Larry Meares (Intusoft), Dr. Richard Redl (ELFI), and Dr. Vatché Vorpérian (JPL) helped me to understand the necessity of describing circuits through analytical analysis. This is the best approach to unmask hidden parasitic elements and find the neutralizing parry.

I had the privilege to form a reviewer team made of worldwide experts who spent time reading—they corrected the work but also polished my English. I warmly thank them for the time they kindly allocated for this book review: Joël Turchi (ON Semi), Dr. José Capilla (ON Semi), Jeff Hall (ON Semi), Yann Vaquette (ON Semi), Nicolas Cyr (ON Semi), Jim Young (ON Semi), Patrick Wang (ON Semi), Dr. Vatché Vorpérian (JPL), Dr. Richard Redl (ELFI), Dhaval Dalal (Innovatech), Analogspiceman, Roland Saint-Pierre (Power Integrations), Steve Sandler (Picotest), Georges Gautier (ESRF), Christopher Merren (International-Rectifier), Arnaud Obin (E-Swin), Dr. Chung-Chuieh Fang (Advanced Analog Technology), Dennis Feucht (Innovatia), Dr. Mike Schutten (General Electric), Dr. Germain Garcia (LAAS-CNRS), and Dr. Didier Balocco (AEGPS).

Last, but not least, I would like to thank Deirdre Byrne and the all the team at Artech House for giving me the opportunity to publish my work with them.

CHAPTER 1

Basics of Loop Control

Often without knowing it, our everyday life utilizes loop control techniques: stretching our muscles to reach a pitcher and pour water into a glass, keeping the bicycle speed constant despite a sudden uphill climb, or maintaining the right pressure on the gas pedal to stay slightly below the maximum speed limit on a long straight road. In all these cases, we have implemented a *feedback control system*: the brain defines a *setpoint* and uses muscles or mechanical power to execute the order. The brain receives information on how the order is executed as the nerves (spinal, optical) permanently feed the information back to it. In the described chains, we have associated a power-amplified control (of which the brain and the muscles represent the *direct path*) with a *return path* (the nervous system). As the information leaves the brain and returns to it via the nervous systems to correct or modify the muscular motion, we say the system operates under *closed-loop* conditions. On the contrary, when the return path is broken, the information confirming that the order has been well or poorly executed is missing. The system is said to run *open loop*. Yes, if you try to ride your bicycle while wearing a blindfold, the biofeedback loop is lost and your body operates in open-loop conditions with all the associated risks!

1.1 Open-Loop Systems

As outlined in the previous section, a system can either run in open-loop or closed-loop conditions. An open-loop system transforms a control signal, the *input*, into an action, the *output*, following a specific relationship that links the output to the input. In this open-loop system, the control input u is independent from the output y. Figure 1.1 shows a simple representation in the time domain of a system where the output of the system relates to the input by a gain factor k. This system is often referred to as *the plant* in the literature and is noted H.

In this diagram, the rectangle represents the transmission chain, whereas the arrows portray the physical input and output variables. Please note the usage of letters u and y to respectively designate the input and the output signals as commonly employed in textbooks. In this drawing, the relationship linking the output y to the input u is simply:

$$y(t) = ku(t) \tag{1.1}$$

As it is assumed that coefficient k does not change with time, the system is said to be *linear time invariant* (LTI).

An application example fitting this model is a person turning a car steering wheel by an angle of θ degrees (the input) to force the wheels turning by an amount of $k\theta$ degrees (the output) on the road. This is what Figure 1.2 depicts. In the early

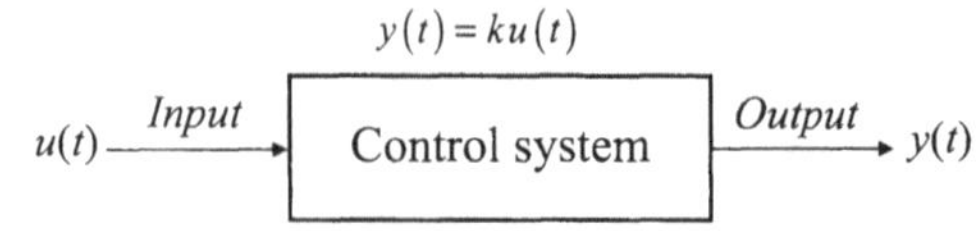

Figure 1.1 A simple representation of a system where the output depends on the input by a factor *k*.

days of cars, the coupling between the steering wheel and the wheels involved mechanical linkages and hydraulic actuators. Turning the wheels at no or low speed, for instance during parking maneuvers, could quickly turn into a physical exercise for the driver, depending on the size of the car. The power or the force available at the output of the system was almost entirely delivered by the control input—the driver's biceps, in this case!

In modern cars, the control technique has evolved into a power-assisted steering system known as electric power steering (EPS). Sensors detect the motion and the torque applied to the steering column and feed a calculator with their signals. The output of the calculator drives a motor via a power module that provides an assistive torque to the steering gear. The driver can then easily rotate the steering wheel and smoothly induce an angle change on the wheels via the amplification chain. Unlike the previous example, the power or the force available at the output no longer derives from the control input but finds its origins from another source of energy. In a car, it can be an electrical battery, for instance. A simplified schematic of the system appears in Figure 1.3: the steering wheel angle is transformed into a voltage (volts, V) by a potentiometer (or a digital encoder). This signal enters the amplifier that delivers the power (watts, W) to efficiently drive the motor. The motor is coupled to the steering gear and delivers the torque (newton-meters, N.m) to change the wheels position. The succession of the various blocks from the input u to the output y is called the *direct path*.

In Figure 1.2, where no amplification chain exists, it is extremely difficult to convey the force over a long distance from the control source to the output without losses or distortion. The exercise becomes even more difficult when geometric shapes are needed to accommodate a complex environment: curvatures, noncollinear shafts, and so on. Fortunately, this is no longer an obstacle when a power-

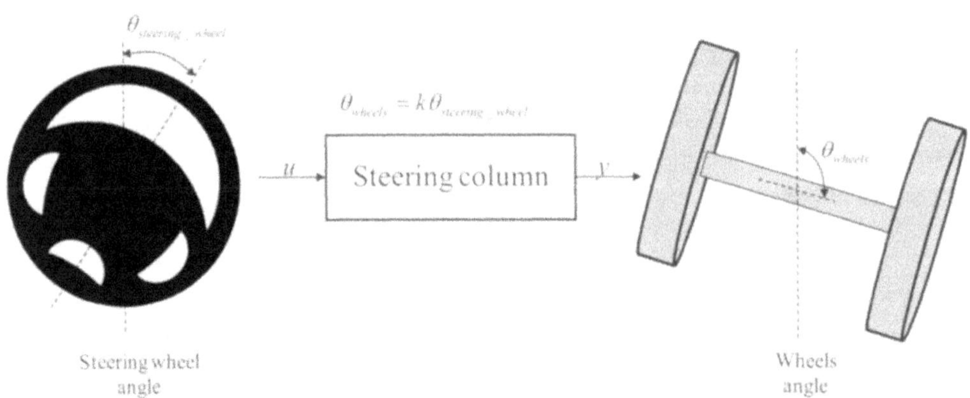

Figure 1.2 The described system directly converts the angle given by a steering wheel into an angle on the car direction.

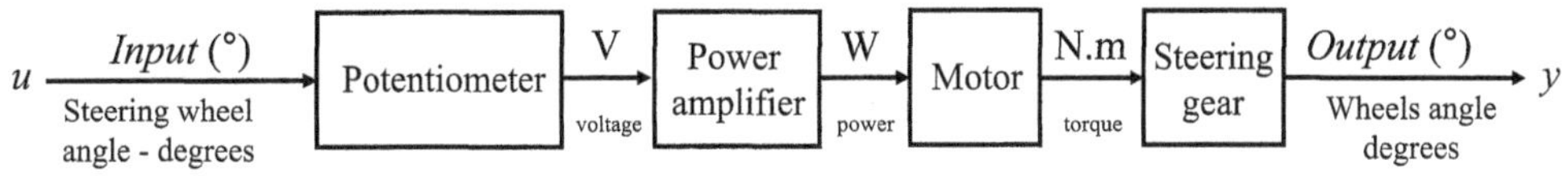

Figure 1.3 A control system featuring an amplification chain. The input variable does not directly control the output but is conveyed through a series of power systems before it is transformed into the final variable.

amplified chain is implemented, as in Figure 1.3. Despite a long distance between the control input and the delivered output, it is easy to transport an electrical signal through a pair of wires and reach a motor or an actuator placed in a remote location. Named *x-by-wire*, this technique nowadays replaces the mechanical and hydraulic links by an electronic control system using electromechanical actuators located close to the point of action: we have seen the steering control for a car (steer-by-wire), but it can be the controls in an aircraft that actuate the flaps or adjust the engine rotation speed (fly-by-wire).

1.1.1 Perturbations

In the described example, we have only considered a single input and a single output. In textbooks, these systems are referred as single-input-single-output (SISO) systems. In reality, any system is affected by several input variables. Therefore, if several input variables are considered, we can envisage as many output variables. In the literature, such a configuration is referred as a multi-input-multi-output (MIMO) system.

Regarding the outputs, we usually select the main output the one that is the most interesting to the designer. In our steer-by-wire system of Figure 1.3, the main output is the angle of the wheels on the road. However, we know that the built-in differential system strives to ensure an evenly distributed torque to each wheel, while allowing them to rotate at different speed. Output variables such as the individual speed of the wheels could then be monitored to deliver information when the car drives along a curve. Another output to consider is the car trajectory, as this is the ultimate goal: forcing the vehicle to follow a curve by acting on the steering wheel without losing car control. Considering all these output variables, the illustration can be updated as Figure 1.4 portrays.

The main input to a control system is usually the one that the output must follow. In our steer-by-wire system, the steering wheel signal obviously represents the main input. However, there could be other inputs that affect the transmission chain. As these inputs are usually undesirable, they are considered as *perturbations*. In our example, assume you try to negotiate a curve in presence of a strong wind. Despite the angle imposed to the car by the steering wheel, you will experience a trajectory deviation. The wind is a perturbation that affects an output, the trajectory. As the

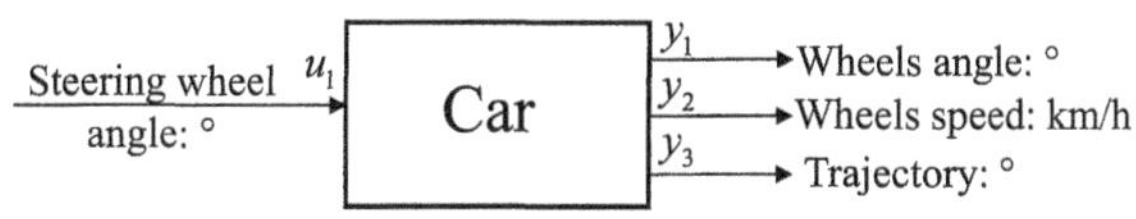

Figure 1.4 A system can have several outputs depending on the designer interest and the goal to fulfill.

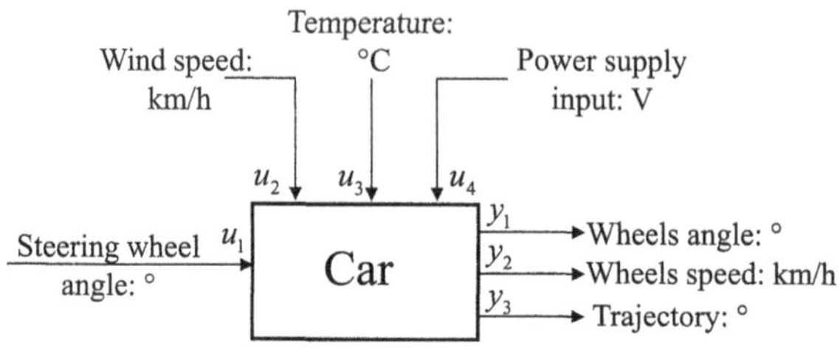

Figure 1.5 Perturbations affect the transmission chain and can be considered as inputs.

assistive torque is derived from an electric motor, the power source supplying the amplifier is getting weak: the high internal temperature lowers the storage inductor value, and the lack of power affects the actuator responsiveness: the imposed angle is not properly replicated on the wheels. The power supply, when its voltage fluctuates, represents a perturbation affecting the transmission chain. All these perturbations can be treated as multiple inputs that have to be accounted for during the design phase. The ultimate goal is to build a rugged device, insensitive to these perturbations. Figure 1.5 represents the system where these individual contributions appear.

1.2 The Necessity of Control—Closed-Loop Systems

In the previous systems, a relationship exists between the input and the output of the system. This is our coefficient k, linking the steering wheel angle and the actual angle applied to the direction system. In reality, because of the numerous perturbations or deficiencies in the transmission chain itself, the final output value may never reach the setpoint imposed by the input. For instance, the coefficient k in Figure 1.1 may have great variability itself, linked to the ambient temperature or process variations. How can we counteract the deficiencies in the chain? In our car example, we could imagine a sensor actually measuring the real angle applied to the wheels based on the steering wheel position. This sensor could be a simple potentiometer or a digital rotary encoder. A calculator could then evaluate the difference, or the *error*, between the imposed setpoint and the obtained angle read by the sensor. In the literature, this error is noted ε (epsilon) and represents the difference between the input setpoint and the final output:

$$\varepsilon = u - y \tag{1.2}$$

From the obtained error, a signal could be generated and injected into the transmission chain to apply a corrective action: if a few degrees are missing or are in excess, the setpoint must be proportionally increased or decreased to compensate the difference. A closed-loop system would therefore be a system where a signal representative of the output is fed back to the control input and acts upon it to ensure the error is kept to a minimum. In such a model, the control signal is no longer the setpoint alone, but the difference signal expressed by (1.2). To upgrade our model representation, we will add a difference box, a circle, illustrating the error signal generation. Figure 1.6 shows the updated sketch. As we deal with electric variables,

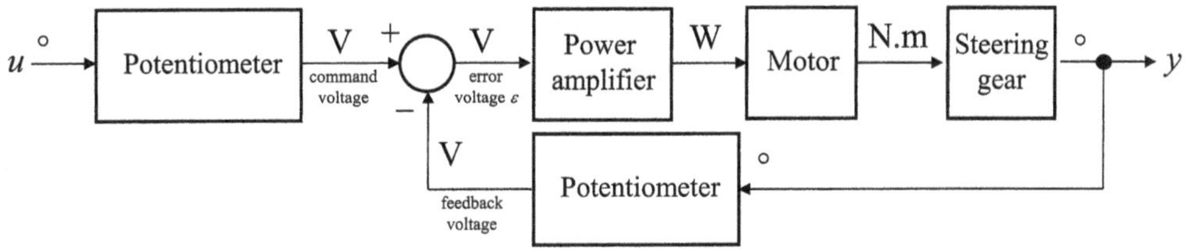

Figure 1.6 The original setpoint is now affected by a return signal to create an error variable. This error variable controls the whole chain.

a second sensor is added in the *return path* (also called the *return chain*) to convert the actual feedback output angle into a voltage and allow the subtraction with the commanded voltage-image of the input variable. The difference with these provides the error signal ε.

If the output y is too large compared to the target imposed by u, the error signal ε will decrease, commanding the system to reduce the output. On the opposite, if the output is too small, ε will grow, commanding an increase of the output. As a first approximation, we can say that the system operates adequately and reaches equilibrium as long as the error signal *opposes* the output variation. If for any reason this relationship is lost (i.e., the negative sign in (1.2) becomes positive), the system can run away and will quickly hit its upper or lower stops. We will come back on this important point in a while.

In the Figure 1.6 example, the control input can be variable or fixed. Imagine the driver is on a road and maintaining the wheels straight for a long period of time. In that case, the system simply maintains the output constant and keeps the wheels in the defined axis, fighting against perturbations (i.e., wind) to maintain the trajectory. The system is told to be a *regulator* or a *regulating system*. A regulator is a control system operating with a constant input or setpoint that maintains a fixed relationship between the setpoint and the output, regardless of what the perturbations are. A voltage regulator is another example: it maintains a constant output voltage despite its operating environment, such as input voltage (ac or dc) and output current changes. In this book, we will mostly deal with power converters delivering a fixed output voltage or current, naturally falling into the regulator area.

If the input permanently changes with time, the control system must ensure that the output precisely tracks the input. The French language uses the term "asservissement" (enslavement) to designate such a system. It literally means that the output must be slave to the input. The goal of such a system is indeed to maintain a relationship between the output and the input regardless of the speed and the amplitude at which the control input changes. Such a system is called a *feedback control system*. Audio amplifiers, the autopilot in aircrafts, or marine navigation systems are good examples of feedback control systems, also denominated *servomechanisms* for the latter, as the controlled variable is a mechanical position. They all use complex feedback architectures to ensure the output perfectly follows the changing setpoint regardless of what the perturbations are (wind speed, stream strength, and so on).

In all the cited examples, an extreme precision is needed in the delivered variable: you cannot afford to have a mismatch of several degrees in the control of an aircraft flap for instance. Such a system must be extremely sensitive to the smallest deviation detected between the setpoint and what the sensor returns. To be that sensitive, it is necessary to *amplify* the error signal ε. When amplified, a small output deviation becomes a variation of higher amplitude that the system can properly treat. The *gain* of a system therefore directly relates to the amount of feedback and hence to its precision: a highly precise system exhibits a high *static* (also called dc) gain in the control path. No gain, no feedback—that is to say you must have gain in the loop to realize a control system.

1.3 Notions of Time Constants

A control system, regardless of its implementation, usually reacts with delay when adjusted to a new setpoint or in response to a perturbation. This delay is inherent to the control chain, as the signal is conveyed through mechanical, physical, or electronic paths. For instance, in our electric power steering system, when you turn the steering wheel, it takes a certain amount of time for the order to propagate as an angle change on the wheels. Another classical example is the heating system in your house: you program a particular temperature setpoint in the presence of other room parameters (e.g., volume, air flow), but you will need to wait tens of minutes if not an hour before the system, via the sensor, considers the read temperature to be adequate. In these examples, the time needed by the system to change from one state to another is called the *time constant*, noted τ, (Greek letter "tau") and relates the system response to a time (s). The time constants can vary from a few milliseconds in the first case to seconds or hours in the given example.

There are many possible ways to represent the time constant of a first-order LTI system; however, they all obey a common differential equation. For electrical engineers, a simple representation is an RC filter as shown in Figure 1.7.

The output voltage $y(t)$ of such a network can be obtained after a few lines of algebra:

$$y(t) = u(t) - Ri(t) \tag{1.3}$$

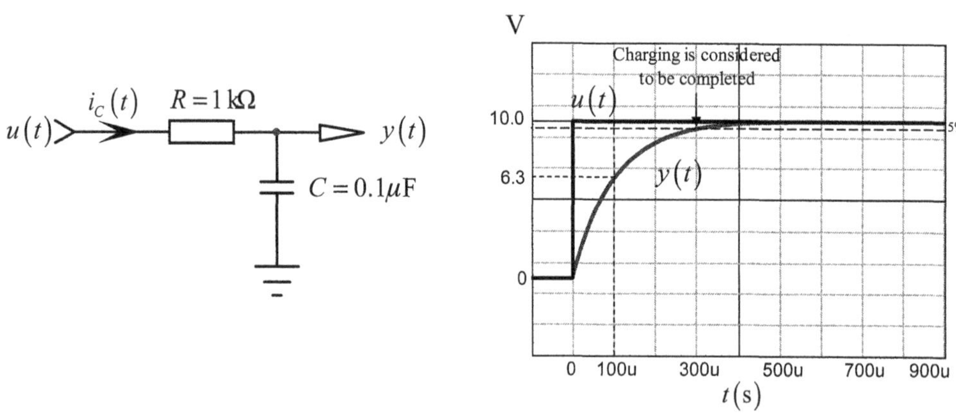

Figure 1.7 An RC low-pass filter response depends on the time constant in response to an input step.

The current in the capacitor depends on the voltage variation across its terminals:

$$i_c(t) = C\frac{dy(t)}{dt} \tag{1.4}$$

Substituting (1.4) into (1.3) and rearranging, we have an equation describing the behavior of a linear time invariant first-order system:

$$y(t) = u(t) - \tau\frac{dy(t)}{dt} \tag{1.5}$$

With $\tau = RC$, the time constant of the system. If R = 1 kΩ and C = 0.1 μF, we have a 100-μs time constant.

The solution to such a differential equation can be found by different means such as the inverse Laplace transform, as we will later see. It can be shown that the solution follows this form:

$$y(t) = A + Be^{-\frac{t}{\tau}} \tag{1.6}$$

where A and B are constant numbers found by solving a 2-unknown/2-equation system for $t = \infty \left(e^{-\frac{\infty}{\tau}} = 0 \right)$ and the *initial condition* at $t = 0 \left(e^{-\frac{0}{\tau}} = 1 \right)$. An initial condition represents the value of the *state* variable at the beginning of the observation. It can be the current in an inductor or the voltage across a capacitor (for instance, when $t = 0$). In Figure 1.7, if we consider the input signal $u(t)$ to be a voltage step of amplitude V_{cc}, then the voltage across the capacitor $v_C(t)$ is simply

$$v_C(t) = V_{cc}\left(1 - e^{-\frac{t}{\tau}}\right) \tag{1.7}$$

This is the familiar exponential curve shown on the right side of Figure 1.7. The time constant can be determined for t equals τ, reached when $e^{-\frac{t}{\tau}} = e^{-1} = \frac{1}{e}$ in (1.7). Solving for the corresponding value of $v_C(t)$, we have:

$$v_C(t) = V_{cc}\left(1 - \frac{1}{e}\right) \approx 63\%\, V_{cc} \tag{1.8}$$

We can apply this definition to our example. As V_{cc} equals 10 V, the capacitor will reach 6.3 V after τ seconds. If we read the x-axis for $v_C(t)$ = 6.3 V, we can determine the system time constant. In this example, it corresponds to 100 μs, the value of the RC product used in the circuit. In a first-order system, the output is within 5 percent of the final target after 3τ have elapsed. In our example, if the input voltage corresponds to a sudden setpoint change and our control system exhibits a 100-μs time constant, the output of the system will be considered within limits after 300 μs.

1.3.1 Working with Time Constants

The time constant represents one of the most troublesome natural parameters in a control system. Why? Because when the loop observes the output while the control input changes, it sees a variable out of range for a certain time: you change the temperature setpoint in the house and no significant temperature variation happens on the sensor before tens of minutes. As a result, the error signal between the setpoint and the returned value is maximum, pushing the system in its upper or lower power limit with all the associated problems (over power, runaway risks, and so on). Is there any wise control strategy, leading to a more reasonable reaction? One widely adopted solution consists of combining the following actions:

- The principle is to link the control voltage amplitude to the difference sensed by the system between the setpoint and the controlled output. Why overamplify a moderate change if a small increase of the control voltage is the right answer? On the other hand, if the change amplitude is large, then it should be compensated with a stronger control voltage. In other terms, it is desirable to have a control voltage amplitude *proportional* to the detected change in the error signal.
- If a drift on the controlled variable or a slow setpoint transition is sensed, why immediately push the control signal to its upper or lower stop? If the perturbation or the change in the operating point is slow, then let's slow down the reaction. On the other hand, if the control chain senses a fast-moving perturbation, let's make it react quickly. The control voltage must thus be sensitive to the slope of the error signal. How do we compute the slope of an error signal? By taking its time *derivative*.
- Finally, we want to obtain a very precise controlled output, exactly matching the setpoint. One way to fulfill this goal is to increase or decrease the control voltage until the detected error signal is null, meaning the output is right on target. In a control system featuring a permanent error between the setpoint and the output, the error voltage is flat: there is nothing the system can do to correct the situation. How do we transform this flat voltage into a growing or decreasing control signal to force the error reduction to zero? By *integrating* the error voltage. This way, we will obtain a ramp, up or down, automatically driving the control voltage until the error becomes zero.

These functions are usually implemented in a *compensating* block taking place after the error signal. The output of this block becomes the new control voltage, v_c. As designers combine a little bit of these functions via the tweaking of their associated coefficients, we call this block a *proportional-integral-derivative (PID)*

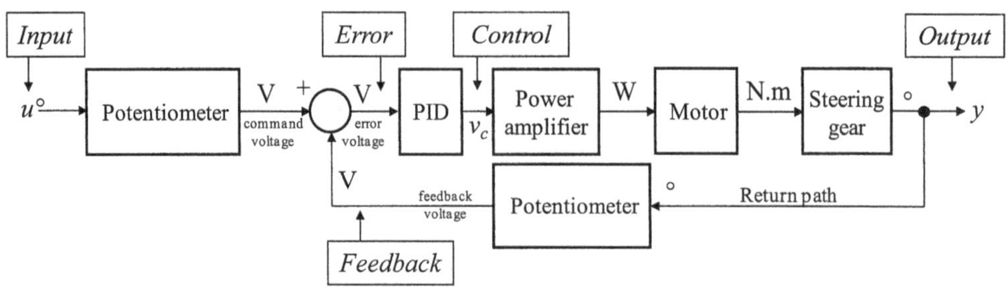

Figure 1.8 A PID block inserted after the error signal offers a way to shape the behavior of the control system.

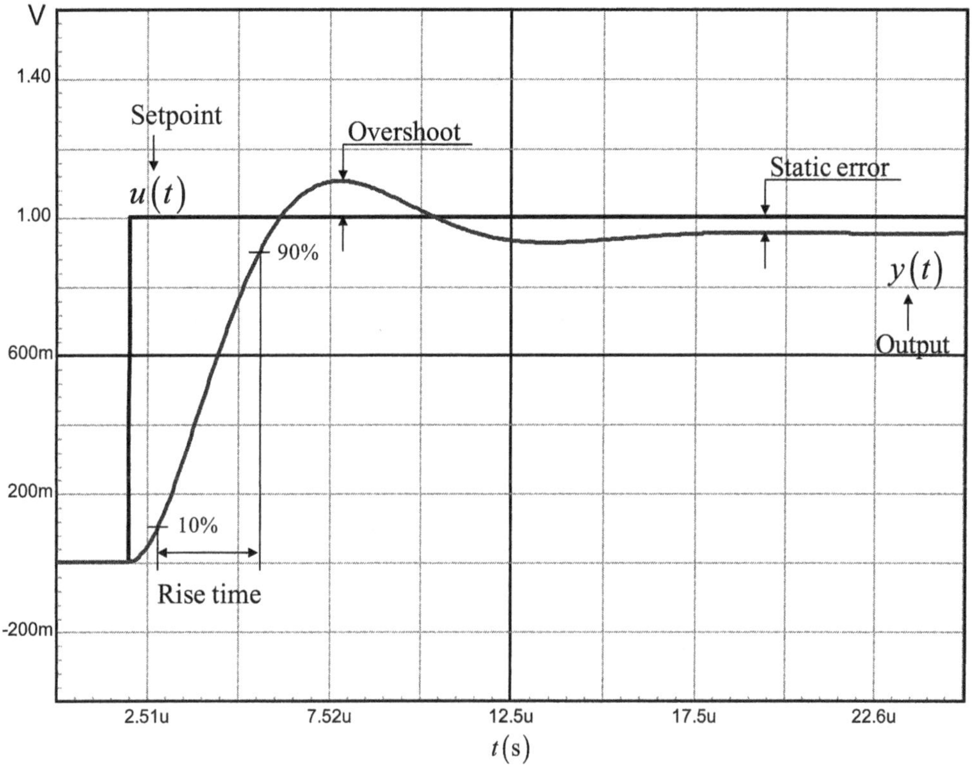

Figure 1.9 The typical response of a control system illustrating some of the aforementioned parameters.

compensator. The term *compensator* means that we want to compensate for system imperfections by purposely tailoring the return chain. Figure 1.8 shows the corresponding update on our control system, including the signal names that we will adopt.

Figure 1.9 shows the typical response of a second-order closed-loop system to a change in the setpoint and illustrates the imperfections we talked about. You can see the output $y(t)$ takes off toward the target and needs a certain time before reaching it. At some point, it even exceeds this target to stabilize to a lower value, giving birth to a permanent mismatch. Implementing PID compensators or correctors will help to minimize these imperfections, leading to control systems exhibiting precision and speed without overshoot. We will study this response in Chapter 2.

Please note that the rise time is measured from 10 to 90 percent of the rising waveform, but the definition can vary. In Chapter 2, it is considered from 0 to 100 percent.

1.3.2 The Proportional Term

The idea is simple: if the deviation or the mismatch with the target is big, the control signal v_c is increased. As the opposite, if the distance from the output variable to the expected target is small, a control signal of low amplitude will be generated. With this approach, a proportional relationship exists between the error signal and the deviation amplitude. This link is implemented with a *proportional gain* introduced

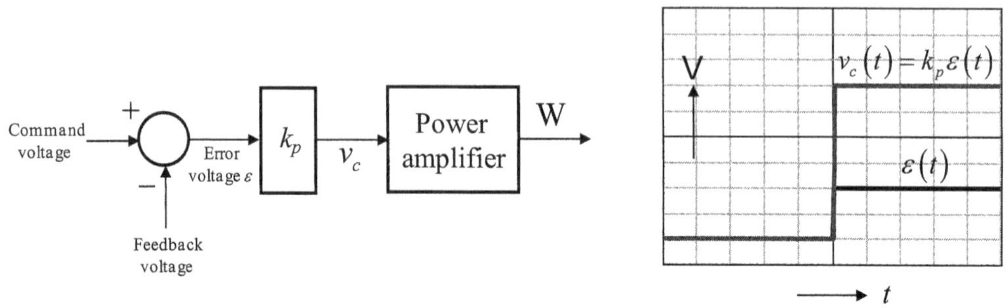

Figure 1.10 The proportional term amplifies the error signal before reaching the power chain.

in the control chain. It is usually noted k_p, as shown in the left side of Figure 1.10. How much gain is needed? In a heating system, if the generated power is high (high k_p value) the output (the heat in our case), will make the temperature rise at a fast pace. On the contrary, if k_p is small, the heat increase will be slower. In case the temperature in the room changes too rapidly, because the heater is pushed to the maximum (k_p is big), there are chances that the target is reached and then exceeded before the loop detects it: an overshoot is created. On the contrary, if you accept a slower but steady reaction (k_p is small), you will limit the overshoot amplitude when reaching the adequate temperature: a proportional gain affects the reaction speed but also the overshoot amplitude.

According to Figure 1.10 drawing, the error signal defined by (1.2) now undergoes a transformation before controlling the amplifier. It becomes

$$v_c(t) = k_p\varepsilon(t) \tag{1.9}$$

If the error voltage $\varepsilon(t)$ is a step, its transformation through (1.9) becomes the scaled-up signal shown on the right side of Figure 1.10.

1.3.3 The Derivative Term

If a gain factor k_p is needed for reaction speed, you could also pay attention to the error signal slew-rate. For a slowly moving error signal, why rush the loop and risk the output overshoot? On the contrary, if the error signal is moving quickly, you must make sure the control signal is sufficiently large to impose a fast-paced change on the output. How do we know if the system requires a small driving signal or a larger one? By looking at the error signal slope. This slope can be assessed by the introduction of a derivative coefficient, noted k_d in Figure 1.11.

According to the drawing, the control voltage v_c becomes

$$v_c(t) = k_d\frac{d\varepsilon(t)}{dt} \tag{1.10}$$

In presence of a fast setpoint change, the control voltage will quickly react thanks to the derivative term in (1.10). This is what the right side of Figure 1.11 shows you. For the opposite, when the change is slow, the control voltage will be of lower amplitude. We will later see that the presence of the derivative term contributes to slowing down the system as it also opposes any output variation, including

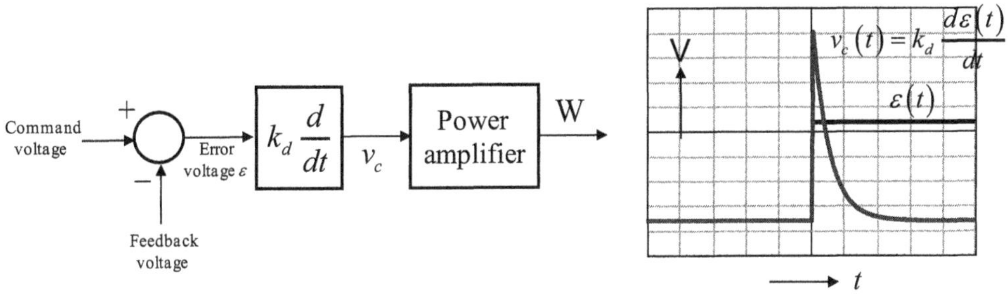

Figure 1.11 The derivative term produces a control voltage sensitive to the slope of the error voltage.

that driven by the loop in response to a perturbation. As a result, the recovery time is affected by the derivative term. Its presence, however, naturally limits the output overshoot.

1.3.4 The Integral Term

What you want from a control system is precision—in other words, the least error between the setpoint and the controlled variable. If the weighted combination of coefficients like k_p and k_d help to reach the target on time while minimizing the overshoot, you need the system to keep its precision over time. In other words, if k_p and k_d clearly affect the transient response, you need another function block that accumulates or *integrates* over time all the long-term errors, eventually canceling the static/dc error. This coefficient is an in *integral* term, noted k_i. It appears in the left side of Figure 1.12.

With the presence of the block, the control voltage in the time domain becomes

$$v_c(t) = k_i \int \varepsilon(t)\,dt \tag{1.11}$$

If you integrate a constant signal of $k_i\varepsilon$ amplitude, you obtain a ramp following $v_c = k_i\varepsilon \cdot t$, where t is the elapsed time. As the resulting control signal increases permanently for a constant error, we assume that the target will be exactly matched after a certain amount of time: systems including an integral term are called null-error systems.

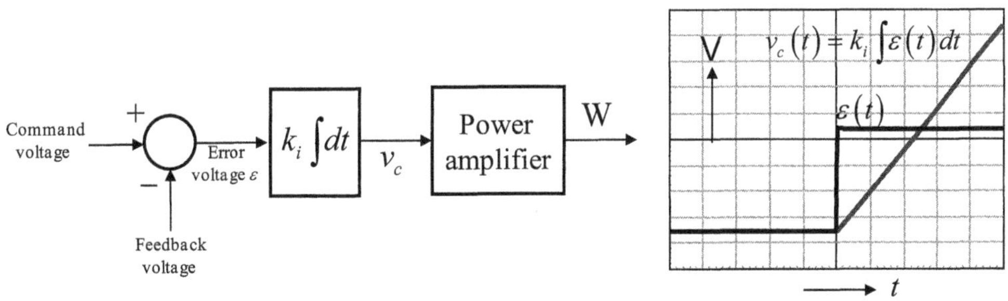

Figure 1.12 The integral term fights all the long-term errors drifts.

1.3.5 Combining the Factors

A well-designed control system reacts quickly to perturbations without excess overshoot or undershoot and exhibits a small static error. However, it is important to understand the necessity of tradeoff between precision (small or zero error) and stability (overshoot amplitude). Addressing this dilemma is partially obtained by combining the above functions through a PID block, where the specific coefficients k_d, k_p and k_i are individually tweaked to reach the desired performance.

This combination appears in Figure 1.13. We can clearly see that the control voltage v_c is actually the sum of all blocks outputs:

$$v_c(t) = k_p\varepsilon(t) + k_i\int \varepsilon(t)\,dt + k_d\frac{d\varepsilon(t)}{dt} \tag{1.12}$$

The fine tuning of each individual coefficient is beyond the scope of this book. However, for the vast majority of power converters (linear or switching), this individual tuning is not directly performed by the designer. Rather, the designer will indirectly control the derivative and integral terms by respectively positioning *zeros* and *poles* in the compensator transfer function to match certain design criteria such as *crossover frequency* and *phase margin*.

It is worth noting that a designer can also favor a term in particular, or two (PI or PD). For instance, it is very possible to stabilize a converter with a proportional term or an integral term alone. Power factor correctors are often designed with a simple integrator in the return chain. We will come back in greater details on this PID type of compensator in Chapter 4.

1.4 Performance of a Feedback Control System

As control inputs and perturbations can be arbitrary by nature, it is extremely difficult to assess the performance of a control system if you ignore the input signal or the perturbation shapes. Furthermore, a control system can cross various operating modes within which its output must stay within known boundaries. These modes can be transient or steady state (i.e., permanent) and must be separately studied to predict the output variations in a variety of situations. In practice, designers judge

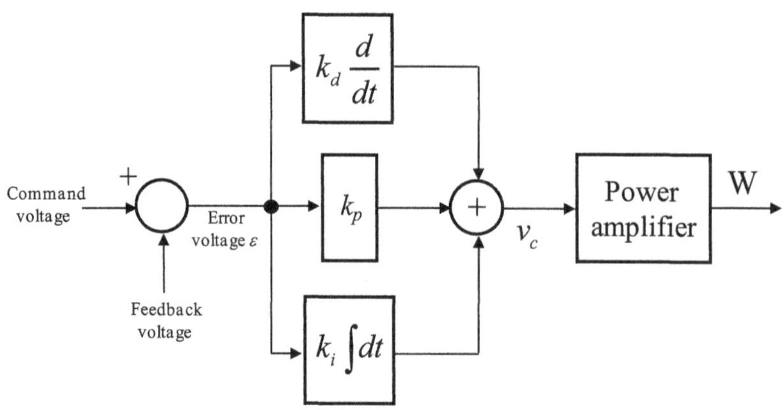

Figure 1.13 A PID system combines all three coefficients to form the control voltage v_c.

the performance of a feedback control system based on its response to a set of test waveforms. There are several of them: the step, the ramp, the Dirac impulse, and the sinusoidal stimulus. As they are among the most commonly used during the analysis of a power converter prototype, we will only look at the step and the sinusoidal input. However, before exploring these signals, it is important to understand the differences between the terms transient and steady state.

1.4.1 Transient or Steady State?

In electrical or electronic systems, the term transient designates a period of time during which the system under study is not at its equilibrium state. For instance, it can be the time necessary to let the system start up and reach its nominal output. When you first power up a 5 V converter, all capacitors are discharged, and its output starts from zero and then rises toward the expected nominal value. Further to a certain time designated as a *transient mode*, the output stabilizes to the regulated value: the converter is in *steady-state operation*. Figure 1.14 illustrates this event for a switching converter.

A transient event can also be seen as a sudden change that makes the system deviate from its equilibrium state: (e.g., the load is steady at 100 mA and you suddenly increase it to 1 A). Observing the output of the converter while this sudden change is applied will give you information on its *transient response*.

During a transient period, the system crosses highly nonlinear states and can no longer be approximated to a linear system for its analysis. The transient response of a converter tells you a lot on the way it has been compensated and will be explored

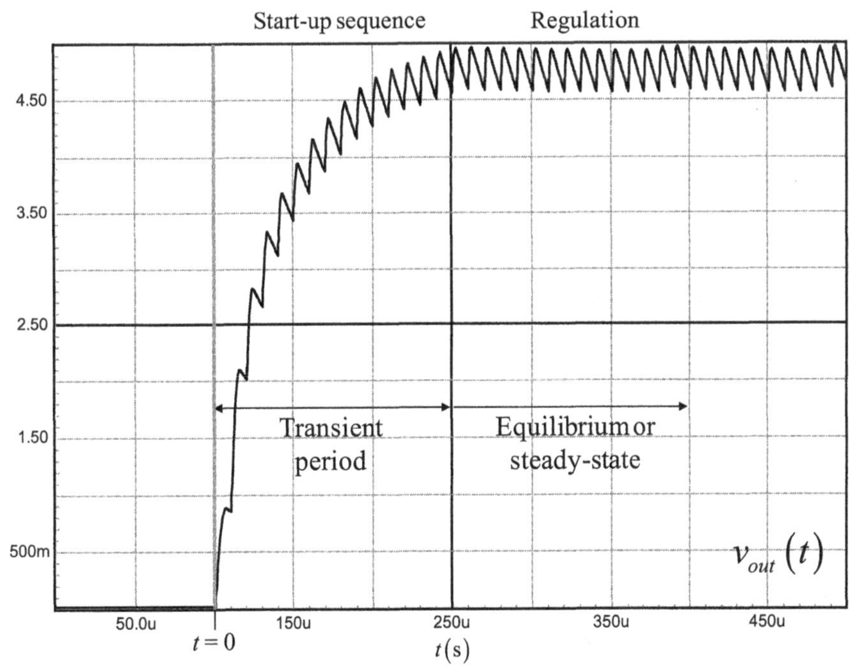

Figure 1.14 The time needed by the converter to stabilize to 5 V is designated as a transient period.

in Chapter 2. Figure 1.15 depicts the typical response of a converter subjected to a transient load step. As you can observe, the current step has been applied after the converter reached equilibrium. The response of Figure 1.15 is typical of a good design: the output slightly deviates from the target and comes back quickly without overshoot or oscillations.

Steady state is the time that the system under study is considered to be at equilibrium: the output has reached a stable operating condition from which it does not deviate. As we will later see, a harmonic analysis can be carried out once the converter reaches steady state. The injected signal perturbs the converter around its *operating point*. Also called *bias point*, it is a working point at which you study the converter. For instance, when you specify a Bode plot or a small-signal analysis, you must indicate the operating conditions at which these data were captured (e.g., V_{in} = 20 V with an output voltage of 5 V delivering 2 A). Further details of the bias point include the duty ratio, the error amplifier output level, and so on. It is important to check these data in a simulation as they indicate a correct dc analysis prior to running the ac sweep. Then, as the deviation brought by the injected harmonic signal around this bias point is small, the converter keeps operating in a linear region, and its response can be analyzed. This is called *small-signal analysis*.

When working with switching converters, observing the current flowing in a capacitor instructs you whether the converter is in steady-state or still in a transient mode. After the converter has stabilized and provided that no external excitations are applied, the time-averaged current in any of the capacitors used in the system must be exactly zero. The same observation can be applied to any inductor where the time-averaged voltage across its terminals must be zero at steady state. Deviations from these values indicate that the converter is not in steady state, suffers instability, or undergoes an ac sweep.

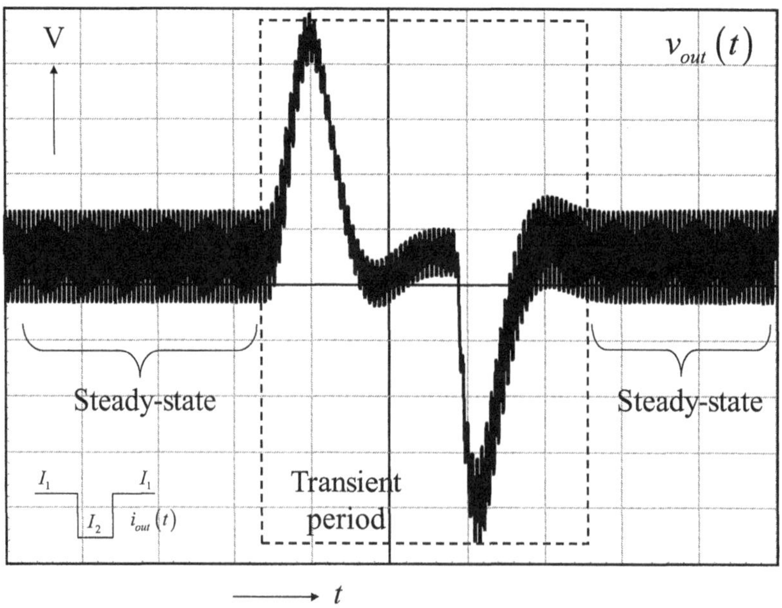

Figure 1.15 This curve combines a steady-state operation before and after the transient event.

1.4.2 The Step

A step function, also named a heaviside function, is a mathematical function whose amplitude is zero for $t < 0$ and equals a constant value for $t \geq 0$. In a closed-loop system, the control input is rarely zero at rest. It can rapidly change from one constant value to another one (for instance, to correct a sudden perturbation). When you study a voltage regulator, the control input is a fixed scaled-up reference voltage V_{ref} you want to replicate on the output (e.g., a 2.5-V reference voltage used to build a 12-V regulator). The perturbations, in this case, are the input variables that can change the operating conditions of the system: the input voltage or the output current. Any of them can thus be stepped to test the system response to a perturbation such as an output current change. Figure 1.16 shows an example where the output voltage of a converter, v_{out}, has been subjected to a steep output current increase.

As Figure 1.16 illustrates, stepping the output of two converters A and B via a current source (or using a resistive load with a switch) displays various things on the performance of these converters and also on the internal loop implementation of their respective control section. Both converters deliver an output voltage of V_0 when loaded by a current I_0. When the current is increased to I_1, converter A output severely dips by a voltage ΔV_A. We say the output *undershoots*, meaning that it passes momentarily below the regulated output level. Then, it recovers by going up quickly, exceeding the output—the system now slightly overshoots—before it stabilizes to V_0, missed by a very small deviation of ΔV_{AA} amplitude: this is the

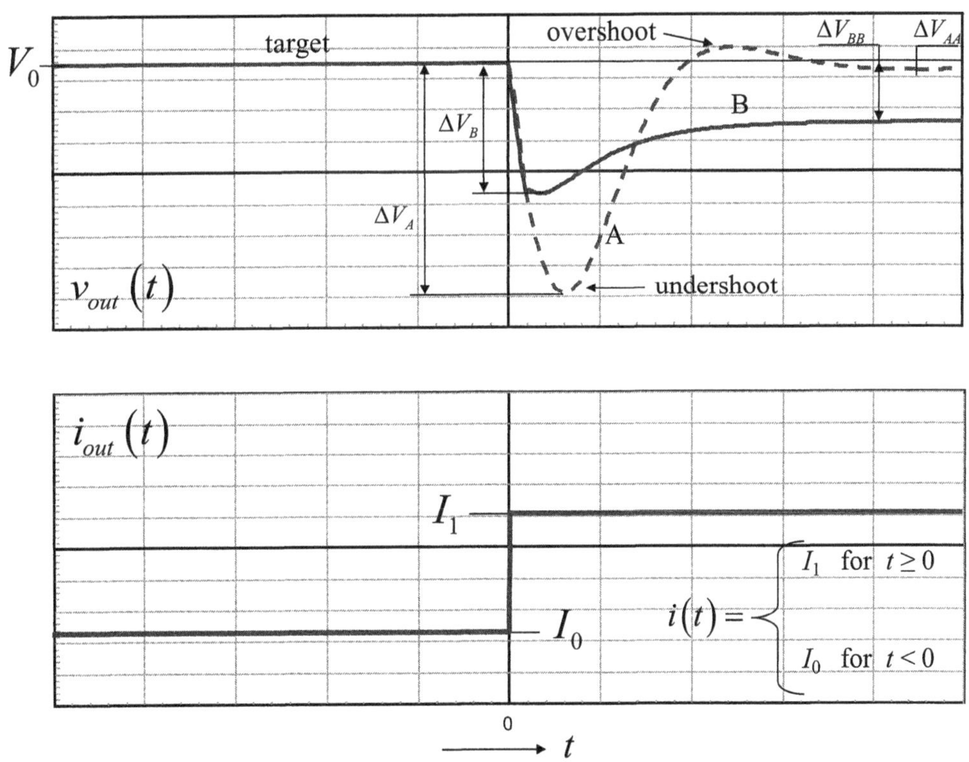

Figure 1.16 The step function can be seen as a switch that suddenly closes (or opens) to apply a sharp discontinuity to the system under study. Here, the output of a converter is suddenly loaded, and the control system tries to correct the perturbation as much as it can.

static error that the system cannot correct. We will later see that a system affected by a large open-loop gain exhibits a very small static error. This theoretical error approaches zero when an integral term (i.e., a pole placed at the origin) is inserted in the control loop. On the second converter, the undershoot ΔV_B is smaller than that of converter A but the static error ΔV_{BB} is larger, almost 10 times the previous one. This teaches us that converter B has a lower gain than converter A, hence a larger static error. As observed, the lower gain system does not generate an overshoot.

1.4.3 The Sinusoidal Sweep

The sinusoidal stimulus offers an alternative to the input step for studying control systems. A sinusoidal signal is used to reveal the *transfer function* of a given system by ac-sweeping one of its inputs while observing one of its outputs. However, as the transfer function study concerns the output response to one particular perturbed input, the other inputs must be biased at a steady-state level during the sweep. For instance, in a converter, the inputs can be the supply voltage, the output current i_{out}, or the control pin v_c. If you study v_{out} to v_c, then the output current and the input voltage are frozen during the ac sweep. A signal of constant amplitude is injected into the selected input and its frequency varied from a starting value (e.g., 10 Hz) to a stop value (e.g., 100 kHz). At each frequency step, the output amplitude is recorded, as well as its phase difference relative to the input. The amplitude of the injected signal must stay within certain limits, a small-signal analysis, to guarantee the system will not be overdriven and remains in a linear zone throughout the sweep. A possible text fixture appears in Figure 1.17, where an oscilloscope can either be used to capture the points of interest or to control the linearity of the output signal. At the end of the sweep, we end up with a series of data points containing the input amplitude V_{in} (kept constant during the sweep), the output amplitude (v_{out}), the phase difference between both signals φ, and, finally, the frequency f at which these points were stored. This series of data points, amplitude phase couples, are representative of the control system transfer function: when a perturbation or a control setpoint change occurs, how does it propagate in terms of amplitude and phase through the system to finally affect the output? This is the answer the study of the transfer function must give us.

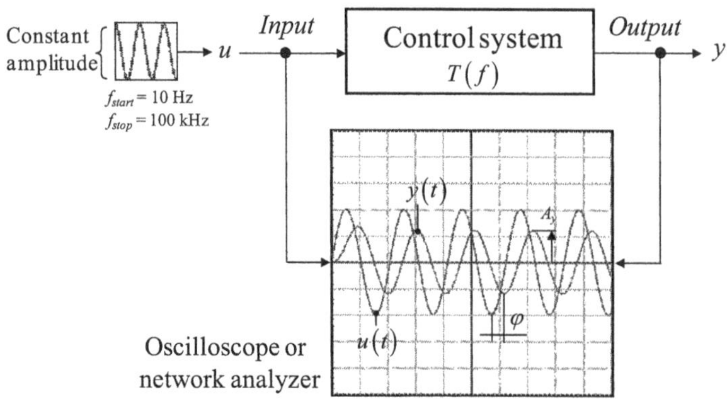

Figure 1.17 The input is ac-swept while the output signal characteristics (amplitude and phase) is recorded at each frequency step.

1.4.4 The Bode Plot

The most common way for plotting the transfer function is to display the magnitude of the ratio V_{out}/V_{in} versus frequency in one graph, while the phase versus frequency appears in a second graph. However, as both the ratio and frequency variations can be quite large, it is usual to logarithmically compress the x and y axis. The final representation becomes a so-called *Bode plot*, after H. Bode, an American engineer working at Bell Labs in the late 1940s. Such a plot is made of two graphs, magnitude and phase, sharing a common horizontal axis graduated in hertz (the log-compressed swept frequency). The upper graph (the magnitude curve) has a vertical scale graduated in decibels (dB), whereas the lower graph simply displays the phase difference in degrees.

The decibel, one tenth of a bell, is a logarithmic unit of measurement commonly used to express the magnitude of a physical quantity (power or current intensity, for instance) relative to a reference level. For instance, when two power levels, P_1 and P_0, need to be compared, the following formula can be used:

$$G_p = 10\log_{10}\left(\frac{P_1}{P_0}\right) \tag{1.13}$$

If a power source P_0 of 10 W is chosen as the reference and you measure a second source P_1 of 30 W, you would say that P_1 is larger than P_0 by 4.8 dB or P_0 is smaller than P_1 by –4.8 dB.

In our case, as we want to compare input and output voltages, the output of our control system to its input stimulus, the formula needs revision. Going back to (1.13), if we consider that power levels P_1 and P_0 are obtained by two rms voltages V_1 and V_0 applied across a common resistor R, then the ratio of powers could be reformulated as follows:

$$G_v = 10\log_{10}\left(\frac{V_1^2/R}{V_0^2/R}\right) = 10\log_{10}\left(\frac{V_1}{V_0}\right)^2 = 20\log_{10}\left(\frac{V_1}{V_0}\right) \tag{1.14}$$

To draw the magnitude curve of our Bode plot, we will simply apply this formula to the collected data points:

$$G_v(f) = 20\log_{10}\left(\frac{V_{out}(f)}{V_{in}(f)}\right) \tag{1.15}$$

For every frequency step, you will thus record and compute the magnitude in decibels plus the phase difference between both input/output signals. Once this information is graphed, you obtain the Bode plot as shown in Figure 1.18 for a first-order system. What kind of frequency step must we select? Usually, to avoid ending up with too many data points, it is recommended to take around 100 points per decade (e.g., 100 points between 10 Hz and 100 Hz and so on). However, if sharp peaks must be observed, there are chances that the 100 points are scattered and the resonance can be masked. In that case, it is recommended to increase the amount of points to, let's say, 1000, to the detriment of the sweep speed of course. By the way, if we take 100 points per decade, what is the resolution of a step then? If we start from f_{start}, the next point will be $f_2 = f_{start} \cdot x$ where x is the ratio increase between the second starting point and the starting point. The third point will be at

$f_3 = (f_{start} \cdot x)x = f_{start} \cdot x^2$. If we select n data points, the equation we need to solve is the following one:

$$f_{start} \cdot x^n = f_{stop} \tag{1.16}$$

Knowing that, over a decade, f_{start} and f_{stop} are linked by a ratio of 10, we have

$$x = 10^{\frac{1}{n}} \tag{1.17}$$

If we start at 10 Hz to end up at 100 Hz, we have $x = 1.02329$ or an increase of 2.33 percent between each point. The second point will therefore be at $f_2 = 10.2329$ Hz, the third at $f_3 = 10.4712$ Hz, and so on.

There are several pieces of information you can extract from the Bode plots represented in Figure 1.18:

- The *cutoff frequency* also called the *corner frequency* is the frequency either below or above which the transfer function magnitude is reduced/increased by 3 dB. In Figure 1.18, we can see that the magnitude is flat in the low frequency area and falls by 3 dB at 1 kHz.
- The cutoff frequency of this first-order system is also the frequency for which the phase lag between the output and the input is 45°. It would be 90° for a second-order system.
- The slope of the magnitude curve is classically given by the vertical displacement divided by the horizontal displacement. In other words,

$$S = \frac{20\log_{10} y_2 - 20\log_{10} y_1}{20\log_{10} x_2 - 20\log_{10} x_1} \tag{1.18}$$

By looking at the graph, for a frequency decade between two points x_1 and x_2, we read that a 20-dB magnitude difference links y_2 and y_1. In other words, using (1.15), $y_2 = 0.1y_1$. Given the decade between x_1 and x_2, we have $x_2 = 10x_1$. Updating (1.18) with these definitions, we have

$$S = \frac{20\log_{10} 0.1y_1 - 20\log_{10} y_1}{20\log_{10} 10x_1 - 20\log_{10} x_1} = \frac{20\log_{10}\left(\frac{0.1y_1}{y_1}\right)}{20\log_{10}\left(\frac{10x_1}{x_1}\right)} = \frac{-20}{20} = -1 \tag{1.19}$$

When using linear-logarithmic (lin-log) scales for the vertical and horizontal axis, it is possible to draw the ac response through asymptotic curves. These curves for both the magnitude and the phase represent a template made of straight lines. In Figure 1.18, we can see the magnitude and phase response of the following first-order transfer function:

$$\frac{V_{out}(s)}{V_{in}(s)} = \frac{1}{1 + s/\omega_p} \tag{1.20}$$

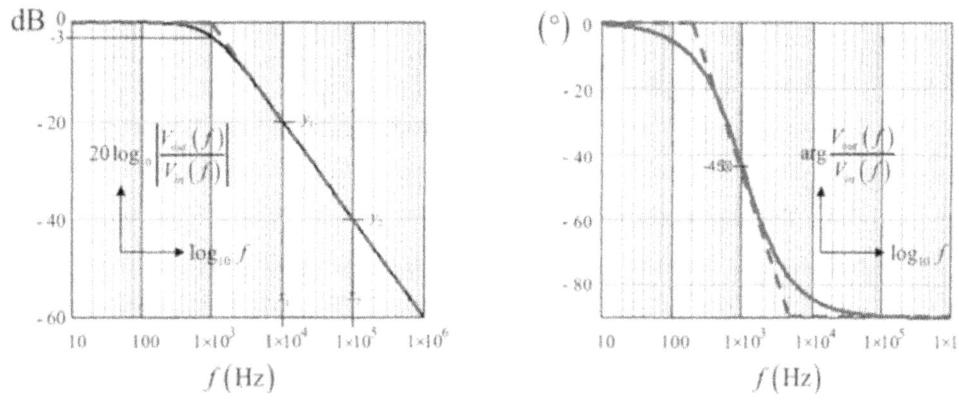

Figure 1.18 The Bode plot is a first-order system exhibiting a –1 slope.

For frequencies well below the pole position (1 kHz), the magnitude curve is almost flat and can be represented by a straight line up to that point. Beyond, the magnitude decreases following a negative slope of –20 dB per decade, also called a –1-*slope*, as given by (1.19). The phase follows almost the same scenario: 0° well below the cutoff frequency and then lags down to 90° when the frequency is far beyond this value. It is usual to draw a 0° flat line up to one fifth of the cutoff point (200 Hz in our example). A falling line then joins a –90° point placed at five times the cutoff value (5 kHz). They are represented as dashed lines in Figure 1.18. As you can see, the magnitude and phase curves deviate from the asymptotes at the corner points. The deviation values at various observation points can be computed as described in the various references given at the end of this chapter.

A first-order system exhibits a down or up slope of 1 or –1, respectively, implying an increase (single zero) or a decrease (single pole) of 20 dB per decade of frequencies. For a second-order system, this slope becomes 2 or –2 (–40 dB per decade) depending if the magnitude increases (double zero) or decreases (double pole) as the frequency is swept.

1.5 Transfer Functions

A control system is characterized by the relationship relating its output, y, to its input u. As our input signals are arbitrary by nature, the *time-domain* analysis offers a known means to study a control system: if, over time, the input $u(t)$ changes by a certain amount, how does it affect the output $y(t)$? Performing this study requires the usage of *differential equations*. A differential equation uses the notion of derivative, a mathematical tool that measures the rate of change of a function when its input varies by a certain quantity. For instance, in the time-domain equation (1.4), we state that the current inside the capacitor depends on the time-derivative of its terminals voltage (actually the *slope* of the voltage applied on the capacitor) and the capacitor value itself. This equation is then substituted into (1.3) to form the final first-order differential equation of the system under study, (1.5). We say first order because we only differentiate once in relationship to a time interval, dt. Should we need to differentiate twice, implying a variable affected by the slope of the slope,

dt^2, we would have a second-order system. Generally speaking, the order of the equation depends on the number of distinct energy storage elements present in the circuit. Should you have one inductor and two independent capacitors (not in parallel or series), this is a third-order equation or a third-order system.

1.5.1 The Laplace Transform

Looking at the output $y(t)$ delivered by these types of equations requires mathematical skills that power electronics engineers, including myself, are often lacking or have forgotten. Rather than solving differential equations, engineers prefer the *Laplace transform*. For our usage in electronics, we can say that the Laplace transform, noted $\mathcal{L}$, is a mathematical tool that converts complex linear differential equations of any order into a simpler set of algebraic expressions. Once the solution of these algebraic equations is found, the expression can be transformed back to the time domain using the *inverse Laplace transform*, denoted $\mathcal{L}^{-1}$.

When used in electrical circuits, the Laplace transform can also be seen as a tool processing periodic or nonperiodic time-domain functions (e.g., $u(t)$ and $y(t)$) to map them into a two-dimensional plan, the complex frequency domain. In this domain, the new expressions, now noted $U(s)$ and $Y(s)$, are a function of a complex argument $s = \sigma + j\omega$ (also known as p in some countries). The resulting function of s now features an argument (the phase) and a magnitude (the amplitude).

The Fourier transform also maps the time-domain function but into a one-dimensional frequency domain, ω. When a linear system is stable and the initial conditions are zero, the Fourier and Laplace transforms can be used interchangeably; they give identical results including transient response. When a system is unstable (i.e., it features a pole in the right half plane), the Fourier transform of the response simply does not exist because the Fourier integral does not converge. The Laplace integral, with complex frequency $s = \sigma + j\omega$, can be made to converge because of the presence of σ, which is at the core of making a stability assessment of a system based on a bounded response. Because part of studying systems is stability assessment, and not just determination of a response, the Laplace transform is the appropriate one to use.

The *unilateral* Laplace transform, meaning that positive time only, is considered (the function is said to be *causal* (e.g., zero before $t = 0$)) looks as follows:

$$U(s) = \mathcal{L}\{u(t)\} = \int_0^\infty u(t)e^{-st}\,dt \tag{1.21}$$

This equation can be used to find the response of a linear circuit to a nonsinusoidal excitation such as a ramp, a step, and so on. By involving the inverse Laplace transform on the output equation that is now a function of s, we can reconstruct the output signal in the time domain. If we are solely interested in a harmonic analysis in steady state, such as what Figure 1.17 shows, s becomes a pure imaginary number equal to $j\omega$: ω is the signal angular frequency in radians per seconds, or $2\pi f$, with f being the waveform frequency, in hertz. In this particular case, the mathematical definition of the Laplace transform becomes that of the unilateral Fourier transform:

$$U(j\omega) = \int_0^\infty u(t)e^{-j\omega t}\,dt \tag{1.22}$$

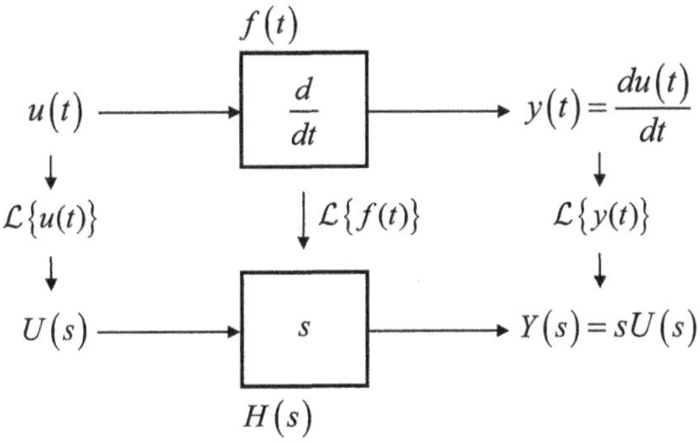

Figure 1.19 The Laplace transform helps to convert a differential equation into a simple algebraic expression. Here, the initial condition u_0 is 0.

In this expression, the term $e^{-j\omega t}$ represents a *phasor*, a reduced mathematical notation including the amplitude and the phase of a sinusoidal signal. The function $U(j\omega)$ will thus carry these characteristics, essential to its description in the complex frequency domain, giving us access to complex parameters such as phase and magnitude. This tool is widely used in the electronic world and in particular in the study of control systems.

There are several interesting properties of the Laplace transform. Among them, the derivative or the integral of a function $u(t)$ are respectively transformed into the Laplace domain by a multiplication or a division by s. The Laplace transform of a derivative is

$$\mathcal{L}\left(\frac{du(t)}{dt}\right) = sU(s) - u_0 \tag{1.23}$$

where u_0 in the expression is the initial state of u at $t = 0$.

Let's imagine, as in Figure 1.19, a box taking the time derivative of the input signal $u(t)$. If we apply a Laplace transform to this expression, considering null initial conditions for u, it becomes a simpler algebraic equation where the multiplication by s now symbolizes the differentiation.

When you write that the voltage across an inductor is defined by

$$V_L(s) = sI_L(s)L \tag{1.24}$$

you simply write that the voltage across the inductor is obtained by differentiating its current I_L further multiplied by the inductor value. In this expression, the term sL is homogenous to the inductor impedance.

The integration follows the same principle and implies a division by s:

$$\mathcal{L}\left(\int u(t)\cdot dt\right) = \frac{U(s)}{s} \tag{1.25}$$

Figure 1.20 illustrates the implementation of this principle using our boxes arrangement.

When you write that the voltage across a capacitor is defined by

$$V_C(s) = \frac{I_C(s)}{sC} \tag{1.26}$$

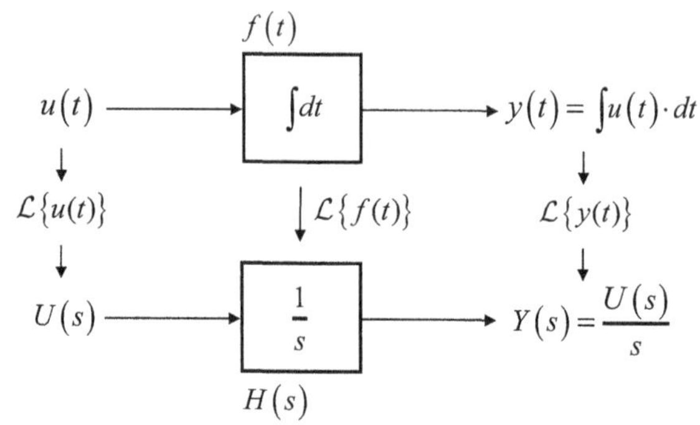

Figure 1.20 The integration term becomes a simple division by s once going through the Laplace transform.

you simply write that the capacitor voltage is obtained by integrating its current further divided by the capacitor value. In this equation, the term $1/sC$ is homogenous to the capacitor impedance.

1.5.2 Excitation and Response Signals

By looking at Figures 1.19 and 1.20, we can see that a relationship now links the output $Y(s)$ to the input $U(s)$ in the frequency domain:

$$Y(s) = H(s)U(s) \tag{1.27}$$

This relationship is called the *transfer function* and is noted $H(s)$ in both figures:

$$\frac{Y(s)}{U(s)} = H(s) \tag{1.28}$$

A transfer function is usually expressed by a quotient made of a numerator and a denominator:

$$H(s) = \frac{N(s)}{D(s)} \tag{1.29}$$

For some values of s, this transfer function can either be zero, $N(s) = 0$, or can go to infinity for $D(s) = 0$. The *roots* of the numerator N are called the *zeros* of the transfer function. The roots of the denominator D are called the *poles* of the transfer function. We will see later on that these roots can be real, complex, or purely imaginary.

As explained in Dr. Vatché Vorpérian's book (see the recommended books list), a transfer function is characterized by an *excitation* signal and a *response* signal. In our previous expressions, $U(s)$ was the excitation and $Y(s)$ the response. Because excitation and response signals can be a current or a voltage, you can easily combine the variables together, as Figure 1.21 details.

There are six possible types of transfer functions. For the voltage and current gains but also for the transadmittance and the transimpedance definitions, the excitation and response signals are collected at different places in the circuit. How-

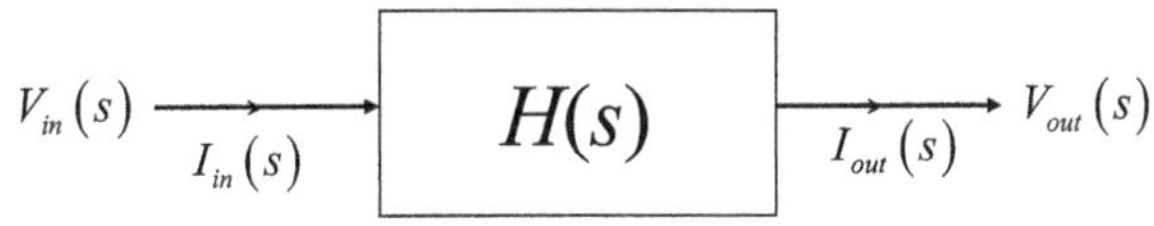

$$H(s)=\frac{V_{out}(s)}{V_{in}(s)} \quad \begin{matrix}\longrightarrow \text{Response}\\ \longrightarrow \text{Excitation}\end{matrix}$$

$$A_v(s)=\frac{V_{out}(s)}{V_{in}(s)} \ \text{voltage gain} \qquad A_i(s)=\frac{I_{out}(s)}{I_{in}(s)} \ \text{current gain}$$

$$Y_t(s)=\frac{I_{out}(s)}{V_{in}(s)} \ \text{transadmittance} \qquad Z_t(s)=\frac{V_{out}(s)}{I_{in}(s)} \ \text{transimpedance}$$

$$\left.\begin{matrix} Y_{in}(s)=\dfrac{I_{in}(s)}{V_{in}(s)} \\ Y_{out}(s)=\dfrac{I_{out}(s)}{V_{out}(s)} \end{matrix}\right| \text{admittance} \qquad \left.\begin{matrix} Z_{in}(s)=\dfrac{V_{in}(s)}{I_{in}(s)} \\ Z_{out}(s)=\dfrac{V_{out}(s)}{I_{out}(s)} \end{matrix}\right| \text{impedance}$$

Figure 1.21 A transfer function implies an excitation and a response signal.

ever, unlike these four transfer functions, the excitation and response signals for the impedance and admittance are observed at a similar location. If you are measuring an impedance, you usually apply a current source—the *excitation*—at the point where you need the impedance value and read the resulting voltage signal—the *response*—at this very point. For an admittance measurement, you apply a voltage—the excitation—and read the resulting current—the response. When you calculate either the impedance or the admittance, you actually compute a transfer function.

These equations will be affected by the frequency of the excitation signal. A Bode plot tells us how the transfer function evolves in gain and phase when studied in the frequency domain, giving the poles and zeros of the transfer function. The construction of the plot can be undertaken by calculating the magnitude of $H(s)$:

$$|H(s)|=\frac{|N(s)|}{|D(s)|} \tag{1.30}$$

but also by evaluating the phase shift it brings:

$$\arg H(s)=\arg N(s)-\arg D(s) \tag{1.31}$$

1.5.3 A Quick Example

The simple RC circuit of Figure 1.7 lends itself very well to a quick application example of the Laplace transform. The time-domain equation of the circuit is given by the following equation:

$$y(t)=u(t)-RC\frac{dy(t)}{dt} \tag{1.32}$$

Considering the capacitor fully discharged at $t = 0$ ($y_0 = 0$), we can apply the Laplace transform to (1.32):

$$Y(s) = U(s) - sRC \cdot Y(s) \tag{1.33}$$

Factoring $Y(s)$ on the left side, we have:

$$Y(s)[1 + sRC] = U(s) \tag{1.34}$$

Rearranging, we have the transfer function we want:

$$H(s) = \frac{Y(s)}{U(s)} = \frac{1}{1 + sRC} \tag{1.35}$$

This is a typical first-order transfer function. The denominator $D(s)$ equals zero for $s_p = -1/RC$. This is a negative root, indicating that the pole is situated in the left-hand portion of the s-plane (see the chapter on poles and zeros). With s equal to $j\omega$, the pole definition is obtained by calculating the magnitude of s_p:

$$\omega_p = |s_p| = \frac{1}{RC} \tag{1.36}$$

As $\omega = 2\pi f$, we can easily extract the cutoff frequency equal to:

$$f_p = \frac{1}{2\pi RC} \tag{1.37}$$

Substituting (1.36) into (1.35), we obtain a slightly different form, often used throughout this book and the literature:

$$H(s) = \frac{1}{1 + s/\omega_p} \tag{1.38}$$

If we now replace s by $j\omega$ and solve for the denominator magnitude, we have:

$$|D(s)| = \left|1 + j\frac{\omega}{\omega_p}\right| = \sqrt{1 + \left(\frac{\omega}{\omega_p}\right)^2} \tag{1.39}$$

Therefore, applying (1.30):

$$|H(s)| = \frac{1}{\sqrt{1 + \left(\frac{\omega}{\omega_p}\right)^2}} \tag{1.40}$$

When ω equals ω_p, the magnitude is $1/\sqrt{2}$. In dB, this number becomes

$$20\log_{10}\left(\frac{1}{\sqrt{2}}\right) \approx -3 \text{ dB} \tag{1.41}$$

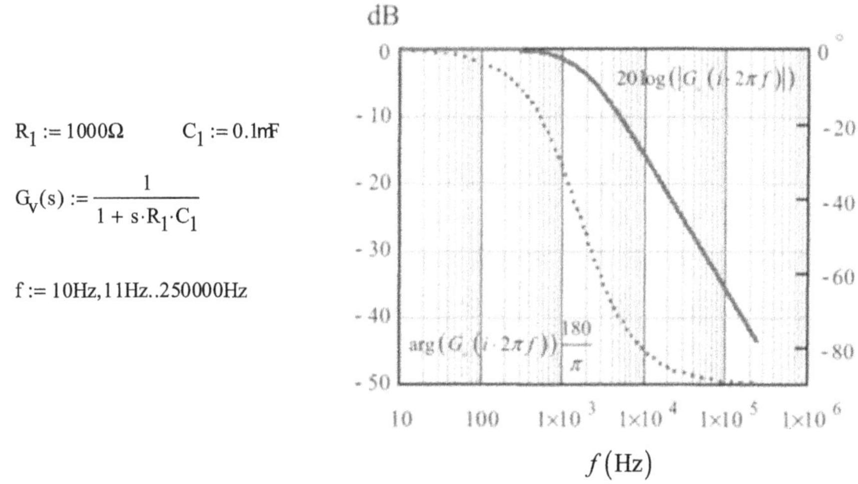

Figure 1.22 The Bode plot of the first-order network. The component values of Figure 1.7 give a cutoff frequency of 1.6 kHz.

The argument of this transfer function is now calculated using (1.31):

$$\arg H(s) = \arg(1) - \arg\left(1 + j\frac{\omega}{\omega_p}\right) \tag{1.42}$$

As the argument of 1 is zero, the argument of $H(s)$ is simply

$$\arg H(s) = -\tan^{-1}\left(\frac{\omega}{\omega_p}\right) \tag{1.43}$$

When ω equals ω_p, the argument reaches –45°.

To plot the transfer function over frequency, we can immediately use (1.40) and (1.43) with mathematical software such as Mathcad®. We then obtain the Bode plot presented in Figure 1.22.

1.5.4 Combining Transfer Functions with Bode Plots

Control systems are often made of cascaded blocks, each offering a particular frequency response. The transfer function of the whole chain is then simply the product of each individual transfer function. An illustrating example appears in Figure 1.23.

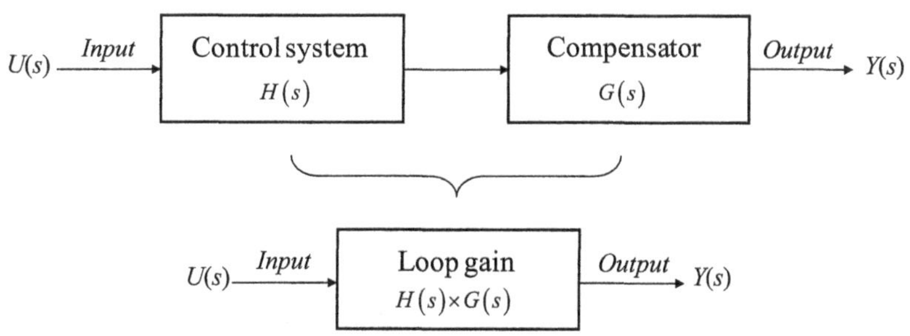

Figure 1.23 With cascaded blocks, the final transfer function is the multiplication of individual transfer functions.

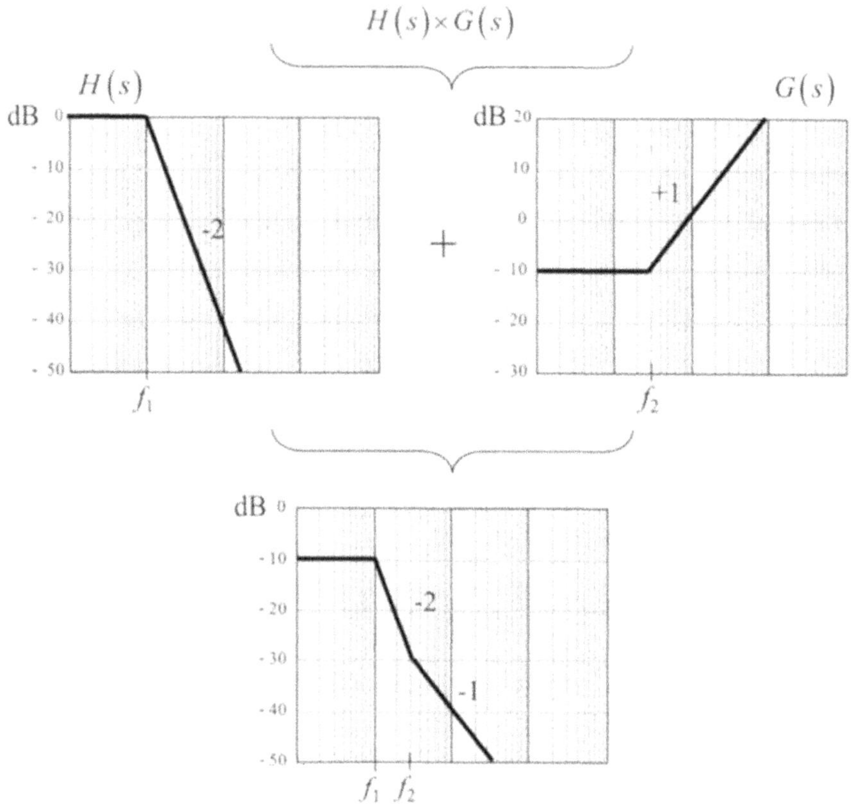

Figure 1.24 When combining asymptotical responses, the curves and their respective slopes add together to form the final ac answer.

If, rather than assessing each transfer function by its Laplace expression, we have access to individual Bode plots, then we can capitalize on the following property of logarithms, independent from their definition base:

$$\log(A \cdot B) = \log A + \log B \tag{1.44}$$

$$\log\left(\frac{A}{B}\right) = \log A - \log B \tag{1.45}$$

Therefore, if you have individually characterized two cascaded blocks $G(s)$ and $H(s)$ through Bode plots, applying (1.44) means that you just need to sum the graphs points by points to obtain the Bode plot of $T(s) = H(s)G(s)$.

$$20\log_{10}\left[G(s)H(s)\right] = 20\log_{10} G(s) + 20\log_{10} H(s) \tag{1.46}$$

When you consider the slopes, they simply add together as shown in Figure 1.24. For instance, assume you combine a second-order low-pass filter for $H(s)$ with a single-zero response for $G(s)$, the second-order filter offers a flat answer until the cutoff frequency f_1 is reached. At this point, its magnitude asymptotically decreases with a –2 slope. The second block magnitude starts with a flat 10-dB attenuation until a frequency f_2 is reached. Beyond this breakout point, the magnitude increases with a +1 slope. The combination of both frequency responses will simply be that presented in Figure 1.24 where the +1 slope opposes the –2 slope at the frequency f_2 to form a –1 slope. The phase characteristics of each block are also summed together, although not represented here:

$$\arg\left[H(s)G(s)\right] = \arg H(s) + \arg G(s) \tag{1.47}$$

1.6 Conclusion

This section ends our quick introduction on the control systems field and its associated terminology. In the coming chapters, we will come back to the topics we've tackled in more detail. Needless to say, the domain is vast and will require effort before mastering it. However, if your interest narrows down to stabilizing simple to moderately complex linear or switching converters, this introduction should get you started.

If you are interested by digging further into the domain of feedback and control systems, the following is a short list of books, articles, and links that will allow you to strengthen your knowledge in that field. Typing in search engines keywords like "modern control theory," "control systems," and so on will lead you to interesting websites and papers.

Selected Bibliography

[1] Stubberud, A., I. Williams, and J. DiStefano, *Schaum's Outline of Feedback and Control Systems*, New York: McGraw-Hill, 1994.

[2] Vorpérian, V. *Fast Analytical Techniques for Electrical and Electronic Circuits*, Cambridge: Cambridge University Press, 2002.

[3] Saucedo, R. *Introduction to Continuous and Digital Control Systems*, New York: Macmillan, 1968.

[4] Basso, C. *Switchmode Power Supplies: SPICE Simulations and Practical Designs*, New York: McGraw-Hill 2008.

[5] Erickson, B., and D. Maksimovic, *Fundamentals of Power Electronics*, New York: Springer, 2001.

[6] "Course on Modeling and Control of Multidisciplinary Systems," http://virtual.cvut.cz/dynlabcourse, last accessed June 2012.

[7] "Welcome To Exploring Classical Control Systems" http://www.facstaff.bucknell.edu/mastascu/eControlHTML/CourseIndex.html, last accessed June 2012.

[8] Astrom, K., and R. Murray, *Feedback Systems: An Introduction for Scientists and Engineers*, Version 2.10b, February 2009, http://www.cds.caltech.edu/~murray/books/AM08/pdf/am08-complete_22Feb09.pdf.

[9] "Colorado Power Electronics Center Publications," http://ecee.colorado.edu/copec/publications.php.

CHAPTER 2
Transfer Functions

We learned in Chapter 1 that a transfer function links a response signal to an excitation signal. A transfer function can be written in a lot of different ways whether it has been derived using brute-force algebra or by implementing a smarter approach (e.g., via known tools such as Thevenin/Norton transformations). What actually matters is the insight you can get just by reading the final equation. By insight, we mean your ability to immediately see where poles or zeros are located, if some gain or attenuation exists, just by reading the equation.

2.1 Expressing Transfer Functions

Linear network theory teaches us that the transfer function H of a network made of capacitors, resistors, and inductors can be expressed the following way:

$$H(s) = \frac{b_0 + b_1 s + b_2 s^2 + b_3 s^3 + ... + b_n s^n}{a_0 + a_1 s + a_2 s^2 + a_3 s^3 + ... + a_m s^m} = \frac{N(s)}{D(s)} \tag{2.1}$$

In this equation, it is important that the order of the denominator m is always greater than or equal to that of the numerator n. When $m > n$ the magnitude of $H(s)$ goes to zero as s goes to infinity. A transfer function satisfying this property is said to be *strictly proper*. The order of the denominator $D(s)$ reflects the order of the network. The roots that cancel this denominator are called the *poles*. On the other hand, the roots that cancel the numerator $N(s)$ are called the *zeros*. The order of the network can be determined by the number of independent storage elements C and L. Should you have two independent capacitors and one inductor, you have three distinct state variables making a third-order circuit: m equals 3 in (2.1) and you have three roots or poles in the denominator.

For instance, if we study a first-order system featuring resistors and a capacitor, its transfer function expression may look like that:

$$H(s) = \frac{b_0 + b_1 s}{a_0 + a_1 s} \tag{2.2}$$

By reading this expression, can you immediately see if the circuit has gain (or attenuation)? Can you identify where the roots of the numerator (the zeros) or the denominator (the poles) are? I cannot. To let you unveil the presence of these elements, you must rearrange that equation into a slightly different format. Factor the terms a_0 and b_0 to obtain

$$H(s) = \frac{b_0}{a_0} \frac{1 + s\dfrac{b_1}{b_0}}{1 + s\dfrac{a_1}{a_0}} = G_0 \frac{1 + s/\omega_{z_1}}{1 + s/\omega_{p_1}} \tag{2.3}$$

Now, from this expression, you can tell that the static gain or the attenuation of the circuit is

$$G_0 = \frac{b_0}{a_0} \tag{2.4}$$

and that a pole and a zero cohabit:

$$\omega_{z_1} = \frac{b_0}{b_1} \tag{2.5}$$

$$\omega_{p_1} = \frac{a_0}{a_1} \tag{2.6}$$

If we apply this factorization to (2.1), we obtain

$$H(s) = \frac{b_0}{a_0} \frac{1 + \dfrac{b_1}{b_0}s + \dfrac{b_2}{b_0}s^2 + \dfrac{b_3}{b_0}s^3 + \ldots + \dfrac{b_n}{b_0}s^n}{1 + \dfrac{a_1}{a_0}s + \dfrac{a_2}{a_0}s^2 + \dfrac{a_3}{a_0}s^3 + \ldots + \dfrac{a_m}{a_0}s^m} \tag{2.7}$$

Then, the next step is to factor the polynomials so that the following form appears:

$$H(s) = G_0 \frac{(1 + s/\omega_{z_1})(1 + s/\omega_{z_2})(1 + s/\omega_{z_3})\ldots}{(1 + s/\omega_{p_1})(1 + s/\omega_{p_2})(1 + s/\omega_{p_3})\ldots} \tag{2.8}$$

This is the preferred factored pole-zero method as Dr. Middlebrook promoted it in his design-oriented analysis course, [1]. What is important is that a familiar structure appears in the final expression. For instance, it can be difficult to obtain a clean factored form as the previous in a second- or a third-order system. In that case, should you write something like

$$H(s) = \frac{1 + s/\omega_{z_1}}{\dfrac{s^2}{\omega_0^2} + \dfrac{s}{\omega_0 Q} + 1} \tag{2.9}$$

then the reader will immediately recognize a frequency response combining a zero and the double poles of a second-order system. In that case, we can show that the roots of the denominator are given by

$$s_1, s_2 = \frac{\omega_0}{2Q}\left(\pm\sqrt{1 - 4Q^2} - 1\right) \tag{2.10}$$

Depending on the quality factor value Q, these roots can be either real or complex conjugates.

2.1.1 Writing Transfer Functions the Right Way

As explained in Chapter 1, a transfer function describes the path from an excitation signal, the input, to the delivered response, the output. For instance, let's consider a voltage transfer function G including an origin pole affected by a coefficient ω_{po}, with a zero and a pole. You can write it this way:

$$G(s) = \frac{1 + s/\omega_{z1}}{\frac{s}{\omega_{po}}\left(1 + s/\omega_{p1}\right)} \tag{2.11}$$

However, the expression does not give you a lot of insight regarding the overall structure of the transfer function. Ideally, it should fit the format recommended by (2.3) where the first term preceding the s-expression is a dc term subscripted with a 0 and sharing a similar dimension as the studied function. Should you write an impedance transfer function, this term should be R_0, homogeneous to ohms. For instance, it could look like

$$Z_{in}(s) = \frac{V_{in}(s)}{I_{in}(s)} = R_0 \frac{1}{1 + s/\omega_{p1}} \tag{2.12}$$

In our case, for a gain, the first term will be called G_0 and will be dimensionless. How do we unveil G_0 in (2.11)? If we factor s/ω_{z1} in the numerator, we obtain

$$G(s) = \frac{\frac{s}{\omega_{z1}}}{\frac{s}{\omega_{po}}} \frac{\frac{\omega_{z1}}{s} + 1}{\left(1 + s/\omega_{p1}\right)} = \frac{s}{\omega_{z1}} \frac{\omega_{po}}{s} \frac{1 + \omega_{z1}/s}{1 + s/\omega_{p1}} \tag{2.13}$$

If we write

$$G_0 = \frac{\omega_{po}}{\omega_{z1}} \tag{2.14}$$

Then (2.13) can be rewritten in the following way:

$$G(s) = G_0 \frac{1 + \omega_{z1}/s}{1 + s/\omega_{p1}} \tag{2.15}$$

With such a formulation, we can see that when the zero and the pole are fixed, changing the position of ω_{po} changes the gain of the system. This gain is often called the *mid-band gain*. What is this term, ω_{po}, by the way?

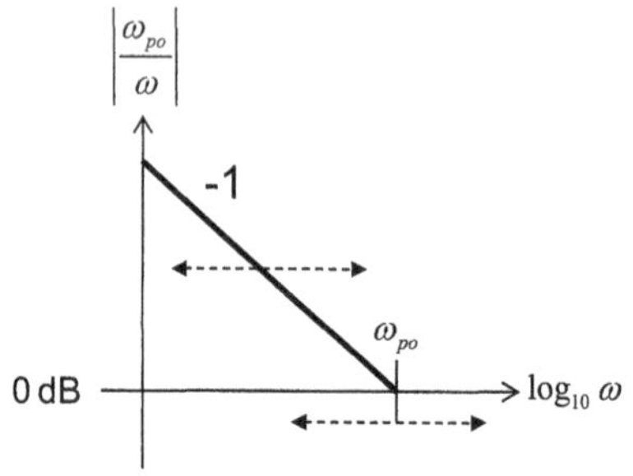

Figure 2.1 The 0-dB crossover pole is a frequency at which the magnitude of ω_{po}/s is 1.

2.1.2 The 0-db Crossover Pole

I named this factor the *0-dB crossover pole*. The origin pole in an expression such as that described by (2.11) is 0, meaning that when $s = 0$, the quotient goes to infinity. However, when s is combined with a coefficient—for instance when you have $1/sRC(1+...)$—it is advantageous to rewrite it as

$$\frac{1}{sRC(1+...)} = \frac{1}{\frac{s}{\omega_{po}}(1+...)} \tag{2.16}$$

in which I call ω_{po} the 0-dB crossover pole. It equals $1/RC$ in (2.16). It correspond to a cutoff frequency at which the magnitude of ω_{po}/s simply equals 1 or 0 dB, hence its name. Figure 2.1 graphically represents the magnitude of ω_{po}/s.

Now, when combined with a zero, as in the G_0 term from (2.14), it creates a break in the slopes and lets you purposely unveil a gain, changing with the position of ω_{po}. This is what Figure 2.2 shows you for two different values of ω_{po}.

The gain G_0 is the mid-band gain defined in (2.14).

2.2 Solving for the Roots

The zeros of a transfer function are the frequency points at which the magnitude of the transfer function is zero: the excitation signal, the input, no longer reaches the output. For the opposite, the poles are the frequency points at which the transfer function goes to infinity. If we consider a transfer function as a fraction made of a

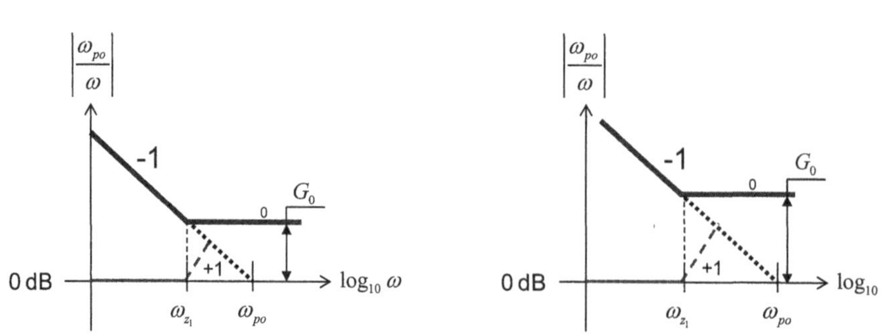

Figure 2.2 When combined with a zero, the gain G_0 changes with the position of ω_{po}.

numerator $N(s)$ over a denominator $D(s)$, then the zeros cancel the numerator while the poles cancel the denominator. In other words, identifying the zeros and the poles of a system is similar to respectively solving for the roots of its numerator and denominator expressions. As all coefficients in (2.7) numerators or denominators are real, the roots can be either purely real or appear in complex conjugate pairs.

Let's go through a few examples where we will use SPICE notations in the upcoming equations: $1k$ = 1000, $1Meg$ = 10^6, $1m$ = 0.001 and $1u$ = 10^{-6}. We assume a transfer function H as follows:

$$H(s) = \frac{s + 5k}{(s + 1k)(s + 30k)} \tag{2.17}$$

The zeros of this transfer function are found when $H(s) = 0$ or when $N(s) = 0$. On the contrary, the poles are the roots of the denominator, $D(s)$ and bring $H(s)$ to infinity. Let's rearrange (2.17) to make it look a little friendlier:

$$H(s) = \frac{(s + 5k)}{(s + 1k)(s + 30k)} = \frac{5k(1 + s/5k)}{1k(1 + s/1k)30k(1 + s/30k)} = \frac{1}{6k}\frac{(1 + s/5k)}{(1 + s/1k)(1 + s/30k)} \tag{2.18}$$

$H(s) = 0$ if we have $N(s) = 0$:

$$(1 + s/5k) = 0 \tag{2.19}$$

Given the factored form, we can identify the following roots, our real zero:

$$s_{z_1} = -5k \tag{2.20}$$

The poles are obtained when making $H(s) = \infty$ or solving for $D(s) = 0$:

$$(1 + s/1k) = 0 \tag{2.21}$$

$$(1 + s/30k) = 0 \tag{2.22}$$

Our real poles can be identified at the following position:

$$s_{p_1} = -1k \tag{2.23}$$

$$s_{p_2} = -30k \tag{2.24}$$

When s equals one of these roots, we encounter either a zero or a pole. The poles or zeros frequencies are obtained by computing the roots magnitude:

$$\omega_{z_1} = |s_{z_1}| = 5000 \text{ rad/s or } f_{z_1} = \frac{\omega_{z_1}}{2\pi} = 796 \text{ Hz} \tag{2.25}$$

$$\omega_{p_1} = |s_{p_1}| = 1000 \text{ rad/s or } f_{p_1} = \frac{\omega_{p_1}}{2\pi} = 159 \text{ Hz} \tag{2.26}$$

$$\omega_{p_2} = |s_{p_2}| = 30000 \text{ rad/s or } f_{p_2} = \frac{\omega_{p_2}}{2\pi} = 4.77 \text{ kHz} \tag{2.27}$$

In these simple examples, the roots are real and you do not need imaginary notation to solve the equations. Let us have a look at a different function:

$$H(s) = \frac{s+4}{(s+0.8)[(s+2.5)^2+4]} \tag{2.28}$$

First, we can massage the expression a little bit to make it easier to work with. Develop the right side of $D(s)$ and factor the terms according to (2.9):

$$H(s) = \frac{4}{0.8 \times 10.25} \frac{(1+s/4)}{(1+s/0.8)\left[\frac{s^2}{10.25} + \frac{s}{\sqrt{10.25}\frac{\sqrt{10.25}}{5}} + 1\right]} = G_0 \frac{1+s/\omega_{z_1}}{(1+s/\omega_{p_1})\left(\frac{s^2}{\omega_0^2} + \frac{s}{\omega_0 Q} + 1\right)} \tag{2.29}$$

We can identify a static gain G_0:

$$G_0 = \frac{4}{0.8 \times 10.25} = 0.488 \tag{2.30}$$

a zero:

$$\omega_{z_1} = 4 \text{ rad/s} \tag{2.31}$$

a pole:

$$\omega_{p_1} = 0.8 \text{rad/s} \tag{2.32}$$

a damped angular frequency ω_0:

$$\omega_0 = \sqrt{10.25} \text{ rad/s} \tag{2.33}$$

and, finally, a quality factor Q:

$$Q = \frac{5}{\sqrt{10.25}} \approx 0.64 \tag{2.34}$$

From (2.10), we can see that a Q greater than 0.5 makes the polynomial within the square root a negative number: the roots are complex. Following the definition given in (2.10), we have

$$s_2 = -2.5 + 2j \tag{2.35}$$

$$s_3 = -2.5 - 2j \tag{2.36}$$

These roots are said to be conjugate. The frequency at which these double poles appear is obtained by calculating the magnitude of either s_2 or s_3:

$$\omega_{p_2} = \omega_{p_3} = |s_{p2}| = |s_{p3}| = \sqrt{2.5^2 + 2^2} = \sqrt{10.25} \text{ rad/s} \tag{2.37}$$

Yes, this is the natural frequency we have evaluated in (2.33). Should we plot (2.29), we would observe the combined action of a pole/zero pair plus a double pole peaking at $\sqrt{10.25}$ radians per second, or 510 mHz.

2.2.1 Poles and Zeros Found by Inspection

In the previous paragraphs, we have learned that zeros and poles could be found by solving for the roots of the considered transfer function. To make our life easier, we have seen that rearranging the equation in a factored form could help us identify the poles/zeros positions in a quicker way. Unfortunately, despite the simplicity brought by rearranging the expression, the starting point still remains the transfer function equation that you must derive from node and mesh analysis of the studied network. Rather than deriving the transfer function through classical analysis techniques, is there a way to actually detect where the poles and zeros are hidden and write the transfer function just by looking at the network arrangement—in other words, by *inspection*? After all, we know that the final result should fit the format given by (2.8). Let us see how we could do that.

If we understand that a transfer function links an output signal (the response) to an input signal (the excitation), then a zero at a certain frequency prevents the excitation from reaching the output. Let us try to apply this theory to the passive filter appearing in Figure 2.3. We can see an ac source delivering a signal through a resistor R_1 to a network made of a series-parallel combination of two resistors and a capacitor. There is one distinct storage element, the capacitor C_1; this is a first-order network. As such, without knowing whether there are poles or zeros, it must fit the format given by (2.3):

$$H(s) = \frac{V_{out}(s)}{V_{in}(s)} = G_0 \frac{1 + s/\omega_{z_1}}{1 + s/\omega_{p_1}} \tag{2.38}$$

The brute-force algebra would be to calculate the expression of impedance Z_1 and apply

$$V_{out}(s) = V_{in}(s) \frac{Z_1}{Z_1 + R_1} \tag{2.39}$$

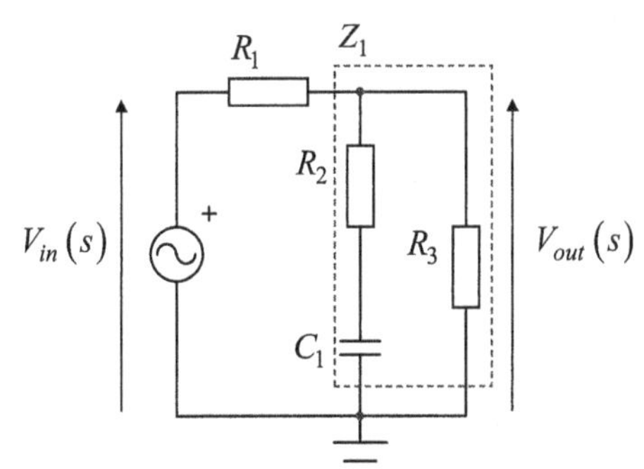

Figure 2.3 A simple first-order system featuring a zero and a pole.

If you go ahead and develop the equation, you will end up with a moderately complicated expression, but chances to make mistakes during the expansions are real. Furthermore, without additional work on the final result, it is unlikely that poles and zeros pop up at the end.

To start the derivation, let us observe the system at dc, when $s = 0$. This is exactly what SPICE does when it starts the simulation: to calculate the dc operating point of the circuit under study, also called the bias point, SPICE opens all capacitors and shorts all inductors. With the equivalent network, it calculates all dc currents and voltages that the simulator will use for the rest of the simulation. In our network, we can do the same. When C_1 is open, we are left with R_1 and R_3. Therefore, the dc attenuation G_0 is simply

$$G_0 = \frac{R_3}{R_1 + R_3} \tag{2.40}$$

Now, if you recall the definition of a zero, it is a frequency point at which the excitation no longer reaches the output. In other words, in Figure 2.3, what element in the input signal path can stop its propagation?

Either an element in series with the signal offers an infinite impedance at a certain frequency or an element linking the signal path to the ground becomes a short circuit, again at a certain frequency point. In our example, the only element that can stop the signal from reaching the output is the series combination of R_2 and C_1. When its resulting impedance is null (short circuit), we have a zero in the transfer function:

$$R_2 + \frac{1}{sC_1} = \frac{1 + sR_2C_1}{sC_1} = 0 \tag{2.41}$$

You can immediately see the zero position is the root of the numerator:

$$\omega_{z_1} = \frac{1}{R_2C_1} \tag{2.42}$$

We can now write the partial transfer function of our network by combining (2.40) and (2.42):

$$H(s) = G_0\frac{N(s)}{D(s)} = \frac{R_3}{R_1 + R_3}\frac{1 + sR_2C_1}{D(s)} \tag{2.43}$$

We are almost there but we lack the denominator expression $D(s)$. This is where the poles hide.

2.2.2 Poles, Zeros, and Time Constants

By definition, a gain is a dimensionless expression. When you say the voltage gain of a system is 20 dB, it is another means to say that the gain is 10 V/V or 10. In other words, if we go back to (2.7), and only consider second-order terms (for the sake of simplicity), we can write the transfer equation of second-order network as follows:

$$H(s) = G_0 \frac{1 + \frac{b_1}{b_0} s + \frac{b_2}{b_0} s^2}{1 + \frac{a_1}{a_0} s + \frac{a_2}{a_0} s^2} \tag{2.44}$$

A term multiplied by s has the dimension of a frequency, Hz. A term multiplied by s^2 has a dimension of a squared frequency, Hz^2. To make sure that all s-terms and s^2-terms lose their dimension when multiplied by a coefficient, these coefficients must have the inverse dimension. Therefore, the terms b_1/b_0 and a_1/a_0 must have a dimension of Hz^{-1}, whereas the terms b_2/b_0 and a_2/a_0 must have a dimension of Hz^{-2}. What offers a dimension of Hz^{-1} or actually seconds? A time constant does. What has a dimension of Hz^{-2} or squared seconds? A product of time constants. And this makes sense when you consider the expression of the zero found in (2.42); it has indeed the dimension of a time constant. How to get the poles then? We need to identify the time constants of our system but in a different manner than what we have shown for the zeros. As we stated, the transfer function denominator $D(s)$ of a linear network does not depend on its excitation or response signals. It only depends on the network structure alone. If you look at transfer functions describing a given network, its output impedance, its input admittance, and so on, then you will see that all these equations share a common denominator $D(s)$.

To study the network alone, we are going to bring its excitation signal to zero. How do we do that? Well, if the excitation signal is a voltage source, we set it to zero: replace it by a short circuit. If the excitation signal is a current source, then open circuit it. Let's apply this technique to Figure 2.3 by shorting to ground the left terminal of R_1:

The time constant is easily calculated by evaluating the resistance "seen" from the capacitor terminals. The first one is obviously R_2, in series with the parallel combination of R_1 and R_3:

$$R = R_2 + R_1 \parallel R_3 \tag{2.45}$$

The pole definition, for this simple first-order system, is simply the inverse of the equivalent time constant:

$$\omega_{p_1} = \frac{1}{\tau} = \frac{1}{RC_1} = \frac{1}{(R_2 + R_1 \parallel R_3)C_1} \tag{2.46}$$

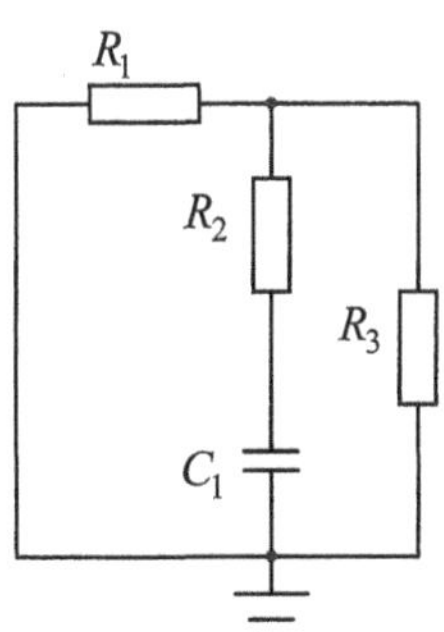

Figure 2.4 By shorting the input voltage, we can reveal the time constants of the system.

The expression of our denominator $D(s)$ is therefore

$$D(s) = 1 + s/\omega_{p1} \tag{2.47}$$

The complete transfer function then becomes

$$H(s) = G_0 \frac{N(s)}{D(s)} = \frac{R_3}{R_1 + R_3} \frac{1 + sR_2C_1}{1 + s(R_2 + R_1 \parallel R_3)C_1} = G_0 \frac{1 + s/\omega_{z1}}{1 + s/\omega_{p1}} \tag{2.48}$$

This is what is called a *low-entropy* equation by analogy to thermodynamic laws. Simply put, the entropy of a system qualifies its degree of internal disorder: to produce the work the system has been designed for, you need to bring less external energy when its entropy is low (elements are well organized and well ordered) than when it is high (elements are in disorder; this is a chaotic organization). For our equations, a low-entropy expression gives you immediate insight, without further work, on the transfer function it realizes. For the opposite, a *high-entropy* equation does not reveal anything and requires further energy through factorizations or expansions before it tells you where poles and zeros are. The analytical technique we just described lets you write low-entropy equations by *inspection*, just by looking at the network schematic and identifying its time constants. Let us check this again via another simple example that appears in Figure 2.5 with an inductive first-order network. Let us try to derive the transfer function applying what we learned.

First, we start from $s = 0$, the dc transfer function. If capacitors are open-circuited at dc, inductors, for the opposite, are considered as short circuits. When shorting L_1, the attenuation G_0 is immediately written as

$$G_0 = \frac{R_3 \parallel R_2}{R_3 \parallel R_2 + R_1} \tag{2.49}$$

The zero is found by identifying a network that prevents the excitation from reaching the output of the circuit under study. A series element admittance can become zero at a frequency point or a network connecting the signal path to ground can have an impedance that drops to zero. In our circuit, the series path is R_1 and offers a fixed value. On the other hand, a network that can potentially shunt the

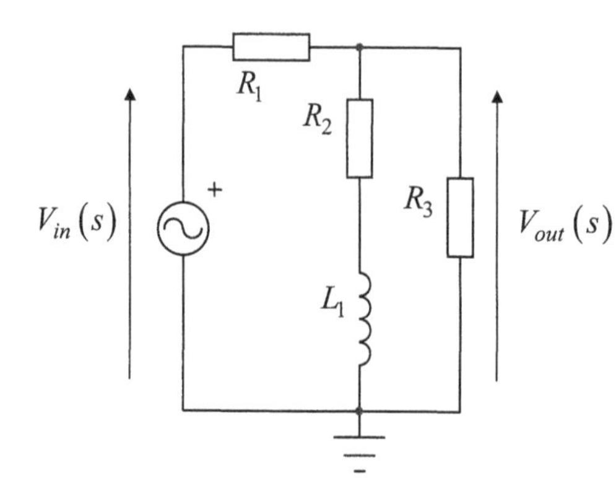

Figure 2.5 The capacitor has been replaced by an inductor.

excitation signal to the ground is R_2 and L_1. This is the place where the zero hides. To unveil it, simply solve

$$R_2 + sL_1 = R_2\left(1 + s\frac{L_1}{R_2}\right) = R_2\left(1 + \frac{s}{\omega_{z_1}}\right) = 0 \tag{2.50}$$

We have our zero position:

$$\omega_{z_1} = \frac{R_2}{L_1} \tag{2.51}$$

If the time constant of the resistor R "driving" the capacitor C is RC, then the time constant of an inductor L driven by a resistor R is L/R. What impedance drives the inductor L_1 in Figure 2.5? To discover it, we set the excitation signal to zero. The impedance is the series combination of R_2 and R_1 in parallel with R_3:

$$R = R_2 + R_1 \,||\, R_3 \tag{2.52}$$

The pole definition, for this first-order system, again, is the inverse of the equivalent time constant:

$$\omega_{p_1} = \frac{1}{\tau} = \frac{R}{L_1} = \frac{(R_2 + R_1 \,||\, R_3)}{L_1} \tag{2.53}$$

The expression of our denominator $D(s)$ is therefore

$$D(s) = 1 + s/\omega_{p_1} \tag{2.54}$$

The complete transfer function then becomes

$$H(s) = G_0\frac{N(s)}{D(s)} = \frac{R_3 \,||\, R_2}{R_3 \,||\, R_2 + R_1}\frac{1 + s\dfrac{L_1}{R_2}}{1 + s\dfrac{L_1}{(R_2 + R_1 \,||\, R_3)}} = G_0\frac{1 + s/\omega_{z_1}}{1 + s/\omega_{p_1}} \tag{2.55}$$

You can see how efficient this fast analytical method can be to express the transfer function of a simple network. The appendix at the end of this chapter shows it at work in a bridge impedance determination. Try to derive it yourself using the classical algebra technique, and you will quickly adopt these fast analytical techniques! Of course, in a small chapter portion, we have just scratched the surface, and as you complicate the network under analysis with more storage elements, you need to apply different techniques such as the extra-element theorem (EET). I encourage you to check [2–6] at the end of this chapter, as they will offer you a means to learn and make this technique efficiently work for you.

2.3 Transient Response and Roots

The stability of a closed-loop system can be assessed in different ways. If an ac-sweep teaches us about phase margin at a given crossover frequency, it does not

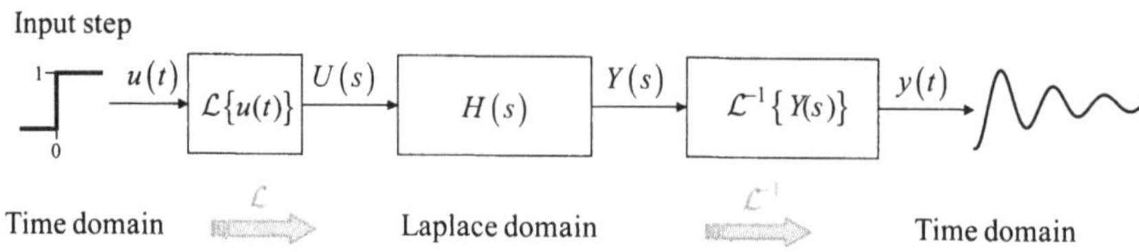

Figure 2.6 You can assess the time-domain response of a Laplace transfer function by exciting its input with a unity step.

readily tell us how the system will react to an incoming perturbation or a sudden change in the input setpoint. A common test consists of exciting the control system input with a given stimulus. As detailed in Chapter 1, there are numerous types of available stimuli: a step, a Dirac pulse, a linear ramp, and so on. Generally, the response to a step is the most popular choice, in particular for regulators such as linear or switching converters. If stepping an electronic load in the laboratory does not require a particular care, applying the technique to a transfer function implies that some precautions will be observed. First, our transfer function is expressed in the Laplace domain, whereas the step belongs to the time domain. A transformation is needed to transit from one domain to the other. Once the step is converted into the Laplace domain, it becomes the excitation signal $U(s)$ to the transfer function $H(s)$ under study. Then, as its output signal $Y(s)$ is also expressed using Laplace notation, an inverse Laplace-transform is necessary to return to the time-domain and see the resulting waveform. Figure 2.6 shows this process.

What is the Laplace-transform of a unity step? Let's have a look at the waveform that appears in Figure 2.7: it is 0 for all negative time values and equals 1 at $t = 0$ and all values beyond. This is a time-domain waveform and, according to Figure 2.6, we must transpose it into the Laplace domain before driving the transfer function of interest. In Chapter 1, we learned the definition of the Laplace transform:

$$U(s) = \mathcal{L}\{u(t)\} = \int_0^{\infty} u(t)e^{-st}\,dt \tag{2.56}$$

As $u(t)$ equals 1 from 0 to ∞, the equation becomes

$$U(s) = \int_0^{\infty} e^{-st}\,dt = \lim_{P\to\infty}\int_0^{P} e^{-st}\,dt = \lim_{P\to\infty}\left[-\frac{e^{-st}}{s}\right]_0^P = \lim_{P\to\infty}\frac{1-e^{-sP}}{s} = \frac{1}{s} \tag{2.57}$$

This is the classical definition of a unity step in the Laplace domain for $s > 0$.

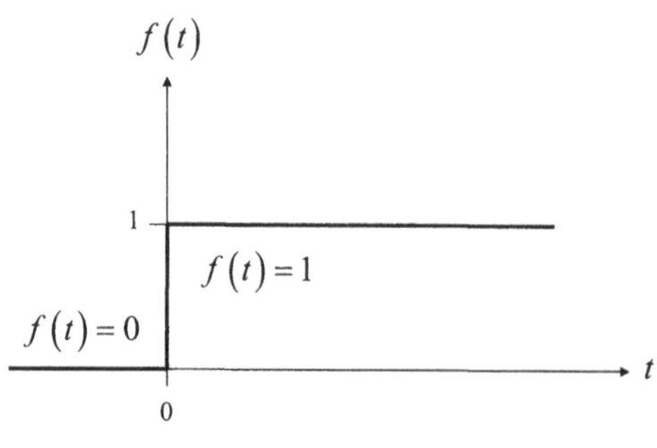

Figure 2.7 A unity step function is 0 for $t < 0$ and jumps to 1 for $t \geq 0$.

If an input signal $1/s$ enters a transfer function $H(s)$, the resulting output signal is nothing else than

$$Y(s) = \frac{1}{s} H(s) \tag{2.58}$$

From this Laplace-domain expression, we need to extract its time-domain correspondence through a reverse Laplace-transform:

$$y(t) = \mathcal{L}^{-1}\{Y(s)\} \tag{2.59}$$

Is it as simple as that? Well, it depends if you use a mathematical solver or try to derive the equation yourself. In the second case, you have to realize that the answer is not the product of the inverse Laplace transform of each individual term: you must rewrite the expression as a sum of individual terms, each having its own inverse Laplace-transform equivalent. Since the Laplace-transform is a linear operator, the inverse-Laplace transform of this sum will be the sum of all individual inverse Laplace-transform terms. However, as most of our transfer functions are often in the form of a rational function $N(s)/D(s)$, you need to split them into a sum of ratios of small polynomials. This technique is called *partial fraction expansion*, and articles on the subject can be found on mathematical textbooks or on the Web. The involved algebra looks simple but you need to be careful when expanding complex transfer functions.

Let's try to apply the technique to our first transfer function (2.17) and see its response to a unity-step stimulus:

$$Y(s) = \frac{1}{s} H(s) = \frac{1}{s} \frac{s+5k}{(s+1k)(s+30k)} \tag{2.60}$$

The technique we are going to use is called the Heaviside cover-up method, named after the English electrical engineer Oliver Heaviside. In the case of (2.60), theory tells us that it can be rewritten or expanded into the following terms:

$$Y(s) = \frac{a_1}{s} + \frac{a_2}{s+1k} + \frac{a_3}{s+30k} \tag{2.61}$$

As you can see, each denominator cancels for the roots we already found in (2.23) and (2.24), but it now also includes $s = 0$. To determine the value of coefficient a_1, a_2, and a_3, the idea is to make s equal the selected root (0 for a_1, $-1k$ for a_2, and so on) while multiplying (2.60) by the denominator including that root (by s for a_1, by $s+1k$ for a_2, and so on). Therefore, the concerned denominator on both sides of the fraction naturally disappears, a bit like if you were covering it up with your finger. It then leaves a simple equation in s, where s takes the value of the selected root. Sounds complicated? Not really as the following details show:

$$a_1 = s\frac{1}{s}\frac{s+5k}{(s+1k)(s+30k)}\bigg|_{s=0} = \frac{0+5k}{(0+1k)(0+30k)} = \frac{1}{6k} = 166.6u \tag{2.62}$$

$$a_2 = (s+1k)\frac{1}{s}\frac{s+5k}{(s+1k)(s+30k)}\bigg|_{s=-1k} = -\frac{1}{1k}\frac{-1k+5k}{-1k+30k} = -\frac{4k}{1k \cdot 29k} = -138u \tag{2.63}$$

$$a_3 = (s+30k)\frac{1}{s}\frac{s+5k}{(s+1k)(s+30k)}\bigg|_{s=-30k} = -\frac{1}{-30k}\frac{-30k+5k}{-30k+1k} = -\frac{25k}{30k\cdot 29k} = -28.7u \tag{2.64}$$

This is it, if we rewrite (2.61) with the right coefficients, we have

$$Y(s) = \frac{166u}{s} - \frac{138u}{s+1k} - \frac{28.7u}{s+30k} \tag{2.65}$$

The inverse Laplace-transform of the previous equation is thus

$$\mathcal{L}^{-1}\{Y(s)\} = \mathcal{L}^{-1}\left\{\frac{166u}{s}\right\} - \mathcal{L}^{-1}\left\{\frac{138u}{s+1k}\right\} - \mathcal{L}^{-1}\left\{\frac{28.7u}{s+30k}\right\} \tag{2.66}$$

We can look at the inverse Laplace-transform tables to get each individual term:

$$\mathcal{L}^{-1}\left\{\frac{166u}{s}\right\} = 166u\cdot\mathcal{L}^{-1}\left\{\frac{1}{s}\right\} = 166u \tag{2.67}$$

$$\mathcal{L}^{-1}\left\{\frac{138u}{s+1k}\right\} = 138u\cdot\mathcal{L}^{-1}\left\{\frac{1}{s+1k}\right\} = 138u\cdot e^{-1k\cdot t} \tag{2.68}$$

$$\mathcal{L}^{-1}\left\{\frac{28.7u}{s+30k}\right\} = 28.7u\cdot\mathcal{L}^{-1}\left\{\frac{1}{s+30k}\right\} = 28.7u\cdot e^{-30k\cdot t} \tag{2.69}$$

Assembling all these terms according to (2.66), we now have our time-domain expression:

$$y(t) = 166u - 138u\cdot e^{-1k\cdot t} - 28.7u\cdot e^{-30k\cdot t} \tag{2.70}$$

The first immediate remark is that the exponents on the exponential terms are the roots of the characteristic equation of $H(s)$. The second remark concerns the zeros. They do influence the time-domain response, but only the poles directly impact the decaying time constants in the response as we just saw. The third remark concerns their signs, negative. It means that when t goes to infinity, since all exponential terms go to zero, the output signal reaches a steady-state value given by the first term: with a 1-V step, the output should reach 166 μV. By the way, we knew it from the start, after we rearranged (2.17) into (2.18) as the dc gain, G_0, is $1/6k$ or 166u. This is confirmed if we plot (2.70) in Figure 2.8.

From these derivation lines, if we try to obtain the step response from a transfer function that contains positive roots in its characteristic equation, the time-domain response will contain exponential terms featuring positive exponents. As t goes to infinity, these exponential terms do not die out to zero but keep increasing, making the output signal severely diverging: the system response is not bounded; you do not control anything. The question that immediately arises is, if my analysis shows that all roots are negative at the considered operating point, can they move and suddenly become positive?

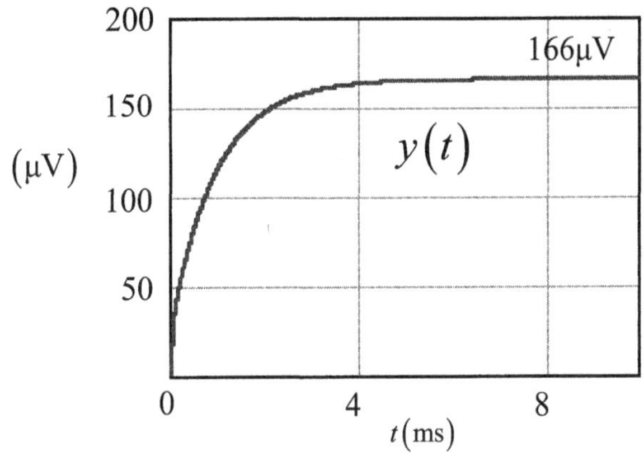

Figure 2.8 Time-domain response of the transfer function *H*(*s*) described by (2.17) when subject to a 1 V input step.

2.3.1 When the Roots Are Moving

In the previous examples, we have fixed-values roots (e.g., –1*k* or –30*k*). They do not depend on other variables, and the transient response to a step won't change as long as the roots remain constant. Now, let us consider a more practical case such as the unity-return regulator appearing in Figure 2.9.

We can see a transfer function *H*(*s*) that, for instance, could be the simplified expression of a voltage-mode switching converter operated in the discontinuous conduction mode (DCM), a buck-boost. We can show that the zero ω_{z_1} is created by the output capacitor C_{out} and its equivalent series resistance (ESR):

$$\omega_{z_1} = \frac{1}{R_{ESR}C_{out}} \tag{2.71}$$

The pole ω_{p_1} depends on the load R_{load} and C_{out}:

$$\omega_{p_1} = \frac{2}{R_{load}C_{out}} \tag{2.72}$$

To stabilize the power supply, we have designed a compensator *G*(*s*) that places a pole ω_{p_2} while offering some gain G_2. This is obviously not the best-known compensation, but we keep it simple for the sake of the example. We will learn in Chapter 3 that a unity feedback system closed-loop response $T_{CL}(s)$ obeys the following

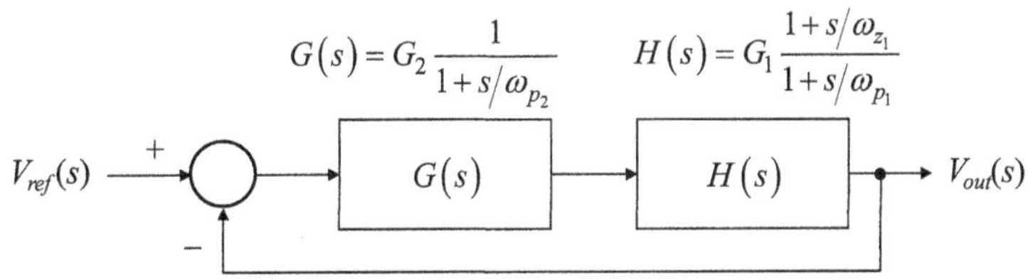

Figure 2.9 This circuit mimics a first-order converter *H* stabilized by a 60-dB gain compensator *G* featuring a single pole response.

expression where the term $1+G(s)H(s)$ represents the characteristic equation of the closed-loop transfer function:

$$T_{CL}(s)=\frac{T_{OL}(s)}{1+T_{OL}(s)}=\frac{G(s)H(s)}{1+G(s)H(s)} \tag{2.73}$$

In this equation, $T_{OL}(s)$ represents the open-loop gain defined as

$$T_{OL}(s)=G(s)H(s)=G_2G_1\frac{1}{1+s/\omega_{p_2}}\frac{1+s/\omega_{z_1}}{1+s/\omega_{p_1}} \tag{2.74}$$

Let us develop (2.73) and see if we can make it fit a familiar format. If we develop and rearrange all the terms, we can show that (2.73) can be rewritten as

$$T_{CL}(s)=\frac{G_2G_1}{1+G_2G_1}\frac{1+s/\omega_{z_1}}{1+s\left(\frac{\frac{1}{\omega_{p_2}}+\frac{1}{\omega_{p_1}}+\frac{G_2G_1}{\omega_{z_1}}}{1+G_2G_1}\right)+s^2\left(\frac{1}{\omega_{p_1}\omega_{p_2}\left(G_2G_1+1\right)}\right)} \tag{2.75}$$

This format fits a second-order equation:

$$T_{CL}(s)=G_{CL}\frac{1+s/\omega_{z_1}}{1+\frac{s}{\omega_0 Q}+\left(\frac{s}{\omega_0}\right)^2} \tag{2.76}$$

In which we can identify the following terms:

$$G_{CL}=\frac{G_2G_1}{1+G_2G_1} \tag{2.77}$$

$$Q=\frac{\omega_0}{\omega_{p_1}+\omega_{p_2}+G_2G_1\omega_{p_1}\frac{\omega_{p_2}}{\omega_{z_1}}} \tag{2.78}$$

$$\omega_0=\sqrt{\omega_{p_1}\omega_{p_2}\left(G_2G_1+1\right)} \tag{2.79}$$

The closed-loop poles of (2.76) are given by the roots of its denominator, $D(s)$. What is worth noting is that the open-loop zero of $H(s)$, ω_{z1}, now appears in the characteristic equation and affects the closed-loop poles. This is always true: the zeros that appear in the plant transfer function or the ones you will place in the compensator G always show up in the characteristic equation. If you place low-frequency zeros to improve the phase margin in a system to be compensated, these zeros will turn into low-frequency poles once the loop is closed. Low-frequency poles imply a slowly reacting system. This is something to keep in mind when evaluating the compensation strategy to follow: is it worth selecting a high crossover frequency requiring the placement of one or two zeros at low frequency, or is it better to adopt a different crossover point and avoid assigning zeros at low frequency?

For stability purposes, what matters now is the expression of (2.76) denominator roots. Since we put our transfer equation under a known second-order form, the roots or the poles of the denominator follow the definition of (2.10):

$$s_{1,2} = \frac{\omega_{p_1} + \omega_{p_2} + G_2 G_1 \omega_{p_1} \dfrac{\omega_{p2}}{\omega_{z1}}}{2} \left(\pm\sqrt{1-4Q^2} - 1 \right) \tag{2.80}$$

From this expression, what matters is the quality factor value:

1. $Q < 0.5$: the expression under the square root is strictly positive; the roots are separate and real.
2. $Q = 0.5$: the expression under the square root is zero; the roots are coincident and real.
3. $Q > 0.5$: the expression under the square root is negative; the roots are imaginary with a real part.

In the expression of Q given by (2.78), there are parameters that can change, while some are less likely to move. This is part of the design discussion to see whether these parameters can be considered real threats or are unlikely to affect the final result despite small variations. It is your responsibility, as a design engineer, to cover the cases where large variations of parameters are likely to happen. Whether they are due to operating conditions (e.g., temperature, bias points) or production spreads (e.g., tolerance, change of component), you must check their impact on the stability. Here, for the sake of the example, we will consider only the power stage zero brought by the output capacitor equivalent series resistance (ESR) described by (2.71). This parasitic element not only moves in relation to the capacitor temperature (the resistance increases as the temperature drops), it is also affected by a wide spread in production. Figure 2.10 shows the equivalent circuit of a typical capacitor. This model can be supplemented with other parasitics, such as a leakage resistance or an equivalent series inductance (ESL), but this simple approach will do for our example.

Let's assume a 470-μF output capacitor. Looking into the manufacturer datasheet, we found that its typical ESR is 90 mΩ at 25°C. However, it can vary

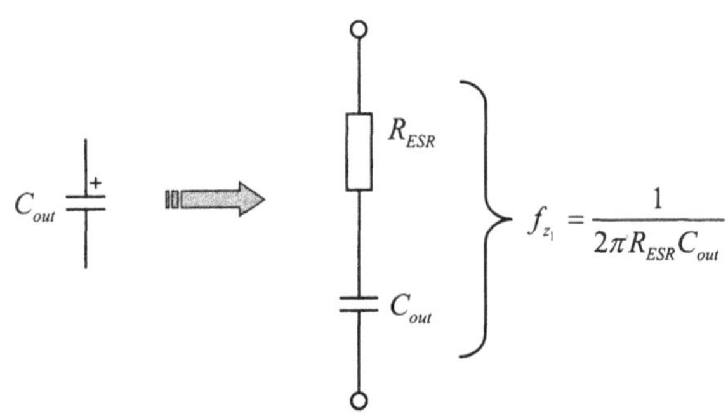

Figure 2.10 A typical capacitor always exhibits stray elements such as an ESR. This parasitic element introduces a zero in the transfer function.

from 50 mΩ to 200 mΩ if we consider a temperature range from –40°C to +105°C and production spreads. The zero given by (2.71) will thus move between

$$f_{z1,low} = \frac{1}{2\pi \times 200m \times 470u} = 1.7 \text{ kHz} \tag{2.81}$$

and

$$f_{z1,high} = \frac{1}{2\pi \times 50m \times 470u} = 6.8 \text{ kHz} \tag{2.82}$$

For the sake of the example, we will assume that the rest of elements constitutive of H and G have the following values:

$$f_{p_1} = 500 \text{ Hz}$$

$$f_{p_2} = 1000 \text{ Hz}$$

$$G_1 = 0.05$$

$$G_2 = 1000$$

Based on these numbers, the quality factor Q described by (2.78) will change as the zero modifies its position in relations to ESR variations. For the lowest zero position, we have

$$Q\big|_{f_{z1}=1.7 \text{ kHz}} = 0.312 \tag{2.83}$$

when the ESR reduces at higher temperature, the quality factor becomes

$$Q\big|_{f_{z1}=6.8 \text{ kHz}} = 0.97 \tag{2.84}$$

The ESR value for which Q equals 0.5 is found by solving for ω_{z_1}

$$\frac{\sqrt{\omega_{p_1}\omega_{p_2}(G_2G_1+1)}}{\omega_{p_1} + \omega_{p_2} + G_2G_1\omega_{p_1}\dfrac{\omega_{p_2}}{\omega_{z_1}}} = 0.5 \tag{2.85}$$

It happens when the zero reaches

$$f_{z_1} = \frac{G_2G_1\omega_{p_1}\omega_{p_2}}{2\sqrt{\omega_{p_1}\omega_{p_2}(1+G_2G_1)} - \omega_{p_1} - \omega_{p_2}}\frac{1}{2\pi} = 2.9 \text{ kHz} \tag{2.86}$$

With a 470-μF capacitor, it corresponds to an ESR of 116 mΩ.

As the ESR varies, it affects the quality factor and changes the nature of the roots. These roots can therefore be purely real; in that case, $Q = 0.312$, and the step response of the closed-loop transfer function is nonringing. According to (2.80), when we have a 1.7-kHz zero, these roots have the following values:

$$s_1 = -90k$$

$$s_2 = -11.1k \tag{2.87}$$

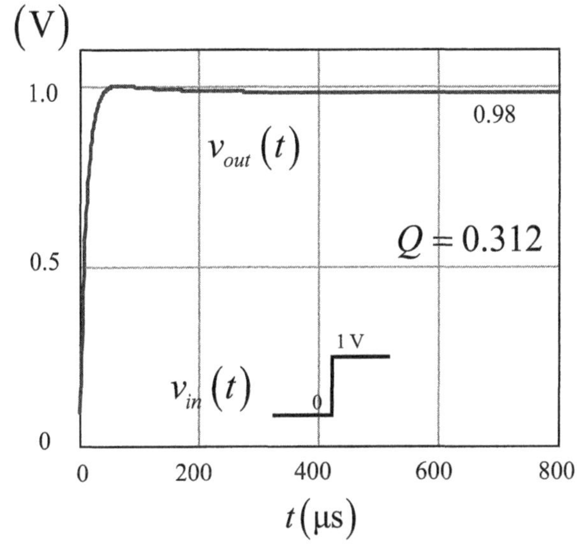

Figure 2.11 The step response for real roots is a nonringing signal with a very minor overshoot.

If we multiply (2.75) by $1/s$ and extract the time-domain response, we obtain, for these conditions, a signal plotted in Figure 2.11. Please note the presence of a light overshoot though, despite a Q less than 0.5. This overshoot is brought by the zero presence that affects the transient response but not the steady-state value. The final value is not 1 V as expected. This is normal; the closed-loop gain defined by (2.77) is less than 1 (0.98 exactly): we have a permanent *static error* of 20 mV.

When the ESR exactly equals 116 mΩ, both roots are coincident and equal:

$$s_1 = s_2 = -32k \tag{2.88}$$

The step response appears in Figure 2.12. Theory tells us that the step response of a second-order system featuring coincident poles should not bring overshoot, but

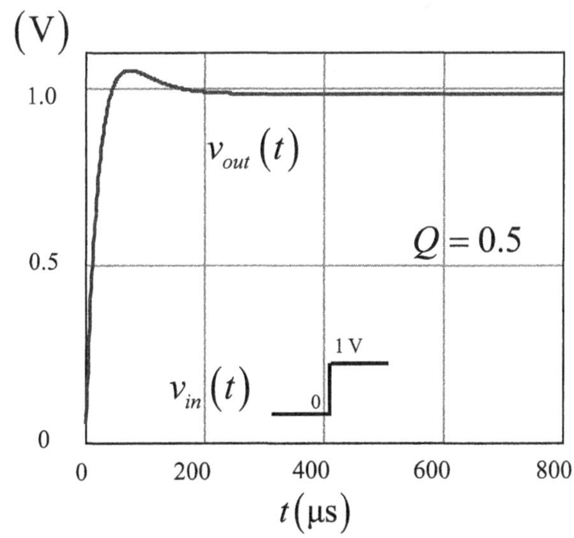

Figure 2.12 With coincident poles, the response is fast, and the overshoot is still weak.

we can see it in the picture. This is because the study considers only a system having only poles, not an extra zero as we have here. Its presence changes the transient response.

Now, when the ESR is really low, the roots become imaginary conjugates:

$$s_1 = -16.3k - j27.2k$$
$$s_2 = -16.3k + j27.2k \tag{2.89}$$

In that case, a more pronounced overshoot starts to appear, as shown in Figure 2.13.

If for any reason the ESR would become negligible (i.e., you chose a multilayer type of capacitor or parallel a lot of low-ESR capacitors), the roots would simply become

$$s_1 = -5.2k - j31.3k$$
$$s_2 = -5.2k + j31.3k \tag{2.90}$$

The corresponding step response exhibits more overshoot than the previous signal but is still stable. Figure 2.14 details the results:

From (2.90), we can compute the natural angular frequency ω_0 as

$$\omega_0 = |s_1| = |s_2| = \sqrt{5.2k^2 + 31.3k^2} = 31.73 \text{ krd/s or } 5.05 \text{ kHz} \tag{2.91}$$

A similar value is obtained if you use (2.79) or (2.89).

As can be observed, by changing the ESR values, the roots exhibit different real and imaginary coefficients. Actually, we can see that a decreasing ESR makes the quality factor grow while the real part of the roots decreases. This makes sense: the real part of the roots expresses losses that damp the second-order system. The Q increase is simply the result of these damping effects going down as the ESR vanishes.

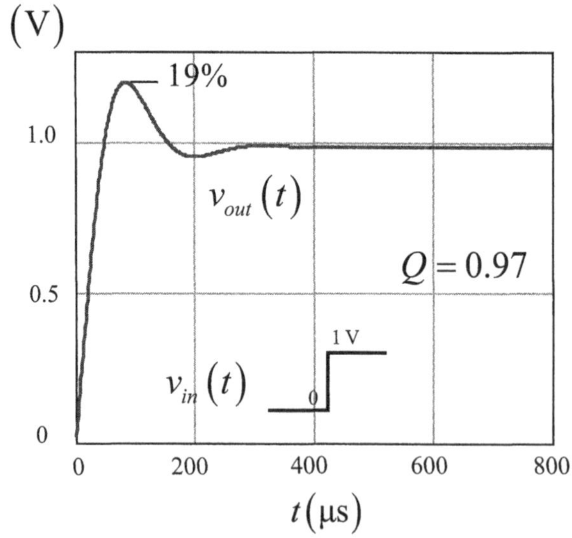

Figure 2.13 As the imaginary coefficients now appear, the closed-loop ac response starts to peak, inducing some ringing in the time-domain response.

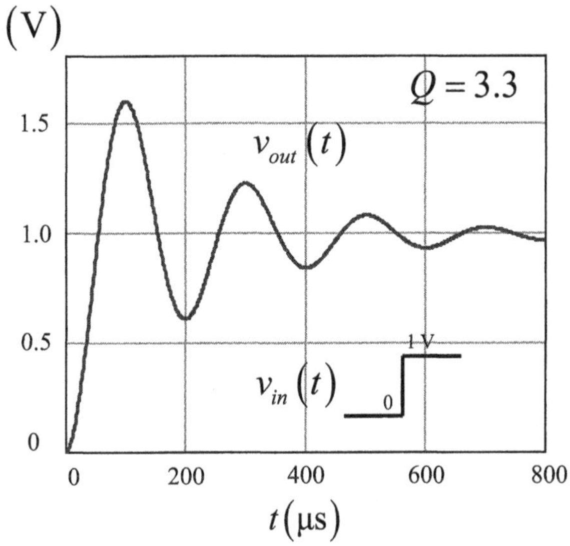

Figure 2.14 Despite a pronounced overshoot, the response is still stable.

2.4 S-Plane and Transient Response

In stability analysis, it is interesting to plot the trajectory of these roots and see how they move in relation to a considered parameter. In the present case, we have chosen the ESR of the output capacitor, but in most of the textbooks the selected parameter is a gain coefficient k. A typical example shown in the literature is that of Figure 2.15.

This is a unity-gain return control system, and it is easy to derive the transfer function from input to output:

$$\frac{V_{out}(s)}{V_{in}(s)} = \frac{kG(s)H(s)}{1+kG(s)H(s)} \tag{2.92}$$

G and H are individual transfer functions, made of a numerator and a denominator:

$$G(s) = \frac{N_G(s)}{D_G(s)} \tag{2.93}$$

$$H(s) = \frac{N_H(s)}{D_H(s)} \tag{2.94}$$

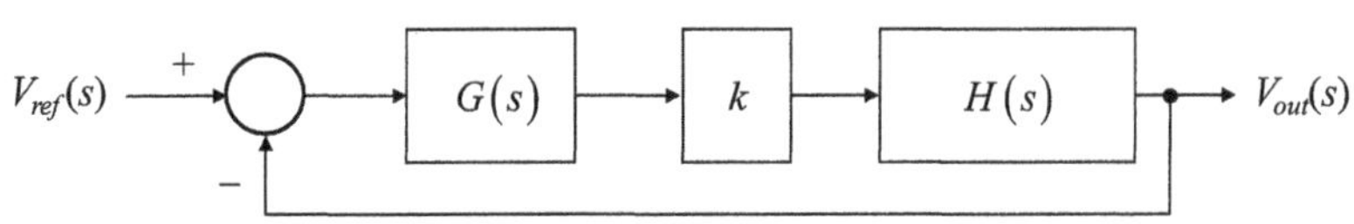

Figure 2.15 In this typical example, the parameter k is a gain inserted in the control loop and subject to wide variations.

If we now reinject these definitions in (2.92), we obtain:

$$\frac{V_{out}(s)}{V_{in}(s)} = \frac{k\frac{N_G(s)}{D_G(s)}\frac{N_H(s)}{D_H(s)}}{1+k\frac{N_G(s)}{D_G(s)}\frac{N_H(s)}{D_H(s)}} = \frac{kN_G(s)N_H(s)}{D_G(s)D_H(s)+kN_G(s)N_H(s)} \tag{2.95}$$

This equation shows that the poles and zeros put in $G(s)$ to shape the open-loop response for adequate crossover frequency and phase margin now appear in the closed-loop transfer equation with the terms $N_G(s)$ and $D_G(s)$. The characteristic equation often noted $\chi(s)$ (Chi, pronounced like "key") can be rewritten as:

$$\chi(s) = D_G(s)D_H(s) + kN_G(s)N_H(s) \tag{2.96}$$

This is an interesting equation because it shows that the zeros you will put in the compensator G (e.g., to boost the phase margin at low frequency) will turn into poles when the loop is closed. A low-frequency open-loop zero turning into a low-frequency closed-loop pole will slow down the response to an incoming perturbation or a setpoint change. If we now look at the parameter k, we can see several cases, depending on whether k is small or high:

- k is small, then the characteristic equation can simplify to $\chi(s) = D_G(s)D_H(s)$: the closed-loop poles are those already present in the open-loop gain equation.
- k now increases, and as the roots are a continuous function of k they move away from their open-loop definition to fully satisfy (2.96).
- If k increases further and becomes really high, then the left side of (2.96) can be neglected and the definition becomes: $\chi(s) = kN_G(s)N_H(s)$. The closed-loop poles are now given by the open-loop zeros!

It is important to check the values taken by the roots as k changes. For instance, are there some conditions where the real parts of the roots dangerously diminish, contributing to eliminate the damping? Can these roots suddenly reverse their negative sign and turn positive, making the system output diverging with all consequences behind? This makes sense, since as we discovered with (2.70) the poles directly influence the time-domain response of our system.

Network theory teaches us that the *complete* or *total* response of a system is generally the sum of its *natural* or *free* response $r_n(t)$ and its *forced* response $r_f(t)$. The free response is obtained by setting the input $u(t)$ to zero and considering nonzero initial conditions. On the contrary, the forced response is obtained by solely considering the input and setting all initial conditions to zero. The forced response in (2.70) was 166 μV, whereas the rest of the terms composed the natural response. The response of a SISO system is thus given by

$$y(t) = r_n(t) + r_f(t) = \sum_{i=1}^{n} C_i e^{p_i t} + r_f(t) \tag{2.97}$$

In this expression, p_i are the poles of the characteristic equation, and C_i are the coefficients of the exponential terms. The number of poles depends on the polynomial degree of the denominator: two poles for a second-order system, three for a third-order system, and so on. The easiest way to study these roots is to place them on a dedicated map called an Argand diagram, commonly denoted as the s-plane. The variable s is classically defined by

$$s = \sigma + j\omega \tag{2.98}$$

The s-plane is thus a two-dimensional graph where the vertical axis represents the imaginary portion of s, $j\omega$, and the horizontal axis, its real part, σ. The area of the plane corresponding to negative roots is called the left half plane (LHP), whereas the area situated to the right of the vertical axis is simply called the right half plane (RHP); the roots placed in this area have positive real parts. The poles are illustrated by a cross, ×, whereas the zeros are designated via a circle, ○. The location of the poles in the s-plane will affect the signal delivered by (2.98) as follows:

1. If the pole is real, $p_i = -\sigma$, it is placed in the LHP; it is called a LHP pole. Its contribution to the response is in the form $Ce^{-\sigma t}$, a decaying exponential component of duration $1/\sigma$. Therefore, if the pole location is far from the origin of the s-plane, then the response is a quickly decaying signal. For the opposite, as the pole's location moves closer to the 0 point, the response will be a slowly decaying signal. Suppose we have in the time-domain expression a term looking like $5e^{-0.1t}$; the signal starting with an amplitude of 5 V will decay and lasts 10 s before dying out to zero.
2. A pole appearing at the origin, $p_i = 0$, defines a term of constant amplitude, defined by the initial conditions: $5e^{-p_i t} = 5e^0 = 5$ V
3. If the pole is real but this time appears in the RHP, it is called a RHP pole and is of the form $p_i = \sigma$. It is a positive root. Its contribution to the output signal is of the form $Ce^{\sigma t}$, a continuously growing component. If you deal with a system featuring a closed-loop RHP pole, it is unstable.
4. When solving the characteristic equation roots, you can find conjugate pairs of roots, leading to a form $p_i = -\sigma \pm j\omega$. In this case, the contribution to the time-domain signal is a decaying sinusoidal signal of frequency ω obeying the form $Ae^{-\sigma t}\sin(\omega t + \varphi)$. A and φ are imposed by the initial conditions. Again, the real part σ represents the losses that damp the response. It is important to note that if the real part is 0, we have imaginary pole pairs of the form $p_i \pm j\omega$. This is an undamped oscillatory component of frequency ω.
5. If the complex pole pair is found to be in the RHP, $p_i = \sigma \pm j\omega$, the response is an exponentially growing sinusoidal signal.

A graphical representation of these various cases appears in Figure 2.16. This is a system featuring two poles, hence a second-order type. These poles could be that of a LC filter in which the quality factor is purposely adjusted for instance by an added resistance. In the first part (a), the poles are separate and purely real. The quality factor is very low; the system is overdamped. The response is similar to that of two cascaded RC filters. In the second plot (b), the poles are coincident; the quality factor is equal to 0.5. The response is faster but there is no overshoot as the roots do not include imaginary parts. In (c), the poles are split and represented

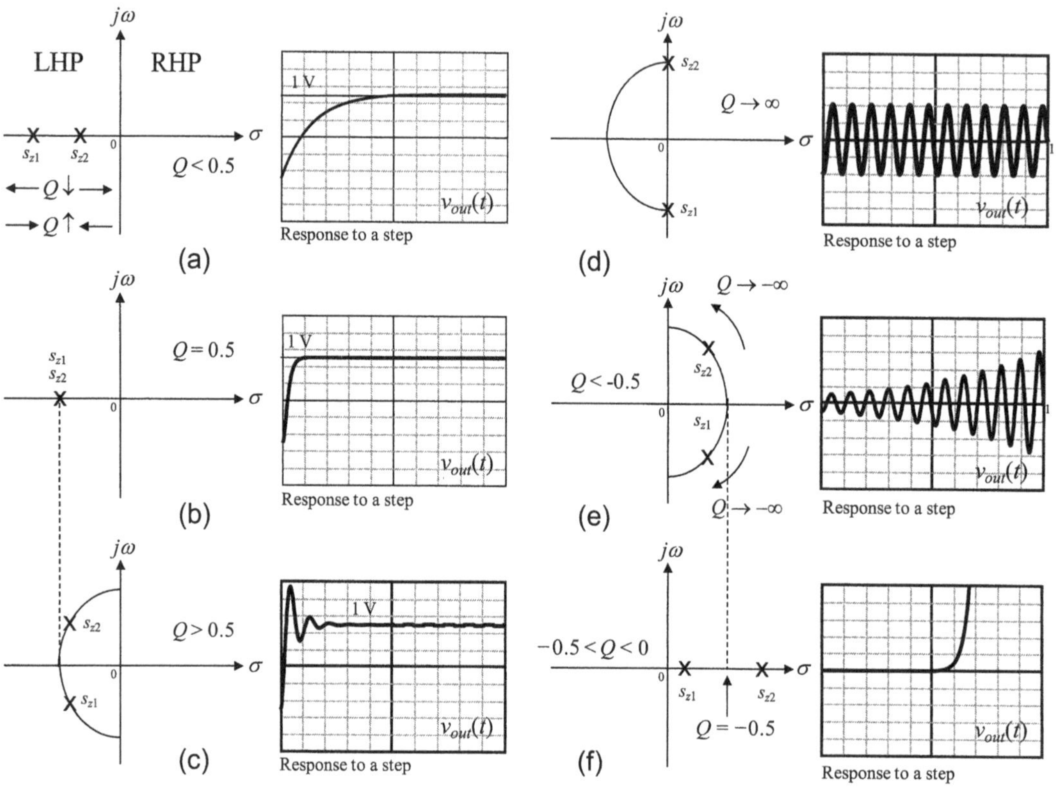

Figure 2.16 Depending on the closed-loop denominator roots position, the output response of a system converges if there are poles occurring in the LHP. If RHP poles appear, the system is unstable.

by conjugate roots. The quality factor is beyond 0.5. An imaginary portion is now present, and you see oscillations. However, the real parts provide the damping and represent the losses in the network that calm down the oscillations. This is a decaying signal. In (d), all losses have disappeared, and the system response is purely oscillatory. The roots are imaginary, and the real parts (the damping) have gone: we have sustained oscillations. In (e), the quality factor is negative and the poles have jumped in the RHP. The exponential exponent is now positive and oscillations grow, the system diverges. In (f), the imaginary portions have disappeared, the system still diverges but without oscillations.

Figure 2.17 shows another example where several poles and zeros are represented. They correspond to the following equation roots:

$$H(s) = \frac{\left[(s+2)^2+4\right]}{\left[(s+1)^2+1\right]}\frac{(s-1)}{(s+3)} \tag{2.99}$$

Please note than one of the zeros, s_{z3}, lies in the RHP; it is a positive root.

Based on what (2.97) taught us, the time-domain response of such a transfer function to a 1-V step is made of three terms (three poles) plus the forced response. The forced response is found by calculating the dc gain of (2.99):

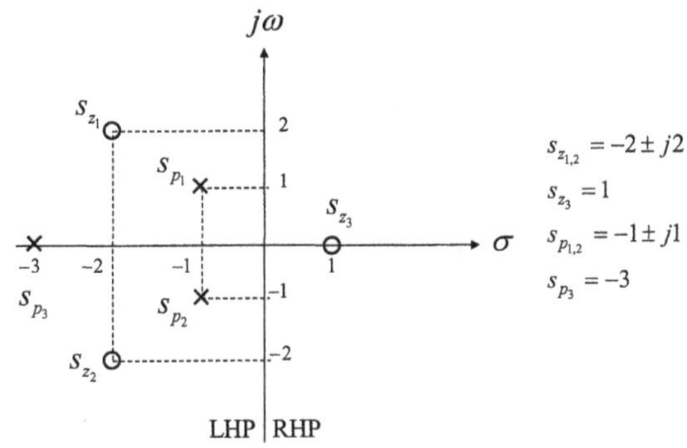

Figure 2.17 The s-plane allows the placement of poles and zeros and helps to check how they move in relationship to a selected parameter.

$$H(0) = \frac{\left[(2)^2 + 4\right](-1)}{\left[(1)^2 + 1\right](3)} = -\frac{4}{3} \tag{2.100}$$

The natural response is made of one pure decaying term (one real pole) and two other terms that are damped sinusoidal signals (conjugate poles pair):

$$\mathcal{L}^{-1}\left\{\frac{1}{s}H(s)\right\} = y_1(t) = \frac{4}{3}e^{-3t} + e^{-t}\cos(t) + 3e^{-t}\sin(t) - \frac{4}{3} \tag{2.101}$$

When all terms have died out to zero, the output is –1.33 V. As you can see, the zeros do not explicitly appear in the various terms, but they play a role in the coefficients—and also in the polarity: the output is negative. Why? We have three zeros and three poles. The two LHP poles and zeros phase lag/lead compensate for each other. The third pole lags by 90°, so what about the RHP zero (RHPZ)? It also lags by 90°, making a total of –180° or a polarity reversal, bringing a negative output for a positive 1-V step! The RHP zero, rather than adding a phase lead as a LHP zero would, further degrades the phase by lagging, just as a pole does. We will see that in more details in a few paragraphs.

Suppose now that s_{z_3} becomes a LHP zero, the response would transform into

$$y_2(t) = \frac{2}{3}e^{-3t} + e^{-t}\cos(t) + e^{-t}\sin(t) + \frac{4}{3} \tag{2.102}$$

The sign is now positive since we have an even number of zeros and poles, making the total phase change to 0°: a positive step gives a positive voltage. Please note that shifting of the zero into the LHP has affected the C coefficients but not the exponential terms that remain exclusively linked to the locations of the poles.

Now, assume the third pole becomes a RHP pole, $s_{p_3} = 3$ in (2.99) and the third zero is still in the LHP as in (2.102). The time-domain response can be expressed as follows:

$$y_3(t) = \frac{116}{51}e^{3t} + \frac{1}{17}e^{-t}\cos(t) - \frac{13}{17}e^{-t}\sin(t) - \frac{4}{3} \tag{2.103}$$

The first term is enough to make the system unstable: the exponent is positive and forces the expression to diverge as t increases. This is the RHP pole effect. If you look at the last term, it is negative again: the two LHP pole/zero pairs compensate each other with a total 0° phase. However, the third LHP zero adds a 90° phase lead that is normally compensated by 90° lag brought by a LHP pole. Here, as this pole is in the RHP, it brings another 90° phase lead, whereas it should be a lag, as with a LHP pole: the total phase lead is now 180°, or another phase reversal. However, as the first positive exponential term immediately dominates the response, you will not see the negative output.

We have graphed all three responses in Figure 2.18. As expected, the RHP pole brings a diverging output. The RHP zero, despite bringing an additional phase lag, does not make the system diverging. As a preliminary conclusion, if a RHP zero forces you to be more careful when compensating the converter, a single RHP pole in the characteristic equation (thus a closed-loop pole) is absolutely insurmountable.

2.4.1 Roots Trajectories in the Complex Plane

The analysis of the characteristic equation (the closed-loop transfer function denominator) is usually carried at a certain operating point. However, we have seen that variables affecting the characteristic equation can move. This is often the case for various gain values, as k in the previous example. At the time scientists studied power electronics in the 1950s, valve-based amplifiers could have wide gain dispersions at warm-up or at various ambient temperatures, and stability could be affected. A significant benefit of feedback was to program the gain via an external set of resistors, making the final chain gain insensitive to the amplifier gain varia-

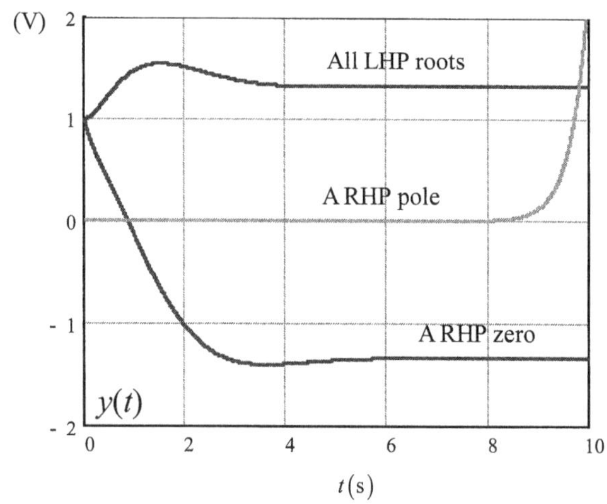

Figure 2.18 Various time-domain outputs depending where the third poles and zeros are in the s-plane.

tions. Besides gain variations, there can be perturbations effects such input voltage or load changes but also open-loop poles or zeros that can slide along the frequency axis when production spreads or loading conditions are involved. In our example from Figure 2.9, sweeping the output capacitor ESR value degraded the transient response. Mathematically, it meant that the roots evaluated at different ESR cases have changed their values. If we would plot these roots in the *s*-plane and link all points together, we would obtain a so-called Evans *root-locus* representation, named after the work of W. R. Evans in the 1950s. By looking at the path taken by the poles, we can check if certain values of the swept parameter bring the roots close to the vertical axis (pure imaginary roots, no damping) or, even worse, make them jump on the right side of the *s*-plane.

Several mathematical programs can compute and plot the poles or zeros on the *s*-plane when a given parameter is swept. Mathcad is one of them with which root locus graphs are very simple to obtain. For instance, Figure 2.19 depicts a root locus plot for (2.80) as the zero s_{z1} is swept from 1.5 kHz to more than 100 kHz. The appendix at the end of this chapter shows how we plotted this picture.

Should the roots approach the imaginary axis or worse, move to the right-half place section, stability would be at stake, as exemplified in Figure 2.16. Fortunately, on the drawing, we see that despite an ESR zero going to almost infinity (the ESR vanishes to 0), the roots always remain in the left plane, guaranteeing some damping. Calculations show that in these conditions, Q does not exceed 3.5 and the step response is stable despite a pronounced overshoot. The trajectory describing the loci of the poles can reveal a lot of information, but the study of the technique is outside the scope of this book. The reader interested in an in-depth analysis of this method will find a lot of information in Chapters 13 and 14 of [6].

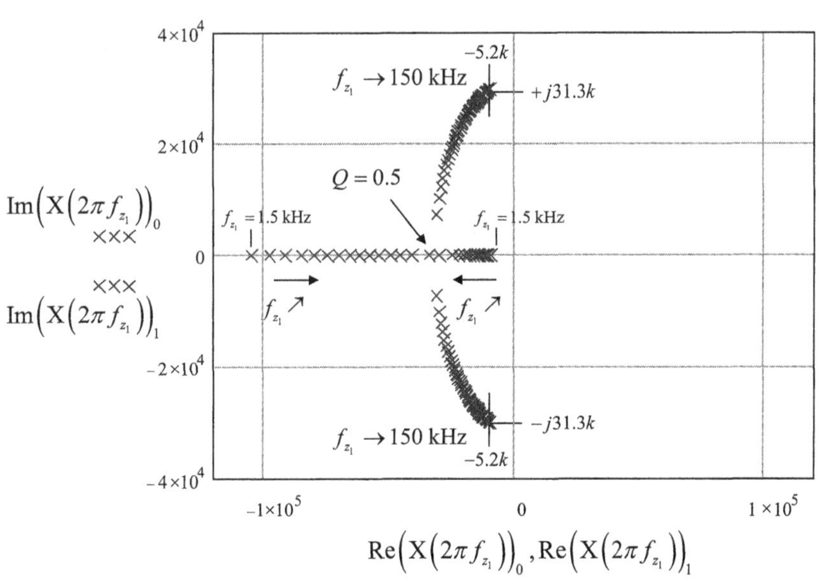

Figure 2.19 In this picture, we can clearly see how the roots move as the ESR zero is swept from 1.5 kHz up to 150 kHz.

2.5 Zeros in the Right Half Plane

Right half plane zeros are usually found in switching converters transferring the energy to the load in two steps: the energy is first stored in the inductor during the on-time and then dumped into the load during the off-time. During the on-time, the load is isolated from the input source while the inductor is energized. During the off-time, the energy stored in the inductor is released to the load: boost, buck-boost, or flyback converters operate this way. These three converters are called indirect energy-transfer converters. They all obey the two-step conversion process we described. For the opposite, the buck converter is a direct energy-transfer converter. You do not need an intermediate step to store energy before transmitting it to the source. This intermediate storing step *actually creates a delay* because the controller must always go through an energy-storing step before answering the increase in the delivered power need. Should it take time to store more energy while the power demand is fast, the converter cannot momentarily keep up its power delivery and the output voltage falls. This event lasts until the stored energy in the inductor has increased.

The output variable momentarily going in the opposite direction compared to what the control expects is the typical signature of a transfer function featuring zero located in the right-half portion of the s-plane, also called a RHP zero or RHPZ.

2.5.1 A Two-Step Conversion Process

Figure 2.20(a) represents a classical boost converter where two switches appear: a power switch SW, usually a MOSFET, and a diode, sometimes called the catch diode. In the continuous conduction mode (CCM) of operation, the inductor current i_L flows in the power switch SW during the on-time or dT_{sw}, where d is the instantaneous duty ratio and T_{sw} the switching period. During the off-time, or $(1 - d)T_{sw}$, the power switch is open and the output diode routes the current to an output network made of the capacitor and the load. Regardless of the control method, voltage or current-mode, this configuration assumes that energy is first stored in the inductor during the on-time and then transferred to the output during the off-time.

Figure 2.20(b) shows an equivalent representation of the boost converter, where the switch/diode network has been replaced by a single pole double throw switch

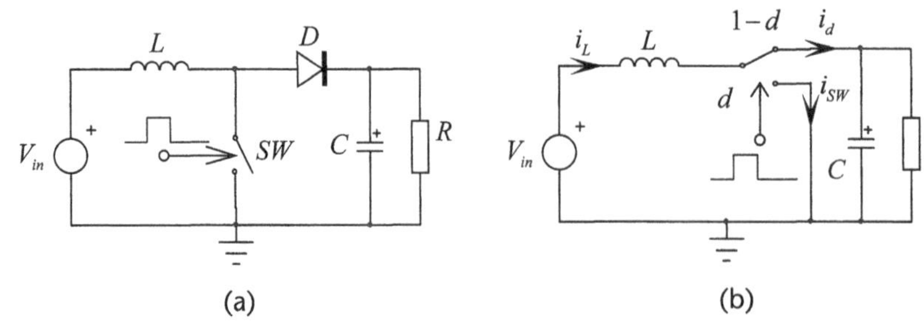

Figure 2.20 A boost converter features two power switches. They can be replaced by a single-pole double-throw switch that represents the operations of the diode and the transistor.

that alternatively routes the inductor current in the two different branches: the power switch or the output diode. If a designer would observe the currents circulating in the output diode, he or she would see a typical waveform shown in bold in Figure 2.21. Our boost converter is designed to deliver power to a given load. The variable of interest, in our case, is thus the available output current I_{out}. This current is actually made of a dc portion on which is superimposed a switching ripple. In theory, the ripple goes into the capacitor and the dc current circulates in the load. The dc current delivered by the boost converter is nothing other than the diode average current I_d. Mathematically, this current can be expressed by

$$I_{out} = I_d = I_L(1 - D) \tag{2.104}$$

where I_d is the average diode current also equal to the dc output current I_{out}. D represents the averaged duty ratio.

If we graph the circulating currents in the diode at the switch opening, we obtain the drawing presented in Figure 2.21. When the switch opens, the current no longer circulates in the power switch but is routed, via the diode, to the output. The average value—or the dc current—circulating in the load is the area of the surface A_0, averaged over a switching cycle T_{sw}.

$$I_{out} = \frac{A_0}{T_{sw}} \tag{2.105}$$

Now, if a sudden current demand occurs on the output, the controller senses the transient and immediately increases the duty ratio by a small value— $\hat{d}$—to build more energy in the inductor. This is what Figure 2.22 shows.

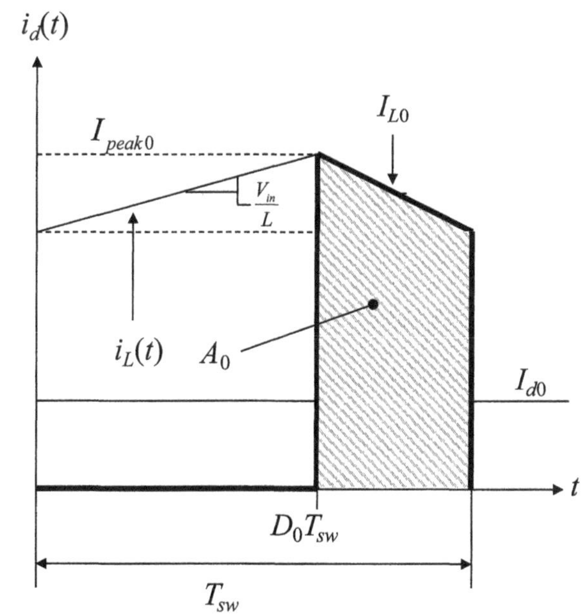

Figure 2.21 The current observed in the output diode at the beginning of the event.

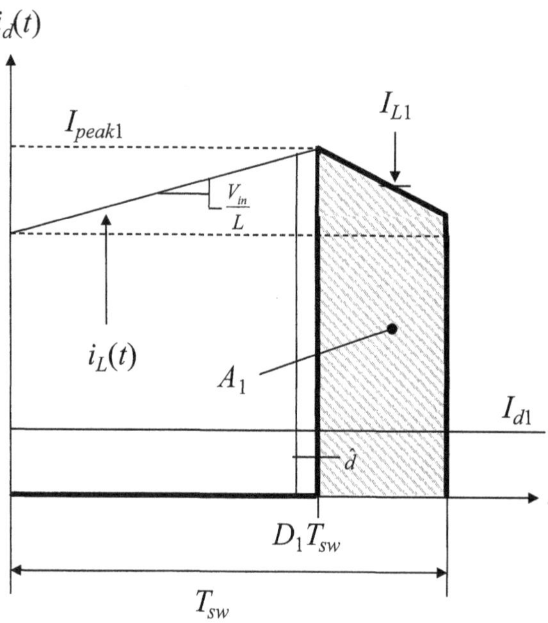

Figure 2.22 As an answer to the output current demand, the controller asks the inductor to store more energy.

In theory, the new surface, A_1, should be larger than A_0 to cope with the output power demand increase. However, as the switching period is fixed, the increase in the on-time duration simply shortens the off-time interval. As this is the time during which the current circulates in the diode to satisfy $A_1 > A_0$, the only condition is that the new peak current I_{peak1} is larger than the first one, I_{peak0}. This is what is sketched in Figure 2.23.

2.5.2 The Inductor Current Slew-Rate Is the Limit

What is the pace at which the average inductor current can change? Lenz's law instructs us that the instantaneous current change rate in an inductor obeys the following formula:

$$\frac{di_L(t)}{dt} = \frac{v_L(t)}{L} \tag{2.106}$$

On average, over a switching cycle, it simply follows

$$\left\langle \frac{di_L(t)}{dt} \right\rangle = \frac{\langle v_L(t) \rangle}{L} = \frac{V_L}{L} \tag{2.107}$$

The exercise now consists of calculating the average value across our inductor. By considering the weighted period of time during which V_{in} or $V_{out} - V_{in}$ are applied across L, we have

$$V_L = V_{in}D - (V_{out} - V_{in})(1 - D) = V_{out}(D - 1) + V_{in} \tag{2.108}$$

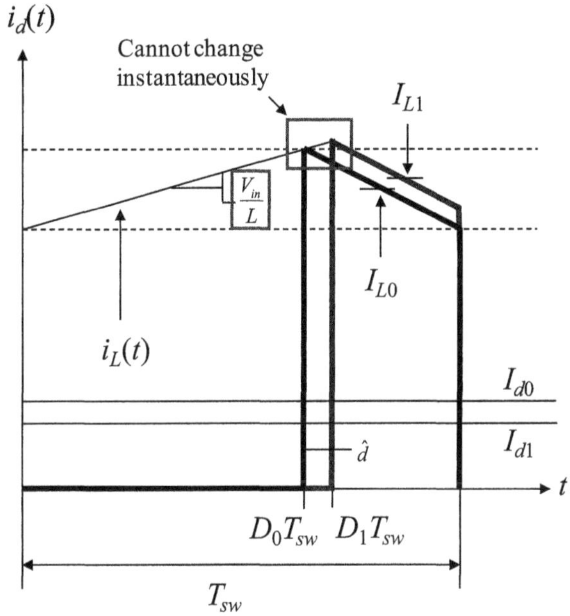

Figure 2.23 As the on-time reduces the off-time duration, the only way to pass more power is to make sure the new peak current is larger than the first one. Unfortunately, the inductor opposes current changes.

Let's assume the following boost operating parameters:

$V_{in} = 10$ V
$D_0 = 0.583$
$V_{out} = V_{in}/(1 - D_0) = 24$ V
$R_{load} = 240\ \Omega$
$L = 1$ mH
$F_{sw} = 100$ kHz

With a 58.3 percent duty ratio, the converter delivers 24 V. We are in steady state and (2.108) gives 0. Now suppose that the duty ratio jumps to $D_1 = 59$ percent or a difference of 0.7 percent. What is the inductor average current slope in this case? Considering a large output capacitor, the output voltage stays constant during the duty ratio change. Applying (2.108) gives a transient average inductor voltage of

$$V_L = V_{out}(D_1 - 1) + V_{in} = 24 \times (0.59 - 1) + 10 = 160 \text{ mV} \tag{2.109}$$

Back to (2.107), the maximum average current slope authorized by the inductor is therefore

$$\left\langle \frac{di_L(t)}{dt} \right\rangle = \frac{V_L}{L} = \frac{160m}{1m} = 160\ \mu\text{A}/\mu\text{s} \tag{2.110}$$

a rather modest value.

When the duty ratio changes from 58.3 percent to 59 percent, it implies an output voltage change of

$$V_{out} = \frac{V_{in}}{1-D} = \frac{10}{1-0.59} = 24.39 \text{ V} \tag{2.111}$$

With a constant 240-Ω load, the output current will increase to

$$I_{out} = \frac{V_{out}}{R_{load}} = \frac{24.39}{240} = 101.6 \text{ mA} \tag{2.112}$$

Brought back to the inductor change, the output current variation given by (2.112) must be accompanied by an average inductor current variation of

$$\Delta I_L = \frac{V_{in}}{R_{load}}\left[\frac{1}{(1-D_1)} - \frac{1}{(1-D_0)}\right] = \frac{10}{240}\left[\frac{1}{(1-0.59)} - \frac{1}{(1-0.583)}\right] = 1.7 \text{ mA} \tag{2.113}$$

Given an average inductor slope 160 μA/μs, this current variation will only be possible within a timeframe of

$$dt = \frac{1.7m}{160u} = 10.6\ \mu\text{s} \tag{2.114}$$

If the duty ratio is swept from 58.3 percent to 59 percent in less than 10.6 μs, the inductor current will not build up at a sufficient pace to make the output current rise at the same speed. As an immediate result, the output current drops rather than increases. On the contrary, if the duty ratio sweep is slow enough, the current can increase in the inductor at sufficient speed to compensate the reduction in $(1 - d)$: the output voltage goes up. How do we make sure the inductor will always have time to build up enough current in case a fast transient occurs? By simply rolling off the crossover frequency or, in other terms, severely limiting the converter bandwidth so that fast transient demands never translates into a fast duty ratio change. If you fail to limit the bandwidth, in a fast output power transient demand, the output voltage will go down despite an increase of duty ratio. In control theory, you have reversed the control loop and oscillations occur. This situation lasts until the current in the inductor builds up to the right value. To prevent this from happening, the RHP zero effect naturally limits the available bandwidth for a given converter. Intuitively, a converter featuring a large inductor, hence operating in a deep CCM, will have a low-frequency RHP zero, severely hampering its possible response time.

2.5.3 An Average Model to Visualize RHP Zero Effects

Average models lend themselves very well to illustrating the effects of a RHP zero. We have built an open-loop boost converter around an auto-toggling model described in [7]. Figure 2.24 portrays the adopted schematic. In this test fixture, a 1-mH inductor driven at a 100-kHz switching period delivers 100 mA to a load (V_{out} = 24 V). The duty ratio is first slowly swept between 58.3 percent and 59 percent. As shown in Figure 2.25, the inductor current nicely follows the demand,

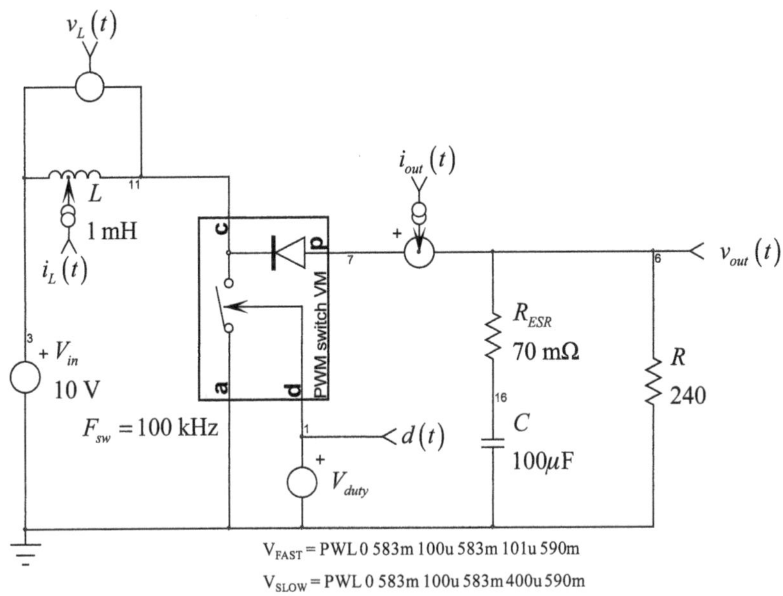

Figure 2.24 A voltage-mode average model whose control input is swept at two different paces can demonstrate the existence of a RHPZ in the CCM-operated boost converter.

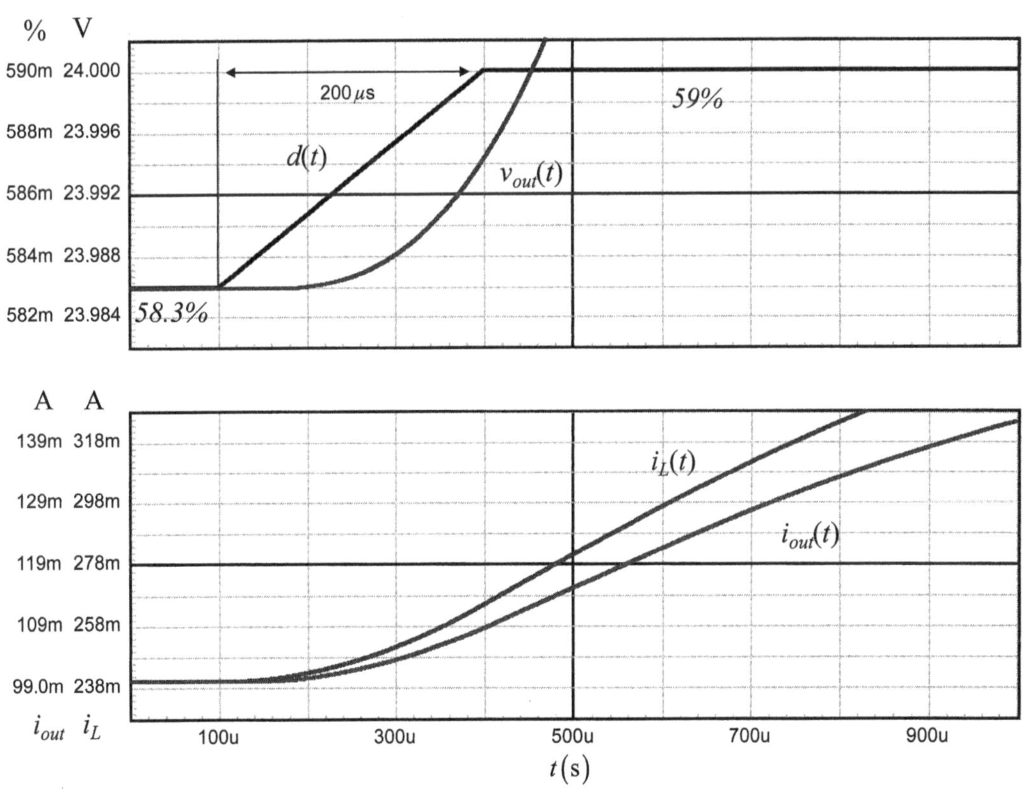

Figure 2.25 When the duty ratio slowly changes and gives time to the inductor current to build up, the output voltage variation is positive, as it should be.

and the output voltage slope is always positive. If we now sweep the duty ratio at a higher speed, Figure 2.26 indicates that despite a regular slope in the inductor current, it does not build up at a sufficient pace to answer the output current demand. As a result, both $v_{out}(t)$ and $i_{out}(t)$ drop. If the system would operate in closed loop, oscillations would occur because the control law is reversed: the duty ratio increases but the output voltage goes down.

How do we prevent this problem from happening? Well, a solution is to clamp the maximum slew rate on the duty ratio control input. In that way, even if a sudden variation is detected on the output, the error voltage will always rise at a speed where the inductor volt-second limit is never reached, giving sufficient time for the inductor current to build up. How do we limit the slew rate? By rolling off the crossover frequency f_c at a position usually well below the worst-case RHP zero position.

2.5.4 The Right Half Plane Zero in the Boost Converter

We have shown the consequences of a RHPZ in a boost converter. It is now interesting to analytically derive its position in the transfer function of the boost converter. To simplify the analysis, we will consider a voltage-mode control. The reader interested in current-mode control and compensation examples can find more informa-

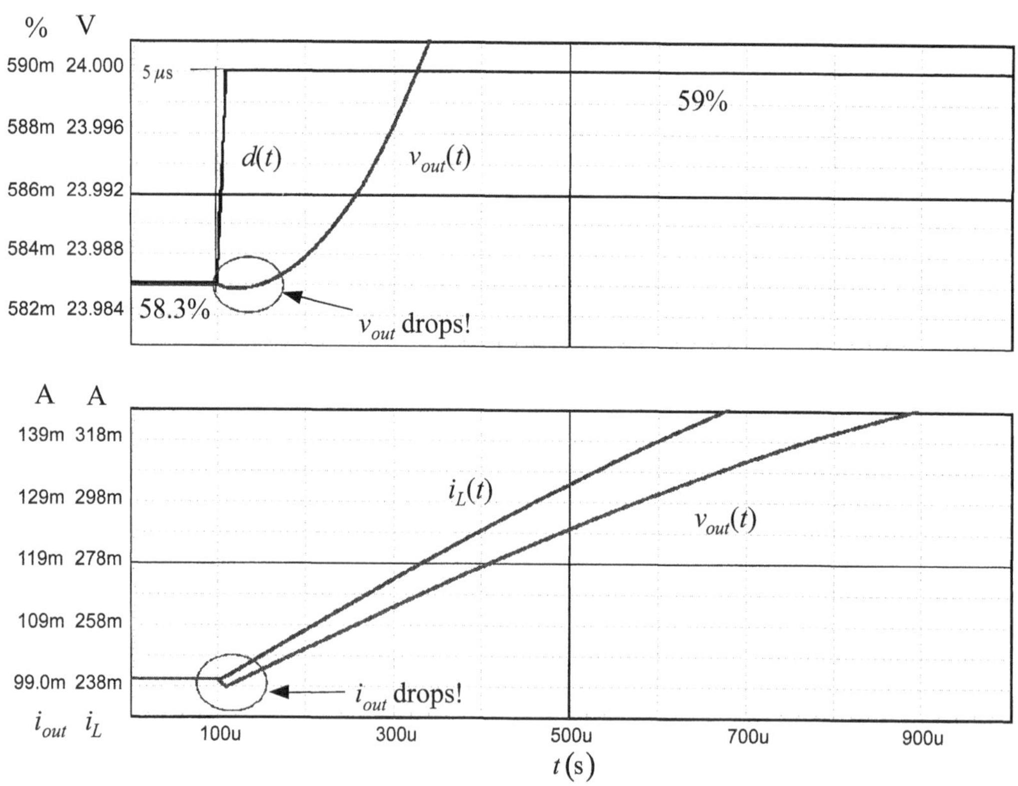

Figure 2.26 In this example, the inductor current builds up too slowly with regard to the output current demand. As a result, the output current drops until the current in the inductor builds up to the right value.

tion in [8]. To derive the transfer function, we can start from the output current expression given in (2.104). This is a large-signal (nonlinear) equation that we must transform into a small-signal expression. The fastest way to do that is to find partial derivatives coefficients for each of the variables, the duty ratio D and the inductor current I_L:

$$\hat{i}_{out} = \left(\frac{\partial I_{out}}{\partial I_L}\hat{i}_L\right)_D + \left(\frac{\partial I_{out}}{\partial D}\hat{d}\right)_{I_L} = \hat{i}_L(1-D) - \hat{d}I_L \tag{2.115}$$

In this equation, the ac inductor current $\hat{i}_L$ appears. What is the expression of an ac inductor current? Simply, it's the ac inductor voltage divided by the inductor impedance. Let us find the expression of the ac inductor voltage by first deriving its average large signal expression, already found in (2.109):

$$V_L = V_{out}(D-1) + V_{in} \tag{2.116}$$

On average, when the converter is at the equilibrium, this equation gives zero. However, under an ac excitation, the average inductor voltage is also ac modulated around zero. By using a partial derivative, we can see that the ac inductor voltage, in this case, is expressed by

$$\hat{v}_L = \hat{v}_{out}(D-1) + \hat{d}V_{out} \tag{2.117}$$

In this equation, the input term V_{in} has disappeared since the input voltage is considered constant during the ac analysis. Furthermore, if we consider a large output capacitor, its impedance at the ac excitation can be considered close to zero. In this case, if we consider $\hat{v}_{out} \approx 0$, we can further simplify the expression:

$$\hat{v}_L \approx \hat{d}V_{out} \tag{2.118}$$

With the ac inductor voltage on hand, it is easy to obtain the ac inductor current we are looking for:

$$\hat{i}_L(s) = \frac{\hat{v}_L(s)}{Z_L} = \frac{\hat{d}(s)V_{out}}{sL} \tag{2.119}$$

Substituting (2.119) in (2.115) gives the final ac output current expression:

$$\hat{i}_{out}(s) = \frac{\hat{d}(s)V_{out}}{sL}(1-D) - \hat{d}(s)I_L \tag{2.120}$$

The average inductor current I_L is the source current I_{in}. Considering a 100 percent efficiency power conversion, we can write

$$V_{in}I_{in} = V_{out}I_{out} = \frac{V_{out}^2}{R} \tag{2.121}$$

From which we have

$$I_{in} = I_L = \frac{V_{out}^{2}}{V_{in}R_{load}} = \frac{V_{out}}{V_{in}}\frac{V_{out}}{R_{load}} = \frac{V_{out}}{(1-D)R_{load}} \tag{2.122}$$

Substituting (2.122) in (2.120), we obtain

$$\frac{\hat{i}_{out}(s)}{\hat{d}(s)} = \frac{V_{out}D'}{sL} - \frac{V_{out}}{D'R_{load}} \tag{2.123}$$

Now factoring the first term and rearranging, we have

$$\frac{\hat{i}_{out}(s)}{\hat{d}(s)} = \frac{V_{out}D'}{sL}\left(1 - \frac{sL}{D'^{2}R_{load}}\right) = \frac{\left(1 - \frac{s}{\omega_{z2}}\right)}{\frac{s}{\omega_0}} \tag{2.124}$$

where

$$\omega_0 = \frac{V_{out}D'}{L} \tag{2.125}$$

$$\omega_{z_2} = \frac{R_{load}D'^{2}}{L} \tag{2.126}$$

This expression links the output current to the duty ratio input. In this equation, we can see a pole at the origin given by the inductor L and a zero featuring a positive root: this is the RHPZ ω_{z_2} we are looking for. Please note that both roots depend on the duty ratio and are moving in relation to the input/output conditions.

If we apply the boost converter numerical values from Figure 2.24, we have the following positions:

$$f_0 = 1.6 \text{ kHz} \tag{2.127}$$

$$f_{z_2} = 6.6 \text{ kHz} \tag{2.128}$$

Our interest now lies in the phase lag brought by this transfer function. The argument of a quotient is the numerator argument minus that of the denominator:

$$\arg\left[\frac{\hat{i}_{out}(\omega)}{\hat{d}(\omega)}\right] = \arg[N(\omega)] - \arg[D(\omega)] = \tan^{-1}\left(-\frac{\omega}{\omega_{z_2}}\right) - \tan^{-1}(\infty) \tag{2.129}$$

In dc, for $\omega = 0$, this expression becomes

$$\lim_{\omega \to 0} \arg\left[\frac{\hat{i}_{out}(\omega)}{\hat{d}(\omega)}\right] = \tan^{-1}(0) - \tan^{-1}(\infty) = -90° \tag{2.130}$$

The presence of the origin pole justifies this permanent phase lag of 90°. Now, with a normal zero, as the frequency increases, we would expect its contributing phase to reach 90°, canceling the origin pole action. Unfortunately, because this is a RHP zero, the total phase lag at $\omega = \infty$ becomes

$$\lim_{\omega \to \infty} \arg\left[\frac{\hat{i}_{out}(\omega)}{\hat{d}(\omega)}\right] = \tan^{-1}(-\infty) - \tan^{-1}(\infty) = -90° - 90° = -180° \tag{2.131}$$

This is the effect of the RHPZ: its phase lag is 90° compared to the 90° phase lead brought by a LHP zero. To obtain a complete picture, we can obtain the transfer function using the average model of Figure 2.24 but also using (2.124) and a calculation software such as Mathcad. Figure 2.27 shows the plots we have obtained. The superposition of both curves confirms the validity of the analytical equation we have derived. Also, as expected, the total phase lag reaches 180°. This is the effect of the origin pole and the RHP zero, which brings an additional –90° rather than 90° as it should for a normal zero. Please note that this RHPZ also exists, at the same location, in a current-mode boost converter.

In this example, should we need to stabilize the converter, the RHPZ position clearly bounds the maximum crossover frequency. To avoid any stability issue, it is recommended to limit the crossover frequency f_c to less than 30 percent of the minimum RHPZ position. In our example, it would imply a crossover frequency of

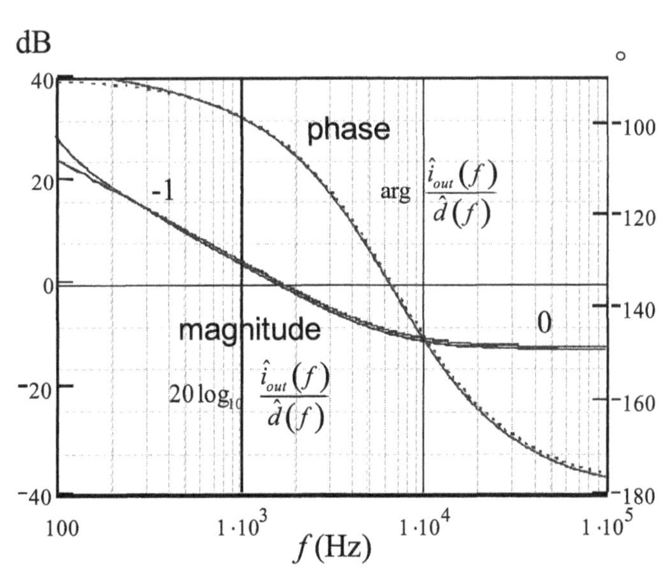

Figure 2.27 The total phase lag reaches 180° whereas it should be 0° if the zero of the denominator would lie in the left-half portion of the s-plane.

$$f_c < 30\% \, f_{z_2} < 0.3 \times 6.6k < 2 \text{ kHz} \tag{2.132}$$

The compensation block G must thus be tailored to force crossover below this value.

One final note on RHPZ: The previous example assumes CCM to illustrate the presence of the RHPZ. It is little known that a RHPZ can also exist in the discontinuous conduction mode (DCM) of operation. However, as it is located in the higher portion of the frequency spectrum, its influence is usually neglected at lower frequencies where crossover takes place.

2.6 Conclusion

Understanding transfer functions is key for the design of fast and stable closed-loop systems. Even if you will never do root locus calculations, it is important to realize that poles positions can move in relation to certain operating parameters. Once these parameters are identified (e.g., our output capacitor ESR), you will know how to efficiently compensate these variations over the power supply lifespan, ensuring a robust design. The RHP zero presence can sometimes hamper the available bandwidth in converters like boost or flyback architectures. Again, being able to analytically locate its worst-case position and visualize its effects on a Bode plot is an important point to let you safely pick a crossover frequency value. Finally, we have seen how fast analytical techniques can dramatically improve your analysis speed and unveil poles and zeros in a few minutes. It requires dexterity and practice but once you master the technique, going back to classical brute force algebra will be difficult!

References

[1] Middlebrook, R. D., "Methods of Design-Oriented Analysis: Low-Entropy Expressions," New Approaches to Undergraduate Education IV, University of California, Santa Barbara, 1992.

[2] Middlebrook, R. D., V. Vorpérian, and J. Lindal, "The *N* Extra Element Theorem," *IEEE Transactions on Circuits and Systems, Fundamental Theory and Applications*, Vol. 45, No. 9, September 1998.

[3] Cochrun, B., and A. Grabel, "A Method for the Determination of the Transfer Function of Electronic Circuits," *IEEE Transactions on Circuit Theory*, Vol. 20, No. 1, January 1973.

[4] Erickson, R. W., "The *n* Extra Element Theorem," http://ecee.colorado.edu/copec/publications.php.

[5] Vorpérian, V., *Fast Analytical Techniques for Electrical and Electronic Circuits*, Cambridge: Cambridge University Press, 2002.

[6] DiStefano, J., A. Stubberud, and I. Williams, *Feedback and Control Systems*, New York: McGraw-Hill, 1990.

[7] Basso, C., *Switch Mode Power Supplies: SPICE Simulations and Practical Designs*, New York: McGraw-Hill, 2008.

[8] Basso, C., "Understanding the RHPZ," Parts I, II, III, and IV, *Power Electronics and Technology*, April, May, June, and July 2009.

Appendix 2A: Determining a Bridge Input Impedance

We are going to use an example given by Dr. Vatché Vorpérian on page 12 of [1]. The circuit diagram appears in Figure 2.28:

The exercise consists of determining the impedance $Z_{in}(s)$ seen from the left side of Figure 2.28:

$$Z_{in}(s) = \frac{V_{in}(s)}{I_{in}(s)} \tag{2.133}$$

If Dr. Vorpérian used the extra element theorem (EET) to obtain the result, we are going to apply the same technique we used in this chapter to find the input impedance. There is one storage element, the capacitor C; this is thus a first-order system. Its general expression can be put under the following form. Analysis will further tell us if a pole or a zero are present:

$$Z_{in}(s) = R_0 \frac{1 + s/\omega_{z1}}{1 + s/\omega_{p1}} \tag{2.134}$$

First, let us derive the input resistance seen at dc, R_0: open-circuit the capacitors and short-circuit the inductors, if any. Our circuit simplifies to that of Figure 2.29.

The input resistance is simply R_1 in series with the series-parallel combination of the remaining elements:

$$R_0 = R_1 + (R_3 + R_4) \,||\, R_2 \tag{2.135}$$

Now, let's see if a zero exists. A zero in this circuit would prevent the excitation signal from reaching the output. As we deal with an impedance expression, the excitation signal is the input current I_{in} and the response is the input voltage V_{in}. What in Figure 2.28 would nullify V_{in}? A short circuit involving the branch where C lies.

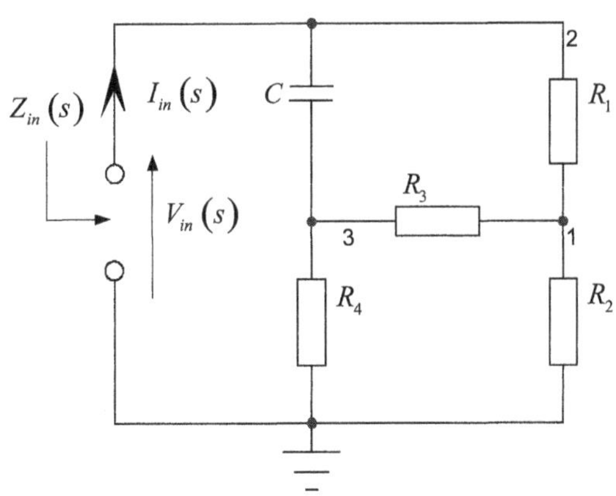

Figure 2.28 The input impedance of this bridge can be derived in a few steps when using analytical techniques.

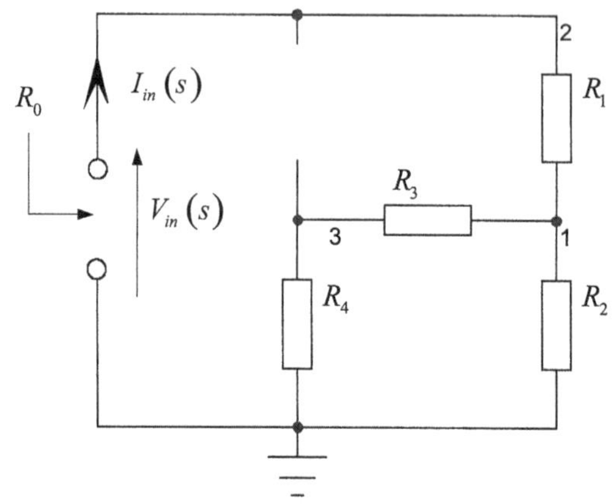

Figure 2.29 The dc input resistance, when the capacitor is removed, is really easy to find.

If we have a short circuit, then $V_{in} = 0$. If $V_{in} = 0$, then node 2 is grounded and R_1 comes in parallel with R_2. The circuit appears in Figure 2.30.

The expression seen from the input impedance port is simply

$$Z_{in}(s) = \frac{1}{sC} + R_4 \parallel (R_3 + R_2 \parallel R_1) = \frac{1 + sCR_4 \parallel (R_3 + R_2 \parallel R_1)}{sC} \tag{2.136}$$

To cancel this expression, we set its numerator to zero and solve for the root. In our case, this is simply

$$1 + sCR_4 \parallel (R_3 + R_2 \parallel R_1) = 0 \tag{2.137}$$

Thus,

$$\omega_{z1} = \frac{1}{sCR_4 \parallel (R_3 + R_2 \parallel R_1)} \tag{2.138}$$

Now that the zero has been found, let's give a look to the pole. To derive the denominator expression $D(s)$, we need to set the excitation to zero and find the time

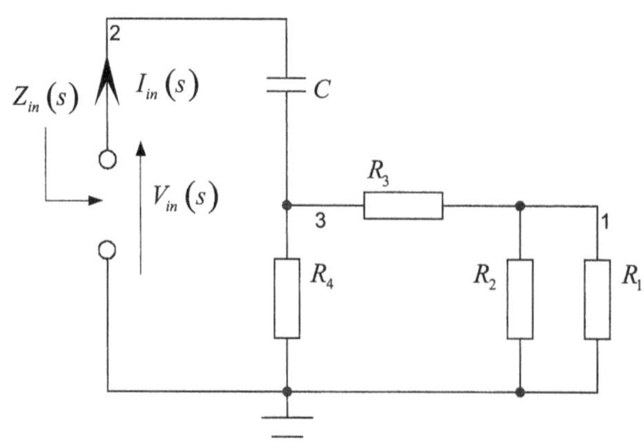

Figure 2.30 The zero expression is found by nullifying the response signal, which is V_{in}.

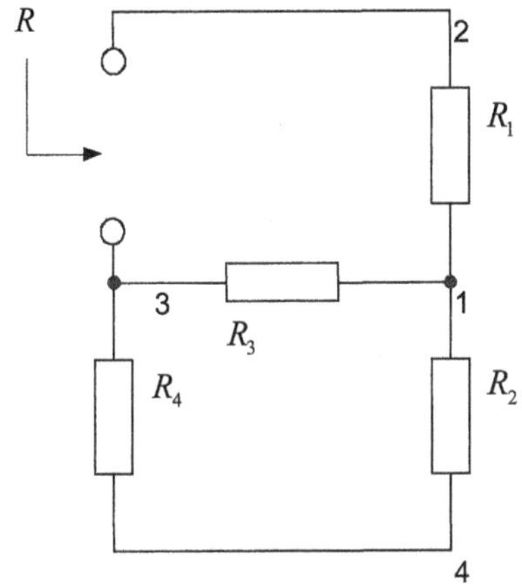

Figure 2.31 The time constant is found by setting the excitation, I_{in}, to zero.

constant of the resulting network. The excitation, in our case, is the input current I_{in}. This is a current generator. When set to zero, it transforms into an open circuit, leading to the above updated schematic.

The next step is to find the resistor R driving the capacitor. Again, looking at the open port in Figure 2.31, a simple resistive arrangement is obtained:

$$R = R_1 + R_3 \,||\, (R_4 + R_2) \tag{2.139}$$

The time constant is therefore

$$\tau = \left[R_1 + R_3 \,||\, (R_4 + R_2)\right]C \tag{2.140}$$

Leading to a pole expression of

$$\omega_{p_1} = \frac{1}{\tau} = \frac{1}{[R_1 + R_3 \,||\, (R_4 + R_2)]C} \tag{2.141}$$

Here we are: we have derived the input impedance expression in less than 10 steps! When gathering equations (2.135), (2.138), and (2.141), we have:

$$Z_{in}(s) = R_1 + (R_4 + R_2) \,||\, R_2 \frac{1 + sC[R_4 \,||\, (R_3 + R_2 \,||\, R_1)]}{1 + sC[R_1 + R_3 \,||\, (R_4 + R_2)]}$$

From this expression, it is easy to identify the dc term and the associated pole and zero.

Reference

[1] Vorpérian, V., *Fast Analytical Techniques for Electrical and Electronic Circuits*, Cambridge: Cambridge University Press, 2002.

Appendix 2B: Plotting Evans Loci with Mathcad

There are several ways to plot Evans roots loci on a computer. One of them uses the popular mathematical tool Mathcad. Let's assume we want to plot the roots loci of a second-order transfer function defined as

$$H(s) = \frac{1}{\left(\frac{s}{\omega_0}\right)^2 + \frac{s}{\omega_0 Q} + 1} \tag{2.142}$$

The denominator of this expression includes the poles of the transfer function. These poles are the roots of the equation

$$\left(\frac{s}{\omega_0}\right)^2 + \frac{s}{\omega_0 Q} + 1 = 0 \tag{2.143}$$

To use Mathcad for the resolution, let's open a new sheet and enter the following equations:

$$\omega_0 := 20008 \quad Q := 0.1 \quad Y(s,Q) := \frac{1}{\left(\frac{s}{\omega_0}\right)^2 + \frac{s}{\omega_0 \cdot Q} + 1}$$

The Q and the resonant frequency ω_0 are arbitrarily selected for this example. As the expression of the denominator follows the form of a second-order polynomial $f(s) = as^2 + bs + c$, we ask the software to identify the individual coefficients by using the function *denom* where Q is the variable. Should we need to also plot the zeros of the transfer function, we would replace *denom* by the keyword *numer* in the following expression:

$$\text{b(Q):=(denom(Y(s,Q))) coeffs,s} \rightarrow \begin{pmatrix} 400320064Q \\ 20008 \\ Q \end{pmatrix} \begin{matrix} \rightarrow c \\ \rightarrow b \\ \rightarrow a \end{matrix}$$

According to this result, the denominator function to solve is in the form of

$$f(s) = Qs^2 + 20008s + 400320064Q = 0 \tag{2.144}$$

Mathcad® can solve this function via the keyword *polyroots* whose results are passed to a two-dimensional vector, X:

$$X(Q) := polyroots(b(Q))$$

We can extract the real and imaginary parts via two dedicated keywords, *Im* and *Re*:

$$\text{Im(X(Q))} = \begin{pmatrix} 0 \\ 0 \end{pmatrix} \qquad \text{Re(X(Q))} = \begin{pmatrix} -1.981 \times 10^5 \\ -2.021 \times 10^3 \end{pmatrix}$$

These two-dimensional vectors include the conjugate roots s_1, s_2 and will be accessed via subscripted notations. Please note that the subscripted notation for a

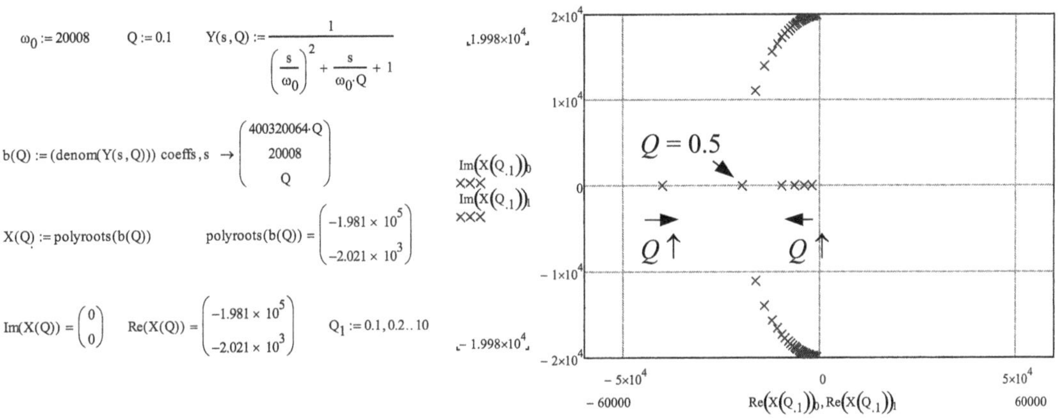

Figure 2.32 This figure gathers all notations and the plot showing how the roots move as a function of *Q*.

vector manipulation in Mathcad is obtained by pressing "[" right after the vector name:

	S_1	S_2
y	$\mathrm{Im}(X(Q))_0$	$\mathrm{Im}(X(Q))_1$
x	$\mathrm{Re}(X(Q))_0$	$\mathrm{Re}(X(Q))_1$

We now have everything to vary *Q* from 0.1 to 10 and plot the points individually made of $\mathrm{Re}(X(Q))_{0 \text{ or } 1}$ as the vertical coordinate and $\mathrm{Im}(X(Q))_{0 \text{ or } 1}$ as the horizontal coordinate. The complete calculation sheet including the roots loci graph appears in Figure 2.32.

To display × for the poles, right click on the graph and select traces/symbols. Pick a cross for both variables and you are done. Should it be zeros instead, pick the ○ as the available symbol, affected by the color of your choice.

We can see the poles moving in the same direction as *Q* increases. When *Q* equals 0.5, both roots are real and coincident. As *Q* continues to grow, the poles split, giving birth to an imaginary portion. As *Q* continues to increase, the real portion (the damping) vanishes and the roots eventually reach the imaginary axis when *Q* reaches infinity.

There are plenty of solutions to plot roots loci in Mathcad. The presented solution is simple and quick to implement. As a drawback, you will find difficult to associate a given set of roots in the plot with the corresponding *Q* value. You can always use the available cursor, but it is not really practical. Searching on the Web will give more comprehensive solutions where *Q* is displayed when a root is identified in the plot but at the expense of an increased sheet complexity.

Appendix 2C: Heaviside Expansion Formulas

We have seen how to obtain the time-domain response of a system whose transfer function is known when its input is subjected to a step function. You have to decompose the quotient into partial fractions and then sum up the inverse Laplace expression of each individual fraction. Despite its simplicity, the process requires two steps: decompose the quotient into partial fractions, and then find the inverse Laplace function of each partial fraction. Heaviside proposed another method that can give you the time-domain response in one single step. It is not really well known, and I could not find a lot of information on the Web. This is the Heaviside development formula described in [1]. It is expressed the following way:

$$\mathcal{L}^{-1}\left\{\frac{N(s)}{D(s)}\right\} = \sum_{k=1}^{n} \frac{N(s_k)}{D'(s_k)} e^{s_k t} \tag{2.145}$$

The exercise consists of identifying the denominator roots s_1, s_2 to s_n if you have n roots. Then, you have to take the derivative of the denominator. Once you have these elements on hand, simply apply the following formula and the resulting time-domain equation shows up:

$$\mathcal{L}^{-1}\left\{\frac{N(s)}{D(s)}\right\} = \frac{N(s_1)}{D'(s_1)} e^{s_1 t} + \frac{N(s_2)}{D'(s_2)} e^{s_2 t} + \ldots + \frac{N(s_n)}{D'(s_n)} e^{s_n t} \tag{2.146}$$

Let's take the transfer function already given in (2.60). It represents the system transfer function multiplied by the step function, $1/s$:

$$\frac{N(s)}{D(s)} = \frac{s+5k}{s(s+1k)(s+30k)} \tag{2.147}$$

We have three roots for $D(s)$. These roots are

$$\begin{aligned} s_1 &= 0 \\ s_2 &= -1k \\ s_3 &= -30k \end{aligned} \tag{2.148}$$

We now develop the denominator:

$$D(s) = s^3 + 31000 \cdot s^2 + 30000000s \tag{2.149}$$

and derive it:

$$\frac{dD(s)}{ds} = 3s^2 + 62000s + 30000000 \tag{2.150}$$

We now calculate

$$\begin{aligned} N(0) &= 5k \\ N(-1k) &= 4k \\ N(-30k) &= -25k \end{aligned} \tag{2.151}$$

and

$$D'(0) = 3 \times 10^7$$

$$D'(-1k) = -2.9 \times 10^7$$

$$D'(-30k) = 8.7 \times 10^8 \tag{2.152}$$

By applying (2.146), we have our time-domain response immediately:

$$\mathcal{L}^{-1}\left\{\frac{N(s)}{D(s)}\right\} = \frac{N(0)}{D'(0)}e^{0t} + \frac{N(-1k)}{D'(-1k)}e^{-1000t} + \frac{N(-30k)}{D'(-30k)}e^{-30000t} \tag{2.153}$$

Developing, we obtain

$$\mathcal{L}^{-1}\left\{\frac{N(s)}{D(s)}\right\} = 166u - 138u \cdot e^{-1000t} - 28.7u \cdot e^{-30000t} \tag{2.154}$$

This is exactly what we found in (2.70) without going through the pain of partial decomposition. Simple, yet elegant, isn't it?

Ok, let's go through another small example. Assume the following transfer function:

$$H(s) = \frac{1+s}{s^2+4} = \frac{N(s)}{D(s)} \tag{2.155}$$

We have two roots for $D(s)$. These roots are

$$s_1 = -2j$$

$$s_2 = 2j \tag{2.156}$$

Let's derive $D(s)$:

$$\frac{dD(s)}{ds} = 2s \tag{2.157}$$

We now calculate

$$N(-2j) = 1 - 2j$$

$$N(2j) = 1 + 2j \tag{2.158}$$

and

$$D'(-2j) = -4j$$

$$D'(2j) = 4j \tag{2.159}$$

By applying (2.146), we have our time-domain response almost immediately:

$$\mathcal{L}^{-1}\left\{\frac{1+s}{s^2+4}\right\} = -\frac{1-2j}{4j}e^{-2jt} + \frac{1-2j}{4j}e^{2jt} \tag{2.160}$$

Rearranging and factoring, we have

$$\mathcal{L}^{-1}\left\{\frac{1+s}{s^2+4}\right\}=\frac{1}{2}\left(\frac{e^{2jt}-e^{-2jt}}{2j}\right)+\left(\frac{e^{2jt}+e^{-2jt}}{2}\right) \tag{2.161}$$

We can identify complex sine and cosine functions. The final expression is

$$\mathcal{L}^{-1}\left\{\frac{1+s}{s^2+4}\right\}=\frac{1}{2}\sin(2t)+\cos(2t) \tag{2.162}$$

Reference

[1] Spiegel, M. R., *Schaum's Outline of Laplace Transforms*, New York: McGraw-Hill, 1965.

Appendix 2D: Plotting a Right Half Plane Zero with SPICE

We have seen that a RHPZ is a numerator root featuring a positive real part. A RHPZ can be put under the following form:

$$H(s)=1-s/\omega_{z1} \tag{2.163}$$

The negative sign indicates the presence of a positive root, located in the right-half portion in the *s*-plane. To show the effect of a RHPZ in a Bode plot, we can try to artificially create one with SPICE. A possible circuit appears in Figure 2.33. The transfer function of such a configuration is quickly obtained:

$$V_{out}(s)=V_{in}(s)-V_{in}(s)\frac{R_1}{\frac{1}{sC_1}}=V_{in}(s)(1-s/\omega_{z1}) \tag{2.164}$$

where $\omega_{z1}=\dfrac{1}{R_1C_1}$.

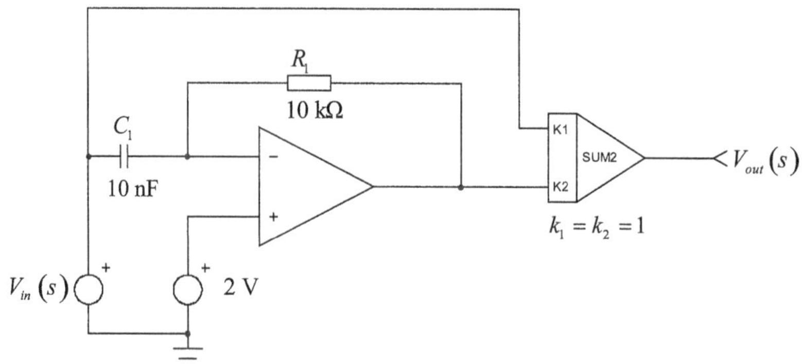

Figure 2.33 A RHPZ is artificially created with an op amp-based differentiator and an adder.

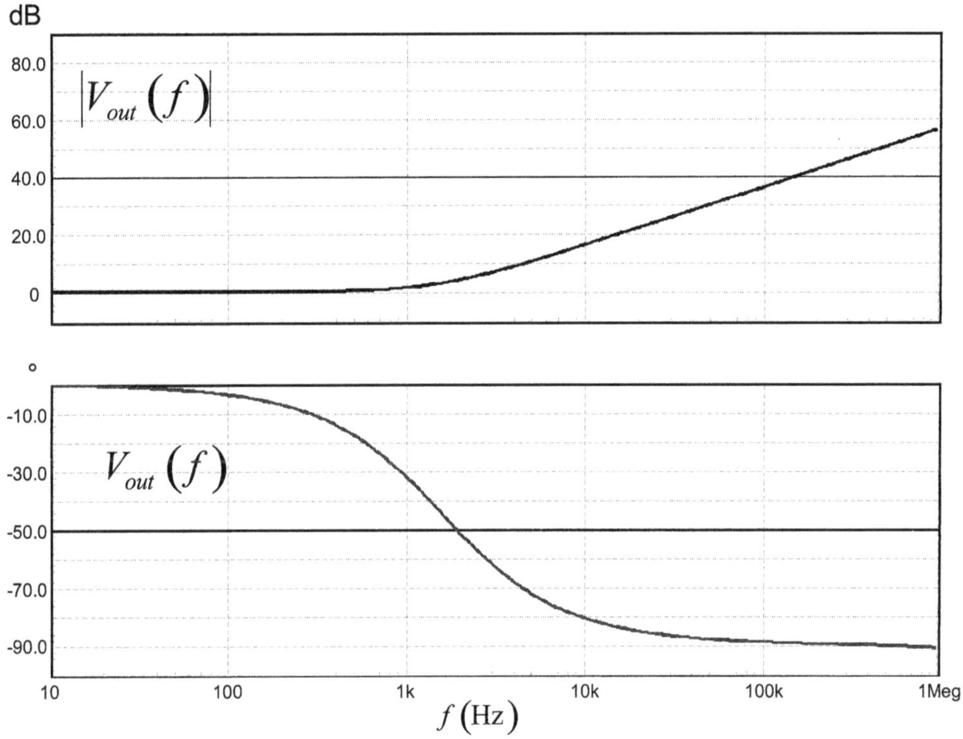

Figure 2.34 As expected, the simulation results reveal a magnitude similar to that of classical zero, but the phase, rather than going up, is actually going down to –90°.

The ac plot of such a transfer function is given in Figure 2.34. As expected, the magnitude is actually that of normal zero: it increases as the frequency increases. But the phase, rather than also heading towards +90°, it actually lags by 90°, as a pole would do.

CHAPTER 3

Stability Criteria of a Control System

In previous chapters, we have learned that a closed-loop system works by permanently comparing its output variable with the setpoint imposed by the control input. The difference between both variables gives birth to an error signal, ε, further processed by the compensator block. The resulting control signal V_c drives the transmission chain to force both input and output variables to match. To operate properly, the control signal must *oppose* the output variations: for instance, if the output signal increases beyond the target, as it subtracts from the control input, it naturally instructs the system to reduce its output, bringing it back within the accepted range. If for any reason the control signal no longer opposes the output but actually amplifies it, the system becomes unstable and goes out of control, with all possible consequences. Understanding stability is key to designing robust and reliable control systems. This third chapter focuses on this subject, introducing parameters such as *phase margin and crossover frequency* but also less known stability criteria such as *modulus* and *delay* margins.

3.1 Building An Oscillator

In the electronic field, an oscillator is a circuit capable of producing a self-sustained sinusoidal signal. In a lot of configurations, cranking up the oscillator involves the noise level inherent to the adopted electronic circuit. As the noise level grows at power-up, oscillations are started and self-sustained. This kind of circuit can be formed by assembling blocks such as those appearing in Figure 3.1. As you can see, the configuration looks very similar to that of our control system arrangement.

In our example, the excitation input is not the noise but a voltage level, V_{in}, injected as the input variable to crank the oscillator. The direct path is made of the transfer function $H(s)$ while the return path consists of the block $G(s)$. To analyze the system, let us write its transfer function by expressing the output voltage versus the input variable:

$$V_{out}(s) = \varepsilon(s)H(s) = [V_{in}(s) - G(s)V_{out}(s)]H(s) \tag{3.1}$$

If we expand this formula and factor $V_{out}(s)$, we have

$$V_{out}(s)[1 + G(s)H(s)] = V_{in}(s)H(s) \tag{3.2}$$

The transfer function of such a system is therefore

$$\frac{V_{out}(s)}{V_{in}(s)} = \frac{H(s)}{1 + G(s)H(s)} \tag{3.3}$$

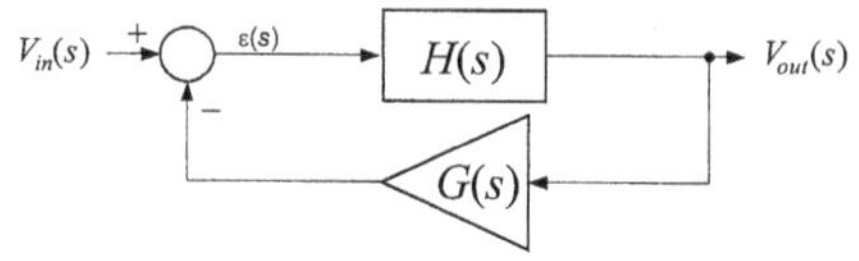

Figure 3.1 An oscillator is actually a control system where the error signal does not oppose the output signal variations.

In this expression, the product $G(s)H(s)$ is called the *loop gain*, also noted $T(s)$. To transform our system into a self-sustained oscillator, an output signal must exist even if the input signal has disappeared. To satisfy such a goal, the following condition has to be met:

$$\lim_{V_{in}(s)\to 0}\left[\frac{H(s)}{1+G(s)H(s)}V_{in}(s)\right]\neq 0 \tag{3.4}$$

To verify this equation in which V_{in} disappears, the quotient must go to infinity. The condition for this quotient to go infinite is that its *characteristic* equation, the denominator $D(s)$, equals zero:

$$1+G(s)H(s)=0 \tag{3.5}$$

To meet this condition, the term $G(s)H(s)$ must equal −1. Otherwise stated, the magnitude of the loop gain must be 1 and its sign should change to minus. A sign change with a sinusoidal signal is simply a 180° phase reversal. These two conditions can be mathematically noted as follows:

$$|G(s)H(s)|=1 \tag{3.6}$$

$$\arg G(s)H(s)=-180^\circ \tag{3.7}$$

When these two expressions are exactly satisfied, we have conditions for steady-state oscillations. This is the so-called Barkhausen criterion, expressed in 1921 by the eponymous German physicist. Practically speaking, in a control loop system, it means that the correction signal no longer opposes the output but returns *in phase* with the exact same amplitude as the excitation signal. In a Bode plot, (3.6) and (3.7) would imply a loop gain curve crossing the 0-dB axis and affected by a 180° phase lag right at this point. In a Nyquist analysis, where the imaginary and real portions of the loop gain are plotted versus frequency, this point corresponds to the coordinates $-1, j0$. Figure 3.2 displays these two curves where conditions for oscillations are met. Should the system slightly deviate from these values (e.g., temperature drift, gain change), output oscillations would either exponentially decrease to zero or diverge in amplitude until the upper/lower power supply rail is reached. In an oscillator, the designer strives to reduce as much as possible the gain margin so that conditions for oscillations are satisfied for a wide range of operating conditions.

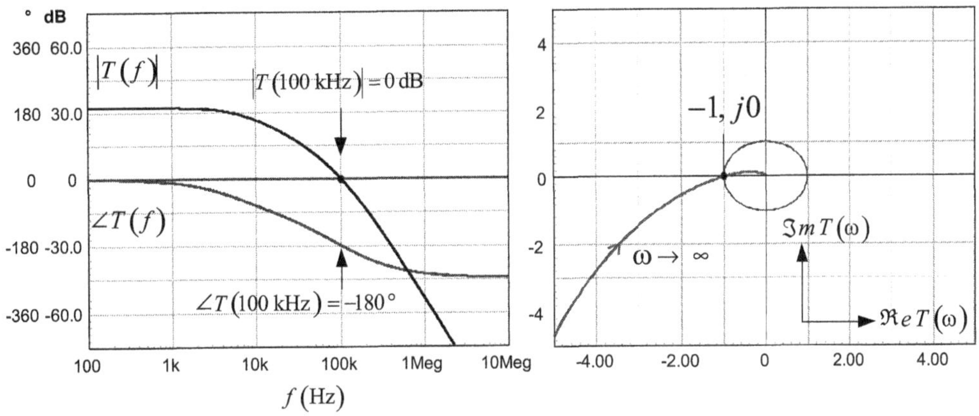

Figure 3.2 Conditions for oscillations can be illustrated either in a Bode diagram or in a Nyquist plot.

3.1.1 Theory at Work

To understand how oscillations are formed, a simulation circuit using SPICE can help us. Figure 3.3 shows our oscillator structure, based on three cascaded RC networks. The ac response of such a third-order configuration appears in Figure 3.4. It shows a flat 0-dB magnitude up to 1 kHz with a slope of –60 dB per decade beyond. The phase starts to drop around 1 kHz and reaches –180° at a 39-kHz frequency. At this point, the magnitude of the RC network transfer function is –29.32 dB. Our experiment will consist of compensating this attenuation via a compensator block $G(s)$ made of a simple gain, independent of frequency. If we cascade a block H

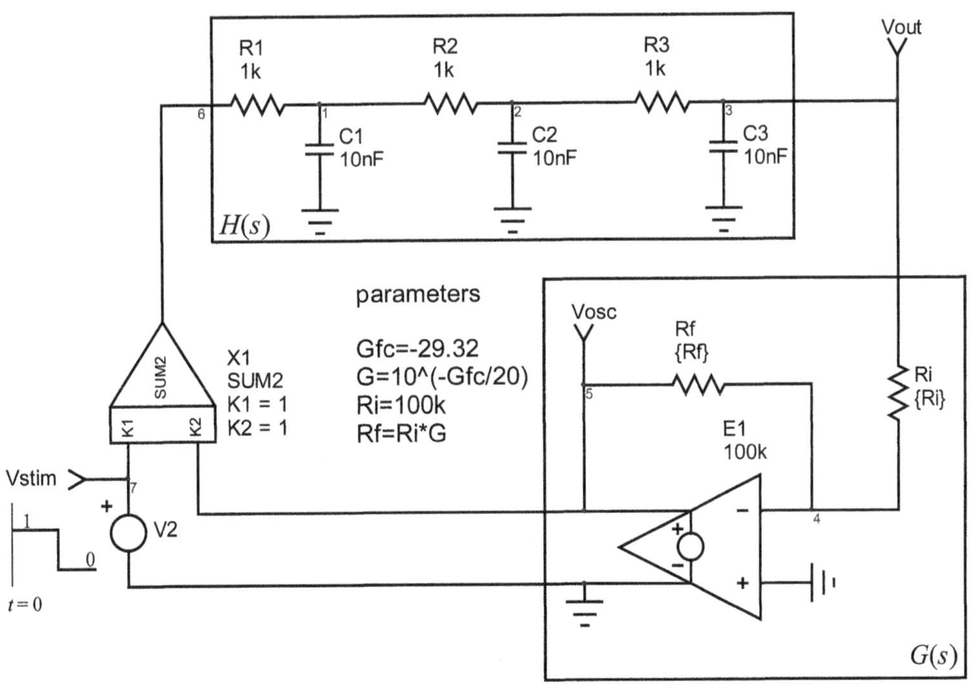

Figure 3.3 We have formed a RC network whose cumulated phase lag goes beyond 180°. An operational amplifier compensates the attenuation to make the total loop gain approaching 1.

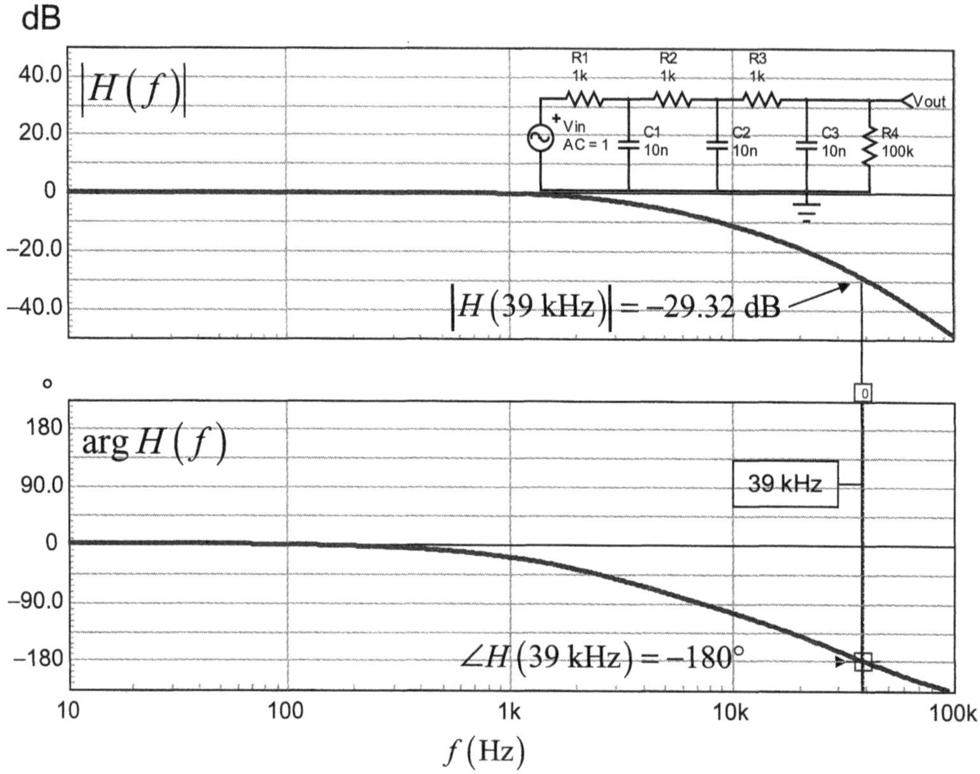

Figure 3.4 The ac response of this network shows a phase lag of 180° at a frequency close to 39 kHz. At this point, the attenuation of the network is 29.32 dB.

featuring an attenuation of –29.32 dB at 39 kHz with another block G showing a permanent gain of 29.32 dB, the resulting loop gain magnitude at 39 kHz will thus be 0 dB. Our conditions for oscillation can therefore be met where $|T(39\text{ kHz})| = 1$ and $\arg T(39\text{ kHz}) = -180°$. The compensation scheme uses a perfect operational amplifier (op amp) wired as an inverter and adjusted in gain by playing with resistor R_f. On the left side of the picture appears our adder subcircuit X_2 used to close the loop and inject the cranking voltage, V_1. This source equals 1 V for a short period of time and immediately goes to 0, obeying what (3.4) describes.

The simulation will then consist of adjusting the compensating gain provided by the op amp to create different operating conditions for the loop gain $T(s)$. The ac responses for three selected cases are plotted in Figure 3.5 at the 0-dB reference point; the phase lag is less than 180° ($G = 27$ dB), exactly 180° ($G = 29.32$ dB), or beyond 180° ($G = 33$ dB).

The simulated transient data covering each compensation scenario appear in Figure 3.6 and reveal different types of responses.

- In the upper section of the picture, the gain was adjusted to 27 dB. In this condition, the magnitude curve crosses the 0-dB axis at a 33.85-kHz frequency where the total phase lag is 171°. The conditions for oscillations are not met. The transient response is a damped oscillatory signal that decays to zero in an exponential envelope.

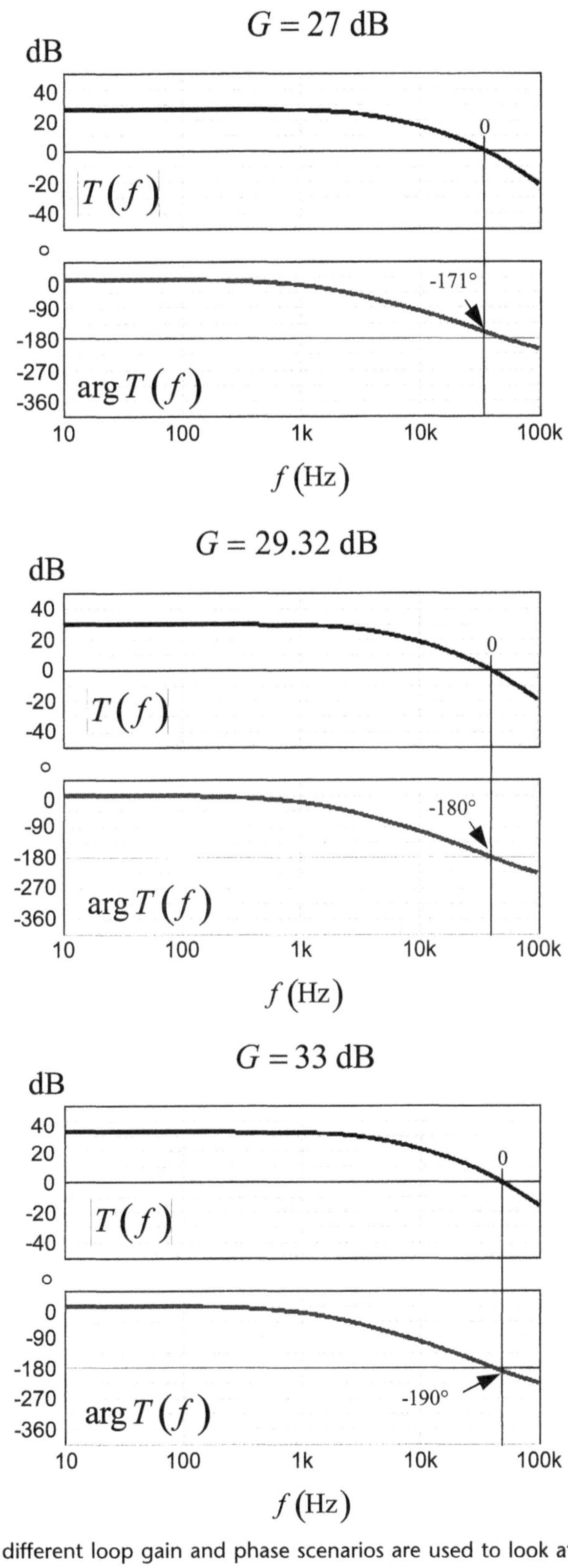

Figure 3.5 Three different loop gain and phase scenarios are used to look at the corresponding transient responses.

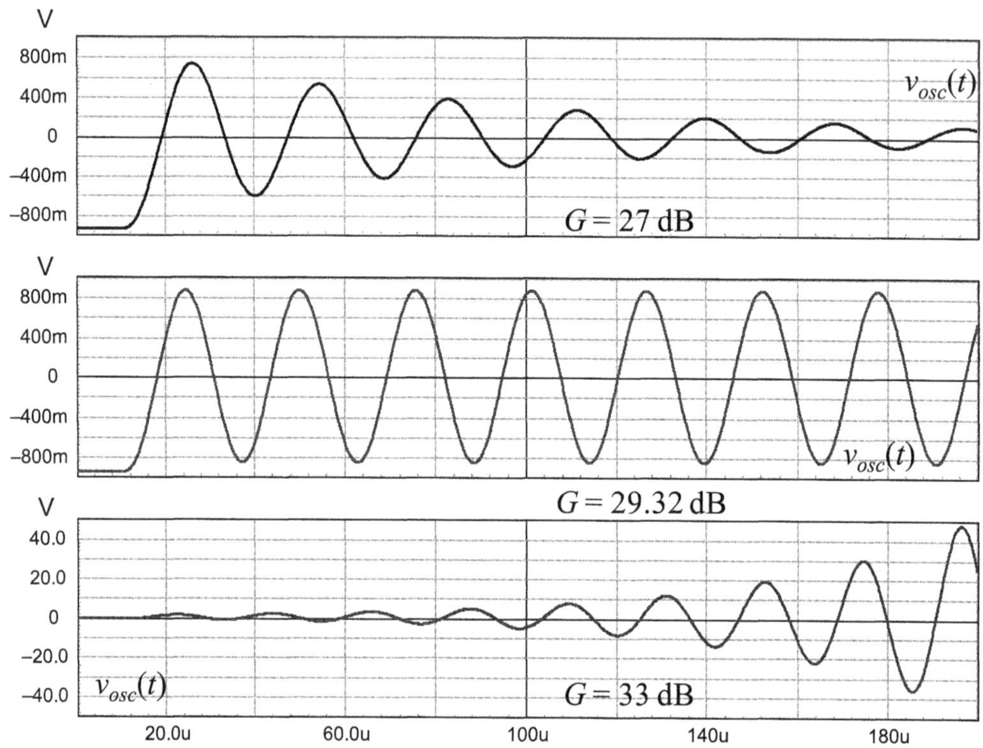

Figure 3.6 Simulation results show a response that is decaying, self-sustained, or diverging, depending on the points at which crossover occurs.

- When the gain is exactly set to 29.32 dB, the 0-dB crossing point is reached at 39 kHz and the phase lag is precisely –180°. The oscillation criteria are now respected, and we have a nice self-sustained sinusoidal waveform at a 39-kHz frequency.
- In the last plot, the op amp gain is pushed to 33 dB, forcing the loop gain to cross over where the phase lag is –190°. The output diverges and exponentially grows until the op amp hits its upper limit voltage range or smoke pops up, preceded by a loud noise.

From these experiments, we can see that oscillations are obtained and self-sustained *only* when the Barkhausen criteria are met. Deviating from this point results in an oscillatory response either decaying or growing.

3.2 Stability Criteria

You understand that our goal with a control system is not to build an oscillator. We want a control system featuring speed, precision, and an oscillation-free response. We must therefore keep away from a configuration where conditions for oscillation or divergence are met. One way is to limit the frequency range within which our system will react. By definition, the frequency range, or the *bandwidth*, corresponds to a frequency value where the closed-loop transmission path from the input to the output drops by 3 dB. The bandwidth of a closed-loop system can be seen as a frequency range where the system is said to satisfactorily respond to its input (i.e., follows the setpoint or efficiently rejects the perturbations). As we will later see, during the design stage, we do not directly control the closed-loop bandwidth but the *crossover frequency* f_c, a parameter pertinent to an open-loop analysis. Both variables are often approximated as equal, and we will see that it is true in one condition only. However, they are not far away from each other, and both terms can be interchanged in the discussion.

We have seen that the open-loop gain represents an important parameter for our control system. When gain exists (i.e., $|T(s)| > 1$), the system works in dynamic closed-loop conditions and can compensate for incoming perturbations or react to setpoint changes. However, there is a limit to the system reaction: the system must offer gain at the frequencies involved in the perturbing signal. If the perturbations or the setpoint changes are too fast, the frequency content of the excitation signal is beyond the bandwidth of the system, implying the absence of gain at these frequencies: the system becomes slow and cannot react, operating as if the loop were unresponsive to varying waveforms. Is an infinite-bandwidth system desirable then? No, because increasing the bandwidth is like widening the diameter of a funnel: you are certainly going to collect more information and react faster to incoming perturbations, but the system will also accept spurious signals such as noise and parasitic data, self-produced by the converter in some cases (the output ripple in switching power supplies, for instance). For this reason, it is mandatory to limit the bandwidth to what your application really requires. Adopting too wide a bandwidth would be detrimental to the *noise immunity* of the system (e.g., its robustness to external parasitic signals).

How do we limit the bandwidth of a control system? By shaping the loop gain curve through the compensator block, G. This block will make sure that after a certain frequency f_c, the loop gain magnitude $|T(f_c)|$ drops and passes below 1 or 0 dB. As we explained, it is roughly the bandwidth of your control system once the loop is closed. The frequency at which this phenomenon occurs is called the *crossover frequency* noted f_c. Is this enough to obtain a robust system? No, we need to ensure another important parameter: the phase of $T(s)$ at the point where its magnitude is 1 must be less than −180°. From our experiments, we have seen that when the loop phase is less than −180° at the crossover frequency, we obtain a response converging toward a stable state. This is obviously a highly desirable characteristic of our control system. To make sure we stay away from the −180° limit at crossover, the compensator $G(s)$ must tailor the loop argument at the selected crossover frequency to build *phase margin*, PM or φ_m. The phase margin can be considered a design or a safety limit ensuring that despite external perturbations or unavoidable production spreads, changes in the loop gain will not put the stability in jeopardy. As we will later see, the phase margin also impacts the transient response of the system. Therefore, its choice does not exclusively depend on stability considerations but also on the type of transient response you want. Mathematically, the phase margin is defined as follows:

$$\varphi_m = 180 + \angle T(f_c) \text{ in degrees} \tag{3.8}$$

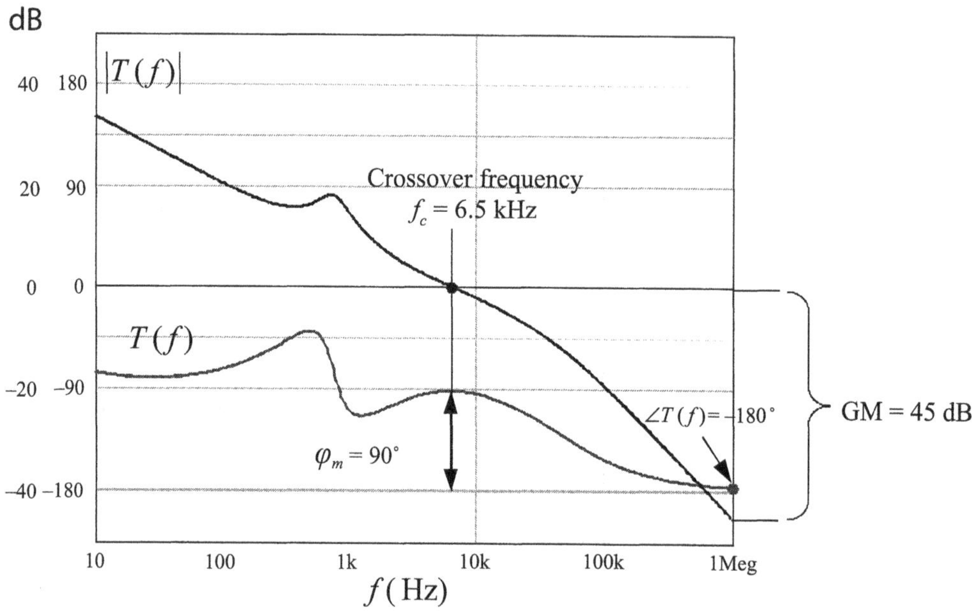

Figure 3.7 In this example, the 0-dB crossover point is located at 6.5 kHz, where the total phase lag offers a phase margin of 90°.

where T represents the open-loop gain made of the cascaded plant H and compensator G gains.

A typical compensated loop gain curve appears in Figure 3.7 and shows a crossover frequency of 6.5 kHz. At this point, the phase of $T(s)$ is –90°. If you start from the –180° line at a 6.5-kHz frequency and positively count the degrees until you cross the argument waveform, you have the phase margin: 90° in this example. This is an extremely robust system that is told to be *unconditionally stable*: despite moderate loop gain variations around the crossover point, there are no possibilities to cross over at a frequency where the phase margin is too small. By too small, we mean a phase margin approaching 30°, a limit below which the system gives an unacceptable ringing response. This is the reason why you have learned at school that 45° was the limit, giving an extra margin compared to 30°. We will later see that there is an analytical origin for these numbers.

3.2.1 Gain Margin and Conditional Stability

Figure 3.8 shows another typical frequency response of a compensated converter, highlighting the 0-dB crossover point as well as the phase margin. We know by experience that the elements constitutive of the converter will exhibit variations along the product life cycle. These variations can be linked to natural production spreads (for instance, resistors or capacitors affected by a lot-to-lot tolerance). The converter environmental operating conditions also have an impact on components. Among these variables, temperature plays an important role and affects passive or active component parameters. It can be capacitors or inductors equivalent series resistors (ESR), the optocoupler current transfer ratio (CTR), or the beta of bipolar transistors for instance. These variations impact the loop gain by shifting it up or down depending on the affected parameters. If the gain curve undergoes a shift, the 0-dB crossover frequency will transition to a new value imposing a different bandwidth to the converter. How can the converter stability be affected under these changes? Well, if the new crossover takes place at a point where the phase margin is weak, you may degrade the transient response so that the overshoot is no longer acceptable. It is thus your responsibility, as a designer, to ensure that these dispersions do not suddenly increase the gain at a frequency where you approach the –180° limit. You need sufficient gain margin as defined by

$$GM = \frac{1}{|T(f_\pi)|} \tag{3.9}$$

where f_π corresponds to the frequency point where $\angle T(s)$ is exactly –180° or $-\pi$ radians (1 MHz in Figure 3.7).

Figure 3.8 portrays typical gain variations of ±10 dB due to production spreads in the selected components. It brings the crossover frequency from 1.5 kHz to 30 kHz. In this area, the phase margin changes from 70° to 45°, safe numbers according to theory. What is the worst case? It is when the new crossover frequency occurs where the total phase lag is 180°, matching the conditions for oscillations. This condition occurs at 1 MHz, implying a positive gain change of 35 dB.

Fortunately, deviations of 35 dB are unlikely to happen in modern electronics circuits. Time ago, when amplifiers or servomechanisms were driven by vacuum tubes-based circuits, warm-up times during the power-on sequence could induce large loop gain variations. Gain provisions were thus necessary to reject a second point where the stability could be in danger. This gain margin, identified on the loop gain curve at the frequency where the total phase lag reaches –180°, is noted GM in Figure 3.7. In modern electronic circuits, gain margins beyond 10 dB are

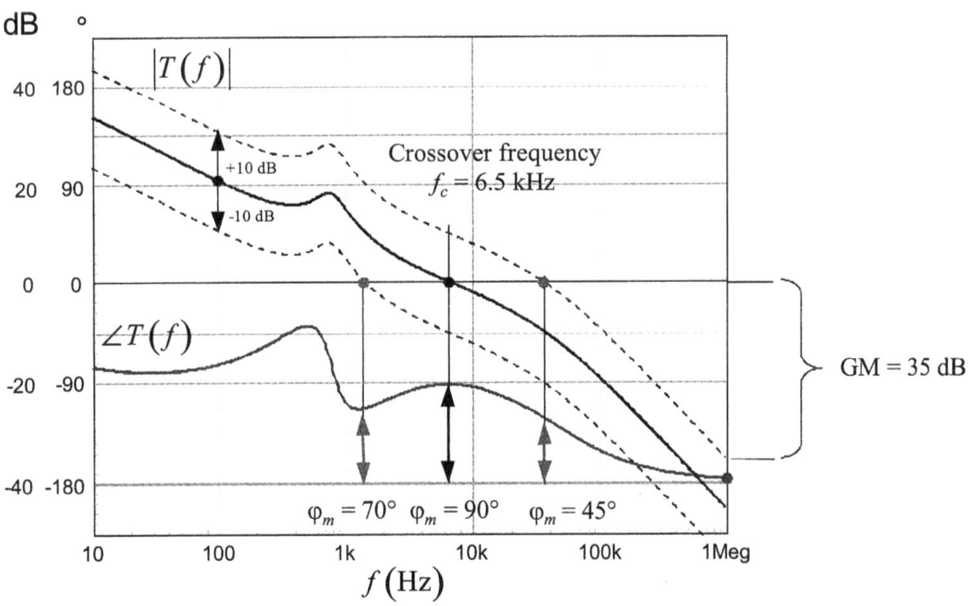

Figure 3.8 The loop gain can show sensitivity to external parameters such as temperature. When a variation occurs, the phase margin must always stay within safe limits.

usually enough, unless your loop gain exhibits extreme sensitivity to an external parameter.

Another example of gain shift appears in Figure 3.9. It shows another compensated converter exhibiting a phase margin of 80° at 10 kHz. Based on what we discussed, we know that gain changes can occur, inducing ups or downs on the gain curve. In our example, we can identify an area around 2 kHz where the phase margin is as small as 18°. If a gain decrease of 20–25 dB occurs, you can end up with a control system showing a dangerously low phase margin around 2 kHz. It would lead to an oscillatory response, perhaps exceeding the overshoot specifications. This kind of system is told to be *conditionally stable*. Fortunately, as already said, a 25-dB variation of gain is unusual and such a system can be considered robust with this gain margin. However, I have seen design cases where the end user (your customer) clearly stated in the specifications that conditional designs were not acceptable, asking for a phase margin greater than 60° at all points below the crossover frequency. In this case, it becomes mandatory to compensate the converter so that no region of reduced phase margins below crossover ever exist whatever the operating conditions are.

It is often believed that a system where the phase curve dips below −180° before crossover is an unstable system. Such a response appears in Figure 3.10. The phase curve quickly drops after 1 kHz and passes the −180° limit at 1.5 kHz for a few kilohertz.

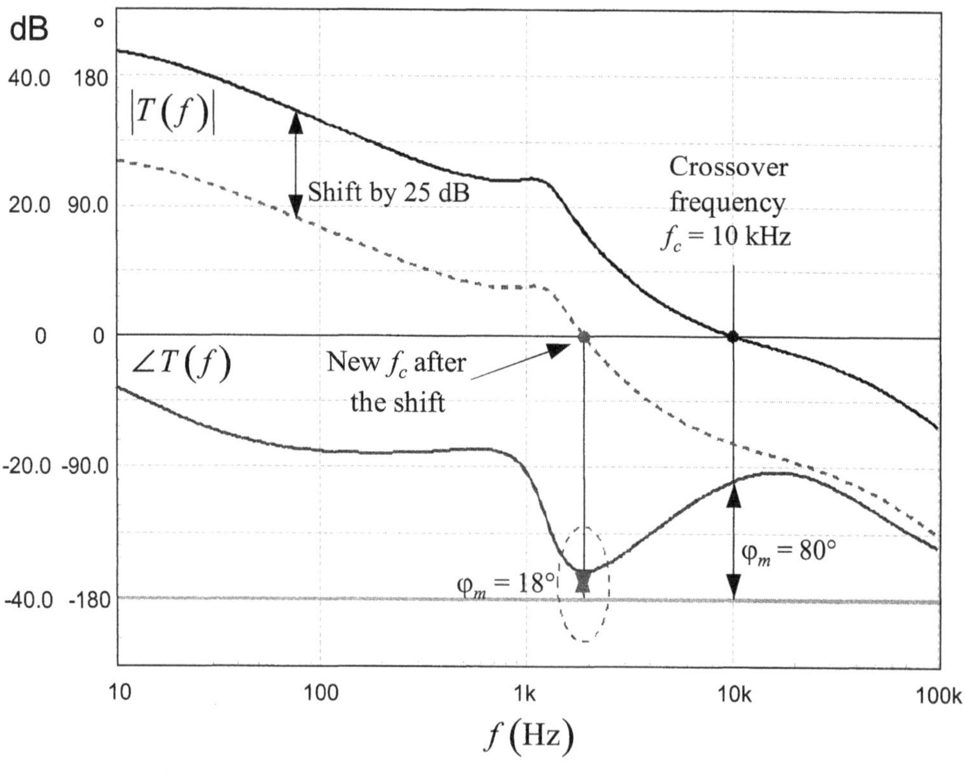

Figure 3.9 In this example, if the gain shifts down by 25 dB, the curve crosses the 0-dB axis at a point where the phase margin is only 18°. Such a low phase margin will give a very oscillatory response, affected by a large overshoot. This is a case for conditional stability.

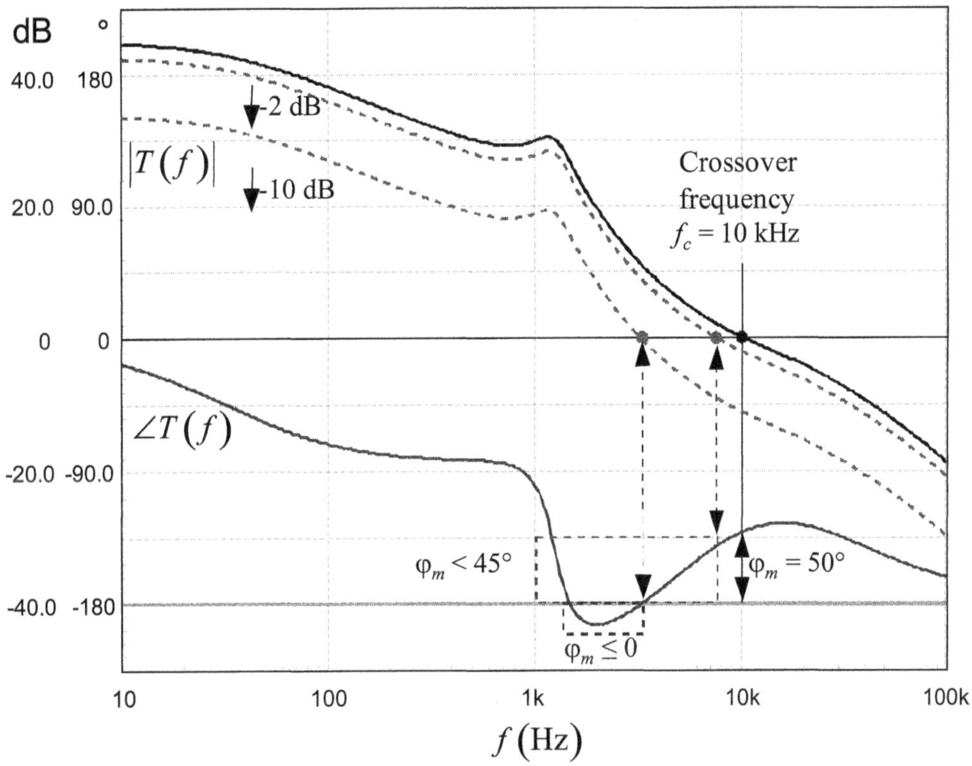

Figure 3.10 The phase lags beyond 180° but in an area where the gain is larger than 1. This is not a problem, and the response is acceptable.

It then goes up again to offer a phase margin of 50° at 10 kHz. Yes, this system is stable simply because we do not satisfy (3.7) at 0 dB. Remember, to cancel the denominator of (3.3), you must have the gain magnitude exactly equal to 1 *and* a phase lag of 180° or beyond. In the graph, we can see that this condition is not satisfied at any point in the picture. However, it is worth noting that the loop is highly conditional. Should the gain reduce by a few decibels and your phase margin becomes less than 45°. Another 10-dB decrease and you enter a dangerous area of zero phase margin where, this time, the oscillation criteria would be met.

3.2.2 Minimum Versus Nonminimum-Phase Functions

In this book, we will essentially use Bode plots for stability analysis. Due to simplicity, the technique is widely adopted in the control engineering industry and, in particular, power conversion. However, the reader must be aware that Bode plots interpretations can sometimes mislead the designer when determining closed-loop stability. Caution must be taken when pure delays or right half plane poles or zeros appear in the transfer function. Here is why: intuitively, if you see a +1 slope in a Bode plot, a zero is in action and you expect the phase to increase and reach +90° at some frequency point. The same reasoning applies to a pole where its –1 slope should bring the phase to –90° at a given frequency. If you combine a pole and a zero, you would see the phase going toward –90° and returning to 0 when the zero starts to kick in. Mathematically, you can write that a minimum phase function has a high-frequency phase asymptote defined as

$$\varphi_{\max} = -90°(n - m) \tag{3.10}$$

where n is the number of poles and m the number of zeros: five poles and four zeros give an asymptote of –90°.

Figure 3.11 shows you two examples. Both functions share similar poles and zeros positions, except that the left curve combines a LHP zero and the right curve implements a RHP zero. If you look at the phase for the left-side function while masking the magnitude graph and knowing that a –90° phase corresponds to a –1 slope and 0° to a 0-slope (flat curve), you could easily reconstruct an asymptotic magnitude graph. On the right side, if you follow the same path, you see a phase curve starting from –90° implying a –1 slope and then touching –180°, meaning that a second –1 slope combines with the first one to become a –2 slope: it can be the result of two cascaded poles, but this is obviously wrong because of the RHP zero presence.

Mathematically, Hendrik Bode demonstrated that it is always possible to link the imaginary and real portions of a transfer function when its poles or zeros sit in the left-half plane: in this case, this transfer function is said to be of *minimum phase*. Practically, it means that you should be able to reconstruct a phase plot from a magnitude plot and vice versa as we shown via the Figure 3.11 left curve. After all, both phase and magnitude are a combination of the real and imaginary parts of the transfer function under study, and a path to reach either one must exist through the computed magnitude or phase values. Now, if RHP zeros or RHP poles appear in the transfer function, they distort the phase information as we showed. The Bode law is violated: this is a *nonminimum phase* function. The remark also applies if a pure delay is part of the transfer function. A delay does not change the magnitude of the function in which it appears but it makes its phase further lag as

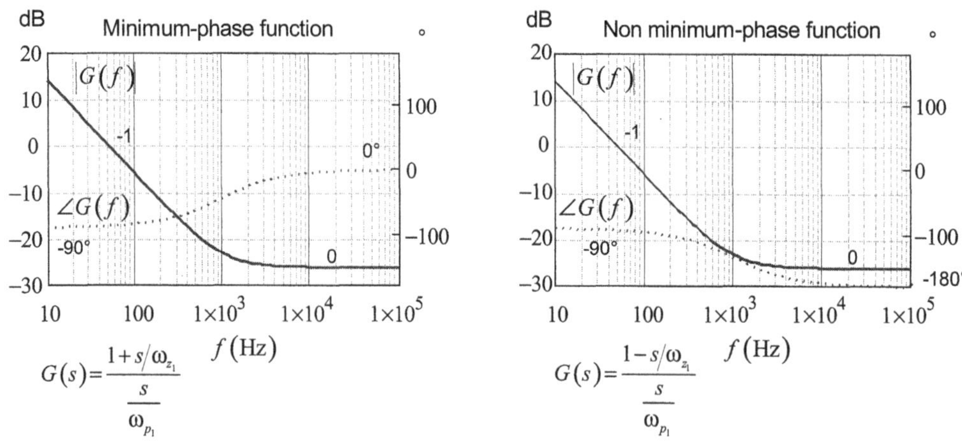

Figure 3.11 With a minimum phase function, you should be able to reconstruct the phase from the magnitude plot and vice versa. On the left side, a –1 slope implies a phase lag of 90° and the 0 slope tells you that the phase has returned to 0°. On the right side, despite similar magnitude plots, the phase curve no longer fits the slopes values: this is a nonminimum-phase function.

the frequency increases, again distorting the phase response and violating the Bode law. In a many theory books, it is clearly stated that you should use Bode plots with caution with the presence of nonminimum-phase functions. In these specific cases, a Nyquist plot must be used, as it does not combine the real and imaginary parts to get magnitude and phase values but individually plots them on a dedicated chart. Even if RHP zeros or poles (also called *unstable* poles or zeros) are present, it does predict the stability or instability without ambiguity.

3.2.3 Nyquist Plots

Though Bode plots are widely used among the design community, it is important to understand Nyquist plots. As we just explained, in some cases, you must apply them to get the right answer, as Bode will fail to predict the instability. This paragraph is a very brief and necessarily incomplete introduction to Nyquist plots, but it should give you the basis in case you need to analyze a linear or switching regulator with this technique. The reader is encouraged to search the available literature and the Web to further learn the vast field of Nyquist plots applications.

A Nyquist graph is a plot where you mark the loci of points in which x,y coordinates are respectively the real and imaginary parts of the function under study as the angular frequency increases from 0 to infinity. Linking all these marks together gives birth to the Nyquist plot. A typical example appears in Figure 3.12. The curve starts from the right where the angular frequency is low (this is your starting frequency in the ac sweep, dc or 0 in theory). Then, as the angular frequency increases, you mark points of x and y coordinates, made of the real and imaginary parts of the loop gain function T, respectively:

$$T(j\omega) = x + jy = \mathrm{Re}\,T(j\omega) + j\,\mathrm{Im}\,T(j\omega) \tag{3.11}$$

A point on the plot corresponds to a certain angular frequency ω. You can reconstruct the transfer function magnitude and phase related to this point by

$$|T(\omega)| = \sqrt{\mathrm{Re}^2\,T(\omega) + \mathrm{Im}^2\,T(\omega)} \tag{3.12}$$

$$\angle T(\omega) = \tan^{-1}\frac{\mathrm{Im}\,T(\omega)}{\mathrm{Re}\,T(\omega)} \tag{3.13}$$

In this graph, there is one particular point where the magnitude of the function is 1 and its argument is $-\pi$. This point, called the "-1" point, appears in the graph at the position $-1,j0$: when the open-loop gain curves exactly passes through this point, we have conditions for oscillations as described by (3.6) and (3.7) with an illustration in Figure 3.2. Coming close to this point also means that the denominator $D(s)$ in (3.3) dangerously approaches zero.

Now suppose you sit on a motorcycle and drive along the path in the prescribed direction as shown in Figure 3.13. All the points you see on your right hand are said to be *enclosed*. For instance, the point positioned at $1,-j$ is enclosed while the "-1" point is not.

For a minimum-phase system, theory teaches us that stability is ensured if the "-1" point is *not* enclosed while sliding the path along the prescribed direction.

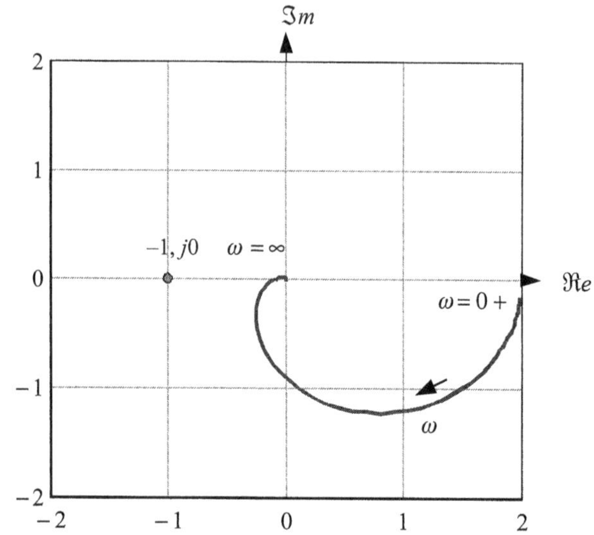

Figure 3.12 A Nyquist plot represents the loci of the loop gain real and imaginary parts as ω increases from 0 to infinity.

Practically speaking, if you leave the "–1" point on your left while going in the prescribed direction, the system is stable. This rule is also known as the Nyquist *left-hand criterion*. An application example is given in Figure 3.14 with two functions *a* and *b*.

When going along the prescribed direction, function *b* encloses the "–1" point and is unstable. On the contrary, function *a* leaves the point on the left while sliding along the path: the function is stable. This simple stability criterion for the Nyquist plot is valid for minimum-phase functions only. For nonminimum phase functions, you will need to apply the more complex Cauchy argument principle whose usage goes beyond this simple introduction.

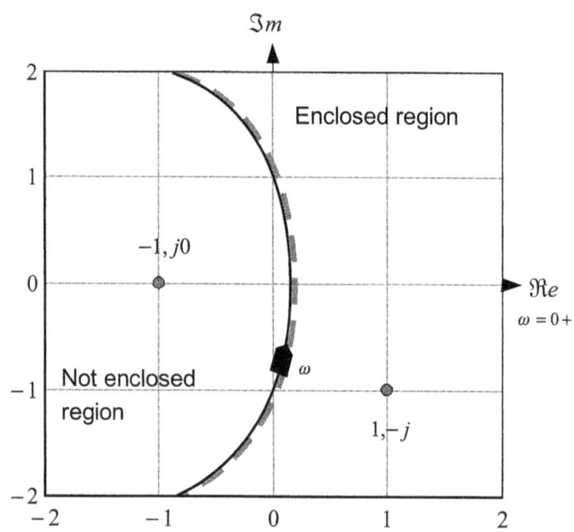

Figure 3.13 All points situated in the right of a path driven along the prescribed direction are said to be enclosed.

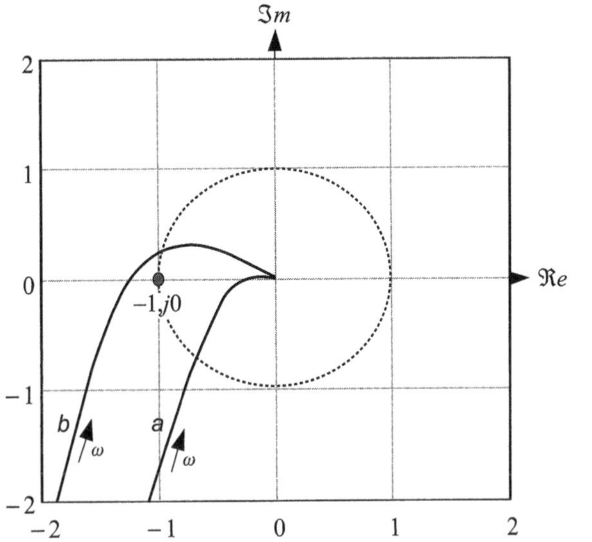

Figure 3.14 In this example, curve b encloses the "–1" and is unstable, while curve a leaves the "–1" on the left when going along the prescribed direction: the function is stable.

3.2.4 Extracting the Basic Information from the Nyquist Plot

From a Nyquist plot, the first information you can extract is the crossover angular frequency, ω_c. Looking at Figure 3.15, we can see that from every coordinates pair $\mathrm{Re}\,T(\omega)$ and $\mathrm{Im}\,T(\omega)$, we can form a triangle whose hypotenuse h starts at the graph origin. Considering triangle geometry and in particular Pythagoras theorem, we can write:

$$h(\omega) = \sqrt{\mathrm{Re}^2\,T(\omega) + \mathrm{Im}^2\,T(\omega)} = |T(\omega)| \tag{3.14}$$

This is the loop gain magnitude definition. When the curve crosses an origin-centered circle of radius 1, the hypotenuse equals 1: we have our crossover angular frequency at $\omega = \omega_c$. The drawback with Nyquist is that the crossover angular frequency is not displayed on the graph. You need to check the point at which the circle crossing occurs. This is admittedly less convenient than with a Bode plot.

Equation (3.8) tells us that the phase margin is the distance from the open-loop gain argument to the –180° or $-\pi$ limit at the crossover frequency (Figure 3.7). On the Nyquist graph, the line –180° or $-\pi$ is the left end of the x-axis when turning clockwise from the right end of this axis. The argument of the open-loop gain $T(s)$ is thus the angle made between the positive section of the x-axis and the hypotenuse h at the crossover angular frequency when turning clockwise. Figure 3.16 details how to form this angle. Finally, the phase margin φ_m is the remaining angle formed by the hypotenuse h and the negative section of the x-axis. The sum of $\angle T(\omega_c)$ and φ_m is well equal to –180°, as defined by (3.8).

The gain margin requires a little more attention. The gain margin concerns the observation of the loop gain magnitude at a frequency f_π where the loop-gain argument is –180° or $-\pi$ (see Figure 3.7 for instance). How much do we need to shift the

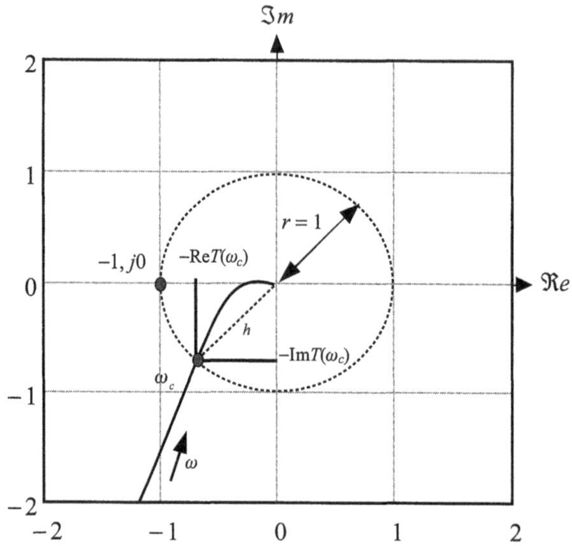

Figure 3.15 When the curve crosses a circle of radius 1, this is the point where the magnitude of *T* is 1.

curve up (or down) to obtain a new 0-dB crossing point and conditions for oscillations? To satisfy this condition, we can write:

$$|T(f_\pi)| \cdot x = 1 \tag{3.15}$$

The multiplicand x is nothing else than our gain margin GM previously defined:

$$\mathrm{GM} = \frac{1}{|T(f_\pi)|} \tag{3.16}$$

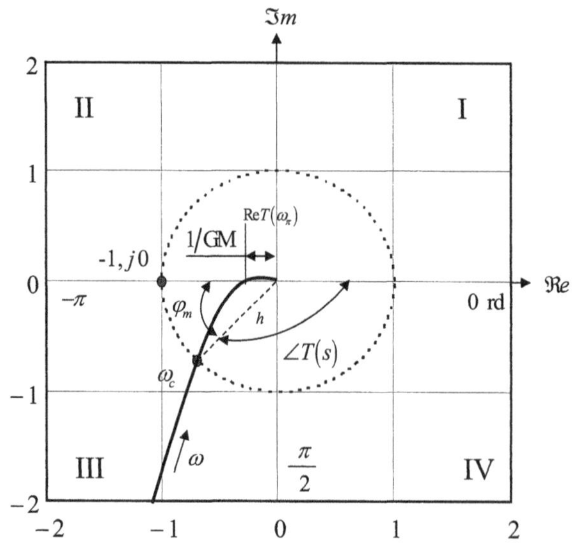

Figure 3.16 The open-loop gain argument is found as the angle measured from the *x*-axis to the triangle hypotenuse at the point where the curve crosses the circle of radius 1.

Where in the Nyquist plot do we see the point where the loop-gain argument is –180°? This is exactly when the curve crosses the x-axis in the left side of the plot in Figure 3.16. At this particular point, the imaginary portion of $T(s)$ is exactly 0. The magnitude expression thus simplifies to:

$$\text{GM} = \frac{1}{\left|\text{Re}\,T(f_\pi)\right|} \tag{3.17}$$

The real part value is directly measurable on the graph as shown in Figure 3.16. Suppose you find the crossing point with the x-axis is 0.25; then, the gain margin is simply:

$$\text{GM} = \frac{1}{0.25} = 4 = 12\text{ dB} \tag{3.18}$$

3.2.5 Modulus Margin

The gain margin alone is not a sufficient condition to qualify the system robustness. By robustness, we mean its ability to efficiently reject the incoming perturbations such as the input voltage or the output current in a voltage regulator case (linear or switching). In the opening paragraph related to Nyquist, we said that the curve must not enclose the "–1" point and should always stay away from it. Our interest is now to check the shortest distance from the curve to the "–1" point at any angular frequency value. This distance is represented by the new hypotenuse h drawn in Figure 3.17.

Applying again Pythagorean geometry and considering the height of the triangle as the imaginary part of the loop gain $\text{Im}\,T(\omega)$, we can write:

$$h = \sqrt{[1 + \text{Re}\,T(\omega)]^2 + [\text{Im}\,T(\omega)]^2} \tag{3.19}$$

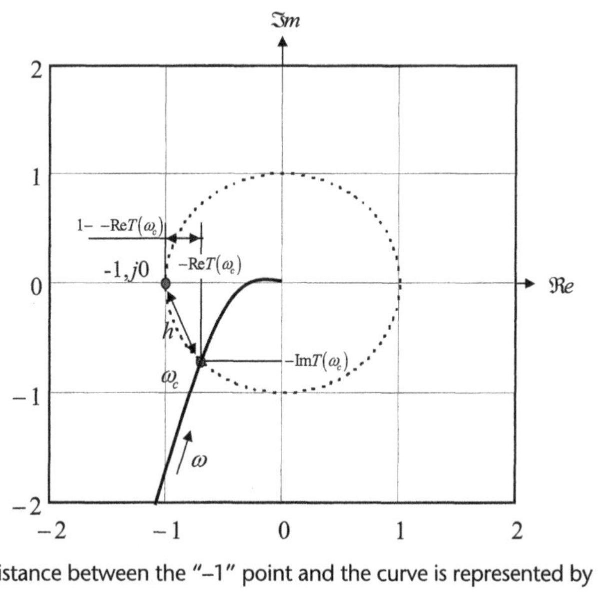

Figure 3.17 The distance between the "–1" point and the curve is represented by the hypotenuse h.

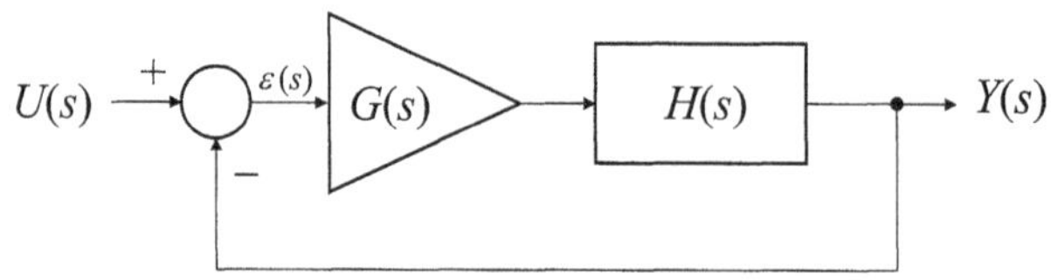

Figure 3.18 The ratio between the error ε and the setpoint U is called the sensitivity function S.

This expression is nothing other than the magnitude of $D(s)$ in our closed-loop gain expression as defined in (3.3):

$$h = |1 + T(s)| \tag{3.20}$$

When this hypotenuse length diminishes, the curve dangerously approaches the "–1" point and the ability to reject the perturbations can be at stake. If we look at Figure 3.18, where a simple unity-return closed-loop system appears, we can derive the relationship linking the error variable ε to the input setpoint u:

$$\frac{\varepsilon(s)}{U(s)} = \frac{1}{1+T(s)} = S \tag{3.21}$$

This function is called the *sensitivity* function, noted S. Ideally, S should be extremely small, meaning that ε, the error, becomes negligibly low and the output matches the setpoint.

Now, assume a perturbation input is inserted into our system, as described by Figure 3.19. This perturbation u_2 could be the input voltage for our converters or the delivered output current.

The transfer function of such a closed-loop system can be derived applying the superposition theorem since we deal with a linear system. First, if we ground the perturbation input U_1, we classically have:

$$Y(s)\big|_{U_1=0} = U_2(s)\frac{T(s)}{1+T(s)} \tag{3.22}$$

Now, if we ground U_2, we obtain the following expression:

$$Y(s)\big|_{U_2=0} = -U_1(s)\frac{1}{1+T(s)} \tag{3.23}$$

The complete expression is thus the sum of (3.22) and (3.23):

$$Y(s) = U_2(s)\frac{T(s)}{1+T(s)} - U_1(s)\frac{1}{1+T(s)} \tag{3.24}$$

As you can read in the previous equation, the perturbation rejection depends on S, which is the inverse of $|1+T(s)|$. As $|1+T(s)|$ becomes smaller, the ability to

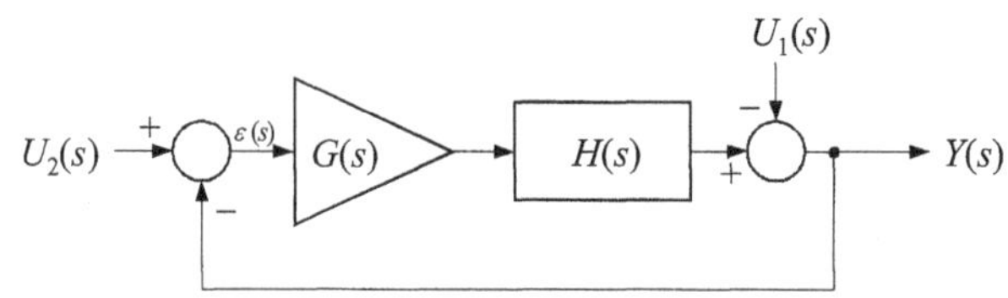

Figure 3.19 When a perturbation is inserted in the control chain, its rejection depends on the sensitivity function.

reject the perturbation weakens and system robustness suffers. Beyond crossover, where the loop gain is below 1, the system loop no longer fights the incoming perturbations. Ideally, the natural gain reduction over frequency must be smooth, without peaking. However, if a hidden resonance suddenly makes $|1+T(s)|$ go to a very small value, much less than one, then the sensitivity function peaks and the perturbation is amplified rather than attenuated. The peak of the sensitivity function or the minimum of $|1+T(s)|$ corresponds to the shortest value of the hypotenuse h in Figure 3.17. It is also the shortest distance between our Nyquist curve and the "–1" point. It is thus always advisable to maintain a minimum value for h. This value is the *modulus margin*, noted ΔM. The term *modulus* can be replaced by the word *magnitude*, if you prefer. It is mathematically defined as the smallest acceptable radius of a circle centered at the point $-1, j0$ and tangent to the loop gain loci. It is usual to adopt a radius of 0.5 for this circle. You can show by calculating the intersection between the 0.5-radius circle and the 1-radius circle that keeping $\Delta M > 0.5$ ensures a gain margin better than 6 dB and a phase margin greater than 30°.

Any points of the loci entering the circle violate the minimum modulus margin we stated. This principle appears as a graphical illustration in the left side of Figure 3.20. As long as no points belonging to the Nyquist loci enter the circle, the modulus margin principle is respected and the gain margin criteria is always fulfilled. As shown on the right side of the same picture, the gain margin can very well be correct; it is not a sufficient condition for stability: a hidden resonance briefly routes the loci into the circle, while returning to its trajectory afterward. You might think the gain margin criterion is satisfied, but the system robustness would be at stake with this fugitive excursion. As a general statement, if the modulus margin is satisfied, the gain margin will also be. In most textbooks, the modulus margin, not the gain margin, is considered a robustness criterion.

For the sake of the practical example, we simulated a buck converter compensated by a structure combining a double pole/double zero arrangement and purposely left the subharmonic poles untreated (weak slope compensation). The corresponding Nyquist plot from the control input to the error amplifier output

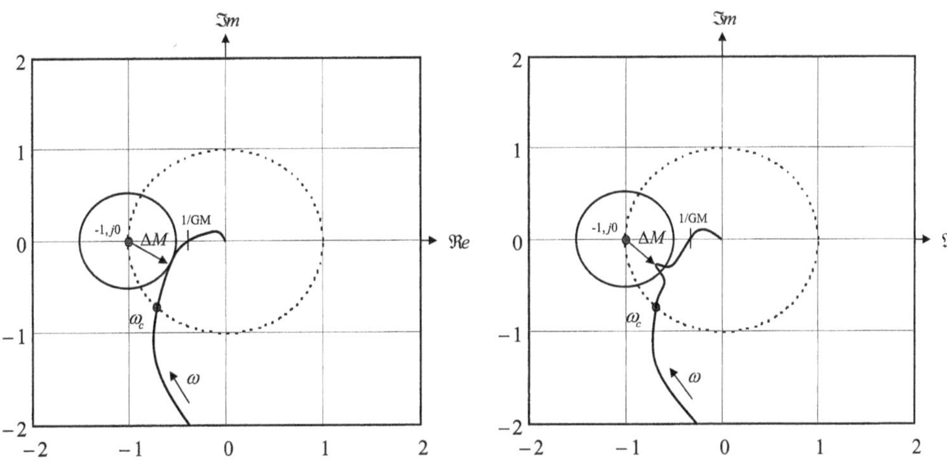

Figure 3.20 Any point entering the 0.5-radius circle violates the 6-dB modulus margin rule.

appears in the upper side of Figure 3.21. We can see the gain margin is within the circle, implying a value less than 6 dB (4.4 dB)—not very good. However, this is not enough to safely characterize the system as the trajectory approaches the "–1" point even closer at a 26.3-kHz frequency, where the distance is further reduced. The modulus margin is way too small here.

A Bode plot can also be used to check the modulus margin criteria. You can derive the sensitivity function S and plot its magnitude in decibels. You can do that in SPICE by combining the real and imaginary parts of the loop gain T:

$$|S| = \frac{1}{|1+T(s)|} = \frac{1}{\sqrt{(1+\operatorname{Re}T(s))^2 + (\operatorname{Im}T(s))^2}} \tag{3.25}$$

Some graphical analysis tools accept programming lines. For instance, with Intusoft's IntuScope graphical viewer, the code could look as follows:

```
* Sensitivity for a transfer function
assertvalid vin vout
phase = phaseextend(phase(vout)–phase(vin)
gain = db(vout) – db(vin)
mag = 10^(gain/20)
real = –mag*cos(phase)
imag = –mag*sin(phase)
```

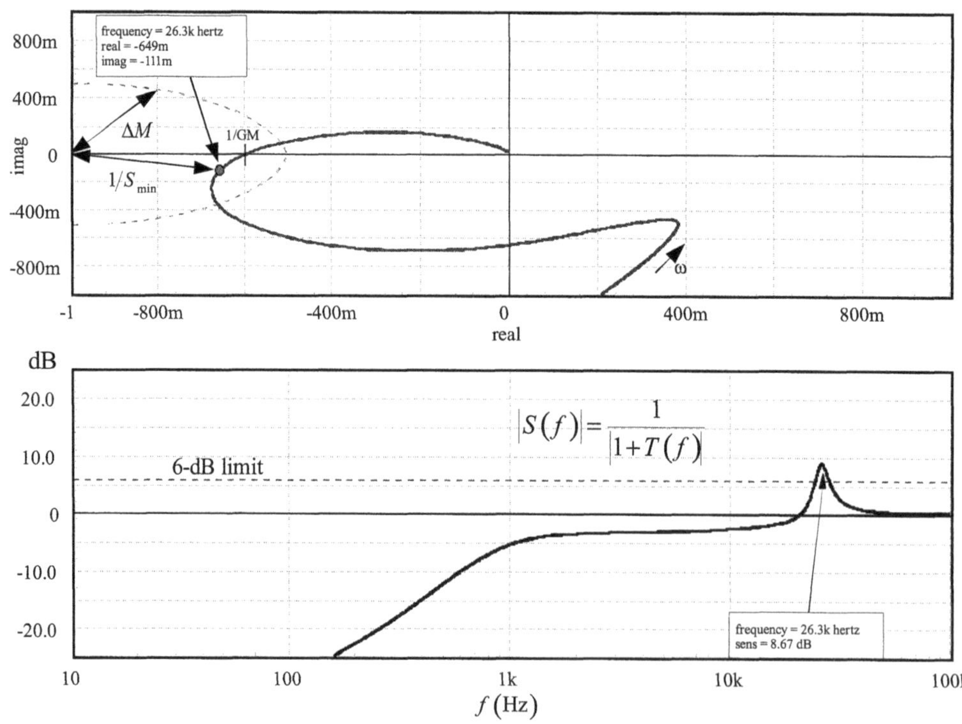

Figure 3.21 The Nyquist plot shows a violation of the 0.5-radius circle, confirmed by the sensitivity function that clearly peaks beyond 6 dB. It confirms the design problem: the modulus margin is too small.

Re2 = (real+1)^2
Imag2 = (imag)^2
D = sqrt(Re2+Imag2)
sens = db(1/D)
plot sens

If a peak exceeds the 6-dB limit, the modulus margin is violated. You can clearly see this peak in the lower side of Figure 3.21.

3.3 Transient Response, Quality Factor, and Phase Margin

We have seen how the gain and modulus margins must be characterized to build robust systems. What design strategy must we now adopt for the other important parameter, the phase margin? In other words, what is the minimum phase margin we should tweak our design to? In most of the available textbooks, and this is what I learned when I was a student, it is often stated that the phase margin should never be less than 45° at the crossover frequency. Period. However, as underlined by Dr. David Middlebrook in his design-oriented analysis course, this is neither the right question nor the correct answer. The phase margin selection depends on the type of transient response you expect from your control system. If you look back to Figure 3.6, in the upper section, you can see a waveform whose amplitude is decaying further to an excitation stimulus. What does this ringing signal remind you? Yes, the response delivered by an *RLC* circuit subjected to a step input. We know that depending on the *quality factor* Q or the *damping ratio* ζ (pronounced "zeta") of the network, the ringing signal will decay more or less quickly. At a certain point, if the damping ratio is cancelled, you have permanent oscillations as in the middle of Figure 3.6. Let us first explore the response delivered by a second-order system, a *RLC* network, to smoothly unveil its relationship to the closed-loop quality factor and the open-loop phase margin.

3.3.1 A Second-Order System, the RLC Circuit

An *RLC* circuit appears in Figure 3.22 where the input is stepped from 0 V to 1 V in a very short time while the output is observed.

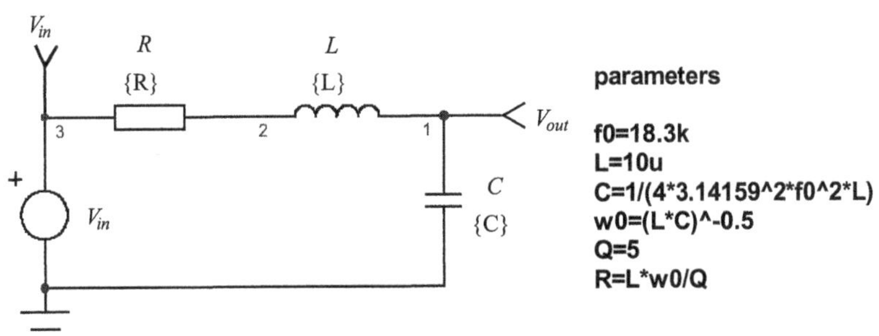

Figure 3.22 A *RLC* network receives an input voltage step and delivers an oscillatory response for high *Q* values.

To have an idea of the time-domain response of such a circuit, we can first evaluate its transfer function. Let's use the classical approach using impedance ratios, as the result is straightforward:

$$V_{out}(s)=V_{in}(s)\frac{\frac{1}{sC}}{R+sL+\frac{1}{sC}} \tag{3.26}$$

Rearranging this equation leads to the following transfer function:

$$\frac{V_{out}(s)}{V_{in}(s)}=\frac{1}{LCs^2+RCs+1} \tag{3.27}$$

This expression can be put in the form of a second-order system using either the *quality factor Q* or the *damping ratio ζ*:

$$\frac{V_{out}(s)}{V_{in}(s)}=\frac{1}{\frac{s^2}{\omega_0^2}+2\zeta\frac{s}{\omega_0}+1}=\frac{1}{\frac{s^2}{\omega_0^2}+\frac{s}{\omega_0 Q}+1} \tag{3.28}$$

Identifying ω_0, Q, and ζ within (3.27), we have the following relationships:

$$\omega_0=\frac{1}{\sqrt{LC}} \tag{3.29}$$

$$f_0=\frac{\omega_0}{2\pi}=\frac{1}{2\pi\sqrt{LC}} \tag{3.30}$$

$$\zeta=\frac{1}{2Q}=\frac{R}{2}\sqrt{\frac{C}{L}}=\frac{RC\omega_0}{2}=\frac{R}{2L\omega_0} \tag{3.31}$$

$$Q=\frac{1}{2\zeta}=\frac{1}{R}\sqrt{\frac{L}{C}}=RC\omega_0=\frac{L\omega_o}{R} \tag{3.32}$$

An LC network can be seen as the linear combination of two storage elements, a capacitor C and an inductor L: this is a second-order system. When a voltage step appears on the input V_{in}, a current circulates in the network. This current induces energy storage in L, while this is the voltage developed across C that dictates the energy it stores. In our example, when the input signal no longer changes, the energy keeps circulating between both elements as a pendulum swinging back and forth: an oscillatory process takes place, implying, without losses, an endless energy transfer between the two storage elements, C and L. If a resistor R is introduced in the path, part of the energy is lost in heat while circulating from C to L and vice versa. As the stored energy diminishes when swinging from one element to the other, the resulting oscillatory signal exponentially decays in amplitude until it completely ceases. In the RLC circuit, the quality factor—or its counter part the damping ratio—quantifies the amount of losses in the network via ohmic paths:

- Q is high or ζ is low → low losses, weak damping, high ringing;
- Q is low or ζ is high → high losses, high damping, low or no ringing.

The circuit is *damped* by the resistor presence in the Figure 3.22 circuit.

To gain further insight into the RLC network response, it is interesting to study the denominator $D(s)$ of (3.28). It is of the following form:

$$D(s) = \frac{s^2}{\omega_0^2} + \frac{s}{\omega_0 Q} + 1 \tag{3.33}$$

The poles or the roots of the equation can be found by solving the 2s values, which satisfy $D(s) = 0$:

$$\frac{s^2}{\omega_0^2} + \frac{s}{\omega_0 Q} + 1 = 0 \tag{3.34}$$

Applying classic algebra formulas, we have the following roots, already introduced in Chapter 2:

$$s_1, s_2 = \frac{\omega_0}{2Q}\left(\pm\sqrt{1-4Q^2} - 1\right) \tag{3.35}$$

Considering the damping ratio instead:

$$s_1, s_2 = \omega_0 \zeta\left(\pm\sqrt{1-\frac{1}{\zeta^2}} - 1\right) \tag{3.36}$$

This expression is close to the one we found in Chapter 2, when looking at the closed-loop compensated converter. Actually, the discussion about Q is similar as we deal with a second-order system:

- $Q < 0.5$ or $\zeta > 1$: the expression below the square root is positive and the roots are negative, real, and separate: there is no imaginary number in the solutions. We have a fully nonringing response; the system is said to be *overdamped*. In that case, theory tells us that (3.28) can be reformulated considering two cascaded RC filter.
- $Q = 0.5$ or $\zeta = 1$: the square root returns 0 and both roots are real, negative, and coincident. The system is said to be *critically damped*. We still have a nonoscillatory response, as no imaginary numbers appear in the roots.

$$s_{1,2} = -\omega_0 \tag{3.37}$$

- $Q > 0.5$: the expression below the square root becomes negative and both roots welcome imaginary numbers: we have an oscillatory response, damped by the presence of the real part, the ohmic losses in the circuit. The poles are complex conjugates with negative real parts:

$$s_{1,2} = -\frac{\omega_0}{2Q} \pm j\omega_0\sqrt{1-\frac{1}{4Q^2}} \tag{3.38}$$

also equal to:

$$s_{1,2} = -\zeta\omega_0 \pm j\omega_0\sqrt{1-\zeta^2} \tag{3.39}$$

In the previous expression, the terms $1/\zeta\omega_0$ or $\frac{2Q}{\omega_0}$ define the time constant of the system.

- As Q approaches infinity, or ζ reaches zero, the real portion fades away (the ohmic losses diminish) until the roots become pure imaginary numbers: we have a fully *undamped* system, also called an oscillator.

To help figure out what is exactly going on, it is interesting to represent the roots loci on the *s*-plan as introduced in the previous chapter. This is what Figure 3.23 represents. If we follow the arrows, we can see that a low-Q condition sets the roots apart, without imaginary portion at all. As the quality factor increases, the roots

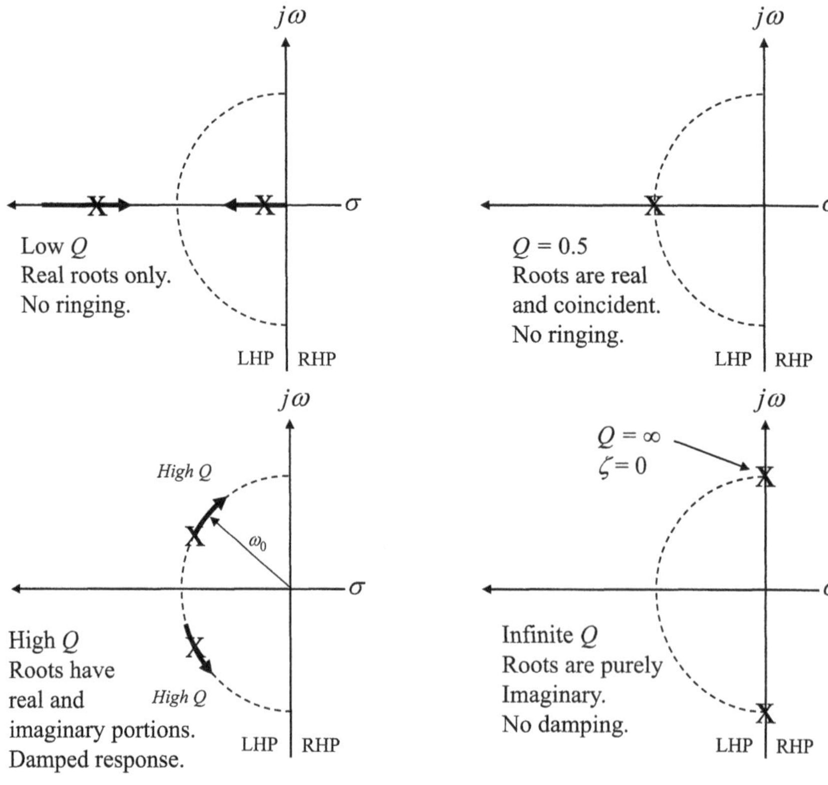

Figure 3.23 The *s*-plane helps to locate the position of the roots and see the contribution of their real or imaginary parts.

move along the x-axis until they meet at Q equals 0.5. At this point, the roots are coincident and still purely real. Then, as Q increases, the imaginary portion starts to show up while the real contribution begins to diminish. The roots slide along a semicircle of radius ω_0. As Q reaches infinity, the real portion—representative of damping or losses—has left and the roots are pure imaginary numbers: the system starts oscillating and never stops.

3.3.2 Transient Response of a Second-Order System

With the transfer function of our RLC network in the s-domain being known, we can derive the delivered response in the time domain. Such an exercise requires the selection of a stimulus applied to the system input. In our case, since we want the response to a step, such a stimulus expression in the Laplace domain is $1/s$. We have

$$v_{out}(t) = \mathcal{L}^{-1}\left\{\frac{1}{s}\frac{1}{\frac{s^2}{\omega_0^2}+2\zeta\frac{s}{\omega_0}+1}\right\} \tag{3.40}$$

The result of this calculus for $\zeta < 1$ appears next:

$$v_{out}(t) = 1 - \frac{1}{\sqrt{1-\zeta^2}}e^{-\zeta\omega_0 t}\sin(\omega_d t + \theta) \tag{3.41}$$

With a damped angular frequency equal to

$$\omega_d = \omega_0\sqrt{1-\zeta^2} \tag{3.42}$$

$$\theta = \cos^{-1}(\zeta) \tag{3.43}$$

Note that both the damped angular frequency (ω_d) and the natural angular frequency (ω_0) are equal when the damping ratio is 0 or the quality factor is infinite. Equation (3.41) can be plotted using a dedicated solver like Mathcad or simply by simulating the Figure 3.22 schematic. The results appear in Figure 3.24 for various Q values.

The right side of (3.41) shows a sinusoidal response in which amplitude is affected by an exponential term. As t increases, provided the exponential exponent remains negative, the sinusoid waveform decays in amplitude to become 0 at $t = \infty$. For certain quality factor values, the response is fully nonoscillatory (e.g., $Q = 0.1$ and $Q = 0.5$). For $Q = 0.1$, the response is extremely slow and slightly faster for $Q =$ 0.5. In both cases, note that there is no overshoot. For Q greater than 0.5, an oscillation superimposes on the signal and brings large over-/undershoots as the system is less damped (Q from 1 to 5). Please note that in spite of large oscillations, the signal eventually stabilizes to 1 V, the input setpoint signal: the response is oscillatory but stable. An unstable system would have a diverging output when responding to an input step (e.g., when the damping disappears and the quality factor becomes infinite as in the lower right side of Figure 3.23).

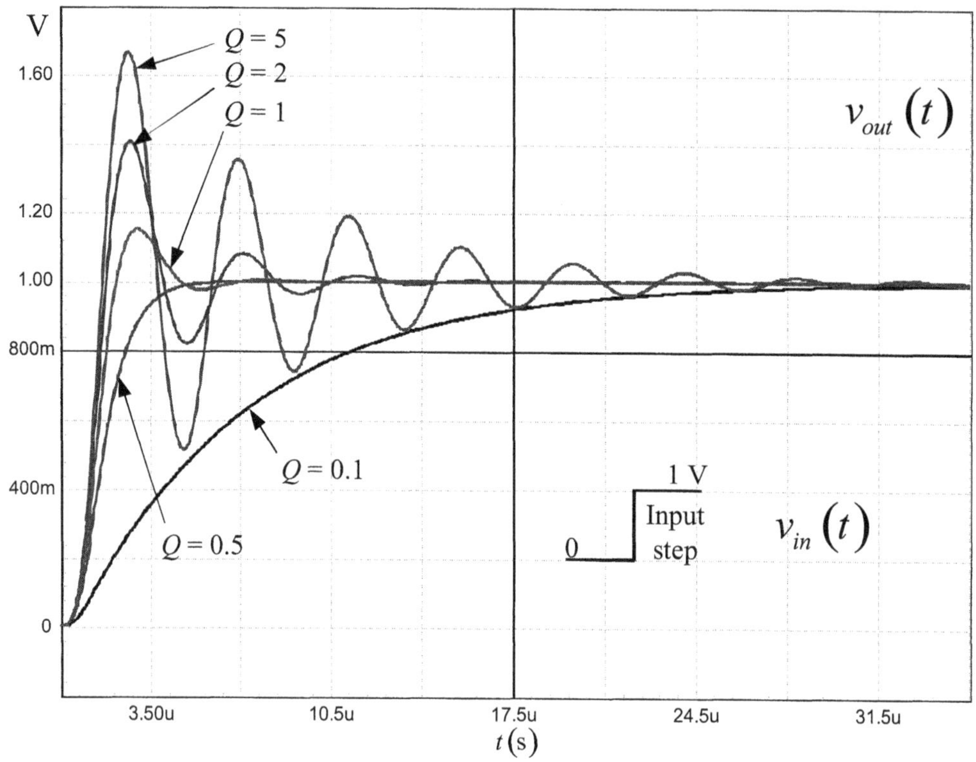

Figure 3.24 Time-domain response of the *RLC* network from Figure 3.22.

Given the impact of the quality factor on the step response, it is important to select the right Q depending on the criteria we want to meet with our control system. Figure 3.25 portrays the typical response of a second-order system where critical parameters affected by Q appear. These parameters are the following:

- The rise time, t_r: this is the time needed for the output to rise from 10 percent to 90 percent of the steady-state value for an overdamped response. For an undamped circuit, whose response appears in Figure 3.25, we consider the time needed to reach 100 percent of the steady-state value, 1 V. After the picture, this time is 17 μs.
- The peak time, t_p: for a Q greater than 0.5, at this moment of time, the signal exceeds the steady-state level and peaks to a maximum before decreasing again.
- The maximum percent overshoot, M_p: this is the maximum value the signal can take at the time t_p, expressed as a percentage of the steady-state level.
- The settling time, t_s: at this moment of time, the output signal is considered to be within the band of tolerance that you or your customer has set. At the time t_s, the output is considered to be steady state.

As detailed in Figure 3.24, depending on the value of Q, different responses are possible. For instance, you might want a response where absolutely no overshoot is accepted. In that case, a Q below 0.5 must be selected. A specification could also be

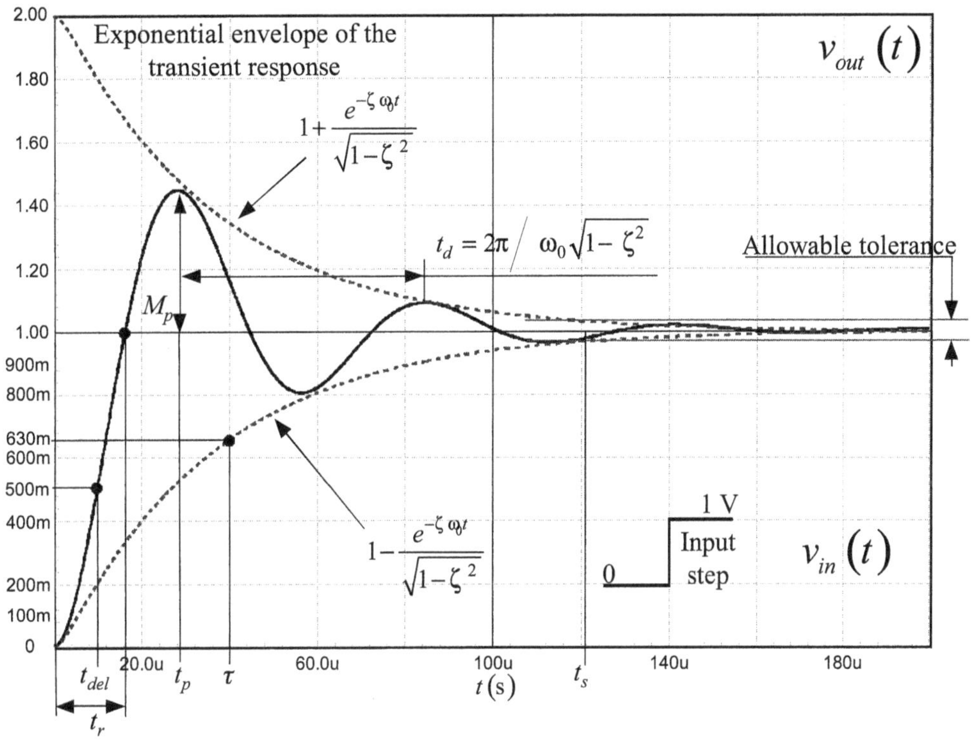

Figure 3.25 This is the classical second-order response found in textbooks where critical parameters appear. In this example, $Q = 2$ or $\zeta = 0.25$.

that a certain amount of overshoot is accepted as long as the signal rises at a pace sufficiently high, specified by the customer. For all of these reasons, it is important to analytically derive some of these important timings. Please note that these equations are derived for a Q greater than 0.5 where the roots are complex (real and imaginary coefficients).

The delay time t_{del} corresponds to the time needed by the output signal to reach 50 percent of the steady-state value, 0.5 V. From (3.41), the equation can be written as follows:

$$0.5 = \frac{1}{\sqrt{1-\zeta^2}} e^{-\zeta\omega_0 t_{del}} \sin(\omega_d t_{del} + \theta) \tag{3.44}$$

Solving this equation symbolically requires a certain amount of mathematical dexterity that I don't have! Happily, an approximated expression has already been derived and appears next:

$$t_{del} \approx \frac{1+0.7\zeta}{\omega_0} \tag{3.45}$$

To plot Figure 3.25, we used a network whose natural frequency was 18.3 kHz with a quality factor of 2. When applying numerical values to (3.45), it gives 10.22 μs, whereas Mathcad® numerical solver delivers 10.064 μs.

For an overdamped system, the rise time t_r is the time needed for the response signal to rise from 10 percent to 90 percent of the final value. For an undamped second-order system, let's say with 10 to 30 percent overshoot, the rise time is defined as the time needed to reach 100 percent of the final value or 1 V in the example. The equation to obtain the rise time definition in this last case is the following:

$$v_{out}(t_r) = 1 = 1 - \frac{1}{\sqrt{1-\zeta^2}} e^{-\zeta\omega_0 t_r} \sin(\omega_d t_r + \theta) \tag{3.46}$$

Otherwise stated:

$$0 = \frac{1}{\sqrt{1-\zeta^2}} e^{-\zeta\omega_0 t_r} \sin(\omega_d t_r + \theta) \tag{3.47}$$

The exponential term $\frac{1}{\sqrt{1-\zeta^2}} e^{-\zeta\omega_0 t_r}$ represents the waveform envelope and cannot be null in the previous equation. Therefore, we need to solve

$$0 = \sin(\omega_d t_r + \theta) = \sin\left(\omega_0 t_r \sqrt{1-\zeta^2} + \cos^{-1}(\zeta)\right) \tag{3.48}$$

Solving for the value of t_r that can cancel the sinus term will give us the result we are looking for:

$$\sin\left(\omega_0 t_r \sqrt{1-\zeta^2} + \cos^{-1}(\zeta)\right) = 0 \tag{3.49}$$

Sin(x) is zero if $x = n\pi$. Considering the case $n = 1$ and solving for t_r, we have

$$t_r = \frac{\pi - \cos^{-1}(\varsigma)}{\omega_0 \sqrt{1-\varsigma^2}} \tag{3.50}$$

For a quality factor of 2 or a damping ratio of 0.25, we find a rise time of 16.4 μs, in agreement with Figure 3.25. For small damping ratios, the previous expression simplifies to

$$\lim_{\zeta \to 0} \frac{\pi - \cos^{-1}(\varsigma)}{\omega_0 \sqrt{1-\varsigma^2}} \approx \frac{\pi - \frac{\pi}{2}}{\omega_0} \approx \frac{1.6}{\omega_0} \tag{3.51}$$

In our case with a 18,300-Hz natural frequency, we have an approximate rise time of

$$t_r \approx \frac{1.6}{2 \times \pi \times 18300} \approx 14\ \mu\text{s} \tag{3.52}$$

Let us now calculate the time t_p at which the output is maximum. To obtain this number, we will derive $v_{out}(t)$ and calculate the time value for which the result equals zero. Mathematically, it means

$$\frac{d}{dt}\left(1 - \frac{1}{\sqrt{1-\zeta^2}} e^{-\zeta\omega_0 t} \sin(\omega_d t + \theta)\right) = 0 \tag{3.53}$$

The derivative of a product *uv* is *vu*'-*uv*', thus,

$$\frac{dv_{out}(t)}{dt} = \frac{\zeta\omega_0 e^{-\zeta\omega_0 t}\sin(\omega_d t+\theta)}{\sqrt{1-\zeta^2}} - \frac{\omega_d e^{-\zeta\omega_0 t}\cos(\omega_d t+\theta)}{\sqrt{1-\zeta^2}} = 0 \tag{3.54}$$

Simplifying, we have

$$\zeta\omega_0 \sin(\omega_d t+\theta) - \omega_d \cos(\omega_d t+\theta) = 0 \tag{3.55}$$

ζ can be extracted from (3.43) to obtain

$$\zeta = \cos(\theta) \tag{3.56}$$

If we substitute this value in (3.42), we obtain a new definition for ω_d, valid for $0 \le \theta \le \pi$:

$$\omega_d = \omega_0\sqrt{1-\cos^2(\theta)} = \omega_0 \sin(\theta) \tag{3.57}$$

Combining (3.56) and (3.57) into (3.55),

$$\sin(\omega_d t+\theta)\cos(\theta)\omega_0 - \omega_0\cos(\omega_d t+\theta)\sin(\theta) = 0 \tag{3.58}$$

If we divide everything by ω_0, we have an equation of the form

$$\sin a\cos b - \cos a\sin b = \sin(a-b) \tag{3.59}$$

Otherwise stated,

$$\sin(\omega_d t + \theta - \theta) = 0 \tag{3.60}$$

Sin (x) can go to zero for $x = 0$ or $x = n\pi$. In our case, $t = 0$ is the starting time; therefore, the next possible value is π:

$$\omega_d t = \pi \tag{3.61}$$

If we solve for t_p, we obtain

$$t_p = \frac{\pi}{\omega_d} = \frac{\pi}{\omega_0\sqrt{1-\zeta^2}} = \frac{1}{2f_0\sqrt{1-\zeta^2}} \tag{3.62}$$

The numerical value for a *Q* of 2 or a damping ratio ζ of 0.25 is $t_p \approx 28.2\ \mu s$. As you can see on Figure 3.25, this value corresponds to half of the oscillating period. The damped period t_d is thus obtained by a simple multiplication by two, in agreement with (3.42):

$$t_d = \frac{2\pi}{\omega_0\sqrt{1-\zeta^2}} = \frac{1}{f_0\sqrt{1-\zeta^2}} \tag{3.63}$$

Now that the time at which the function peaks is known, we can substitute its value into (3.41) while replacing ω_d by its definition with (3.42):

$$v_{out}(t_p) = 1 - \frac{1}{\sqrt{1-\zeta^2}} e^{-\zeta\omega_0\left(\frac{\pi}{\omega_0\sqrt{1-\zeta^2}}\right)} \sin\left(\omega_0\sqrt{1-\zeta^2}\left(\frac{\pi}{\omega_0\sqrt{1-\zeta^2}}\right)+\theta\right)$$

$$= 1 + \frac{e^{-\frac{\zeta\pi}{\sqrt{1-\zeta^2}}}}{\sqrt{1-\zeta^2}} \sin(\theta) \tag{3.64}$$

From (3.56), the right term in the denominator of (3.64) can be reformulated, again considering $0 \le \theta \le \pi$:

$$\sqrt{1-\zeta^2} = \sqrt{1-\cos^2\theta} = \sqrt{\sin^2\theta} = \sin\theta \tag{3.65}$$

Therefore, (3.64) can be further simplified to

$$v_{out}(t_p) = 1 + e^{-\frac{\zeta\pi}{\sqrt{1-\zeta^2}}} \tag{3.66}$$

The mathematical definition for the maximum percent overshoot M_p is the following one:

$$M_p = \frac{V_{out}(t_p) - V_{out}(\infty)}{V_{out}(\infty)} \tag{3.67}$$

With our steady-state value being 1 V, the right-side extra term in (3.66) is simply the overshoot:

$$M_p(\%) = e^{-\frac{\zeta\pi}{\sqrt{1-\zeta^2}}} \times 100 \tag{3.68}$$

If we express the damping ratio in relationship to the quality factor Q, the previous formula becomes

$$M(\%) = e^{-\frac{\pi}{\sqrt{4Q^2-1}}} \times 100 \tag{3.69}$$

From (3.68) and (3.69), we can plot the relationship between the overshoot and the quality factor (or the damping ratio) of our second-order system. It appears in Figure 3.26.

For the quality factor of 2 used to plot Figure 3.25, we have a 44 percent overshoot. What is the maximum overshoot value the output can be affected by? Well, when the damping ratio reaches zero or when the quality factor goes to infinity, we have

$$\lim_{Q\to\infty} e^{-\frac{\pi}{\sqrt{4Q^2-1}}} = 100\% \tag{3.70}$$

In this case, the input step is simply doubled. This is what was already shown in Figure 3.25, where the top of the envelope is 2 V, corresponding to a 100 percent overshoot.

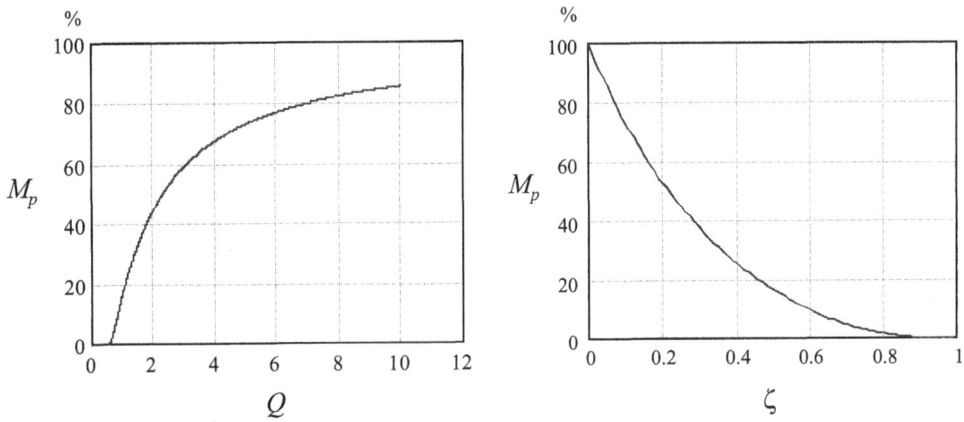

Figure 3.26 The overshoot depends on the selected quality factor when above 0.5 or on the damping ratio when lower than 2.

By applying these formulas, we can find the various peaks values. The first one is equal to

$$V_{peak1} = \left(1 + e^{-\frac{\pi}{\sqrt{4Q^2-1}}}\right) \times 1 = (1 + 0.444) \times 1 = 1.44 \text{ V} \tag{3.71}$$

in line with Figure 3.25 results.

The rest of the peaks can be found by selecting different $n\pi$, with $n = 2$ (negative), $n = 3$ (positive), and so on. For instance, the second peak, which is an undershoot, can be computed as follows given the 1-V step input:

$$V_{peak2} = \left(1 - e^{-\frac{2\pi}{\sqrt{4Q^2-1}}}\right) \times 1 = (1 - 0.197) \times 1 = 802 \text{ mV} \tag{3.72}$$

This is what we can read in Figure 3.25. The third peak is found by letting $n = 3$:

$$V_{peak3} = \left(1 + e^{-\frac{3\pi}{\sqrt{4Q^2-1}}}\right) \times 1 = (1 + 87.7m) \times 1 = 1.087 \text{ V} \tag{3.73}$$

This is exactly the displayed value.

It is sometimes interesting to derive the damping ratio (or the quality factor) and the resonant frequency from an observed waveform such as the one presented in Figure 3.27. The best way to obtain this parameter is to measure the attenuation level between two consecutive peaks as those represented in Figure 3.27. From the picture, the ratio α between the two positive overshoots is

$$\alpha = \frac{V(t_0)}{V(t_0 + t_d)} = \frac{e^{-\frac{\pi}{\sqrt{4Q^2-1}}}}{e^{-\frac{3\pi}{\sqrt{4Q^2-1}}}} \tag{3.74}$$

From the picture, we extracted a value of 2.413. Please make sure you considered only the overshoot; do not forget to remove the final value from your readings

(5 V here). From this equation, if we take the natural logarithm of both sides, we have a definition for δ (delta) known as the *logarithmic decrement*:

$$\delta = \ln(\alpha) = \ln\left(\frac{e^{-\frac{\pi}{\sqrt{4Q^2-1}}}}{e^{-\frac{3\pi}{\sqrt{4Q^2-1}}}}\right) = \ln\left(e^{-\frac{\pi}{\sqrt{4Q^2-1}}}\right) - \ln\left(e^{-\frac{3\pi}{\sqrt{4Q^2-1}}}\right)$$

$$= \frac{1}{\sqrt{4Q^2-1}}(-\pi + 3\pi) = \frac{2\pi}{\sqrt{4Q^2-1}} \tag{3.75}$$

Otherwise written:

$$\delta = \frac{2\pi}{\sqrt{4Q^2-1}} \tag{3.76}$$

From this value, the quality factor Q is a few lines of algebra away:

$$Q = \sqrt{\frac{4\pi^2 + \delta^2}{4\delta^2}} = \sqrt{\left(\frac{\pi}{\delta}\right)^2 + \frac{1}{4}} \tag{3.77}$$

If you prefer the damping ratio instead, the formula becomes

$$\zeta = \frac{1}{\sqrt{\left(\frac{2\pi}{\delta}\right)^2 + 1}} \tag{3.78}$$

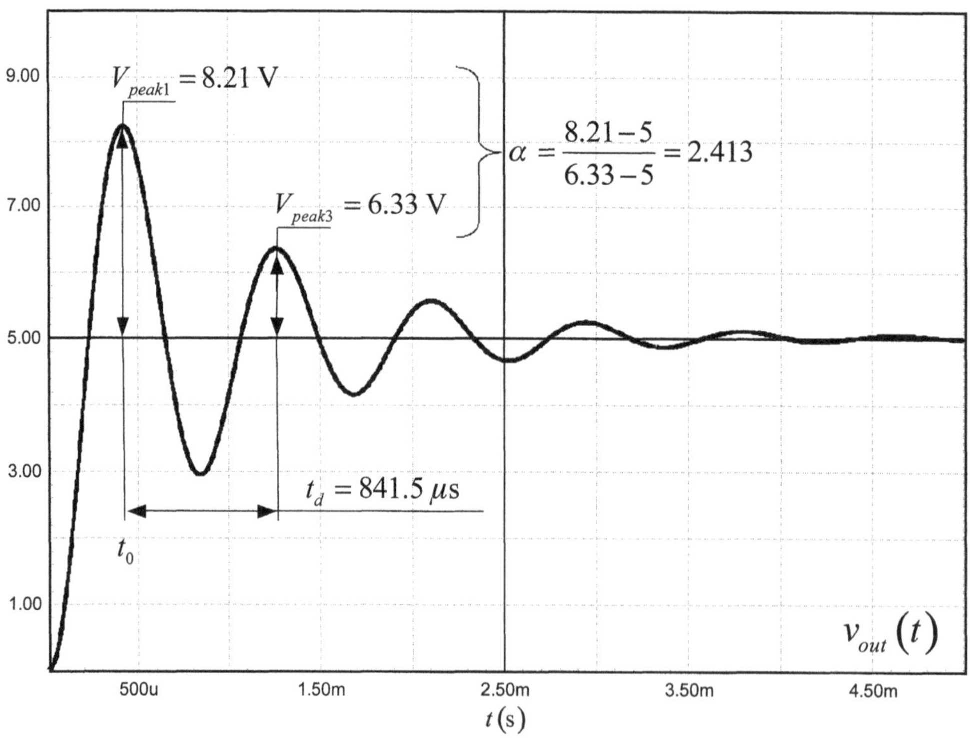

Figure 3.27 The typical response of a second-order filter to a 5-V step with a quality factor of 3.6.

If we apply the numerical value found in Figure 3.27, we have

$$Q = \sqrt{\left(\frac{3.14}{\ln(2.413)}\right)^2 + 0.25} = 3.6 \tag{3.79}$$

and

$$\zeta = \frac{1}{\sqrt{\left(\frac{6.28}{\ln(2.413)}\right)^2 + 1}} = 0.139 \tag{3.80}$$

The resonant frequency is easily obtained via (3.63) once the time distance between two peaks has been measured. We have 841.5 μs from the graph

$$f_0 = \frac{1}{t_d\sqrt{1-\zeta^2}} = \frac{1}{841.5u\sqrt{1-0.139^2}} = 1.2 \text{ kHz} \tag{3.81}$$

As the calculation is based on the identification of peaks of two consecutive positive amplitudes, you realize that the parameter extraction can become approximated if not impossible when the response is really damped (e.g., for quality factors below 1). When Q is 0.5, in any case, the two poles are coincident and the overshoot is gone.

We almost have all our timing definitions except the settling time. The settling time t_s represents the time needed for the output to be within its allowable tolerance. Let's say this tolerance is 2 percent of V_{out}. According to (3.41), the transient response is a sinusoidal signal whose envelope amplitude is a decaying exponential. This envelope can be put under the following expression:

$$1 \pm e^{-\zeta\omega_0 t} \tag{3.82}$$

At $t = 0$, the exponential term is 1 and the signal starts from 0 as confirmed in Figure 3.25. At $t = t_s$, the envelope must have reduced to a point where it stays within the "tunnel" defined by the allowable tolerance. If we consider a 2 percent tolerance and a 1-V signal, we can write

$$1 \pm e^{-\zeta\omega_0 t} = 1 \pm 0.02 \tag{3.83}$$

which, further to simplification, leads to solving for t when

$$e^{-\zeta\omega_0 t} = 0.02 \tag{3.84}$$

By using logs, we have

$$t_s = -\frac{\ln 0.02}{\omega_0\zeta} \approx \frac{3.9}{\omega_0\zeta} \tag{3.85}$$

If we apply the numerical values of the circuit used to plot Figure 3.25 (ω_0 = 114.4 krad/s):

$$t_s = \frac{3.9}{0.25 \times 114.4k} = 136\ \mu\text{s} \tag{3.86}$$

Table 3.1 Summary of Parameter Definitions for an Under-Damped Second-Order System

Parameter	*Definition*	*Value*
ω_d	The damped angular frequency in relationship to the natural angular frequency ω_0	$\omega_d = \omega_0\sqrt{1-\zeta^2}$
t_{del}	Delay time, defined for $v_{out}(t)$ equals 50 percent of the final value	$t_{del} \approx \frac{1+0.7\zeta}{\omega_0}$
δ	Delta, the logarithmic decrement	$\delta = \ln\left(\frac{x(t_0)}{x(t_0+t_d)}\right) = \frac{2\pi}{\sqrt{4Q^2-1}}$
t_r	Rise time, defined when $v_{out}(t)$ reaches 100 percent of the final value	$t_r = \frac{\pi - \cos^{-1}(\varsigma)}{\omega_0\sqrt{1-\varsigma^2}}$
t_p	Peak time, the time at which $v_{out}(t)$ reaches a maximum	$\frac{\pi}{\omega_0\sqrt{1-\zeta^2}}$
$M_p(\%)$	Overshoot value	$e^{-\frac{\zeta\pi}{\sqrt{1-\zeta^2}}} \times 100$
t_s	Settling time, the time needed for $v_{out}(t)$ to be within x percent of the final value	$-\frac{\ln x}{\omega_0\zeta}$

From these equations, we can see the quality factor (or the damping ratio) affects a lot of parameters. Its choice clearly depends on the final set of specifications you have: is overshoot acceptable, what settling time is acceptable, and so on.

Table 3.1 offers a summary of the variables we have derived for an underdamped system ($Q > 0.5$).

We have seen in Figure 3.6 that the phase margin changes the response of the system, exactly as the Q of the *RLC* system does. Therefore, there must be a relationship between the phase margin as measured on an *open-loop* system and the quality factor observed after the loop is *closed*. If we can find it, we will be able to pick a phase margin design goal based on the type of transient response we expect.

3.3.3 Phase Margin and Quality Factor

Assume a simple converter as the one appearing in Figure 3.28. We can see a plant featuring a transfer function $H(s)$ and a compensator $G(s)$ processing the error signal ε.

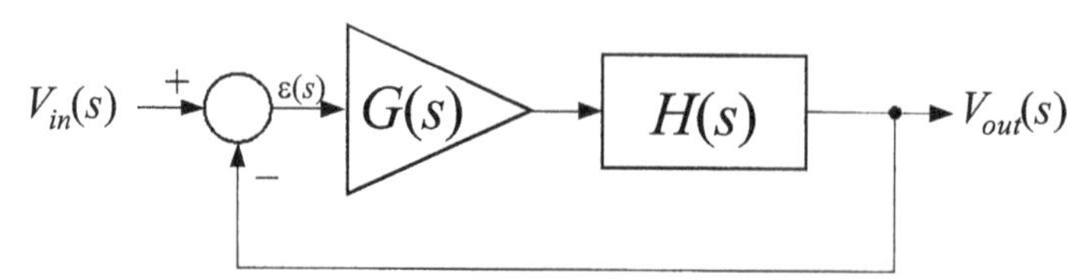

Figure 3.28 A compensated converter featuring a return path of a unity gain.

The closed-loop transfer function of such a system can be easily derived with a few lines of algebra:

$$V_{out}(s) = \left[V_{in}(s) - V_{out}(s)\right]G(s)H(s) \tag{3.87}$$

As we have seen, the loop gain $G(s)H(s)$ is also noted $T(s)$. Substituting and rearranging (3.87), we have:

$$\frac{V_{out}(s)}{V_{in}(s)} = \frac{T(s)}{1+T(s)} \tag{3.88}$$

This is the closed-loop gain expression of Figure 3.28 unity feedback system. Now, let's assume that its open-loop response $T(s)$ is plotted in Figure 3.29. We know that a control system reacts to incoming perturbations as long as some gain exists at the frequencies of concern. We also know that stability is ensured if we limit the bandwidth by forcing the gain to roll off as the frequency increases. To limit the phase lag at crossover, and thus obtain a good phase margin, we bend the loop gain to cross the 0-dB axis with a –1 slope, implying a single pole response slightly before and after f_c. We also learned that at some point, when the phase lag reaches 180°, the loop gain must be sufficiently low to ensure a good gain margin. To speed-up the gain decrease beyond f_c and ensure sufficient gain margin, a second pole is generally installed after the crossover point. If we use a magnifier to observe

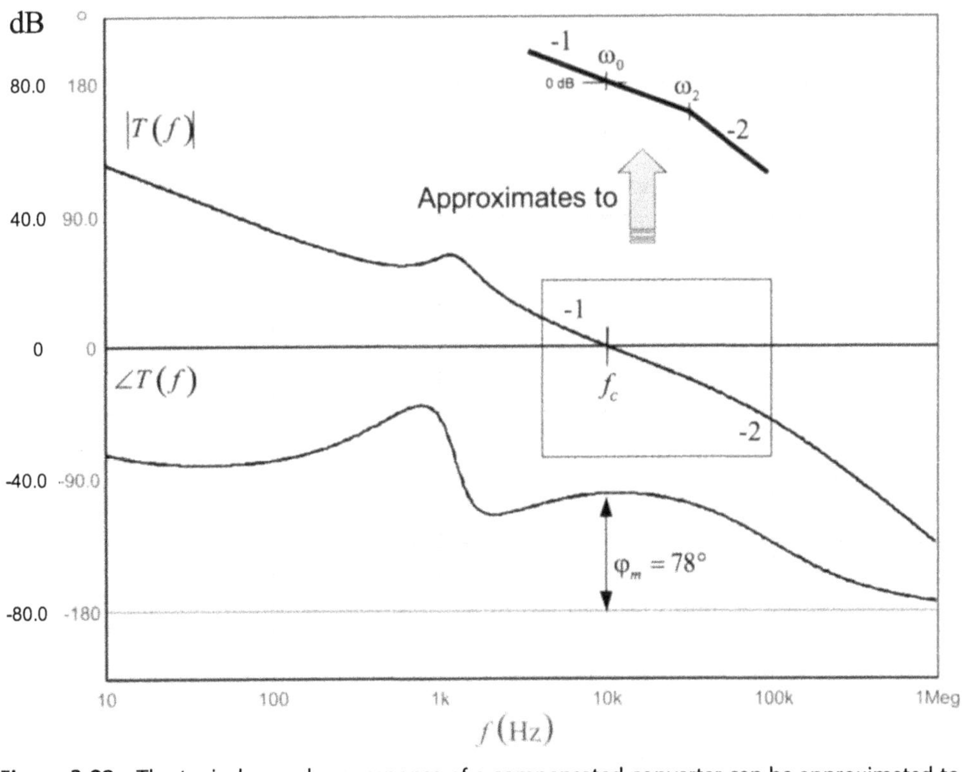

Figure 3.29 The typical open-loop response of a compensated converter can be approximated to a second-order system in the vicinity of the crossover frequency.

the curve around the crossover frequency (see the framed area in Figure 3.29), we can see a two-pole configuration. This two-pole configuration is constructed with a 0-dB crossover pole, ω_0, and a high-frequency pole, ω_2. The transfer function of these cascaded poles can be put under the following form:

$$T(s) \approx \frac{1}{\left(\frac{s}{\omega_0}\right)\left(1+\frac{s}{\omega_2}\right)} \tag{3.89}$$

In this approximated expression, we clearly do not consider extra poles and zeros away from f_c, naturally eliminating their impact on the transfer function. However, our interest lies in the approximate response the converter is going to deliver once its loop is closed. In other terms, let us identify the closed-loop transfer function derived from (3.89) and rearranged according to (3.88):

$$\frac{T(s)}{1+T(s)} = \frac{\frac{1}{\left(\frac{s}{\omega_0}\right)\left(1+\frac{s}{\omega_2}\right)}}{1+\frac{1}{\left(\frac{s}{\omega_0}\right)\left(1+\frac{s}{\omega_2}\right)}} = \frac{1}{\frac{s^2}{\omega_0\omega_2}+\frac{s}{\omega_0}+1} \tag{3.90}$$

The right term of (3.90) looks familiar. It actually recalls the form already derived in (3.28) from which we can establish a relationship with the coefficients of (3.90):

$$\frac{1}{\frac{s^2}{\omega_0\omega_2}+\frac{s}{\omega_0}+1} = \frac{1}{\frac{s^2}{\omega_r^2}+\frac{s}{\omega_r Q_c}+1} \tag{3.91}$$

The identification of the closed-loop quality factor Q_c and the resonant frequency ω_r is straightforward:

$$Q_c = \sqrt{\frac{\omega_0}{\omega_2}} \tag{3.92}$$

$$\omega_r = \sqrt{\omega_0\omega_2} \tag{3.93}$$

We now have an equation that describes the approximated closed-loop response of our converter and it includes a quality factor Q_c. The next step is to derive a relationship between Q_c and the key design parameter, the open-loop phase margin φ_m. First, based on (3.89), let us calculate the crossover frequency brought by the location of the 0-dB crossover pole ω_0 and its associated high frequency pole ω_2. At the crossover point f_c, we know that the magnitude of $T(s)$ equals 1 or 0 dB. Therefore, if we assume a harmonic excitation at the crossover frequency, we can replace s by $j\omega_c$ in (3.89) and write:

$$\left|\frac{1}{\left(\frac{j\omega_c}{\omega_0}\right)\left(1+\frac{j\omega_c}{\omega_2}\right)}\right| = 1 \tag{3.94}$$

Otherwise stated:

$$\left|\left(\frac{j\omega_c}{\omega_0}\right)\left(1+\frac{j\omega_c}{\omega_2}\right)\right| = \left|\left(\frac{j\omega_c}{\omega_0}\right)\right|\left|\left(1+\frac{j\omega_c}{\omega_2}\right)\right| = 1 \tag{3.95}$$

Extracting the individual magnitudes gives us

$$\frac{\omega_c}{\omega_0}\sqrt{1+\left(\frac{\omega_c}{\omega_2}\right)^2} = 1 \tag{3.96}$$

To get rid of the square root, we square both sides of the expression:

$$\left(\frac{\omega_c}{\omega_0}\right)^2\left(1+\left(\frac{\omega_c}{\omega_2}\right)^2\right) = 1 \tag{3.97}$$

From (3.92), we extract ω_0 and plug it in (3.97):

$$\frac{\omega_c^2\left(\omega_2^2+\omega_c^2\right)}{\left(Q_c\omega_2\right)^4} = 1 \tag{3.98}$$

Solving for ω_c, we obtain

$$\omega_c = \frac{\omega_2\sqrt{\left(\sqrt{1+4Q_c^4}-1\right)}}{\sqrt{2}} \tag{3.99}$$

Equation (3.99) shows us how the closed-loop quality factor and the open-loop crossover frequency are linked. It is important for this remark to be well understood: Q_c represents the resulting *closed-loop* response quality factor based on the pole/zero arrangement describing the approximated *open-loop* compensated transfer function $T(s)$ near the crossover frequency in (3.89).

To continue further with our analysis, we can evaluate the argument of $T(s)$ defined in (3.89) at the crossover point ω_c:

$$\arg T(\omega_c) = -\left(\tan^{-1}\frac{\omega_c/\omega_0}{0}+\tan^{-1}\frac{\omega_c}{\omega_2}\right) = -\frac{\pi}{2}-\tan^{-1}\frac{\omega_c}{\omega_2} \tag{3.100}$$

Looking at Figure 3.7, the phase margin φ_m represents the distance between the total phase lag at crossover and the –180° limit. We can write

$$\arg T(\omega_c) = -180+\varphi_m \tag{3.101}$$

Rearranging and introducing π rather than 180°, we obtain

$$\varphi_m = \pi + \arg T(\omega_c) \tag{3.102}$$

Substituting (3.100) in (3.102), we obtain

$$\varphi_m = \pi - \tan^{-1}\frac{\omega_c}{\omega_2} - \frac{\pi}{2} = \frac{\pi}{2} - \tan^{-1}\frac{\omega_c}{\omega_2} \tag{3.103}$$

Remembering our (far away) trigonometric classes, we know that

$$\tan^{-1} x + \tan^{-1}\frac{1}{x} = \frac{\pi}{2} \tag{3.104}$$

Replacing $\pi/2$ in (3.103) by its expression in (3.104), we obtain

$$\varphi_m = \tan^{-1}\frac{\omega_c}{\omega_2} + \tan^{-1}\frac{\omega_2}{\omega_c} - \tan^{-1}\frac{\omega_c}{\omega_2} \tag{3.105}$$

Collecting the remaining terms, we finally obtain a simpler definition based on the position of the second pole in relationship to the crossover frequency:

$$\varphi_m = \tan^{-1}\frac{\omega_2}{\omega_c} \tag{3.106}$$

We have already defined the crossover angular frequency versus the closed loop quality factor in (3.99). If we capitalize on this definition in (3.106) we have

$$\varphi_m = \tan^{-1}\left(\sqrt{\frac{2}{\sqrt{\left(1+4Q_c^4\right)}-1}}\right) \tag{3.107}$$

The next step is to extract the closed-loop quality factor from (3.107) and simplify the result:

$$Q_c = \frac{\sqrt[4]{1+\tan^2(\varphi_m)}}{\tan(\varphi_m)} \tag{3.108}$$

Again, looking back at what we learned at school, we remember that

$$1+\tan^2\varphi_m = \frac{1}{\cos^2\varphi_m} \tag{3.109}$$

Replacing the expression under the root in (3.108), we have

$$Q_c = \frac{\sqrt[4]{\dfrac{1}{\cos^2(\varphi_m)}}}{\tan(\varphi_m)} = \frac{1}{\sqrt{\cos\varphi_m}}\frac{\cos\varphi_m}{\sin\varphi_m} = \frac{\sqrt{\cos\varphi_m}}{\sin\varphi_m} \tag{3.110}$$

This expression also works backward if you already have measured the closed-loop quality factor:

$$\varphi_m = \cos^{-1}\left(\frac{\sqrt{4Q_c^4+1}-1}{2Q_c^2}\right) \tag{3.111}$$

This is it! We now have a relationship between our main design criterion, the open-loop phase margin φ_m, and the closed-loop quality factor Q_c. Figure 3.30

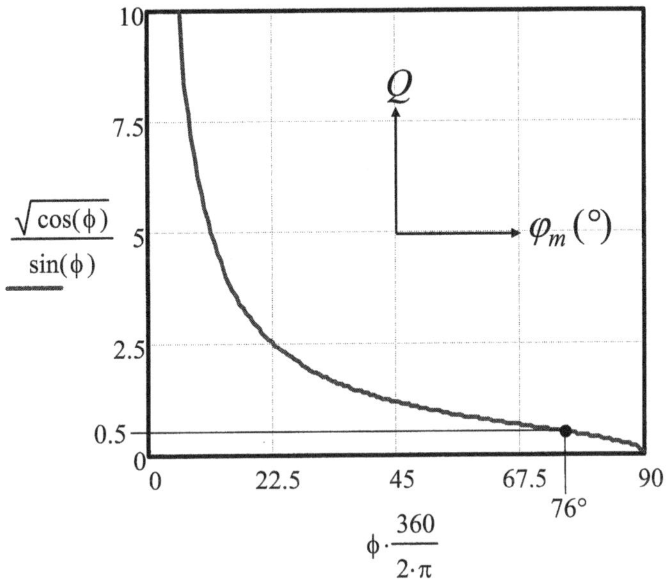

Figure 3.30 The graph shows the evolution of the closed-loop quality factor Q as you select different phase margins φ_m.

graphs the quality factor your closed-loop system will exhibit depending on the phase margin you have selected at the design stage.

We have drawn the ac closed-loop response of a second-order system compensated to give different open-loop phase margins. The result appears in Figure 3.31 and clearly show the peaking as with a *RLC* filter.

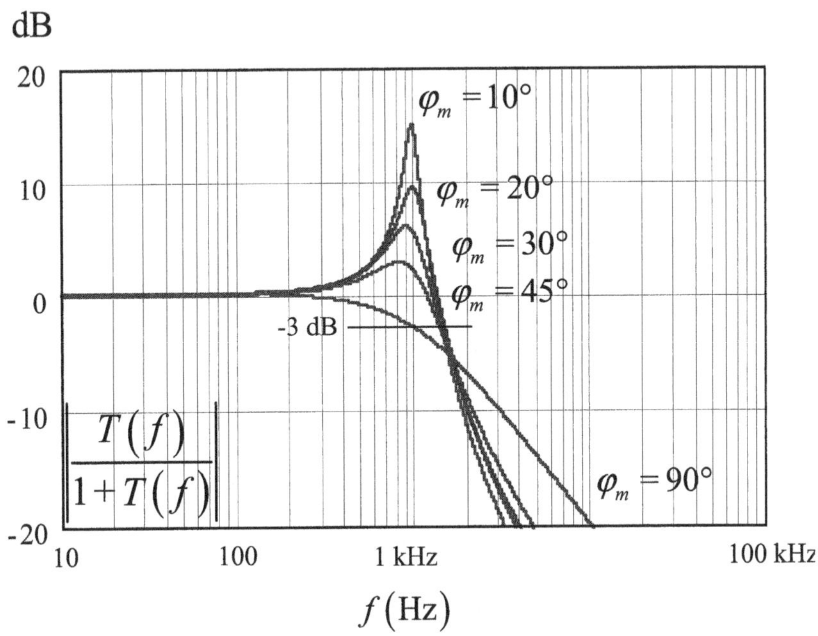

Figure 3.31 By selecting different open-loop phase margin values, we modify the ac response of the closed-loop system.

The peaking amplitude noted M_m can also be analyzed. Going back to the transfer function of a second-order system as defined by (3.28), we first express its magnitude by replacing s with $j\omega$:

$$T(j\omega) = \frac{1}{\frac{(j\omega)^2}{\omega_0^2} + 2\zeta\frac{j\omega}{\omega_0} + 1} = \frac{1}{1 - \left(\frac{\omega}{\omega_0}\right)^2 + j2\zeta\frac{\omega}{\omega_0}} \tag{3.112}$$

The magnitude can be extracted as follows:

$$|T(j\omega)| = \frac{1}{\sqrt{\left[1 - \left(\frac{\omega}{\omega_0}\right)^2\right]^2 + \left(2\zeta\frac{\omega}{\omega_0}\right)^2}} \tag{3.113}$$

To obtain the frequency at which the expression peaks, the resonant frequency, we derive the previous function and solve for ω, which makes the result go to zero:

$$\frac{d|T(j\omega)|}{d\omega} = 0 \tag{3.114}$$

Solving for the equation leads to

$$\omega_m = \omega_0\sqrt{1 - 2\zeta^2} \tag{3.115}$$

defined for $\zeta < \frac{1}{\sqrt{2}}$.

The value of the overshoot itself is obtained by plugging (3.115) into (3.113). Once done, we obtain

$$M_m = \frac{1}{2\zeta\sqrt{1 - \zeta^2}} \tag{3.116}$$

If you want to combine speed and lack of overshoot, Figure 3.23 suggests a Q of 0.5 where both poles are coincident, without imaginary contributions. Reading the corresponding phase margin in Figure 3.30, we can see that a recommended value of 76° satisfies this request for such a Q. We are far away from the 45° target found in the majority of textbooks! What does it mean then? It means that you select the phase margin based on the transient response you need. In the response to a load step, once the loop is closed, the open-loop phase margin mostly affects the recovery speed and a little the undershoot depth. If a fast recovery is needed and some overshoot accepted, then reducing the phase margin can be an option. On the contrary, if absolutely no overshoot is tolerated, you have no other choice than to increase the phase margin to the detriment of the recovery speed. Whatever solution you select, you must ensure that, despite changing operating conditions such as input/output voltage or load, ambient temperature, and production spreads (equivalent series resistors of capacitors, for instance), the phase margin never goes below 45°: the obtained ringing would be unacceptable. Some customers like military or space agencies ask for a minimum phase margin of 90° and must be backed by Monte Carlo analysis to prove that dispersions are under control. For the general case, aiming for a typical value around 70° should become a good design practice.

3.3.4 Opening the Loop to Measure the Phase Margin

The ac response of a closed-loop system can be deduced from its transient response but it is complex to extract parameters such as crossover or phase margin. To study a closed-loop system, we must physically open the loop at some appropriate point to access the transfer functions of interest: $H(s)$, the plant transfer function, $G(s)$ the compensator we designed, and $T(s)$ the open-loop gain. The first transfer function, $H(s)$, is the starting point of control loop study: we must know the frequency response of the system we want to stabilize. The compensator block G is then tailored based on the collected data and the desired response: we want to force the loop gain crossover at the selected frequency while ensuring enough phase margin at this salient point. The study of the loop gain on the hardware prototype will eventually tell us if G was well designed, revealing the obtained crossover frequency and the associated phase margin. As we will later see, the study of the whole system can also be undertaken analytically or by using a SPICE simulator.

A typical loop opening appears in Figure 3.32: this is the classical example found in textbooks. A signal is injected in the left input while its effects are observed on outputs A or B depending on what you want to plot, $H(s)$, $G(s)$, or $T(s)$. The drawing represents a control system where the output must follow the input signal. We have seen in Chapter 1 that other types of control systems exist and are called regulators. These systems work with a fixed control input and ensure a constant

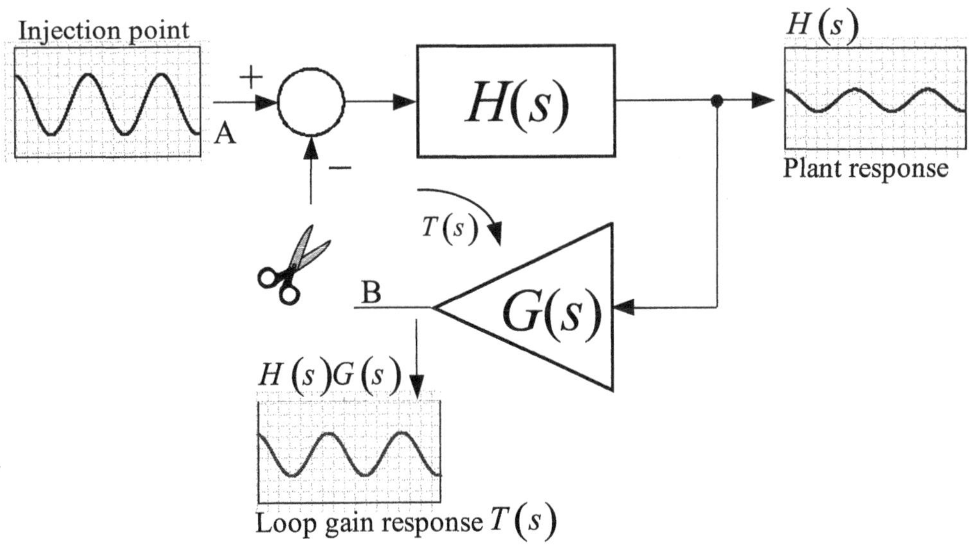

Figure 3.32 The loop is opened just prior the inverted input on the adder. Please note that the negative sign is not included in the path under study.

output variable regardless of the operating conditions. A dc-to-dc converter is a regulator: it delivers a constant voltage even if the input level changes or if the current drawn by the load varies.

In such a regulator, a reference source, often noted V_{ref}, imposes the setpoint while the system strives to ensure a constant output (usually a multiple of V_{ref}) whatever the perturbations. The perturbations are the input voltage changes, the delivered current or the temperature if it needs to be accounted for. In such an arrangement, the loop gain $T(s)$ is given by

$$T(s) = H(s)G(s) \tag{3.117}$$

As you can see, the reversal sign brought by the inverting input is not included in the formula since we open the loop at point B. The system becomes unstable when the signal injected in A becomes lagged by 180° in B. The phase margin is thus measured by reference to the −180° distance and the condition for instability is noted:

$$\angle T(j\omega) = -180° \tag{3.118}$$

To reflect an implementation closer to reality, the Figure 3.32 sketch must be updated by what is proposed in Figure 3.33.

In this drawing, the varying setpoint is replaced by a fixed reference voltage V_{ref}. The output is subtracted to this voltage level and the error voltage ε enters the compensation block G. It then further controls the power H to deliver the right output level. As explained, this level can be affected by perturbations such as the input voltage and the output current. In this representation, these perturbations are modeled as sources that subtract from the plant output. We will later see, in particular for the output impedance, that different arrangements can be made. In

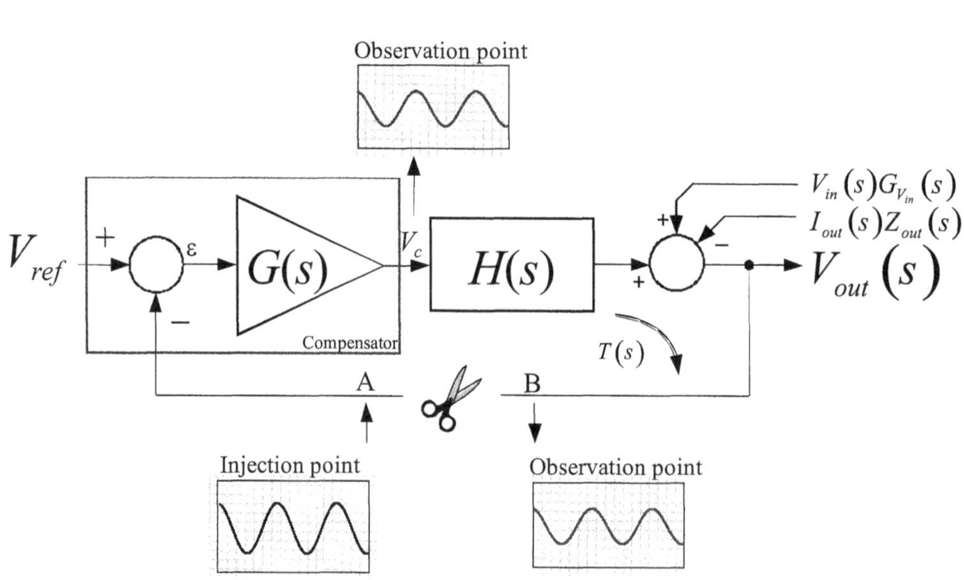

Figure 3.33 A regulator maintains a constant output regardless of external perturbations. The opening of the loop can take place in the return path, and several signals can be observed.

the example, the return path is broken and used to inject the modulating signal. If we inject in A, we can observe V_c and check that the transfer function of G is the one we were looking for:

$$G(s) = \frac{V_c(s)}{V_A(s)} \tag{3.119}$$

Now, if we inject in A but observe V_c and V_{out}, we have the plant transfer function $H(s)$:

$$H(s) = \frac{V_{out}(s)}{V_c(s)} \tag{3.120}$$

Finally, if we inject in A and observe B, we have the complete open-loop gain, $T(s)$:

$$T(s) = \frac{V_B(s)}{V_A(s)} \tag{3.121}$$

In this particular configuration, the reversal action of the compensator is now included and the loop gain definition becomes

$$T(s) = -H(s)G(s) \tag{3.122}$$

The instability condition is now that the signal injected in A returns in phase in B (with equal amplitude, of course). This complete lag implies an instability condition redefined as

$$\angle T(j\omega) = -360° \text{ or } 0° \tag{3.123}$$

The phase margin is thus measured by reference to –360° or 0°. For this reason, all plots in Figure 3.34 refer to a similar phase margin. Network analyzers or SPICE graphical tools usually display the open-loop phase as in the upper graph.

It is important to note that this ac analysis makes sense only if the perturbation such as the input voltage or the input current is kept constant during the ac sweep. You select them to fix the operating point at which the converter must be analyzed, and they no longer change during the harmonic sweep.

In the given examples, the loop has been physically open, meaning the operating point of the closed-loop system is lost. In order to put the system in conditions representative of its closed-loop working mode, you will need to recreate the operating point during the ac sweep. If the converter is compensated to deliver 19 V/3 A while the input voltage is 100 V, then physically opening the loop implies the addition of an external bias keeping a 19 V output while it delivers 3 A. If it can sometimes work to extract the plant transfer function of a given converter, it is almost impossible for the open-loop gain. As the dc gain of the loop is purposely made extremely high, a very small variation in the dc signal (e.g., noise, thermal drift of the source) causes the output to jump into its lower or upper bounds. When manipulating high-power systems, this is not something you want to undertake. Fortunately, alternative methods exist and authorize open-loop measurements without opening the loop. We will come back to the technique when studying practical cases in Chapter 9.

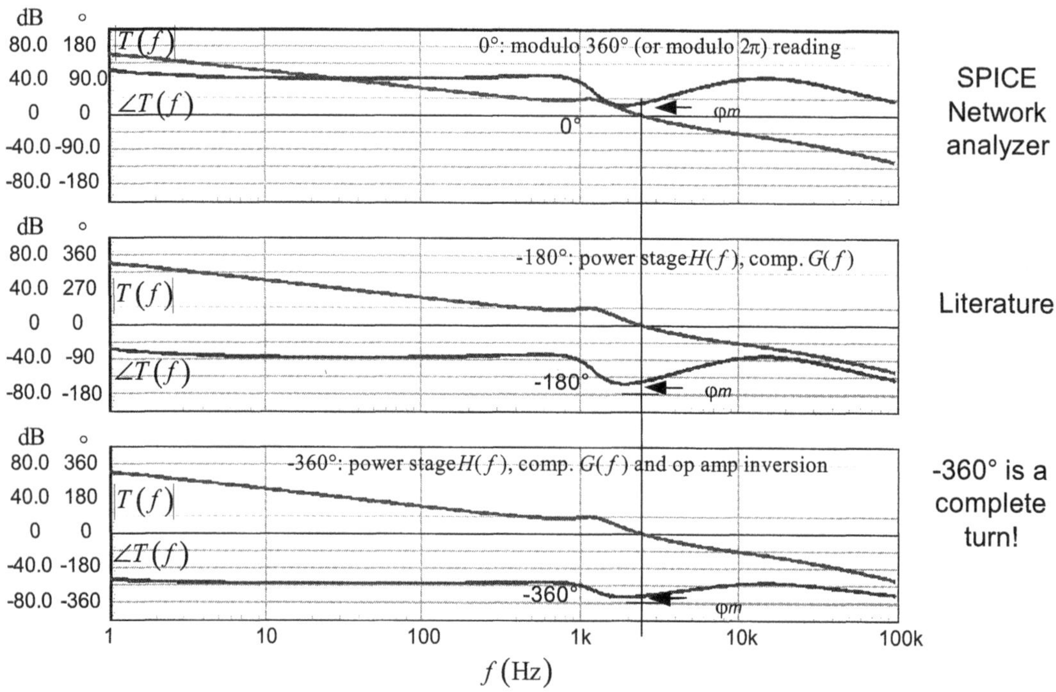

Figure 3.34 All these curves refer to the same phase margin value.

3.3.5 The Phase Margin of a Switching Converter

Further to these theoretical derivations, it is time to look at a working example. Figure 3.35 shows an average model described in [1], wired in a voltage-mode buck configuration. The compensation elements around the operational amplifier are automatically evaluated using the k-factor technique, a method that will be described in an upcoming chapter. Thanks to this automated template, we can easily select the phase margin of our choice at a constant crossover value (10 kHz) and check for the transient response to a sudden load change.

The output is subjected to a current step ranging from 1 A to 2 A in 1 μs. The results appear in Figure 3.36. The 76° phase margin gives a little overshoot of 0.05 percent whereas the 49° margin triples that overshoot, still reasonable though, given the vertical axis scale of 20 mV per division. However, you can observe a faster recovery in the 49° phase case (70 μs) versus the 76° case (227 μs). Why do we still have overshoot with 76° when theory states there should be none? It is because (3.89) is a simplified view of the transfer function, captured only in the vicinity of the crossover frequency, with a two-pole configuration, without zeros. If you have extra zeros in the transfer function or a lower open-loop gain, the Q factor approximation we have been through does not work anymore and extra work is required as detailed in [2]. So what is the interest of the derivation we just went through? It is to analytically show how the phase margin in the open-loop transfer function affects the transient response once the loop is closed. Its choice is capital to meet the specifications. As confirmed by Figure 3.36, a small phase margin leads to a peaky closed-loop response and a large phase margin implies a sluggish but nonovershooting response. As usual, you will need to trade one parameter versus

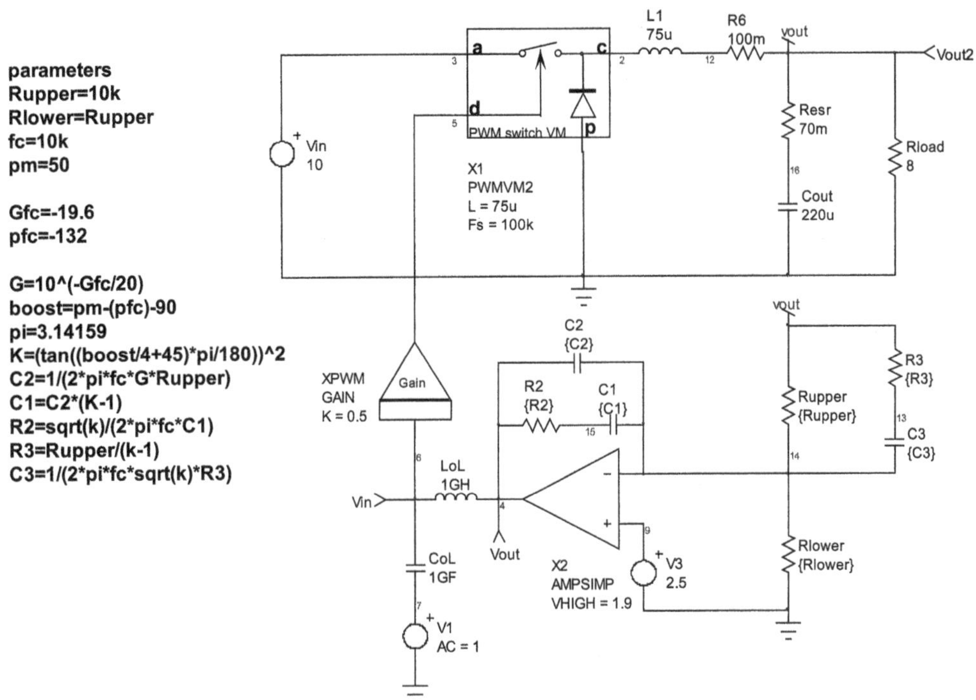

Figure 3.35 A buck converter operated in voltage mode is used to illustrate the impact of the phase margin on the transient response.

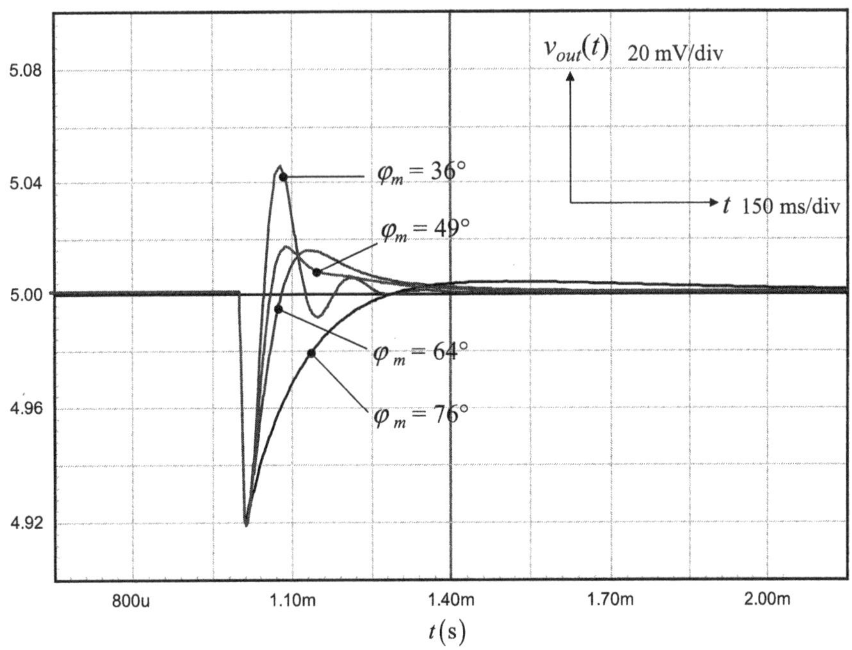

Figure 3.36 The phase margin has been adjusted at different values—f_c is constant—and it clearly affects the transient response in both the recovery time and the overshoot above the 5-V target.

the other one: a sluggish but nonringing response or a fast settling time with some overshoot?

3.3.6 Considering a Delay in the Conversion Process

Phase margin is an important parameter in the design of control systems. Not only because it must be selected based on the desired transient response as we just explained, but also because the transfer function ac response will change with time and production spreads. Changes can be imputed to temperature drifts, aging, production shifts, or simply a component replacements. It is your responsibility as a design engineer that these variations do not jeopardize the stability of the converter. In other words, you must give the total loop lag some freedom to safely move within a certain range, hence the term margin.

A loop is made of various components, regardless of whether they belong to the compensator G, the plant H, or the sensors. Some components are passive and can undergo the deterioration/variation of some of their stray elements. The equivalent series resistor of a capacitor is a typical example, as it creates a zero in the transfer function that affects the small-signal response of the converter. As the ESR moves, the zero position will be affected by temperature or lots-to-lots variations. Therefore, provisions must be made to shield the control loop against these variations. Some other blocks are active like comparators, analog-to-digital converters, logic gates, and so on. A simple example of such an active function is the pulse width modulator (PWM) or the PWM block in a switching converter. Its purpose is to drive the power switch conduction time in relationship to the power demand. A typical circuit implementation appears in Figure 3.37.

The circuit works by comparing the error voltage $v_{err}(t)$ to a ramp of a period T_{sw}. When the error voltage (pin +) is above the ramp signal (pin –), the comparator is in a high state and drives the switch on. This high state is called the on-time

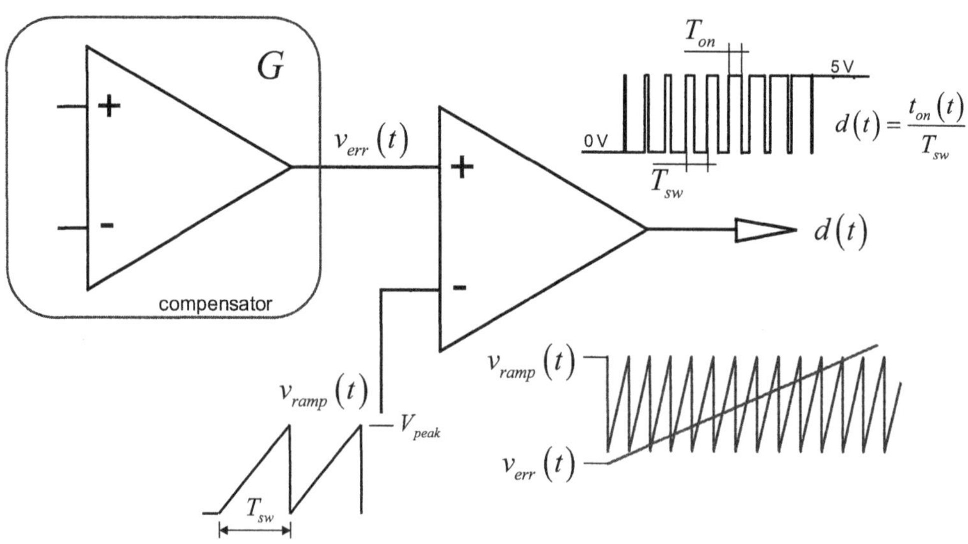

Figure 3.37 A pulse-width modulator uses a simple comparator that toggles every time the error signal crosses an artificial ramp.

(noted T_{on}) and lasts until the ramp signal exceeds the error voltage, where the comparator returns to its low state. The duration T_{on} over the switching period T_{sw} is called the duty ratio and is noted D:

$$D = \frac{T_{on}}{T_{sw}} \tag{3.124}$$

As T_{on} changes over time, the duty ratio can also be expressed as an instantaneous variable:

$$d(t) = \frac{t_{on}(t)}{T_{sw}} \tag{3.125}$$

By controlling the duty ratio via the error voltage, the control system has a means to adjust the power transfer of the converter. In order to perform ac analysis, this modulating block must be modeled to study the loop gain of our converter, $V_c(s)$ to $V_{out}(s)$. The time-domain equation is simple: the transition occurs when the sawtooth and the error levels meet. The sawtooth equation is that of a line going up to V_{peak} in a time duration of T_{sw}:

$$v_{ramp}(t) = V_{peak}\frac{t}{T_{sw}} \tag{3.126}$$

When t equals T_{on}, v_{err} has crossed the sawtooth level, hence

$$v_{err}(t) = V_{peak}\frac{T_{on}(t)}{T_{sw}} = d(t)V_{peak} \tag{3.127}$$

where $T_{on}(t)$ is the on-time value at the instant $v_{err}(t)$ equals the sawtooth value. Rearranging gives us

$$d(t) = \frac{v_{err}(t)}{V_{peak}} \tag{3.128}$$

If we average this equation over a switching period, we obtain a large-signal equation:

$$D(V_{err}) = \frac{V_{err}}{V_{peak}} \tag{3.129}$$

The ac transfer function of this equation is simply obtained by a differentiation: you will differentiate $D(V_{err})$ with respect to V_{err} by checking its sensitivity to V_{err}. This sensitivity is actually the small-signal gain:

$$\hat{d} = \frac{dD(V_{err})}{dV_{err}}\hat{v}_{err} = \frac{1}{V_{peak}}\hat{v}_{err} \tag{3.130}$$

We obtain the small-signal gain of the PWM block noted K_{PWM}:

$$K_{PWM} = \frac{1}{V_{peak}} \tag{3.131}$$

Suppose the peak value of the sawtooth is 2 V. Then the small-signal gain of the modulator is 0.5 or –6 dB. When studying the transfer function of a power stage (a buck, for instance), you will have to add the pulse-width modulator small-signal

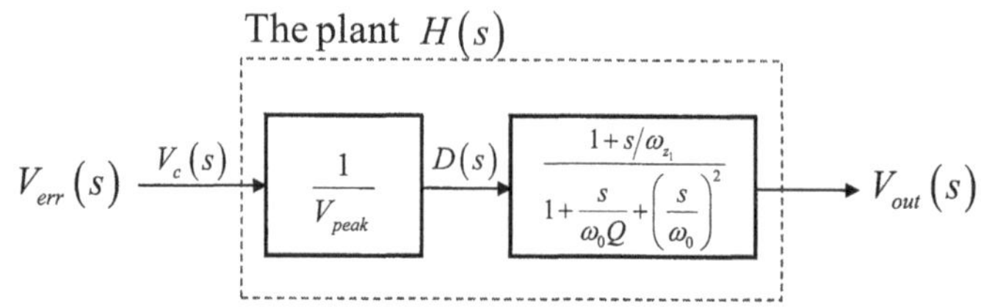

Figure 3.38 The plant transfer function $H(s)$ also includes the PWM modulator gain.

gain to the control input to account for its own ac transfer function. This is what has been done in Figure 3.35 with the insertion of the subcircuit XPWM. If we want to model the power stage alone, including the modulator gain, we obtain the drawing shown in Figure 3.38.

As we all know, the comparator is affected by a response time. When a voltage difference exists between both inputs, the decision to actually toggle the output requires a processing time: charging parasitic capacitances, switching branches, and so on. For a comparator like the LM311, it can be as high as 250 ns. Practically, it means that the compensator instructs a change in the duty ratio because the operating conditions impose it, but this change will only occur 250 ns later; this is a *delay*, also called a *transport* or *propagation* delay, in the control loop. Intuitively, if a delay appears in slow-bandwidth system (e.g., 1 kHz), 250 ns of delay is a relatively small contribution to the control chain. Expand the bandwidth to 100 kHz (very common in miniature high-frequency dc-dc converters for wireless applications), and these 250 ns start to trouble the whole control loop.

Figure 3.39 depicts a signal $u(t)$ entering a block affected by a delay τ.

The output signal $y(t)$ appears after a delay of τ seconds has elapsed. When you observe the output signal $y(t)$, you can say that what you see is actually a signal that occurred τ seconds before (i.e., at the instant t_2-τ on the picture). Mathematically, this observation can be written as follows:

$$y(t) = u(t - \tau) \tag{3.132}$$

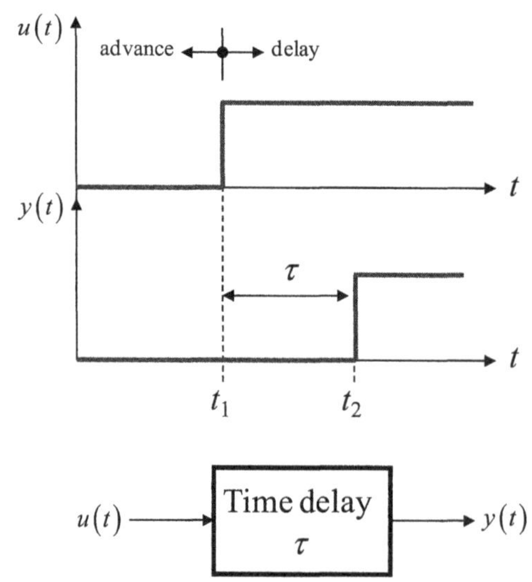

Figure 3.39 A delay can be represented by a time-domain shift to the input signal.

Now, to include this delay block in the Laplace domain and thus be able to run small-signal analysis, we will first derive the Laplace transform of (3.132):

$$\mathcal{L}[y(t)] = \mathcal{L}[u(t-\tau)] \tag{3.133}$$

At first glance, we do not know the result of this equation, but we can think of a different approach to solve it. Assume $u(t)$ is a sinusoidal waveform with an amplitude A. Thanks to Euler, we can express this signal as

$$u(t) = Ae^{j\omega t} \tag{3.134}$$

Once this signal passes through the delay block, its amplitude is unaffected, but it becomes delayed as (3.132) describes. The term t in (3.134) simply becomes t-τ:

$$y(t) = Ae^{j\omega(t-\tau)} = Ae^{j\omega t - j\omega\tau} \tag{3.135}$$

This is of the form $e^a e^b = e^{a+b}$. Otherwise stated:

$$y(t) = Ae^{j\omega(t-\tau)} = Ae^{j\omega t}e^{-j\omega\tau} \tag{3.136}$$

The first term is $u(t)$, and the second term is the delay affecting the input signal. If we take the Laplace transform of the previous equation, we obtain

$$Y(s) = U(s)e^{-s\tau} \tag{3.137}$$

The transfer function of the time delay block is thus

$$\frac{Y(s)}{U(s)} = e^{-s\tau} \tag{3.138}$$

Going back to the frequency domain, the ac response of the delay block is

$$\frac{Y(j\omega)}{U(j\omega)} = e^{-j\omega\tau} = e^{j\varphi} \tag{3.139}$$

The form $Ae^{j\varphi}$ is the Euler expression of a sinusoidal waveform under the form of

$$Ae^{j\varphi} = Ae^{j\omega t} = A(\cos\omega t + j\sin\omega t) \tag{3.140}$$

Because of the negative sign in the exponent, $\varphi = -\omega\tau$. Therefore,

$$\arg e^{-j\omega\tau} = \tan^{-1}\left(\frac{\sin(-\omega\tau)}{\cos(\omega\tau)}\right) = -\omega\tau \tag{3.141}$$

and

$$\left|e^{-j\omega\tau}\right| = \sqrt{\cos^2\omega\tau + \sin^2\omega\tau} = 1 \tag{3.142}$$

To check the frequency response of the block, we can implement a SPICE simulation using an arrangement suggested in [3]. The setup appears in Figure 3.40 and shows a buffered delay line, loaded by its characteristic impedance (50 Ω in our example).

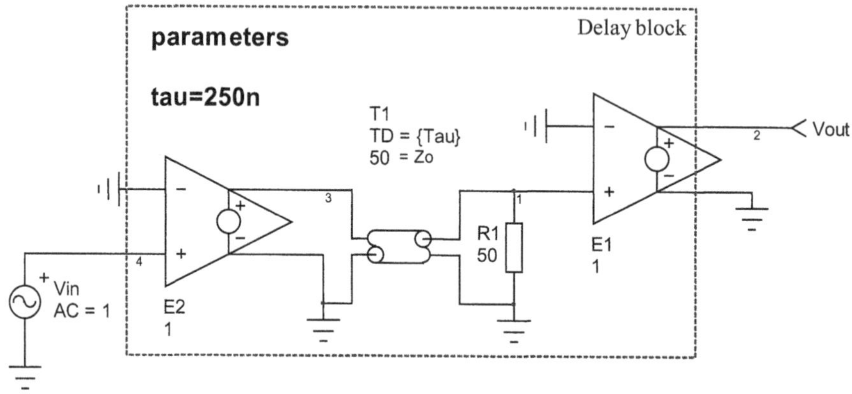

Figure 3.40 A simple delay line helps to model a delay that appears in a power converter.

The ac response of such a subcircuit with a 250-ns delay is given in Figure 3.41.

The magnitude is 1 over the whole frequency sweep, but the phase lags as we go down the horizontal axis. At 1 kHz, the phase lag contributed by the delay is negligible. At 100 kHz, (3.141) predicts a phase lag of

$$\varphi = -\omega\tau = -100k \times 6.28 \times 250n \times \frac{180}{\pi} = -9° \tag{3.143}$$

This is what the graph reads, while it goes down to −18° at 200 kHz. A 250-ns delay does not seem much, but if we consider the whole transmission chain, it can

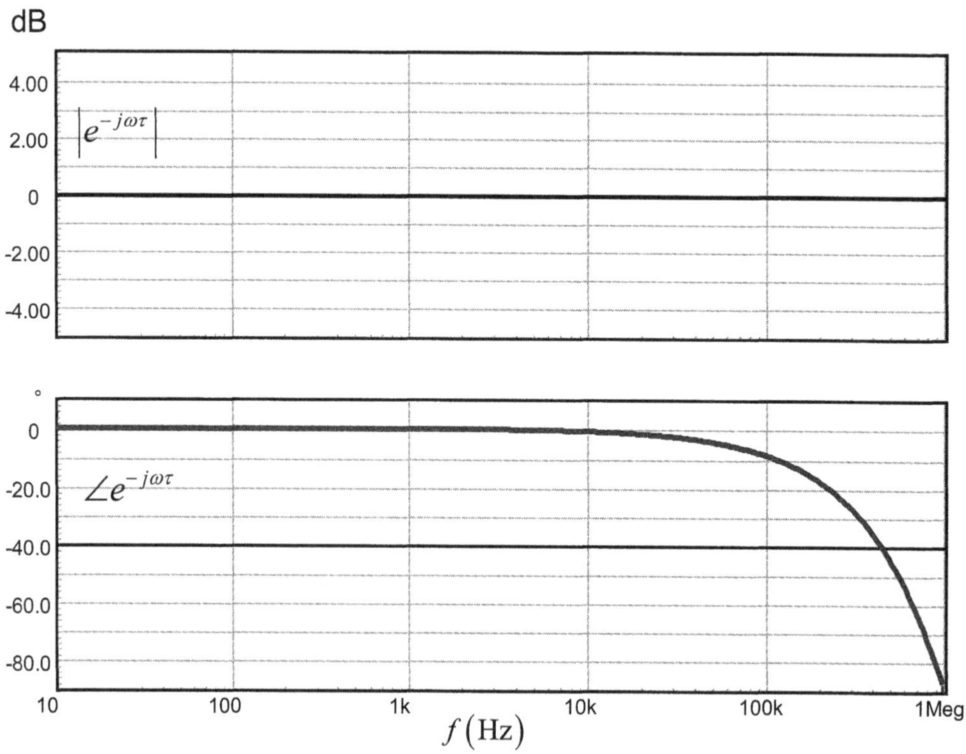

Figure 3.41 As expected, the magnitude is 0 dB all over the spectrum, but the phase lags as the frequency increases.

be much longer. The delay does not only depend on the PWM block but also on the internal logic arrangement, including the way you turn off the power switch. The logic propagation delay can be very small, but if you slowly turn off the MOSFET for EMI considerations, this is another delay. A total transport delay of 300–400 ns is thus not uncommon at all.

3.3.7 The Delay in the Laplace Domain

Now that we know a delay exists in the transmission chain, we can update Figure 3.38 representation to make it appear in Figure 3.42:

The new plant transfer function for our voltage-mode buck converter can thus be expressed as

$$H(s) = \frac{e^{-s\tau}}{V_{peak}} \frac{1 + s/\omega_{z_1}}{1 + \frac{s}{\omega_0 Q} + \left(\frac{s}{\omega_0}\right)^2} \tag{3.144}$$

The question now is how to run pole/zero analysis or root locus tests with the extra term $e^{-s\tau}$? We clearly need to replace this expression with a pole/zero combination capable of reproducing the Figure 3.41 ac response. What do we see there? A phase lag increasing as we go down the x-axis. What transfer function brings phase lag as frequency increases? A pole, indeed. Unfortunately, the magnitude of a pole is not flat over frequency. To counteract the magnitude decrease versus frequency, why not including a zero then? A zero would certainly cancel the magnitude decrease of the pole but would also cancel its phase lag. Unless we purposely insert a RHP zero! This zero will have the same magnitude as a left half plane zero, but the phase will lag, cumulating with that of the pole. Amplitude will be flat, but not the phase. The simplified equality could thus look as follows:

$$e^{-st} \approx \frac{1 - s/\omega_\tau}{1 + s/\omega_\tau} \tag{3.145}$$

What value shall we assign to ω_τ to match the delay block? As both left- and right-side expressions arguments must be equal, we can write

$$\arg(e^{-s\tau}) = \arg\left(\frac{1 - s/\omega_\tau}{1 + s/\omega_\tau}\right) \tag{3.146}$$

According to (3.141), we can further write

$$-\omega\tau = \arg(1 - s/\omega_\tau) - \arg(1 + s/\omega_\tau) \tag{3.147}$$

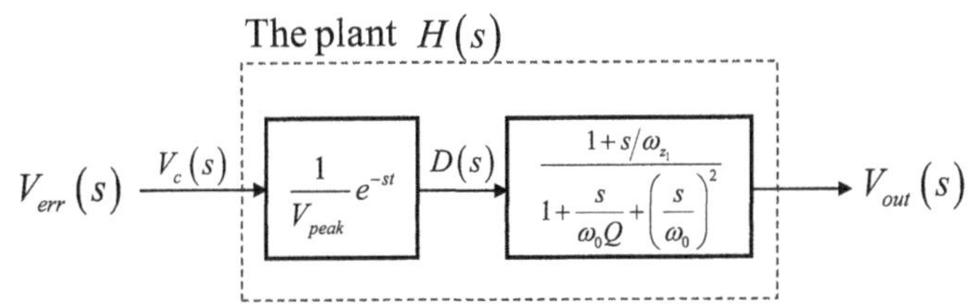

Figure 3.42 The PWM delay now appears in the whole transmission chain.

Replacing s by $j\omega$ and applying complex number formulas:

$$-\omega\tau = \tan^{-1}\left(-\frac{\omega}{\omega_\tau}\right) - \tan^{-1}\left(\frac{\omega}{\omega_\tau}\right) = -2\tan^{-1}\left(\frac{\omega}{\omega_\tau}\right) \tag{3.148}$$

The Taylor series of $\tan^{-1}(x)$ is $x - \frac{x^3}{3} + \frac{x^5}{5} + \ldots$. Applied to the right side of (3.148), we have

$$-\omega\tau \approx -2\left[\frac{\omega}{\omega_\tau} - \frac{\left(\frac{\omega}{\omega_\tau}\right)^3}{2} + \frac{\left(\frac{\omega}{\omega_\tau}\right)^5}{5}\right] \tag{3.149}$$

If we consider that $\omega_\tau \gg \omega$ along our ac analysis, then all the terms in cube and above can be neglected. The above equation simplifies to

$$\omega\tau \approx 2\frac{\omega}{\omega_\tau} \tag{3.150}$$

From which we can easily extract ω_τ:

$$\omega_\tau = \frac{2}{\tau} \tag{3.151}$$

Equation (3.145) can now be updated as

$$e^{-s\tau} \approx \frac{1 - \frac{s\tau}{2}}{1 + \frac{s\tau}{2}} \tag{3.152}$$

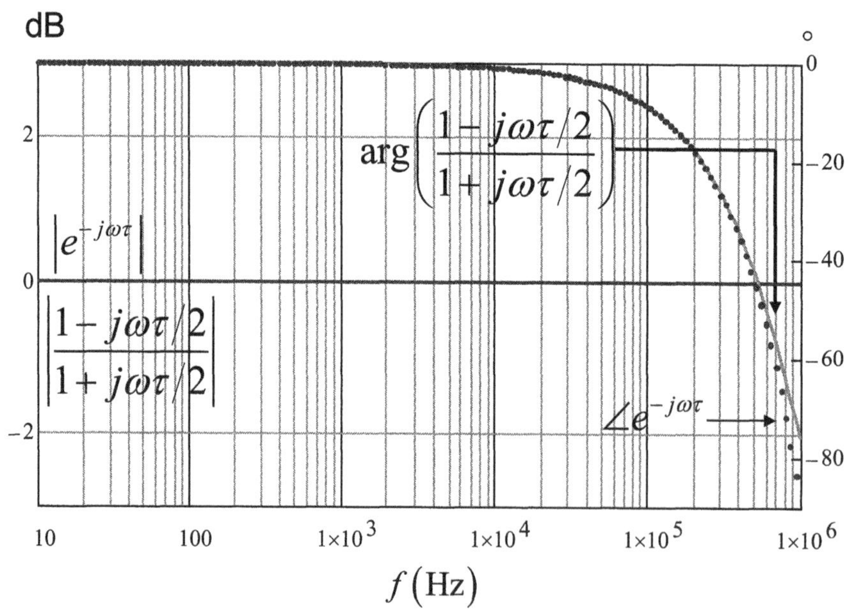

Figure 3.43 The ac response of the exponential expression versus its simplified version is very good. τ is 250 ns in this example.

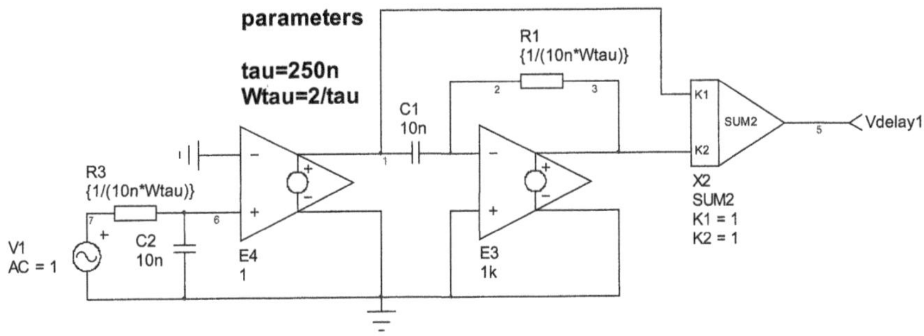

Figure 3.44 An op amp realizing the RHPZ function to which a pole is added makes a simple and efficient delay block.

This expression is nothing more that the first-order Padé approximation for the exponential

$$e^x \approx \frac{1+\dfrac{x}{2}}{1-\dfrac{x}{2}} \tag{3.153}$$

Given the approximation that we made, $\omega_\tau \gg \omega$, the simplified formula will deviate from the original expression as the frequency increases. If the deviation is too wide, a higher-order Padé expression must be selected. However, the discussion goes well beyond the scope of this book. For those interested by the subject, a comprehensive coverage of the subject appears in [4].

As shown in Figure 3.40, the delay could be built in SPICE using a delay line. A delay line in SPICE usually increases the computational time and can sometimes

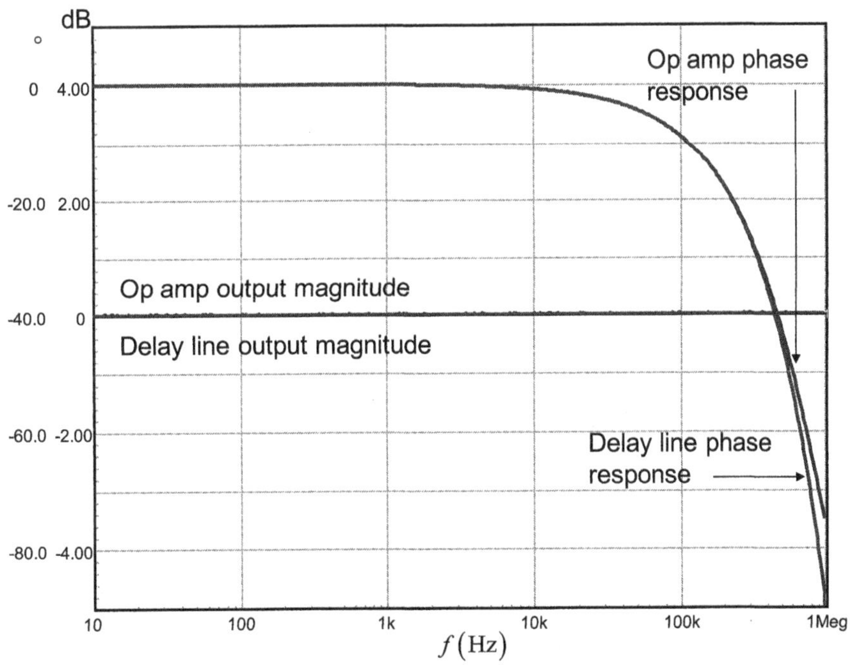

Figure 3.45 The ac response of the op amp-based delay agrees quite well with that of the delay line. It still remains a first-order model though.

lead to convergence issues. Also, some simple simulators do not include delay lines in their primitives list. No problem, Chapter 2 taught us how to build a RHP zero with an op amp and an adder. If we add a simple pole coincident with the zero, we have our delay!

The ac response of the block appears in Figure 3.45 for a 250-ns transport delay. It is in excellent agreement with the response plotted in Figure 3.43 but still remains a first-order approximation.

3.3.8 Delay Margin versus Phase Margin

The definition we gave for the delay block shows that its insertion in the loop gain equation does not affect its magnitude, only its phase. For stability analysis in a unity feedback system, the classical characteristic equation $1+T(s)=0$ must be updated to account for the delay block presence:

$$\chi(s)=1+e^{-s\tau}T(s)=0 \tag{3.154}$$

What matters now is to check how much the delay τ can safely vary without jeopardizing the system stability. If we call the maximum delay τ_{max}, (3.154) can be rewritten:

$$\chi(s)=1+e^{-s\tau_{max}}T(s)=0 \tag{3.155}$$

The conditions for which $\chi(s)$ equals zero are still the same. At crossover, we have

$$\left|e^{-s\tau_{max}}T(\omega_c)\right|=1 \tag{3.156}$$

since $\left|e^{-s\tau_{max}}\right|=1$, (3.156) is similar to

$$\left|T(\omega_c)\right|=1 \tag{3.157}$$

Regarding the argument, this is where the change takes place:

$$-\pi=\arg(e^{-s\tau_{max}})+\arg T(\omega_c)=-\omega\tau_{max}+\arg T(\omega_c) \tag{3.158}$$

Now, remembering the definition for the phase margin given in (3.8):

$$\varphi_m=\pi+\arg T(\omega_c) \tag{3.159}$$

we extract T from the previous expression and substitute it into (3.158), solving for $\omega\tau_{max}$:

$$\tau_{max}=\frac{\varphi_m}{\omega_c} \tag{3.160}$$

where ω_c represents the crossover frequency of the compensated open-loop gain, and φ_m the phase margin (in radians) measured without delay. If the system works well with the actual delay τ, then the delay margin $\Delta\tau$ is defined as

$$\Delta\tau=\tau_{max}-\tau \tag{3.161}$$

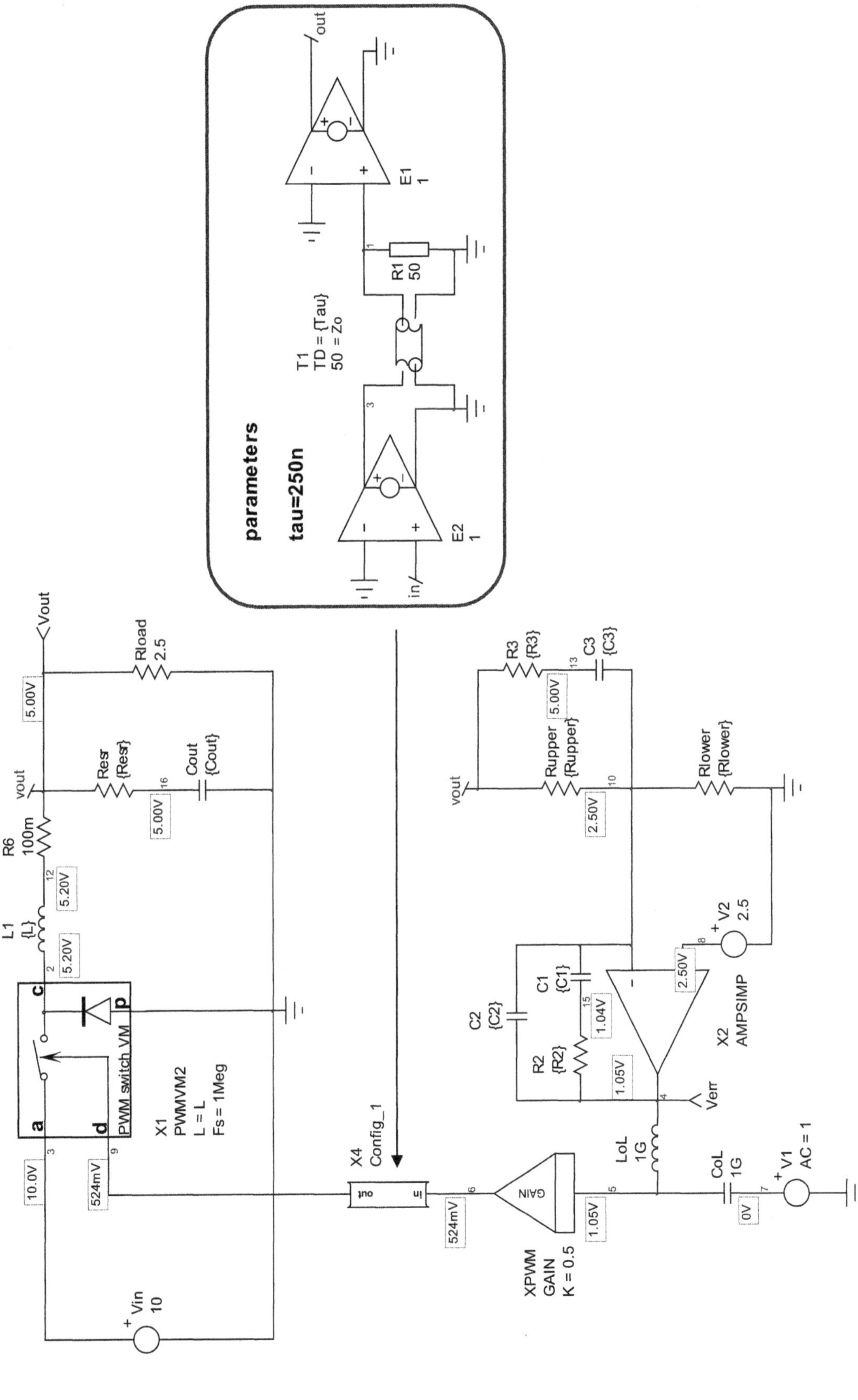

Figure 3.46 The modulating section shows the addition of the delay block in series with the PWM subcircuit.

Let's take an example of a buck converter where we now consider a delay in the PWM block. The previous buck converter schematic shown in Figure 3.35 has been updated by inserting the delay block in series with the PWM block. The whole circuit appears in Figure 3.46. The initial conversion delay is 250 ns and the switching frequency is 1 MHz.

The loop ac response $T(s)$ is given in Figure 3.47. The compensation imposes a 100-kHz crossover frequency. This value seems high but is not uncommon in small dc-dc converters used by cell phones makers. The compensation we made leads to a 49.5° phase margin at the 100-kHz crossover frequency. If we apply (3.160), as we already have a 250-ns delay when we measure $\arg T(s)$, we can calculate the delay we could further accept in the modulating chain or elsewhere in the conversion process:

$$\Delta\tau = \frac{\varphi_m}{\omega_c} = \frac{49.5}{2\pi \times 100k}\frac{\pi}{180} = 1.375\ \mu\text{s} \tag{3.162}$$

Given the 250-ns original delay, the total delay would therefore be

$$\tau_{\max} = \tau + \Delta\tau = 1.375u + 250n = 1.625\ \mu\text{s} \tag{3.163}$$

To check this result, we have purposely increased the delay to the 1.625 μs and the updated Bode plot appears in Figure 3.48. As expected, the phase margin has gone down to 0: the system is completely unstable.

As a conclusion on the delay margin, we must always compare the phase margin to the crossover frequency. The previous lines show that when the crossover point is high, a small added delay may perturb the system and make it unstable.

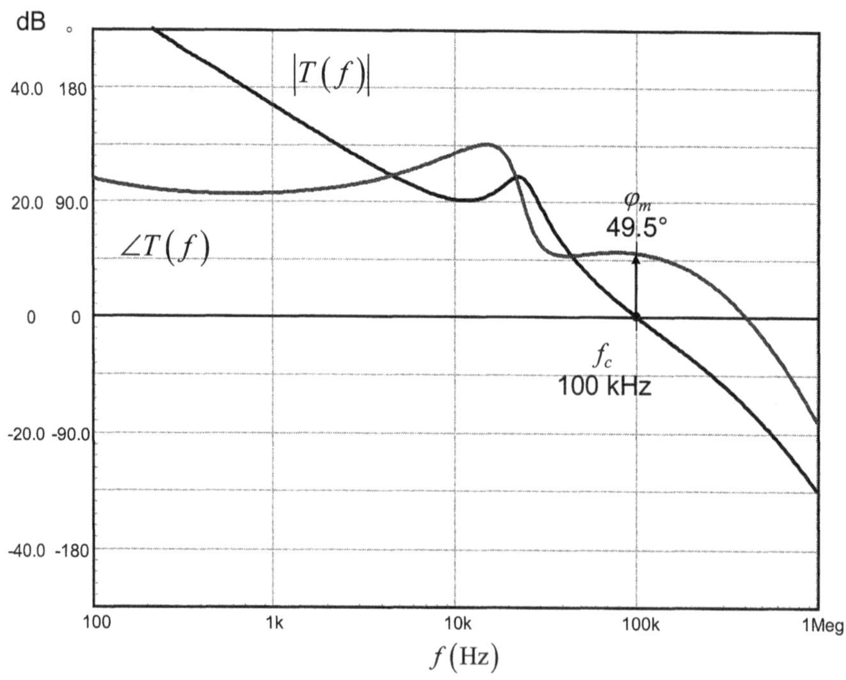

Figure 3.47 Once compensated, this 1-MHz switching frequency buck converter exhibits a 100-kHz crossover frequency with a 49° phase margin. The PWM delay is 250 ns.

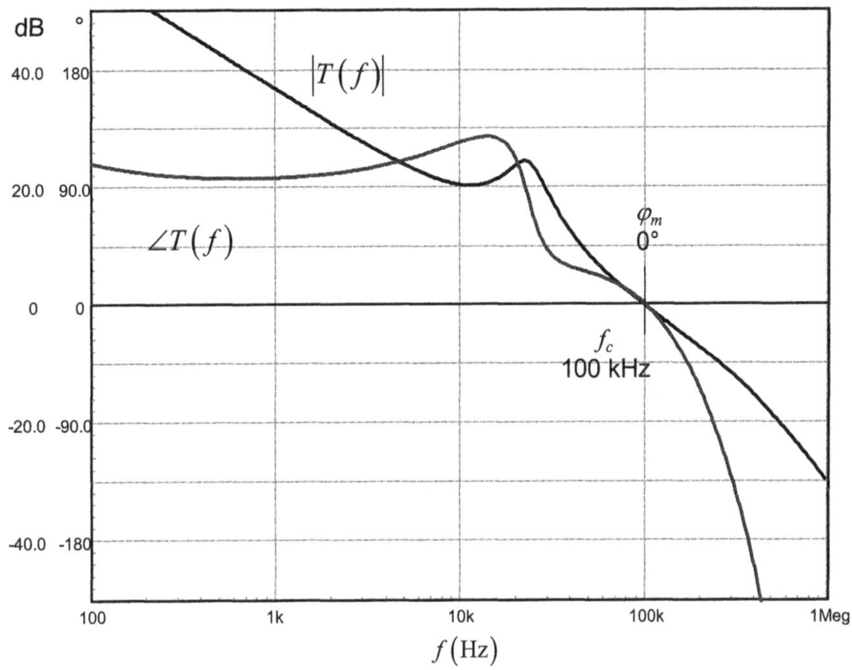

Figure 3.48 With a delay pushed to the maximum value (1.62 μs), the phase margin vanishes to 0 degree as expected.

Phase margin alone can sometimes lead to an erroneous conclusion on the system robustness, especially in high-bandwidth systems. Delay margin will thus be preferred to phase margin in high-speed dc-to-dc converters.

3.4 Selecting the Crossover Frequency

With a closed-loop control system such as the one presented in Figure 3.28, we expect the output to exactly follow the input setpoint with a gain of 1 in this example. In a unity-gain system, we should have both input and output signals exactly matching each other in magnitude and phase. Can this matching be maintained at any frequencies? No, there is a physical limit beyond which the system reaction time starts to increase. This physical limit is called the *bandwidth*: if the setpoint signal changes too rapidly, the control system will not be able to follow it. Similarly, if the frequency of the perturbation or its spectrum content is higher than the system bandwidth, correction will no longer be ensured, and drift or distortion will appear in the output. A solution would be to extend the bandwidth as much as we can, but it would make the system sensitive to incoming noises as well as to its self-generated noise if we consider a switching converter output ripple. There must be a limit imposed to the closed-loop bandwidth, and this limit is set by the open-loop crossover frequency f_c.

We have seen that, in certain conditions, the closed-loop response of the system under study could be approximated to a second-order transfer function whose Q depends on the open-loop phase margin. As with any transfer function, the system is affected by a bandwidth, exactly like a filter. By bandwidth, we mean a reduction

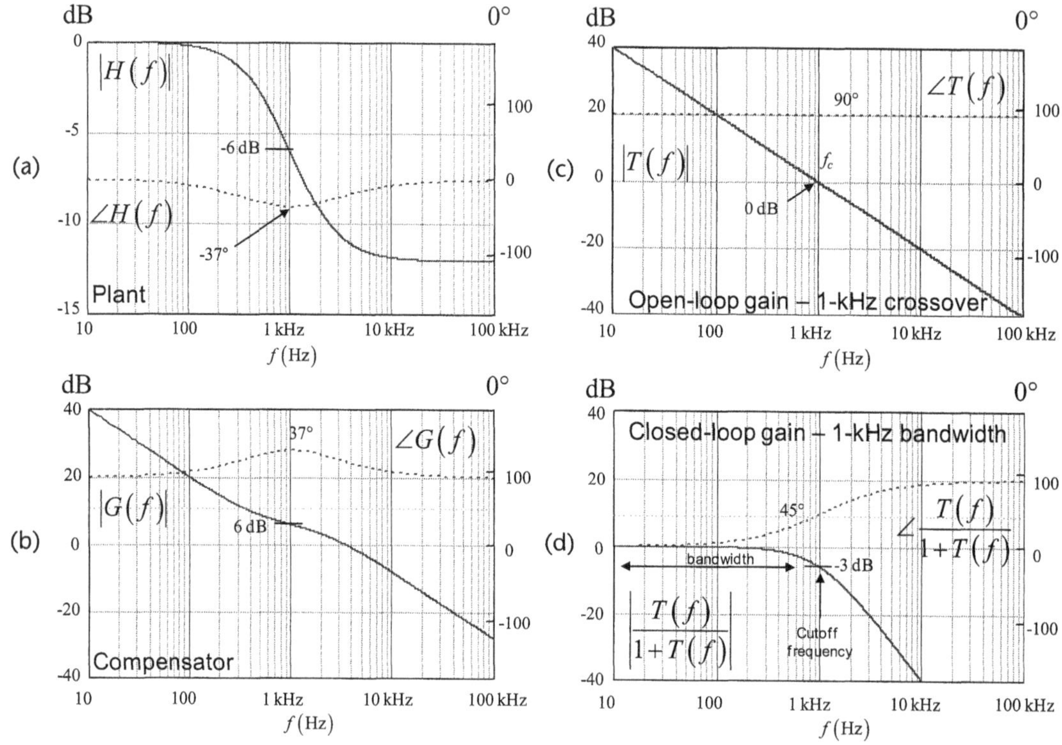

Figure 3.49 The crossover is often approximated to the bandwidth of the system. They exactly match with each other when the phase margin is 90°.

by 3 dB from its zero-frequency magnitude value. If this value is 1 or 0 dB, as expected in the example, then the frequency at which the gain magnitude has fallen to 0.707 or – 3 dB is also called the cutoff frequency of the control system. How do we set the bandwidth or the cutoff frequency of our closed-loop system? By choosing a crossover frequency f_c during the study of the open-loop transfer function. This crossover frequency will then be obtained by tailoring the transfer function of the compensator G (i.e., placing poles, zeros, and gain to compensate the deficiencies of the plant transfer function H). This is exactly what Figure 3.49 shows you.

We start from the plant whose transfer function $H(s)$ appears in the Figure 3.49(a). As we want a 1-kHz crossover frequency, we observe the magnitude and the phase of the transfer function at this frequency point. We read an attenuation of –6 dB, together with a phase lag of 37°. The compensator frequency response appears in Figure 3.49(b) and shows a positive translation of +6 dB and a 37° phase increase at 1 kHz. When we plot $T(s) = H(s)G(s)$ in Figure 3.49(c), we measure a crossover frequency of 1 kHz, together with a phase margin of 90°. If we graph in Figure 3.49(d) the closed-loop gain as expressed by (3.90), we observe a flat response until 1 kHz, where the gain drops by 3 dB: this is our cutoff frequency. As you could see in Figure 3.31, changing the phase margin affects the cutoff frequency and the bandwidth of the system. The case where the crossover frequency exactly equals the cutoff frequency occurs only when the phase margin is 90°. For the rest of the cases, we can say that for undamped systems, the closed-loop bandwidth is roughly equal to 1.5 times the open-loop crossover frequency. This is confirmed by Figure 3.31 for phase margins different than 90°.

In most switching converters design examples, it is common to arbitrarily place the crossover frequency to one-fifth or one-tenth of the switching frequency. However, it is little known that the crossover frequency actually affects other parameters of the converter, such as its output impedance: a relationship exists between both variables. Therefore, once the output capacitor has been selected (based on its operating parameters, such as rms current, temperature, or acceptable voltage ripple), the designer can analytically select his crossover frequency to match the desired output undershoot. In the same way we learned how the phase margin did affect the transient response (recovery time and overshoot), we will explore the link between the crossover frequency and the output impedance.

For linear converters, the crossover frequency also affects the output impedance. If the output undershoot is the design criteria, it can be interesting to select the crossover frequency to minimize it. On the contrary, if the settling time matters, the crossover frequency can be chosen to match a certain design goal. We will come back to the linear case in some of the design examples.

3.4.1 A Simplified Buck Converter

A buck converter is a switching system that takes a voltage and decreases it to a lower, regulated value (e.g., a 5-V output obtained from a 10-V input). Basically, a buck converter can be modeled as a low-impedance square-wave generator followed by an LC network. Such a simplified representation appears in Figure 3.50.

The output impedance of the network can be derived by shorting the input excitation. In this case, we have the parallel combination of the inductor and the capacitor networks:

$$Z_{out} = (sL_{out} + r_L) \,||\, \left(r_C + \frac{1}{sC_{out}} \right) \tag{3.164}$$

By inspection, we can see that the inductor resistive path r_L dominates the impedance in dc (L_{out} is shorted and C_{out} is open). The inductor then enters the picture as the frequency increases. In the upper frequency portion of the spectrum, the capacitor impedance starts to take over the inductive section until it becomes a short circuit and leaves the impedance value to its series loss r_C. If we ac-sweep the

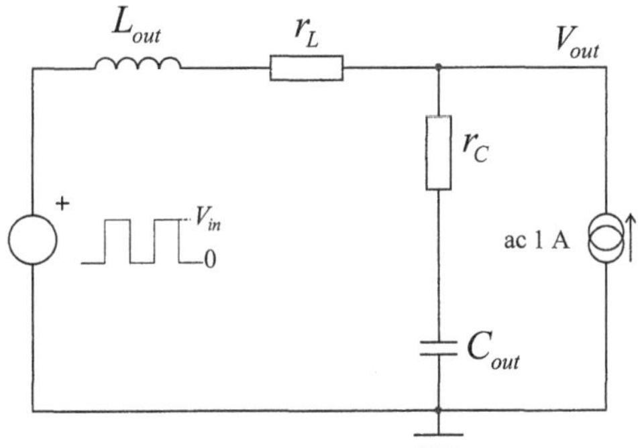

Figure 3.50 A simplified buck representation where the current source ac sweeps the output impedance.

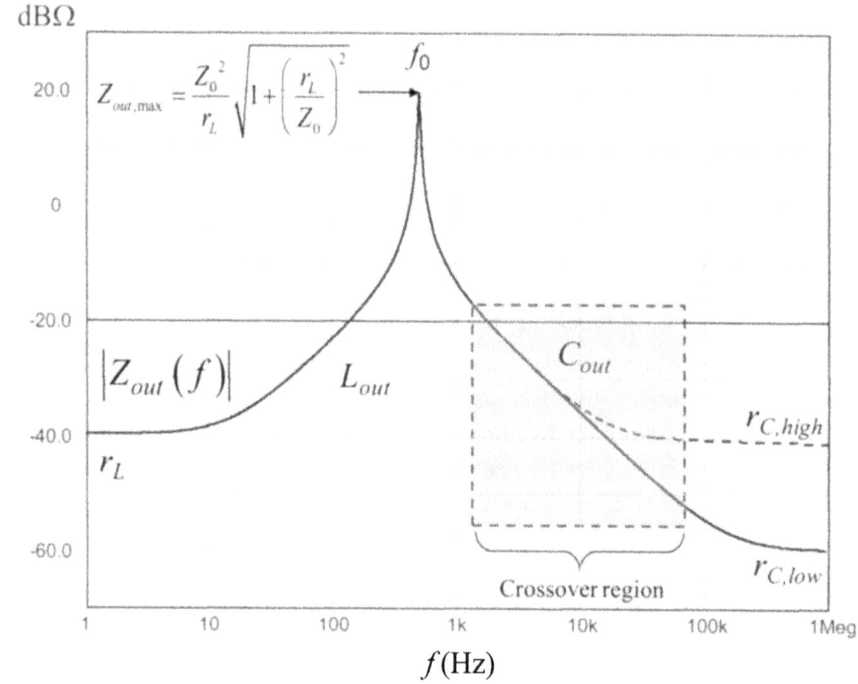

Figure 3.51 As shown by (3.164), the ohmic losses dominate the output impedance at both extremes of the graph ($f = 0$ and $f = \infty$).

output impedance of this passive network using SPICE, we obtain a graph as the one appearing in Figure 3.51.

As observed, a peaking occurs at the resonant frequency f_0. The maximum of this peaking can be analytically derived if we neglect the capacitor ESR contribution r_C, as shown in [1]:

$$Z_{out,\max} = \frac{Z_0^2}{r_L}\sqrt{1+\left(\frac{r_L}{Z_0}\right)^2} \tag{3.165}$$

where $Z_0 = \sqrt{\frac{L_{out}}{C_{out}}}$ is the characteristic impedance of the filter and r_L is the inductor series resistor. Such peaking is typical of a buck output impedance behavior where the LC filter has been optimized to minimize the losses (as losses damp the filter). This situation induces a high quality factor, hence a severe peaking in the impedance graph. One of the feedback aims is to minimize the output impedance so that the output voltage drop is kept minimum when a load step occurs. On this plot, the natural output impedance of the filter dramatically peaks at the resonant frequency. Therefore, if we select a crossover frequency below the LC filter resonance, we will not have enough gain to get rid of the resonance, and, despite a good phase margin, the system will severely ring. If we want to obtain a good transient response, we have to make sure the loop gain remains high enough to tame the peaking when it occurs. In other words, the crossover frequency f_c must be selected well above f_0 so that some gain exists when the peaking appears. Usually, a ratio of five is enough, but a closed-loop impedance is important to reveal the presence of a peaking somewhere. If it still exists, you will have to increase the gain at the resonant frequency or select a higher crossover point.

In Figure 3.51, if we select a crossover region beyond the resonance, we can see an impedance graph dominated by the output capacitor impedance C_{out} and its resistive loss r_C, unless it is a very low value. At the crossover frequency, we consider an output impedance combining both elements:

$$|Z_{out}(f_c)| \approx \sqrt{\left(\frac{1}{2\pi f_c C_{out}}\right)^2 + r_C^2} \tag{3.166}$$

With this approach, we purposely ignored a more comprehensive model of the capacitor where its series parasitic inductance, the equivalent series inductor (ESL), is considered. This element plays a role in large di/dt load steps and, given the extremely sharp transient it brings, the loop cannot fight it. The only way to reduce its effects is to select/combine low-ESL types of capacitors such as multilayer devices. In dc-dc converters supplying motherboards, current slopes of several tens of amperes per microsecond are not uncommon. In these cases, it is crucial to account for the ESL presence when computing the converter undershoot.

Figure 3.52 shows the typical response of a power supply submitted to a brutal load step. The output capacitor features both stray elements, ESR and ESL. The first drop is due to the ESL presence. Its amplitude is classically $L\frac{di_{out}(t)}{dt}$, where L is the

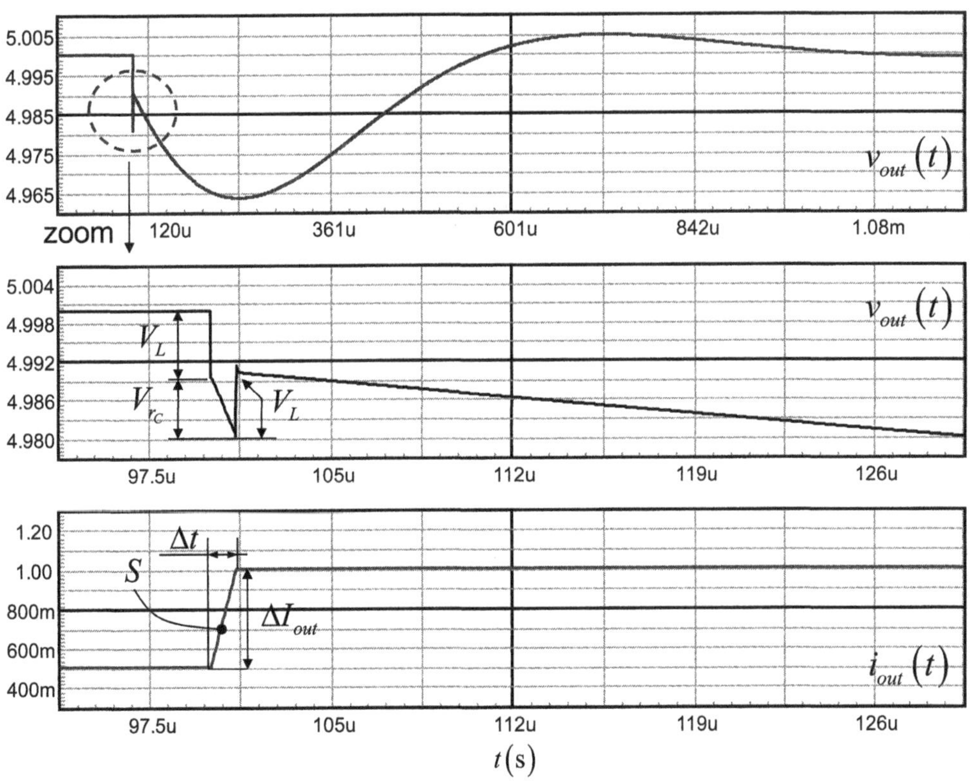

Figure 3.52 When all parasitic elements are present, the response to a current step reveals different areas where each stray element plays a role.

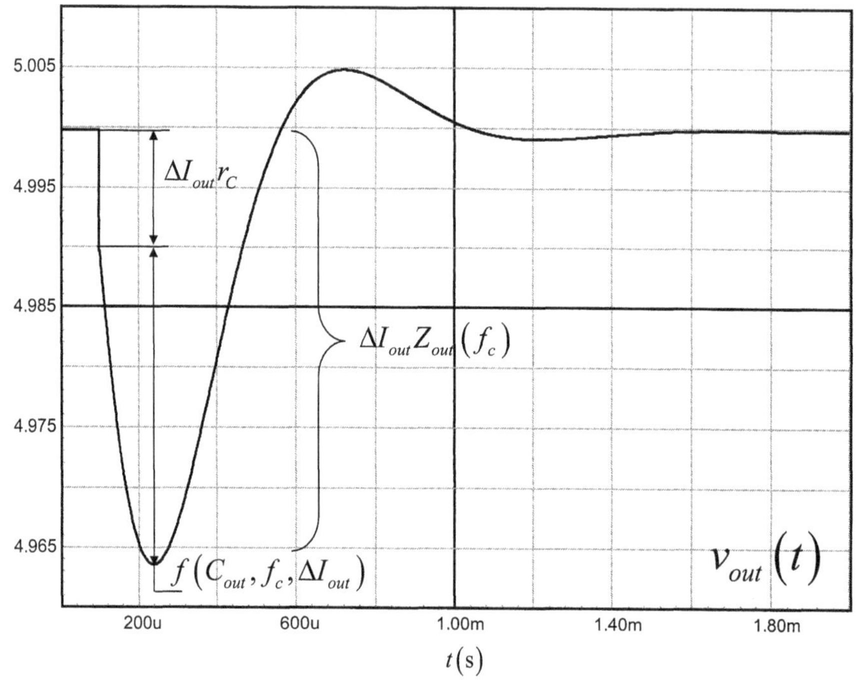

Figure 3.53 The drop is made of a capacitive undershoot topping a resistive drop. The bandwidth can play on the capacitive contribution, but not on that from the ESR.

ESL of the output capacitor and $di_{out}(t)/dt$ is the output current slope S. On top of the ESL voltage, you have the resistive voltage brought by the ESR and simply equal to $r_C \Delta I_{out}$. The variation slope is that of the current. When the current has reached its maximum value and flattens, the ESL drop disappears. The voltage continues to undershoot until the loop eventually takes over. As we will see, the undershoot value depends on the output capacitor and the crossover frequency. Reference [5] studies in detail the implications of the stray elements in the design of dc-dc converters aimed at powering motherboards specifically.

If we now consider much slower slopes than before (e.g., slopes of 1 A/μs or so, a typical test value for offline converters and general-purpose power supplies), the ESL contribution is usually weak and can be neglected. The voltage response will thus combine a capacitive and a resistive contribution only as shown in Figure 3.53.

This time, in the absence of ESL, the first drop is inherent to the resistive component of the capacitor. It looks straight, but it has the same shape as in Figure 3.52 zoom. Again, besides selecting a low-ESR type, there is nothing you can do to counteract it. The second drop is the pure capacitive contribution. Its amplitude depends on several variables, such as the crossover frequency, the output capacitor itself, and the current step. This is the component on which we can play to make it reach the value we want via crossover frequency selection. Let us see how.

3.4.2 The Output Impedance in Closed-Loop Conditions

A power supply can always be represented by its Thevenin-equivalent circuit featuring a dc generator, V_{th}, accompanied by its output impedance, R_{th}. To obtain

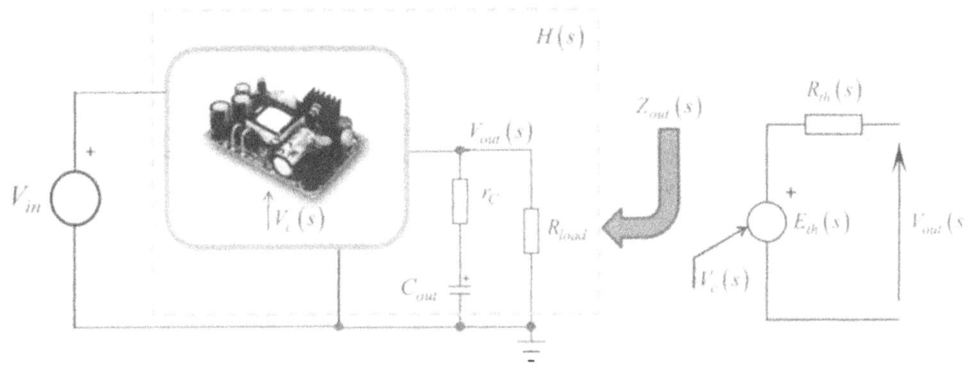

Figure 3.54 A converter can always be replaced by its Thevenin-equivalent model.

these characteristics, we can take a converter operated in open-loop conditions and extract its parameters as suggested by Figure 3.54. The converter is controlled by a voltage V_c, and the power supply delivers current to a load. There is an output capacitor C_{out} affected by an ESR r_C.

A simpler representation appears in Figure 3.55 where we can see the control voltage V_c driving the power stage H. Please note that H encompasses the total plant transfer function captured at a certain V_{in}, while delivering current to its load. The power stage output impedance is externally modeled via the addition of an output impedance—the R_{th} equivalence—denominated $Z_{out,OL}$. OL stands for open loop.

If we write a few lines of algebra, without surprise, we can express the output voltage as

$$V_{out}(s) = V_c(s)H(s) - I_{out}(s)Z_{out,OL}(s) \tag{3.167}$$

The second term expresses the drop incurred to the output impedance $Z_{out,OL}$. Now, let's assume that we want to transform this open-loop system into a closed-loop converter. We need to insert a subtraction block and a compensator box $G(s)$. The new circuit appears in Figure 3.56, featuring a unity-gain return path. This is a different representation—more physical—than that of Figure 3.19, but they are equivalent. The new control variable is no longer V_c but V_{ref}.

Following the signals path, we can derive the transfer function $V_{out}(s)/V_{ref}(s)$ of this simple architecture:

$$V_{out}(s) = \left[V_{ref}(s) - V_{out}(s)\right]G(s)H(s) - Z_{out,OL}(s)I_{out}(s) \tag{3.168}$$

$$V_{out}(s)\left[1 + G(s)H(s)\right] = V_{ref}(s)G(s)H(s) - Z_{out,OL}(s)I_{out}(s) \tag{3.169}$$

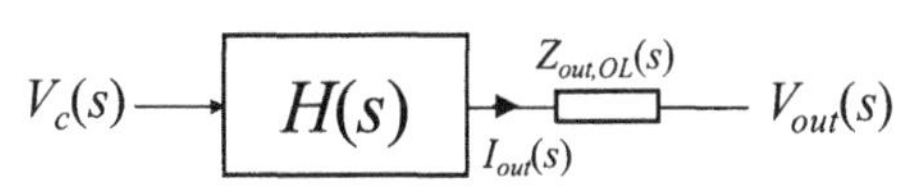

Figure 3.55 This simplified schematic represents a converter affected by an output impedance $Z_{out,OL}$ which generates a voltage drop.

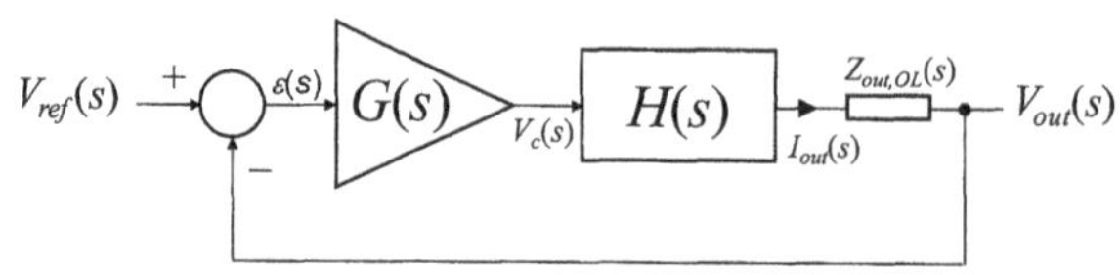

Figure 3.56 This simplified schematic represents a converter affected by an output impedance $Z_{out,OL}$, which generates a voltage drop.

$$V_{out}(s) = V_{ref}(s)\frac{G(s)H(s)}{1+G(s)H(s)} - \frac{Z_{out,OL}(s)}{1+G(s)H(s)}I_{out}(s) \tag{3.170}$$

Identifying the loop gain $T(s)$ as $G(s)H(s)$, we can reformulate the closed-loop expression as

$$V_{out}(s) = V_{ref}(s)\frac{T(s)}{1+T(s)} - \frac{Z_{out,OL}(s)}{1+T(s)}I_{out}(s) \tag{3.171}$$

In this expression, we have two right-side members. The first one shows that, despite the precision you put in V_{ref}, the output voltage will never exactly match the setpoint even with a zero-output impedance converter. There will always be a small error, the dc *static error*, between the output voltage and the setpoint you want to reach. If we neglect the output impedance dc drop in closed-loop conditions, then the deviation between the theoretical setpoint (V_{ref}) and the measured output could be reformulated as follows:

$$\varepsilon_0 \approx V_{ref} - V_{ref}\frac{T_{(0)}}{1+T_{(0)}} = V_{ref}\left(1-\frac{T_{(0)}}{1+T_{(0)}}\right) = V_{ref}\left(\frac{1}{1+T_{(0)}}\right) \tag{3.172}$$

If the dc gain $T_{(0)}$ goes to infinity, the error ε_0 is nullified and the output exactly matches the setpoint V_{ref}. Therefore, as a good design practice, it is recommended to grow the dc gain to minimize the static error of your converter. A known means to bring the loop gain to infinity for $s = 0$ is to integrate the error voltage as seen in Chapter 1. What is the Laplace equivalent of integration? You introduce the term $\frac{1}{s}$ in the transfer function of $G(s)$. We will later see that the s alone in the denominator is called an *origin pole* since the quotient goes to infinity in dc, when s equals 0. Exactly what we are looking for!

A large dc gain has another good impact if you look at the second right-side member of (3.171). The open-loop impedance has been transformed and is now divided by the loop gain $T(s)$. Again, this second term is similar to that found in (3.24): the output current is a perturbation that is fought by the loop gain in the denominator. Practically, the expression of the new output impedance, the closed-loop output impedance $Z_{out,CL}$, is then

$$Z_{out,CL}(s) = Z_{out,OL}(s)\frac{1}{1+T(s)} \tag{3.173}$$

Calculating the magnitude of this expression, we obtain

$$|Z_{out,CL}(s)| = |Z_{out,OL}(s)|\left|\frac{1}{1+T(s)}\right| \tag{3.174}$$

The closed-loop output impedance is then the open-loop output impedance affected by a corrective term, dependent upon the loop gain value. If we have installed a pole at the origin in our compensating path $G(s)$, we know that the loop gain will go to infinity in dc (for $s = 0$), bringing the closed-loop output impedance to almost 0. In our *RLC* filter of Figure 3.50, the dc closed-loop output impedance will thus be much smaller than r_L, due to the control system operation. As the frequency increases, we learned that $T(s)$ must be tailored through $G(s)$ to roll off the gain at a certain point, forcing a 0-dB crossover at a selected frequency, f_c. If $T(s)$ now drops in magnitude as the frequency is increased, its action on the output impedance becomes weaker, until unity is reached at crossover. At this point, the closed-loop output impedance has come back to its open-loop value:

$$\left|Z_{out,CL}(f_c)\right| \approx \left|Z_{out,OL}(f_c)\right| \tag{3.175}$$

To illustrate this theory, we have inserted the buck converter featuring the open-loop output impedance curve of Figure 3.51 in closed-loop configuration. The compensated loop gain $T(s)$ and the resulting plots (open-loop and closed-loop) are displayed in Figure 3.57.

In this picture, we clearly see the impact of the decreasing loop gain, which eventually reaches unity at a 10-kHz frequency. At this point, the output impedance is almost that without feedback. Let's try to derive what hides behind the term *almost*.

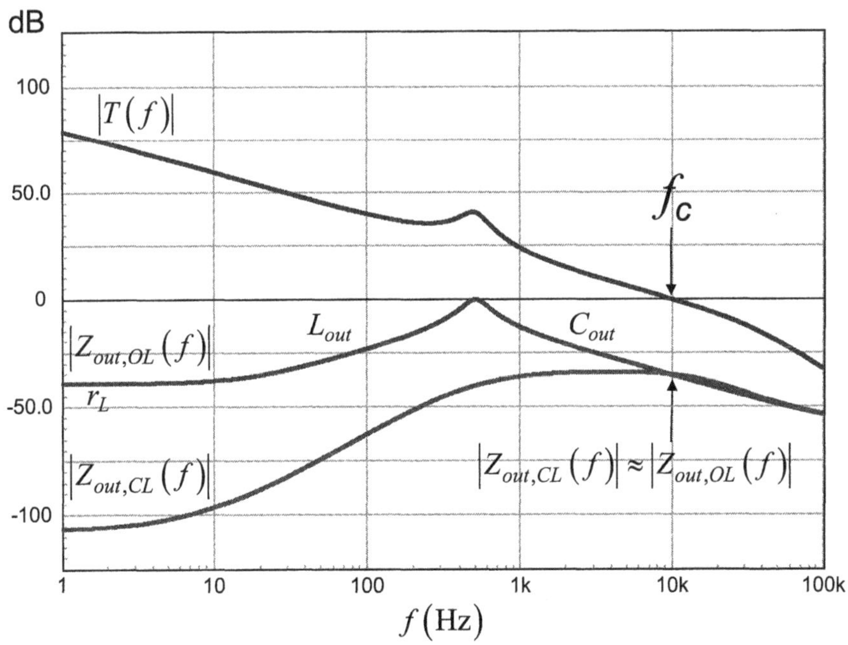

Figure 3.57 The loop gain reduces in magnitude as the frequency increases. It clearly impacts the output impedance.

3.4.3 The Closed-Loop Output Impedance at Crossover

To obtain the exact definition of the output impedance at crossover, we will derive the magnitude of the right term in (3.174). Please note that we keep observing the loop gain in the vicinity of the crossover frequency, as already highlighted in Figure 3.29.

$$\left|\frac{1}{1+T(j\omega_c)}\right| = \left|\frac{1}{\dfrac{1}{j\dfrac{\omega_c}{\omega_0}\left(1+j\dfrac{\omega_c}{\omega_2}\right)}+1}\right| = \sqrt{\frac{(\omega_c\omega_2)^2+\omega_c^4}{(\omega_c\omega_2)^2+(\omega_c^2-\omega_0\omega_2)^2}} \tag{3.176}$$

In this equation, we can now substitute ω_c by its definition from (3.99) and ω_0 from (3.92):

$$\begin{aligned}\left|\frac{1}{1+T(j\omega_c)}\right| &= \sqrt{\frac{Q^4\omega_2^2}{\omega_0^2+Q^4\omega_2^2+\omega_0\omega_2-\omega_0\omega_2\sqrt{4Q^4+1}}}\\ &= \frac{1}{2Q\sqrt{\dfrac{2Q^2+1-\sqrt{1+4Q^4}}{\left(1+\sqrt{1+4Q^4}\right)\left(\sqrt{1+4Q^4}-1\right)}}}\end{aligned} \tag{3.177}$$

Now replace Q by its definition in (3.110) and have fun simplifying the result:

$$\left|\frac{1}{1+T(j\omega_c)}\right| = \frac{1}{\sqrt{\dfrac{1}{\cos(\varphi_m)}\left(2\cos(\varphi_m)+\left[\dfrac{1+\cos^2(\varphi_m)}{\sin^2(\varphi_m)}\right]\left[\cos^2(\varphi_m)-1\right]+1-\cos^2(\varphi_m)\right)}} \tag{3.178}$$

If everything goes well, you should find

$$\left|\frac{1}{1+T(j\omega_c)}\right| = \frac{1}{\sqrt{2-2\cos(\varphi_m)}} \tag{3.179}$$

This is the expression of the closed-loop gain magnitude. For a phase margin of 90°, we obtain 0.707 or exactly –3 dB. In that case, both the open-loop crossover frequency and the closed-loop cutoff frequency are equal. For different phase margin values, the closed-loop gain starts to peak and the cutoff frequency deviates from the open-loop crossover frequency. This is what has been shown in Figure 3.31.

Back to our output impedance equation in (3.166), the exact definition thus becomes

$$|Z_{out}(f_c)| = \sqrt{\left(\frac{1}{2\pi f_c C_{out}}\right)^2+r_C^2}\,\frac{1}{\sqrt{2-2\cos(\varphi_m)}} \tag{3.180}$$

From this expression, we are now able to link the selection of the crossover frequency with a design criterion, the output impedance. You could also link the crossover frequency with the closed-loop settling time if you wish. However, the voltage drop in a converter represents an important parameter, and we are going to use it in the following example to set the crossover frequency. Let's assume we have selected a 1000-μF capacitor based on several parameters such as its rms current capability at high temperature, its size, and also its cost. From the manufacturer data sheet, we read that its ESR is 30 mΩ. Now, assume the specification imposes a maximum undershoot V_p of 90 mV when the output undergoes a current step of 2 A. First, let's calculate the contribution of the ESR alone:

$$V_{ESR} = \Delta I_{out} r_C = 2 \times 30m = 60 \text{ mV} \tag{3.181}$$

It is obvious that if this drop approaches the maximum undershoot, there is nothing you can do beyond selecting a bigger capacitor featuring a smaller ESR or associate capacitors in parallel. Here, the approach is different: we will try to select the crossover frequency to match the undershoot specifications. At this point, from the simplified impedance definition value, we can extract the crossover frequency so that the 1000-μF addition to the 60-mV ESR drop makes the whole undershoot stay below 90 mV. With a 2-A step, the closed-loop output impedance must stay below

$$Z_{out}(f_c) < \frac{V_p}{\Delta I_{out}} < \frac{90m}{2} < 45 \text{ m}\Omega \tag{3.182}$$

Equation (3.180) is really too comprehensive to be used as is. Keep in mind that capacitors and ESRs are affected by dispersions going up to ±30 percent or ±40 percent, so it is really meaningless to run calculations down to the third decimal. Experience shows that the original equation (3.175) gives acceptable practical results. Adopting this approach, the way to derive the crossover frequency is to solve for f_c in the following equation:

$$\sqrt{\left(\frac{1}{2\pi f_c C_{out}}\right)^2 + r_C^2} \approx Z_{out}(f_c) \tag{3.183}$$

From which we can extract the crossover frequency f_c target:

$$f_c > \frac{1}{2\pi C_{out}\sqrt{Z_{out}(f_c)^2 - r_C^2}} > \frac{1}{6.28 \times 1m \times \sqrt{45m^2 - 30m^2}} > 4.7 \text{ kHz} \tag{3.184}$$

Obviously, this is an approximation and practical experiments in the bench will be needed to confirm this choice. However, experience shows that the final result is often not too far from what is expected. The previous derivation is a good starting point for the crossover frequency selection.

3.4.4 Scaling the Reference to Obtain the Desired Output

In most of our drawing examples for the sake of simplicity (e.g., in Figure 3.56), the output is directly compared to the setpoint. In reality, this setpoint is made of a reference source whose value can be quite low compared to the observed output. For instance, the vast majority of stable sources are built on a bandgap voltage reference whose level is 1.25 V. Therefore, in high-voltage applications where the output voltage can be much higher than 1.25 V (400 V in power factor correction circuits), we do not directly observe Vout but a reduced image of it. This is done by installing a resistive divider between the observed variable and the input of the error amplifier. Figure 3.58 shows this typical arrangement with an op amp.

This divider introduces a scaling term equal to

$$\alpha = \frac{R_{lower}}{R_{lower} + R_{upper}} \tag{3.185}$$

In the previous example, the op amp undergoes a local feedback via the element Z_f. The presence of this element brings a well-known characteristic called a virtual ground. It means that considering an infinite or very large op amp open-loop gain A_{OL}, the voltage ε between the plus and minus input is null, or extremely small: both input levels must be equal at steady state. If it is not the case, the op amp fights the difference along its dynamic range. If it cannot ensure this equality, it is stuck in its upper or lower stop. The presence of this virtual ground has an impact on the ac analysis of this block. In ac, as we consider a voltage regulator, we are interested in the converter's ability to reject the incoming perturbations such as the input voltage or the input current. During these tests, the reference voltage does not change: its ac modulation is 0. As both op amp inputs have equal levels because of the virtual ground, there is no ac voltage modulation across the lower-side resistor R_{lower}, and it simply disappears from the ac analysis. We can quickly go through a few equations to show that.

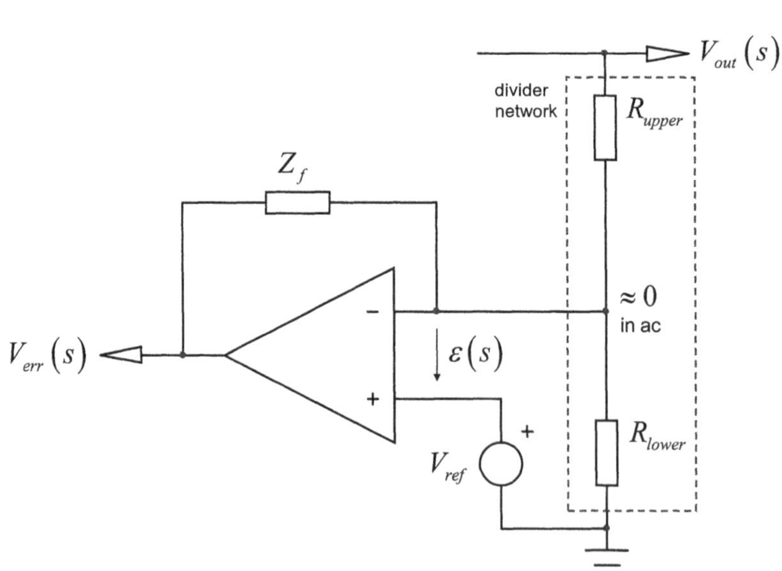

Figure 3.58 In power converters, a divider is inserted between the observed output voltage and the reference voltage.

The output voltage of Figure 3.58 configuration can be obtained by applying the superposition theorem:

$$V_{err}(s)\big|_{V_{out}=0} = V_{ref}(s)\left(\frac{Z_f}{R_{upper} \parallel R_{lower}} + 1\right) \tag{3.186}$$

$$V_{err}(s)\big|_{V_{ref}=0} = -V_{out}(s)\frac{Z_f}{R_{upper}} \tag{3.187}$$

Adding both expressions leads to the error voltage equation:

$$V_{err}(s) = V_{ref}(s)\left(\frac{Z_f}{R_{upper} \parallel R_{lower}} + 1\right) - V_{out}(s)\frac{Z_f}{R_{upper}} \tag{3.188}$$

If we now derive the previous expression with respect to V_{out} and keeping V_{ref} constant, we obtain the ac small-signal response in which the lower-side resistor plays no role:

$$V_{err}(s) = -\frac{Z_f}{R_{upper}} \tag{3.189}$$

Please note that the resistor plays a role in dc as it fixes, together with R_{upper} and V_{ref}, the operating point in (3.188). But owing to the virtual ground, only R_{upper} plays a role in poles/zeros calculations.

Is this always the case? No, when there is no virtual ground, as with an operational transconductance amplifier (OTA), for instance, the virtual ground is lost and both R_{upper} and R_{lower} play a role in the ac analysis. We will see it in the chapter dedicated to compensating with an OTA. This remark also applies if Z_f is not a resistance but an impedance made of a series resistor-capacitor combination. In this case, we have well a virtual ground in ac (the capacitor offers a certain impedance) and R_{lower} is off the picture for the calculation. However, in dc, at s equals 0, if you try to evaluate the output static error, the capacitor impedance is infinite and the virtual ground is lost: the divider comes back in the calculation and you must account for its presence together with the op amp open-loop gain. Fortunately, in the vast majority of cases, because of the origin pole we insert via the series combination of a resistor and a capacitor for Z_f, and also because we perform an ac sweep where the op amp virtual ground is effective, we simply ignore R_{lower}.

If we place ourselves in this latter case, we can transform Figure 3.56 representation to now include the divider ratio, solely affecting the reference voltage and not the rest of the chain. The new drawing appears in Figure 3.59.

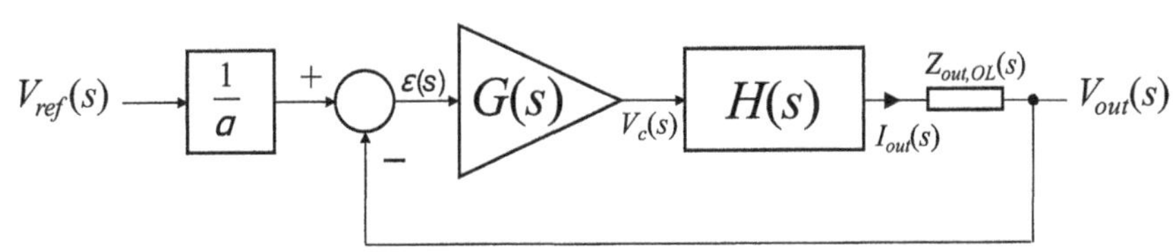

Figure 3.59 The resistive divider does not play a role in the output impedance expression; it only affects V_{ref}.

We are still in a unity feedback system; all the equations we have derived so far still apply. If you rederive the output voltage expression, you should be able to show that the updated expression obeys

$$V_{out}(s) = \frac{V_{ref}}{\alpha}\frac{G(s)H(s)}{1+G(s)H(s)} - \frac{Z_{out,OL}}{1+G(s)H(s)}I_{out}(s) \tag{3.190}$$

In this equation, the first term, again, is the theoretical dc output value you would expect from the converter if we neglect its output impedance contribution:

$$V_{out} \approx \frac{V_{ref}}{\alpha}\frac{G_{(0)}H_{(0)}}{1+G_{(0)}H_{(0)}} \tag{3.191}$$

In presence of a large dc open-loop gain $G_{(0)}H_{(0)}$, this equation simplifies to

$$V_{out} \approx \frac{V_{ref}}{\alpha} \tag{3.192}$$

From this expression, we can see that a divider value α allows us to scale the output voltage to any combination of the reference voltage independently from the other loop parameters. This is what feedback is all about: we want the output to match the input setpoint by a factor made independent from the forward path of the control system. With a ratio of 0.5 and a 2.5-V reference voltage, we expect the output voltage to equal 5 V. However, this is a theoretical value. Equation (3.191) shows that a correction factor must be added to account for the open-loop gain. This open-loop gain, as large as it can be, always introduces a small deviation between the expected target and the final measurement. In presence of extremely large open-loop gains (90 dB or more), as when an origin pole exists in the transfer function $G(s)$, the error becomes extremely small. Neglecting the closed-loop impedance effect, the static output error between the theoretical value given by (3.192) and that obtained via (3.191) becomes

$$\varepsilon_0 \approx \frac{V_{ref}}{\alpha} - \frac{V_{ref}}{\alpha}\frac{G_{(0)}H_{(0)}}{1+G_{(0)}H_{(0)}} = V_{ref}\left(\frac{1}{\alpha} - \frac{1}{\alpha\left(1+\dfrac{1}{T_{(0)}}\right)}\right) \tag{3.193}$$

From this equation, we can see that despite having an extremely precise reference voltage, when embedded into a low gain loop, the output will always deviate from the expected target defined by (3.192).

To exercise these new equations and results, we have built a buck converter delivering 5 V from a 10 V dc source.

Its linear representation is given in Figure 3.60, using an averaged circuit, the PWM switch model, which is extremely useful in the ac analysis of switching converters. To understand how this model works, please check [1] at the end of this chapter. As confirmed by the bias points in Figure 3.60, the switching regulator delivers the right output when loaded by a 2-Ω resistor (2.5 A).

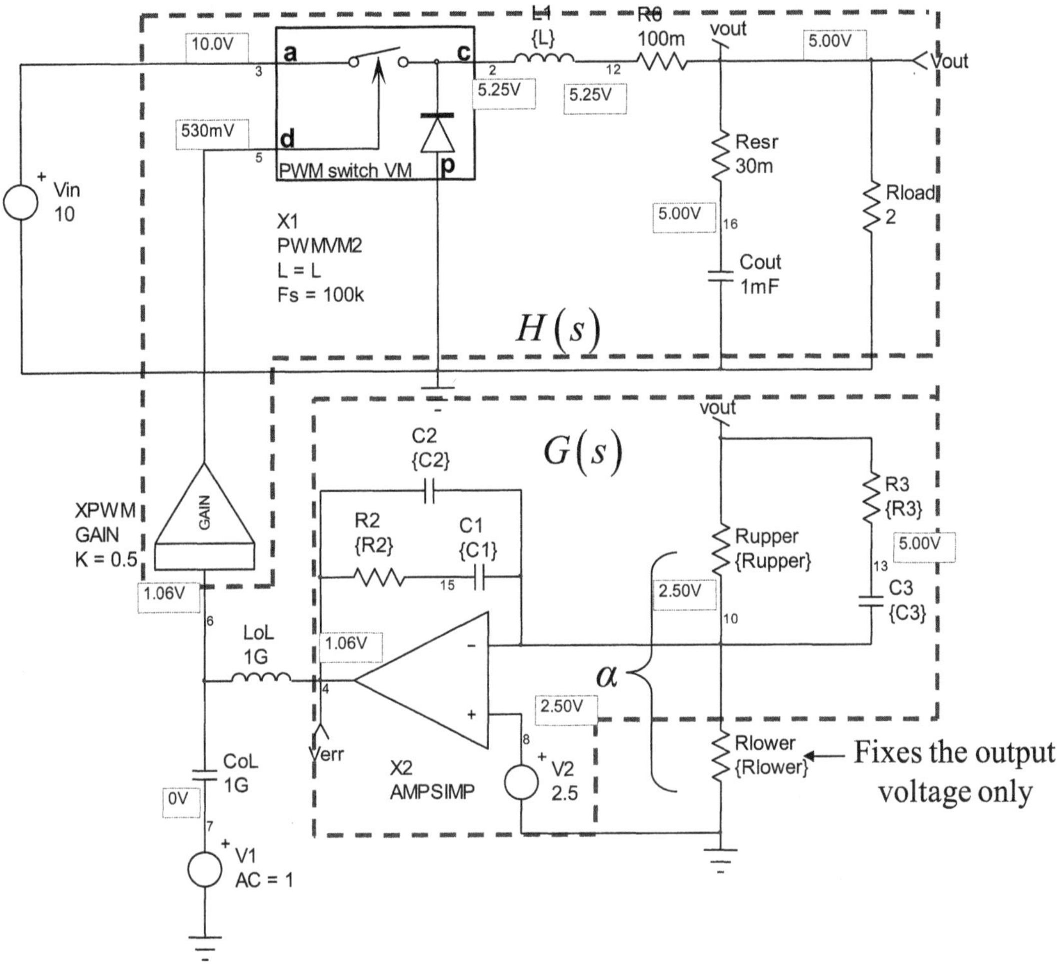

Figure 3.60 A simple linear buck averaged model can be used as an example to check the validity of our approach.

Suppose we have the following values in our system:

$V_{ref} = 2.5$ V, the reference voltage
$H_0 = 4.7$, the power stage dc-gain
$G_0 = 2000$, the error amplifier dc-gain

In that case, (3.193) would give the following output error:

$$\varepsilon_0 \approx 2.5 \times \left(\frac{1}{0.5} - \frac{1}{\left(\frac{1}{4.71 \times 2000} + 1 \right) \times 0.5} \right) = 530\ \mu\text{V} \tag{3.194}$$

Leading to an output voltage of $5 - 0.53m = 4.9995$ V. Figure 3.60 output file (.OUT) gives us a level of 4.9993 V, very close to this number.

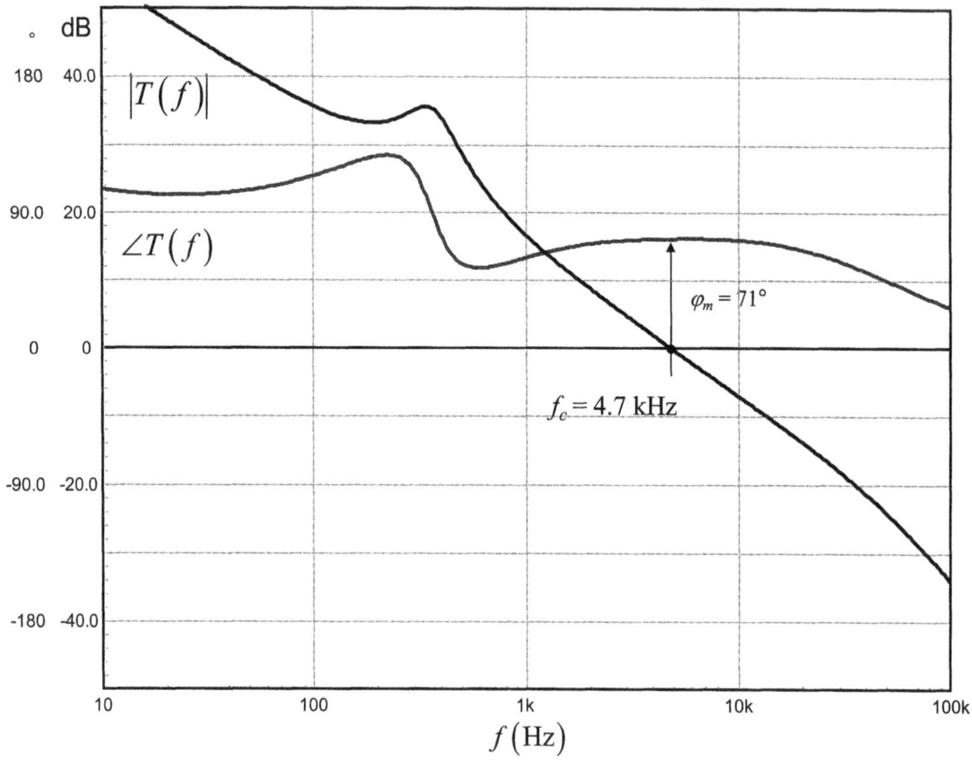

Figure 3.61 The compensated regulator exhibits a 4.7-kHz bandwidth together with a comfortable phase margin of 71°.

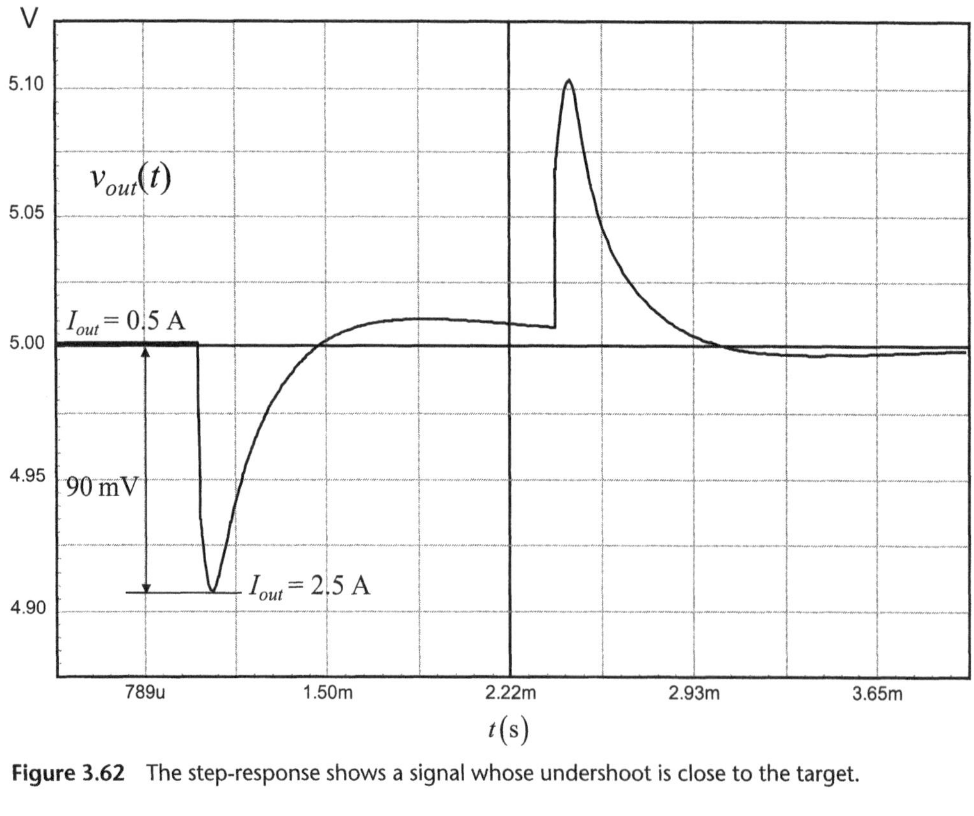

Figure 3.62 The step-response shows a signal whose undershoot is close to the target.

Without disclosing the operating details yet, we have compensated the regulator to offer a 4.7-kHz bandwidth together with a 71° phase margin, following the recommendations of (3.184). The resulting Bode plot appears in Figure 3.61.

The load resistor R_{load} is now been replaced by a current source stepping the output from 0.5 A to 2.5 A with a slope of 1 A per microsecond. When we observe the output, the voltage deviation reaches 90 mV as shown in Figure 3.62. This is very close to the values we have targeted in the design example. Of course, a real experiment has to be carried out with the selected capacitor to check if the breadboard results match the calculations. This is a general practice that any serious designer must follow: always verify on a bench prototype if the theoretical assumptions were correct and lead to an acceptable practical result.

3.4.5 Increasing the Crossover Frequency Further

In the given example, we have selected the crossover frequency to reduce the capacitive contribution and make the total drop fit the specifications. This method works well for slow converters such as ac-dc adapters for notebooks or netbooks, where the crossover frequency often lies between 1 and 5 kHz. The output capacitor is selected for its rms current capabilities and the acceptable output ripple. Given the available ESR, the designer can apply this strategy to reduce the capacitive drop. In high-speed converters such as dc-dc switchers for the telecom market, the drop must really be reduced to a minimum value that is the ESR contribution. The only possibility is then to increase the crossover frequency until the capacitive drop vanishes to a negligible value, the ESR term remaining alone. This is what we did in Figure 3.63 where the converter crossover frequency was purposely increased while the open-loop phase margin was purposely kept to 60°.

As the crossover frequency increases, the undershoot brought by the capacitor is reduced as the converter reacts faster. A possible compensation strategy, in that case, would be to select the crossover frequency so that the capacitive contribution becomes negligible, leaving the ESR drop alone. It would allow us to push the crossover frequency just enough to reduce the undershoot brought by the capacitor and not beyond what is really necessary. This naturally reduces the risks of picking up parasitic noises and making the loop unstable. As an arbitrary value, let's assume we want to reduce the capacitive contribution to a maximum of 20 percent of r_C. The output impedance equation could thus be updated so that its value gives $1.2r_C$ at the crossover frequency value we seek:

$$\sqrt{\left(\frac{1}{2\pi f_c C_{out}}\right)^2 + r_C^2} = 1.2r_C \tag{3.195}$$

Solving for fc gives us

$$f_c \approx \frac{0.24}{C_{out} r_C} \tag{3.196}$$

Figure 3.63 curves were obtained with a 1000-μF capacitor featuring a 20-mΩ ESR. If we apply the previous formula to reduce the undershoot to almost the ESR

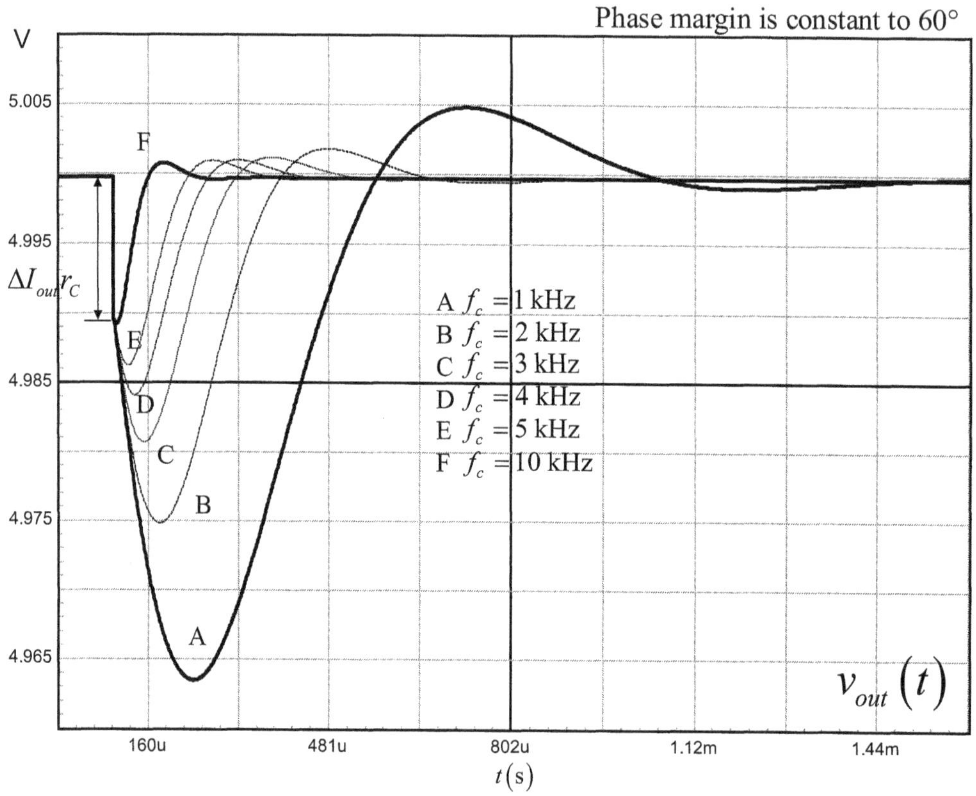

Figure 3.63 When the crossover frequency increases, the capacitive undershoot reduces until the drop becomes dictated by the ESR only.

contribution alone, we obtain a crossover frequency of 12 kHz, in line with what curve F brings. We will later use this formula in our LDO compensation example described in Chapter 9.

3.5 Conclusion

In this chapter, we have tried to show that phase margin and crossover frequency could be analytically selected to match the project specifications. The simple formula linking the open-loop phase margin to the closed-loop quality factor gives the designer a relationship between a design criteria and a final response. Rather than pulling the minimum phase margin criteria out of thin air, as it often the case in textbooks, the proposed method explores an analytical path showing its relationship to the transient behavior you will get in closed loop. The designer can thus tweak his system, depending on the needed performance in terms of response speed or overshoot. Finally, the output impedance derivation for a switching converter highlights the link between the crossover frequency and the voltage drop in response to a current step. You will no longer select your crossover frequency based on a rule of thumb with the risk of pushing it too far: check the voltage drop specifications, find the output capacitor you have, and adjust the crossover to the needed value.

Among the design criteria, if the phase and gain margins are the most popular variables, we have shown that they are not a sufficient condition for system robustness. Modulus and delay margins should be used instead, especially if hidden resonances exists in the plant transfer function and if a high-bandwidth system is desirable.

It is also important to keep in mind that the analytic descriptions we went through, phase margin and transient response followed by the output impedance prediction, are approximations. They have to be used as rules of thumb during the design phase. For instance, should you want to obtain the exact transient signature of a system, you must calculate its response to a step excitation in Laplace and then convert the result to its time-domain equivalent as shown in [2]. Needless to say, most of the designers never undertake such a complex work when several poles and zeros are at stake. Furthermore, this is a small-signal result and we know that nonlinear elements are at play in this mode. The resulting signal is then an approximation as well.

References

[1] Basso, C., *Switch Mode Power Supplies: SPICE Simulations and Practical Designs*, New York: McGraw-Hill, 2008.

[2] Peretz, M. M., and S. Ben-Yaakov, "Revisiting the Closed Loop Response of PWM Converters Controlled by Voltage Feedback," APEC 2008, Austin, TX.

[3] Adar, D., and S. Ben-Yaakov, "Generic Average Modeling and Simulations of Discrete Controller," APEC 2001, Anaheim, CA.

[4] Özbay, H., *Feedback Control Theory*, Boca Raton, FL: CRC Press, 2000.

CHAPTER 4

Compensation

The performance of a control system depends on numerous parameters, among which the compensator transfer function $G(s)$ plays a central role. This is the place where you combine poles/zeros to shape the open-loop frequency response you are looking for but also make sure that robustness is ensured over the converter lifetime. This shape is motivated by the transient performance you want once the system operates in closed loop. For instance, in one case, precision and an absolute absence of overshoot is key. In another one, you want the fastest response to a set-point change or an incoming perturbation, sacrificing the precision of the output variable. Having these elements in mind will actually dictate the way you design your compensator. This chapter will explore the possible compensation strategies applied to practical examples such as linear or switching converters.

4.1 The PID Compensator

Literature and the Web in particular abound on the subject of PIDs. If you browse the available pages, such as those in [1], you will learn that the first attempts to use this type of compensator date to governor designs, back at the end of the nineteenth century. Industrial applications to navy ships start to take off in the 1920s. Whereas mechanical/pneumatic systems make an extensive use of PID-based systems, power electronics engineers prefer to place poles and zeros.

Rather than again exploring the processing block, we will quickly show what its peculiarities are and how to bridge it with our power conversion world. Already introduced in Chapter 1, a PID is made of three distinct blocks, each processing the error signal with a mathematical treatment: proportional, integral, and derivative. By individually adjusting the coefficient of each block, the designer can tune the behavioral parameters of a control system such as rise time, damping ratio, or response time.

The combination of several mathematical processing blocks can be written in different forms. The control law in its standard form obeys the following equation:

$$v_c(t) = k_p\left(\varepsilon(t) + \frac{1}{\tau_i}\int_0^t \varepsilon(t)dt + \tau_d\frac{d\varepsilon(t)}{dt}\right) \tag{4.1}$$

where $v_c(t)$ is the control signal delivered by the compensator G, and ε is the error level between the output y and the input variable u.

- The first term $P = k_p\varepsilon(t)$ is the proportional term. It generates a control signal proportional to the error amplitude. The idea is to generate a corrective signal in proportion to the input/output mismatch amplitude: if the error is large, you expect an energetic correction; if you have a small error, a small corrective action is necessary. k_p is the parameter that lets you tweak the proportional term. If k_p is high, you have a fast response, but risks of overshoot exist. If k_p is small, you have a slow system exhibiting a sluggish response, but overshoots are reduced.

- The second term $I = \frac{1}{\tau_i}\int_0^t \varepsilon(t)dt$ is linked to the error signal integral. It can be seen as an accumulation or an integration over time of the long-term errors or drifts: you generate a corrective signal as long as you sense an error between the input and the output. In theory, the integral term cancels the static error between the setpoint and the output. In transient, the integral term slows down the response and increases the overshoot. You adjust the integral contribution via the coefficient τ_i. It has the dimension of a time constant. Sometimes, as it integrates or accumulates the errors, the integral block can saturate. To prevent this problem, it is possible to add an anti-windup system.
- The third term $D = \tau_d \frac{d\varepsilon(t)}{dt}$ is relative to the differentiation of the error signal—in other words, to the slope of the incoming perturbation or the setpoint change. Should the slope be very steep, this block delivers a large amplitude. For the opposite, a slow-moving perturbation will generate a lower-amplitude corrective signal. You adjust the derivative contribution through the coefficient τ_d that also has the dimension of a time constant. It can be seen as an anticipation factor by accounting for the variation of the perturbation. In static, as the derivative of the signal is zero, the derivative term has no effect on the output signal. Dynamically, it helps stabilize the output and increases the response speed.

Figure 4.1 shows the general architecture of this PID compensator.

Should you want to develop (4.1), you'll end up with a different expression known as the *parallel PID* formula:

$$v_c(t) = k_p\varepsilon(t) + k_i\int_0^t \varepsilon(t)dt + k_d\frac{d\varepsilon(t)}{dt} \tag{4.2}$$

In this expression, the parameters *ki* and *kd* no longer have the dimension of a time constant:

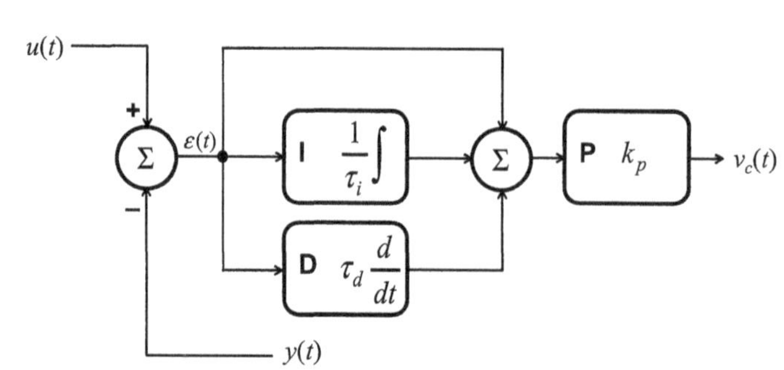

Figure 4.1 A typical PID implementation showing the three distinct blocks.

$$k_i = \frac{k_p}{\tau_i} \tag{4.3}$$

$$k_d = k_p \tau_d \tag{4.4}$$

The distribution of the k_p term implies a slightly different practical implementation, as shown in Figure 4.2.

In some cases, the whole chain is not necessary and some blocks can disappear. This is the case if you associate a proportional block with an integral block: you create a PI compensator as shown in Figure 4.3.

In this particular case, (4.2) simplifies to

$$v_c(t) = k_p \varepsilon(t) + k_i \int_0^t \varepsilon(t) dt \tag{4.5}$$

Different associations remain possible, like a simple proportional block: you just need a gain k_p for the compensation. Similarly, if you need an integrator, a single integral block is what you will pick.

4.1.1 The PID Expressions in the Laplace Domain

The PID control laws given in (4.1) and (4.2) are a continuous-time expression. Small-signal analysis is usually carried in the frequency domain. We will thus apply the Laplace transform to (4.1), remembering that a multiplication by s implies differentiation and division by s implies integration:

$$G(s) = \frac{V_c(s)}{\varepsilon(s)} = k_p \left(1 + \frac{1}{s\tau_i} + s\tau_d\right) \tag{4.6}$$

In this equation, the term $s\tau_d$ is physically improper. If we isolate this derivative term, we have a continuous +1 slope as the frequency increases. The output voltage goes to infinity as s also goes. If V_D is the derivative block output voltage, we have

$$V_D(s) = \tau_d s = \frac{s}{\omega_{z_1}} \tag{4.7}$$

in which $\omega_{z_1} = 1/\tau_d$.

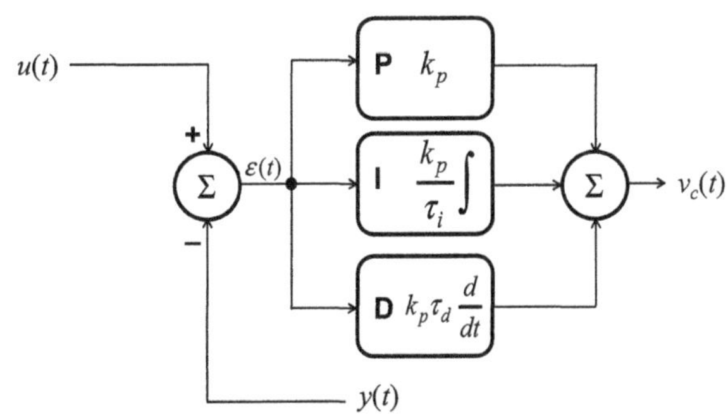

Figure 4.2 The parallel form of the PID implies the development of (4.1).

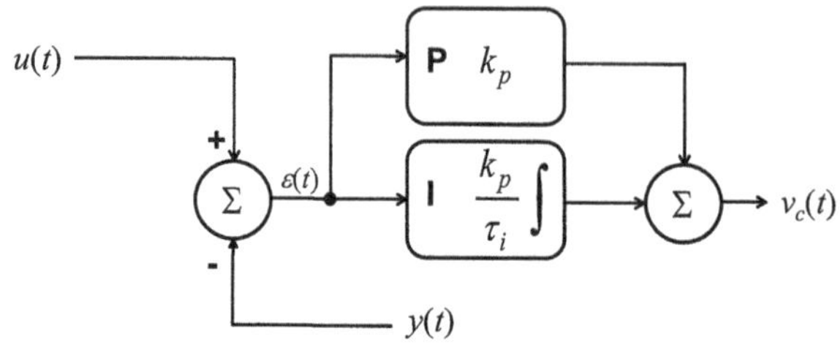

Figure 4.3 A PI block does not implement the derivative term.

Implementing the derivative term as recommended in (4.6) would bring an extreme sensitivity to incoming noises or perturbations that would corrupt the return chain. To prevent high-frequency noise from coming in, it is common practice to include a pole that safely excludes the upper portion of the frequency spectrum. Figure 4.4 represents both approaches where a pole placed five times above the zero stops the gain excursion ($N = 5$). Including this pole in the PID equation, (4.6) gives birth to the *filtered-PID* equation:

$$G(s) = \frac{V_c(s)}{\varepsilon(s)} = k_p\left(1 + \frac{1}{s\tau_i} + \frac{s\tau_d}{1 + \frac{s\tau_d}{N}}\right) \tag{4.8}$$

If we develop this equation by letting k_p enter the parenthesis, we have again the parallel form, already represented in Figure 4.2 and updated in Figure 4.5.

It is identical in terms of the earlier transfer function of G but introduces individual tuning parameters such as k_i and k_d:

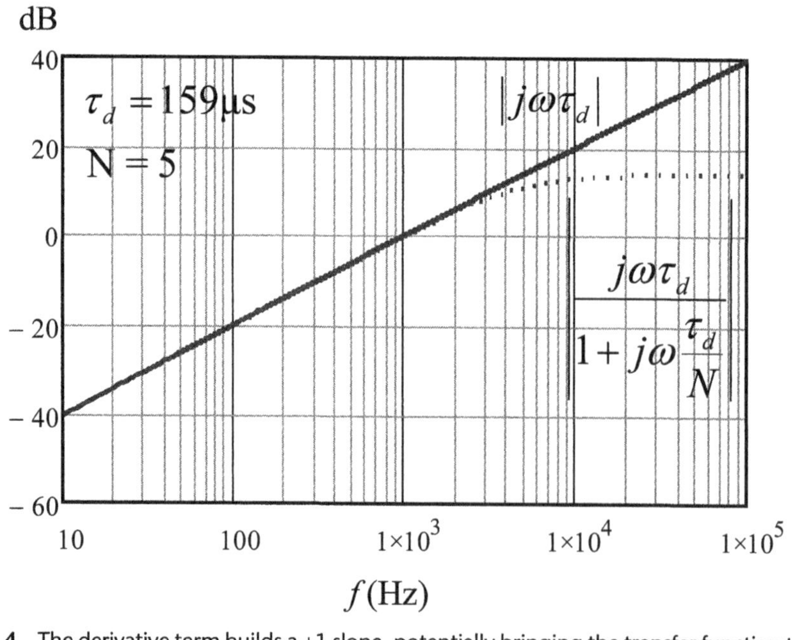

Figure 4.4 The derivative term builds a +1 slope, potentially bringing the transfer function to infinity as *s* also goes to infinity. The addition of a high-frequency pole safely clamps the gain excursion.

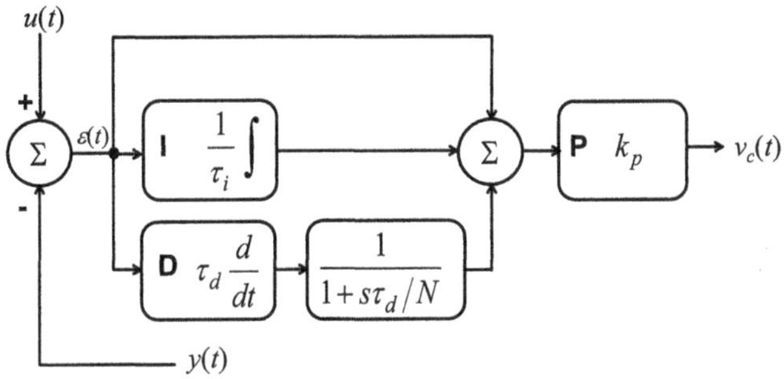

Figure 4.5 The parallel form filtered PID structure now includes a high-frequency pole.

$$G(s) = \frac{V_c(s)}{\varepsilon(s)} = k_p + \frac{k_i}{s} + \frac{sk_d}{1+s\dfrac{k_d}{Nk_p}} \tag{4.9}$$

In that case, the links between the coefficients is easy as k_p is unchanged:

$$k_i = \frac{k_p}{\tau_i} \tag{4.10}$$

$$k_d = k_p \tau_d \tag{4.11}$$

Let us now develop (4.8) a little more and factor the terms in a familiar way. We obtain the following equation:

$$G(s) = k_p\left(1+\frac{1}{s\tau_i}+\frac{s\tau_d}{1+s\dfrac{\tau_d}{N}}\right) = \frac{1+s\left(\dfrac{\tau_d}{N}+\tau_i\right)+s^2\left(\dfrac{\tau_d\,\tau_i}{N}+\tau_d\,\tau_i\right)}{s\dfrac{\tau_i}{k_p}\left(1+\dfrac{\tau_d}{N}s\right)}$$
$$= \frac{\left(1+s/\omega_{z1}\right)\left(1+s/\omega_{z2}\right)}{\dfrac{s}{\omega_{po}}\left(1+\dfrac{s}{\omega_{p1}}\right)} \tag{4.12}$$

This is it—a filtered PID is nothing more than a compensator setting a double zero, an origin pole, and a high-frequency pole! Now it sounds more friendly to us engineers than (4.6), doesn't it? What matters is the relationship to go from the time constant definitions to the poles/zeros placements. Some algebra is necessary to do that.

4.1.2 Practical Implementation of a PID Compensator

We, power electronics engineers, place poles and zeros, not individual coefficients as they appear in (4.12). However, for the sake of the study, it is interesting to understand how to bridge a PID compensator to an op amp-based circuitry and learn how to go from one configuration to the other. First, we can redevelop the right side of (4.12) as follows:

$$G(s) = \frac{(1+s/\omega_{z1})(1+s/\omega_{z2})}{\dfrac{s}{\omega_{po}}\left(1+\dfrac{s}{\omega_{p1}}\right)} = \frac{1+s\left(\dfrac{1}{\omega_{z1}}+\dfrac{1}{\omega_{z2}}\right)+s^2\left(\dfrac{1}{\omega_{z1}\,\omega_{z2}}\right)}{\dfrac{s}{\omega_{po}}\left(1+\dfrac{s}{\omega_{p1}}\right)} \tag{4.13}$$

The right side of this equation should match the developed terms of the PID compensator as it appears in (4.12). In other words,

$$\frac{1+s\left(\dfrac{\tau_d}{N}+\tau_i\right)+s^2\left(\dfrac{\tau_d\tau_i}{N}+\tau_d\tau_i\right)}{s\dfrac{\tau_i}{k_p}\left(1+\dfrac{\tau_d}{N}s\right)} = \frac{1+s\left(\dfrac{1}{\omega_{z1}}+\dfrac{1}{\omega_{z2}}\right)+s^2\left(\dfrac{1}{\omega_{z1}\,\omega_{z2}}\right)}{\dfrac{s}{\omega_{po}}\left(1+\dfrac{s}{\omega_{p1}}\right)} \tag{4.14}$$

From this expression, we can identify the individual terms by solving this set of four equations featuring four unknowns:

$$\frac{\tau_d}{N}+\tau_i = \frac{1}{\omega_{z1}}+\frac{1}{\omega_{z2}} \tag{4.15}$$

$$\frac{\tau_d\tau_i}{N}+\tau_d\tau_i = \frac{1}{\omega_{z1}\,\omega_{z2}} \tag{4.16}$$

$$\frac{\tau_i}{k_p} = \frac{1}{\omega_{po}} \tag{4.17}$$

$$\frac{\tau_d}{N} = \frac{1}{\omega_{p1}} \tag{4.18}$$

To go from known poles and zeros to the PID coefficients, we found:

$$\tau_d = \frac{(\omega_{p1}-\omega_{z1})(\omega_{p1}-\omega_{z2})}{(\omega_{p1}\,\omega_{z1}+\omega_{p1}\,\omega_{z2}-\omega_{z1}\,\omega_{z2})\,\omega_{p1}} \tag{4.19}$$

$$N = \frac{\omega_{p1}^2}{(\omega_{p1}\,\omega_{z1}+\omega_{p1}\,\omega_{z2}-\omega_{z1}\,\omega_{z2})\,\omega_{p1}}-1 \tag{4.20}$$

$$\tau_i = \frac{\omega_{z1}+\omega_{z2}}{\omega_{z1}\,\omega_{z2}}-\frac{1}{\omega_{p1}} \tag{4.21}$$

$$k_p = \frac{\omega_{po}}{\omega_{z1}}-\frac{\omega_{po}}{\omega_{p1}}+\frac{\omega_{po}}{\omega_{z2}} \tag{4.22}$$

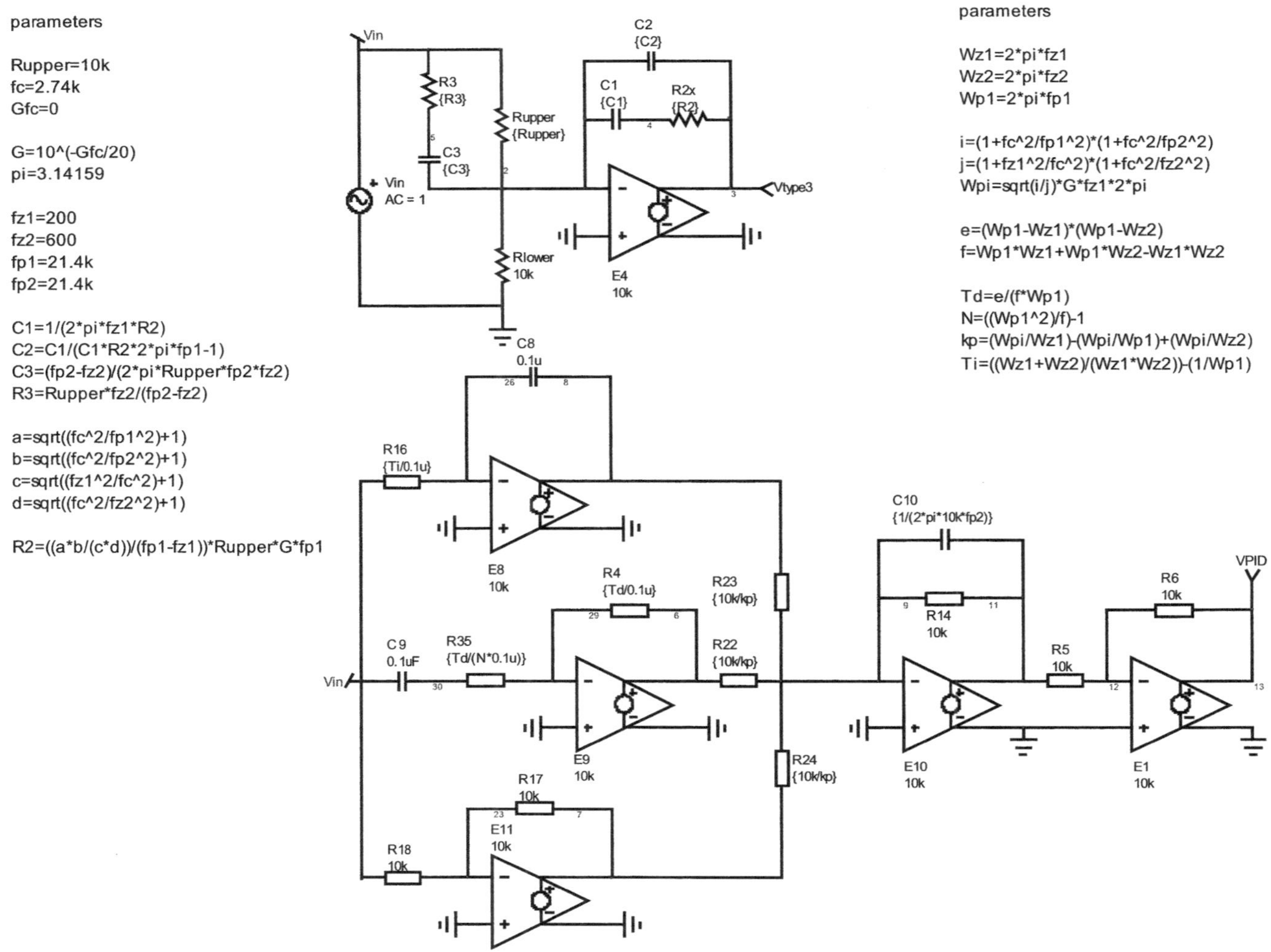

Figure 4.6 We compare the frequency response of a type 3 compensator classically built with an op amp with that of the equivalent filtered-PID circuit featuring a second high-frequency pole.

To check the validity of these calculations, we have drawn two filters. One is the classical type 3 based on an op amp. A type 3, as we will later see, is made of two zeros, two poles, and an origin pole. To compare the responses, the other filter implements the filtered-PID equation to which a second high-frequency pole f_{p2} has been added. Its location is the same as in the type 3 configuration. Its purpose is to force the gain roll-off at high frequencies, beyond the crossover point. It helps to build a good gain margin but also ensures noise immunity by filtering high-frequency spurious noises.

The comparison SPICE schematic appears in Figure 4.6. Voltage-controlled sources have been used to mimic an op amp behavior, and there is no bias point for this pure ac simulation.

The type 3 compensator organized around E_4 places a double zero, an origin pole, and two poles, with one usually at high frequencies. The PID is made of the integral block with source E_8, while the filtered derivative term uses source E_9. To maintain the inverting sign after the proportional block built with E_{10}, a final inverter is added with E_1. The polarity of the PID output signal is thus the same as that of the upper circuit using source E_4. Please note that all elements are automatically computed in the dedicated parameters windows. An ac signal is injected on the left (V_{in} node), and we can now compare the transfer functions. They appear in Figure 4.7 and confirm the validity of our derivations: the curves perfectly superimpose.

The other way around is also possible, starting from PID coefficients to rebuild a poles-/zeros-based transfer function. The values we have derived are the following ones:

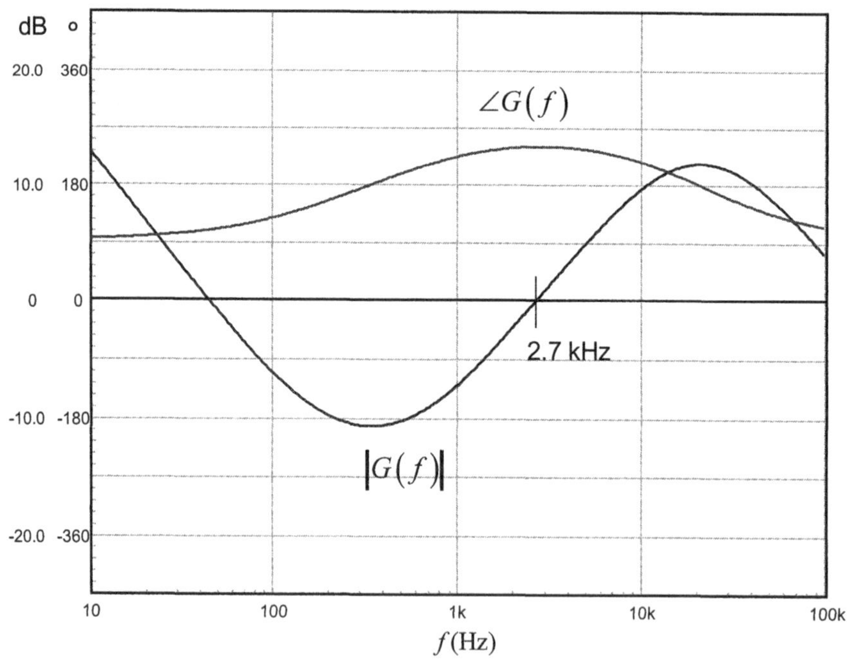

Figure 4.7 The transfer function from the filtered PID exactly matches that from the op amp-based type 3 compensator. Both curves are perfectly superimposed, confirming the validity of our calculations.

$$f_{z1} = \frac{\tau_d - \sqrt{-4N^2\tau_d\tau_i + N^2\tau_i^2 - 2N\tau_d\tau_i + \tau_d^2} + N\tau_i}{2\tau_d\tau_i(1+N)2\pi} \tag{4.23}$$

$$f_{z2} = \frac{\tau_d + \sqrt{-4N^2\tau_d\tau_i + N^2\tau_i^2 - 2N\tau_d\tau_i + \tau_d^2} + N\tau_i}{2\tau_d\tau_i(1+N)2\pi} \tag{4.24}$$

$$f_{p1} = \frac{N}{2\pi \cdot \tau_d} \tag{4.25}$$

$$f_{po} = \frac{k_p}{2\pi \cdot \tau_i} \tag{4.26}$$

Please note that (4.23) and (4.24) give real zeros if the following conditions are met [2]:

$$N\tau_i - \tau_d \geq 2N\sqrt{\tau_i\tau_d} \tag{4.27}$$

If you find imaginary roots for the zero positions given by (4.23) and (4.24), it simply means that the PID you study implements complex zeros. We will study their effects later in this chapter.

4.1.3 Practical Implementation of a PI Compensator

In the case you have built a PI compensator, as presented in Figure 4.3, the compensator formula excludes the derivative term and transforms into a simpler Laplace expression:

$$G(s) = \frac{V_c(s)}{\varepsilon(s)} = k_p + \frac{k_p}{s\tau_i} = k_p\left(\frac{1+s\tau_i}{s\tau_i}\right) = \frac{1+s\tau_i}{s\frac{\tau_i}{k_p}} = \frac{1+s/\omega_z}{s/\omega_{po}} \tag{4.28}$$

If you factor the term s/ω_z and rearrange the equation, you obtain a familiar format:

$$G(s) = G_0\left(1+\frac{\omega_z}{s}\right) \tag{4.29}$$

Unlike in the complex full PID expression, the gain, the zero, and the pole are immediately identified as:

$$G_0 = \frac{\omega_{po}}{\omega_z} \tag{4.30}$$

$$\omega_z = \frac{1}{\tau_i} \tag{4.31}$$

$$\omega_{po} = \frac{k_p}{\tau_i} \tag{4.32}$$

In the literature, this PI compensator is referred as a type 2a compensator.

The PI expression in (4.28) reveals a pole at the origin associated with a single zero. A possible implementation of such expression appears in Figure 4.8. The PI

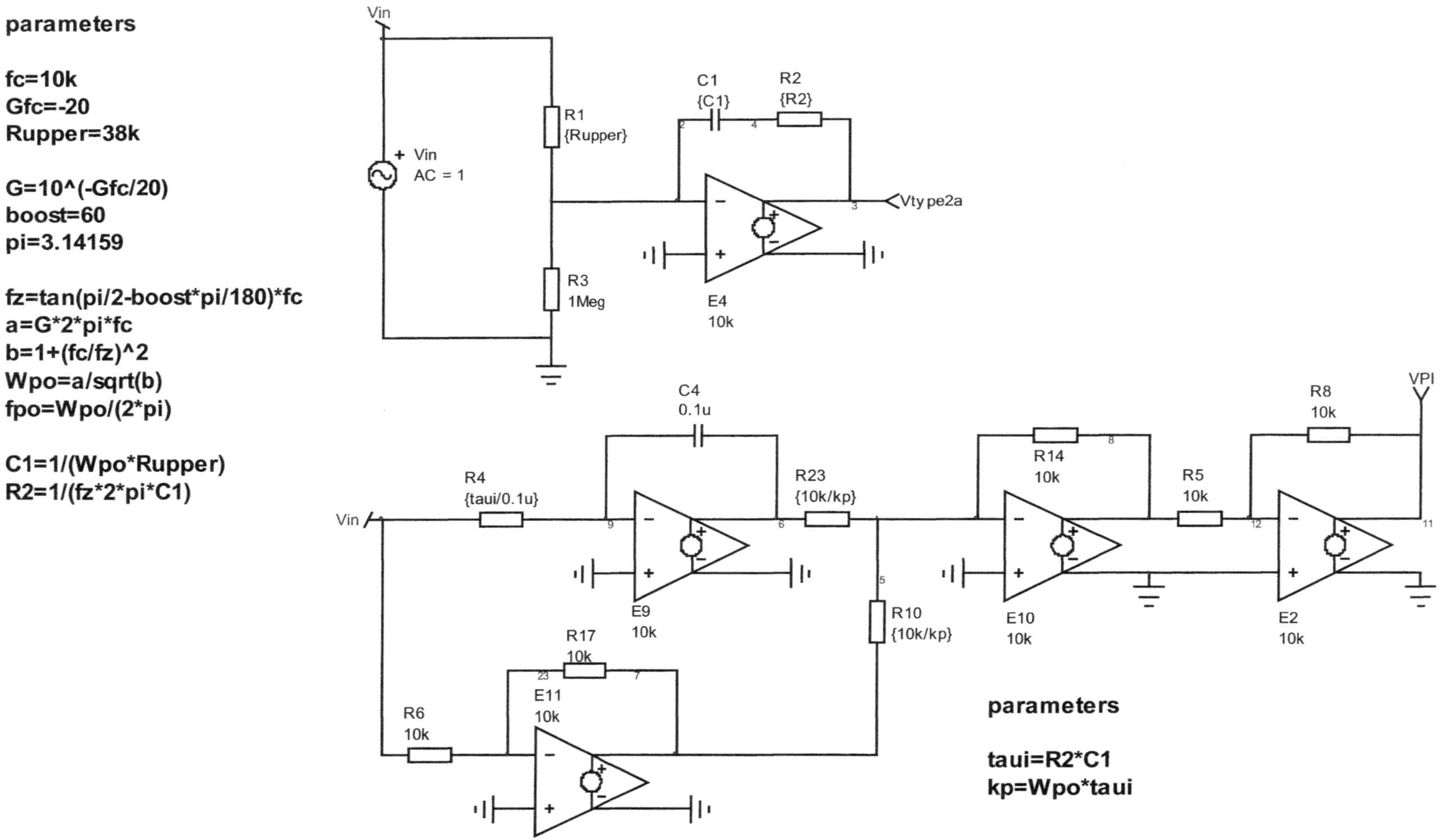

Figure 4.8 The electrical construction of the PI compensator is simplified due to the absence of the derivative term.

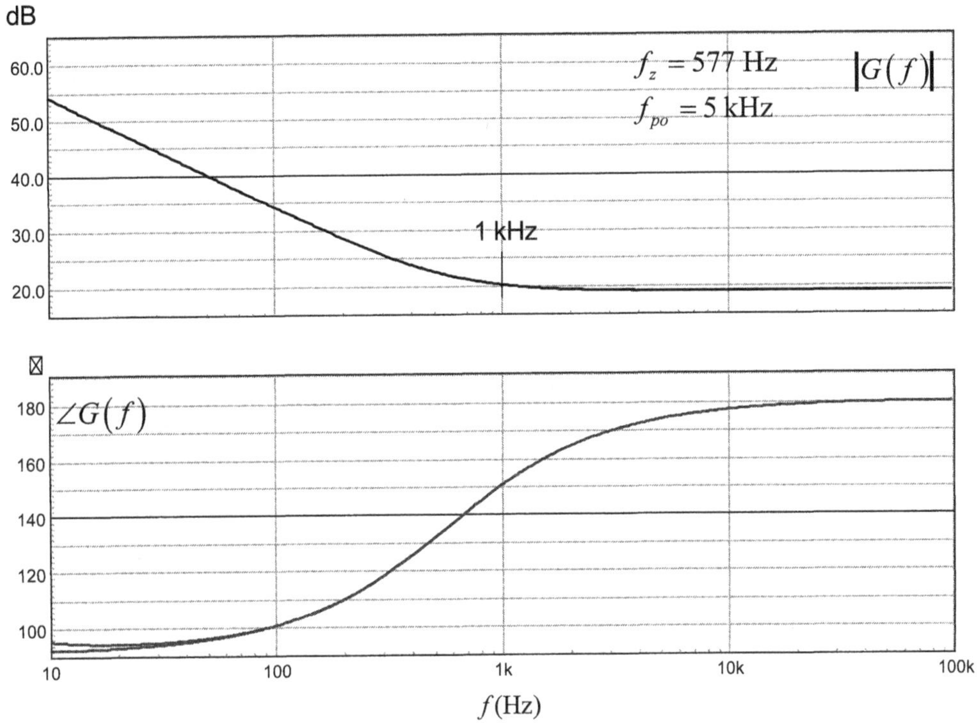

Figure 4.9 The ac response of the compensator, regardless of whether it is made around a type 2a or a PI compensator, is similar.

compensator coefficients are easily derived and automated in the simulation sheet. The ac response is given in Figure 4.9 and confirms the good agreement between the reference compensator (a type 2a) and the PI circuit.

4.1.4 The PID at Work in a Buck Converter

Now that we understand how to build a PID compensator, it is time to put it to work. The idea is to use a buck converter operated in voltage-mode. The schematic of such a switching converter appears in Figure 4.10. The principle of operation is well known. A power switch SW is operated on and off at a pace fixed by an internal clock generator. Typical operating frequencies are between 50 kHz to 1 or 2 MHz for the vast majority of converters. The switch control pattern is produced by the pulse width modulator (PWM) block, just as we studied in Chapter 3. In this technique, the error voltage directly controls the duty ratio and continuously adjusts the switch on-time t_{on} in relation to the operating conditions. Usually, when the error voltage increases, so does the duty ratio, instructing SW to remain closed longer. If SW is permanently closed, you have V_{in} at the diode cathode. Inversely, if SW is permanently open, you have 0. The voltage at the diode cathode is thus a square-wave signal toggling between V_{in} and 0. As we want a dc output voltage, a low-pass filter made of L and C attenuates all unwanted harmonics and the output delivers a dc level equal to the average value of the square-wave signal. Please note that both L and C are affected by parasitic resistive elements, noted r_L and r_C,

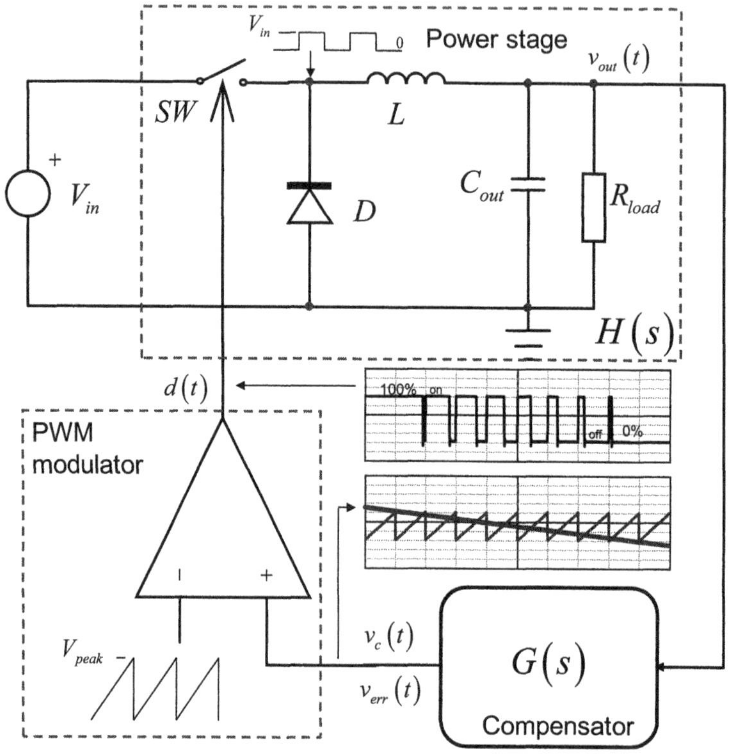

Figure 4.10 A buck converter meets the output voltage requirement—here 5 V—by adjusting the duty ratio of the power switch.

respectively. They do not appear in this schematic for the sake of clarity. We can show that the output voltage obeys the following relationship:

$$V_{out,avg} = \frac{1}{T_{SW}} \int_{0}^{t_{on}} V_{in} dt = dV_{in} \tag{4.33}$$

If you control the duty ratio d from 0 to 100 percent, you have a means to adjust the output voltage between 0 and V_{in}.

The first thing to do when you want to compensate a converter is to get its small-signal transfer function from the control input V_c to the output V_{out}. In Figure 4.10, this is $H(s)$, driven by the PWM block. This transfer function can be measured in the lab, derived using small-signal analysis or simply simulated using an average model. We used this last option to plot the ac response of the power stage. The test fixture appears in Figure 4.11, where the capacitor and the inductor are now in series with their respective parasitic elements. We have two distinct storage elements (two state variables), so this is a second-order system. As already demonstrated, the small-signal model of the PWM modulator is simply the inverse of the sawtooth peak amplitude. If you have a 2-V amplitude, the PWM gain is 0.5 or −6 dB. The operating point is selected by biasing the control input to a level that gives you the correct output voltage: 1.05 V applied to v_c gives 5 V in the example. The ac response is that of a second-order system, as shown in Figure 4.12.

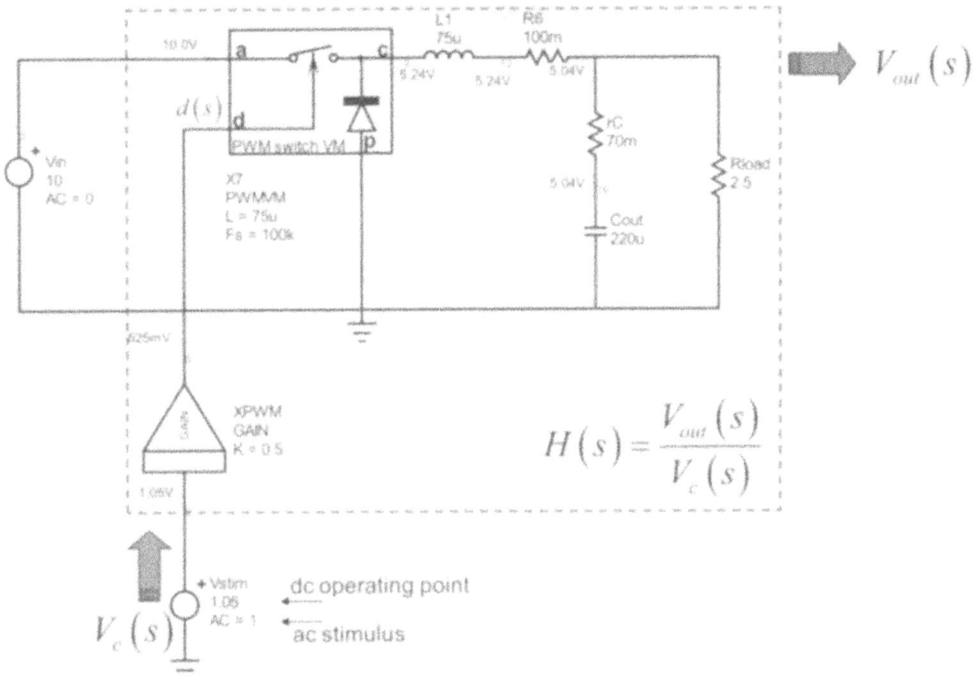

Figure 4.11 This text fixture uses an average model to extract the small-signal response of the voltage-mode buck converter.

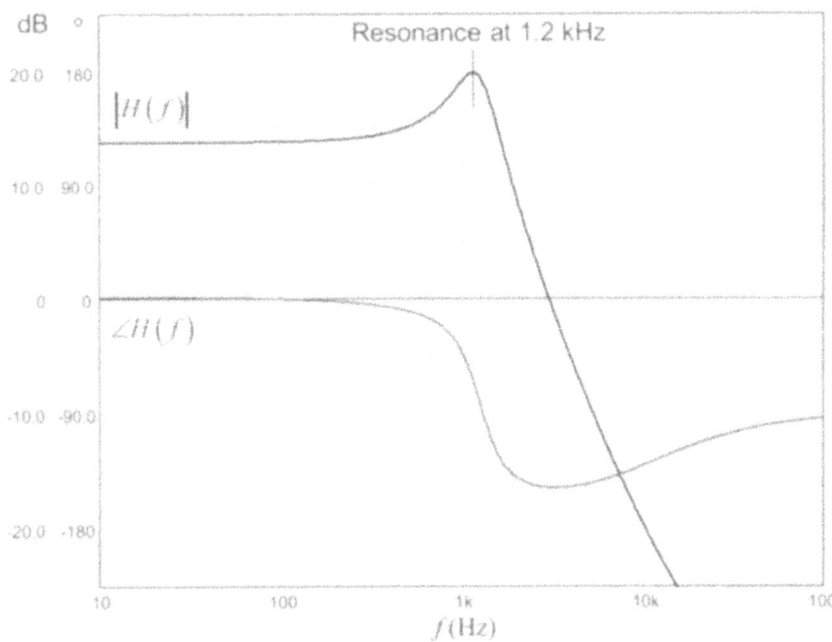

Figure 4.12 The *LC* filter introduces a resonance in the ac transfer function.

It can be shown that the small-signal analysis carried on the buck converter operated in voltage mode leads to the following equations:

$$H(s) = \frac{V_{out}(s)}{V_c(s)} = H_0 \frac{1 + s/\omega_{z_1}}{\left(\frac{s}{\omega_0}\right)^2 + \frac{s}{Q\omega_0} + 1} \tag{4.34}$$

In which we have

$$\omega_{z_1} = \frac{1}{r_C C} \tag{4.35}$$

$$\omega_0 \approx \frac{1}{\sqrt{LC}} \tag{4.36}$$

$$Q \approx R\sqrt{\frac{C}{L}} \tag{4.37}$$

$$H_0 = \frac{V_{in}}{V_{peak}} \tag{4.38}$$

where R, L, and C are, respectively, the load resistor (R_{load}), the output inductor (L_1), and the output capacitor (C_{out}) in Figure 4.12. The ESR is designated by r_C. V_{in} is the input voltage, and V_{peak} the PWM block sawtooth peak amplitude.

Ideally, we would like a fast transient response, almost no overshoot, and a precise output. In other words, a flat nonpeaking closed-loop gain from input to output is the goal to reach, with a bandwidth of 10 kHz for the sake of the example. To meet this requirement, we must add a PID compensator to the Figure 4.11 converter. The updated schematic using the average model appears in Figure 4.13. In such a representation, the transfer function we study is classically defined as

$$\frac{Y(s)}{U(s)} = \frac{V_{out}(s)}{V_{ref}(s)} \tag{4.39}$$

To obtain a flat ac closed-loop response, let us first study the open-loop expression of our compensated buck. It is the multiplication of the power stage transfer function $H(s)$ derived in (4.34) multiplied by the filtered-PID transfer function $G(s)$ given in (4.12):

$$T_{OL}(s) = H(s)G(s) = H_0 \frac{1 + s/\omega_{z_1}}{\left(\frac{s}{\omega_0}\right)^2 + \frac{s}{Q_0\omega_0} + 1} \frac{1 + s\left(\frac{\tau_d}{N} + \tau_i\right) + s^2\left(\frac{\tau_d \tau_i}{N} + \tau_d \tau_i\right)}{s\frac{\tau_i}{k_p}\left(1 + \frac{\tau_d}{N}s\right)} \tag{4.40}$$

This is a rather complex expression that we can try to simplify. As suggested in [3], if we purposely adjust the double zeros present in the numerator of $G(s)$ to match the position of the double poles that appear in the denominator of $H(s)$, then the expression greatly simplifies. In other words, if we have

$$\left(\frac{s}{\omega_0}\right)^2 + \frac{s}{Q_0\omega_0} + 1 = 1 + s\left(\frac{\tau_d}{N} + \tau_i\right) + s^2\left(\frac{\tau_d \tau_i}{N} + \tau_d \tau_i\right) \tag{4.41}$$

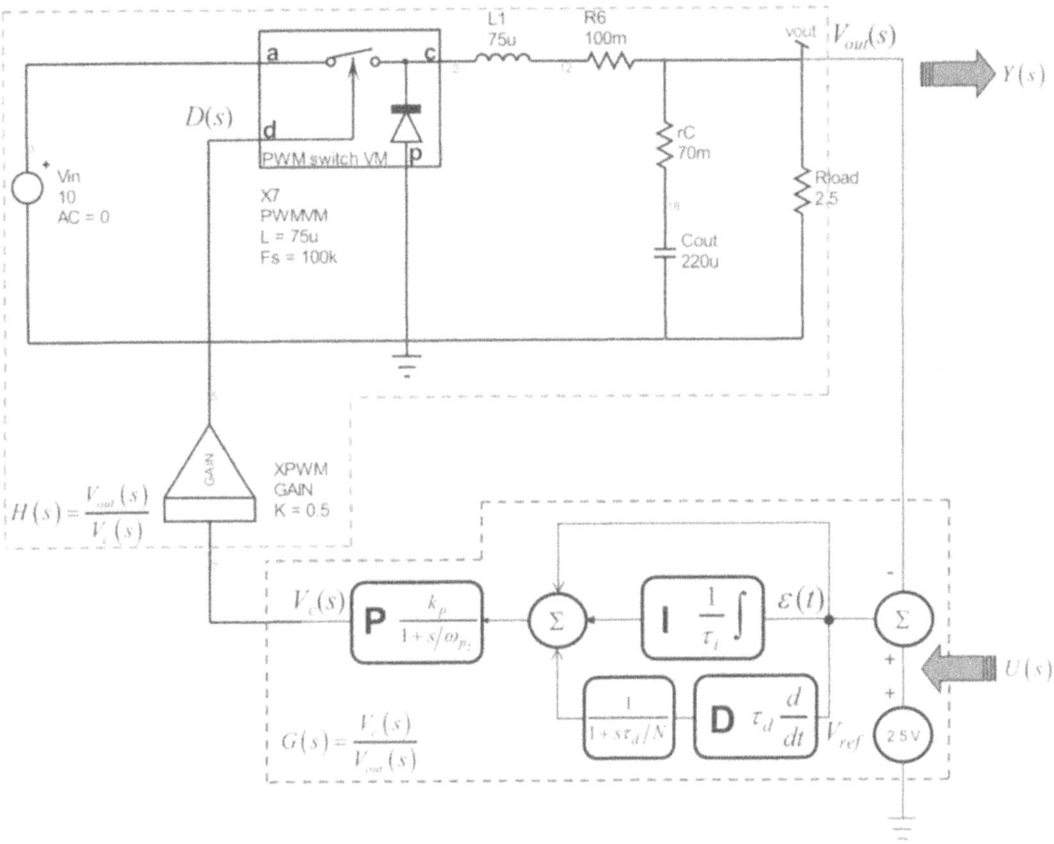

Figure 4.13 Once the compensator has been added, we obtain a complete buck converter maintaining an output voltage of 5 V. This is obviously a simplified representation.

then the loop gain expression simplifies to

$$T_{OL}(s) = H_0 \frac{1+s/\omega_{z1}}{s\frac{\tau_i}{k_p}\left(1+\frac{\tau_d}{N}s\right)} \tag{4.42}$$

We know that the closed-loop expression of a unity-return control system is the following:

$$T_{CL}(s) = \frac{T_{OL}(s)}{1+T_{OL}(s)} \tag{4.43}$$

Substituting (4.42) in (4.43) gives

$$T_{CL}(s) = \frac{H_0 \dfrac{(1+s/\omega_{z1})}{s\frac{\tau_i}{k_p}\left(1+\frac{\tau_d}{N}s\right)}}{1+H_0 \dfrac{(1+s/\omega_{z1})}{s\frac{\tau_i}{k_p}\left(1+\frac{\tau_d}{N}s\right)}} = \frac{1+s/\omega_{z1}}{1+s\left(\frac{1}{\omega_{z1}}+\frac{\tau_i}{H_0k_p}\right)+s^2\left(\frac{\tau_d\tau_i}{N\cdot H_0k_p}\right)} \tag{4.44}$$

This is our closed expression once the double poles have been neutralized by the double zeros of the PID. If you observe the denominator, you can see that it is following the form of a second-order transfer function:

$$1+s\left(\frac{1}{\omega_{z1}}+\frac{\tau_i}{H_0 k_p}\right)+s^2\left(\frac{\tau_d\tau_i}{N\cdot H_0 k_p}\right)=1+\frac{s}{\omega_c Q_c}+\left(\frac{s}{\omega_c}\right)^2 \tag{4.45}$$

Since we want a flat ac response from this second-order transfer function, we can adjust our PID coefficients to match a closed-loop quality factor of 0.5 (coincident poles, no peaking) with a bandwidth of 10 kHz $\omega_c = 62.8$ krad/s. We have four unknown, τ_d, τ_i, k_p and N. We thus need four equations. The first two come from the pole/zero neutralization in (4.41). The rest are coming from (4.45) since Q_c and ω_c are the objectives to reach:

$$\frac{\tau_d}{N}+\tau_i=\frac{1}{\omega_0 Q_0} \tag{4.46}$$

$$\frac{\tau_d\tau_i}{N}+\tau_d\tau_i=\frac{1}{\omega_0^2} \tag{4.47}$$

$$\frac{1}{\omega_{z1}}+\frac{\tau_i}{H_0 k_p}=\frac{1}{\omega_c Q_c} \tag{4.48}$$

$$\frac{\tau_d\tau_i}{N\cdot H_0 k_p}=\frac{1}{\omega_c^2} \tag{4.49}$$

The symbolic extraction gives birth to four quite ugly equations. Sorry, I did not have the courage to give them a little "massage" and make them look more friendly. Mathcad® is to blame!

$$\tau_d=\frac{Q_0Q_c^2\omega_{z1}^2\omega_0^2+Q_c^2\omega_{z1}\omega_0\omega_c^2+Q_0Q_c^2\omega_c^4-Q_c\omega_{z1}^2\omega_0\omega_c-2Q_0Q_c\omega_{z1}\omega_c^3+Q_0\omega_{z1}^2\omega_c^2}{\omega_0\omega_c(Q_c\omega_c-\omega_{z1})(Q_c\omega_c^2-\omega_{z1}\omega_c+Q_0Q_c\omega_{z1}\omega_0)}$$
$$=1.116\text{ ms} \tag{4.50}$$

$$\tau_i=-\frac{Q_c\omega_c^2-\omega_{z1}\omega_c+Q_0Q_c\omega_{z1}\omega_0}{Q\,\omega_{z1}\omega_0\omega_c-Q_0Q_c\omega_0\omega_c^2}=14.6\ \mu\text{s} \tag{4.51}$$

$$N=-\frac{Q_0Q_c^2\omega_{z1}^2\omega_0^2+Q_c^2\omega_{z1}\omega_0\omega_c^2+Q_0Q_c^2\omega_c^4-Q_c\omega_{z1}^2\omega_0\omega_c-2Q_0Q_c\omega_{z1}\omega_c^3+Q_0\omega_{z1}^2\omega_c^2}{\omega_0Q_c\omega_{z1}(Q_c\omega_c^2-\omega_{z1}\omega_c+Q_0Q_c\omega_{z1}\omega_0)}$$
$$=72.4 \tag{4.52}$$

$$k_p=\frac{Q_c\omega_{z1}(\omega_{z1}\omega_c+Q_cQ_0\omega_{z1}\omega_0-Q_c\omega_c^2)}{H_0Q_0\omega_0(\omega_{z1}-Q_c\omega_c)^2}=0.178 \tag{4.53}$$

Now that we have our PID coefficients, we can plot the compensator transfer function and the plants for reference. This is what Figure 4.14 shows. The compensator response is rather unusual. We can clearly see an origin pole (infinite gain for $s = 0$) but then a kind of notch appears, exactly at the LC resonant frequency.

With the help of (4.23) through (4.26), we can check the resulting poles and zeros corresponding to the calculated PID. If we look at the zeros placed by the adopted compensation strategy, we have

$$f_{z1} = \frac{\tau_d - \sqrt{-4N^2\tau_d\tau_i + N^2\tau_i^2 - 2N\tau_d\tau_i + \tau_d^2} + N\tau_i}{2\tau_d\tau_i(1+N)2\pi} = 144.7 - j1.23k \quad (4.54)$$

$$f_{z2} = \frac{\tau_d + \sqrt{-4N^2\tau_d\tau_i + N^2\tau_i^2 - 2N\tau_d\tau_i + \tau_d^2} + N\tau_i}{2\tau_d\tau_i(1+N)2\pi} = 144.7 + j1.23k \quad (4.55)$$

No wonder a notch appears in the transfer function of $G(s)$: the two zeros are conjugate complex numbers. This actually makes sense since we wanted to neutralize a pair of coincident complex poles located at ω_0. How do you neutralize them? By placing a pair of coincident complex zeros at ω_0. If a double complex pole induces a peak in the transfer function, a double complex zero is seen as a notch in the ac plot. This is exactly what Figure 4.14 shows.

Regarding the additional poles brought by G, they are located at

$$f_{p1} = \frac{N}{2\pi \cdot \tau_d} = 10.3\text{ kHz} \quad (4.56)$$

$$f_{po} = \frac{k_p}{2\pi \cdot \tau_i} = 1.9\text{ kHz} \quad (4.57)$$

Again, it makes perfect sense. The compensator must force the loop gain decrease so that the selected crossover point is reached at the right frequency. What in the plant transfer function opposes a magnitude decrease as the frequency increases? A zero. This is ω_{z1} found to be at 10.3 kHz with (4.35). Naturally, the zero (real in this case, a LHP zero) is neutralized by placing a pole f_{p1} at its frequency. Finally, the 0-dB crossover pole f_{po} is adjusted to make sure the curve is perfectly flat until 10 kHz are reached. The final result appears in Figure 4.15.

The open-loop transfer function is perfect and offers a phase margin of 80°. The closed-loop response is also excellent and shows a flat transmission up to 11.8 kHz, the cutoff frequency. There is absolutely no peaking; we expect an excellent transient response.

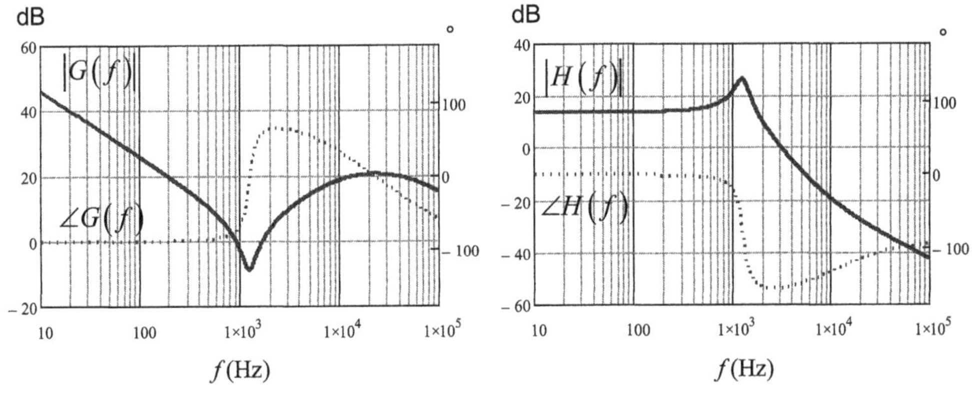

Figure 4.14 The compensator and the plant transfer functions are plotted using Mathcad.

4.1.5 The Buck Converter Transient Response with the PID Compensation

The transient response of our compensated converter can be evaluated in several ways. The simplest one is to use the closed-loop transfer function obtained with (4.43). It describes the relationship between the reference voltage v_{ref} and the output voltage v_{out}. To plot the response with Mathcad we apply a step to (4.43) and we take the inverse Laplace transform. However, as this equation is an ac small-signal equation, it does not convey a dc operating point. Suppose we have a 5-V output and we want to check the resulting impact of a 300-mV voltage step on the reference voltage V_{ref}. The equation we should plot is thus

$$v_{out}(t) = 5 + \mathcal{L}^{-1}\left(\frac{0.3}{s}\frac{T_{OL}(s)}{1+T_{OL}(s)}\right) \tag{4.58}$$

The resulting waveform appears in Figure 4.16. As expected, it is a perfect nonringing answer, in agreement with Figure 4.15 predictions. The gain is 1 (or 0 dB), justifying the 300-mV output step for a 300-mV input.

This is a purely analytical prediction, and it is interesting to see how it compares to our SPICE model implementing a buck average model. After all, in a regulator application, such as dc-dc/ac-dc converters or linear regulators, the reference voltage is never stepped since we want a constant output voltage. Rather, the output current or the input voltage are perturbed and impose the regulator to reject them in relationship to operating conditions.

The updated simulation schematic appears in Figure 4.17. You can see the average model whose input receives a signal from the B2 source. This source simply clamps the PID error voltage so that the duty ratio voltage does not exceed 100 percent. The PID implementation slightly differs from that in Figure 4.6 but the result is similar. The additional pole ω_{p2} added through subcircuit X6 does not impact the response because it is placed at high frequencies. This second pole—present in a type 3 compensator—forces frequency roll-off in the upper portion of the spectrum. Indeed, if you look at (4.13), the equation magnitude does not reach 0 as s goes to infinity. With the addition of this extra pole, we make sure the denominator degree is greater than that of the numerator: the function is said to be *proper*.

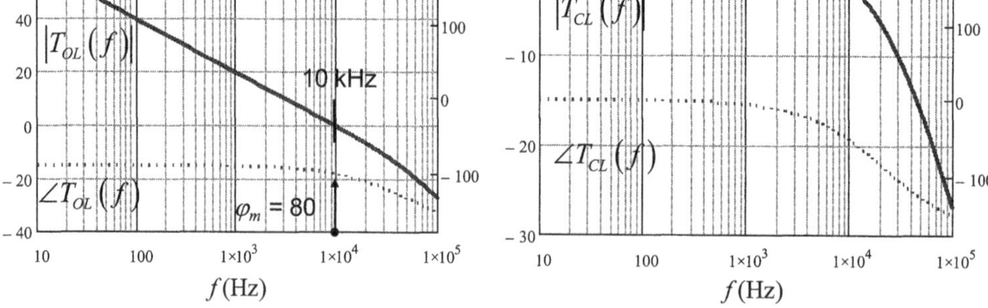

Figure 4.15 The compensated control system offers a flat response with a 10-kHz crossover frequency.

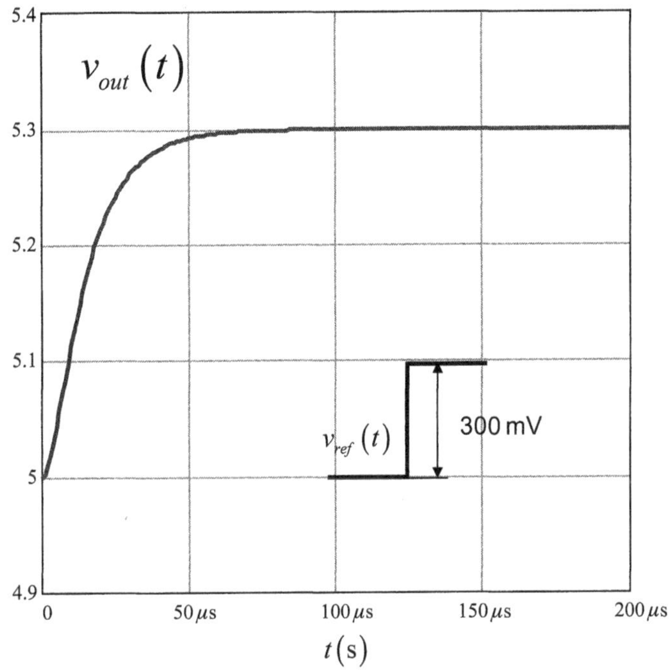

Figure 4.16 The output transient response further to a 300-mV step on the reference voltage confirms the excellent compensation scheme.

To check the transient response, the load has been replaced by a current source, stepping the current from 1 A to 2 A in 1 μs. The results are displayed in Figure 4.18 and show a stable but oscillatory waveform. It really contradicts the Figure 4.15 plot showing a low-Q flat closed-loop transfer function. From the transient result, we can derive the quality factor value using the formula we have seen in Chapter 3:

$$Q = \sqrt{\left(\frac{\pi}{\ln k}\right)^2 + \frac{1}{4}} = \sqrt{\left(\frac{\pi}{\ln\frac{21.5}{48.4}}\right)^2 + \frac{1}{4}} = 3.9 \tag{4.59}$$

It is far from our 0.5 design target! Also, if you look carefully, the oscillations do not correspond to a 10-kHz signal (our crossover point) but to the 1.2-kHz LC network resonant frequency. So where is the problem?

4.1.6 The Setpoint Is Fixed: We Have a Regulator!

In Chapter 1, we detailed the definition of a control system where the output must faithfully follow a setpoint change. In a switching or linear converter case, the setpoint is the reference voltage V_{ref} that the output must match, scaled up or down by a divider ratio. When operating, V_{ref} never changes (unless it is an adjustable output, of course) and the system must deliver a stable and precise output despite incoming perturbations. In that case, we talk about a regulator as introduced in Chapter 1. Such a system can be modeled in a way where perturbations appear. This model has already been presented before in its simplified form. The updated version that is shown in Figure 4.19 now includes the input voltage as another perturbation.

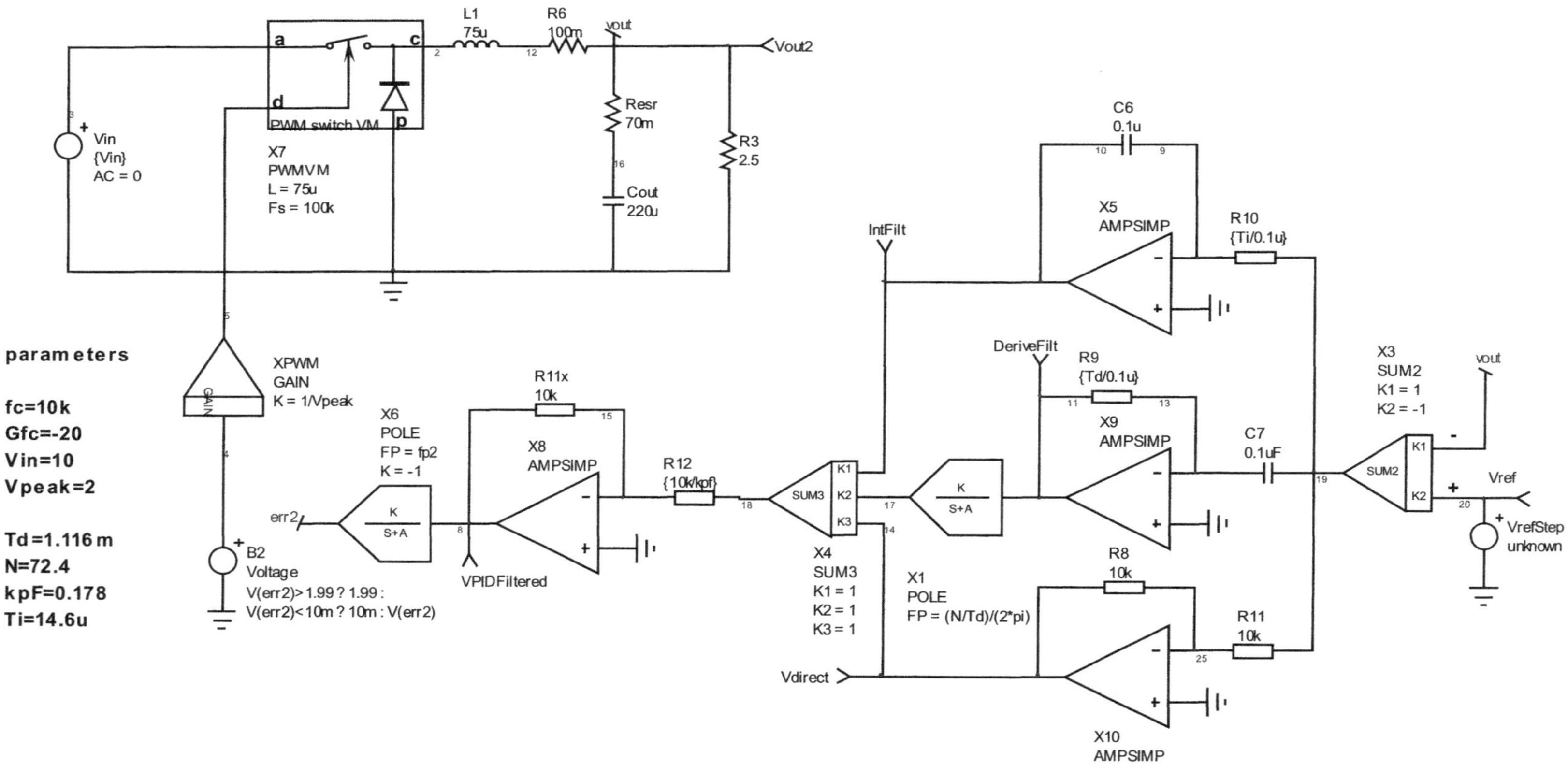

Figure 4.17 The complete simulation schematic implements the buck model with the full PID chain. The coefficients are those we already computed.

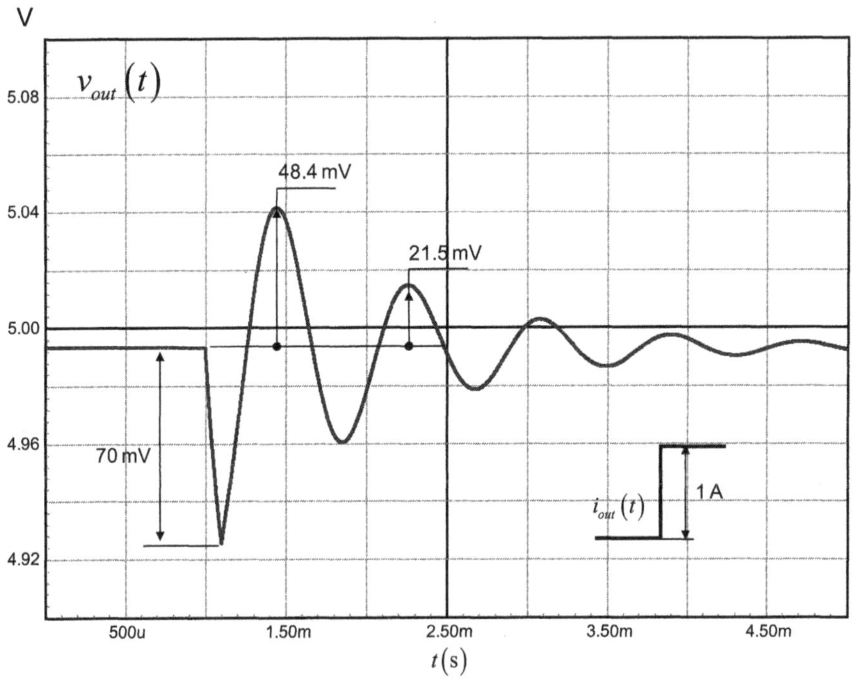

Figure 4.18 The transient response to an output current step is stable but oscillatory.

In a previous chapter, we have seen that in a system like that of Figure 4.19 all perturbations were multiplied by the sensitivity function S. Therefore, without applying the superposition theorem, we can directly write the output voltage equation of Figure 4.19 system:

$$V_{out}(s) = V_{ref}(s)\frac{T(s)}{1+T(s)} - Z_{out}(s)I_{out}(s)\frac{1}{1+T(s)} + V_{in}(s)G_{V_{in}}\frac{1}{1+T(s)} \tag{4.60}$$

In reality, as the reference voltage V_{ref} is fixed, its ac value $\hat{v}_{ref}$ is 0. The same comment applies to the input voltage V_{in}: when we step the output, we assume that the input voltage does not change. The previous equation defining the output voltage can thus be rewritten as

$$V_{out}(s)\Big|_{\hat{v}_{ref}=0,\hat{v}_{in}=0} = -Z_{out}(s)I_{out}(s)\frac{1}{1+T(s)} \tag{4.61}$$

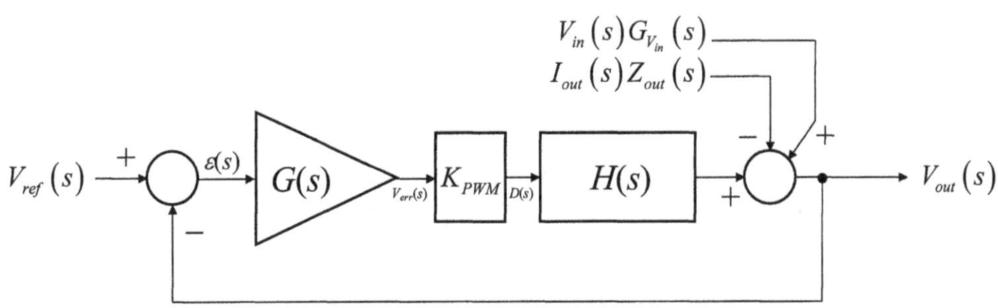

Figure 4.19 A regulator operates with a fixed setpoint constantly fights the incoming perturbations.

This is the small-signal deviation obtained when a step load is applied to the output. As you can read, the output impedance plays the main role, together with the sensitivity function. The transfer function from V_{ref} to V_{out} is not pertinent in this case. Despite Bode predicting a good answer, we just missed the main point in our regulator example: the output impedance Z_{out} fixes the transient response to a load step. Studying the transfer function $V_{out}(s)/V_{ref}(s)$ is simply not enough; we must always check the magnitude of $Z_{out}(s)$.

4.1.7 A Peaky Output Impedance Plot

With the help of the SPICE circuit from Figure 4.17, we can easily plot the output impedances in an open- or closed-loop condition. We just add a 1-A ac current source in parallel with the load, and we directly obtain the output impedance by plotting the output voltage vector. This is what Figure 4.20 shows you.

As you can see, despite the PID action, the output impedance natural peaking ($Z_{out,OL}$) is not tamed at all. Can we check the resulting quality factor from this plot? It is certainly difficult to measure it directly from the graph. However, by looking at Appendix 4B, we can see that a relationship links the quality factor Q to the group delay τ_g. To use it, we must first display the closed-loop output impedance argument and calculate the corresponding group delay. The results appear in Figure 4.21 and show a group delay of 1.015 ms. Applying (4.206) derived in Appendix 4B, we estimate the quality factor to a value of:

$$Q = \tau_g \pi f_0 = 1200 \times 1.015m \times 3.14159 = 3.82 \tag{4.62}$$

This is in good agreement with the measured value from the transient step in Figure 4.18. The difference is that (4.62) is obtained from a small-signal graph where the system is linear. Equation (4.59) comes from the transient response of a system perhaps operated in a nonlinear zone, thus potentially leading to corrupted results.

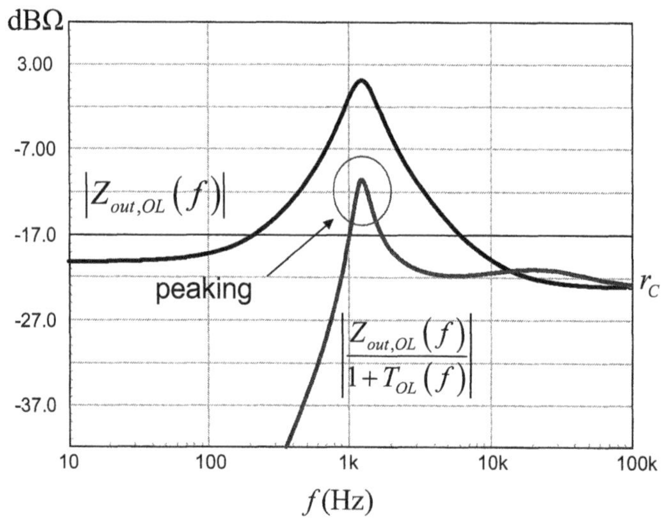

Figure 4.20 The closed-loop output impedance is still peaky despite the PID presence.

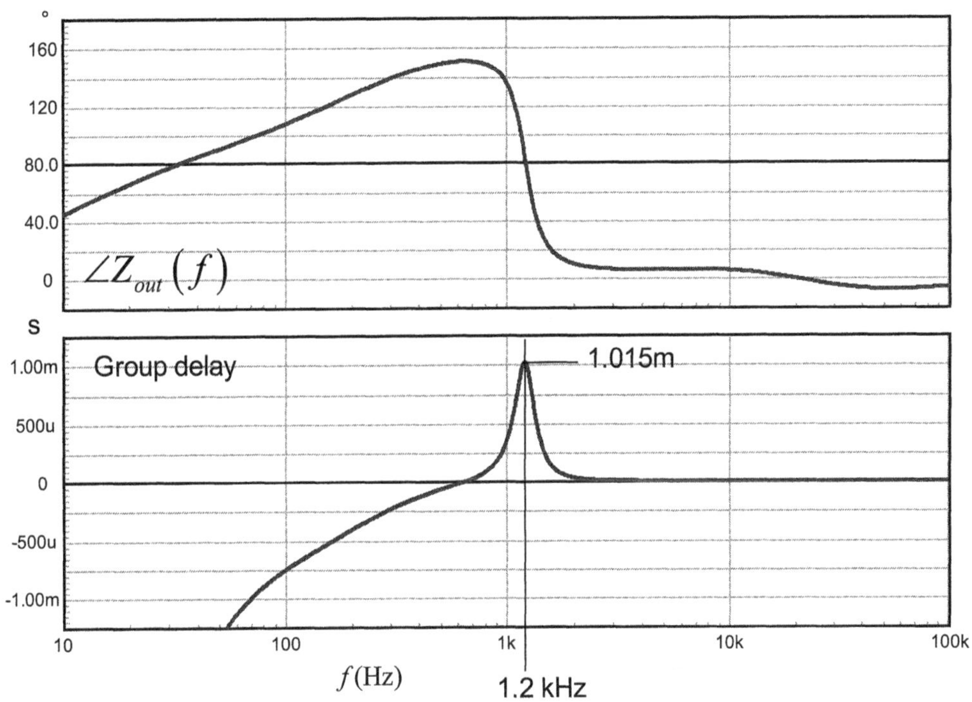

Figure 4.21 By ac sweeping the output impedance and determining its group delay, we have a means to determine the quality factor.

These measurements confirm that the output impedance of our closed-loop converter is that of a second-order system featuring a Q of 3.9. No wonder it gives an oscillatory response! We have compared the response of the PID-stabilized buck of Figure 4.17 with that of a RLC circuit also featuring a Q of 3.9. Figure 4.22 shows that both resulting waveforms perfectly agree. The PID compensator we

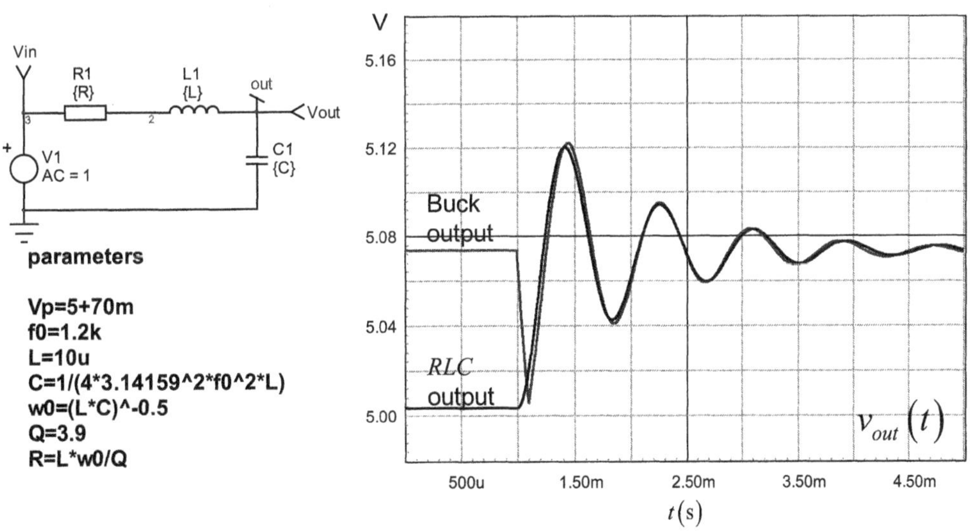

Figure 4.22 The compensated buck behavior is similar to that of an undamped *RLC* network affected by a quality factor of 3.9.

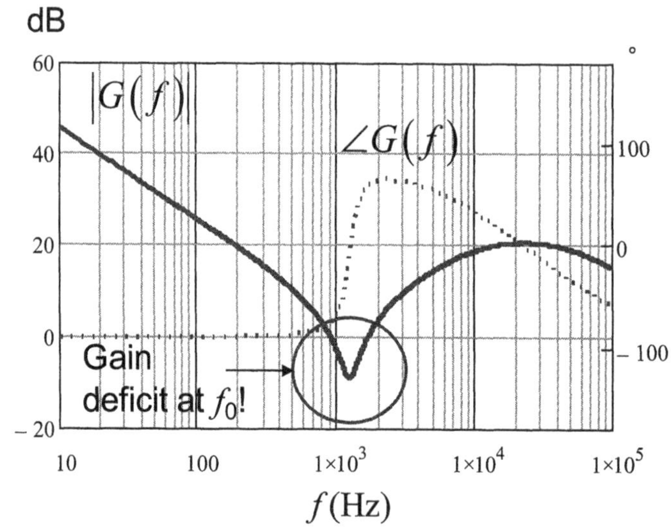

Figure 4.23 In the compensator block, the presence of a conjugate pair of zero reduces the gain at resonance—the opposite of what should be done!

selected did not tame the naturally peaking open-loop impedance of the buck converter. Let's see why and how to remedy that.

From (4.61), we learned that the transient response of our switching converter directly depends on its closed-loop output impedance characteristics. To make them as low as possible, it is in our interest to grow the loop gain $T(s)$ so that the perturbations—the output current step, in our case—are efficiently rejected. To be more precise, we must ensure that sufficient gain exists at the resonant frequency so that efficient damping occurs. Unfortunately, in our PID compensator, to perfectly compensate the double pole at f_0, the equations system leads us to also place a double zero at this frequency via the compensator G. However, because of the quality factor greater than 0.5, the complex poles conjugate pair asked for a complex conjugate pair of zeros for a complete cancelation. If a complex conjugate pair of poles peaks, a complex conjugate pair of zeros works as a rejector: the gain dips at the resonance, whereas it would need to increase! This is the region we highlighted in Figure 4.23.

It is clear from this graph that a classical PID approach applied *stricto sensu* to the case of a buck converter does not give the right results (i.e., a fast, nonringing response). Let's see a more general method that will give better results.

4.2 Stabilizing the Converter with Poles-Zeros Placement

PID control with individual coefficients calculation is often used for the control of complex plants (e.g., by using Ziegler and Nichols experimental processes, as described in [4]). In switching or linear regulators designs, engineers rarely use the PID method we described. If computing the PID compensator individual coefficients leads to positioning poles and zeros as shown in (4.54) to (4.57), designers prefer to directly place these elements in order to tailor the system response to their needs. For instance, without going through a complex polynomial analysis as we did in the

PID example, you can simply select the crossover frequency of your choice and the exact phase margin you expect at this point by properly placing the poles and zeros. You can then shift one zero or one pole and see the effect on the response speed or the recovery time, for instance. At the end, you will end up with a configuration that a PID algorithm could have recommended, but the steps to get there, in the author's opinion, are faster and simpler.

4.2.1 A Simple Step-by-Step Technique

As a starting point, you need the plant transfer function $H(s)$ to learn about the system you want to stabilize. We have seen how to get it from the buck converter in a previous paragraph (e.g., by applying the method described in Figure 4.11). Other methods exist, such as analytical analysis (equation-based small-signal model), laboratory experiments with a network analyzer, or even a Ziegler-Nichols approximation (good results for well-damped systems). From this graph, you need two pieces of data: the plant magnitude and the argument at the selected crossover frequency, $|H(f_c)|$ and $\angle H(f_c)$. These elements will tell you how to shape the compensator frequency response so that you obtain the desired crossover frequency while phase and gain margins values are within the limits you have fixed. However, this is not enough. Observe the magnitude and phase plots to detect the presence of peaks or notches and make sure the selected crossover point is far away from a resonance or in a zone where the phase lag does not excessively degrade. Failure to damp resonating peaks leads to an oscillatory response as we experienced with the PID-compensated buck.

The compensator is the place where you will arrange poles, zeros, and gains/attenuations to shape the loop gain T. This compensator uses an active element, usually an operational amplifier—an op amp—but a lot of industrial projects can implement different types of amplifiers, such as the TL431 (open-collector self-contained op amp with a reference voltage), a shunt regulator (a kind of active Zener circuit), or even an operational transconductance amplifier (OTA). Very often, in industrial applications, these processing blocks transmit the error signal via an isolating device, an opto-coupler, adding complexity to the final transfer equation. We will cover all these structures in detail in dedicated chapters.

Despite the usage of different types of amplifiers, the compensator ac shaping process follows a common agenda:

1. Identify on the plant transfer function $H(s)$ the magnitude and argument at the crossover frequency f_c you have selected.
2. In the compensator, place a pole at the origin to offer a high gain at dc and efficiently reject the perturbations. Input voltage and output current are perturbations in a switching or linear regulator. A high dc gain ensures a good audio or input rejection, while it also guarantees the lowest (theoretically null) dc static error on the output. This is the integrating block in our PID expression. This origin pole is found in most of compensators. However, we will later see a compensation scheme in which there is no origin pole. For instance, output impedance shaping for high-speed dc-dc regulators implements a pure proportional architecture without an origin pole.

3. As a complement to the origin pole, identify the position of poles, zeros, and gain/attenuation points to (a) force crossover at the selected frequency f_c, (b) correct the plant phase response by locally building phase boost, and (c) ensure the targeted open-loop phase margin. Depending on the needed correction, you can deal with three types of compensators, bringing a phase boost from 0° to 180°. They are designated as types 1, 2, and 3 and can be declined with op amps, TL431, and so on.
4. Once the compensator has been designed, plot the loop gain response $T(s)$ and verify that the crossover frequency associated with margins (phase and gain) are within acceptable limits. Check that sweeping the input voltage and the load does not degrade the margins. Do the same check with parasitic element changes, such as capacitor ESRs. Finally, a load step will tell you if the response looks like what you were expecting.

4.2.2 The Plant Transfer Function

Obtained by a SPICE simulation, two examples of plant transfer functions appear in Figure 4.24 and Figure 4.25. The first one is a damped current-mode flyback converter operated in continuous conduction mode (CCM). The first thing is to identify a crossover frequency f_c. We have seen some guidelines in a previous chapter where f_c was chosen based on the transient response to a load step. In this example, we have selected 5 kHz. Looking at the graph, we do not see resonating peaks on the magnitude curve, and the phase gently stays above –90° up to a fairly high frequency. No hidden traps then and the 5 kHz number does not seem to be

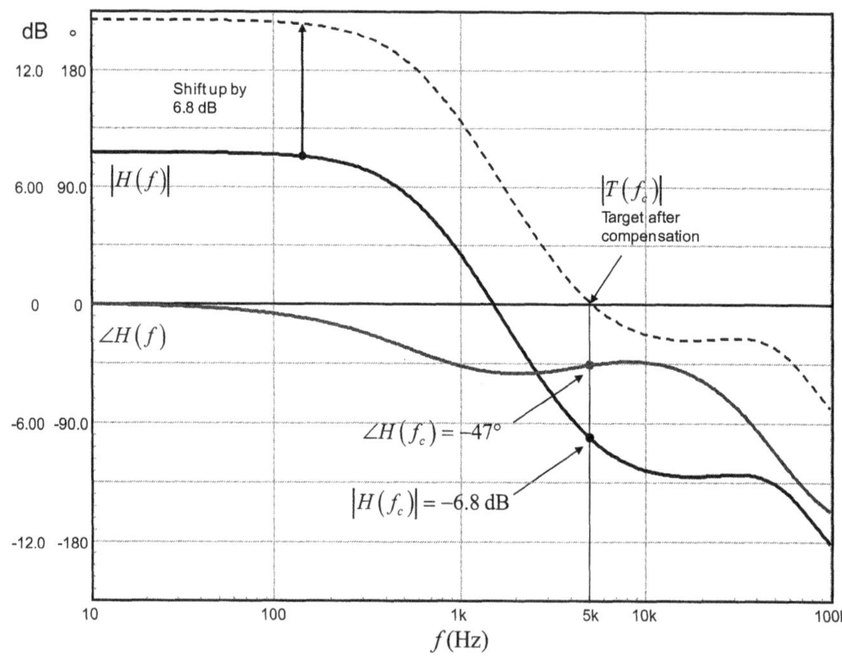

Figure 4.24 This is a current-mode flyback CCM converter transfer function. To obtain a 5-kHz crossover frequency, G must compensate a gain deficit at this frequency.

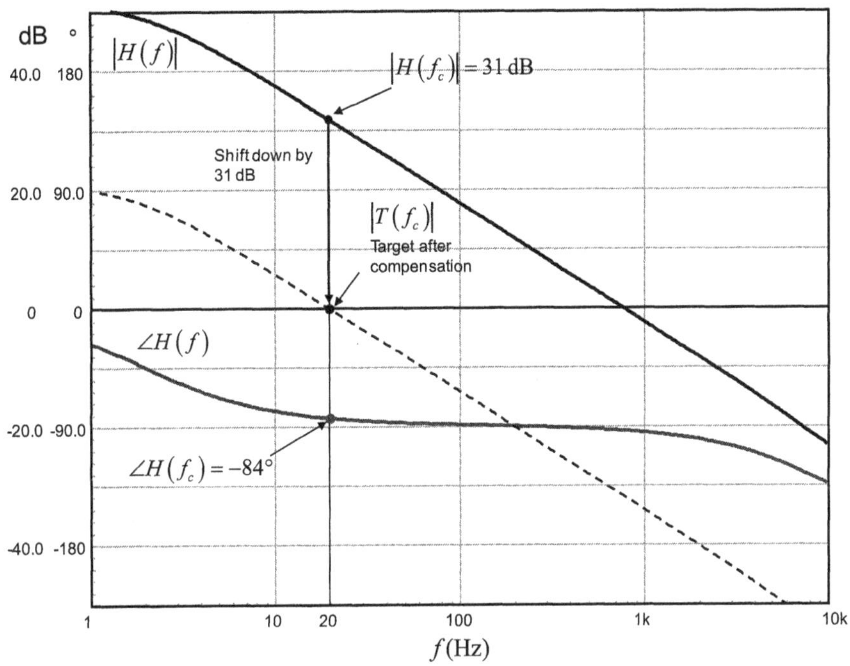

Figure 4.25 With this transfer function of a BCM power factor correction boost converter, the designer must shape the compensator G to compensate an excess of gain at the selected 20-Hz crossover frequency.

a difficult goal to accomplish. What is the plant magnitude value at 5 kHz? From the graph, we read a gain of –6.8 dB. Since we want a 5-kHz crossover frequency, it means that the compensator magnitude $|G|$ at 5 kHz must exactly be 6.8 dB so that $|H(5\text{ kHz})\cdot G(5\text{ kHz})|$, the open-loop gain T, is exactly 1 or 0 dB at this point. In this particular example, we will shift the curve up to compensate the gain deficit at the selected crossover point.

However, there are situations where you observe an excess of gain at crossover rather than a deficit. This is what happens in Figure 4.25 for a power factor corrector (PFC) operated in borderline conduction mode (BCM). The crossover frequency is usually selected at a low value to reject the 100/120-Hz output ripple that would pollute the error signal otherwise. A 20-Hz crossover frequency can be a number PFC designers deal with. From the graph, we observe a 31-dB gain excess at 20 Hz. Therefore, we must shift the curve down by 31 dB: the compensator magnitude $|G|$ at 20 Hz must be –31 dB so that $|H(20\text{ Hz})\cdot G(20\text{ Hz})|$ is 1 or 0 dB.

4.2.3 Canceling the Static Error with an Integrator

We have learned that the loop gain phase lag at the crossover frequency $\angle T(f_c)$ must keep away from the 360° limit; otherwise, an oscillatory response with a pronounced overshoot is obtained. The distance to this limit is the phase margin noted φ_m. Very often, the phase margin represents a design target (e.g., 70° could be asked by your customer or your project manager). In fact, 90° is not an unusual target in space/military designs. Sometimes, conditional stability can be banned, and your customer will ask for bench measurements to show compliance with his demand. It is thus your duty as a design engineer to shape the compensator response and reach the crossover frequency goal together with other requirements such as phase margin. To obtain these numbers, we must first understand what a classical compensator is made of. As the lowest static error is usually wanted, an integrating block is necessary. Please note it is not always the case as sometimes a static error is accepted to get rid of the integrator block. As seen in the PID section, this block is identified in the compensator transfer function as the origin pole. An origin pole is a simple division by s in the compensator transfer function:

$$\frac{V_{out}(s)}{V_{in}(s)} = G(s) = \frac{1}{s} \tag{4.63}$$

At dc (or at the origin of the x-axis if you prefer), when s equals 0, the gain is simply infinite. With an infinite gain at dc, theory tells us that the static error (the dc or steady-state difference between the output and the setpoint) is canceled. In reality, there is a limit brought by the op amp open-loop gain A_{OL} (80–90 dB, for instance). This limit will surely induce a static error but with high open-loop gain values, this error is usually negligible. As we said, some designers actually want a static error and do not place an origin pole. This is the case for high-speed dc-dc converters for motherboards, as we will later see in an example. In the case of our origin pole, the phase lag brought by (4.63) is computed as follows:

$$\lim_{\omega \to 0} \arg \frac{V_{out}(j\omega)}{V_{in}(j\omega)} = \lim_{\omega \to 0} \arg\left(-j\frac{1}{\omega}\right) = \arctan(-\infty) = -\frac{\pi}{2} \tag{4.64}$$

In this equation, we can see that an origin pole brings a permanent phase lag of 90°, independent from frequency. This is an important result: every time you identify a pole at the origin in a transfer function, you have a phase lag of 90° that will cumulate with that of other poles and zeros when present. Should you have two origin poles $\left(G(s) = \frac{1}{s^2}\right)$, the phase lag will be 180°.

You can see an integrator in Figure 4.26 made around a simple voltage-controlled voltage source. If we consider an infinite gain, the transfer function of such a configuration is simply

$$G(s) = -\frac{Z_f}{Z_i} = -\frac{\frac{1}{sC_1}}{R_1} = -\frac{1}{sR_1C_1} \tag{4.65}$$

If we consider the cutoff angular frequency ω_{po} to be $1/R_1C_1$, then (4.65) can be rewritten as

$$G(s) = -\frac{\omega_{po}}{s} \tag{4.66}$$

Replacing s with $j\omega$, we can extract the magnitude of this expression:

$$|G(j\omega)| = \left|0 + j\frac{\omega_{po}}{\omega}\right| = \frac{\omega_{po}}{\omega} = \frac{f_{po}}{f} \tag{4.67}$$

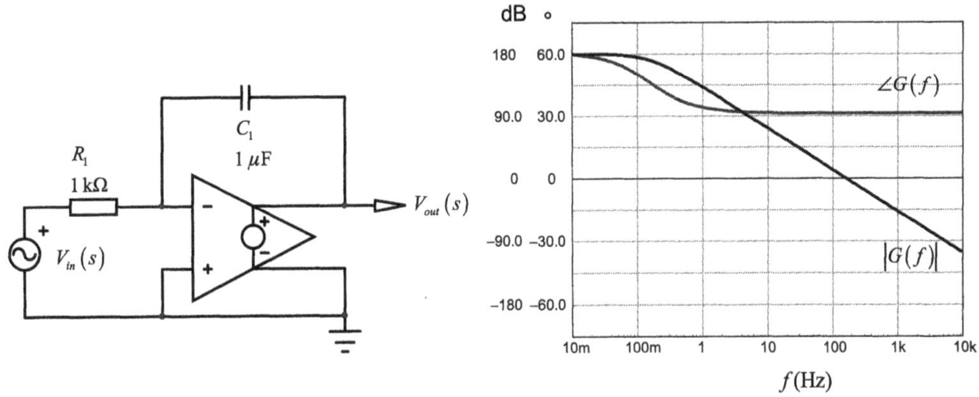

Figure 4.26 A simple integrator built with an op amp in an inverting configuration brings a total phase lag of 270° or a phase lead of 90°.

The argument is obtained from (4.66); now considering the inverting sign brought by the op amp:

$$\arg G(j\omega) = \arg\left(0 + j\frac{\omega_{po}}{\omega}\right) = \tan^{-1}(\infty) = \frac{\pi}{2} \tag{4.68}$$

When ω approaches zero, the gain goes infinite. When ω reaches ω_{po}, the so-called 0-dB crossover pole, the gain is 1 or 0 dB. However, in reality, the op amp exhibits a finite open-loop gain usually noted A_{OL}. If we now consider it, we show in Appendix 4D that the transfer function becomes

$$G(s) \approx \frac{A_{OL}}{1 + s/\omega_{p_1}} \tag{4.69}$$

where A_{OL} is the open-loop gain of the op amp (1,000 or 60 dB in the magnitude graph) and ω_{p_1} defined by

$$\omega_{p_1} = \frac{\omega_{po}}{A_{OL}} \tag{4.70}$$

The magnitude of the integrator is computed by replacing s with $j\omega$:

$$|G(j\omega)| = \frac{A_{OL}}{\sqrt{1 + \left(\dfrac{\omega}{\omega_{p_1}}\right)^2}} \tag{4.71}$$

In dc, when s approaches 0, the integrator gain does not reach infinity but is clamped to A_{OL}. This is what is displayed in the right side of Figure 4.26. In the same figure, you can observe a phase lead of 90° or a lag of 270°. Both angles are equal, as we can add or subtract $\pm 2k\pi$ to any angle without affecting its value. If we subtract 360° from 90°, we obtain –270°, which physically makes better sense. In effect, a negative phase, by convention, illustrates a delay: in the time domain, the output of an op amp can appear only later than its input stimulus. We will therefore use this latter approach to manipulate the various stability limits: an inverting integrator will delay the phase by 270°. The simulator, however, applies a mathematical treatment to the numbers it manipulates and ignores whether 360° must be added or subtracted. You just need to keep in mind that both –270° and 90° refer to the same angle when analyzing phase charts.

These tangent calculations can sometimes be tricky; this is why I wrote a dedicated appendix. Tangent lovers, please proceed to Appendix 4C.

4.2.4 Adjusting the Gain with the Integrator: The Type 1

This is the notion of 0-dB crossover pole already tackled in a previous chapter. In (4.67), the term f_{po} relates to a frequency at which the gain magnitude is 1 or 0 dB. By changing its value, you have a means to modify the gain you need at another frequency point (e.g., the crossover frequency). Suppose you will use a simple integrator to close your converter loop. Assume the crossover frequency is 20 Hz and the gain excess at this point is 23 dB. You must adjust f_{po} so that a 20-Hz modulation brings an attenuation of –23 dB or $10^{-\frac{23}{20}} = 70.8\ m$. In other words, you have to solve

$$\frac{f_{po}}{20} = 0.0708 \tag{4.72}$$

It implies placing the 0-dB crossover pole at

$$f_{po} = 20 \times 0.0708 \approx 14 \text{ Hz} \tag{4.73}$$

This is typically a compensation scheme that could fit power factor correction circuits where the crossover frequency is purposely put low to avoid reacting on the output ripple. Its schematic appears in Figure 4.27. In the technical literature, it is referred as a type 1 compensator: no phase boost, an origin pole for a static large gain, and a certain amount of gain or attenuation at a selected frequency. As already indicated, this circuit introduces a permanent phase lag of 270°.

The test fixture in Figure 4.27 implements an auto-bias circuit made of the voltage-controlled voltage source E_1. Its purpose is to bias the error amplifier at a level where it is operated in a linear way, far away from its lower or upper stops. You adjust this level via the source V_2. It is put to around 2.5 V in this example. E_1 thus biases the divider network at the calculated level. Since we automated our calculations for a 400 V output, the bias point on R_1 right terminal confirms it. The capacitor value is obtained from (4.65) and is simply

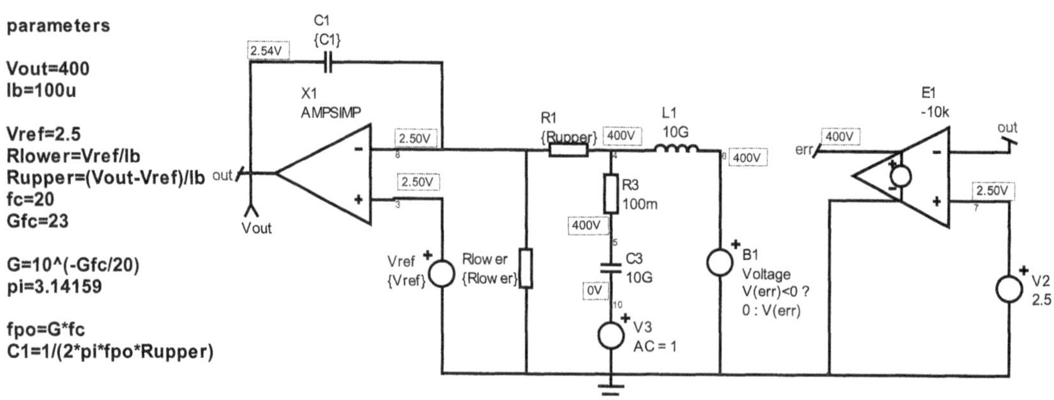

Figure 4.27 This circuit automates a type 1 compensator where the divider network is calculated based on bias current requirements. The cutoff frequency is adjusted through capacitor C_1.

$$C_1 = \frac{1}{2\pi R_{upper} f_{po}} \tag{4.74}$$

With an upper resistor calculated at 4 MΩ, the capacitor value is 28 nF or 33 nF for the next normalized value. The ac plot from Figure 4.28 confirms the validity of our calculations. The phase is fixed to 90° or –270°, no phase boost at all.

In this example, the 0-dB crossover pole position is a means to adjust the gain to any value at the selected frequency. We will see in the following examples how this can be coupled to other expressions.

4.2.5 Locally Boosting the Phase at Crossover

We know that an origin pole is needed if we want the lowest static error on the output. However, as observed in Figure 4.26, the phase brought by this arrangement is absolutely flat. It means that if you add an integrator to an existing design, you simply bring an additional phase lag of 270° and an infinite gain at dc—that is all. Should you need to improve the phase at some particular point (crossover, for instance), the integrator alone is of no help at all. A poor phase margin is due to an excessive phase lag of the transmission chain around the crossover frequency. To improve the situation, we must counteract the phase lag by adding some phase lead somewhere in the chain: the compensator G is the place to do it.

The upper side of Figure 4.29 shows an argument example of the plant we want to stabilize. At the selected 4-kHz crossover frequency, the plant lags by 71°.

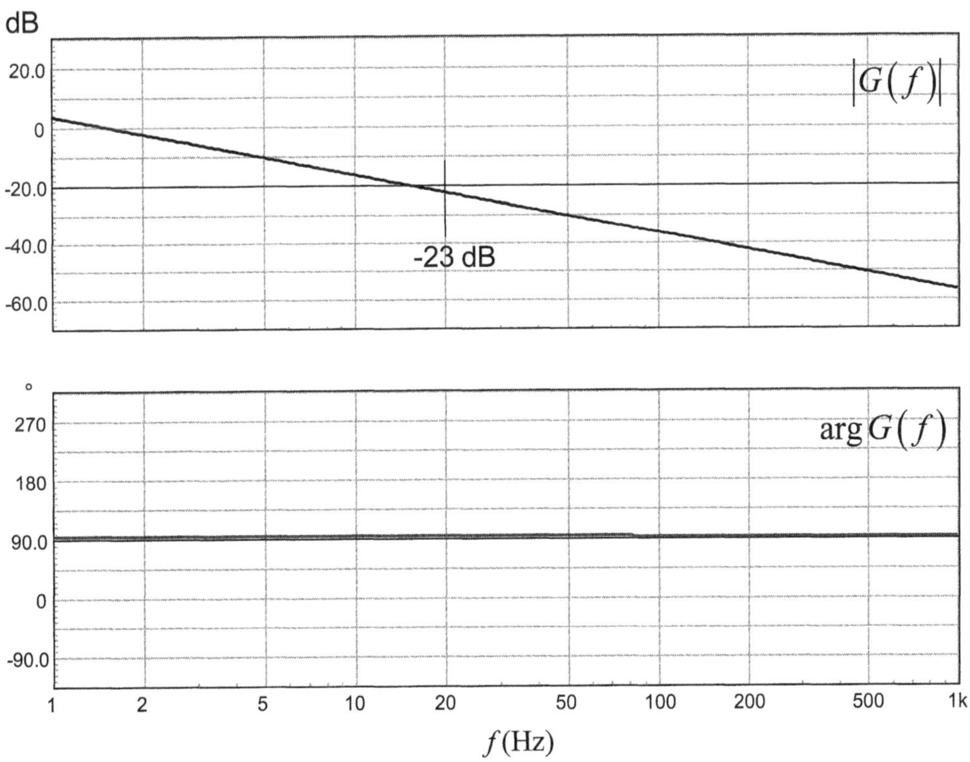

Figure 4.28 The ac plot confirms an attenuation of 23 dB at the selected 20-Hz frequency.

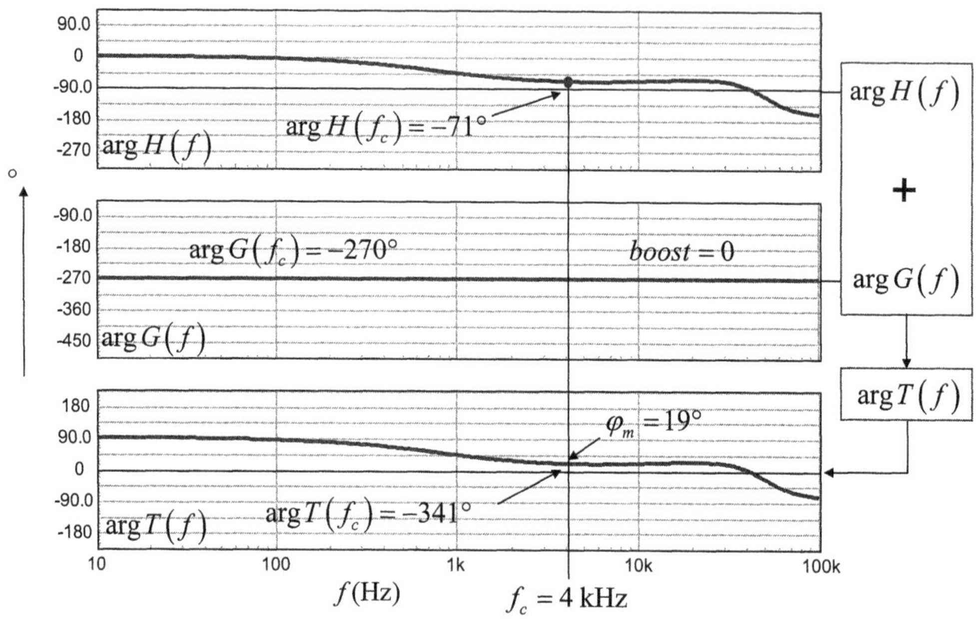

Figure 4.29 Boosting the phase means locally decreasing the plant phase lag by providing phase lead at the crossover point: this is the so-called phase boost.

To obtain the smallest static error, we insert a pole at the origin. We know from the previous paragraph that the addition of an origin pole via an inverting integrator lags the phase by –270°. Therefore, once this integrator is put in series with the plant, we have a total phase lag of

$$\arg T(4\text{ kHz}) = -71° - 270° = -341° \tag{4.75}$$

If we would keep the integrator and close the loop, we would obtain a phase margin equal to

$$\varphi_m = 360° - 341° = 19° \tag{4.76}$$

This is what Figure 4.29 shows you in its lowest part as you simply sum the plant argument to that of the compensator. Please note that the phase margin is measured as the distance between the loop gain argument and the –360° limit or 0° as –360° is a complete turn. We measure 19°.

If our design target is 70°, we are far from this number. To correct this figure, we need to reduce the loop gain phase lag at crossover so that its distance to the –360° line (or 0°) is equal to our phase margin. The necessary increase, or *phase boost*, is part of the following equation:

$$\arg T(f_c) - 270° + boost = -360° + \varphi_m \tag{4.77}$$

Solving for the *boost* value, we have

$$boost = -360° + \varphi_m - \arg T(f_c) + 270° = \varphi_m - \arg T(f_c) - 90° \tag{4.78}$$

In our example, we would need a phase boost of

$$boost = \varphi_m - \arg T(f_c) - 90° = 70 + 71 - 90 = 51° \tag{4.79}$$

It means that the compensator G must be arranged so that its argument at crossover (4 kHz) is no longer –270° (as the sole integrator would provide) but should be

$$\arg G(4\text{ kHz}) = -270° + 51° = -219° \tag{4.80}$$

Once the compensator is tailored to exhibit this phase, new simulation results showed in Figure 4.30 confirm the calculations and the proper phase margin: we have a robust design.

4.2.6 Placing Poles and Zeros to Create Phase Boost

In the integrator transfer function, the argument is flat to –270° along the frequency axis. To change this argument and actually reduce the phase lag at some point, what can we do? We can place a zero in the compensator. We know that the +1-slope of a zero is accompanied by a phase starting from 0 at dc and linearly increasing to 90°. Indeed, this phase lead, well placed on the frequency axis, will bend the integrator phase lag and make it decrease as wished. A zero is defined by the following expression:

$$G(s) = 1 + \frac{s}{\omega_z} \tag{4.81}$$

The magnitude of such transfer function is found by replacing s by $j\omega$:

$$|G(j\omega)| = \left|1 + s\frac{\omega}{\omega_z}\right| = \sqrt{1 + \left(\frac{\omega}{\omega_z}\right)^2} \tag{4.82}$$

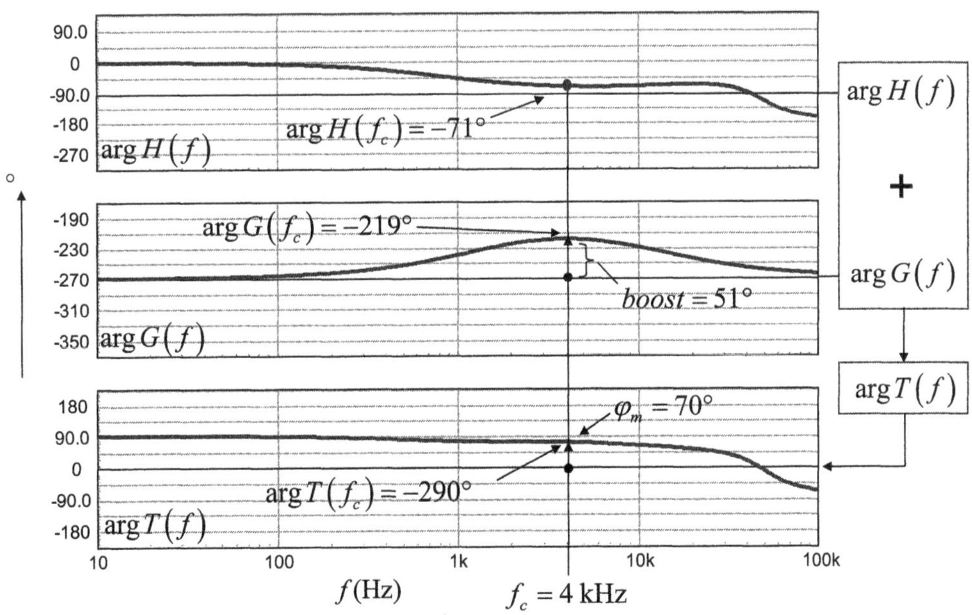

Figure 4.30 By locally decreasing the phase lag or boosting the phase at the crossover, we obtain the required phase margin.

The zero argument is derived as

$$\arg G(j\omega) = \tan^{-1}\left(\frac{\omega}{\omega_z}\right) \tag{4.83}$$

If we plot (4.82) and (4.83), Figure 4.31 displays the ac response of a zero, placed at 1.4 kHz in this example.

As observed on the graph, the zero is able to bring a phase lead that will oppose the phase lag of the plant, to the benefit of stability. However, we want the phase boost to be calibrated to a certain amount: 51° in the design example of Figure 4.30. Therefore, at some point on the frequency axis, when the wanted boost is obtained, the phase lead must be brought to zero. Also, the zero alone would bring the compensator gain to infinity as frequency increases. We limit its action by adding a pole.

The expression of a pole is the following one:

$$G(s) = \frac{1}{1 + s/\omega_p} \tag{4.84}$$

To compute its magnitude, we replace s with $j\omega$:

$$|G(j\omega)| = \frac{1}{\sqrt{1 + \left(\frac{\omega}{\omega_p}\right)^2}} \tag{4.85}$$

The pole argument is derived as

$$\arg G(j\omega) = \arg(1) - \arg\left(1 + j\frac{\omega}{\omega_p}\right) = -\tan^{-1}\left(\frac{\omega}{\omega_p}\right) \tag{4.86}$$

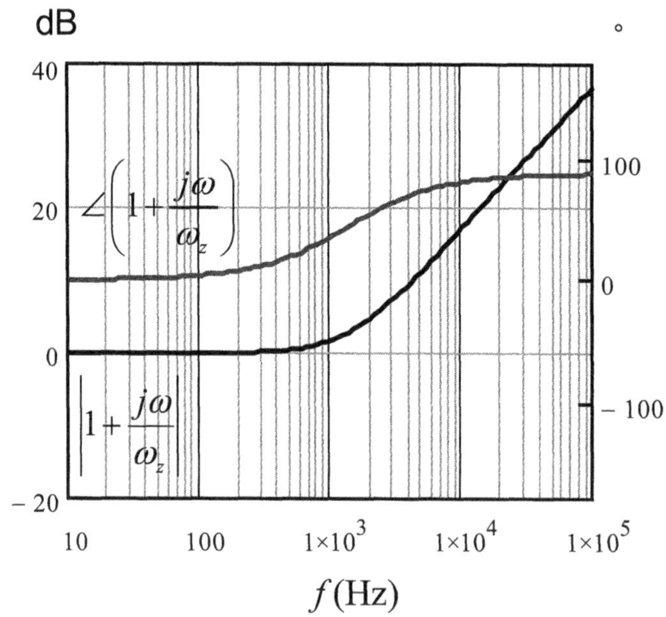

Figure 4.31 The ac response of a zero placed at 1.4 kHz: the phase starts to increase before the zero position and continues toward 90°.

If we now plot (4.85) and (4.86), we obtain the drawing of Figure 4.32. We can see a phase starting from 0 and linearly decreasing toward the –90° asymptote.

In the first case, with the zero, the phase is only growing positively (phase lead), while the pole only brings negative phase (phase lag). As we want a certain amount of phase lead—actually, our phase boost—we can combine a zero and a pole together to exactly calibrate the amount of phase boost we want.

When we associate a pole and a zero, also called a single pole/zero pair, the total argument before either one starts to kick in is 0°. If the zero is placed before the pole, the phase starts to lead as we progress along the frequency axis. This what we showed in Figure 4.31. If the pole now enters in action, its phase lag counteracts that of the zero and the phase starts to drop (Figure 4.32). Ultimately, as the zero argument asymptote is 90° and that of the pole –90°, the phase contribution of the combined pole-zero is 0°. The combined transfer function of the pole-zero combination is as follows:

$$G(s) = \frac{1 + s/\omega_z}{1 + s/\omega_p} \tag{4.87}$$

The magnitude is extracted by replacing s with $j\omega$ and is the quotient of the numerator magnitude by that of the denominator:

$$|G(j\omega)| = \frac{\sqrt{1 + \left(\frac{\omega}{\omega_z}\right)^2}}{\sqrt{1 + \left(\frac{\omega}{\omega_p}\right)^2}} \tag{4.88}$$

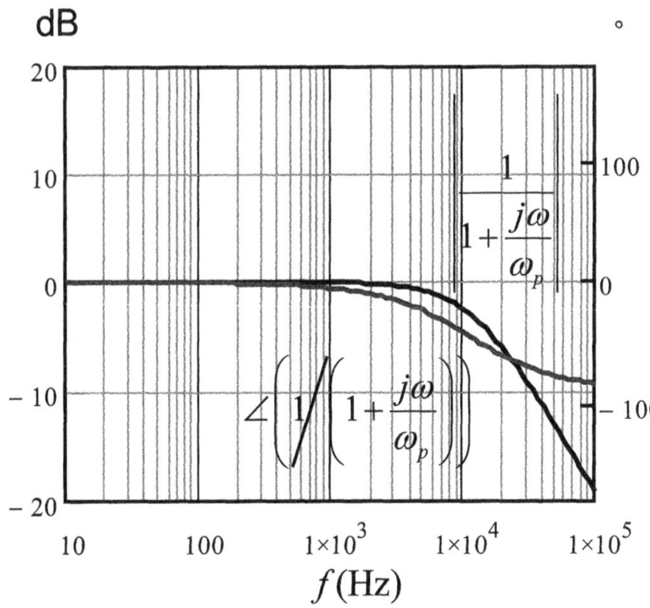

Figure 4.32 The ac response of a pole placed at 11.4 kHz: the phase starts to lag before the pole position and decreases toward –90°.

The argument is simply the difference between (4.83) and (4.86):

$$\arg G(j\omega) = \tan^{-1}\left(\frac{\omega}{\omega_z}\right) - \tan^{-1}\left(\frac{\omega}{\omega_p}\right) \tag{4.89}$$

We have plotted (4.87) in Figure 4.33 where the zero was placed at 1.4 kHz and the pole at 11.4 kHz. As expected, the combined action of the pole and the zero creates a localized phase lead at a certain frequency. If the pole and the zero are coincident, they perfectly neutralize each other: flat 0-dB magnitude over frequency and 0° contribution. As you split both pole and zero, you start to create phase lead in between. When they are split apart so that the pole kicks in after the zero argument has reached its asymptotic value, the boost is maximum to 90°. In Figure 4.34, we have placed a pole at x times the zero position and looked at the total argument for various values of x:

$$f_p = f_z \cdot x \tag{4.90}$$

You can see the boost building up as the pole moves away from the zero.

Now, where to position the pole and the zero so that the phase boost occurs exactly at the crossover frequency? In other words, when we place a pole and a zero, at what frequency does the phase boost peak, before it returns to zero? We can easily obtain that answer by deriving (4.89) and looking at the frequency that cancels the result:

$$\frac{d}{df}\left(\tan^{-1}\left(\frac{f}{f_z}\right) - \tan^{-1}\left(\frac{f}{f_p}\right)\right) = \frac{1}{f_z\left(\frac{f^2}{f_z^2}+1\right)} - \frac{1}{f_p\left(\frac{f^2}{f_p^2}+1\right)} = 0 \tag{4.91}$$

Solving this equation shows that the phase boost peaks at the geometric means of the pole/zero frequencies:

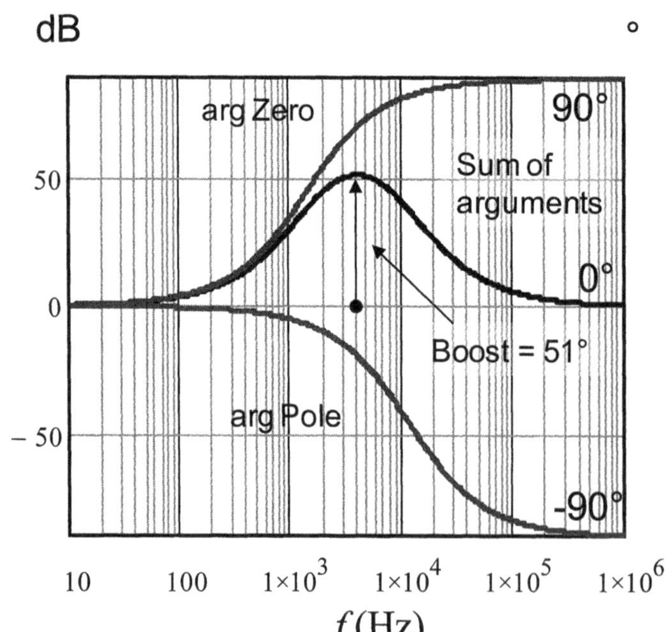

Figure 4.33 When a pole and a zero are combined, we create a calibrated amount of phase lead.

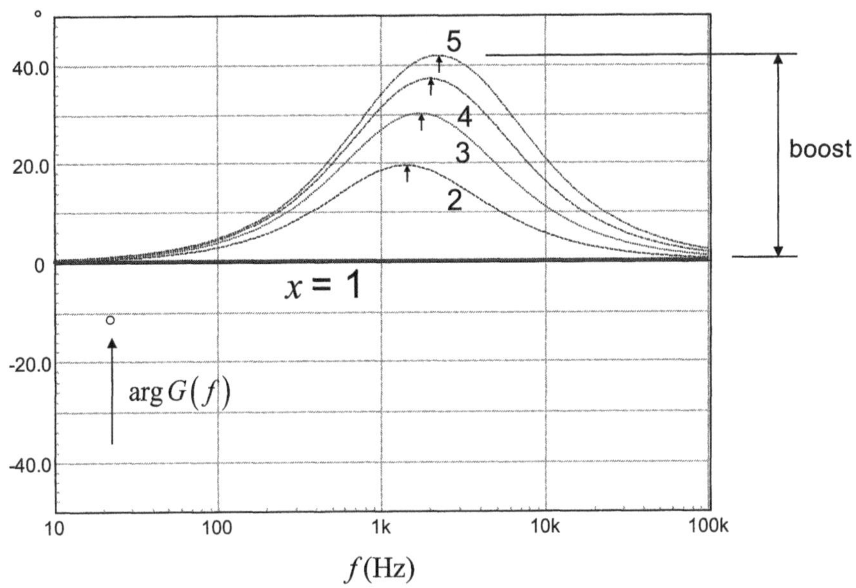

Figure 4.34 By splitting the pole and the zero apart, you can adjust the phase lead to your needs.

$$f_{max} = \sqrt{f_p f_z} \tag{4.92}$$

With our zero at 1.4 kHz and the pole at 11.4 kHz, the peak occurs at:

$$f_{max} = \sqrt{1.4k \times 11.4k} = 4 \text{ kHz} \tag{4.93}$$

as confirmed by Figure 4.33.

4.2.7 Create Phase Boost up to 90° with a Single Pole/Zero Pair

We have learned that combining a pole and a zero provides a phase boost adjustable from 0° to 90°. As we want this boost to occur at the crossover frequency to locally compensate a phase margin deficit, we need equations to properly place the pole and the zero. As both are currently unknown, we actually need two equations:

$$boost = \tan^{-1}\left(\frac{f_c}{f_z}\right) - \tan^{-1}\left(\frac{f_c}{f_p}\right) \tag{4.94}$$

and

$$f_c = \sqrt{f_p f_z} \tag{4.95}$$

From (4.95), we can extract the zero definition:

$$f_z = \frac{f_c^2}{f_p} \tag{4.96}$$

We can substitute this definition in (4.94):

$$boost = \tan^{-1}\left(\frac{f_p}{f_c}\right) - \tan^{-1}\left(\frac{f_c}{f_p}\right) \tag{4.97}$$

To help solve this equation, we can introduce a coefficient k and write $k = f_p/f_c$ and rewrite (4.97):

$$boost = \tan^{-1}(k) - \tan^{-1}\left(\frac{1}{k}\right) \tag{4.98}$$

We must now remember a trigonometric formula involving the arctangent:

$$\tan^{-1}(k) + \tan^{-1}\left(\frac{1}{k}\right) = \frac{\pi}{2} \tag{4.99}$$

We can extract $\tan^{-1}(1/k)$ from this equation and reinject the result into (4.98). Once it is done, we obtain

$$2\tan^{-1}(k) = boost + \frac{\pi}{2} \tag{4.100}$$

Solving for k, we simply have

$$k = \tan\left(\frac{boost}{2} + \frac{\pi}{4}\right) \tag{4.101}$$

This expression is nothing else than the "k factor," introduced by Dean Venable in the 1990s [5].

Knowing that $k = f_p/f_c$, we obtain our pole definition:

$$f_p = k \cdot f_c = \tan\left(\frac{boost}{2} + \frac{\pi}{4}\right)f_c \tag{4.102}$$

And the zero is derived from (4.96):

$$f_z = \frac{f_c}{k} = \frac{f_c}{\tan\left(\frac{boost}{2} + \frac{\pi}{4}\right)} \tag{4.103}$$

When the crossover frequency is identified together with the necessary phase boost, you can select the pole and the zero position using these derivations. The limit given by the method lies in the placement of the crossover frequency at the exact geometric means between the pole and the zero. What if you need to place the pole to counteract an identified zero in the transfer function? In that case, you can no longer apply the previous synopsis. No need to panic; (4.94) is still valid to place the remaining pole or zero at the right position. Let's assume you have to place a zero f_z at 800 Hz and the crossover frequency f_c is 8 kHz. As the needed phase boost is 55°, where do you place the pole? You simply extract f_p from the rearranged equation (4.94), and you are done:

$$\tan^{-1}\left(\frac{f_c}{f_p}\right) = \tan^{-1}\left(\frac{f_c}{f_z}\right) - boost \tag{4.104}$$

To solve this equation, we can use the following trigonometric relationship:

$$\tan(A - B) = \frac{\tan A - \tan B}{1 + \tan A \tan B} \tag{4.105}$$

If we apply the formula to (4.104), we have an equation in the form of

$$\frac{f_c}{f_p} = \frac{\frac{f_c}{f_z} - \tan(boost)}{1 + \frac{f_c}{f_z}\tan(boost)} \tag{4.106}$$

Solving for f_p gives us

$$f_p = \frac{f_z f_c + \tan(boost) f_c^2}{f_c - f_z \tan(boost)} \tag{4.107}$$

If we place the zero at 800 Hz with a crossover frequency at 8 kHz, and we need a 55° phase boost, then the pole should be placed at

$$f_p = \frac{800 \times 8k + \tan(55°) \times 8k^2}{8k - 800 \cdot \tan(55°)} = 14.2 \text{ kHz} \tag{4.108}$$

The result is confirmed by the plot that appears in Figure 4.35. Should you fix the pole instead, the zero would be located at the following value:

$$f_z = \frac{f_c f_p - \tan(boost) f_c^2}{f_c + f_p \tan(boost)} \tag{4.109}$$

4.2.8 Mid-Band Gain Adjustment with the Single Pole/Zero Pair: The Type 2

We know how to place the single pole/zero pair to meet the phase boost requirement at crossover. How do we now adjust the compensator to provide the needed gain or attenuation at crossover? We simply marry the single pole/zero pair with an origin pole such as the one described by (4.66). In that case, (4.87) becomes:

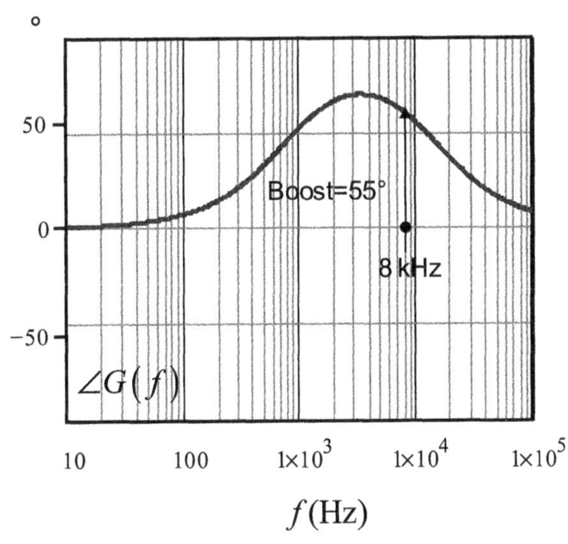

Figure 4.35 The crossover frequency is no longer in the geometric means of the pole/zero position, but the phase boost is what we wanted.

$$G(s) = -\frac{1+s/\omega_z}{\frac{s}{\omega_{po}}(1+s/\omega_p)} \tag{4.110}$$

Please note the "–" sign presence as we now deal with an inverting op amp-based compensator.

This equation does not fit the transfer function format described in Chapter 2. Let's rework it by factoring s/ω_z in the numerator:

$$G(s) = -\frac{s}{\omega_z}\frac{\left(\frac{\omega_z}{s}+1\right)}{\frac{s}{\omega_{po}}(1+s/\omega_p)} \tag{4.111}$$

The 0-dB crossover pole ω_{po} can go to the numerator, and a simplification by s is possible:

$$G(s) = -\frac{s}{\omega_z}\frac{\omega_{po}}{s}\frac{\left(\frac{\omega_z}{s}+1\right)}{(1+s/\omega_p)} = -\frac{\omega_{po}}{\omega_z}\frac{1+\omega_z/s}{1+s/\omega_p} = -G_0\frac{1+\omega_z/s}{1+s/\omega_p} \tag{4.112}$$

In this expression, the term G_0 is called the mid-band gain and is equal to

$$G_0 = \frac{\omega_{po}}{\omega_z} \tag{4.113}$$

As ω_z is fixed by the amount of needed phase boost, you will place ω_{po} depending on the desired gain/attenuation you want at crossover. We have built a type 2 compensator. The design flow is rather simple: you select the pole and zero based on the needed phase boost at the crossover frequency, and you adjust the gain or attenuation at f_c using (4.113). Figure 4.36 shows how it works on an asymptotic construction.

A type 2 can be assembled in a lot of different ways, whether you use op amps, TL431, shunt regulators, or so on. The classical type studied in textbooks is made around an op amp and appears in Figure 4.37.

We won't spend more time on this structure, as Chapter 5 is dedicated to it and goes into the details of numerous configurations. Let's familiarize ourselves with the type 2 in a quick design example.

4.2.9 Design Example with a Type 2

Let's assume we want to stabilize a power supply that has a gain deficit of 18 dB at a 5-kHz selected crossover frequency. The necessary phase boost is 68°. From (4.101) and (4.102), we place a pole at

$$f_p = \tan\left(\frac{boost}{2}+\frac{\pi}{4}\right)f_c = \tan\left(\frac{68^\circ}{2}+45^\circ\right)\times 5k = 25.7 \text{ kHz} \tag{4.114}$$

The zero is placed at

$$f_z = \frac{f_c}{\tan\left(\frac{boost}{2}+\frac{\pi}{4}\right)} = \frac{5k}{\tan\left(\frac{68^\circ}{2}+45^\circ\right)} = 972 \text{ Hz} \tag{4.115}$$

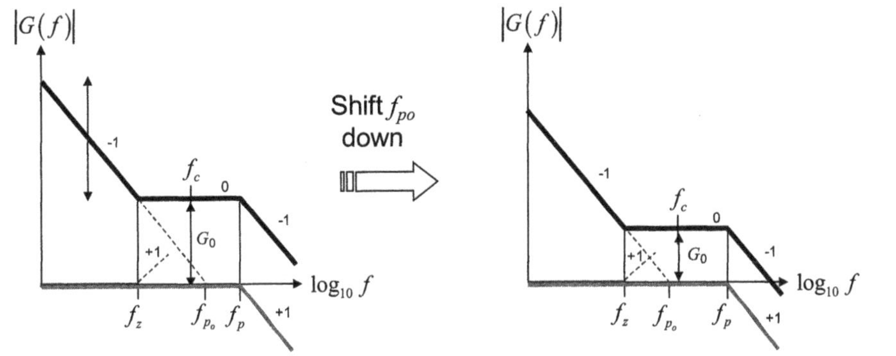

Figure 4.36 Adjusting the 0-dB crossover pole position (while the pole/zero pair is untouched) gives a way to tweak the mid-band gain G_0. Here the mid-band gain reduction is achieved by shifting down the 0-dB crossover pole.

Now the 0-dB crossover pole: we need an 18-dB gain at 5 kHz. From (4.113) and (4.115), it must be placed at

$$f_{po} = f_z \cdot 10^{G_0/20} = 972 \times 8 \approx 7.8\,\text{kHz} \tag{4.116}$$

If we now asymptotically construct the resulting response, we have the diagram of Figure 4.38.

We have simulated this type 2 configuration, and Figure 4.39 plots the ac response typical of that type of architecture.

The calculation methodology for Figure 4.37 components is treated in detail in Chapter 5 and will not be detailed here.

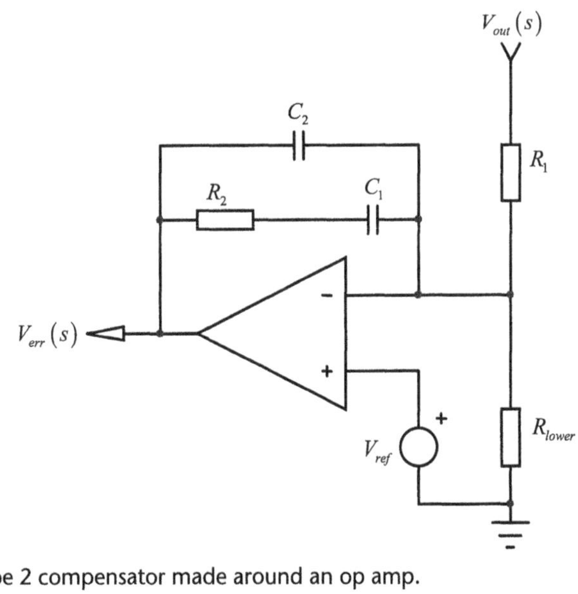

Figure 4.37 A type 2 compensator made around an op amp.

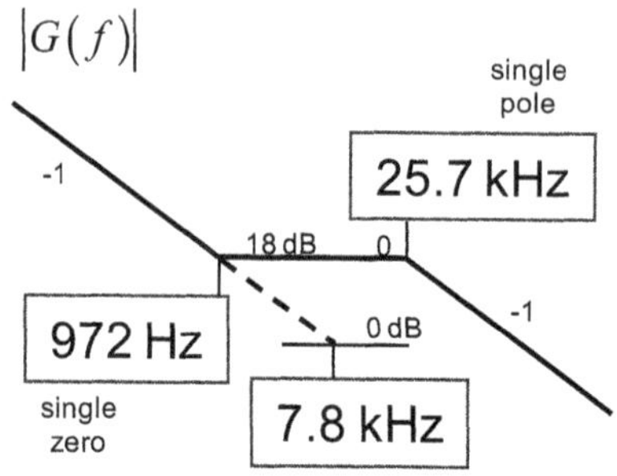

Figure 4.38 The asymptotical response of the designed type 2 compensator.

4.2.10 Create Phase Boost up to 180° with a Double Pole/Zero Pair

The maximum boost brought by a pole/zero combination is 90°. This is easily seen in Figure 4.33, and it occurs when the pole and zero are split far away from each other. There are designs, however, that require more than a 90° phase boost. In that case, you will have to combine two zeros and two poles, also called a double pole/zero pair. The principle is the same as with the single pole/zero pair except that the double zero phase asymptote is now 180° and that of the double pole is –180°. When they are far away from each other, the maximum phase boost becomes 180°.

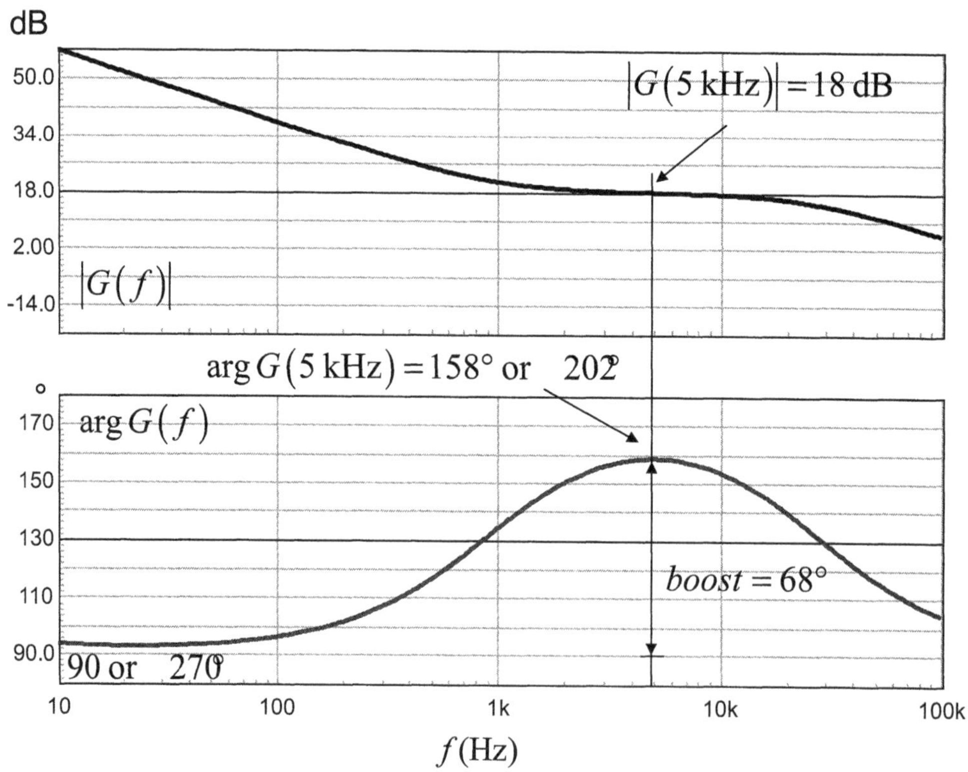

Figure 4.39 A SPICE simulation of the designed type 2 confirms the targets are reached.

The combined transfer function of the double pole-zero combination is as follows:

$$G(s) = \frac{(1 + s/\omega_{z_1})(1 + s/\omega_{z_2})}{(1 + s/\omega_{p_1})(1 + s/\omega_{p_2})} \tag{4.117}$$

The magnitude is extracted by replacing s with $j\omega$ and is the quotient of the numerator magnitude by that of the denominator:

$$|G(j\omega)| = \frac{\sqrt{1+\left(\frac{\omega}{\omega_{z1}}\right)^2}\sqrt{1+\left(\frac{\omega}{\omega_{z2}}\right)^2}}{\sqrt{1+\left(\frac{\omega}{\omega_{p1}}\right)^2}\sqrt{1+\left(\frac{\omega}{\omega_{p2}}\right)^2}} \tag{4.118}$$

The argument is simply the difference between numerator argument and that of the denominator:

$$\arg G(j\omega) = \tan^{-1}\left(\frac{\omega}{\omega_{z1}}\right) + \tan^{-1}\left(\frac{\omega}{\omega_{z2}}\right) - \tan^{-1}\left(\frac{\omega}{\omega_{p1}}\right) - \tan^{-1}\left(\frac{\omega}{\omega_{p2}}\right) \tag{4.119}$$

If we assume that the two zeros and the two poles are coincident (e.g., a double pole and a double zero), then (4.91) can be reformulated with f rather than ω:

$$\frac{d}{df}\left(2\tan^{-1}\left(\frac{f}{f_{z1,2}}\right) - 2\tan^{-1}\left(\frac{f}{f_{p1,2}}\right)\right) = \frac{2}{f_z\left(\frac{f^2}{f_{z1,2}^2}+1\right)} - \frac{2}{f_p\left(\frac{f^2}{f_{p1,2}^2}+1\right)} = 0 \tag{4.120}$$

Solving for f shows that the phase boost also peaks at the geometric means of the double pole-double zero location:

$$f_{\max\, boost} = \sqrt{f_{z1,2} f_{p1,2}} \tag{4.121}$$

As we did for the pole/zero pair, we can calculate the phase boost brought by the combination. The phase boost is obtained from (4.119), which is rearranged considering the double pair:

$$boost = 2\tan^{-1}\left(\frac{f_c}{f_{z1,2}}\right) - 2\tan^{-1}\left(\frac{f_c}{f_{p1,2}}\right) = 2\left(\tan^{-1}\left(\frac{f_c}{f_{z1,2}}\right) - \tan^{-1}\left(\frac{f_c}{f_{p1,2}}\right)\right) \tag{4.122}$$

This expression has already been solved in (4.99) and (4.100). The difference is simply the factor 2 in the right term, leading to the final expression of k for the double pole/zero pair:

$$k = \tan\left(\frac{boost}{4} + \frac{\pi}{4}\right) \tag{4.123}$$

The definition derived in [5] features a squared right term in (4.123), probably chosen for convenient expressions using k in the compensator components calculations. The poles and zeros positions with respect to k are, however, similar:

$$f_{p_{1,2}} = k \cdot f_c \tag{4.124}$$

$$f_{z_{1,2}} = \frac{f_c}{k} \tag{4.125}$$

In the previous example, a single pole/zero pair brought a phase boost of 51° at 8 kHz. We have now placed a double zero at 1.4 kHz and a double pole at 11.4 kHz. As confirmed by Figure 4.40, the phase boost is doubled to 102°.

There are some cases where the double zero is fixed, and you must adjust one pole to match the needed phase boost. It can, for instance, be the case for a voltage-mode buck converter where the designer usually places the double zero $f_{z_{1,2}}$ at the LC filter resonant frequency and a pole f_{p_2} at half the switching frequency for noise immunity. The remaining pole is adjusted to meet the phase boost. Extracted from (4.119), its position is simply

$$f_{p_1} = \frac{f_c}{\tan\left(2\tan^{-1}\left(\frac{f_c}{f_{z_{1,2}}}\right) - boost - \tan^{-1}\left(\frac{f_c}{f_{p_2}}\right)\right)} \tag{4.126}$$

Assume a crossover frequency of 10 kHz, a double zero located at 1.2 kHz, and a high-frequency pole placed at 50 kHz. Suppose we need a phase boost of 130°; then, if we follow (4.126), this second pole must be placed at

$$f_{p_1} = \frac{10k}{\tan\left(2\tan^{-1}\left(\frac{10k}{1.2k}\right) - 120° - \tan^{-1}\left(\frac{10k}{50k}\right)\right)} = 14.3 \text{ kHz} \tag{4.127}$$

We have entered (4.119) in Mathcad and plotted the argument response based on the previous poles/zeros locations. The result appears in Figure 4.41.

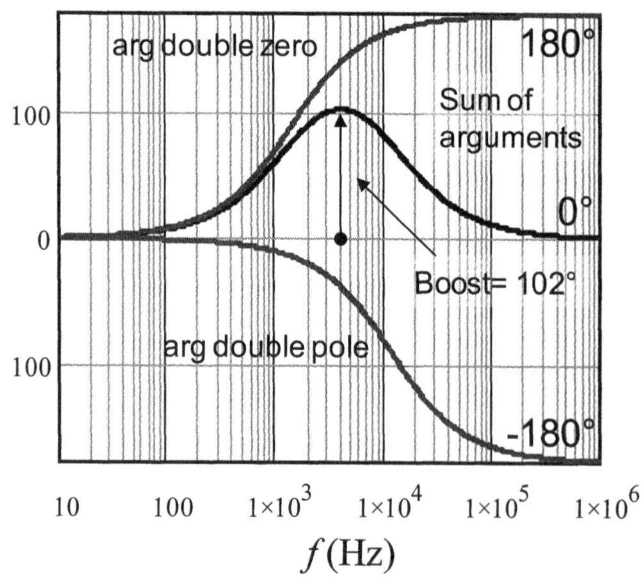

Figure 4.40 The double pole/zero pair doubles the phase boost of that given by a single pole/zero pair.

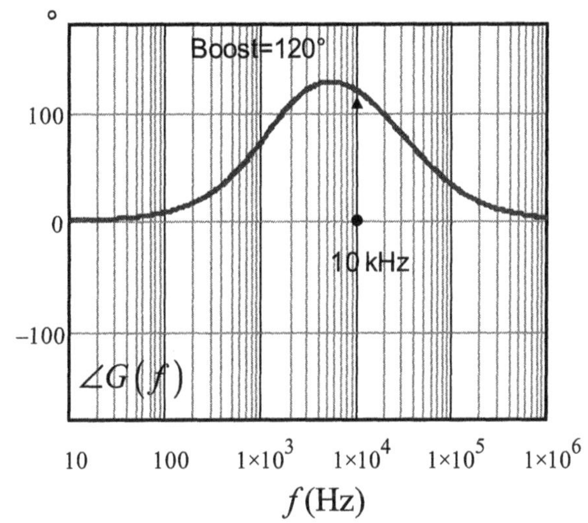

Figure 4.41 The phase boost at 10 kHz confirms the target at 120°.

4.2.11 Mid-Band Gain Adjustment with the Double Pole/Zero Pair: The Type 3

The mid-band gain adjustment in the type 2 compensator was made using the 0-dB crossover pole position. In our double pole/zero pair, the same method is used: an origin pole is inserted in (4.117) to create the type 3 compensator:

$$G(s) = -\frac{(1+s/\omega_{z_1})(1+s/\omega_{z_2})}{\frac{s}{\omega_{po}}(1+s/\omega_{p_1})(1+s/\omega_{p_2})} \tag{4.128}$$

Please note the "–" sign presence as we now deal with an inverting op amp-based compensator.

This equation does not fit the transfer function format described in Chapter 2. Let's rework it by factoring s/ω_{z_1} in the numerator:

$$G(s) = -\frac{s}{\omega_{z_1}}\frac{\left(\frac{\omega_{z_1}}{s}+1\right)(1+s/\omega_{z_2})}{\frac{s}{\omega_{po}}(1+s/\omega_{p_1})(1+s/\omega_{p_2})} \tag{4.129}$$

The 0-dB crossover pole ω_{po} can go to the numerator and a simplification by s is possible:

$$G(s) = -\frac{\omega_{po}}{s}\frac{s}{\omega_{z_1}}\frac{\left(\frac{\omega_{z_1}}{s}+1\right)(1+s/\omega_{z_2})}{(1+s/\omega_{p_1})(1+s/\omega_{p_2})} = -G_0\frac{\left(\frac{\omega_{z_1}}{s}+1\right)(1+s/\omega_{z_2})}{(1+s/\omega_{p_1})(1+s/\omega_{p_2})} \tag{4.130}$$

in which

$$G_0 = \frac{\omega_{po}}{\omega_{z_1}} \tag{4.131}$$

Figure 4.42 shows an asymptotic construction of a type 3 magnitude.

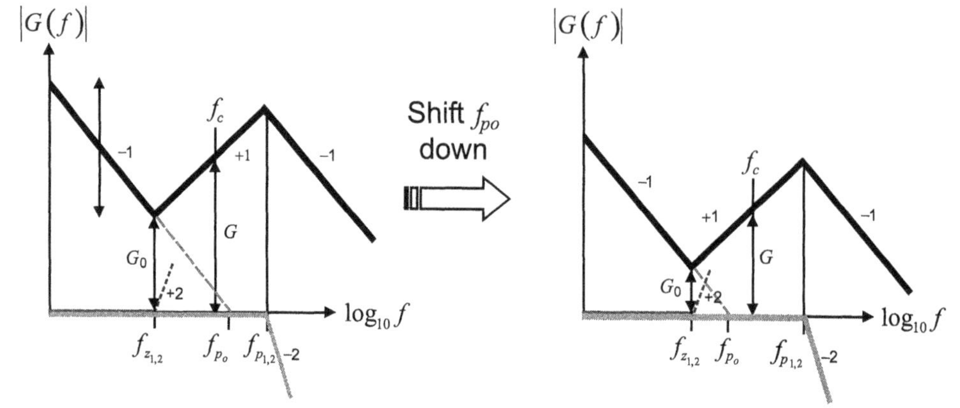

Figure 4.42 By changing the position of the 0-dB crossover pole, you have a means to adjust the mid-band gain value to crossover at the right frequency.

As ω_{z_1} is fixed by the amount of needed phase boost, you will place ω_{po} depending on the required gain/attenuation you want at crossover. We have built a type 3 compensator. The design flow is rather simple: you select the poles and zeros based on the needed phase boost at the crossover frequency, and you adjust the gain or attenuation at f_c. Unfortunately, (4.131) cannot be used alone to calculate the 0-dB crossover pole position. Why? Because, as you can see in Figure 4.42, the crossover occurs after the two zeros have kicked in, in the middle of a +1 slope. The case differs from that of the type 2 where the slope was unchanged (0 slope) at the point the zero was positioned. The gain G at this position shown in Figure 4.42 is not G_0 and must account for the position of the double pole/zero pair. The formula derives from (4.130):

$$\omega_{po} = G \cdot \omega_{z_1} \frac{\sqrt{1+\left(\frac{\omega_c}{\omega_{p_1}}\right)^2}\sqrt{1+\left(\frac{\omega_c}{\omega_{p_2}}\right)^2}}{\sqrt{1+\left(\frac{\omega_{z_1}}{\omega_c}\right)^2}\sqrt{1+\left(\frac{\omega_c}{\omega_{z_2}}\right)^2}} \tag{4.132}$$

If we consider a double coincident poles/zeros pair, the formula becomes

$$\omega_{po} = G \cdot \omega_{z_{1,2}} \frac{\left(\omega_{p_{1,2}}^2 + \omega_c^2\right)}{\omega_{p_{1,2}}^2 \sqrt{\left(\frac{\omega_{z_{1,2}}}{\omega_c}\right)^2 + 1}\sqrt{\left(\frac{\omega_c}{\omega_{z_{1,2}}}\right)^2 + 1}} \tag{4.133}$$

In this expression, G is the wanted gain or attenuation at crossover.

A type 3 can be assembled in a lot of different ways, regardless of whether you use op amps, TL431, shunt regulators, or so on. The classical type studied in textbooks is made around an op amp and appears in Figure 4.43.

The calculation of the components values appears in Chapter 5, dedicated entirely to the op amp architectures, and will not be repeated here.

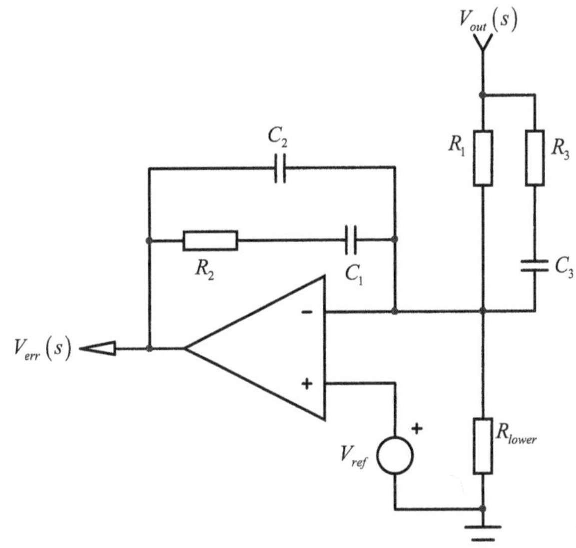

Figure 4.43 A type 3 compensator uses the same basis as a type 2 to which an *RC* network is added in parallel with the upper resistor R_1.

4.2.12 Design Example with a Type 3

In this example, we assume a converter whose crossover frequency must be set to 5 kHz. At this frequency, the plant shows a gain deficit of 10 dB. The required phase boost is 158°. From the formula given in (4.124), we can calculate the position of the double pole:

$$f_{p1,2} = \tan\left(\frac{boost}{4} + \frac{\pi}{4}\right) f_c = \tan\left(\frac{158°}{4} + 45°\right) \times 5k \approx 52 \text{ kHz} \tag{4.134}$$

The double zero position comes easily with (4.125):

$$f_{z1,2} = \frac{f_c}{\tan\left(\frac{boost}{4} + \frac{\pi}{4}\right)} = \frac{5k}{\tan\left(\frac{158°}{4} + 45°\right)} \approx 480 \text{ Hz} \tag{4.135}$$

The gain G at 5 kHz must be 10 dB. Applying (4.132) or (4.133) will tell us to position the 0-dB crossover pole at the following value:

$$f_{po} = G \cdot f_{z1,2} \frac{\left(f_{p1,2}^2 + f_c^2\right)}{f_{p1,2}^2 \sqrt{\left(\frac{f_{z1,2}}{f_c}\right)^2 + 1} \sqrt{\left(\frac{f_c}{f_{z1,2}}\right)^2 + 1}}$$

$$= 10^{\frac{10}{20}} \times 480 \times \frac{52k^2 + 5k^2}{52k^2 \times \sqrt{1 + \left(\frac{480}{5k}\right)^2} \times \sqrt{1 + \left(\frac{5k}{480}\right)^2}} \approx 146 \text{ Hz} \tag{4.136}$$

The asymptotic construction of the compensator is given in Figure 4.44.

We simulated this compensator and the ac response is shown in Figure 4.45. It confirms the calculations we made.

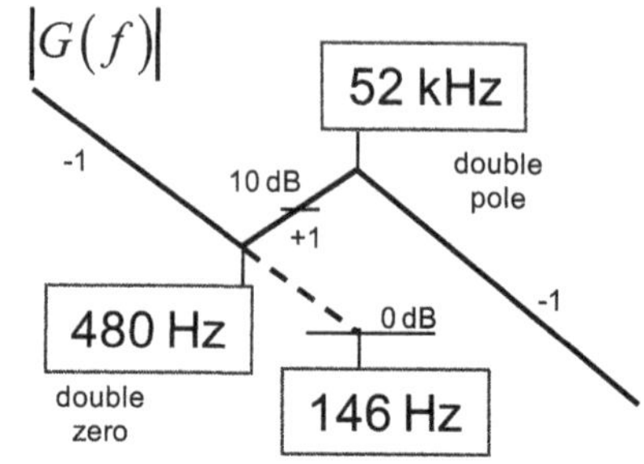

Figure 4.44 The asymptotic construction of the type 3 shows the positions of the poles and zeros pairs.

4.2.13 Selecting the Right Compensator Type

Figure 4.46 depicts the possible architectures with their respective ac responses. You select the compensator type based on the converter you have to stabilize.

- Type 1: as its phase boost is zero, it can be used only in converters where the plant phase lag is minimal at the selected crossover frequency. Assume your plant exhibits a phase lag of 40° at the crossover frequency. Then, if you add another –270° from the type 1, you end with a total phase lag of 310°, giving a phase margin of 50°. However, as with any integral term, it will bring a response with overshoot that can sometimes be too severe. This type is widely used in PFC applications.
- Type 2: this is the most popular structure in current-mode control ac-dc or dc-dc converters. With a phase boost up to 90°, it perfectly suits the compensation needs of current-mode converters such as flyback, forward, boost, and buck-boost. It will also work for voltage-mode control types of converters,

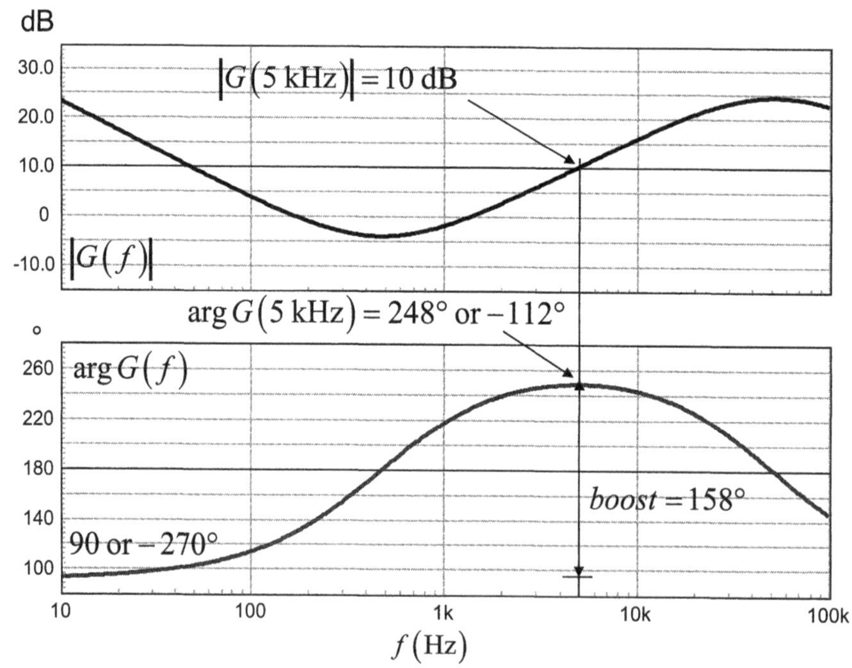

Figure 4.45 The SPICE simulation of the compensator shows the right gain at 5 kHz and the correct phase boost.

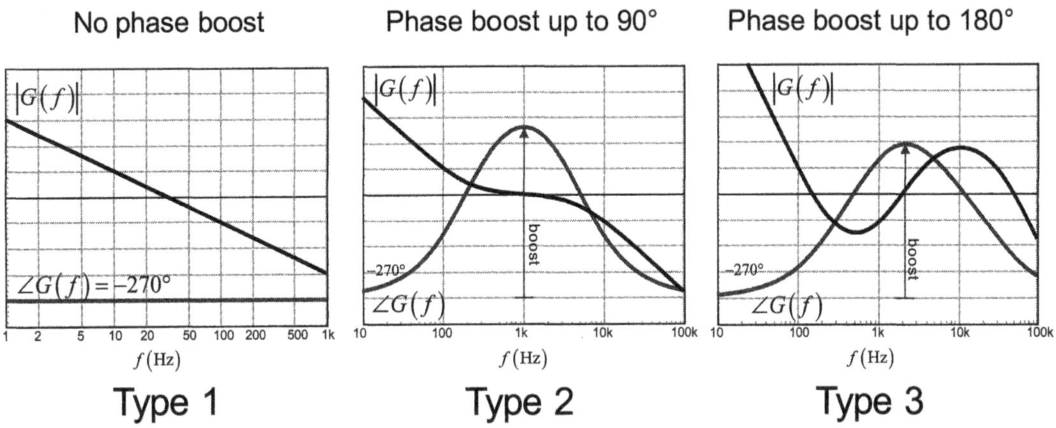

Figure 4.46 This is a brief summary of the possible architectures with their respective ac responses.

but the limited boost will restrict its usage to discontinuous mode only. The type 2 is sometimes declined in 2a or 2b, building PI or filtered-proportional compensators, respectively. They are covered in Chapter 5.

- Type 3: with its phase boost capability up to 180°, the type 3 is often used in voltage-mode or direct duty ratio converters. Easily declined in op amp-based architectures, it becomes more complicated to build with OTA- or TL431-based compensators.

4.2.14 The Type 3 at Work with a Buck Converter

Now that we have seen how to place poles and zeros independently from tweaking individual PID parameters, it is time to apply the technique to the buck converter already presented in Figure 4.11. The main problem with the PID compensator example was that the cancellation of the resonating double poles led to the placement of a conjugate zeros pair, giving rise to a gain decrease in the compensator at exactly the resonant frequency (the notch in Figure 4.23). If the control-to-output response was good (see Figure 4.16), the perturbation rejection showed an oscillatory response because of an undamped output impedance. To ensure proper damping of the output impedance, it is important to always verify that sufficient gain exists at the resonant frequency (if there is one in the plant function, of course) so that the resulting closed-loop output impedance quality factor is less than one. With a voltage-mode converter, the classical compensation method works as follows:

1. Select a crossover frequency f_c that is at least three to five times away from the resonating peak f_0 (never before the peak or a similar problem would occur by gain deficiency at resonance).
2. Place a real zero pair at the resonance frequency. This time, the zeros are real and not conjugate. We will see that these zeros can be split to increase the stability in light load conditions.
3. If the ESR-linked zero appears before crossover, neutralize it as we did with the PID by placing a pole right at its location. If the zero appears far away from the bandwidth, simply place the pole at half of the switching frequency. This first pole can then be moved to adjust the phase margin at the wanted value if necessary.

4. To force gain decrease at high frequency and ensure gain margin exists, place a second pole sufficiently high so that its presence does not hamper phase margin. It is usually placed at half of the switching frequency.

The converter transfer function in Figure 4.12 shows a resonant frequency at 1.2 kHz. To respect what we said from the previous point 1, we have to select a crossover frequency beyond 5 kHz. Let us adopt 10 kHz in this example, without considering undershoot specifications for the sake of simplicity. As the phase lag from the plant is not far from 180°, we are going to need an amount of phase boost greater than 90° if we want a decent phase margin. This is the role of the type 3 compensator we have already described.

The transfer function of such a compensator is as follows:

$$G(s) = -G_0 \frac{\left(1+\frac{\omega_{z_1}}{s}\right)\left(1+\frac{s}{\omega_{z_2}}\right)}{\left(1+\frac{s}{\omega_{p_1}}\right)\left(1+\frac{s}{\omega_{p_2}}\right)} \tag{4.137}$$

in which $G_0 = \frac{\omega_{po}}{\omega_{z_1}}$.

To apply the compensation strategy to our voltage-mode buck converter, we will place the double zeros at 1.2 kHz and will adjust the first pole around the plant zero position—as already recommended by (4.56) as a matter of fact—to tweak the phase margin to our target. With a 100-kHz operating frequency, the second pole will be arbitrarily placed at 50 kHz. This pole ensures further gain decrease at high frequencies and improves noise immunity. Now, where do we place the 0-dB crossover pole? Its position depends on the selected crossover frequency and the plant gain (or deficiency) at this frequency: ω_{po} will be placed to exactly force crossover at f_c. The principle is to shift up (or down) the plant magnitude H at f_c plot by the compensator magnitude G at f_c so that $|G(f_c)H(f_c)| = 1$. We have seen that in Figure 4.24 and Figure 4.25.

To force crossover at 10 kHz, we look at the plant transfer function $H(s)$, unveiled in Figure 4.14. From this curve, we can extract the gain deficiency at 10 kHz: around −20 dB. The phase lag is also extracted to be in the vicinity of 130°. We can also precisely derive these values from (4.34), for example, in an automated sheet:

$$|H(10k)| = H_0 \frac{\sqrt{1+\left(\frac{f_c}{f_{z_1}}\right)}}{\sqrt{\left(1-\frac{f_c^2}{f_0^2}\right)^2+\left(\frac{f_c}{f_0 Q}\right)^2}}$$

$$= \frac{10}{2.5} \frac{\sqrt{1+\left(\frac{10k}{10.3k}\right)}}{\sqrt{\left(1-\frac{10k^2}{1.24k^2}\right)^2+\left(\frac{10k}{1.24k\times 1.45}\right)^2}} = 0.108 \text{ or } -19.3\text{ dB} \tag{4.138}$$

$$\angle H(10k) = \tan^{-1}\left(\frac{f}{f_{z1}}\right) - \tan^{-1}\left[\frac{f_c}{f_0 Q}\frac{1}{\left(1-\frac{f_c^2}{f_0^2}\right)}\right]$$

$$= \tan^{-1}\left(\frac{10k}{10.3k}\right) - \tan^{-1}\left[\frac{10k}{1.24k\times 1.45}\frac{1}{\left(1-\frac{10k^2}{1.24k^2}\right)}\right] = -134° \quad (4.139)$$

Since the plant attenuation at 10 kHz is –19.3 dB, the compensator G must provide an amplification of exactly 19.3 dB at this frequency. As we want a phase margin of 70° at 10 kHz, what phase boost must we design the compensator for? We will use (4.78) for that purpose:

$$boost = \varphi_m - \arg T(f_c) - 90° = 70 + 134 - 90 = 114° \quad (4.140)$$

Following the compensation strategy unveiled a few lines earlier, we are going to place a double zero at the LC network resonant frequency (1.2 kHz):

$$f_{z_1} = f_{z_2} = 1.2\text{ kHz} \quad (4.141)$$

One pole will be placed at 50 kHz for noise immunity purposes:

$$f_{p_2} = 50\text{ kHz} \quad (4.142)$$

The remaining pole can be placed straight at the ESR zero (10.3 kHz), and you will check if the resulting phase margin suits your needs. You can also compute the position of this pole to exactly meet the 114° phase boost suggested by (4.140). If we stick to the second option, we use the boost formula given by (4.119):

$$\arg G(f_c) = boost = \tan^{-1}\left(\frac{f_c}{f_{z1}}\right) + \tan^{-1}\left(\frac{f_c}{f_{z2}}\right) - \tan^{-1}\left(\frac{f_c}{f_{p1}}\right) - \tan^{-1}\left(\frac{f_c}{f_{p2}}\right) \quad (4.143)$$

$$\tan^{-1}\left(\frac{10k}{f_{p_1}}\right) = 2\tan^{-1}\left(\frac{10k}{1.2k}\right) - \tan^{-1}\left(\frac{f_c}{f_{p_2}}\right) - 114° = 166.3° - 11.3° - 114° = 41° \quad (4.144)$$

From which we can obtain the pole position:

$$f_{p_1} = \frac{f_c}{\tan(41°)} = \frac{10k}{0.87} = 11.5\text{ kHz} \quad (4.145)$$

We are going to use the template we already tested in Figure 4.17. This time, the PID parameters are tweaked to place the previous real poles and zeros we have derived. The 0-dB crossover pole is computed in the integral term k_i, which naturally depends on the wanted gain at crossover, 19.3 dB in our case. In this particular case, as we have split the poles, (4.136) needs to be updated to account for this fact:

$$f_{po} = G \cdot f_{z_1} \frac{\sqrt{\left(\frac{f_c}{f_{p_1}}\right)^2 + 1}\sqrt{\left(\frac{f_c}{f_{p_2}}\right)^2 + 1}}{\sqrt{\left(\frac{f_{z_1}}{f_c}\right)^2 + 1}\sqrt{\left(\frac{f_c}{f_{z_2}}\right)^2 + 1}}$$

$$= 10^{\frac{20}{20}} \times 1.2k \times \frac{\sqrt{1+\left(\frac{10k}{11.3k}\right)^2} \times \sqrt{1+\left(\frac{10k}{50k}\right)^2}}{\sqrt{1+\left(\frac{1.2k}{10k}\right)^2} \times \sqrt{1+\left(\frac{10k}{1.2k}\right)^2}} \approx 1.9 \text{ kHz} \tag{4.146}$$

The complete template appears in Figure 4.47. This time, the PID coefficients are derived from the poles/zeros positions we fix, and we use definitions given in (4.19) through (4.22) to parameterize the PID blocks.

The ac responses brought by the compensator and the compensated loop gain are, respectively, given in Figure 4.48 and Figure 4.49.

It is now interesting to look at the closed-loop output impedance since we know that the transient response to a load step will depend upon this figure. It is given in Figure 4.50. As you can see, the peaking found in the classical PID compensation has gone. The curve is very smooth, and we expect a good transient response without significant overshoot.

We have stepped the buck converter output from 1 A to 2 A in a 100-μs slew-rate. The response is given in Figure 4.51. The undershoot is small (0.74 percent), and the spurious oscillations are gone. This is a well-compensated design.

Until now, the transient response and ac analysis have been carried on a converter operated in the continuous conduction mode (CCM). This mode simply states that the inductor current does not return to zero within one switching cycle. When the load is getting lighter, the inductor fully demagnetizes within a switching cycle and the converter is told to operate in the discontinuous conduction mode (DCM). This new operating mode, for a voltage-mode converter, drastically changes its ac response. This is what is shown in Figure 4.52. The gain is reduced and the transition frequency drops from 3 kHz in CCM to around 150 Hz in DCM. Regarding the asymptotic phase lag, it approaches 180° for a CCM converter but remains below 90° for the DCM operation. What is the problem then? Well, we have compensated our buck converter in the worst load condition case (at minimum input voltage) so that the phase margin meets the design target. What happens if the converter now transitions into DCM? SPICE is extremely helpful as the models derived in [6] are auto-toggling: they automatically transition between CCM and DCM and deliver the correct ac response. The loop gain ac response of the CCM-compensated buck converter now operated in DCM is given in Figure 4.53.

From this picture, we can see that the phase margin degrades significantly when the buck converter enters DCM. For a 500-Ω load (I_{out} = 10 mA), the phase margin drops to 40°. How can we improve the situation? By splitting the zero pair located at the resonant frequency. We are going to push one zero above the resonant frequency—at 3 kHz—while the other one will shift down to 300 Hz. These values can be analytically derived

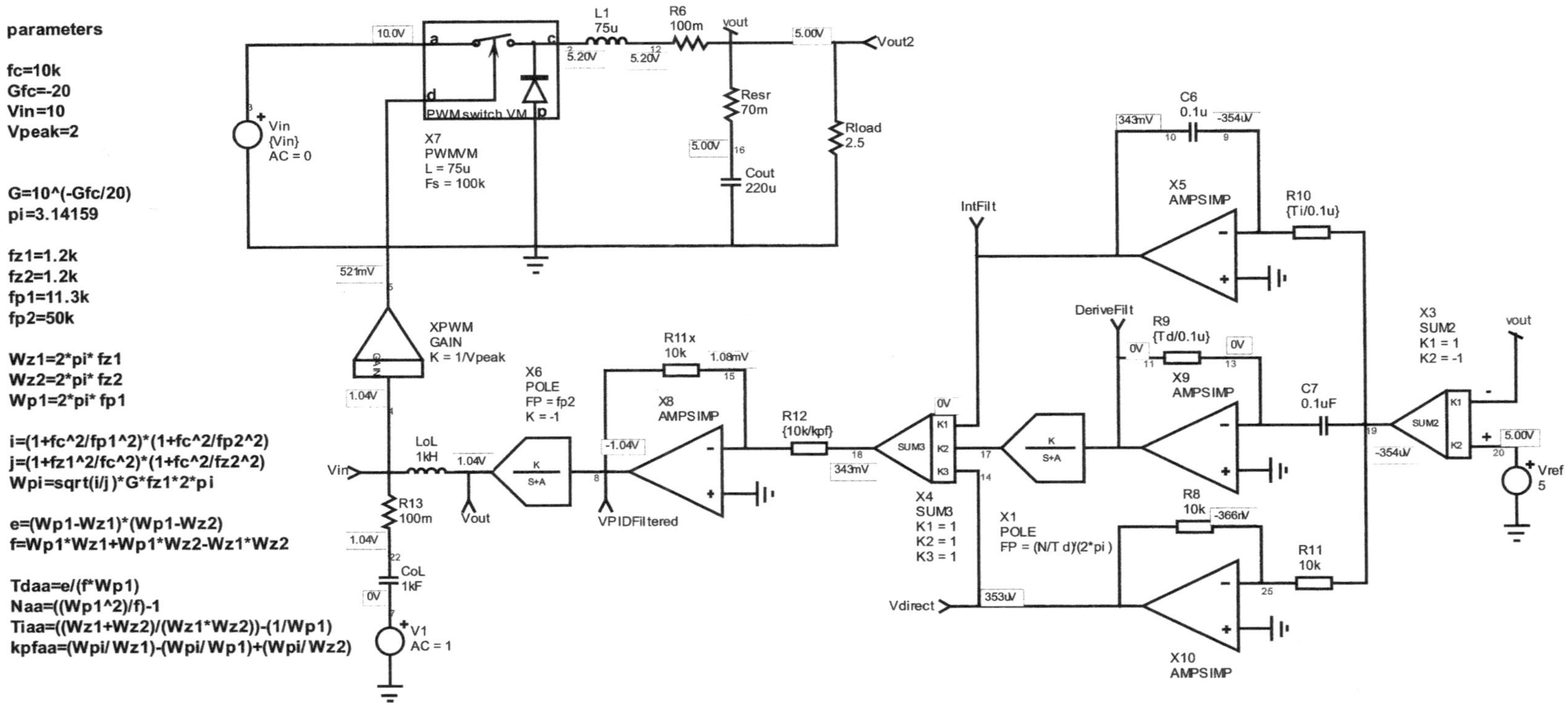

Figure 4.47 The simulation template shows the filtered-PID compensator whose coefficients are now computed based on the wanted real poles and zeros.

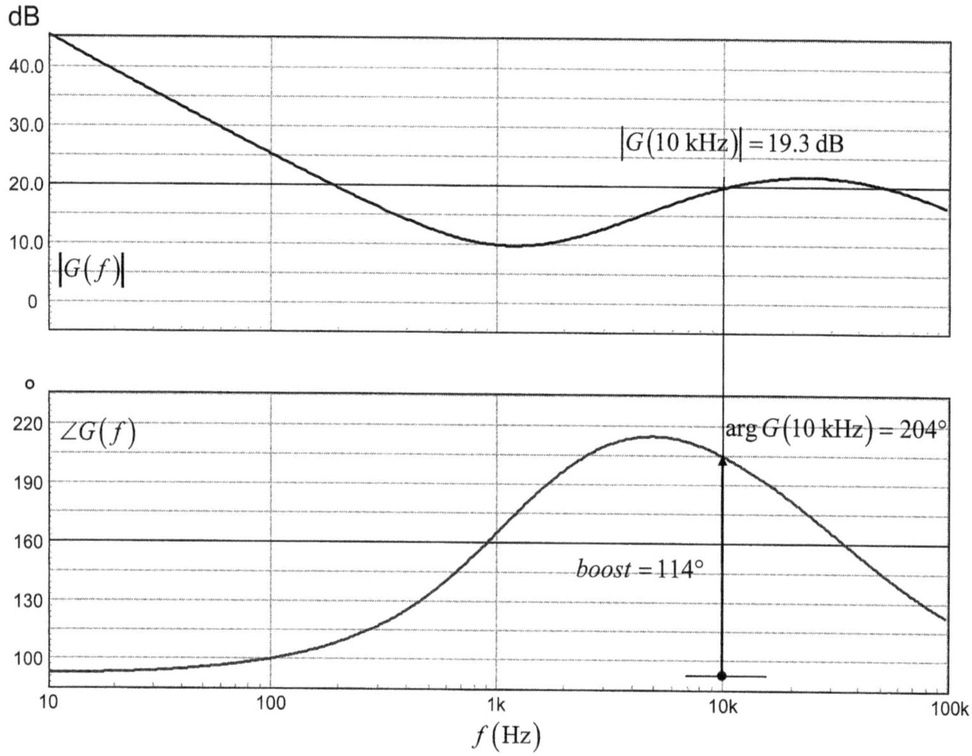

Figure 4.48 The compensator boosts the phase by 114°, as expected.

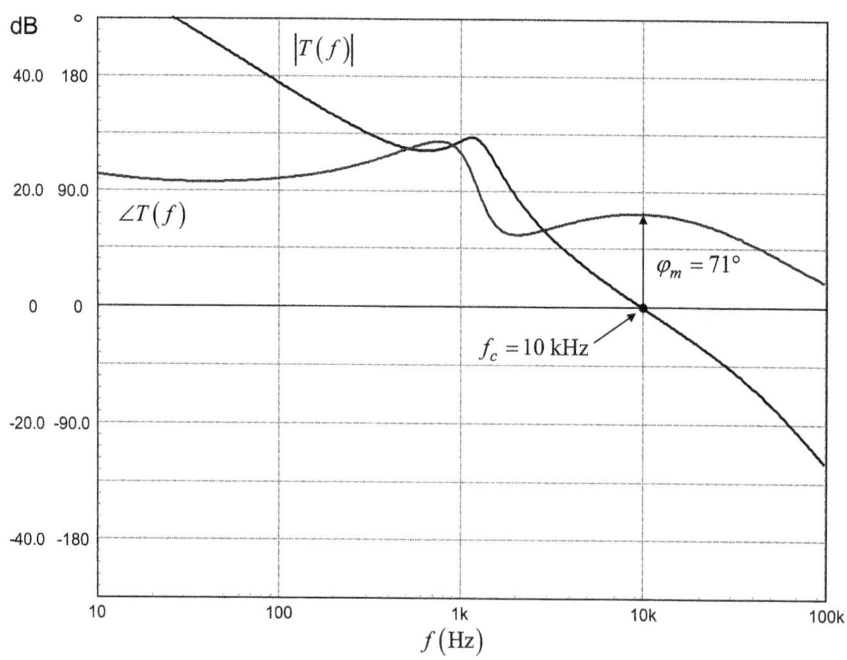

Figure 4.49 The phase margin is 70° as expected at a 10-kHz crossover frequency.

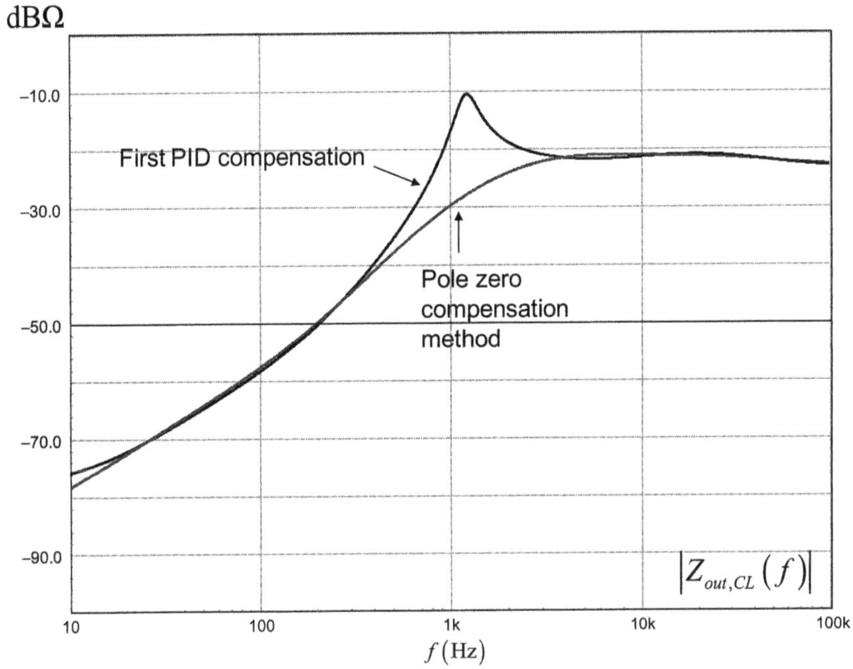

Figure 4.50 The closed-loop output impedance of the compensated buck converter does not reveal any peak.

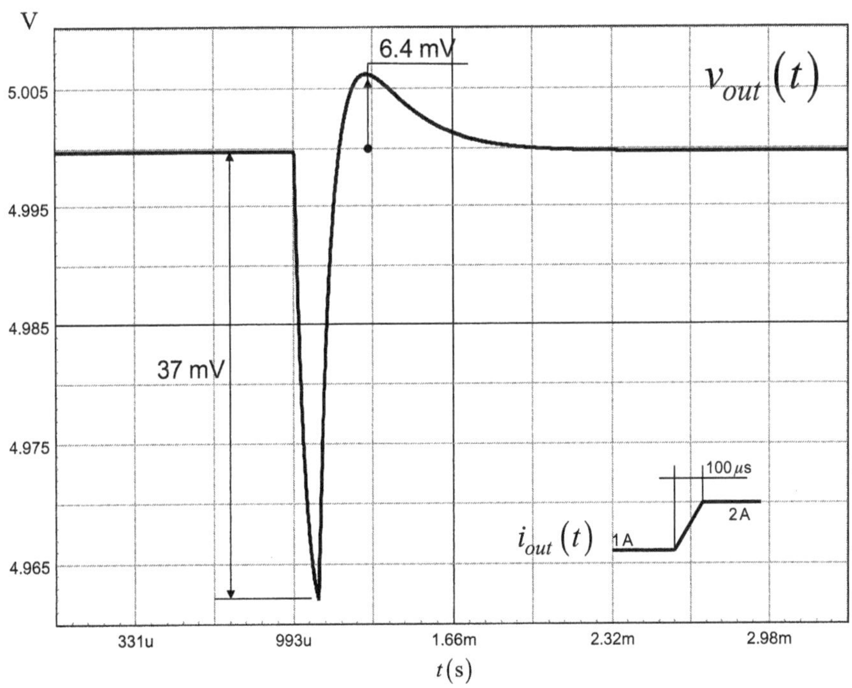

Figure 4.51 The overshoot keeps extremely small when the compensated buck converter is subjected to a load step.

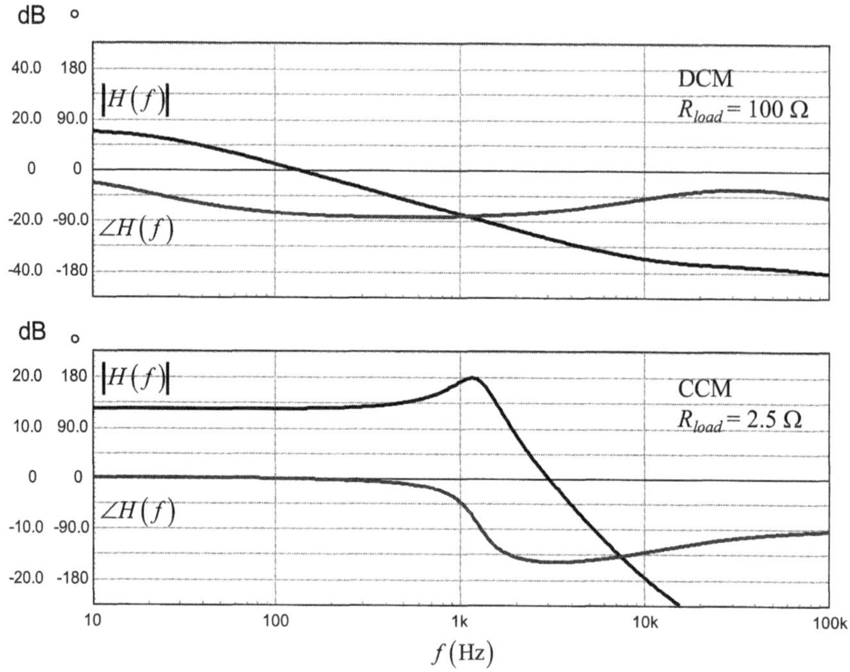

Figure 4.52 When operated in DCM, the second-order CCM operated buck converter turns into a first-order converter.

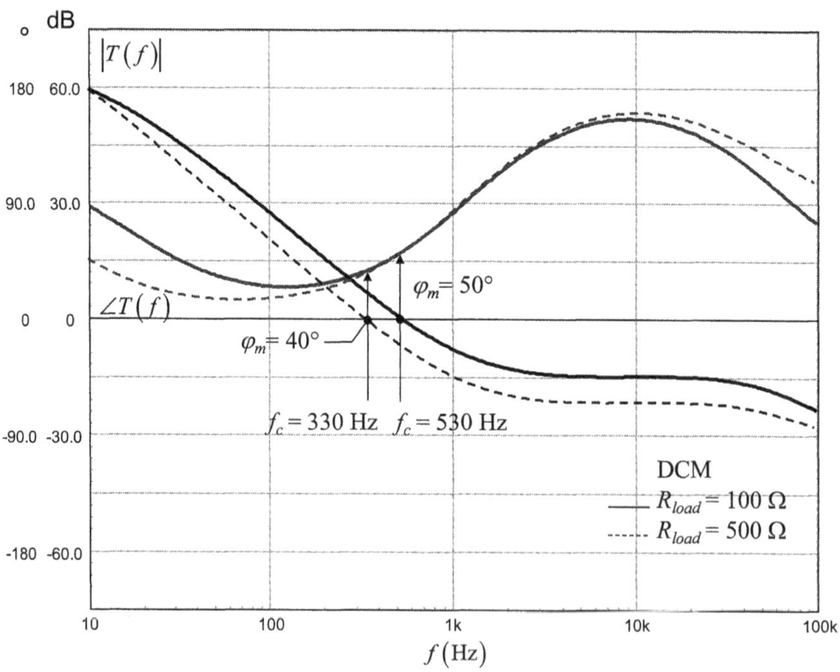

Figure 4.53 When transitioning in DCM, because of the drastic plant gain change, the crossover frequency is reduced and the phase margin at this new point becomes questionable.

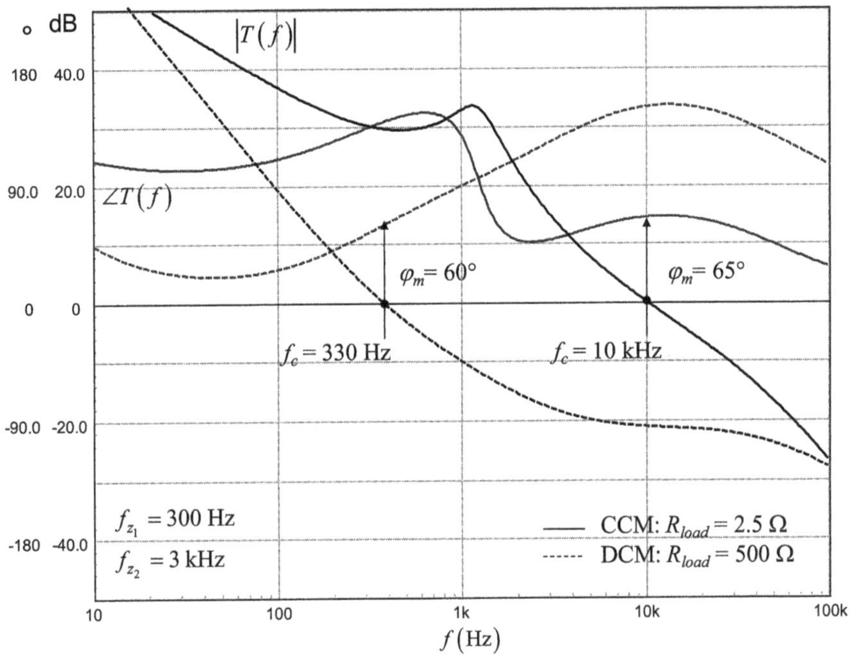

Figure 4.54 By splitting the zeros, you improve the phase margin in DCM.

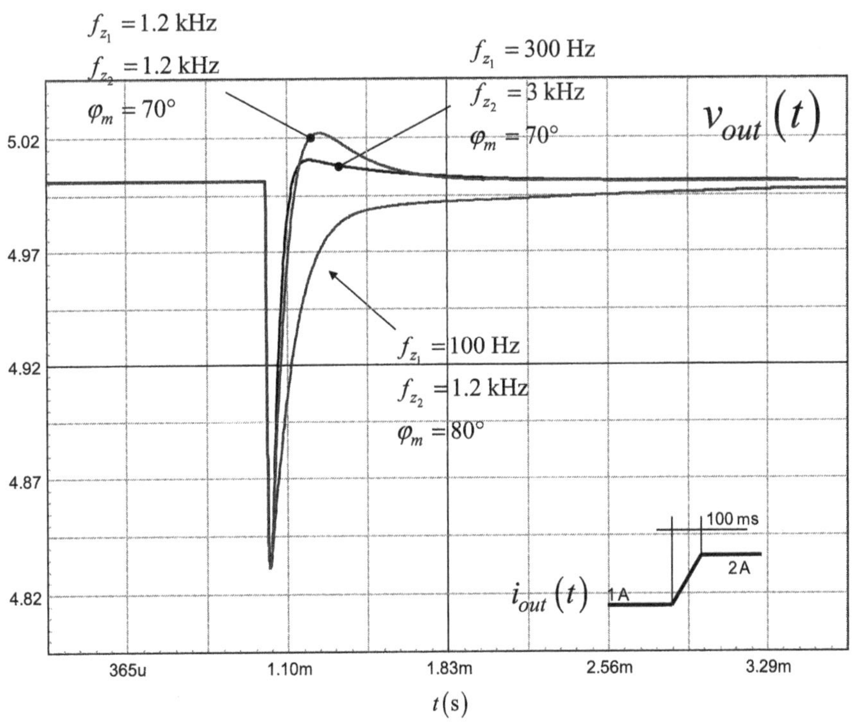

Figure 4.55 Splitting the zeros affects the transient response in CCM, mainly recovery time and the overshoot.

	Frequency	Overshoot	Settling time	Phase margin
f_{z_1}	↗	↗	faster	↘
	↘	↘	slower	↗
f_{z_2}	↗	↗	faster	↘

Figure 4.56 This array shows how splitting the zeros can change the transient response of the affected converter.

but the exercise is iterative by nature since splitting these zeros affects both the CCM and DCM ac responses: should you tweak the phase margin in DCM, you must verify what it becomes in CCM. A SPICE simulator is actually the best to quickly run the exercise and check both responses in a few milliseconds.

We have simulated the converter with the new zeros position (300 Hz and 3 kHz), and Figure 4.54 shows you the results: the phase margin in DCM is much better than before, unfortunately to the detriment of ac response in CCM: the phase margin drops by 6° (see Figure 4.49) but remains at a comfortable value of 65° though.

In Figure 4.55, we have tested three different transient responses to the 1A load step and compared them to the reference one, the 1.2-kHz zero pair. When a zero is going down the frequency axis, it improves the phase margin and reduces the overshoot. However, it increases the recovery time. On the contrary, if this zero is shifted up, the converter recovers faster but overshoots. This is normal, as we learned in Chapter 2 that the open-loop zeros (those you put in the compensator, f_{z1}, f_{z2} in this example) turn into closed-loop poles. If you put a zero at a low frequency to boost the phase, you will naturally slow down the transient response. The array in Figure 4.56 tells you how moving the zeros affects the closed-loop performance.

The transient response resulting from this compensation scheme differs from that obtained with the PID calculation methodology: the response is nonoscillatory but there is a rather deep overshoot, typical of systems including an integral term. Certain applications do not tolerate these undershoots and ask for stringent regulation limits. This is the case for high-speed dc-dc converters for motherboards. For this type of application, the two methods we described cannot be implemented.

4.3 Output Impedance Shaping

In the previous methods, we have seen how the converter fights the sudden current increase or release: an undershoot and an overshoot are typical of an inductive-shaped output impedance. If you look back to Figure 4.50, you can indeed see an impedance growing as the frequency also increases, confirming an inductive-like output impedance. Capitalizing on this shape, [7] offers a simplified representation of a closed-loop buck converter impedance, associating an equivalent inductance and the output capacitor ESR. The schematic appears in Figure 4.57.

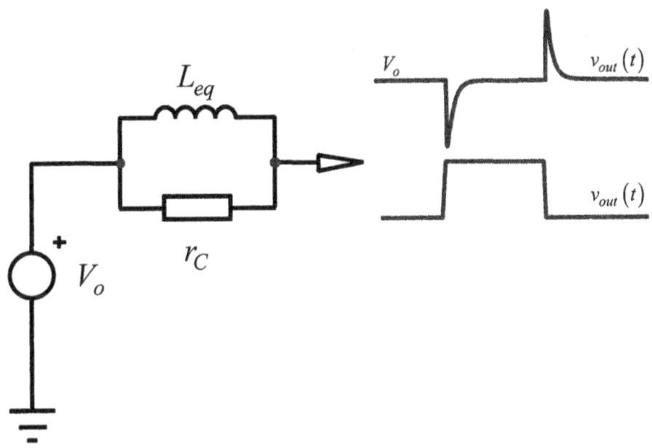

Figure 4.57 The simplified equivalent model of the buck converter associates an inductor and a resistor.

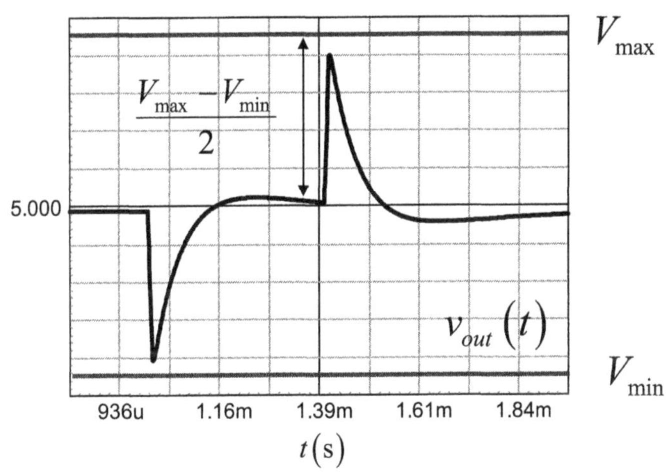

Figure 4.58 The designer must limit both the undershoot and the overshoot to stay within the specifications.

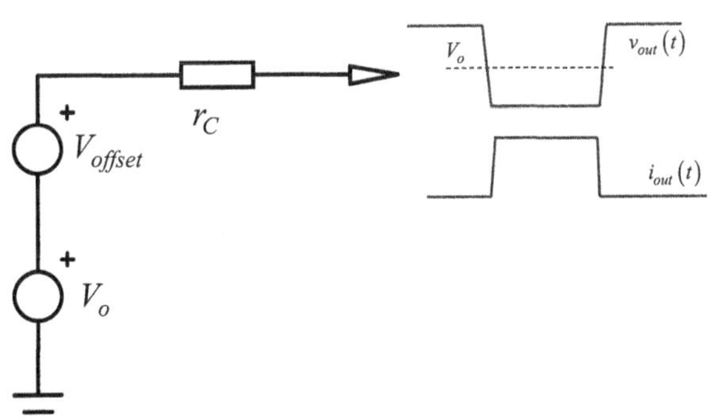

Figure 4.59 If the output impedance becomes purely resistive, the output signal is no longer affected by large under-/overshoots.

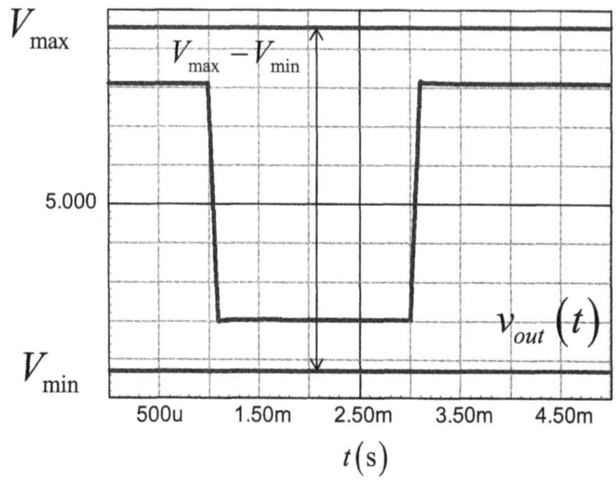

Figure 4.60 A way to limit the voltage excursions is to shape the output impedance and make the inductive term disappear.

The combined action of the inductor and the resistor generates the typical signal represented on the output. The shape is typical of a converter involving an integral term in its compensation chain. As observed, the undershoot and the overshoot can be quite large and exceed the allowable band. This what we shown in Figure 4.58, where margin barely exists.

To get rid of these large excursions around the target output voltage, the best would be suppress the inductive term from Figure 4.57. This is the approach proposed in Figure 4.59.

In this configuration, the targeted level is purposely shifted by an offset to position the output voltage exactly in the middle of the peak-to-peak excursion. This way, the allowable excursion doubles compared to the previous approach. Figure 4.60 portrays the resulting waveform. The method is called *adaptative voltage positioning*.

4.3.1 Making the Output Impedance Resistive

The closed-loop output impedance $Z_{out,CL}$ of the voltage-mode converter is defined by its open-loop output impedance $Z_{out,OL}$ divided by the open-loop gain T_{OL}:

$$Z_{out,CL}(s) = \frac{Z_{out,OL}(s)}{1+T_{OL}(s)} \tag{4.147}$$

The open-loop output impedance has already been derived in Appendix 4A and obeys the following expression:

$$Z_{out,OL}(s) = R_0 \frac{\left(1+\frac{s}{\omega_{z1}}\right)\left(1+\frac{s}{\omega_{z2}}\right)}{\left(\frac{s}{\omega_0}\right)^2+\frac{s}{\omega_0 Q}+1} \tag{4.148}$$

The loop gain T_{OL} is made of the plant transfer function $H(s)$ and the compensator transfer function $G(s)$:

$$T_{OL}(s) = H(s)G(s) \tag{4.149}$$

The buck transfer function is that of a second-order system with a zero linked to the output capacitor ESR:

$$H(s) = H_0 \frac{1+\dfrac{s}{\omega_{z_1}}}{\left(\dfrac{s}{\omega_0}\right)^2 + \dfrac{s}{\omega_0 Q} + 1} \tag{4.150}$$

The idea is to shape the closed-loop output impedance and make it look like a resistance over the frequency spectrum. To fulfill this goal, there is one circuit upon which we can act: the compensator G. This is the place where we will arrange poles or zeros to make the closed-loop output impedance look resistive. What resistance, by the way? The output capacitor equivalent series resistance, r_C, is a good choice, as it is the value toward which the impedance heads at high frequency (Figure 4.61). Equation (4.147) can now be updated to reflect this choice:

$$r_C = \frac{Z_{out,OL}(s)}{1+T_{OL}(s)} = \frac{Z_{out,OL}(s)}{1+H(s)G(s)} \tag{4.151}$$

Substituting (4.148) and (4.150) into (4.151), we have

$$r_C = R_0 \frac{\left(1+\dfrac{s}{\omega_{z_1}}\right)\left(1+\dfrac{s}{\omega_{z_2}}\right)}{\left(\dfrac{s}{\omega_0}\right)^2 + \dfrac{s}{\omega_0 Q} + 1} \frac{1}{1+H_0 \dfrac{1+\dfrac{s}{\omega_{z_1}}}{\left(\dfrac{s}{\omega_0}\right)^2 + \dfrac{s}{\omega_0 Q} + 1} G(s)} \tag{4.152}$$

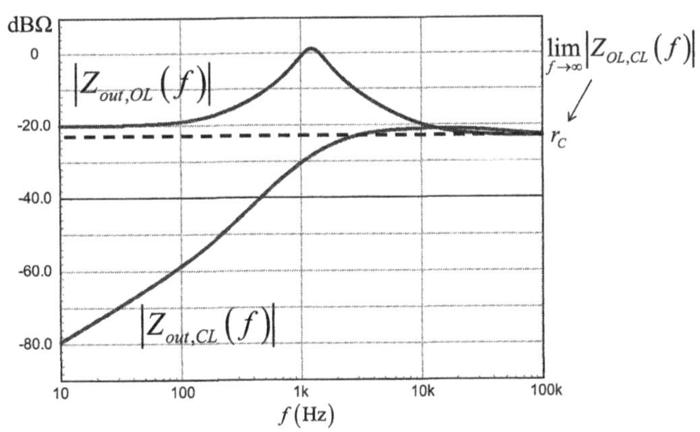

Figure 4.61 When you compare the buck converter open and closed-loop output impedances, they both converge to a fixed value at high frequency, the output capacitor ESR, r_C.

In which $G(s)$ must be shaped to meet the resistive output impedance goal. If we extract the compensator transfer function from (4.152), we obtain

$$G(s) = \frac{R_0\left(1+\frac{s}{\omega_{z2}}\right)}{r_C H_0} - \frac{\left(\frac{s}{\omega_0}\right)^2 + \frac{s}{Q\omega_0} + 1}{H_0\left(1+\frac{s}{\omega_{z_1}}\right)} \tag{4.153}$$

At first glance, I am not able to guess the frequency response of such a transfer function. To gain insight on this expression, we can plot it with Mathcad®. To run this exercise, let us consider the following component values:

$L = 75\ \mu\text{H}$

$r_L = 300\ \text{m}\Omega$

$r_C = 30\ \text{m}\Omega$

$C_{out} = 220\ \mu\text{F}$

$R_{load} = 2.5\Omega$

$\omega_{z_1} = r_L/L = 4\ \text{krad/s so } f_{z_1} = 636\ \text{Hz}$

$\omega_{z_2} = 1/(r_C C_{out}) = 151\ \text{krad/s so } f_{z_2} = 24.1\ \text{kHz}$

$R_0 = r_L \parallel R_{load} = 268\ \text{m}\Omega$

$\omega_0 \approx 1/\sqrt{LC_{out}} = 7.78\ \text{krad/s so } f_0 \approx 1.2\ \text{kHz}$

$Q \approx \sqrt{L/C_{out}}/(r_L + r_C) = 1.77$

$H_0 = V_{in}/V_{peak} = 10/2 = 5$

The plot appears in Figure 4.62.

I confess that I did not expect this magnitude shape when looking at (4.153), but this is it. You can see a dc gain less than 5 dB and a zero kicking around 500 Hz. Then the 1-slope breaks into a 0-slope as the pole appears around 20 kHz. The shape fits the following equation:

$$G(s) = K_0 \frac{1+\frac{s}{\omega_{zG}}}{1+\frac{s}{\omega_{pG}}} \tag{4.154}$$

The fun part now is to identify the pole and the zero definition, as well as the gain K_0 with the elements part of (4.153). From these definitions, the authors of [8] did a great job and came up with the following relationships:

$$K_0 = \frac{r_L - r_C}{H_0 r_C} = 1.8 \tag{4.155}$$

$$a = \frac{r_L}{\omega_{z_1}\omega_{z_2}} - \frac{r_C}{\omega_0^2} = 47.7p \tag{4.156}$$

$$b = r_L\left(\frac{1}{\omega_{z_1}} + \frac{1}{\omega_{z_2}}\right) - \frac{r_C}{Q\omega_0} = 74.2u \tag{4.157}$$

$$c = r_L - r_C = 0.27 \tag{4.158}$$

$$\omega_{p_G} = \omega_{z_1} = 4\,\text{krad/s},\ f_{p_G} = f_{z_1} = 24.1\,\text{kHz} \tag{4.159}$$

$$\omega_{z_G} = \frac{b - \sqrt{b^2 - 4ac}}{2a} = 3.64\,\text{krad/s},\ f_{z_G} = 580\,\text{Hz} \tag{4.160}$$

If we plot (4.154) fed with the previous gain, pole, and zero positions and compare it to the original plot from Figure 4.62, the agreement is good, as seen in Figure 4.63.

Now that the transfer function for the compensator is known, we need to find an implementation. The op amp architecture offered in Figure 4.64 seems like a good fit.

To calculate the component values, we need to derive the transfer function of this compensator. Given the inverting configuration, the first expression is

$$G(s) = \frac{V_c(s)}{V_{out}(s)} = -\frac{R_1}{Z_1(s)} \tag{4.161}$$

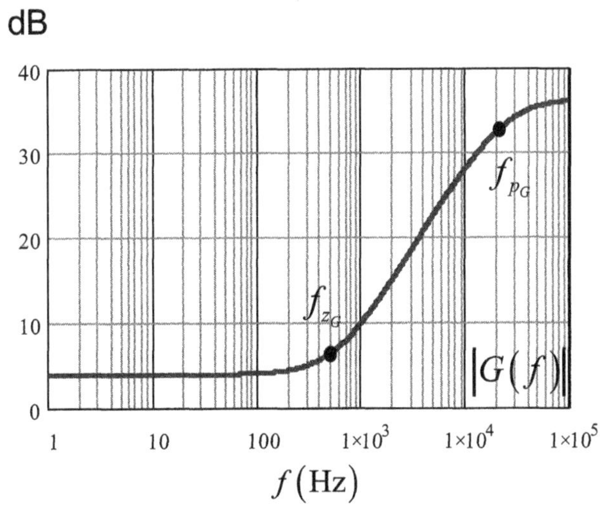

Figure 4.62 The required transfer function combines a pole and a zero but no origin pole: no integral term.

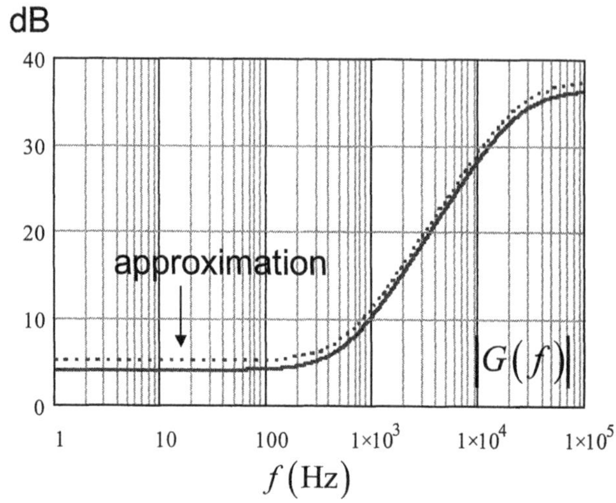

Figure 4.63 The agreement between the original curve and the approximation is good.

The impedance Z_1 is made of a series-parallel combination:

$$Z_1 = \frac{R_2\left(R_3 + \frac{1}{sC_1}\right)}{R_2 + \left(R_3 + \frac{1}{sC_1}\right)} \tag{4.162}$$

If we substitute this definition in (4.161), then develop and rearrange the final expression, we obtain

$$G(s) = -\frac{R_1}{R_2}\frac{[1 + sC_1(R_2 + R_3)]}{1 + sR_3C_1} = -K_0\frac{1+\frac{s}{\omega_{zG}}}{1+\frac{s}{\omega_{pG}}} \tag{4.163}$$

in which we have

$$K_0 = \frac{R_1}{R_2} \tag{4.164}$$

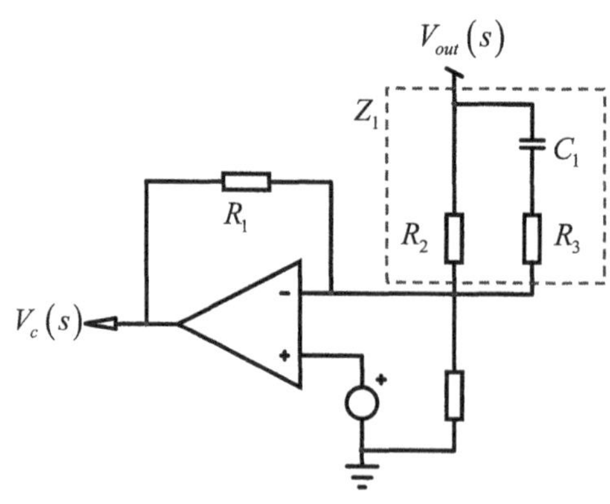

Figure 4.64 The simple pole/zero combination can be easily implemented with an op amp.

$$\omega_{zG} = \frac{1}{C_1(R_2 + R_3)} \tag{4.165}$$

$$\omega_{pG} = \frac{1}{R_3 C_1} \tag{4.166}$$

If you fix R_1 to 10 kΩ (for instance) and solve for the three remaining unknowns C_1, R_2, and R_3, you should obtain the following results:

$$R_3 = \frac{R_1 \omega_{zG}}{K_0(\omega_{pG} - \omega_{zG})} = 137\ \Omega \tag{4.167}$$

$$R_2 = \frac{R_1}{K_0} = 5.5\ \text{k}\Omega \tag{4.168}$$

$$C_1 = \frac{K_0(\omega_{pG} - \omega_{zG})}{R_1 \omega_{pG} \omega_{zG}} = 48\ \text{nF} \tag{4.169}$$

If we assign these values to (4.163), we obtain the Bode plot presented in Figure 4.65, a copy of that displayed in Figure 4.63.

Having the compensator transfer function on hand, we can now plot the final output impedance as defined by (4.147). This graph is shown in Figure 4.66. As observed, the closed-loop output impedance is perfectly resistive along the frequency axis and shows a value of –30.4 dBΩ, which is around 30 mΩ, our output capacitor ESR. The proposed scheme works well on paper and has transformed our inductive open-loop output impedance into a resistive closed-loop output impedance.

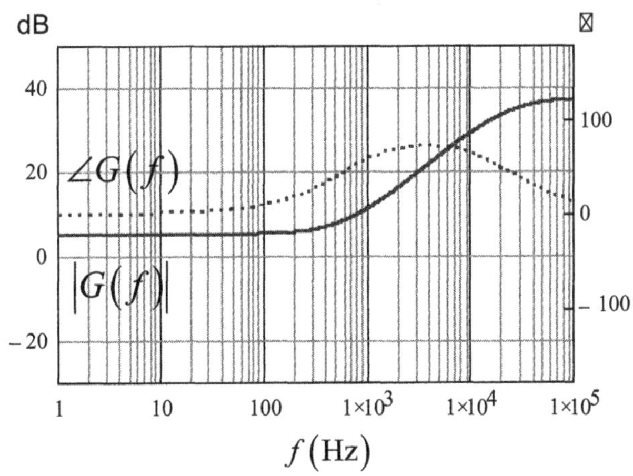

Figure 4.65 The compensator Bode plot does not include an integral term (no origin pole).

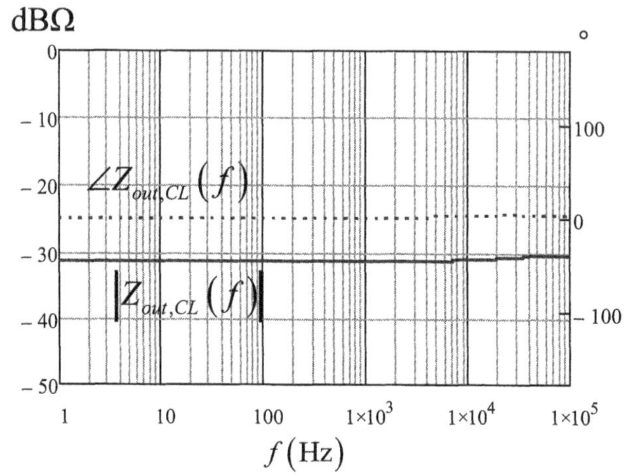

Figure 4.66 The output impedance is perfectly resistive as expected!

We can run a simulation to check the validity of our calculations. The simulated schematic appears in Figure 4.67. On the left, the compensator values are automatically calculated to place the pole and zero recommended by (4.159) and (4.160). Let us look at the ac transfer function to check the crossover frequency and the phase margin at this point. The ac response is given in Figure 4.68. We can see a crossover frequency slightly above 20 kHz and a phase margin of 90°: this is a rock-solid design!

To check the effects of this compensation on the closed-loop impedance, we are going to add a 1-A ac source in parallel with the load, while the loop opening network made of *LoL* and *CoL* is removed. The plot proposed in Figure 4.69 confirms the resistive nature of the output impedance. It is time to test the transient response of the whole converter. The response to a 1-A step is given in Figure 4.70. It is a perfect square wave whose amplitude is exactly the 30-mΩ ESR multiplied by the 1 A current step.

Is the solution a panacea? Well, if we take a closer look at the open-loop ac response from Figure 4.68, we can see a rather low dc gain, below 20 dB. We learned that the dc gain directly impacts the output static error. If we program the op amp divider network to get 5V, despite the precision of the reference voltage, an open-loop dc gain of 10 will not guarantee a 5-V output. To compensate for this problem, we have purposely shifted the target by altering resistor R_7 on the simulation schematic. It purposely shifts up the output target to reach 5 V, and the dc bias shows this to work. If you change the input conditions, this low gain will not shield the output from moving. This voltage-mode implementation of a resistive output impedance can work only if you have a stable input voltage. This is the case for high-speed dc-dc converters for motherboards where the designers generate a 3.3 V or a 1.5 V output directly from the 12V rail. In applications where the input voltage widely varies, this voltage mode solution does not work, and you will need to consider the current-mode version. This is the subject of [8, 9]. The founding equations differ from that of the voltage-mode structure, but the principle remains the same.

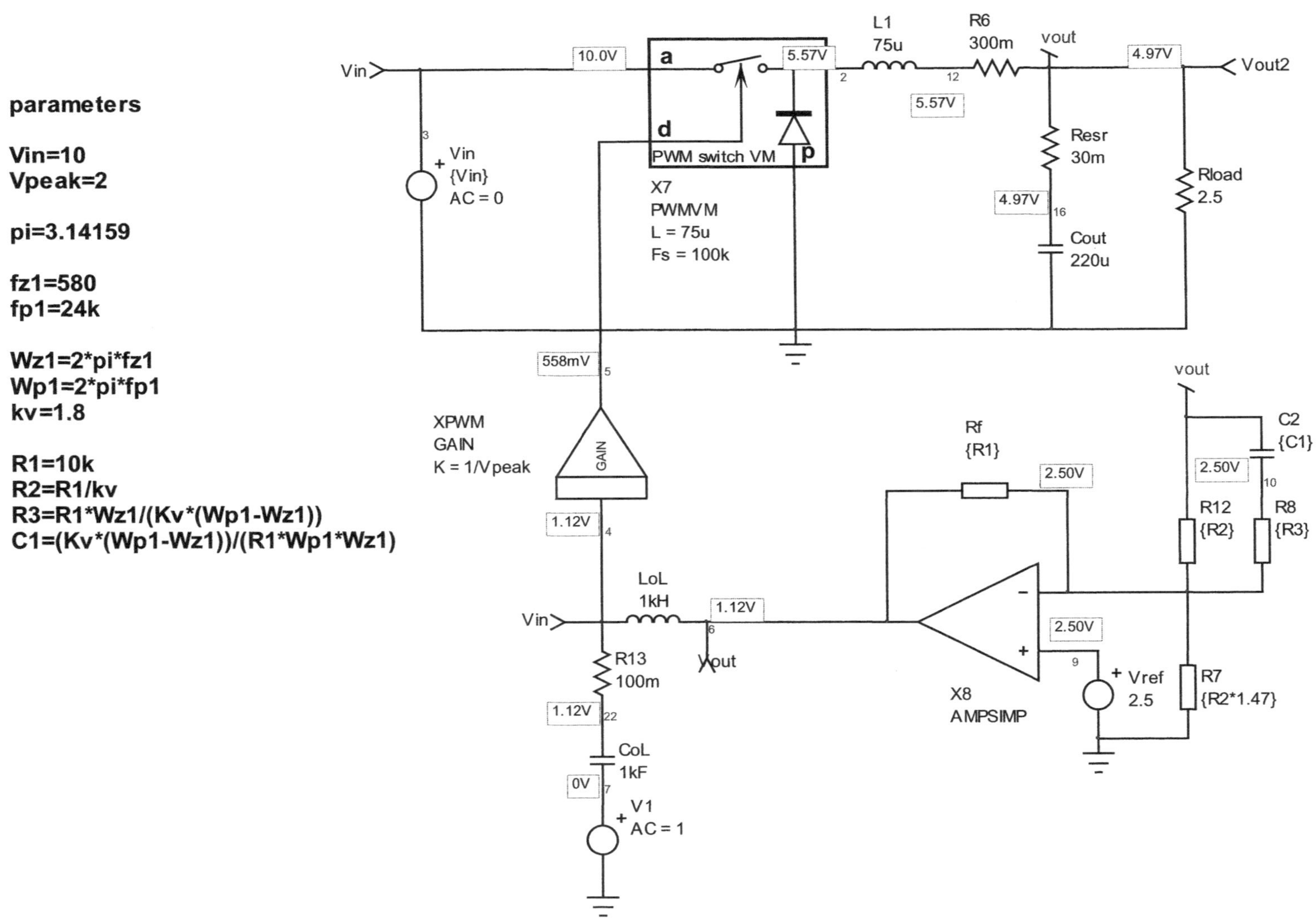

Figure 4.67 The simulation circuit shows the buck converter whose loop is now closed with the Figure 4.64 circuit.

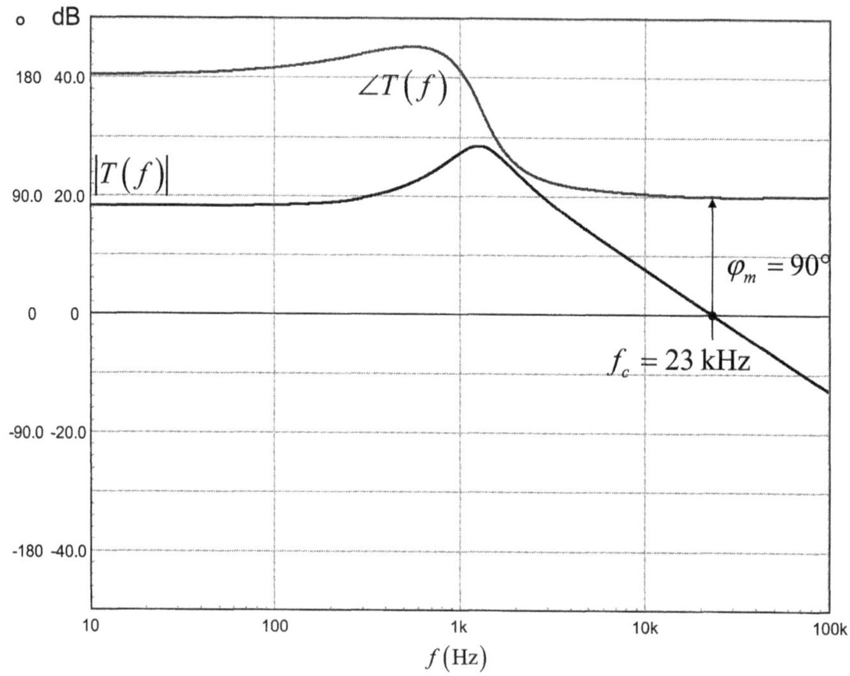

Figure 4.68 The ac response shows a good phase margin despite a high crossover frequency.

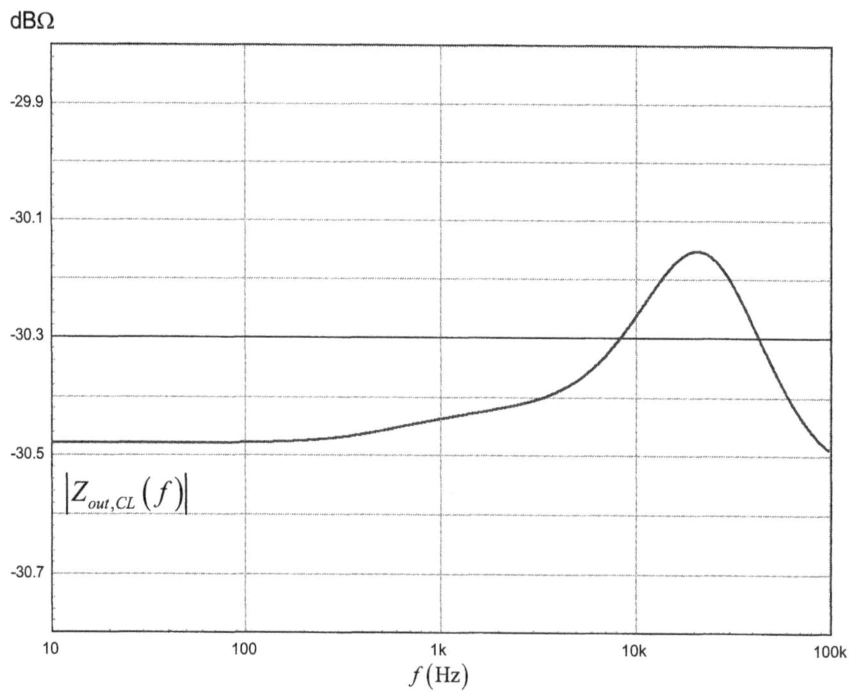

Figure 4.69 The ac simulation of the output impedance confirms our analytical calculations: it is almost constant at −30 dBΩ.

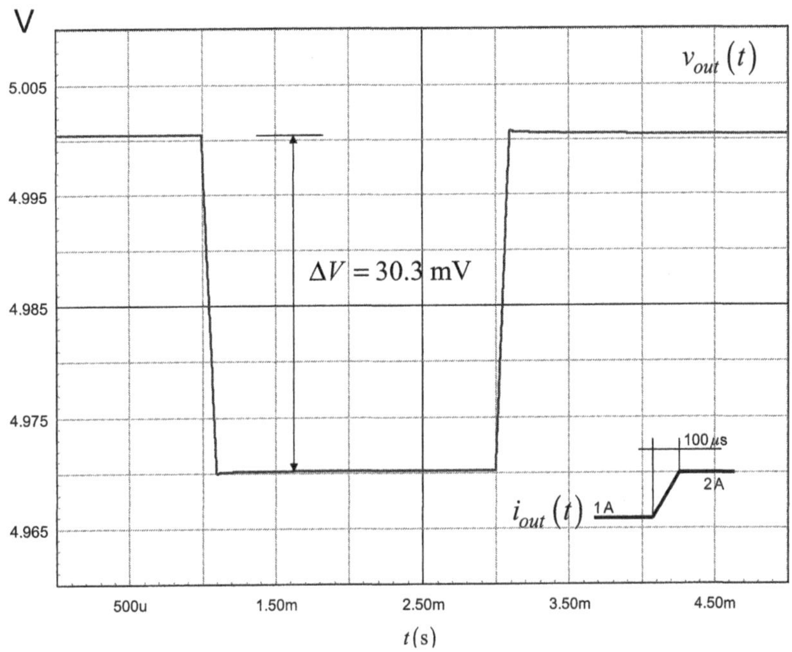

Figure 4.70 As expected, there is no over- or undershoot on this waveform. The total deviation is exactly the current multiplied by the ESR value.

4.4 Conclusion

I was taught PID at university in a rather complex way. There were no connections to pole-zero placement that I was more familiar with. In this chapter, after several algebraic manipulations, we found that a filtered-PID was actually a type 3 compensator, a well-known circuitry for those who compensate converters. Compensating a switching converter by focusing only on the setpoint-to-output transfer function is offering an incomplete view of the transmission chain. Most of our converters are actually regulators, and the setpoint (the reference voltage) is fixed when the converter undergoes a load change. What matters to the transient response is thus the output impedance expression. This is the element that dictates the response and will tell you if your design will deliver a nonringing response. We discovered this fact by blindly applying a PID-based compensation method to the voltage-mode buck converter under study. The result was not convincing, leading to an oscillatory response. We obtained a nonpeaking output impedance with the classical pole/zero placement technique, and the step response was perfectly damped. In both methods, we concentrated our efforts on stabilizing the setpoint-to-output transfer function, without specifically looking at the output impedance. In the third example, we explicitly looked at its expression and worked the compensator to make it fully resistive. The result is a nice square-wave signal, without over-/undershoots. This technique is widely used in high-speed dc-dc converters for personal computers and can be applied in situations where a stringent specification concerns the load-step response.

Now that we reviewed a few of the available compensation techniques, we are going to discover how to build compensators with operational amplifiers.

References

[1] Wikipedia contributors, "PID Controller," http://en.wikipedia.org/wiki/PID_controller, last accessed June 3, 2012.

[2] Besançon-Voda, A., and S. Gentil. "Régulateurs PID Analogiques et Numériques," *Techniques de l'ingénieur*, R7416, 1999.

[3] Retif, J.-M. "Automatique Regulation," 4ième année GE, Institut National des Sciences Appliquées, 2008.

[4] Ziegler, J. G., and N. B. Nichols. "Optimum Settings for Automatic Controllers," *Trans. ASME*, Vol. 64, 1942, pp. 759–768.

[5] Venable, D. "The k-Factor: A New Mathematical Tool for Stability Analysis and Synthesis," *Proceedings of Powercon 10*, 1983, pp. 1–12.

[6] Basso, C. *Switch-Mode Power Supplies: SPICE Simulations and Practical Designs*, New York: McGraw-Hill, 2008.

[7] Erisman, B., and R. Redl. "Modify Your Switching Supply Architecture for Improved Transient Response," *EDN Magazine*, November 11, 1999.

[8] Yao, K., M. Xu, Y. Meng, and F. Lee. "Design Considerations for VRM Transient Response Based on the Output Impedance," *IEEE Proceedings*, Vol. 18, No. 6, November 2003.

[9] Erisman, B., and R. Redl. "Optimizing the Load Transient Response of the Buck Converter," *APEC'1998 Proceedings*, Vol. 1, pp. 170–176.

Appendix 4A: The Buck Output Impedance with Fast Analytical Techniques

An output impedance is a transfer function. It is made of an excitation signal—a current source—that produces a voltage across the impedance it sweeps, the response signal. This known principle is found in Figure 4.71.

As already indicated, the output impedance expression must follow the form indicated in the picture: a static ohmic expression followed by an *s*-quotient without dimension. First, we start with the dc term, R_0. It is found by putting all storage elements in their dc states: the capacitors are open and the inductors short circuited. You obtain the simple circuit of Figure 4.72.

From that drawing, we see immediately that the dc term is simply:

$$R_0 = r_L \parallel R_{load} \tag{4.170}$$

Now, observing Figure 4.71, do we see branches that would prevent the excitation from reaching the output? In other terms, what could prevent the excitation current from generating a voltage across R_{load}? Well, if R_{load} was simply short

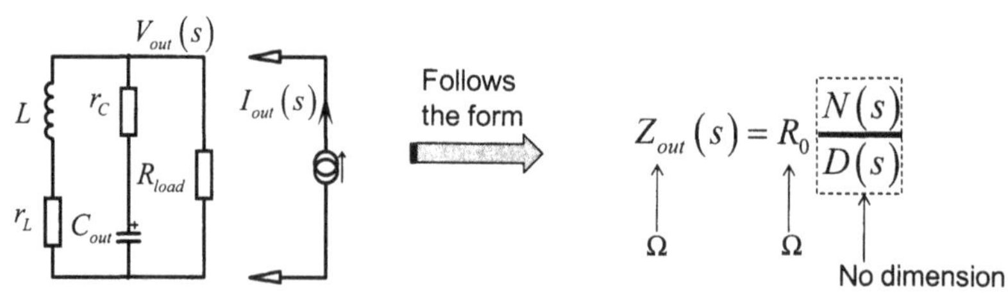

Figure 4.71 An output impedance is one of the six possible transfer functions.

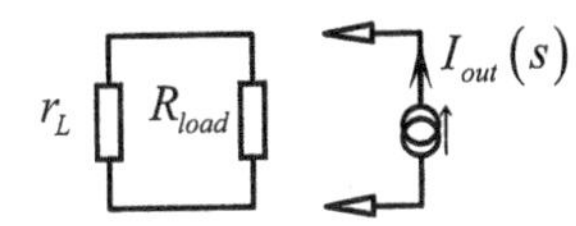

Figure 4.72 When all capacitors are open and inductors short circuited, the circuit greatly simplifies.

circuited by the parallel networks, the voltage would be zero. What, in this parallel network, could be a short circuit? Potentially the RL and RC branches. Mathematically, you can then write two equations as follows:

$$sL + r_L = r_L\left(1 + s\frac{L}{r_L}\right) = 0 \tag{4.171}$$

and

$$r_C + \frac{1}{sC} = \frac{sr_cC + 1}{sC} = 0 \tag{4.172}$$

Solving these simple equations gives us two zeros expressions:

$$\omega_{z_1} = \frac{r_L}{L} \tag{4.173}$$

$$\omega_{z_2} = \frac{1}{r_C C} \tag{4.174}$$

We are now halfway to our impedance definition:

$$Z_{out}(s) = R_0 \frac{(1 + s/\omega_{z_1})(1 + s/\omega_{z_2})}{D(s)} \tag{4.175}$$

The most complex part of the process is the derivation of denominator $D(s)$. Theory tells us that this denominator solely depends on the network structure, not on the excitation. If the excitation signal does not play a role when you derive the denominator, put it to zero in the network you study: short the excitation voltage or open the excitation current. For our impedance case, the current source just leaves the picture. Once $D(s)$ is derived, it appears as such in the remaining transfer functions like input impedance, input to output gain, and so on. Hey, we have it already from (4.34) then! That's right, but for the sake of the exercise, let us try to find it using the fast analytical techniques described in [1]. What does $D(s)$ look like? This is a two-storage element circuit—thus, a second-order network. It must obey the following expression:

$$D(s) = 1 + a_1 s + a_2 s^2 \tag{4.176}$$

As explained in the referenced work, the general form of the coefficient of s^k in the previous equation (a_1, a_2) must be of dimension $(Hz)^{-k}$ so that the denominator remains dimensionless. These coefficients sum all the combined network time constants equal to R_xC or L/R_y, where resistors R_x or R_y are the respective resistances seen at the capacitive or the inductive ports under some particular configurations we will see. These configurations involve so-called dc and high-frequency states defined as follows:

- Dc state: the capacitor is simply open (its impedance is infinite at $s = 0$), while the inductor is a short circuit.
- High-frequency state: in this case, the capacitor is now a short-circuit, and the inductor becomes open.

According to [1], the possible terms for a second-order network denominator are

- $\tau_1 + \tau_2$ for a_1. Dimension is time (s) or Hz^{-1}.
- $\tau_1 \tau_2'$ or $\tau_1' \tau_2$ are the two possible choices for a_2, where τ_1 or τ_2 are the time constants from earlier. Dimension is squared time (s^2) or Hz^{-2}. τ_1' τ_2' will be defined in a few lines.
- In all cases, the time constants τ are of the form RC or L/R.

Applying definitions from [1], the denominator formula for a second-order system is given by

$$D(s) = 1 + s(\tau_1 + \tau_2) + s^2(\tau_1 \tau_2') \tag{4.177}$$

equivalent to

$$D(s) = 1 + s(\tau_1 + \tau_2) + s^2(\tau_1' \tau_2) \tag{4.178}$$

For a_1, we are first looking at the resistance R seen from the inductor port when the capacitor is put in its dc state (open circuited). The sketch appears in Figure 4.73. The resistance seen at the inductor port is straightforward and equal to

$$R = r_L + R_{load} \tag{4.179}$$

Leading to

$$\tau_1 = \frac{L}{r_L + R_{load}} \tag{4.180}$$

Now, we look at the capacitor port resistance when the inductor is in its dc state. This is what Figure 4.74 represents. In this case, the equivalent resistance seen at the capacitor port is simply

$$R = r_C + r_L \parallel R_{load} \tag{4.181}$$

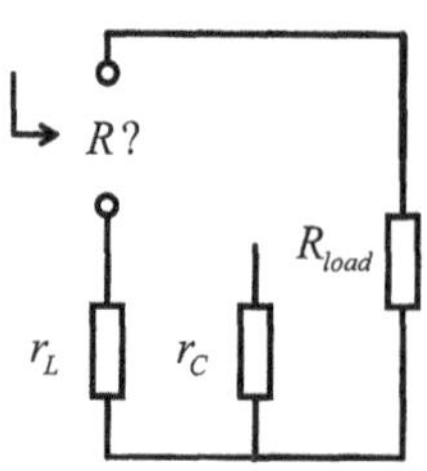

Figure 4.73 The capacitor is open to find the inductor driving resistance.

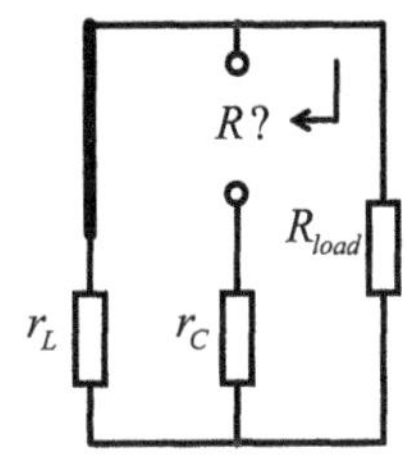

Figure 4.74 An inductor in its dc state is a short circuit.

thus

$$\tau_2 = C(r_C + r_L \parallel R_{load}) \tag{4.182}$$

Following the definition of (4.177), the first coefficient a_1 is defined by

$$a_1 = \tau_1 + \tau_2 = \left(\frac{L}{r_L + R_{load}} + C[(r_L \parallel R_{load}) + r_C] \right) \tag{4.183}$$

The coefficients for a_2 require a little more attention. We have seen that we have to evaluate $\tau_1\tau_2'$ or $\tau_1'\tau_2$, where τ_1 and τ_2 are the same time constants evaluated for a_1. Therefore, we can either decide to find how τ_1 (involving L) combines with τ_2' (involving C) or see how τ_2 (involving C) combines with τ_1' (involving L). Either combination must lead to a similar result for a_2. Let's see the first option where we select τ_1 and must find the resistance seen at port C when L is in its high-frequency state (open circuited). The equivalent schematic appears in Figure 4.75.

The resistance seen at the capacitor port in this case is simply

$$R = r_C + R_{load} \tag{4.184}$$

therefore,

$$\tau_2' = (r_C + R_{load})C \tag{4.185}$$

If we choose the second option, where we select τ_2, we must find the resistance seen at port L while C is in its high-frequency state (short circuit). The equivalent schematic appears in Figure 4.76.

The resistance is defined by a simple series-parallel association:

$$R = r_L + r_C \parallel R_{load} \tag{4.186}$$

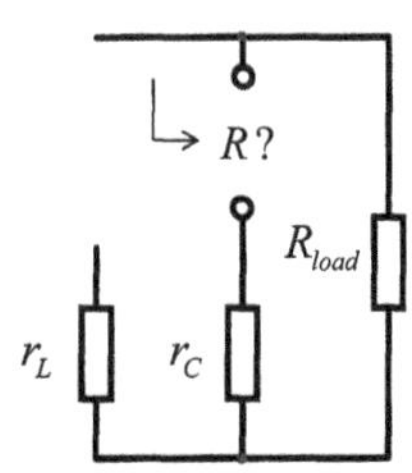

Figure 4.75 The inductor in its high-frequency state is an open circuit and leaves the picture.

which gives us

$$\tau_1' = \frac{L}{r_L + r_C \,||\, R_{load}} \tag{4.187}$$

Following (4.177), we have our expression for coefficient a_2:

$$a_2 = \tau_1 \tau_2' = \frac{L}{r_L + R_{load}} C(r_c + R_{load}) = LC\frac{r_c + R_{load}}{r_L + R_{load}} \tag{4.188}$$

which is equivalent to

$$a_2 = \tau_1' \tau_2 = \frac{L}{r_L + R_{load} \,||\, r_C} C[(r_L \,||\, R_{load}) + r_C] = LC\frac{r_L \,||\, R_{load} + r_C}{r_L + R_{load} \,||\, r_C} = LC\frac{r_c + R_{load}}{r_L + R_{load}} \tag{4.189}$$

If you develop this expression, it is similar to that in (4.188). We are all set; the complete denominator function is thus:

$$D(S) = 1 + s\left(\frac{L}{r_L + R_{load}} + C[(r_L \,||\, R_{load}) + r_C]\right) + s^2 LC\frac{r_c + R_{load}}{r_L + R_{load}} \tag{4.190}$$

We now can update (4.175). We obtain our final output impedance definition:

$$Z_{out}(S) = (r_L \,||\, R_{load}) \frac{\left(1 + s\frac{r_L}{L}\right)(1 + s r_C C)}{1 + s\left(\frac{L}{r_L + R_{load}} + C[(r_L \,||\, R_{load}) + r_C]\right) + s^2\left(LC\frac{r_C + R_{load}}{r_L + R_{load}}\right)} \tag{4.191}$$

It can be put under the familiar form

$$Z_{out}(S) = R_0 \frac{(1 + s/\omega_{z_1})(1 + s/\omega_{z_2})}{1 + \frac{s}{\omega_0 Q} + \left(\frac{s}{\omega_0}\right)^2} \tag{4.192}$$

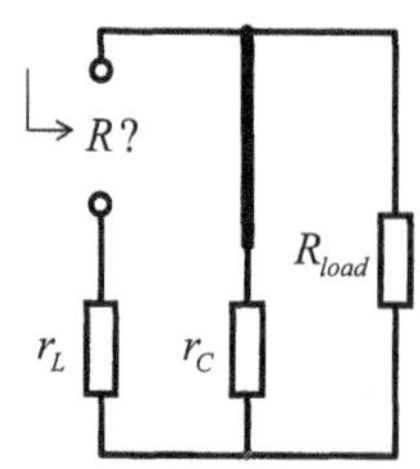

Figure 4.76 In this choice, we keep τ_2 and look at the resistance seen from the inductive port while C is a short circuit.

in which we have

$$R_0 = r_L \parallel R_{load} \tag{4.193}$$

$$\omega_{z_1} = \frac{r_L}{L} \tag{4.194}$$

$$\omega_{z_2} = \frac{1}{r_C C} \tag{4.195}$$

$$\omega_0 = \frac{1}{\sqrt{LC}}\sqrt{\frac{r_L + R_{load}}{r_C + R_{load}}} \tag{4.196}$$

$$Q = \frac{LC\omega_0 (r_C + R_{load})}{L + C[r_L r_C + R_{load}(r_L + r_C)]} \tag{4.197}$$

If we now consider r_C and r_L equal to zero, (4.196) and (4.197) simplify to (4.36) and (4.37).

Reference

[1] Erickson, R. W. "The n Extra Element Theorem," CoPEC, http://ecee.colorado.edu/~ecen5807/course_material/EET, last accessed June 3, 2012.

Appendix 4B: The Quality Factor from a Bode Plot with Group Delay

It can be sometimes difficult to extract the quality factor from a Bode plot whose magnitude weakly peaks. This is the case of the graph in Figure 4.77, where the transient response overshoot indicates a quality factor beyond 0.5. However, the flat magnitude plot makes it difficult to measure.

Let us now have a look at the Bode plot generated from a second-order network where Q has been purposely swept from a low value of 0.6 up to 10. This is shown in Figure 4.78 in which the magnitude starts to peak when Q exceeds 1. In all cases, the phase starts from 0° in dc down to –180° reached in the upper portion of the frequency spectrum. However, you can notice that the phase rate of change depends on the quality factor value: at low Qs, the phase smoothly drifts to –180°, whereas it moves in a much sharper way as Q increases.

In other terms, we can clearly see that the phase curve slope increases with Q. Therefore, there must be a relationship between this slope and the quality factor. This relationship is the *group delay* noted τ_g and defined as

$$\tau_g = -\frac{d\varphi(\omega)}{d\omega}[S] \tag{4.198}$$

The group delay has a dimension of time and is a well-known parameter in optics or in audio. In audio, the group delay evaluates the phase nonlinearity as a signal propagates through a filter, a pair of wires, a loudspeaker, and so on. If

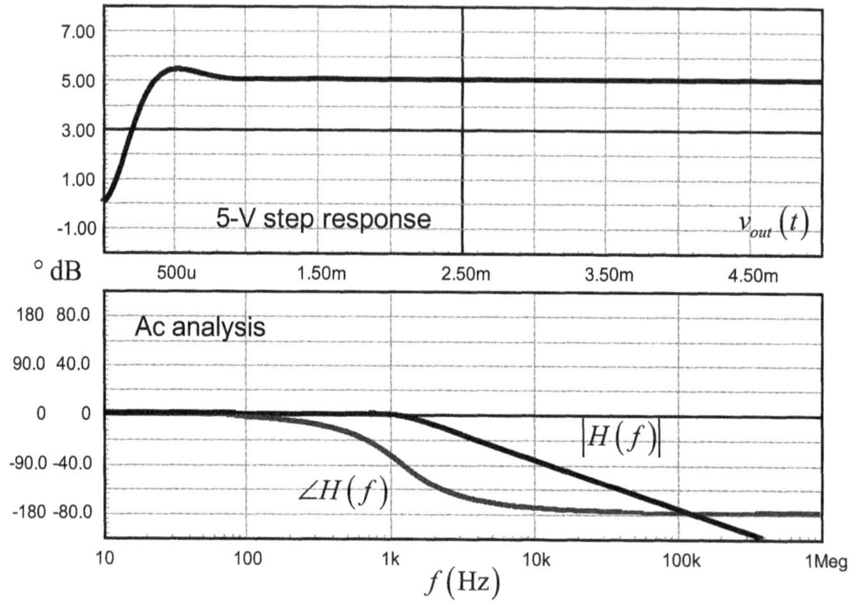

Figure 4.77 In this graph, it is difficult to extract the quality factor value, despite a value above 0.5 given the overshoot in the time-domain response.

the phase variation is linear over a certain band of frequencies, the group delay is constant or uniform and the frequency components of the signal within the considered band are equally treated: the signal is simply shifted in time and its integrity preserved. For the opposite, if the phase suddenly changes at certain frequencies, the group delay peaks, indicating phase distortion at these points. The ears or a

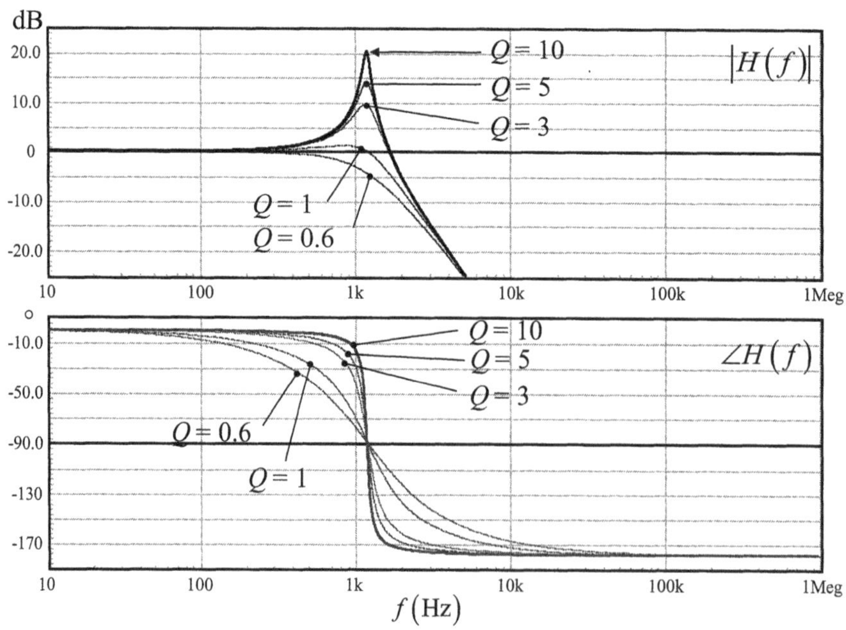

Figure 4.78 As you can see in this plot, the phase rate of change increases as *Q* does.

measurement instrument will then reconstruct the signal by combining frequency components affected by different delays. This is obviously a situation detrimental to the signal integrity, resulting in audio distortion. Of course, if you play (loud) Motorhead's "The Ace of Spades," none will notice it and more distortion will actually be enjoyable!

To see if a relationship links the quality factor to group delay, let us study the transfer function of a second-order network:

$$H(s) = \frac{1}{1 + \frac{s}{\omega_0 Q} + \left(\frac{s}{\omega_0}\right)^2} \tag{4.199}$$

If we replace s by $j\omega$, we can rewrite this function, separating the real and imaginary parts:

$$H(j\omega) = \frac{1}{\left(1 - \frac{\omega^2}{\omega_0^2}\right) + j\frac{\omega}{\omega_0 Q}} = \frac{1}{x + jy} \tag{4.200}$$

From this definition, the magnitude and the argument come easily:

$$|H(\omega)| = \frac{1}{\sqrt{x^2 + y^2}} = \frac{1}{\sqrt{\left(1 - \frac{\omega^2}{\omega_0^2}\right)^2 + \left(\frac{\omega}{\omega_0 Q}\right)^2}} \tag{4.201}$$

$$\angle H(\omega) = \tan^{-1}\left(\frac{y}{x}\right) = \tan^{-1}\left[\frac{\omega}{\omega_0 Q}\frac{1}{\left(1 - \frac{\omega^2}{\omega_0^2}\right)}\right] \tag{4.202}$$

We can now apply the group definition stated in (4.198) through (4.202):

$$\tau_g(\omega) = -\frac{d}{d\omega}\tan^{-1}\left[\frac{\omega}{\omega_0 Q}\frac{1}{\left(1 - \frac{\omega^2}{\omega_0^2}\right)}\right] = \frac{Q\omega_0(\omega^2 + \omega_0^2)}{Q^2\omega^4 - 2Q^2\omega^2\omega_0^2 + Q^2\omega_0^4 + \omega^2\omega_0^2} \tag{4.203}$$

At the resonance, where $\omega = \omega_0$, the phase change is the sharpest. This equation simplifies to

$$\tau_g = \frac{2Q}{\omega_0} \tag{4.204}$$

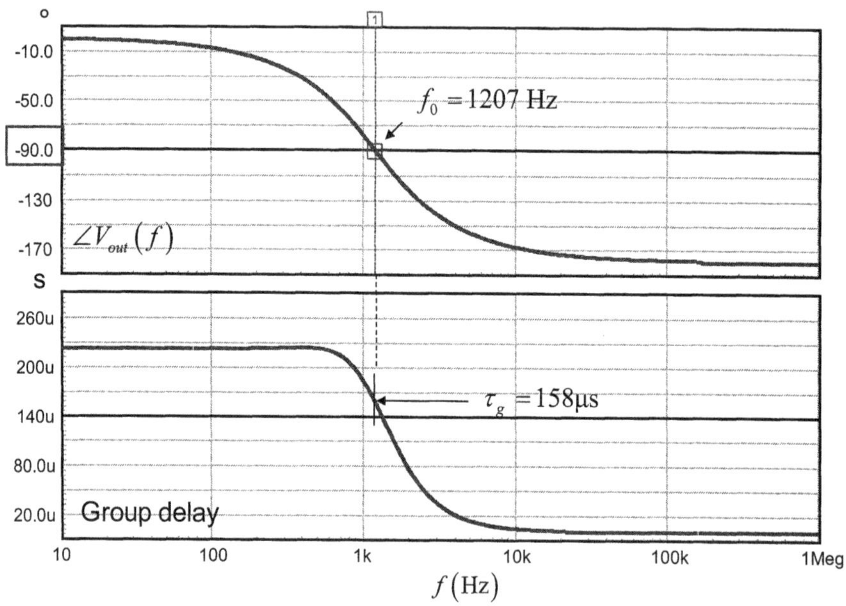

Figure 4.79 A 0.6-Q brings a rather flat magnitude response without peaking. Group delay measurement helps you to compute the *Q* quite easily.

From this expression, we have our relationship linking *Q* to the group delay at the resonance:

$$Q = \frac{\tau_g \omega_0}{2} = \tau_g \pi f_0 \tag{4.205}$$

Let us see if we can easily apply the concept to a Bode plot. We have simulated the transfer function of an *RLC* filter tuned to 1.2 kHz, affected by a quality factor of 0.6. As shown in Figure 4.79, the transfer function argument crosses the –90° line and indicates the resonant frequency value. We have plotted the group delay by the execution of a simple script in Intusoft's Intuscope graphical tool. Under Cadence's Probe, you will use the mathematic function *G*. Type *VG(3)* and Probe will display the group delay of node 3, for instance. From the graph, we measure a group delay of 158 μs at 1.2 kHz. Applying the formula derived in (4.205), we have

$$Q = \tau_g \pi f_0 = 1207 \times 158u \times 3.14159 = 0.599 \tag{4.206}$$

This is exactly the value we have chosen for the *RLC* filter *Q*.

Appendix 4C: The Phase Display in Simulators or Mathematical Solvers

The way a simulator or a mathematical solver displays the phase can sometimes puzzle people who are not used to manipulating complex numbers. As a reminder, a complex number is defined as

$$z = x + jy \tag{4.207}$$

where x and y are, respectively, the real and imaginary parts of the complex number z. Such a number and its conjugate $\bar{z}$ are represented as vectors in Figure 4.80's so-called Argand plane. The x-axis represents the real part, while the y-axis is the imaginary part. Please note that the angle α is positive in quadrant I (we turn counterclockwise, or positively) whereas in quadrant IV, this angle is considered negative as we turn clockwise, or negatively.

From this drawing, we can obtain the magnitude and the argument of the considered number. The *magnitude* of z, also found in textbooks as *modulus*—"module" in French—is obtained by applying Pythagorean geometry. We say magnitude and not *amplitude* because the length of a vector can only be positive:

$$|z| = \sqrt{x^2 + y^2} \tag{4.208}$$

In loop control theory, when we cascade transfer function blocks expressed in Laplace notation, we also manipulate complex numbers since we replace s with $s = j\omega$ in harmonic analysis. If we have cascaded blocks G and H, then the loop gain magnitude $|T|$ is the product of the individual magnitudes:

$$|T(s)| = |G(s)H(s)| = |G(s)| \cdot |H(s)| \tag{4.209}$$

Of course, if the individual amplitudes G and H are converted in decibels, like in a Bode plot, you would simply sum them to obtain that of T since $\log(ab) = \log(a) + \log(b)$.

For a quotient in a transfer function, the magnitude is the quotient of the individual magnitudes. For instance, if we have

$$G(s) = \frac{N(s)}{D(s)} \tag{4.210}$$

then

$$|G(s)| = \frac{|N(s)|}{|D(s)|} \tag{4.211}$$

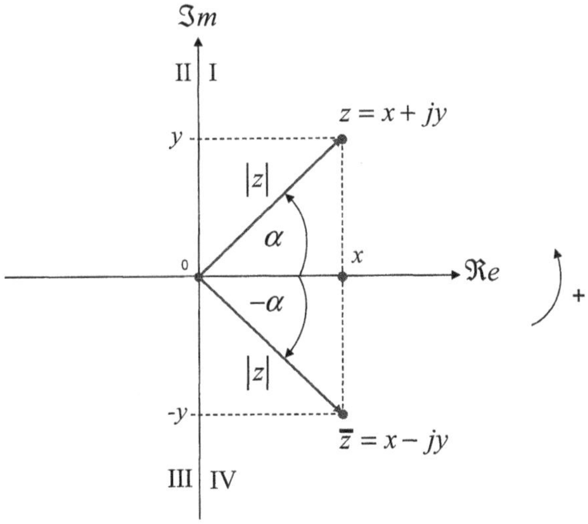

Figure 4.80 The complex number z and its conjugate expression.

The argument is obtained in a similar way if we look at the angles shown in the graph. The tangent of α is classically obtained by dividing the angle opposite side (the sinus amplitude) by the angle cosinus:

$$\tan(\alpha) = \frac{\sin\alpha}{\cos\alpha} = \frac{y}{x} \tag{4.212}$$

As α is the argument of z that we want, we extract it by using the arctangent function:

$$\arg(z) = \operatorname{atan}\left(\frac{y}{x}\right) \text{ or } \tan^{-1}\left(\frac{y}{x}\right) \tag{4.213}$$

If we apply these definitions to our cascaded gain blocks H and G, the argument of T is simply the sum of these arguments:

$$\arg T(s) = \arg\left[G(s)H(s)\right] = \arg G(s) + \arg H(s) \tag{4.214}$$

For the quotient in (4.210), the argument is the difference in arguments between that of the numerator and that of the denominator:

$$\arg G(s) = \arg N(s) - \arg D(s) \tag{4.215}$$

These properties are used extensively throughout the chapters to compute and evaluate various transfer functions magnitude and arguments.

Calculating the Tangent

The tangent function works on an angle and returns a height. This is the height of a straight line starting from the x-axis, tangent to the unit circle, and intersecting with a line starting from the plan origin while forming the angle of interest. Such a simple representation appears in Figure 4.81.

In this picture, you see the plane divided in four subdivisions called quadrants. Each corresponds to a certain combination of x and y signs in the Argand plane. For instance, in quadrant I, x and y are both positive, while in quadrant II y is still

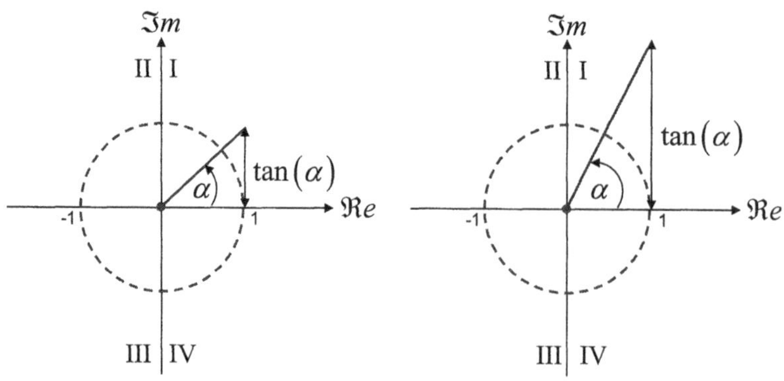

Figure 4.81 The tangent function is defined for an angle α such that $-90° < \alpha < 90°$ or $-\frac{\pi}{2} < \alpha < \frac{\pi}{2}$.

positive but x is negative. In quadrant I, as the angle α widens and approaches 90° or $\pi/2$, the tangent function goes to infinity as $\cos(\pi/2)$ in (4.212) equals 0. The same occurs when the angle approaches –90° or $-\pi/2$ in quadrant IV. Now what value does the tangent function return for angles appearing in quadrants II or III, for instance, 120° or 225°? To understand the answer, it is interesting to plot the sine/cosine functions and compute the tangent according to (4.212). This is what Figure 4.82 shows. The left side of the picture describes the sine and cosine functions, while the angle changes from 0° to 360° at which point, a complete revolution has been made since 0° and 360° refer to the same location on the circle. The right side of the picture displays the trigonometric circle on which some typical angles are reported. α_1 to α_5 are positive angles, since they start from 0 and increase while turning counter-clockwise.

In this picture, we can see that the tangent exists for the angle of α_1 and returns a height h_1 located in quadrant I. When α_2 is reached (90° or $\pi/2$), we are at the sine peak while the cosine is 0. Obviously, at this point, we have a division by zero and the tangent is infinite or undefined: this is a discontinuity in the function. Now, when we progress further on the trigonometric circle and pass the 90° angle, we reach the angle α_3. At this point, the cosine function has become negative, making the tangent jump to a negative value calculated as h_2 and now located in quadrant IV. If we further increase the angle and eventually reach π, the tangent returns 0,

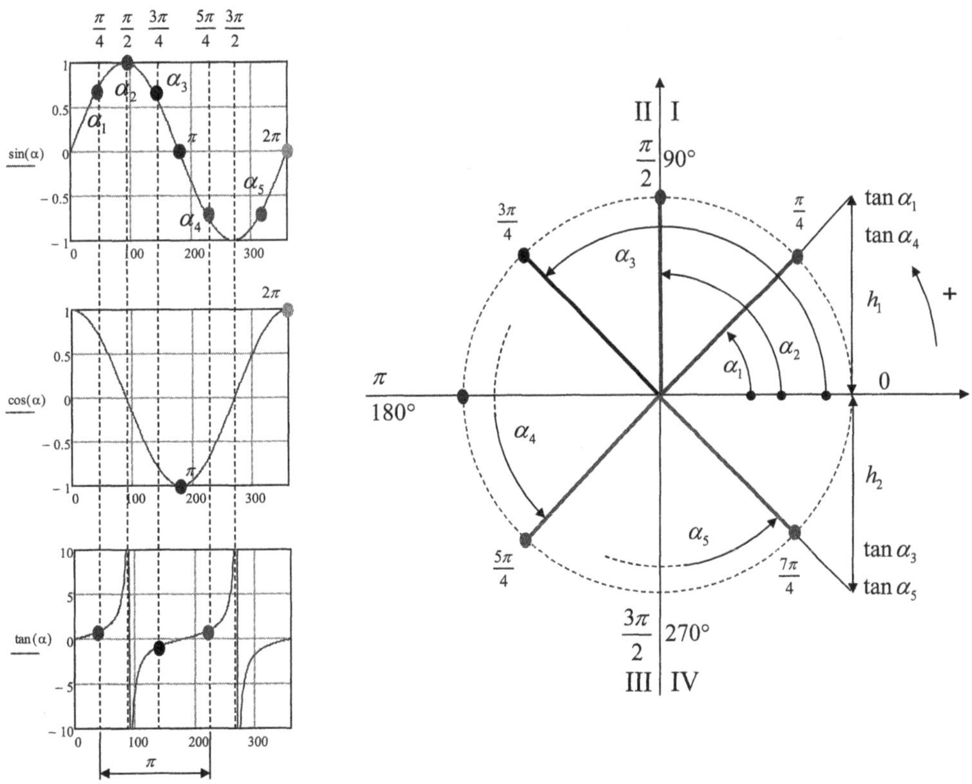

Figure 4.82 When the angle approaches 90°, the tangent goes to infinity, and as the angle continues to widen, the tangent becomes negative: we have jumped to the other side of the circle.

$\sin(\pi) = 0$. As the angle positively progresses in quadrant III, the tangent builds up positively again in quadrant I and gives h_1 at the angle of $\frac{5\pi}{4}$ or 225°. The tangent increases until an angle of $\frac{3\pi}{2}$ or 270° is reached, leading to an infinite value. Beyond this point, the tangent is negative again and as the angle opens to $\frac{7\pi}{4}$ or 315°, the tangent returns a height h_2 located in quadrant IV.

Accounting for the Quadrant

To illustrate the ambiguity with a more practical case, we can feed our calculator with the angles values we used in degrees and compute the tangent height h. This is what we will get:

$\tan(45°) = 1$

$\tan(135°) = -1$

$\tan(225°) = 1$

$\tan(315°) = -1$

For two different angles, 45°/225° or 135°/315°, distant from 180°, the results returned by the tangent function are similar. Since the arctangent is the reciprocal of the tangent and returns an angle, we have an ambiguity: is $\tan(\alpha) = 1$ coming from an angle of 45° or 225°? Arctangent will always return an angle comprised between –90° and 90°. If your complex number or vector lies in quadrants I or IV, the arctangent will return the correct value. If the vector lies in quadrant II or III, there is a problem and arctangent will return the wrong angle. Mathematically, we say that the tangent function is not bijective: for a given tangent value, two

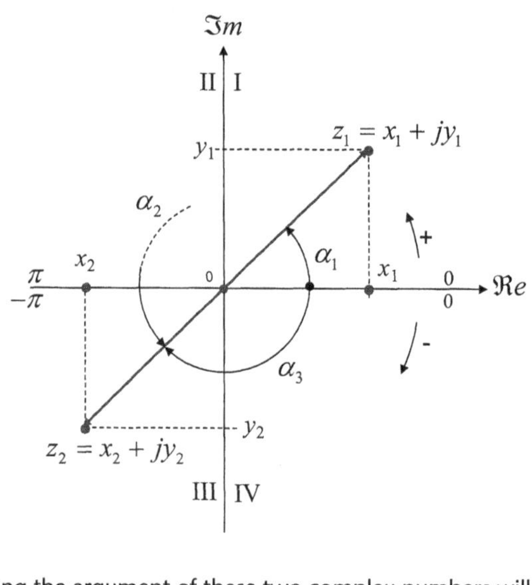

Figure 4.83 Computing the argument of these two complex numbers will return a similar angle if no precautions are taken.

possible angles exist. A simple illustration appears in Figure 4.83. If you compute the argument of these complex numbers where $x_1 = -x_2 = 1$ and $y_1 = -y_2 = 1$, (4.213) will incorrectly return the same angle for both vectors while they are obviously different:

$$\arg(z_1) = \tan^{-1}\left(\frac{1}{1}\right) = 45° \quad (4.216)$$

$$\arg(z_2) = \tan^{-1}\left(\frac{-1}{-1}\right) = 45° \quad (4.217)$$

This can also be easily verified as the periodicity (or the modulo) of the tangent is π. We can add π to any angle, as it won't affect its tangent calculation:

$$\tan(\alpha + \pi) = \frac{\sin(\alpha + \pi)}{\cos(\alpha + \pi)} = \frac{-\sin\alpha}{-\cos\alpha} = \tan\alpha \quad (4.218)$$

In Figure 4.83, we have considered angles belonging to the closed-open interval $[0°,360°)$ or $[0,2\pi)$. These are the angles α_1 and α_2 in Figure 4.83. In the field of complex numbers, however, the so-called *principal argument* of a complex number is defined in the open-closed interval $(-\pi, \pi]$ or $(-180°,180°]$. Therefore, we will consider positive arguments from 0 to π in quadrants I and II, $y > 0$, while turning counter-clockwise. For the opposite, we will consider negative arguments from 0 to less than $-\pi$ in quadrants III and IV, $y < 0$, while turning clockwise. Thus, in Figure 4.83, assuming $\alpha_1 = 45°$, the correct answer for the argument of z_2 is found by subtracting π (we turn clockwise) to α_1: $\alpha_3 = \alpha_1 - 180° = -135°$.

The reason why 360° or 2π is excluded from the angle definition range comes from the fact that 0° and 360° refer to identical positions on the circle. So when an angle opens beyond 359.99°, the next position is ... 0°. The same applies to the range $(-\pi, \pi]$ or $(-180°,180°]$ as $-\pi$ and π (or −180° and 180°) refer to a similar position on the circle. Applying this convention, if y_2 is 0 in Figure 4.83, the magnitude of z_2 becomes x_2 (−1, a negative real number) and its argument is π or 180°.

This technique will be widely used when calculating the argument of inverting compensators. It is important that you feel comfortable with it. Assume you have a zero and an origin pole wired in an inverting configuration. The transfer function is (please note the negative sign):

$$G(s) = -\frac{\omega_{po}}{\omega_z}\left(1 + \frac{\omega_z}{s}\right) \quad (4.219)$$

The argument of the compensator is

$$\arg G(j\omega) = \arg\left(-1 - \frac{\omega_z}{j\omega}\right) = \arg\left(-1 + j\frac{\omega_z}{\omega}\right) = \tan^{-1}\left(-\frac{\omega_z}{\omega}\right) = -\tan^{-1}\left(\frac{\omega_z}{\omega}\right) \quad (4.220)$$

As you can see, the complex number lies in quadrant II ($x = -1$); therefore, the correct answer is obtained by adding π to (4.220):

$$\arg G(\omega) = \pi - \tan^{-1}\left(\frac{\omega_z}{\omega}\right) \tag{4.221}$$

This formula can be further tweaked. We need to use a formula also popular in this book:

$$\tan^{-1}(x) + \tan^{-1}\left(\frac{1}{x}\right) = \frac{\pi}{2} \tag{4.222}$$

From which we can state

$$\tan^{-1}\left(\frac{1}{x}\right) = \frac{\pi}{2} - \tan^{-1}(x) \tag{4.223}$$

Substituting (4.223) in (4.221), we have

$$\arg G(\omega) = \pi - \tan^{-1}\left(\frac{\omega_z}{\omega}\right) = \pi - \left[\frac{\pi}{2} - \tan^{-1}\left(\frac{\omega}{\omega_z}\right)\right] = \frac{\pi}{2} + \tan^{-1}\left(\frac{\omega}{\omega_z}\right) \tag{4.224}$$

At dc, the argument is $\frac{\pi}{2}$ or −270°, and, for high frequency, the argument is π or −180°. A similar result is obtained from (4.221).

Improving the Arctangent Function

To return the right angle, solvers and simulators must know the quadrant in which the evaluated complex number resides. One known function is *atan*2 and is implemented in Mathcad®. The function observes the signs of y and x and, depending on the quadrant, applies the following scaling factors as detailed in [1]:

$$\operatorname{atan2}(x,y) = \begin{cases} \tan^{-1}\left(\dfrac{x}{y}\right) & x > 0 \\ \pi + \tan^{-1}\left(\dfrac{x}{y}\right) & y \geq 0, x < 0 \\ -\pi + \tan^{-1}\left(\dfrac{x}{y}\right) & y < 0, x < 0 \\ \dfrac{\pi}{2} & y > 0, x = 0 \\ -\dfrac{\pi}{2} & y < 0, x = 0 \\ \text{undefined} & y = 0, x = 0 \end{cases} \tag{4.225}$$

The *atan*2 function will return angles moving from π to $-\pi$ with a discontinuity as the angles approach −180° or $-\pi$. When $x = 0$, where the tangent has an infinite positive or negative height, the function properly returns the angle of $\pm\frac{\pi}{2}$. A possible implementation of *atan*2 is proposed by [1] and gives adequate results:

$$\text{atan}\,2(x,y) = 2\tan^{-1}\left(\frac{y}{\sqrt{x^2+y^2}+x}\right) \tag{4.226}$$

To compare these functions, we have plotted *atan* and *atan*2 from the sine and cosine waveforms already used in Figure 4.82. The plots appear in Figure 4.84.

The upper right corner shows the result delivered by *atan* and gives an angle bounded between $-\frac{\pi}{2}$ and $\frac{\pi}{2}$: the result is displayed *modulo* π or with a periodicity of π as also shown in Figure 4.82.

The lower-right corner shows the results computed by the *atan*2 function. As expected, the phase nicely increases to 180° and suddenly jumps to −179° (−180 or $-\pi$ is excluded). This time, the result is displayed *modulo* 360 or 2π. Some simulators display the phase in this mode. However, in some cases, you would prefer to map these results in the segment $[0,360)$ or $[0,2\pi)$ for an easier reading. A possible way to program this in Mathcad® is as follows:

$$360_{\text{map}}(x,y) := \left|\begin{array}{l}\text{atan2}(x,y)+2\pi \text{ if atan2}(x,y)<0\\ \text{atan2}(x,y) \text{ otherwise}\end{array}\right.$$

The result must then be multiplied by $360/2\pi$ for a proper display in degrees.

If you do not want to type these lines, the built-in function *angle* gives similar results as shown in Figure 4.85 in which we used Figure 4.84's sine and cosine functions. The discontinuity has gone, and the graph shows a linearly increasing angle. When you approach 360°, the function jumps back to zero, indicating a complete circle revolution.

Phase Display in a SPICE Simulator

In some simulation packages, the graphical analysis tool does the 0°–360° mapping automatically for you. This is the case for Intusoft's graphical analysis tool Intuscope, where the function *phaseextend* is implemented. This software routine observes the phase discontinuity in the *atan*2 function and applies the strategy of the 360_{map} function. There is a slight difference, though, as it authorizes the phase to extend beyond −360° if necessary. Let us see how this technique applies to different circuits.

As a first example, we have selected a simple op amp-based inverter drawn in Figure 4.86.

The gain of such a simple cell is easily derived, and we find

$$|G(s)| = \left|\frac{V_{out}(s)}{V_{in}(s)}\right| = -\frac{R_f}{R_i} \tag{4.227}$$

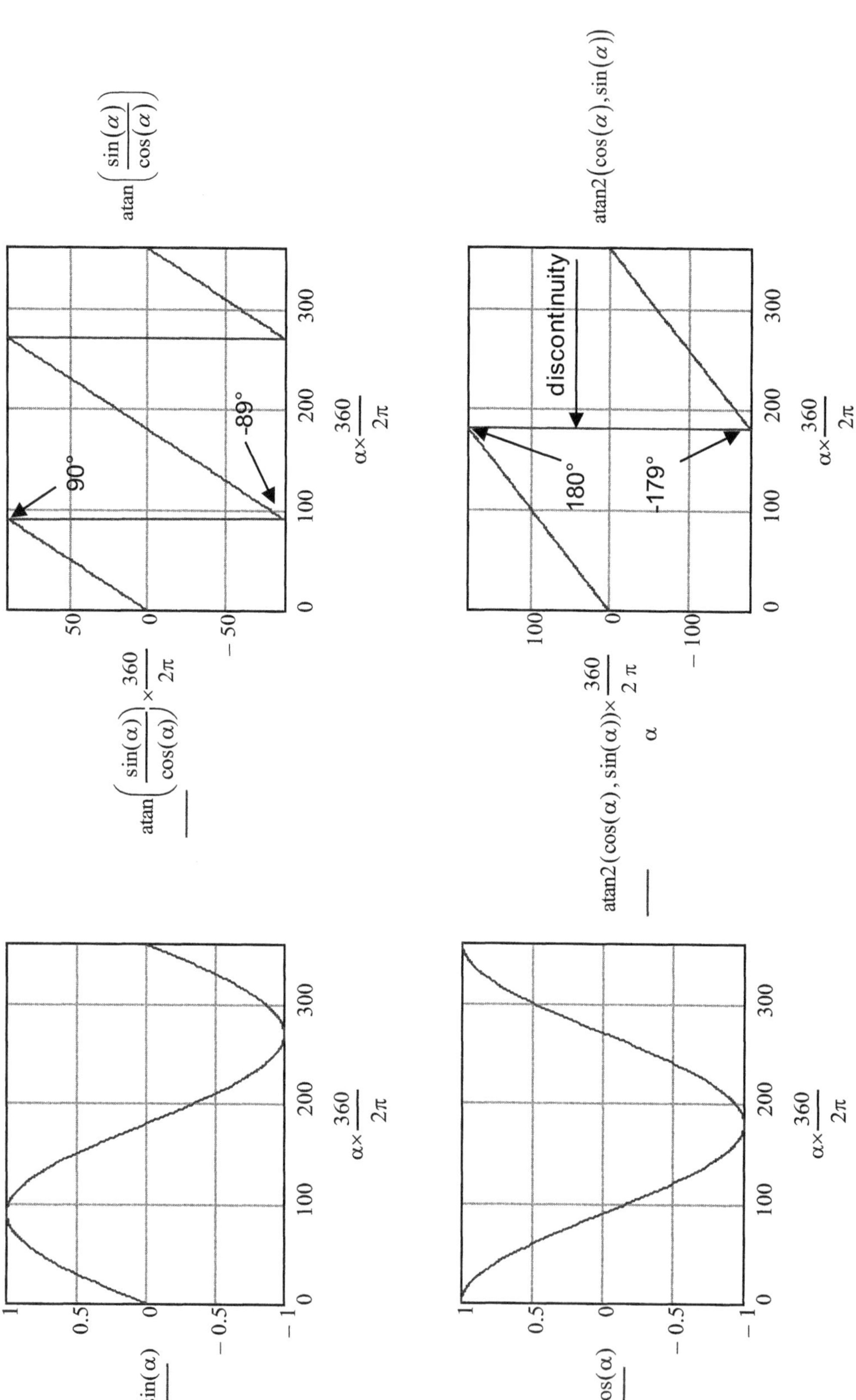

Figure 4.84 If the *atan* expression returns an angle between –90° and 90°, *atan2* accounts for the quadrant and returns an answer corrected as expressed in (4.225).

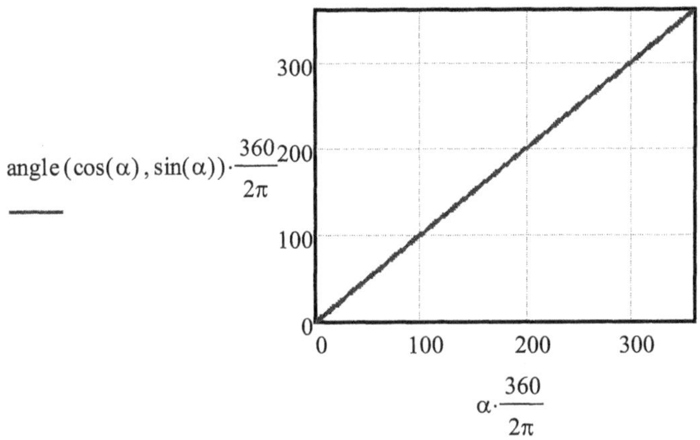

Figure 4.85 The function *angle* automatically maps angles into the $[0,2\pi)$ segment.

The mathematical argument of this transfer function is the argument of a negative real (no imaginary part as when $y_2 = 0$ in Figure 4.83) and equals

$$\arg\left(-\frac{R_f}{R_i}\right) = \pi \tag{4.228}$$

The simulation results for equal resistors values (–1 gain) are given in Figure 4.87. The upper section of the figure confirms the calculation with a displayed 180° phase.

However, at school, we learned that the phase brought by an inverting circuit was –180° and not 180° as expressed by (4.228). Could the simulator be wrong?

Mathematically, talking about an angle of 180° or –180° is the same: it refers to a similar angle in the polar plane, as an angle is always defined $\pm k2\pi$ or $\pm k \times 360°$: should you add or subtract a complete circle revolution to an angle, you return to the same starting point on the circle. If you add –360° to 180°, you obtain –180° as shown in the lower section of Figure 4.87. If this option gives more physical sense to the representation of $G(s)$ in (4.227), both diagrams in Figure 4.87 are, however, mathematically equal.

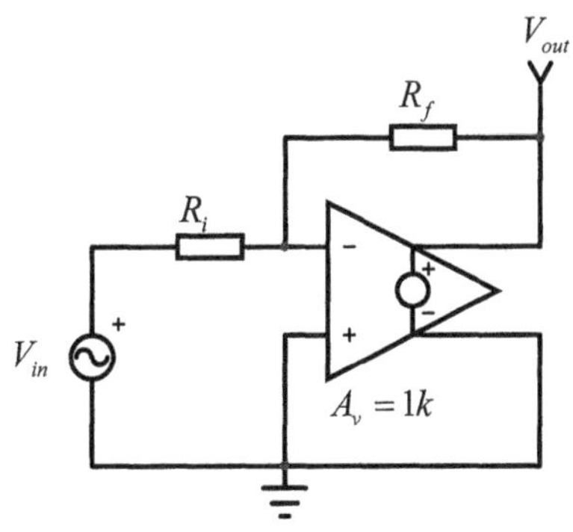

Figure 4.86 A simple inverter made around a voltage-controlled control source.

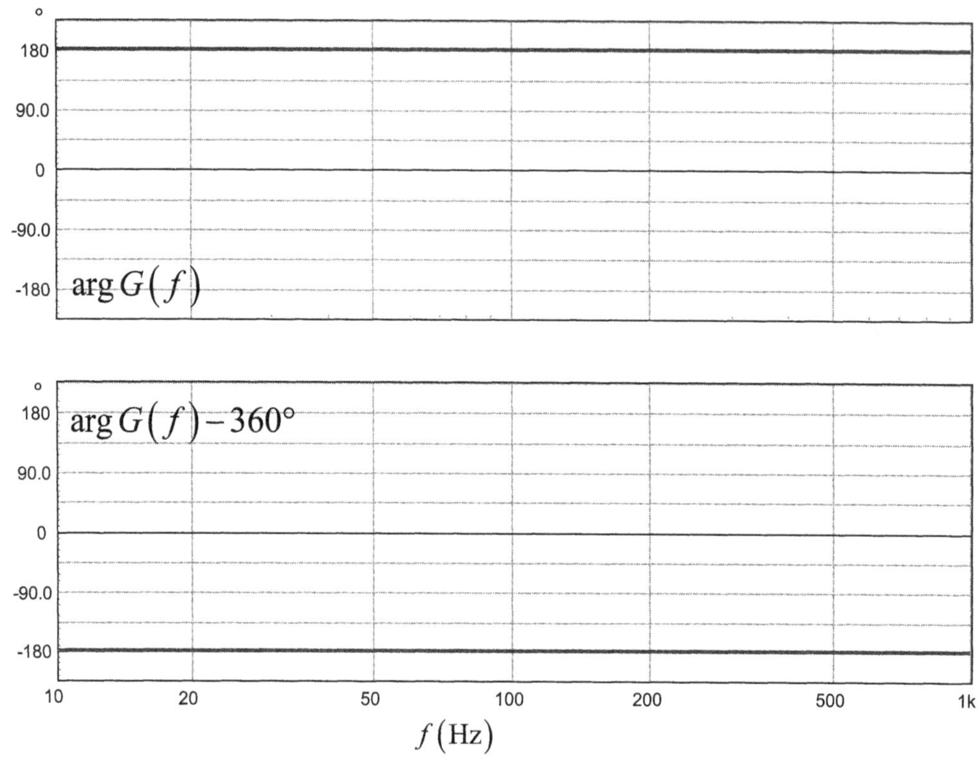

Figure 4.87 The simulator returns the mathematical value of the argument: 180° or π.

In Figure 4.89, we plotted the signals coming out from an inverting integrator. We can say that w_2 lags w_1 by 270°, but if we select a period on the right side of the picture, we can also say that w_2 leads w_1 by 90°! Again, both statements are mathematically equal: $90° = -270° + 360°$. If you plot the phase response of this inverting integrator, what will SPICE display? Well, the inverter argument is 180° according to (4.228) and the argument of the origin pole is –90°. Summing these two values will give a displayed phase of 90°, equivalent to an argument of –270°, as confirmed by Figure 4.89. We will use the second notation in our phase calculations for the sake of compliance with most of the theory books. The negative phase notation emphasizes, after all, the fact that the time-domain response of the considered network is always delayed compared to the excitation signal.

Most of simulation packages use the function *atan*2 to display an argument signed between $-\pi$ (excluded) and π. Some graphical investigation tools, such as Intuscope, enhance the *atan*2 function with a software routine known as *phaseextend* that computes the phase down to –360° and beyond. It works more or less like the 360_{map} routine we detailed. When you start an ac analysis, the simulation engine calculates the argument of the transfer function from the very first frequency points. The returned argument is that of the function *atan*2, signed between $-\pi$ (excluded) and π. As the ac sweep proceeds, if the function senses an argument discontinuity between adjacent simulation points—for instance, when a jump occurs from –179° to 180° (or vice versa)—*phaseextend* does its job and smoothes the discontinuity,

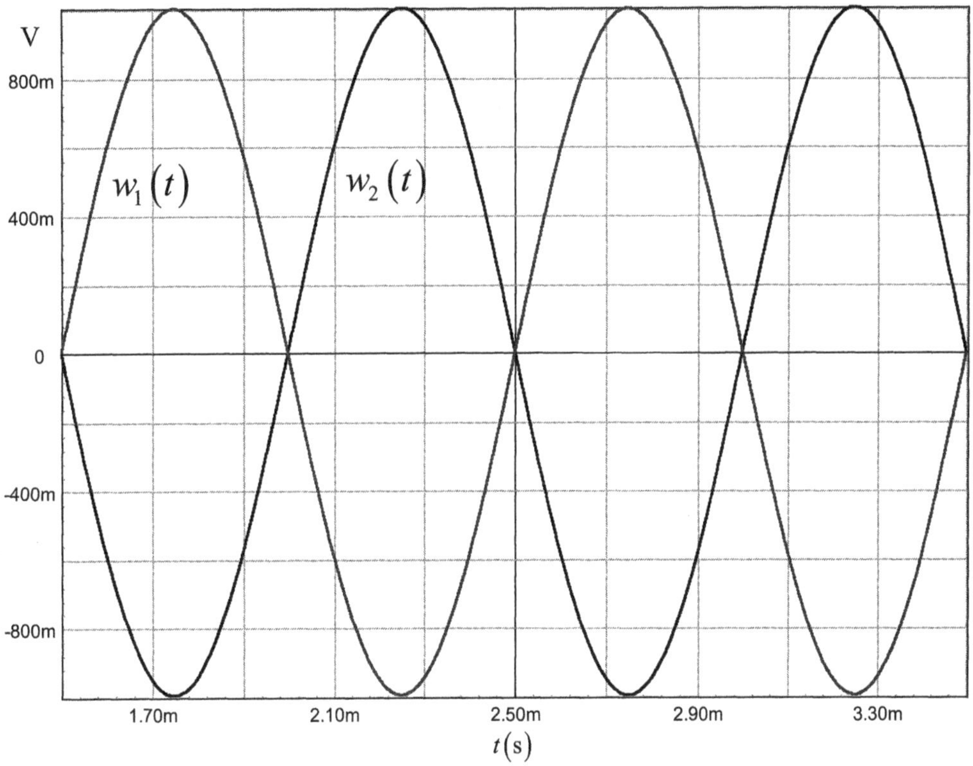

Figure 4.88 These two signals are out of phase. However, can you say that waveform 1 precedes waveform 2 by 180° or waveform 2 lags waveform 1 by 180°?

returning an argument down to –360°. Let us check this theory on the Figure 4.90 third-order RC network on which we launched an ac sweep.

From the simulation data, we can use the Intuscope built-in calculation engine and apply macros to the real and imaginary portions of the V_{out} vector. This is what we did to compare the results returned by the classical arctangent function and the function described by (4.226). The results are plotted in Figure 4.91. In the lower section, the *atan* function returned results signed between $-\frac{\pi}{2}$ and $\frac{\pi}{2}$, as expected. In the middle section, we implemented (4.226), and it nicely returned an argument signed between $-\pi$ (excluded) and π. You can observe that the phase toggling point in the returned value occurred a little before 400 Hz. Finally, in the upper section of the figure, the phase lags from 0° down to 270° (we have three poles) using the *phaseextend* function that efficiently smoothes the reported discontinuity around 400 Hz. Please note that the phase starts from 0 since at 1 Hz, where the ac sweep started, the returned argument was 0.

In a new simulation, we did not start the ac sweep from 1 Hz, but from 400 Hz. At this frequency, the simulation engine evaluated the argument returned by *atan*2 to be already beyond 180°. This explains why Intuscope now displays the phase from 180° rather than 0° as in the previous case. This is what you can see in Figure 4.92, and it is correct.

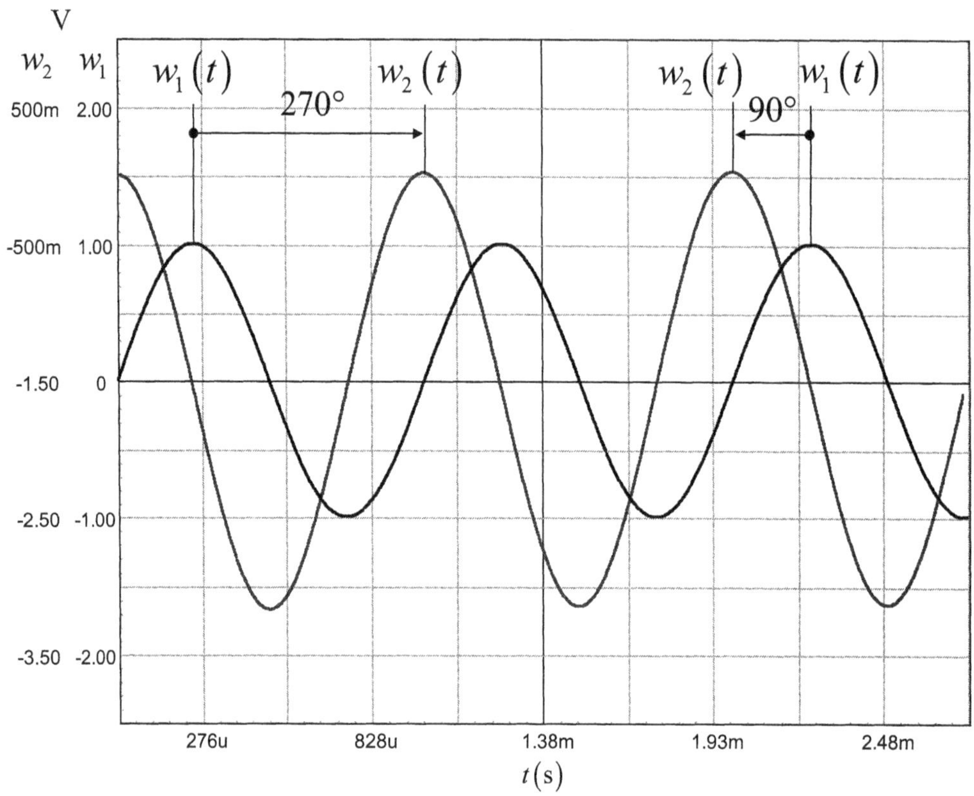

Figure 4.89 In this picture, you can say w_2 leads w_1 by 90° or w_2 lags w_1 by 270°; it is a similar statement!

Conclusion

An engineer who is not familiar with the intricacies of circuit simulation can often be puzzled by phase or argument evaluations. We have seen that engineering judgment is necessary to interpret the results displayed by the graphical viewer. The change in the start frequency is a typical example where attention is required to understand the answer.

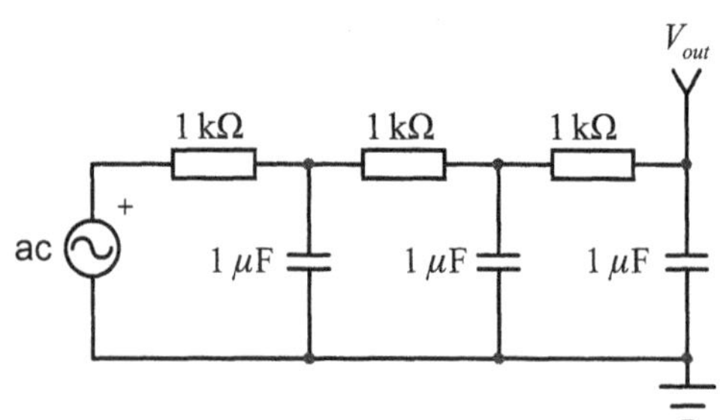

Figure 4.90 This third-order *RC* network has a 0 argument at dc and its phase smoothly shifts to –270°.

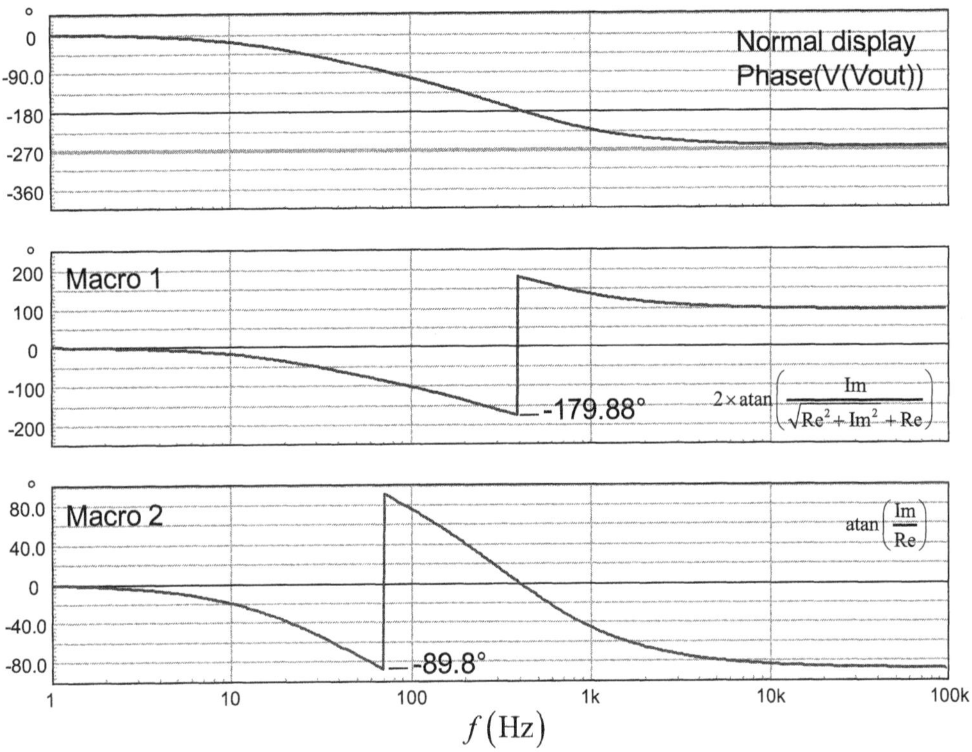

Figure 4.91 The *phaseextend* function built in Intuscope displays the phase down to –270°. Using macros, we can nicely reproduce the results predicted in Figure 4.84.

Reference

[1] Wikipedia contributors, "atan2," http://en.wikipedia.org/wiki/Atan2, last accessed June 3, 2012.

Appendix 4D: Impact of Open-Loop Gain and Origin Pole on Op Amp-Based Transfer Functions

An operational amplifier is often considered a perfect device, featuring an infinite open-loop gain and infinite bandwidth. In reality, the open-loop gain is finite, and designers place a pole at low frequency for stability purposes. The typical ac response of such op amp, a μA741, for instance, appears in Figure 4.93 and shows a typical open-loop gain A_{OL} of 106 dB together with a low-frequency pole f_p placed below 10 Hz.

When we consider an op amp to build a compensator or a simple amplifier, we usually do not consider the origin pole and the open-loop gain, assuming they can be neglected. However, it is always interesting to understand how these parameters can affect the final performance, especially if they are moving from lot to lot in production or if they change with temperature. Data-sheet indications showing the wide spread of a parameter should always trigger your interest in investigating

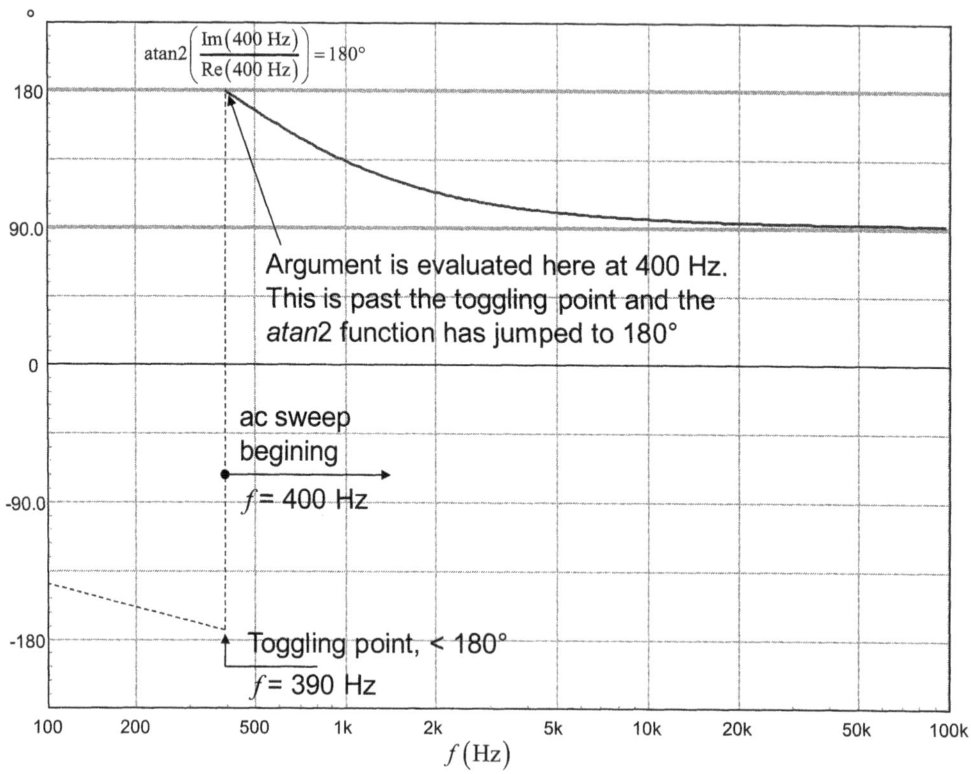

Figure 4.92 The displayed phase starts from 180° because the simulator evaluates the argument at a frequency point (400 Hz), where the *atan2* function has already jumped to 180°.

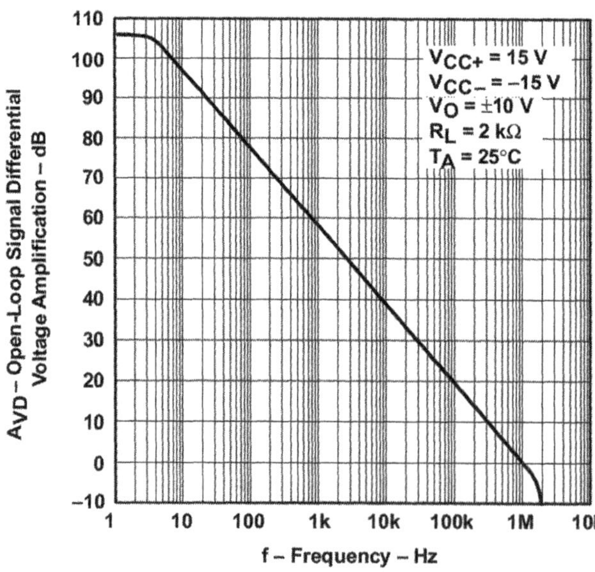

Figure 4.93 The open-loop gain is around 106 dB typically for this μA741 from TI, and the low frequency pole appears below 10 Hz.

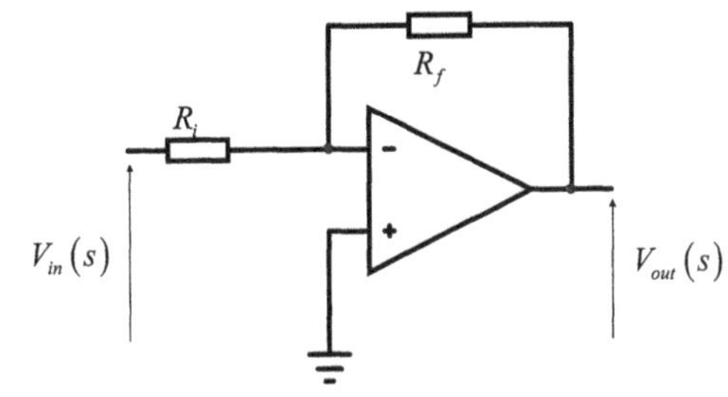

Figure 4.94 A simple inverter built around an op amp.

the consequences of its variations upon the final design performance. For instance, the μA741 gain is typically indicated to be 200k (106 dB) but can potentially go down to 20k (86 dB) in some cases. To understand how this variation will affect the design, you must run the analytical analysis and assess the final impact. From the derived equations, you will, as design engineer, decide to either include and compensate for the effects or simply ignore them.

Let us consider the circuit proposed in Figure 4.94. You recognize a simple inverter whose gain depends on R_f and R_i. We know that the gain of such a circuit is $-R_f/R_i$, considering an infinite open-loop gain and no low-frequency pole. To see the effects of these elements, we need to make them appear in the op amp equivalent circuit. A simplified representation including these new comers appears in Figure 4.95.

To calculate the transfer function of such an arrangement, we need a few equations. Let's start with the output voltage, V_{out}:

$$V_{out}(s) = A_{OL} \cdot \varepsilon(s) \frac{1}{1 + \frac{s}{\omega_p}} \tag{4.229}$$

The error voltage, ε, is equal to the difference between the negative and positive inputs:

$$\varepsilon = V_{(+)} - V_{(-)} \tag{4.230}$$

As the positive input is grounded, let's apply the superposition theorem to this linear network and derive the negative input node voltage. The negative input voltage

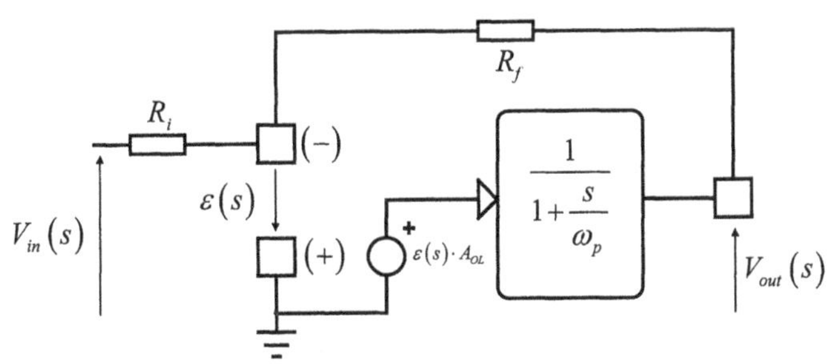

Figure 4.95 When including the open-loop gain and the origin pole, the circuit slightly complicates.

is the sum of the voltage seen at this pin when V_{out} is grounded plus the voltage seen at the same pin when V_{in} is grounded. Therefore, we have

$$\varepsilon(s) = -\left[V_{out}(s)\frac{R_i}{R_i + R_f} + V_{in}(s)\frac{R_f}{R_i + R_f}\right] \tag{4.231}$$

If we substitute this value into (4.229), re-arrange the expression, and extract the ratio V_{out}/V_{in}, we obtain the following equation:

$$G(s) = \frac{V_{out}(s)}{V_{in}(s)} = -\frac{R_f}{\frac{R_f + R_i}{A_{OL}}\left(1 + \frac{s}{\omega_p}\right) + R_i} \tag{4.232}$$

This transfer expression does not really conform to the format introduced in Chapter 2. To gain insight into the function, it must be rearranged to fit the following equation:

$$G(s) = G_0 \frac{1}{1 + \frac{s}{\omega_{peq}}} \tag{4.233}$$

where G_0 is the low-frequency or dc gain and ω_{peq} the new pole brought by the structure.

Let us develop and rearrange the (4.232) denominator:

$$D(s) = R_i + \frac{R_f + R_i}{A_{OL}} + s\left(\frac{R_i + R_f}{A_{OL}\omega_p}\right) \tag{4.234}$$

From this expression, we can factor the left term $R_i + \frac{R_f + R_i}{A_{OL}}$:

$$D(s) = \left(R_i + \frac{R_f + R_i}{A_{OL}}\right)\left[1 + s\left(\frac{R_i + R_f}{A_{OL}\omega_p}\right)\frac{1}{\left(R_i + \frac{R_f + R_i}{A_{OL}}\right)}\right] \tag{4.235}$$

If we develop and factor the right term, we have

$$D(s) = \left(R_i + \frac{R_f + R_i}{A_{OL}}\right)\left[1 + s\frac{R_i + R_f}{\omega_p(R_f + R_i + A_{OL}R_i)}\right] \tag{4.236}$$

The complete expression can be rewritten as follows:

$$G(s) = -\frac{R_f}{R_i + \frac{R_f + R_i}{A_{OL}}}\frac{1}{1 + s\frac{R_i + R_f}{\omega_p(R_f + R_i + A_{OL}R_i)}} \tag{4.237}$$

in which

$$G_0 = -\frac{R_f}{R_i + \frac{R_f + R_i}{A_{OL}}} \tag{4.238}$$

and the equivalent pole is

$$\omega_{peq} = \omega_p \frac{R_f + R_i + A_{OL} R_i}{R_i + R_f} \tag{4.239}$$

In these expressions, when the open-loop gain is really high, the inverter gain G_0 is solely set by R_f and R_i and equals $-R_f/R_i$. This is, by the way, the benefit of feedback where, despite variations of the open-loop gain A_{OL}, the transmission gain depends solely on the external elements R_f and R_i. With high open-loop gains, the equivalent pole is relegated to infinity where it can be neglected.

Let's go through a quick example to see if these statements are always valid. Suppose we want to build a 30-kHz bandwidth inverter featuring a gain of 10. For this purpose, we will choose resistors such as R_f = 10 kΩ and R_i = 1 kΩ. With a μA741 dc-gain of 200k and a low-frequency pole located at 10 Hz, we obtain the following values from (4.238) and (4.239):

$$G_0\big|_{A_{OL}=200k} = -9.999 \tag{4.240}$$

$$f_{peq}\big|_{A_{OL}=200k} \approx 182 \text{ kHz} \tag{4.241}$$

From these values, we can see that we will have a flat response up to 182 kHz with a gain of 10, as expected. The ac response appears in Figure 4.96(a), and it is flat at 30 kHz: the buffer works as expected.

If we now use the lower op amp open-loop gain A_{OL} of 20k, the new values for the transmission gain and the pole are the following ones:

$$G_0\big|_{A_{OL}=20k} = -9.995 \tag{4.242}$$

$$f_{peq}\big|_{A_{OL}=20k} \approx 18.2 \text{ kHz} \tag{4.243}$$

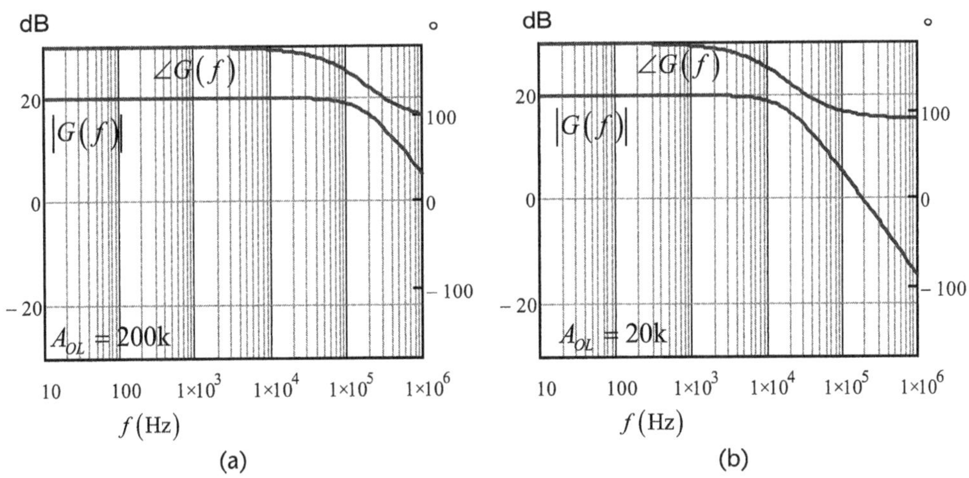

Figure 4.96 When the op amp open-loop gain falls to 20k, the inverter gain is almost unchanged, but the ac response is seriously affected. The pole has shifted below 20 kHz, and the expected flat ac response up to 30 kHz is lost.

The ac response is given in the Figure 4.96(b) where you can see that the buffer ac response now suffers beyond 10 kHz. Your design is no longer able to pass the specification. The phase lag can bring instability if this buffer is included in a compensator chain, for instance.

What can you do then? You can select an op amp with lower open-loop gain variations, but it is not unusual to find ratios of 1 to 3 or even more, as we have seen. You can also choose a high-speed type of op amp. This device features a moderate open-loop gain but the designers pushed the low-frequency pole way up the frequency axis. For instance, National-Semiconductor LM6132BI data sheet says the typical open-loop gain is 15k, which is much lower than the μA741 minimum value. To calculate the position of the low-frequency pole, we can use another parameter given by op amp manufacturers, the gain bandwidth product GBW. It is defined as

$$\text{GBW} = A_{OL} \times f_p \tag{4.244}$$

where A_{OL} is the op amp open-loop gain and f_p is the point where the ac magnitude drops by 3 dB. This number is constant, given the single-pole response of the compensated op amp: when the amplitude drops by 10 (–20 dB), the frequency has increased also by 10 (a decade). For the LM6132BI, the typical GBW product is 10 MHz. From this value, we can extract the position of the low-frequency pole:

$$f_p = \frac{\text{GBW}}{A_{OL}} = \frac{10Meg}{15k} = 667\text{ Hz} \tag{4.245}$$

If we plug the new open-loop gain of 15k and the 667-Hz pole position in (4.239), our inverter circuit with a gain of 10 will have a new pole situated at 910 kHz. When the open-loop gain drops to 6k, as indicated in the data sheet, together with a minimum GBW of 7 MHz (f_p = 1.1 kHz), the new pole will be located at 600 kHz. We are safe with our 30-kHz bandwidth requirements.

The Integrator Case

If you now consider an integrator as drawn in Figure 4.26, the resistance R_f in (4.232) is replaced by the capacitor impedance:

$$Z_f = \frac{1}{sC_1} \tag{4.246}$$

while R_i remains in place. Developing (4.232) accounting for these changes gives

$$G(s) = \frac{\dfrac{1}{sC_1}}{\dfrac{\dfrac{1}{sC_1} + R_1}{A_{OL}}\left(1 + \dfrac{s}{\omega_p}\right) + R_1} = \frac{A_{OL}\omega_p}{s + \omega_p + s^2 R_1 C_1 + sR_1C_1\omega_p + sA_{OL}R_1C_1\omega_p} \tag{4.247}$$

Rearranging this equation by factoring ω_p and considering $\omega_{po} = 1/R_1C_1$ leads to

$$G(s) = \frac{A_{OL}}{1 + s\left(\dfrac{1}{\omega_p} + \dfrac{1}{\omega_{po}} + \dfrac{A_{OL}}{\omega_{po}}\right) + s^2\left(\dfrac{1}{\omega_p \omega_{po}}\right)} \tag{4.248}$$

This is a second-order equation. In the denominator second term, if we consider a large open-loop gain A_{OL}, the expression simplifies to

$$G(s) \approx \frac{A_{OL}}{1+s\left(\frac{A_{OL}}{\omega_{po}}\right)+s^2\left(\frac{1}{\omega_p\omega_{po}}\right)} \tag{4.249}$$

We can rewrite this expression and make it fit the classical second-order polynomial form:

$$G(s) = \frac{1}{1+\frac{s}{\omega_0 Q}+\left(\frac{s}{\omega_0}\right)^2} \tag{4.250}$$

We need to identify the terms ω_0 and Q with (4.249) and (4.250):

$$\omega_0 = \sqrt{\omega_{po}\omega_p} \tag{4.251}$$

Q is obtained by solving the simple equation

$$\frac{A_{OL}}{\omega_{po}} = \frac{1}{Q\omega_0} \tag{4.252}$$

which gives

$$Q = \frac{\omega_{po}}{A_{OL}\omega_0} \tag{4.253}$$

Knowing that the open-loop gain A_{OL} is large, Q is naturally a very small value.

The roots or the poles of the denominator are obtained by using a formula introduced in Chapter 3, (3.35):

$$s_1, s_2 = \frac{\omega_0}{2Q}\left(\pm\sqrt{1-4Q^2}-1\right) \tag{4.254}$$

From (4.253) we extract ω_0:

$$\omega_0 = \frac{\omega_{po}}{A_{OL}Q} \tag{4.255}$$

If we substitute this definition in (4.254), we obtain the first root:

$$s_1 = \frac{\omega_{po}}{A_{OL}}\frac{\left(\sqrt{1-4Q^2}-1\right)}{2Q^2} \tag{4.256}$$

Considering first order terms only, we know that

$$(1+x)^a \approx 1+ax \tag{4.257}$$

(4.256) can be rewritten as

$$s_1 \approx \frac{\omega_{po}}{A_{OL}} \frac{1 - \frac{4Q^2}{2} - 1}{2Q^2} = -\frac{\omega_{po}}{A_{OL}} \tag{4.258}$$

The pole ω_1 is located at

$$\omega_1 \approx \frac{\omega_{po}}{A_{OL}} \tag{4.259}$$

The second pole is calculated with s_2:

$$s_2 = \omega_0 \frac{\left(-\sqrt{1-4Q^2} - 1\right)}{2Q} \tag{4.260}$$

Applying (4.257) and considering Q as a small quantity, we have

$$s_2 \approx -\omega_0 \frac{\left(1 - \frac{4Q^2}{2}\right) - 1}{2Q} = \omega_0 \left(\frac{2Q^2 - 2}{2Q}\right) \approx -\frac{\omega_0}{Q} \tag{4.261}$$

If we substitute in this equation the definitions from (4.251) and (4.253), we have

$$s_2 = -\frac{\sqrt{\omega_{po}\omega_p}}{\frac{\omega_{po}}{A_{OL}\sqrt{\omega_{po}\omega_p}}} \tag{4.262}$$

Extracting the magnitude of this root leads to the second pole definition:

$$\omega_2 \approx \omega_p A_{OL} \tag{4.263}$$

ω_1 is nothing but the 0-dB crossover pole divided by the op amp open-loop gain A_{OL}, whereas the second pole is the gain bandwidth product GBW defined in (4.244). If we neglect the high-frequency pole, the transfer function simplifies to

$$G(s) \approx \frac{A_{OL}}{1 + \frac{s}{\omega_1}} \tag{4.264}$$

Let's assume the following component values were used to build Figure 4.26, which now includes the op amp open-loop gain and a low-frequency pole located at 30 Hz:

$R_1 = 10\ \text{k}\Omega$

$C_1 = 0.1\ \mu\text{F}$

$A_{OL} = 10000$

$f_p = 30\ \text{Hz}$

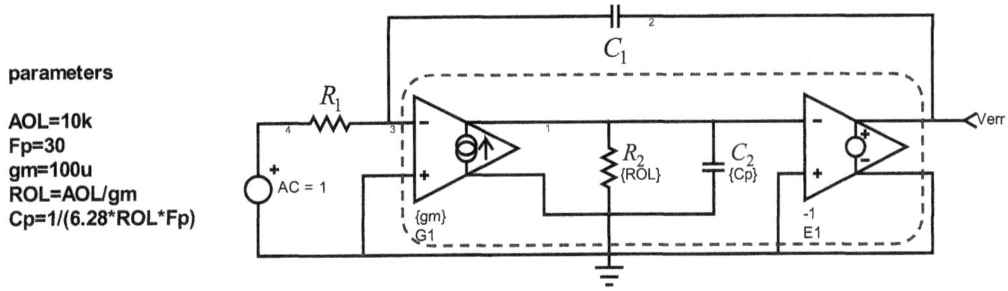

Figure 4.97 A simple setup will tell us if our equations are correct.

If we plot the ac response from such an integrator, then from (4.259) and (4.263) we should see inflexion points at the following frequencies:

$$f_{po} = \frac{1}{2\pi R_1 C_1} = \frac{1}{6.28\times 10k\times 0.1u} = 159\text{ Hz} \quad (4.265)$$

$$f_1 = \frac{\omega_{po}}{2\pi A_{OL}} = \frac{1}{2\pi R_1 C_1 A_{OL}} \approx 16\text{ mHz} \quad (4.266)$$

$$f_2 = f_p A_{OL} = 30\times 10000 = 300\text{ kHz} \quad (4.267)$$

To verify these numbers, we have built a simple ac setup as shown in Figure 4.97. The first stage is a voltage-controlled current source affected by a transconductance g_m of 100 μS. Multiplied by the resistor R_2, it gives the open-loop gain of 10,000. The low-frequency pole is contributed by C_2 and set to 30 Hz.

The input source performs an ac sweep, and the results appear in Figure 4.98. The poles are exactly at the predicted positions.

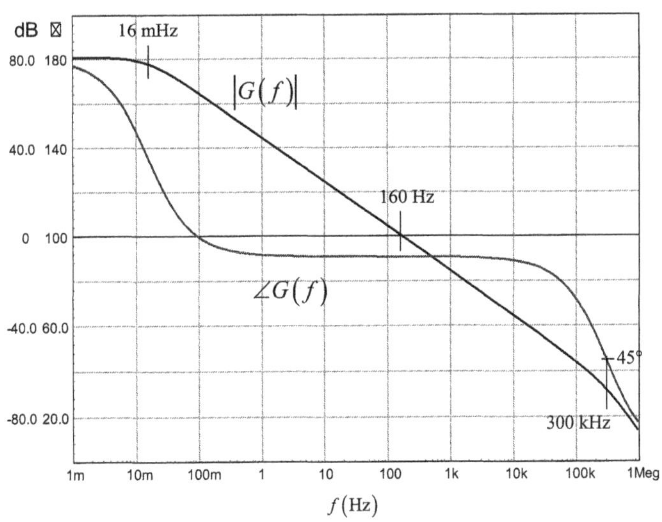

Figure 4.98 This plots confirms that the op amp-based integrator featuring a low-frequency pole and a finite open-loop gain is a second-order system.

This appendix shows that you must look at the ac op amp characteristics if you plan to design high-speed power converters where a high crossover frequency is needed. In this case, you will need to consider the combination open-loop gain/low-frequency pole and check whether or not they impact your design. For that purpose, a dispersion parameter can be affected to the op amp open-loop gain, especially if you run a Monte Carlo analysis in SPICE.

Appendix 4E: Summary of Compensator Configurations

Table 4E.1 gives the correspondence between P, I, and D individual blocks with op amp-based compensators.

Table 4E.1 Summary of Compensator Configurations and Their Transfer Functions

Action Mode	Basic Element	Transfer Function	Implementation	Bode Plot $\lvert G(s)\rvert$	Type
Proportional	P	$G(s)=-k_p$	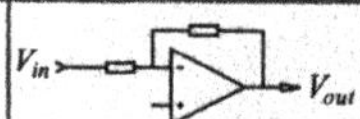	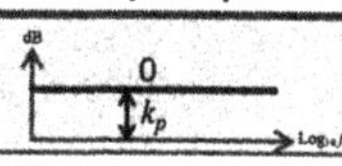	
Integral	I	$G(s)=-k_i\frac{1}{s}$	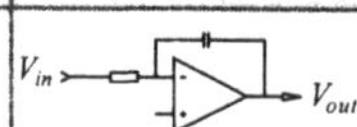	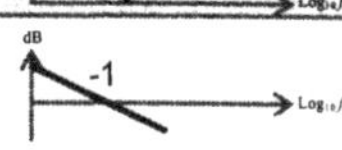	1
Derivative	D	$G(s)=-sk_d$	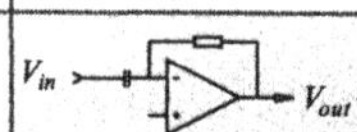	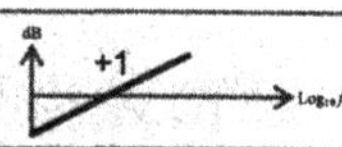	
Proportional Integral	PI	$G(s)=-G_0\frac{1+s/\omega_{z_1}}{s}$	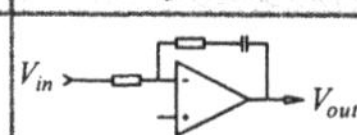	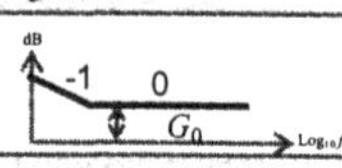	2a
Proportional Integral + 1st-order lag	PI_1	$G(s)=-G_0\frac{1+s/\omega_{z_1}}{s}\frac{1}{1+s/\omega_{p_1}}$	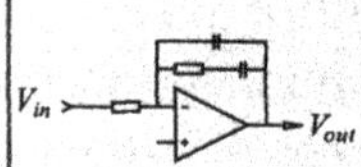	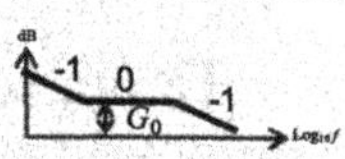	2
Proportional Integral Derivative	PID	$G(s)=-G_0\frac{(1+s/\omega_{z_1})(1+s/\omega_{z_2})}{s}$	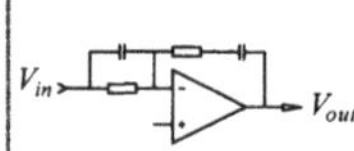	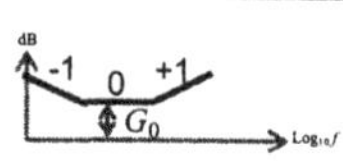	3a
Proportional Integral Derivative + 2nd order lag	PID_2	$G(s)=-G_0\frac{(1+s/\omega_{z_1})(1+s/\omega_{z_2})}{s(1+s/\omega_{p_1})(1+s/\omega_{p_2})}$	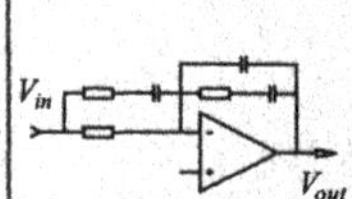	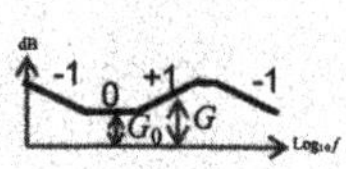	3

CHAPTER 5

Operational Amplifiers-Based Compensators

The operational amplifier, commonly abbreviated as *op amp*, represents the basis of the closed loop system (see [1, 2]). Its function, in a feedback system, is to amplify the error detected between a fixed and stable reference level (e.g., a current or a voltage) and the monitored state variable (e.g., an output voltage or a current). Besides its monitoring purposes, the op amp lends itself very well to building the compensation stage. Its role is to shape the loop gain transfer function $T(s)$ of the system we need to stabilize by compensating the plant deficiencies such as lack of gain, phase lag, and resonating peaks. By wiring passive components such as capacitors and resistors across the op amp, we will build a transfer function $G(s)$, which, once inserted in the return path, will adjust the loop gain crossover frequency to the targeted value and provide the necessary phase and gain margins [3]. In this section, we will cover the previously introduced three basic types, 1, 2, and 3, plus some extra configurations where an isolated feedback is necessary—in offline power converters, for instance. Figure 5.1 shows the typical nonisolated configuration on the left, whereas the isolated version appears on the right side.

Nonisolated structures are often found in so-called point of load (POL) regulators where a voltage is locally created by decreasing or increasing an available dc rail (12 V to 5 V, for instance). As source and load share a common ground, the converter is said to be nonisolated. If you now plan to power a DVD player or simply a cell phone from a mains outlet, unless you want to reiterate Luigi Galvani experiments and risk a lethal electric shock, it is strongly advised to isolate the output ground (touched by the end user) from the live wire. The high-frequency power transformer performs this isolation task by having primary and secondary sides physically separated. However, the output monitoring data must be returned to the control section—the pulse width modulator—usually placed on the nonisolated side. As shown in the right side of Figure 5.1, an optocoupler is used in the wide majority of cases to return this output voltage error information. Other means to convey the information exist, such as high-frequency low-power transformers or even piezoelectric devices. They will not be described here. As the optocoupler is one of the most important elements in an isolated power supply, an appendix is entirely dedicated to this device in the end of this chapter.

Let's start with the simplest compensating structure, type 1.

5.1 Type 1: An Origin Pole

Type 1 is the typical compensator found in circuits where no phase boost at the selected crossover frequency is necessary. Also called an integrator, it includes a single

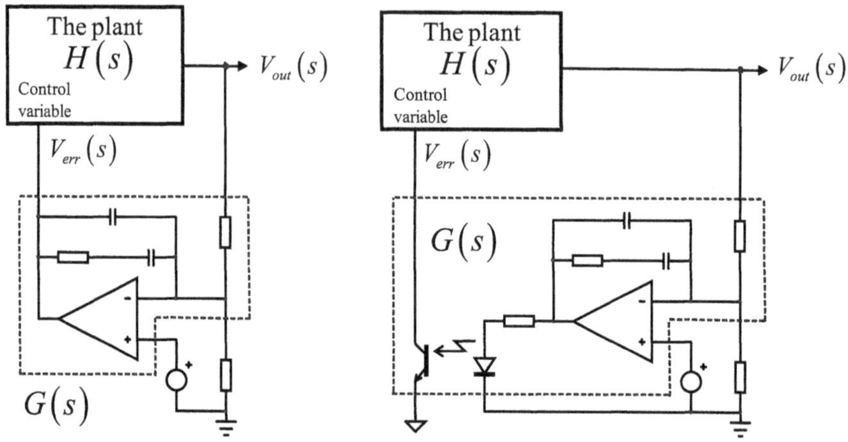

Figure 5.1 The basic op amp configuration and its isolated counterpart.

pole at the origin and offers a permanent phase lag of 270° or a phase lead of 90°. The simplest form of the compensator appears in Figure 5.2.

Throughout the text, the transfer function of the op amp-based circuit $G(s)$ is defined by

$$G(s) = \frac{V_{err}(s)}{V_{out}(s)} \tag{5.1}$$

In this type 1 configuration, you derive G by dividing the capacitor impedance by the upper resistor R_1:

$$G(s) = \frac{V_{err}(s)}{V_{out}(s)} = -\frac{Z_f}{Z_i} = -\frac{\frac{1}{sC_1}}{R_1} = -\frac{1}{sC_1R_1} \tag{5.2}$$

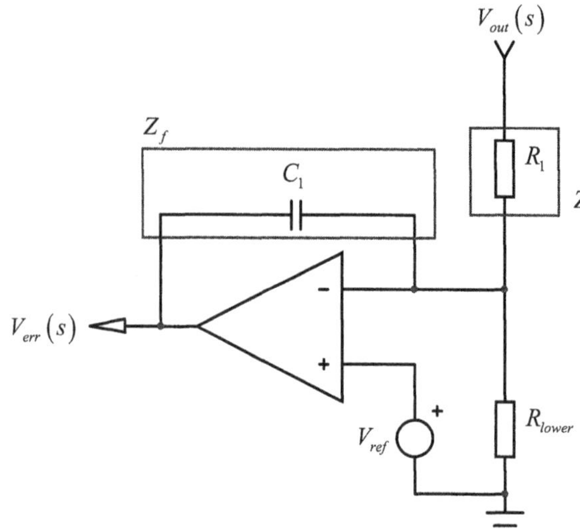

Figure 5.2 A capacitor C_1 connected across the output and the inverting input of the op amp creates an origin pole together with the upper resistor R_1.

In which we unveil the so-called 0-dB crossover pole angular frequency:

$$\omega_{po} = \frac{1}{R_1 C_1} \tag{5.3}$$

As we explained before, (5.2) features an origin pole for $s = 0$. However, the time constant of (5.3) can be seen as a gain, pushing up or down the −1 slope. When the frequency reaches the value defined by (5.3), the gain of the type 1 compensator is 1 or 0 dB—hence its name, the 0-dB crossover pole.

By reworking (5.2) with (5.3), we obtain the following form:

$$G(s) = -\frac{1}{\dfrac{s}{\omega_{po}}} \tag{5.4}$$

The magnitude of (5.4) is given by

$$|G(j\omega)| = \left| j\frac{\omega_{po}}{\omega} \right| = \sqrt{\left(\frac{\omega_{po}}{\omega}\right)^2} = \frac{\omega_{po}}{\omega} \tag{5.5}$$

Whereas the argument of this simple integrator is obtained via

$$\arg G(j\omega) = \arg\left(0 + j\frac{\omega_o}{\omega}\right) = \tan^{-1}(\infty) = \frac{\pi}{2} \tag{5.6}$$

also equal to $\frac{\pi}{2} - 2\pi = -\frac{3\pi}{2}$ or $-270°$.

According to (5.5), to adjust the gain or the attenuation we want at a certain frequency, our crossover choice, for instance, we just need to calculate where to position the 0-dB crossover pole f_{po}.

5.1.1 A Design Example

Let us assume we want to compensate a gain deficiency G_{f_c} of −20 dB at a 1-kHz frequency. To do that, we will define the position at which the 0-dB crossover pole must be located so that the magnitude of $G(s)$ is exactly 20 dB when the frequency reaches 1 kHz. A 20-dB gain translates to

$$G = 10^{\frac{-G_{f_c}}{20}} = 10 \tag{5.7}$$

According to (5.5), to attenuate the input signal by 20 dB at 1 kHz, we must place the 0-dB crossover pole at

$$f_{po} = f_c G = 10 f_c = 10\text{ kHz} \tag{5.8}$$

Assuming the upper resistor R_1 is 10 kΩ, then capacitor C_1 can be easily calculated according to (5.3):

$$C_1 = \frac{1}{2\pi R_1 f_{po}} = \frac{1}{6.28 \times 10k \times 10k} = 1.6\text{ nF} \tag{5.9}$$

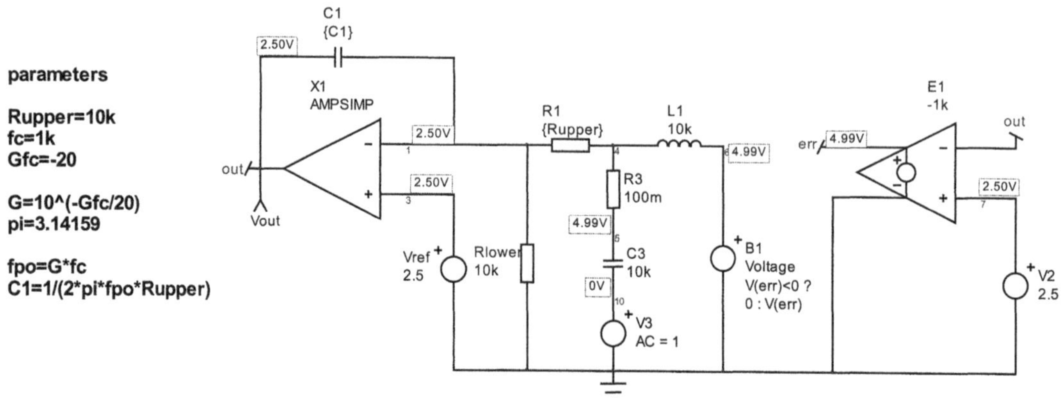

Figure 5.3 The automated simulation of the type 1 example.

The SPICE simulation of such a compensator appears in Figure 5.3.

As you can see in the schematic, the operating points have been reflected and show that the op amp output is well biased, away from its lower or upper stops. This situation is obtained thanks to the voltage-controlled voltage source E_1, which automatically biases the type 1 network (here a 5-V output) to make the op amp work within its linear region ($V_{err} = 2.5$ V). In case you change the op amp and replace this model by another one, or adopt a different upper resistor value, the bias point will automatically adjust to the value imposed by V_2. No further tweak is necessary. The L_1/C_3 network (also labeled *LoL*/*CoL* in other fixtures) makes this auto-adjustment possible: when SPICE calculates the dc point—the circuit bias point—it shorts the inductors and opens

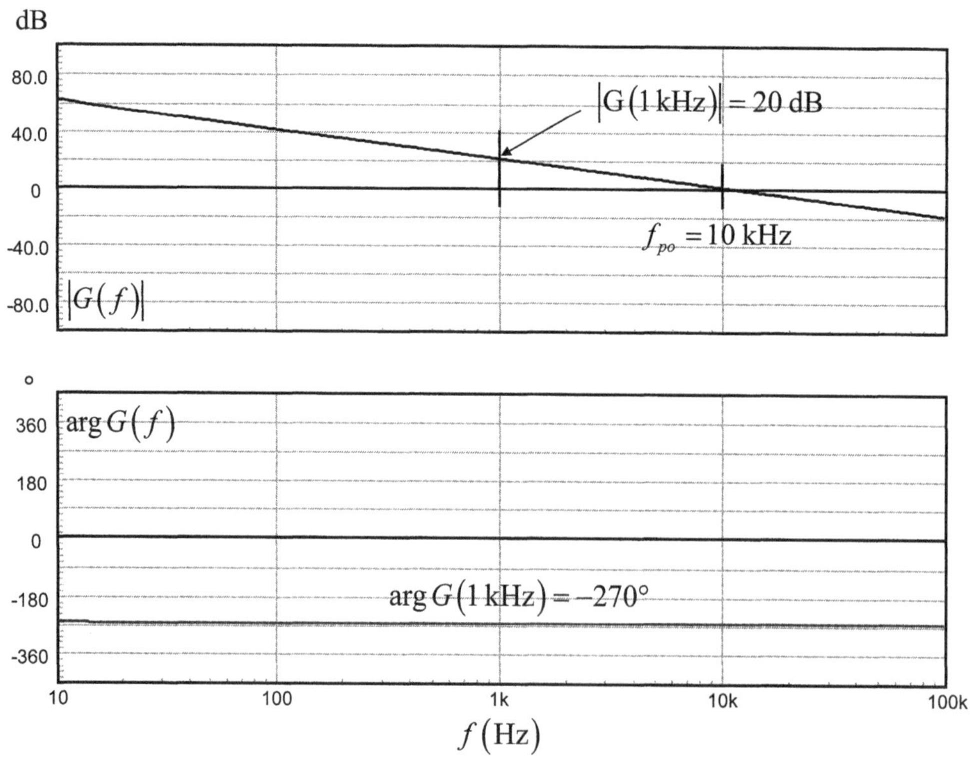

Figure 5.4 the resulting plot shows a +20-dB gain at 1 kHz, as expected.

all the capacitors. As L_1 is a short circuit and C_3 is off, the loop is closed and E_1 adjusts to have the *out* node equal to 2.5 V. Once the ac simulation starts, L_1 being of high value, it blocks the modulation while it easily passes through C_3. This way, any change in the circuit leads to a new bias point calculation every time you start the simulation.

The simulation results appear in Figure 5.4 and confirm the right gain at 1 kHz: +20 dB. The phase is permanently lagging at 270°. Please note that most simulators are displaying the phase with a 2π modulo:

$$\theta = \frac{\pi}{2} \pm k2\pi \tag{5.10}$$

If $k = 0$, then the displayed angle will be $\pi/2$ or 90°. To ease the representation, we purposely added –360° to the phase curve in order to offer a more common value of –270°.

Please see Appendix C in Chapter 4 for explanations on phase display in simulators.

5.2 Type 2: An Origin Pole, plus a Pole/Zero Pair

Offering an origin pole, one zero, and one higher frequency pole, the type 2 compensator provides a phase boost up to 90°. Its electrical configuration appears in Figure 5.5. The transfer function is obtained by calculating the impedance offered by the network placed in the op amp feedback path, Z_f, and dividing it by the upper resistor, R_1:

$$\frac{Z_f}{Z_i} = \frac{\left(\frac{1}{sC_1} + R_2\right)\frac{1}{sC_2} \Big/ \left(\left(\frac{1}{sC_1} + R_2\right) + \frac{1}{sC_2}\right)}{R_1} \tag{5.11}$$

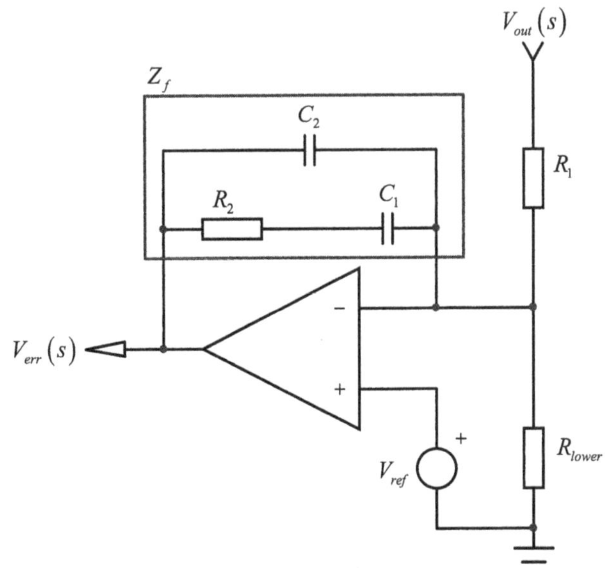

Figure 5.5 By adding an extra pole/zero pair to type 1, type 2 offers a possible phase boost up to 90°.

Developing and rearranging (5.11), we have

$$G(s) = -\frac{Z_f}{Z_i} = -\frac{1 + sR_2C_1}{sR_1(C_1 + C_2)\left(1 + sR_2\left[\frac{C_1C_2}{C_1 + C_2}\right]\right)} \tag{5.12}$$

From there, we can factor sR_2C_1 and have the equation fit a more common format:

$$G(s) = -\frac{R_2}{R_1}\frac{C_1}{C_1 + C_2}\frac{1 + 1/sR_2C_1}{\left(1 + sR_2\left[\frac{C_1C_2}{C_1 + C_2}\right]\right)} = -G_0\frac{1 + \omega_z/s}{1 + s/\omega_p} \tag{5.13}$$

where

$$G_0 = \frac{R_2}{R_1}\frac{C_1}{C_1 + C_2} \tag{5.14}$$

$$\omega_z = \frac{1}{R_2C_1} \tag{5.15}$$

$$\omega_p = \frac{1}{R_2\frac{C_1C_2}{C_1 + C_2}} \tag{5.16}$$

In these definitions, ω_z and ω_p are known because of the phase boost we want for our converter (see Chapter 4 on compensation strategies). To adjust the crossover point, that is to say provide the exact gain or attenuation at the selected crossover frequency f_c, we need to extract R_2 from (5.13). Thanks to its right-side form, we can express the magnitude of (5.13) at the crossover frequency f_c involving the pole and zero, independently from the passive elements that fix them:

$$|G(f_c)| = G_0\frac{\sqrt{1 + \left(\frac{f_z}{f_c}\right)^2}}{\sqrt{1 + \left(\frac{f_c}{f_p}\right)^2}} \tag{5.17}$$

By substituting C_1 and C_2 extracted from (5.15) and (5.16) into (5.17) and then solving for R_2, we obtain the following value:

$$R_2 = \frac{R_1f_pG}{f_p - f_z}\frac{\sqrt{\left(\frac{f_c}{f_p}\right)^2 + 1}}{\sqrt{\left(\frac{f_z}{f_c}\right)^2 + 1}} \tag{5.18}$$

where G represents the needed magnitude at the crossover frequency as defined by (5.7). The rest of the component values come out easily:

$$C_1 = \frac{1}{2\pi R_2 f_z} \tag{5.19}$$

$$C_2 = \frac{C_1}{2\pi f_p C_1 R_2 - 1} \tag{5.20}$$

A possibility exists to simplify these formulas. In a lot of practical cases, the capacitance value of C_2 is much smaller than that of C_1. As a result, if we now consider $C_2 \ll C_1$, the overall transfer function simplifies to

$$G(s) \approx -\frac{R_2}{R_1}\frac{1/sR_2C_1 + 1}{(1 + sR_2C_2)} = -G_0\frac{1+\omega_z/s}{1+s/\omega_p} \tag{5.21}$$

Following the same path as before, we can extract the component values we are looking for:

$$G_0 = \frac{R_2}{R_1} \tag{5.22}$$

$$\omega_p = \frac{1}{R_2C_2} \tag{5.23}$$

$$\omega_z = \frac{1}{R_2C_1} \tag{5.24}$$

$$R_2 = R_1 G \tag{5.25}$$

$$C_2 = \frac{1}{2\pi f_p R_2} \tag{5.26}$$

$$C_1 = \frac{1}{2\pi R_2 f_z} \tag{5.27}$$

The argument of a type 2 configuration where an origin pole and a zero/pole pair cohabit is given by

$$\arg G(\omega) = \arg\left(-1 + j\frac{\omega_z}{\omega}\right) - \arg\left(1 + j\frac{\omega}{\omega_p}\right) \tag{5.28}$$

The numerator complex number lies in quadrant II; we need to add π to correct the angle returned by $\tan^{-1}$ (see Appendix 4C for a refresher on complex numbers):

$$\arg G(f) = \pi - \tan^{-1}\left(\frac{f_z}{f}\right) - \tan^{-1}\left(\frac{f}{f_p}\right) \tag{5.29}$$

In dc, because of the origin pole and the op amp inversion, the phase lags by

$$\lim_{f\to 0} \arg G(f) = \pi - \tan^{-1}(\infty) - \tan^{-1}(0) = \pi - \frac{\pi}{2} = \frac{\pi}{2} \text{ or } -\frac{3\pi}{2} \tag{5.30}$$

As seen in Chapter 4, the phase boost at the crossover frequency f_c is $\arg G(f_c) - \frac{\pi}{2}$. Otherwise expressed, the phase boost of a type 2 compensator is

$$boost = \pi - \tan^{-1}\left(\frac{f_z}{f_c}\right) - \tan^{-1}\left(\frac{f}{f_p}\right) - \frac{\pi}{2} = \left[\frac{\pi}{2} - \tan^{-1}\left(\frac{f_z}{f_c}\right)\right] - \tan^{-1}\left(\frac{f_c}{f_p}\right) \tag{5.31}$$

The term $\frac{\pi}{2} - \tan^{-1}\left(\frac{f_z}{f_c}\right)$ is simply $\tan^{-1}\left(\frac{f_c}{f_z}\right)$ so

$$boost = \tan^{-1}\left(\frac{f_c}{f_z}\right) - \tan^{-1}\left(\frac{f_c}{f_p}\right) \tag{5.32}$$

5.2.1 A Design Example

Let's assume we need to provide a 15-dB gain at a selected 5-kHz frequency together with a phase boost of 50°. Where do we place our pole and zero pair? As we have detailed in the pole-zero section, the pole can be placed at the following location:

$$f_p = \left[\tan(boost) + \sqrt{\tan^2(boost) + 1}\right] f_c = 2.74 \times 5k = 13.7 \text{ kHz} \tag{5.33}$$

As the phase peaks at the geometric mean between the pole and the zero, this latter is placed at

$$f_z = \frac{f_c^{\,2}}{f_p} = \frac{(5k)^2}{13.7k} \approx 1.8 \text{ kHz} \tag{5.34}$$

Applying the design equations (5.18) through (5.20), we can calculate the component values we need, considering the upper resistor R_1 fixed to 10 kΩ in this example:

$$R_2 = \frac{R_1 f_p G}{f_p - f_z} \frac{\sqrt{\left(\frac{f_c}{f_p}\right)^2 + 1}}{\sqrt{\left(\frac{f_z}{f_c}\right)^2 + 1}} = 64.8 \text{ k}\Omega \tag{5.35}$$

$$C_1 = \frac{1}{2\pi R_2 f_z} = 1.3 \text{ nF} \tag{5.36}$$

$$C_2 = \frac{C_1}{2\pi f_p C_1 R_2 - 1} = 206 \text{ pF} \tag{5.37}$$

The SPICE simulation of the type 2 compensator appears in Figure 5.6 and makes use of the automated bias point adjustment previously described. The ac sweep results are given in Figure 5.7 and confirm our calculations.

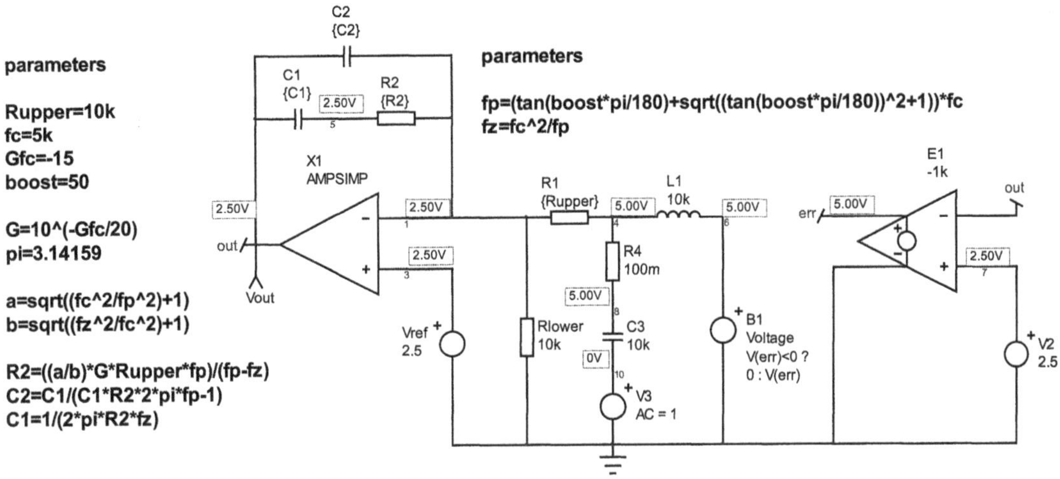

Figure 5.6 In this test fixture, the E_1 source automatically adjusts the bias point of the compensator to place the op amp output within its regulation range (2.5 V). Should you change the divider network, the bias point would automatically be recomputed.

In this particular example, as we deal with a rather high crossover frequency, the simplified formulas given by (5.25) through (5.27) would give almost similar results because C_2 is small compared to C_1. In case the crossover frequency would be much lower, in the vicinity of a few tens of hertz as with a PFC, for instance, then you might need to use the full formulas to obtain the exact gain and phase boost you are looking for.

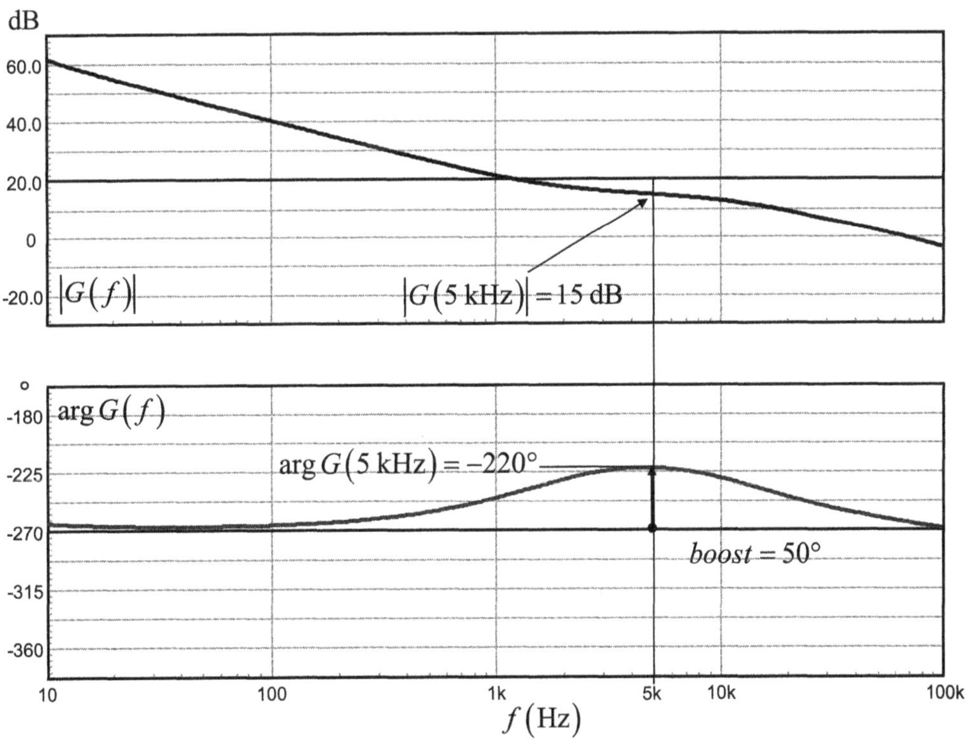

Figure 5.7 The Bode plot confirms the right 50° phase boost and the gain at the selected 5-kHz crossover frequency.

5.3 Type 2a: An Origin Pole plus a Zero

There are applications where the high-frequency pole on the op amp is not necessary. For instance, if you associate the op amp with an optocoupler, you might want to place the pole after the optocoupler to improve the noise immunity of the whole chain. Figure 5.8 shows you the adopted configuration in its simplest form. This is a PI compensator.

Following the previous design derivation method, we quickly obtain the transfer function:

$$G(s) = -\frac{R_2 + \frac{1}{sC_1}}{R_1} = -\frac{1 + sR_2C_1}{sR_1C_1} = -\frac{1 + s/\omega_z}{s/\omega_{po}} \tag{5.38}$$

We can factor s/ω_z and rearrange the equation:

$$G(s) = -\frac{s}{\omega_z}\frac{\omega_{po}}{s}\left(1 + \frac{\omega_z}{s}\right) = -G_0\left(1 + \frac{\omega_z}{s}\right) \tag{5.39}$$

From which the zero, the 0-dB origin pole and the mid-band gain can be extracted:

$$\omega_z = \frac{1}{R_2C_1} \tag{5.40}$$

$$\omega_{po} = \frac{1}{R_1C_1} \tag{5.41}$$

$$G_0 = \frac{\omega_{po}}{\omega_z} = \frac{R_2}{R_1} \tag{5.42}$$

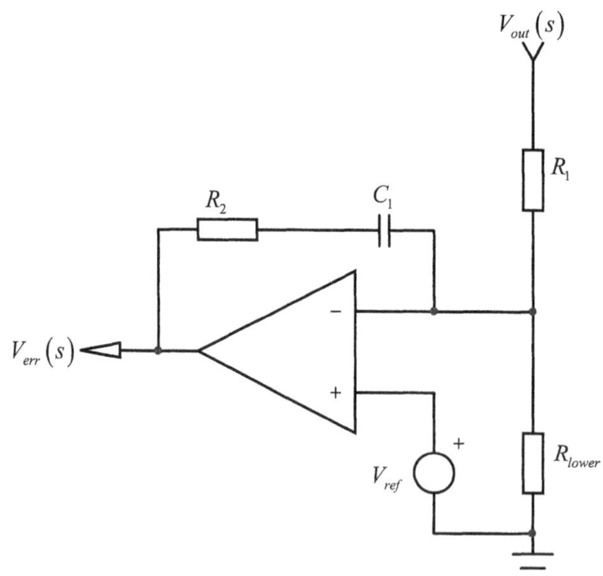

Figure 5.8 Type 2a does not implement the high-frequency pole.

As the frequency increases, C_1 becomes a short circuit and the gain reduces to a ratio of resistors as expressed by

$$\lim_{s\to\infty}|G(s)| = G_0 = \frac{R_2}{R_1} \tag{5.43}$$

The argument is calculated using (5.39). Because of the negative sign on G_0 (the inverting configuration), we are in quadrant II, and π must be added to the arctangent result:

$$\arg G(j\omega) = 180° + \arg\left(1 - j\frac{\omega_z}{\omega}\right) = 180° + \tan^{-1}\left(-\frac{\omega_z}{\omega}\right) = 180° - \tan^{-1}\left(\frac{\omega_z}{\omega}\right) \tag{5.44}$$

When ω reaches infinity, the right term is 0, and we have the phase of a simple inverter. When displayed under the SPICE graphical viewer, it can show up as 180° or –180°, as both angles are similar.

5.3.1 A Design Example

Suppose we need to compensate a circuit whose gain excess at a 10-Hz crossover frequency is 20 dB. The needed phase boost to stabilize the device is 45°. How do we place our zero and the 0-dB crossover pole to meet these criteria? We know that in dc, the argument of the inverting type 2a is 90°. The phase boost is thus

$$boost = \arg G(j\omega_c) - 90° \tag{5.45}$$

If we substitute (5.44) in the previous equation, we have

$$boost = 180° - \tan^{-1}\left(\frac{\omega_z}{\omega_c}\right) - 90° = 90° - \tan^{-1}\left(\frac{\omega_z}{\omega_c}\right) \tag{5.46}$$

Solving for ω_z and converting to f_z, we have

$$f_z = f_c \tan\left(\frac{\pi}{2} - boost\right) = 10 \cdot \tan(90 - 45) = 10 \text{ Hz} \tag{5.47}$$

We have the zero position, where do we place the 0-dB crossover pole now? From (5.38), we calculate the magnitude at the selected crossover frequency:

$$|G(f_c)| = \frac{\sqrt{1+\left(\frac{f_c}{f_z}\right)^2}}{f_c} f_{po} \tag{5.48}$$

We will now place the 0-dB crossover pole to realize an attenuation G of –20 dB at 10 Hz. From (5.48), we have

$$f_{po} = \frac{G \cdot f_c}{\sqrt{1+\left(\frac{f_z}{f_c}\right)^2}} = \frac{10^{-\frac{20}{20}} \times 10}{\sqrt{1+\left(\frac{10}{10}\right)^2}} = \frac{1}{\sqrt{2}} \approx 0.7 \text{ Hz} \tag{5.49}$$

The simulation circuit appears in Figure 5.9.

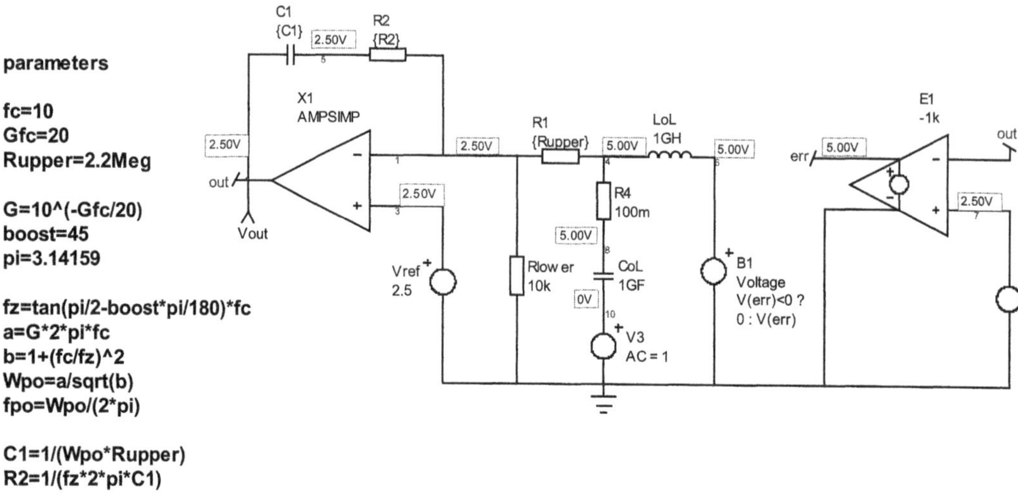

Figure 5.9 The automated simulation circuit for the type 2a compensator.

The ac simulation are given in Figure 5.10 and confirm the validity of our equations.

5.4 Type 2b: Some Static Gain plus a Pole

There are some applications where a simple proportional compensation combined with a gain rolloff is the adequate solution. In this situation, type 2b can be

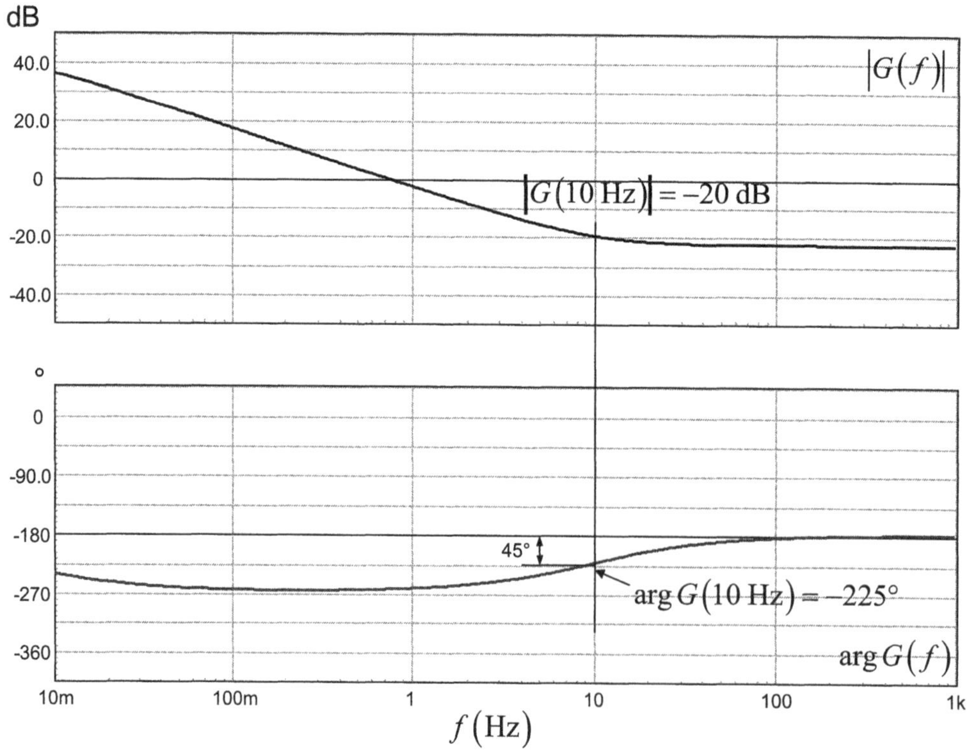

Figure 5.10 The simulations results confirm the 10-Hz zero and the phase lag up to 180°.

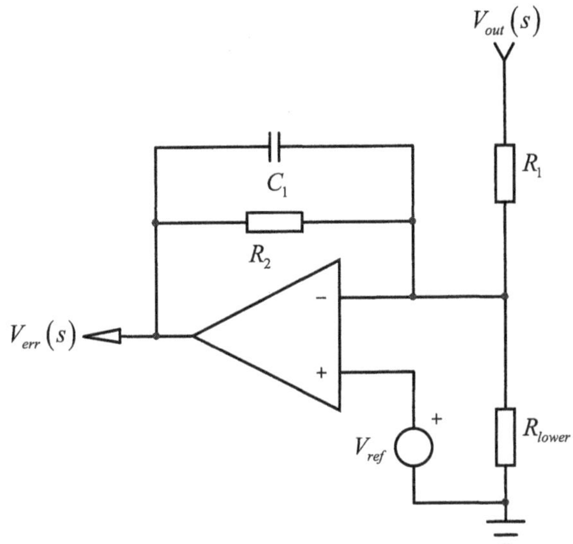

Figure 5.11 Type 2b inserts a pole together with some static gain.

envisaged. Its architecture appears in Figure 5.11. The transfer function equation is straightforward:

$$G(s) = -\frac{\frac{1}{sC_1}R_2}{\frac{1}{sC_1}+R_2}\frac{1}{R_1} = -\frac{R_2}{R_1}\frac{1}{1+sR_2C_1} = -G_0\frac{1}{1+s/\omega_p} \tag{5.50}$$

with

$$G_0 = \frac{R_2}{R_1} \tag{5.51}$$

and

$$\omega_p = \frac{1}{R_2C_1} \tag{5.52}$$

At high-frequency, the gain goes to

$$\lim_{s\to\infty}|G(s)| = 0 \tag{5.53}$$

In dc, the gain reaches a plateau, also called the static gain:

$$\lim_{s\to 0}|G(s)| = G_0 \tag{5.54}$$

The argument of type 2b is derived as follows:

$$\arg G(j\omega) = \pi - \tan^{-1}\left(\frac{\omega}{\omega_p}\right) \tag{5.55}$$

At dc, when the capacitor is considered an open circuit, the argument is that of a simple inverter and is 180° or –180°. When the frequency increases toward infinity, the phase reaches

$$\lim_{s\to\infty} \arg G(s) = \pi - \tan^{-1}\left(\frac{\omega}{\omega_p}\right) = \pi - \tan^{-1}(\infty) = \pi - \frac{\pi}{2} = \frac{\pi}{2} \text{ or } -270° \tag{5.56}$$

5.4.1 A Design Example

Let's assume a 50-dB static gain is needed, followed by a pole installed at 10 kHz. R_1 is supposed to be 10 kΩ. Using (5.54), we have

$$R_2 = R_1 10^{\frac{50}{20}} = 10k \times 316 \approx 3.2\ \text{M}\Omega \tag{5.57}$$

The capacitor is obtained using (5.52):

$$C_1 = \frac{1}{2\pi R_2 f_p} = \frac{1}{6.28 \times 3.2M \times 10k} = 4.7\ \text{pF} \tag{5.58}$$

The simulator has been fed with these calculated values and the commented results show up in Figure 5.12.

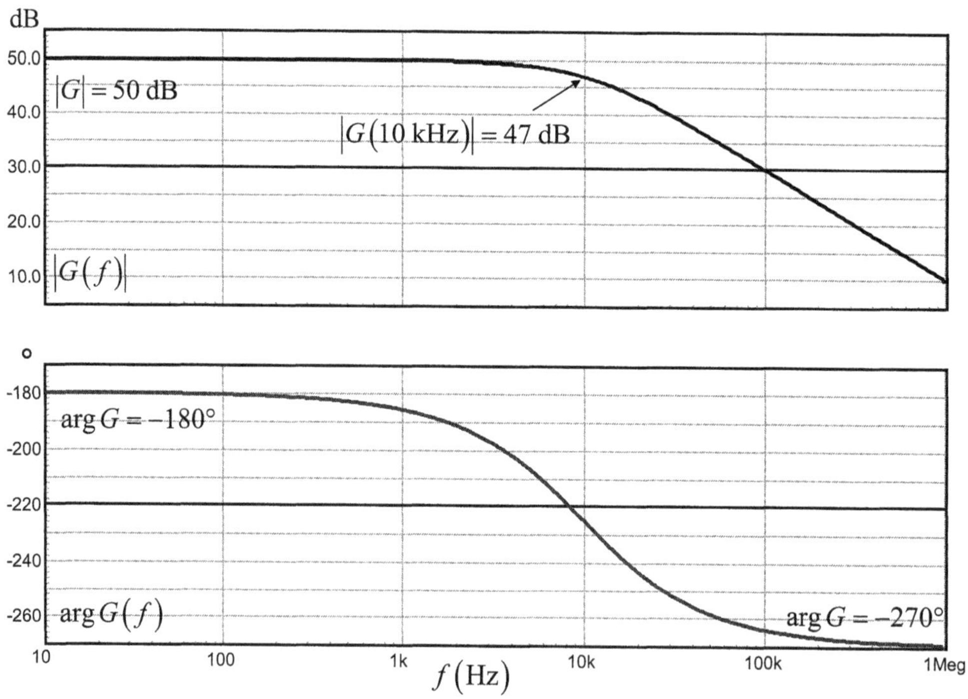

Figure 5.12 The phase starts from 180° down to 90° or from –180° down to –270°; this is the type 2a argument over frequency.

5.5 Type 2: Isolation with an Optocoupler

In the majority of offline applications (e.g., ac-dc adapters), the secondary side information is carried back to the primary via an opto-isolator, also known as an optocoupler. This element plays an important role, and you must understand how it works to successfully stabilize a converter that uses one. The simplified representation of an opto-isolator appears in Figure 5.13. In a nutshell, the set of parameters needed to stabilize a loop are the following ones:

- *Current transfer ratio (CTR)*: this is the amount of current I_C flowing in the collector when the LED is forward biased by a current I_F. Its definition is simply

$$\mathrm{CTR} = \frac{I_C}{I_F} \tag{5.59}$$

- C_{opto}: this is the equivalent parasitic capacitor seen across the collector and the emitter of the device. It actually sets the pole frequency f_{pole} inherent to the optocoupler when wired in either a common collector ($R_{pulldown}$) or common emitter (R_{pullup}) configuration:

$$f_{pole} = \frac{1}{2\pi C_{opto} R_{pullup}} \tag{5.60}$$

- You cannot get rid of that pole but you can shift it by adding an extra capacitor across the optocoupler collector-emitter terminals.
- V_f: this is the LED forward voltage. It usually equals roughly 1 V and stays rather constant given the low forward currents in play.
- $V_{CE,sat}$: this is the minimum voltage appearing across the collector-emitter junction when the LED is forward biased. It obviously depends on the collector current and the LED biasing level. In general, this saturation voltage is in the vicinity of 300 mV.

The equivalent simplified schematic we have used appears next and shows the presence of the parasitic capacitor, which, when combined with the pull-up resistor,

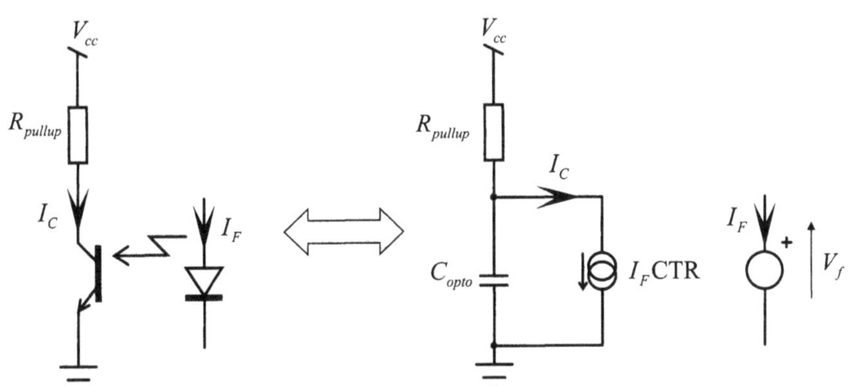

Figure 5.13 The simplified ac model of the optocoupler that we will use in the following lines.

introduces a pole. Any capacitor connected from the collector to ground will be supplemental to the parasitic capacitor already present. This is something to keep in mind for future calculations.

There are basically three possible ways to connect an optocoupler to the op amp output. They appear in Figure 5.14. In the first option, noted (a), an op amp directly drives the optocoupler LED with a simple resistor in series with the op amp output.

The second option (b) shows the optocoupler LED wired as a pull-up load. Please note that in this configuration, the LED current not only depends on the op amp output but also on the observed voltage, V_{out} in this case. Finally, solution (c) is a variation of (b) in the sense that the LED current now solely depends on the op amp output current. We will explore these configurations in several design examples.

In solution (a), the op amp output is made of $V_{out}(s)G(s)$ alone, and the optocoupler chain provides an additional gain block, featuring a pole brought by C_{opto} combined with the pull-up resistor. A type 2 compensator already features one pole. If we directly cascade another stage also exhibiting a pole within our frequency zone of interest, we will surely degrade the op amp transfer function and, in particular, the phase boost we are looking for. For that reason, why not getting rid of the pole in the op amp chain (like with a type 2a) and only keep the pole brought by the optocoupler? This would ensure the inclusion of the optocoupler in the calculation chain, thus accounting for its effects (on the pole but also on the gain with its CTR). We know that the optocoupler pole can be shifted by adding an extra capacitor from collector to ground, a capacitor that will also help improve

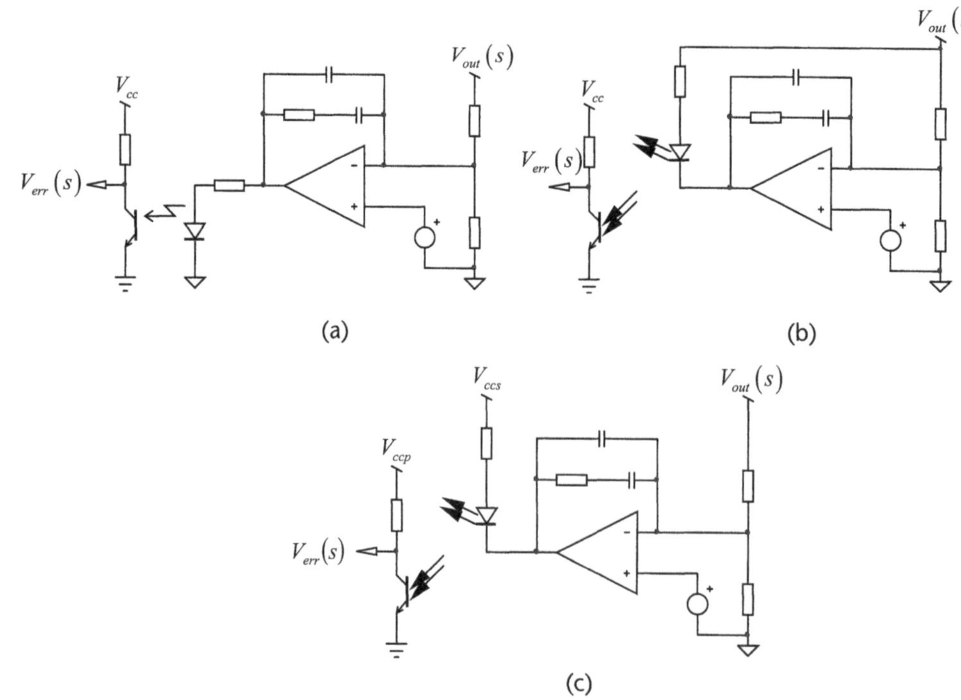

Figure 5.14 Three possibilities exist to wire an op amp output to an optocoupler.

the noise immunity. Indeed, the connection from the optocoupler to the feedback pin of the considered controller is often long and subject to noise pickup. Adding an extra capacitor for the pole and placing it exactly between the feedback pin and the controller ground (close to the integrated circuit) is the best way to avoid spurious noise injection. This is the adopted scheme for all our compensation examples featuring an optocoupler.

5.5.1 Optocoupler and Op Amp: the Direct Connection, Common Emitter

The type 2 compensator connected to the optocoupler appears in Figure 5.15. The current flowing through the LED is made of a dc component (the bias point) and an ac component. In ac and neglecting the diode dynamic resistance, the LED current is simply the op amp output voltage divided by the LED series resistor R_{LED}:

$$I_{LED}(s) = \frac{V_{op}(s)}{R_{LED}} \tag{5.61}$$

Note the position of C_2 that has moved on the nonisolated optocoupler side. C_2 represents the total capacitance on the collector node, including C_{opto}.

The ac output voltage $V_{err}(s)$ is generated by the collector current flowing in the loading network consisting of the pull-up resistor in parallel with the total capacitance seen from the collector to ground, C_2. The capacitor, C_2, as explained, will be the combination of the added external capacitor C_{col} (placed close to the controller feedback pin) and the optocoupler parasitic element, C_{opto}. They both appear in parallel:

$$C_2 = C_{opto} \mid\mid C_{col} \tag{5.62}$$

Now, as the LED current expression is known, we can easily link it to the output voltage with the help of (5.59). Please note the presence of the minus sign given the inversion brought by the common-emitter configuration:

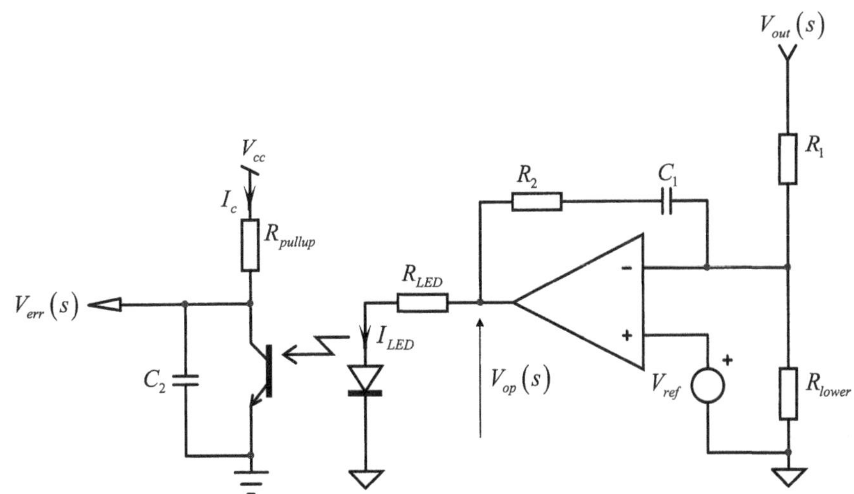

Figure 5.15 The op amp is directly driving the LED via a series resistor. Note the position of C_2 that has moved on the non-isolated optocoupler side. C_2 represents the total capacitance on the collector node, including C_{opto}.

$$V_{err}(s) = -I_{LED}(s)\text{CTR}(R_{pullup} \parallel C_2) = -V_{op}(s)\frac{R_{pullup}}{R_{LED}}\frac{1}{1+sR_{pullup}C_2}\text{CTR} \quad (5.63)$$

The op amp solely contributes to a type 2a transfer function already seen with (5.38). Therefore, combining (5.63) with (5.38) and rearranging, we have

$$G(s) = \text{CTR}\frac{R_{pullup}}{R_{LED}}\frac{R_2}{R_1}\frac{1+1/sR_2C_1}{1+sR_{pullup}C_2} = G_0\frac{1+\omega_z/s}{1+s/\omega_p} \quad (5.64)$$

where

$$G_0 = \text{CTR}\frac{R_{pullup}}{R_{LED}}\frac{R_2}{R_1} = G_1G_2 \quad (5.65)$$

$$G_1 = \text{CTR}\frac{R_{pullup}}{R_{LED}} \quad (5.66)$$

$$G_2 = \frac{R_2}{R_1} \quad (5.67)$$

$$\omega_z = \frac{1}{R_2C_1} \quad (5.68)$$

$$\omega_p = \frac{1}{R_{pullup}C_2} \quad (5.69)$$

It is important to note that the negative sign has disappeared since two inverting circuits are in series, the op amp and the optocoupler in a common-emitter arrangement. This polarity can fit certain types of controllers, but it has to be checked before adopting this configuration.

In (5.65), the mid-band gain is actually composed of two terms: the optocoupler network ($G_1 = R_{pullup}\text{CTR}/R_{LED}$) cascaded with the op amp contribution ($G_2 = R_2/R_1$). The optocoupler network features elements usually imposed or fixed by the controller and the optocoupler. This is the case for the pull-up resistor and the CTR. One adjustment variable for the mid-band gain can therefore be the LED resistor. However, this resistor has a limit beyond which the proper bias point is no longer ensured. When the op amp output is to its highest level (VOH), the current injected in the LED and seen on the collector multiplied by CTR has to be sufficiently strong to bring the collector to the saturation voltage, $V_{CE,sat}$ (≈300 mV, depending on R_{pullup} and the V_{cc} level). If not, the output voltage on the collector will limit the dynamics of the allowed feedback signal, hampering the converter operation in some conditions. As a result, an upper limit for R_{LED} has to be derived. The maximum current in the collector is equal to

$$I_{C,\max} = \frac{V_{cc} - V_{CE,sat}}{R_{pullup}} \quad (5.70)$$

Reflected back to the LED, it becomes

$$I_{LED,\max} = \frac{I_{c,\max}}{\text{CTR}_{\min}} = \frac{V_{cc} - V_{CE,sat}}{R_{pullup}\text{CTR}_{\min}} \tag{5.71}$$

This current must be reached when the op amp delivers its maximum output voltage at a given supply level:

$$I_{LED,\max} = \frac{\text{VOH} - V_f}{R_{LED}} \tag{5.72}$$

As both (5.71) and (5.72) must be equal, we can extract the maximum value of the LED resistor:

$$R_{LED,\max} = \frac{R_{pullup}(\text{VOH} - V_f)\text{CTR}_{\min}}{(V_{cc} - V_{CE,sat})} \tag{5.73}$$

5.5.2 A Design Example

We want to reuse the compensation parameters already exposed in the type 2 design example: a 15-dB gain at 5 kHz together with a 50° phase boost at this point. Equations (5.33) and (5.34) recommended we place a pole at 13.7 kHz and a zero at 1.8 kHz, respectively. Suppose we have the following parameters:

VOH = 10 V; the op amp maximum output voltage.
V_f = 1 V; the LED forward voltage.
$V_{CE,sat}$ = 0.3 V; the optocoupler saturation voltage.
V_{cc} = 5 V; the pull-up V_{cc} level.
R_{pullup} = 1 kΩ; the optocoupler pull-up resistor.
$\text{CTR}_{\min}$ = 0.8; the minimum optocoupler current transfer ratio.
R_1 = 10 kΩ; the upper resistor in the resistor bridge observing the output variable.
f_{opto} = 15 kHz; the optocoupler pole that has been characterized with R_{pullup}.

With these elements on hand, let us calculate with (5.73) the maximum LED series resistor we can select:

$$R_{LED,\max} = \frac{R_{pullup}(\text{VOH} - V_f)\text{CTR}_{\min}}{(V_{cc} - V_{CE,sat})} = \frac{1k(10-1)}{5-0.3}0.8 = 1.5\ \text{k}\Omega \tag{5.74}$$

Adopting a 20 percent safety margin and choosing a normalized value, we finally have R_{LED} = 1.2 kΩ. With this resistor now known, we can evaluate the gain brought by the optocoupler chain alone and defined by

$$G_1 = \frac{R_{pullup}}{R_{LED}}\text{CTR}_{\min} = \frac{1k}{1.2k}0.8 = 0.666 \tag{5.75}$$

Given the needed 15-dB gain G at the crossover point (5 kHz), the second gain G_2 is simply

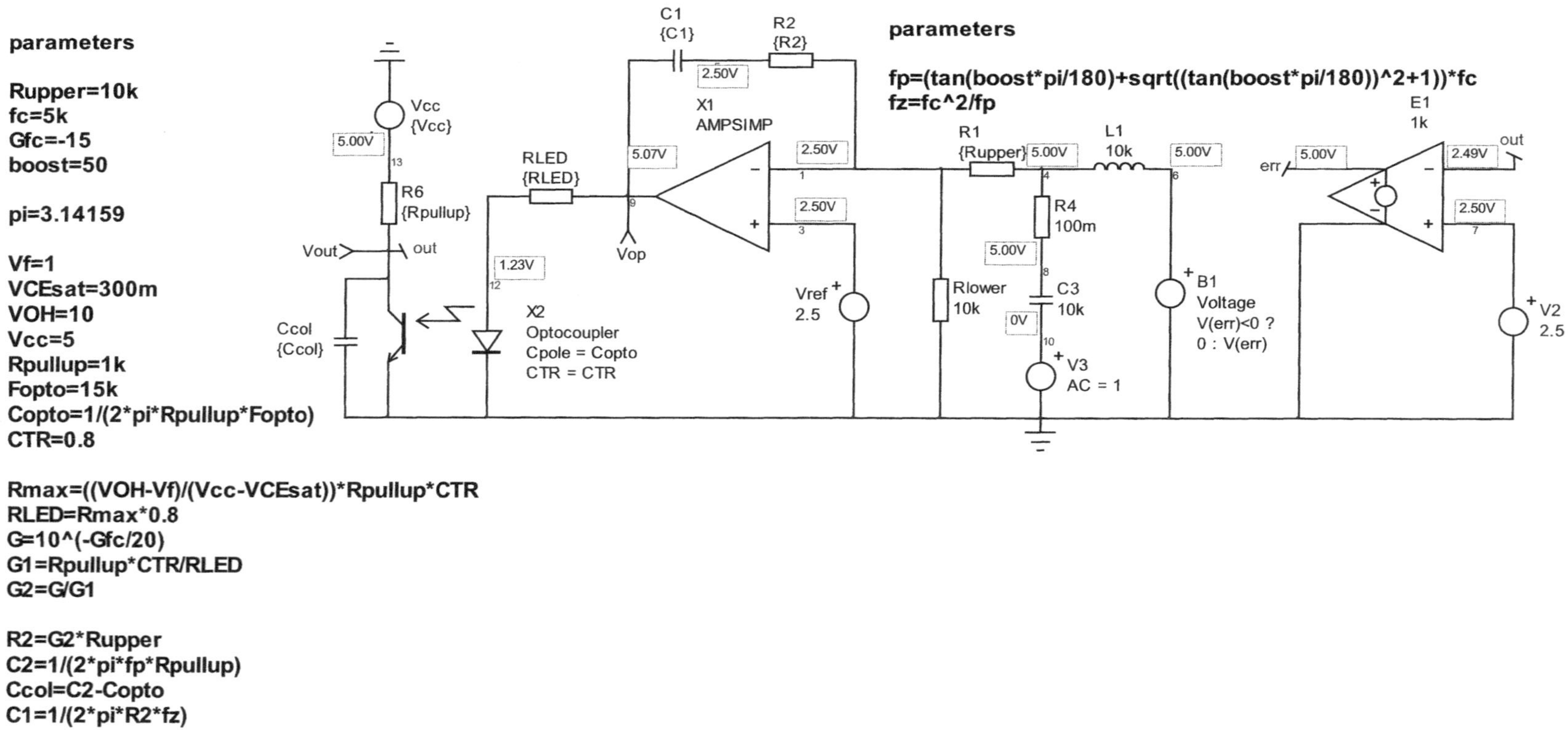

Figure 5.16 The ac simulation of the type 2 compensator directly driving the optocoupler.

$$G_2 = \frac{G}{G_1} = \frac{10^{\frac{15}{20}}}{0.666} = 8.44 \tag{5.76}$$

From (5.67) we can immediately extract the value of R_2:

$$R_2 = R_1 G_2 = 10k \times 8.44 = 84.4\ \text{k}\Omega \tag{5.77}$$

The zero depends on R_2, which [due to (5.68)], leads to C_1:

$$C_1 = \frac{1}{2\pi f_z R_2} = \frac{1}{6.28 \times 1.8k \times 84.4k} \approx 1\ \text{nF} \tag{5.78}$$

The pole is obtained thanks to C_2, the total capacitance seen across the optocoupler collector to ground. For a 13.7-kHz location and a 1-kΩ pull-up resistor, we have

$$C_2 = \frac{1}{2\pi R_{pullup} f_p} = \frac{1}{6.28 \times 1k \times 13.7k} = 11.6\ \text{nF} \tag{5.79}$$

As explained, C_2 is actually made of the external capacitor C_{col} that we will place in parallel with the internal optocoupler parasitic capacitor C_{opto}. Given the characterized optocoupler pole at 15 kHz, the parasitic capacitor is found to be

$$C_{opto} = \frac{1}{2\pi f_{opto} R_{pullup}} = \frac{1}{6.28 \times 15k \times 1k} = 10.6\ \text{nF} \tag{5.80}$$

As the total capacitance is 11.6 nF, the added capacitor C_{col} is simply the difference between (5.79) and (5.80):

$$C_{col} = C_2 - C_{opto} = 11.6n - 10.6n = 1\ \text{nF} \tag{5.81}$$

Sometimes, this capacitor can be negative, meaning that the optocoupler pole position limits the placement of the high-frequency pole. In this case, try to reduce the crossover frequency and compute the pole position until (5.81) gives a minimum capacitance of 100 pF. Try also to change the crossover frequency where less phase boost is needed. This capacitor, placed across the feedback and the ground pins, very close to the controller, will not only play its ac stabilization role but will also ensure a good noise immunity.

Figure 5.16 depicts the simulation template that we used for this example.

Further to a simulation, the results appear in Figure 5.17 and show the correct values at 5 kHz. Please note that the phase lag of the whole chain in now 90° and no longer 270° since $G(s)$ does not reverse the input signal owing to the optocoupler configuration.

5.5.3 Optocoupler and Op Amp: The Direct Connection, Common Collector

In some cases, because the control law polarity is not suitable, the direct connection with the optocoupler wired in a common-emitter configuration does not work. An

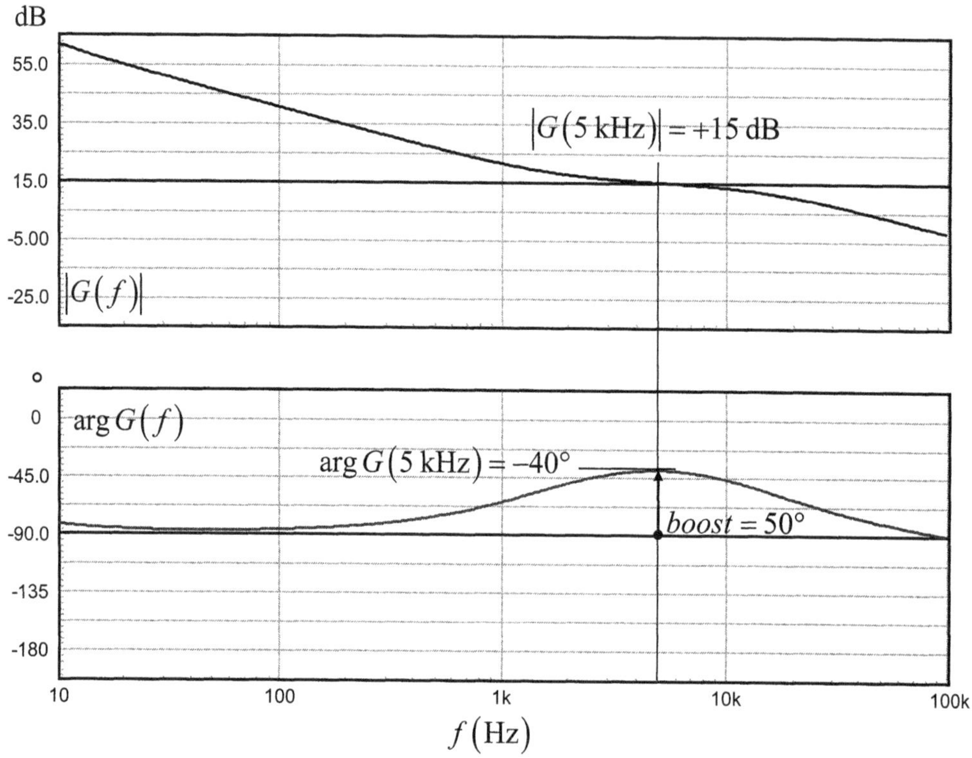

Figure 5.17 The simulation results show an excellent agreement between the input data and the final results.

option consists of placing the optocoupler loading resistor in the emitter rather than in the collector: common collector versus common emitter. This option appears in Figure 5.18.

The transfer function is exactly the same as in (5.64), except that the minus sign is coming back and R_{pullup} is replaced by $R_{pulldown}$:

$$G(s) = -\mathrm{CTR}\frac{R_{pulldown}}{R_{LED}}\frac{R_2}{R_1}\frac{1+1/sR_2C_1}{1+sR_{pulldown}C_2} = -G_0\frac{1+\omega_z/s}{1+s/\omega_p} \tag{5.82}$$

where:

$$G_1 = \mathrm{CTR}\frac{R_{pulldown}}{R_{LED}} \tag{5.83}$$

$$G_1 = \mathrm{CTR}\frac{R_{pulldown}}{R_{LED}} \tag{5.84}$$

$$G_2 = \frac{R_2}{R_1} \tag{5.85}$$

$$\omega_z = \frac{1}{R_2C_1} \tag{5.86}$$

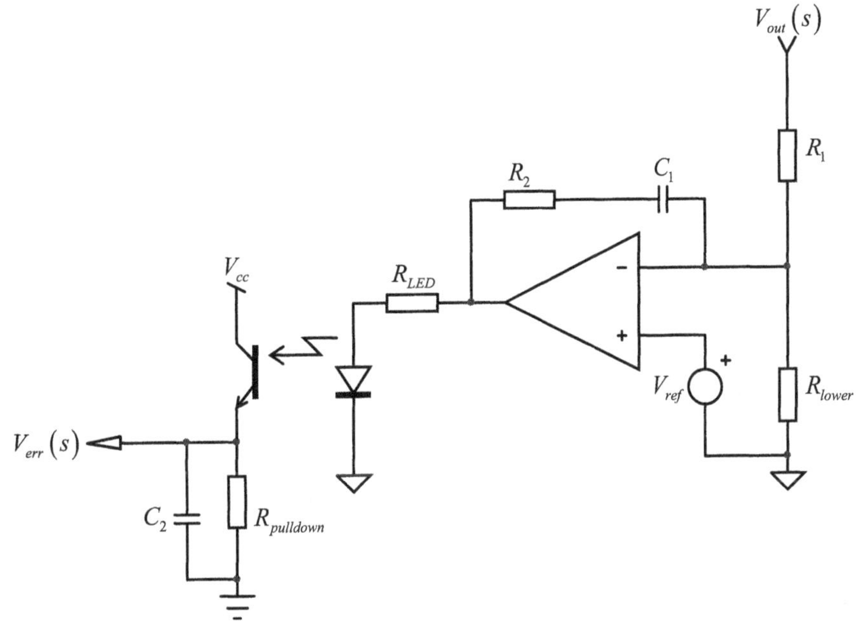

Figure 5.18 The loading resistor is now in the emitter of the optocoupler, rather than its collector, restoring the negative sign to $G(s)$.

$$\omega_p = \frac{1}{R_{pulldown}C_2} \tag{5.87}$$

The design methodology exposed in the previous design example remains the same. The maximum LED resistor now depends upon the pull-down resistor:

$$R_{LED,\max} = \frac{R_{pulldown}(\text{VOH} - V_f)\text{CTR}_{\min}}{(V_{cc} - V_{CE,sat})} \tag{5.88}$$

5.5.4 Optocoupler and Op Amp: The Direct Connection Common Emitter and UC384X

Within the power supply controllers, the UC384X still occupies an important part of modern designs: low cost, flexibility, and ease of availability are among its characteristics. Despite the presence of an internal op amp, it is easy to connect it to one of the previous secondary-side type 1 or type 2 configurations. Indeed, in most cases, it is easier and more effective to have the intelligence—the reference and the compensator—placed on the secondary side and feed the information back to the primary side via an optocoupler. This is what Figure 5.19 suggests. In this example, the op amp is wired as a unity gain inverter, bringing the right polarity from the optocoupler collector. Make sure the op amp resistors are sufficiently high compared to R_{pullup}—otherwise, both the gain and the pole position will be affected. The design procedure exposed in the previous lines is unchanged, given the unity gain of the UC384X op amp configuration. Please note that this configuration requires an auxiliary supply on the secondary side to power the op amp independently from V_{out}; otherwise, the converter cannot start up.

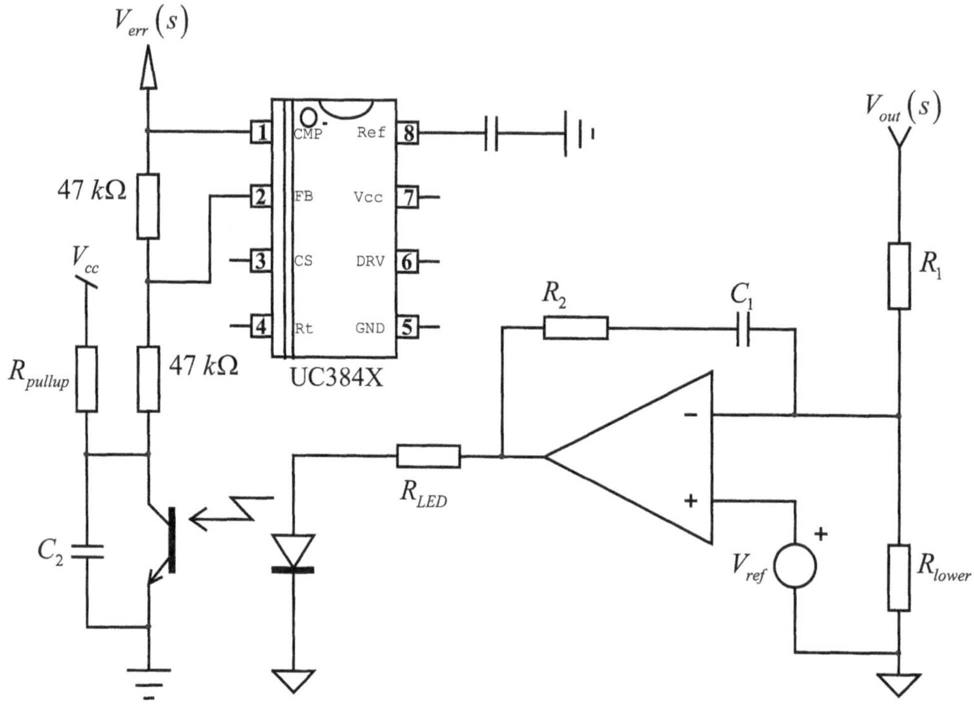

Figure 5.19 Connecting our type 2 chain to a UC384X controller is an easy exercise.

5.5.5 Optocoupler and Op Amp: Pull Down with Fast Lane

New generation controllers, such as the NCP120X from ON Semiconductor, offer a simple feedback pin internally pulled up by a resistor. To reduce the power flow, the feedback level must decrease as shown in Figure 5.20.

If we take the op amp example from Figure 5.15, for instance, we find the polarity is not correct: when the output voltage V_{out} approaches the target, V_{op} decreases. This leads to an increase of V_{err}, rather than the opposite needed by Fig-

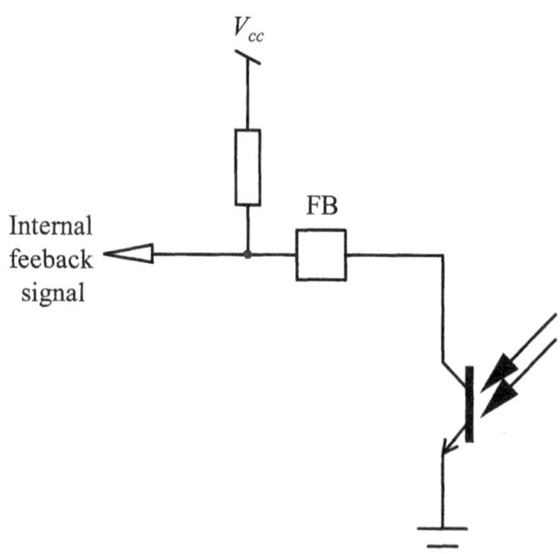

Figure 5.20 Most modern PWM controllers feature a feedback input that must be pulled down to reduce the peak current setpoint.

ure 5.20. Rather, it is more effective to arrange the op amp in a different manner, as proposed by Figure 5.21.

In this configuration, the LED current is now dependent on the op amp output voltage $V_{op}(s)$ as well as directly on the converter output voltage, $V_{out}(s)$. When C_1 is a short circuit at high frequency, the op amp output is ac flat (there is no coming through modulation, since C_1 is a short), and its dc output maintains the bias point at the right level. Therefore, if V_{out} is modulated, it is not seen at the op amp output. However, if V_{out} is modulated, since the cathode of the LED is now fixed by the op amp output, there is an ac current flowing in the LED:

$$\underset{s \to \infty}{I_{LED}(s)} = \frac{V_{out}(s)}{R_{LED}} \tag{5.89}$$

This characteristic, also seen with the TL431, is imputed to the so-called *fast lane*, meaning that V_{out} has a direct access to the LED current without passing through the op amp. This particular arrangement leads to the following equation:

$$I_{LED}(s) = \frac{V_{out}(s) - V_{op}(s)}{R_{LED}} \tag{5.90}$$

The op amp output is that of a type 1 compensator already described by (5.2). By substituting $V_{op}(s)$ of (5.2) in (5.90), we have

$$I_{LED}(s) = \frac{V_{out}(s) + \left(V_{out}(s)\frac{1}{sR_1C_1}\right)}{R_{LED}} = \frac{V_{out}(s)}{R_{LED}}\left(1 + \frac{1}{sR_1C_1}\right) = \frac{V_{out}(s)}{R_{LED}}\left(\frac{1 + sR_1C_1}{sR_1C_1}\right) \tag{5.91}$$

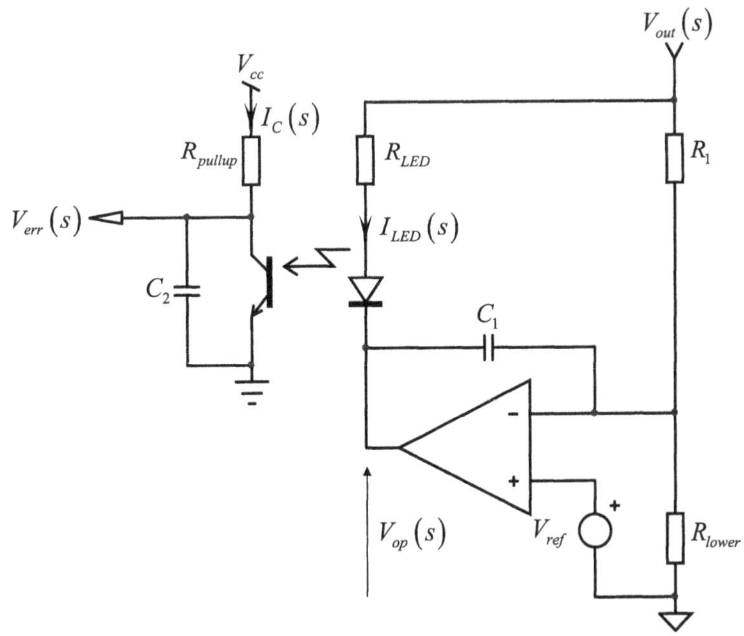

Figure 5.21 The LED drive is no longer ground referenced but directly hooked to the monitored output, creating another ac path. Despite C_1 alone across the op amp, we really have a type 2 configuration with a pole/zero pair.

In this equation, we can see the effect of the fast lane. Rather than having a simple type 1 as Figure 5.21 suggests at first glance, we have an origin pole and a zero: a type 2a! To finish our type 2 configuration, we can add a high-frequency pole as we did before across the collector-emitter of the optocoupler.

$$V_{err}(s) = -I_C(s)(R_{pullup} \parallel C_2) = -I_C(s)R_{pullup}\frac{1}{1 + sR_{pullup}C_2} \tag{5.92}$$

Substituting (5.91) multiplied by the optocoupler CTR (to get the collector current) in (5.92), we have

$$G(s) = -\frac{R_{pullup}}{R_{LED}}\text{CTR}\frac{1 + sR_1C_1}{sR_1C_1(1 + sR_{pullup}C_2)} \tag{5.93}$$

By factoring sR_1C_1 in the numerator, we obtain

$$G(s) = -\frac{R_{pullup}}{R_{LED}}\text{CTR}\frac{1 + \dfrac{1}{sR_1C_1}}{1 + sR_{pullup}C_2} = -G_0\frac{1 + \omega_z/s}{1 + s/\omega_p} \tag{5.94}$$

This is our complete type 2 error amplifier, where

$$G_0 = \frac{R_{pullup}}{R_{LED}}\text{CTR} \tag{5.95}$$

$$\omega_p = \frac{1}{R_{pullup}C_2} \tag{5.96}$$

$$\omega_z = \frac{1}{R_1C_1} \tag{5.97}$$

$$C_2 = \frac{1}{2\pi f_p R_{pullup}} \tag{5.98}$$

$$C_1 = \frac{1}{2\pi R_1 f_z} \tag{5.99}$$

In the circuit presented in Figure 5.21, the maximum LED current is linked to the minimum op amp output swing VOL, the converter output voltage V_{out}, and the LED series resistor. Again, as explained in the previous design example, the LED resistor features an upper value limit to ensure that the optocoupler collector can always bring $V_{err}(s)$ to ground. Otherwise, the loop might have a regulation problem if its dynamics are hampered in some way. The maximum collector current is given by

$$I_{C,\max} = \frac{V_{cc} - V_{CE,sat}}{R_{pullup}} \tag{5.100}$$

Reflected back to the LED, it becomes

$$I_{LED,\max} = \frac{I_{C,\max}}{\text{CTR}_{\min}} = \frac{V_{cc} - V_{CE,sat}}{R_{pullup}\text{CTR}_{\min}} \tag{5.101}$$

The LED current depends on the series resistor value of course, but also on V_{out} and the minimum op amp output voltage VOL:

$$I_{LED,\max} = \frac{V_{out} - V_f - \text{VOL}}{R_{LED}} \tag{5.102}$$

Equating (5.101) and (5.102), we can extract the LED resistor limit:

$$R_{LED,\max} = \frac{R_{pullup}(V_{out} - V_f - \text{VOL})\text{CTR}_{\min}}{(V_{cc} - V_{CE,sat})} \tag{5.103}$$

In light of (5.95), we can see that the mid-band gain is adjusted by calculating the LED resistor value. Unfortunately, as confirmed by (5.103), there is an upper limit for this resistor. An upper limit for the LED resistor implies a lower limit for the mid-band gain:

$$G_0 \geq \frac{R_{pullup}}{R_{LED,\max}}\text{CTR} \tag{5.104}$$

This is the tragedy inherent to the fast lane presence, also affecting the TL431 circuitry. If you'd like to attenuate at the selected crossover frequency rather than amplify, you are stuck with a minimum gain imposed by (5.104). Let us see the implications with a design example.

5.5.6 A Design Example

For the sake of simplicity, we will reuse the compensation parameters already exposed in the type 2 design example, slightly changed for the gain: a 5-dB amplification at 5-kHz together with a 50° phase boost at this point. Equations (5.33) and (5.34), respectively, recommended we place a pole at 13.7 kHz and a zero at 1.8 kHz. Suppose we have the following parameters:

VOL = 0.2 V; the op amp minimum output voltage.
$V_f = 1$ V; the LED forward voltage.
$V_{CE,sat} = 0.3$ V; the optocoupler saturation voltage.
$V_{cc} = 5$ V; the pull-up V_{cc} level.
$V_{out} = 5$ V; the monitored output voltage.
$R_{pullup} = 1$ kΩ; the optocoupler pull-up resistor.
$\text{CTR}_{\min} = 0.8$; the minimum optocoupler current transfer ratio.
$R_1 = 10$ kΩ; the upper resistor in the resistor bridge observing the output variable.
$f_{opto} = 15$ kHz; the optocoupler pole that has been characterized with R_{pullup}.

With these elements on hand, let us calculate with (5.103) the maximum LED series resistor we can use:

$$R_{LED,\max} = \frac{R_{pullup}(V_{out} - V_f - \text{VOL})\text{CTR}_{\min}}{(V_{cc} - V_{CE,sat})} = \frac{1k(5-1-0.2)}{5-0.3}0.8 \approx 647\ \Omega \tag{5.105}$$

When working with (5.95) to adjust the mid-band gain, we will have to select a LED resistor value below what (5.105) recommends. Let's see what (5.95) suggests when we extract R_{LED} out of it:

$$R_{LED} = \frac{R_{pullup}}{G_0} \mathrm{CTR}_{\min} = \frac{1k}{10^{\frac{5}{20}}} 0.8 = 450\ \Omega \tag{5.106}$$

We are below 647 Ω by ≈30 percent, so it is safe to select that resistor value. The next component is the zero capacitor C_1. Dependent upon the upper resistor R_1, its calculation is straightforward:

$$C_1 = \frac{1}{2\pi f_z R_1} = \frac{1}{6.28 \times 1.8k \times 10k} = 8.8\ \mathrm{nF} \tag{5.107}$$

The pole is realized with the pull-up resistor and the compound capacitor C_2 equal to C_{col} and C_{opto} in parallel:

$$C_2 = \frac{1}{2\pi R_{pullup} f_p} = \frac{1}{6.28 \times 1k \times 13.7k} = 11.6\ \mathrm{nF} \tag{5.108}$$

Given the extracted optocoupler pole of 15 kHz, the parasitic capacitor C_{opto} is found to be

$$C_{opto} = \frac{1}{2\pi f_{opto} R_{pullup}} = \frac{1}{6.28 \times 15k \times 1k} = 10.6\ \mathrm{nF} \tag{5.109}$$

As the total capacitance is 11.6 nF, the added capacitor C_{col} is simply the difference between (5.108) and (5.109):

$$C_{col} = C_2 - C_{opto} = 11.6n - 10.6n = 1\ \mathrm{nF} \tag{5.110}$$

We are now all set and can start a simulation to verify our calculations. This is what appears in Figure 5.22, showing the proper operating points.

The Bode plot response of the previous compensation circuit is shown in Figure 5.23.

Equation (5.104) teaches us that the gain cannot be smaller than

$$G_0 \geq 20\log_{10}\left(\frac{1k}{647} 0.8\right) = 1.84\ \mathrm{dB} \tag{5.111}$$

Should we need a 0-dB gain at 5 kHz rather than the adopted 5 dB, we would violate (5.105), and the loop could not properly bias the LED and the optocoupler. This is the inherent limitation of the fast lane and can often be a problem where attenuation rather than gain are needed. To get rid of this issue, why not trying to suppress the fast lane?

5.5.7 Optocoupler and Op Amp: Pull-Down with Fast Lane, Common Emitter, and UC384X

Given the return of the negative sign in the transfer equation of our compensator, we can connect this circuit to either a NCP120X type of controller or a UC384X. The

parameters

Vout=5
Rupper=(Vout-2.5)/250u
fc=5k

Gfc=-5
boost=50

G=10^(-Gfc/20)
pi=3.14159

VOL=0.2
VCEsat=0.3
Vdd=5
Vf=1
A=Vout-Vf-VOL
B=Vdd-VCEsat
Rmax=0.8*(A/B)*Rpullup*CTR

Rpullup=1k
RLED=CTR*Rpullup/G
C1=1/(2*pi*fz*Rupper)
C2=1/(2*pi*fp*Rpullup)

Ccol=C2-Copto

Fopto=15k
Copto=1/(2*pi*Fopto*Rpullup)
CTR = 0.8

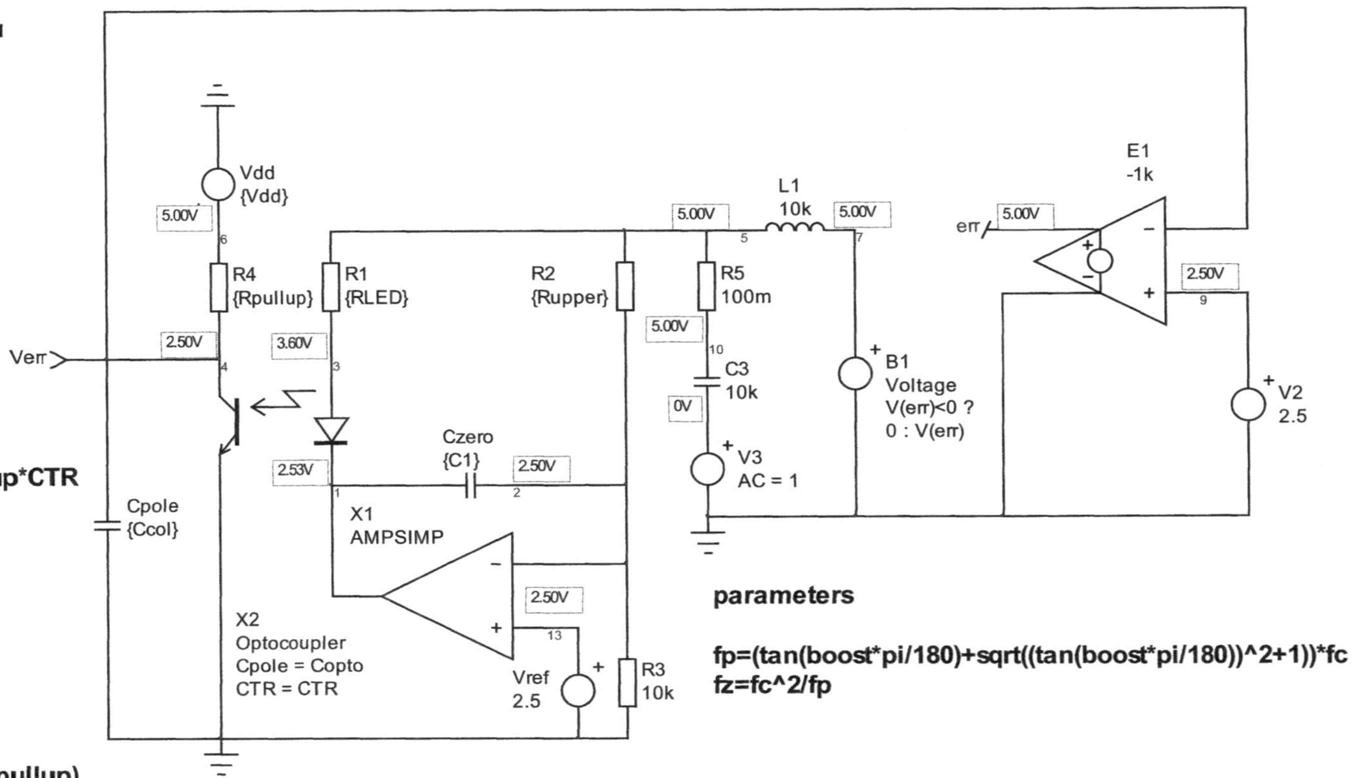

parameters

fp=(tan(boost*pi/180)+sqrt((tan(boost*pi/180))^2+1))*fc
fz=fc^2/fp

Figure 5.22 The simulation fixture shows a type 2 associating an op amp and an optocoupler. This time, the op amp drives the LED from its cathode rather its anode as seen before.

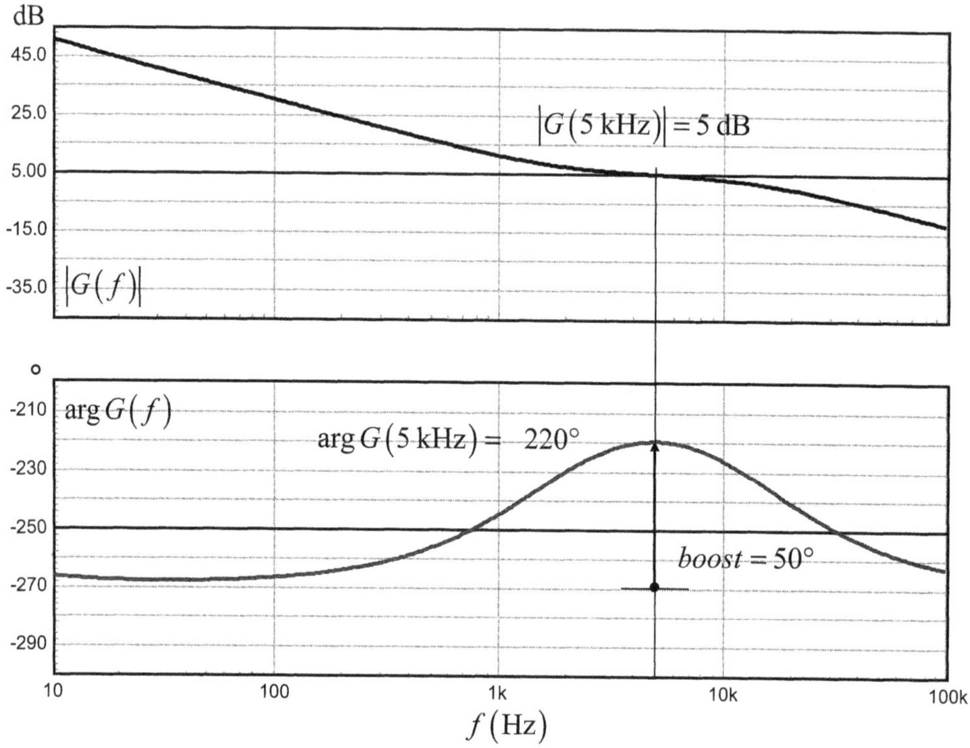

Figure 5.23 The Bode plot response of the type 2 return chain confirms the 5-dB gain at 5 kHz with the right phase boost.

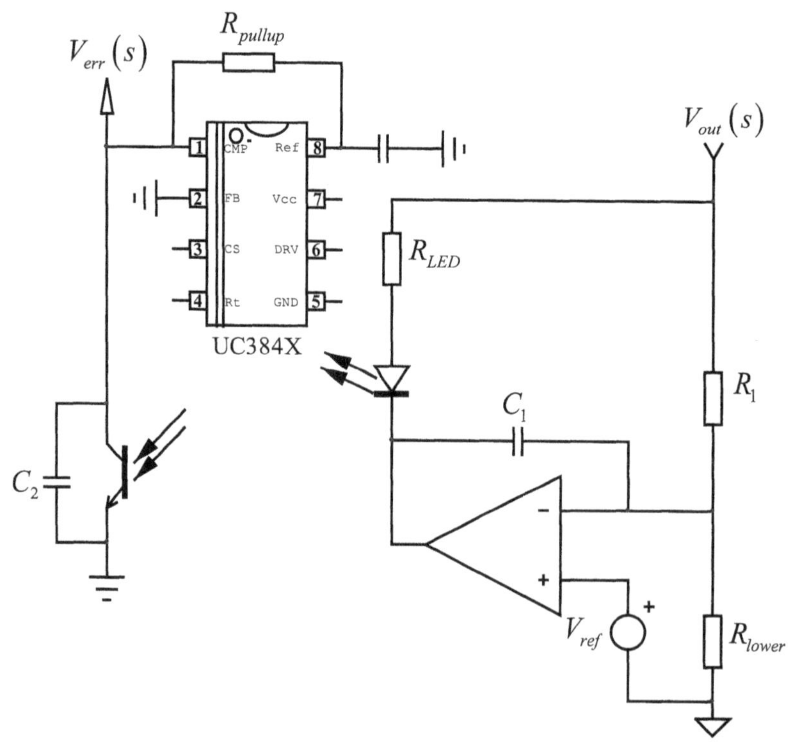

Figure 5.24 The UC384X can also be implemented without using its internal operational amplifier. A simple pull-up resistor to the 5-V reference is enough to drive it from the optocoupler collector.

error amplifier inside this chip cannot deliver more than 1 mA. Therefore, a solution is to hook its output (pin 1, CMP) to the 5-V reference voltage via a pull-up resistor and ground the feedback input (pin 2, FB). Then, connect the optocoupler collector to pin 1 and you are all set! This configuration is described in Figure 5.24.

The design procedure is the same that we just have seen in the previous design example.

Compared to what was shown in Figure 5.19, when V_{out} is absent at startup, the op amp does not impose a current into the LED and the optocoupler collector is high, naturally imposing the maximum peak current: the converter can start without problems, even if the op amp is not supplied at the beginning.

5.5.8 Optocoupler and Op Amp: Pull Down Without Fast Lane

In applications where more flexibility is needed in the choice of the compensated crossover gain, a solution consists in suppressing the fast lane. The fast lane actually comes from a direct ac relationship between the LED current and the monitored variable. To get rid of it, either connect the LED anode to a regulated dc voltage, independent from the output level, or isolate it via a Zener diode and a capacitor. This is what Figure 5.25 suggests. This solution is actually very close to the direct LED drive disclosed in Figure 5.15, except that we now drive the optocoupler diode cathode. Also, as we will see, the error signal polarity is now compatible with pull-down feedback inputs such as the one already presented in Figure 5.20. A similar regulation architecture is found on the popular NCP120X series from ON Semiconductor, for instance. Capitalizing on what has already been derived, we can immediately express the LED ac current:

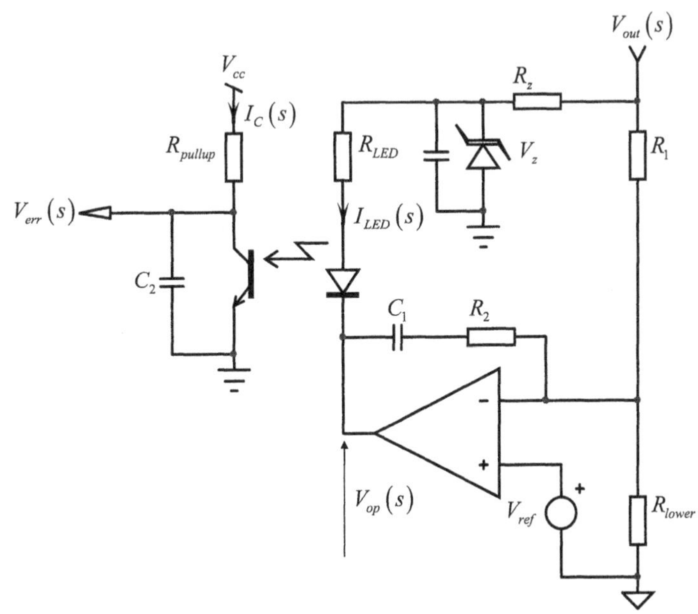

Figure 5.25 By inserting an ac-decoupling network between the output voltage and the LED anode, it becomes possible to get rid of the fast lane.

$$I_{LED}(s) = -\frac{V_{op}(s)}{R_{LED}} \tag{5.112}$$

The op amp ac output $V_{op}(s)$ is that of a type 2a whose expression was derived in (5.39):

$$\frac{V_{op}(s)}{V_{out}(s)} = -\frac{R_2 + \frac{1}{sC_1}}{R_1} = -\frac{1 + sR_2C_1}{sR_1C_1} = -\frac{sR_2C_1}{sR_1C_1}\left(\frac{1}{sR_2C_1} + 1\right) = -\frac{R_2}{R_1}\left(1 + \frac{1}{sR_2C_1}\right) \tag{5.113}$$

The second stage is made of the optocoupler path and classically involves the pull-up resistor and the capacitor from collector to ground, C_2:

$$V_{err}(s) = -I_{LED}(s)R_{pullup}\frac{1}{1 + sR_{pullup}C_2}\text{CTR} \tag{5.114}$$

Combining (5.112), (5.113), and (5.114), we obtain the complete transfer function:

$$G(s) = -\text{CTR}\frac{R_{pullup}}{R_{LED}}\frac{R_2}{R_1}\frac{1 + 1/sR_2C_1}{1 + sR_{pullup}C_2} = -G_0\frac{1 + \omega_z/s}{1 + s/\omega_p} \tag{5.115}$$

where

$$G_0 = \text{CTR}\frac{R_{pullup}}{R_{LED}}\frac{R_2}{R_1} = G_1G_2 \tag{5.116}$$

$$G_1 = \text{CTR}\frac{R_{pullup}}{R_{LED}} \tag{5.117}$$

$$G_2 = \frac{R_2}{R_1} \tag{5.118}$$

$$\omega_z = \frac{1}{R_2C_1} \tag{5.119}$$

$$\omega_p = \frac{1}{R_{pullup}C_2} \tag{5.120}$$

As we have shown before, the LED resistor must be sized to let the optocoupler properly pull the feedback node down to ground. A sufficient current must therefore flow through R_{LED} to let the op amp fully control the feedback voltage. However, the limit on the LED resistor fixes only one gain expression, G_1, but as we have a type 2a with the op amp, we can freely set R_2 without limitation at all. The equation for $R_{LED,\max}$ is similar to the one already derived:

$$R_{LED,\max} = \frac{R_{pullup}(V_Z - V_f - \text{VOL})\text{CTR}_{\min}}{(V_{cc} - V_{CE,sat})} \tag{5.121}$$

In this expression, V_Z represents the Zener voltage derived from the output voltage. This voltage must also be carefully selected. If too low, you limit the voltage excursion over the op amp output, and if too high, you reduce the ac decoupling effects you are looking for—that is to say, ac-separating the LED anode voltage from the output voltage $V_{out}(s)$. Setting the Zener voltage around two thirds of the output voltage looks like a reasonable choice.

Once the LED resistor has been selected, we need to calculate the Zener diode biasing resistor R_Z. This resistor supplies the current not only for the LED branch but also for the Zener diode to make it operate far enough from its knee (the lower its dynamic impedance, the better ac isolation from V_{out} it provides). The LED maximum current is actually reached when the feedback voltage V_{FB} (or V_{err}) is pulled down to $V_{CE,sat}$. In this case, the LED current is simply

$$I_{LED,\max} = \frac{V_{cc} - V_{CE,sat}}{R_{pullup}\text{CTR}_{\min}} \tag{5.122}$$

The Zener dropping resistor R_Z is therefore immediately derived through

$$R_Z = \frac{V_{out} - V_Z}{I_{LED,\max} + I_{Zbias}} \tag{5.123}$$

Substituting (5.122) in (5.123), we obtain

$$R_Z = \frac{(V_{out} - V_Z)R_{pullup}\text{CTR}_{\min}}{V_{cc} - V_{CE,sat} + I_{Zbias}R_{pullup}\text{CTR}_{\min}} \tag{5.124}$$

5.5.9 A Design Example

Let's assume we still want the same crossover frequency of 5 kHz associated with a 50° phase boost, but rather than amplifying at this frequency, we want a –10-dB attenuation. The variables are as follows:

VOL = 0.2 V; the op amp minimum output voltage.
$Vf = 1$ V; the LED forward voltage.
$V_{CE,sat} = 0.3$ V; the optocoupler saturation voltage.
$V_{cc} = 5$ V; the pull-up V_{cc} level.
$R_{pullup} = 1$ kΩ; the optocoupler pull-up resistor.
$\text{CTR}_{\min} = 0.8$; the minimum optocoupler current transfer ratio.
$V_{out} = 12$ V; the converter output voltage.
$R_1 = 38$ kΩ; the upper resistor in the resistor bridge observing the output variable.
$V_{out} = 12$ V; the monitored output voltage.
$f_{opto} = 15$ kHz; the optocoupler pole that has been characterized with R_{pullup}.
$V_Z = 8.2$ V; the Zener diode breakdown voltage.
$I_{Zbias} = 1$ mA; the Zener diode bias current.

With these elements on hand, let us calculate with (5.121) the maximum LED series resistor we can select:

$$R_{LED,\max} = \frac{R_{pullup}(V_Z - V_f - \text{VOL})\text{CTR}_{\min}}{(V_{cc} - V_{CE,sat})} = \frac{1k \times (8.2 - 1 - 0.2) \times 0.8}{5 - 0.3} \approx 1.2\ \text{k}\Omega \tag{5.125}$$

Adopting a 20 percent safety margin and choosing a normalized value, we finally have $R_{LED} = 910\ \Omega$. With its resistor now known, we can evaluate the gain brought by the optocoupler chain alone defined by

$$G_1 = \frac{R_{pullup}}{R_{LED}}\mathrm{CTR}_{\min} = \frac{1k}{910}0.8 = 0.88 \tag{5.126}$$

Given the needed –10-dB attenuation at the crossover point (5 kHz), the second gain G_2 is simply

$$G_2 = \frac{10^{\frac{-10}{20}}}{0.88} = 0.328 \tag{5.127}$$

From (5.118) we can immediately extract the value of R_2:

$$R_2 = R_1 G_2 = 38k \times 0.328 \approx 12.5\ \mathrm{k}\Omega \tag{5.128}$$

The zero depends on R_2, which, after (5.34) and (5.119), leads to C_1:

$$C_1 = \frac{1}{2\pi f_z R_2} = \frac{1}{6.28 \times 1.8k \times 12.5k} \approx 7\ \mathrm{nF} \tag{5.129}$$

The pole is obtained owing to C_2, the total capacitance seen across the optocoupler collector to ground. For a 13.7-kHz upper pole location—see (5.33) and (5.120)—and a 1-kΩ pull-up resistor, we have

$$C_2 = \frac{1}{2\pi R_{pullup} f_p} = \frac{1}{6.28 \times 1k \times 13.7k} = 11.6\ \mathrm{nF} \tag{5.130}$$

As explained, C_2 is actually made of an external capacitor C_{col} in parallel with the internal optocoupler parasitic capacitor C_{opto}. Given the extracted optocoupler pole of 15 kHz, the parasitic capacitor is found to be

$$C_{opto} = \frac{1}{2\pi f_{opto} R_{pullup}} = \frac{1}{6.28 \times 15k \times 1k} = 10.6\ \mathrm{nF} \tag{5.131}$$

As the total capacitance is 11.6 nF, the added capacitor C_{col} is simply the difference between (5.130) and (5.131):

$$C_{col} = C_2 - C_{opto} = 11.6n - 10.6n = 1\ \mathrm{nF} \tag{5.132}$$

The ac design is finished, so let's see what resistance value we can select for R_Z. Applying (5.124), we have

$$R_Z = \frac{(V_{out} - V_Z) R_{pullup} \mathrm{CTR}_{\min}}{V_{cc} - V_{CE,sat} + I_{Zbias} R_{pullup} \mathrm{CTR}_{\min}} = \frac{(12 - 8.2) \times 1k \times 0.8}{5 - 0.3 + 1m \times 1k \times 0.8} = 552\ \Omega \tag{5.133}$$

We can also apply a safety margin on this value and select a 470-Ω resistor. This design is now over, and we can feed our simulator with these values. This is what Figure 5.26 shows you with the proper bias point values.

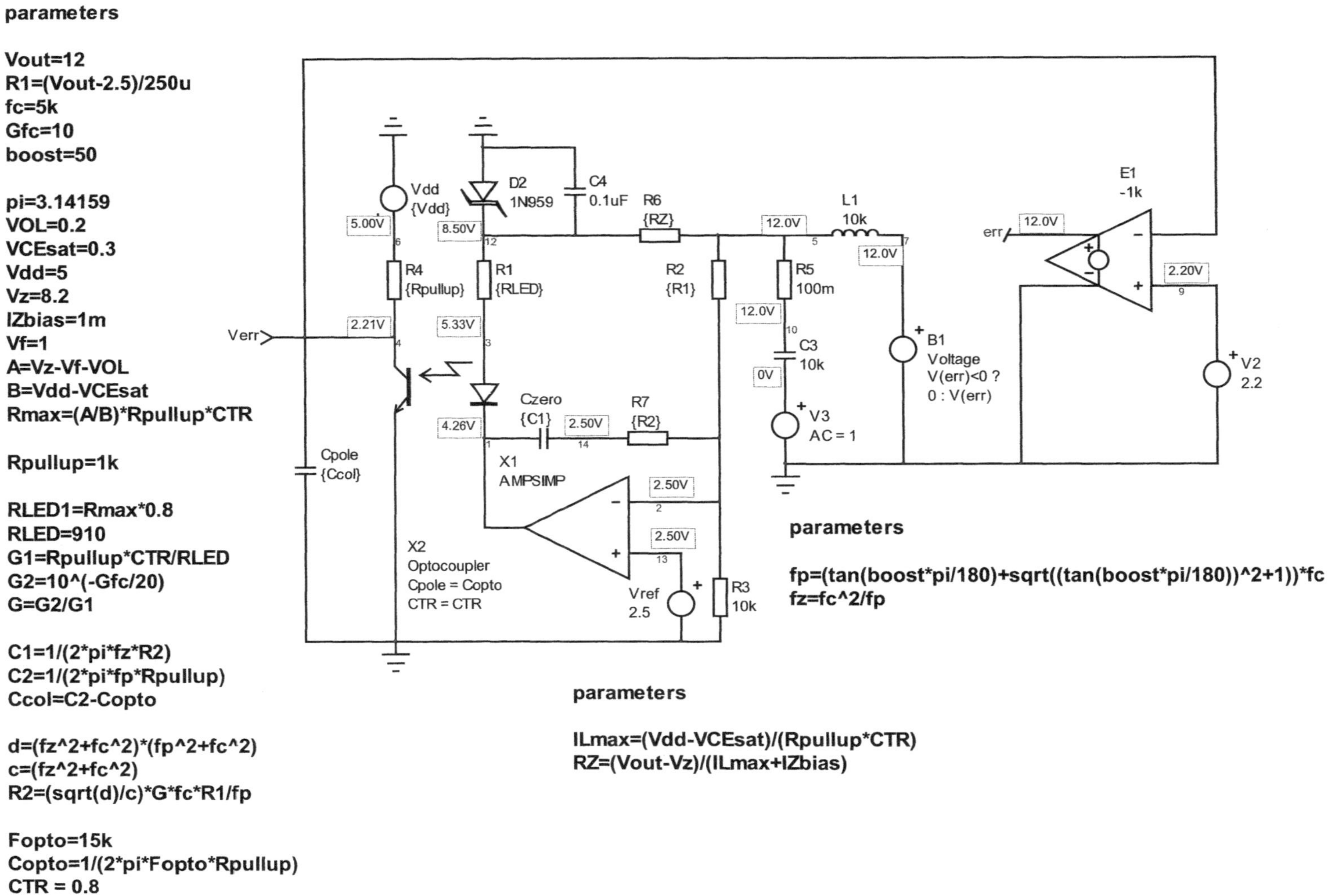

Figure 5.26 An ac simulation test fixture for the type 2 compensator where the fast lane on the secondary side has been removed.

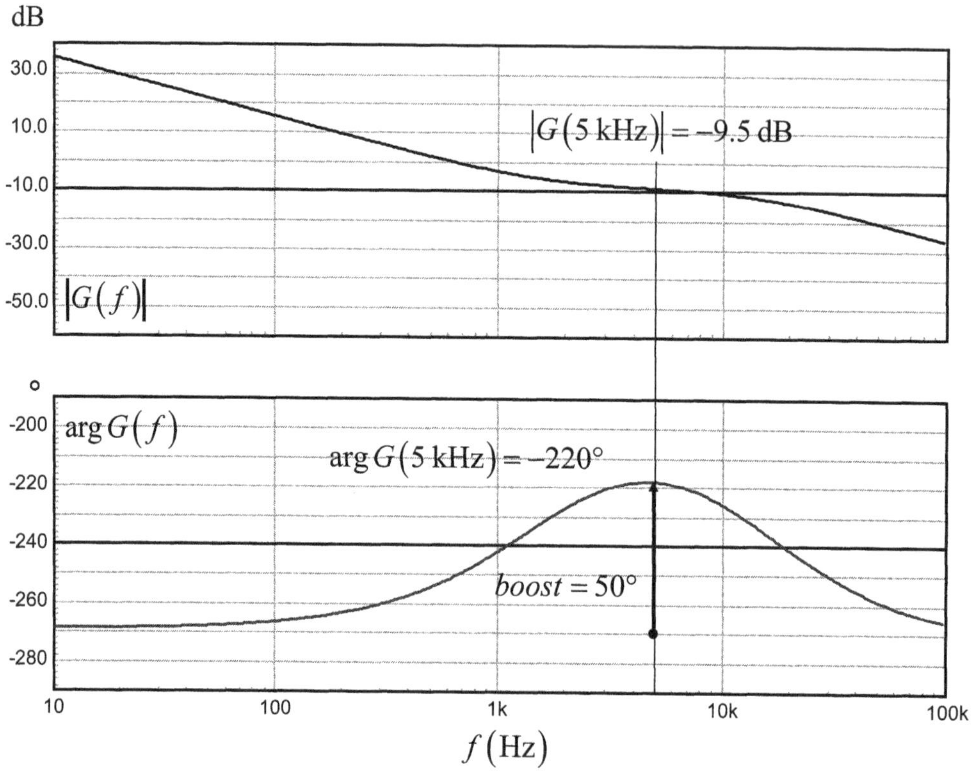

Figure 5.27 The ac sweep shows the attenuation slightly less than our initial target of –10 dB. This is linked to the LED dynamic resistor and the imperfect ac decoupling provided by the Zener network.

The ac simulation results appear in Figure 5.27 and show a slight discrepancy between the –10-dB target and the –9.5-dB result read on the graph. This is mainly linked to the imperfect decoupling brought by the Zener network. If this dispersion between the target and the result is considered too wide, it can easily be compensated by increasing the attenuation/gain target by the observed difference. We increase the attenuation target to –10.5 dB in this case.

In this example, we have shown a Zener-based network. It could of course be replaced by a linear regulator, either integrated or using a bipolar transistor. We could even think of an auxiliary winding dedicated to this purpose. With this latter, the designer must ensure the lack of ac coupling between the auxiliary voltage feeding the optocoupler LED and the regulated rail (Figure 5.28).

5.5.10 Optocoupler and Op Amp: A Dual-Loop Approach in CC-CV Applications

In certain applications, it is necessary to control the output voltage V_{out} or the delivered current I_{out} and ensure one of them remains constant. This is a called a constant-current constant-voltage converter (CC-CV). Even in pure CC applications, a voltage-loop is always present to prevent output voltage runaways in light-load conditions. Widely used in notebook or cell-phone chargers, their typical

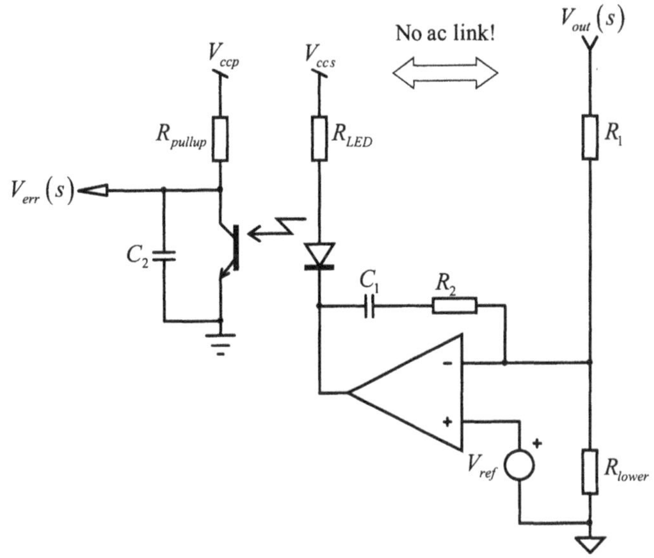

Figure 5.28 The Zener-based network can be replaced by a dc source, which must be ac-decoupled from V_{out}.

output characteristic appears in Figure 5.29: one control section operates at a time, either the voltage loop or the current loop.

The voltage implementation requires a sensing network made of two resistors similar to those used in the previous examples. The current regulation loop needs a sensing element, usually a few resistors, plus a low-voltage reference voltage.

Figure 5.30 portrays the typical secondary side of a CC-CV control circuitry, where the sensing ground is actually the converter output ground. Please note the insertion of a sense resistor in series with the secondary-side winding. The voltage developed across this element is, however, negative when referenced to the load ground (see the sensed voltage arrow in Figure 5.30). Since the reference voltage, usually a 2.5-V source, is linked to the output ground, a means has to be found to authorize the current loop implementation. A zoom of this portion appears in Figure 5.31.

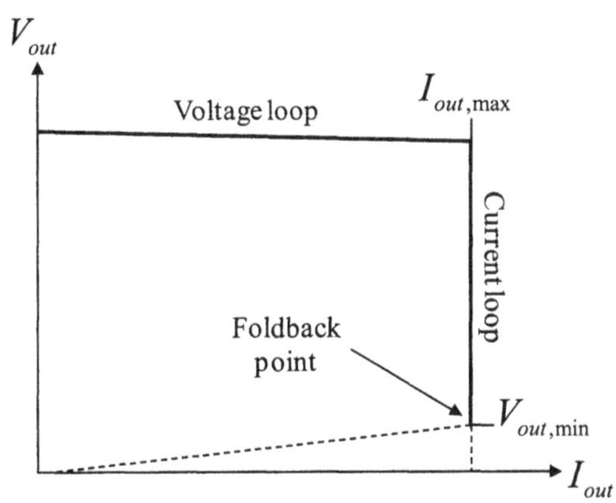

Figure 5.29 This is the typical charging profile of a CC-CV converter.

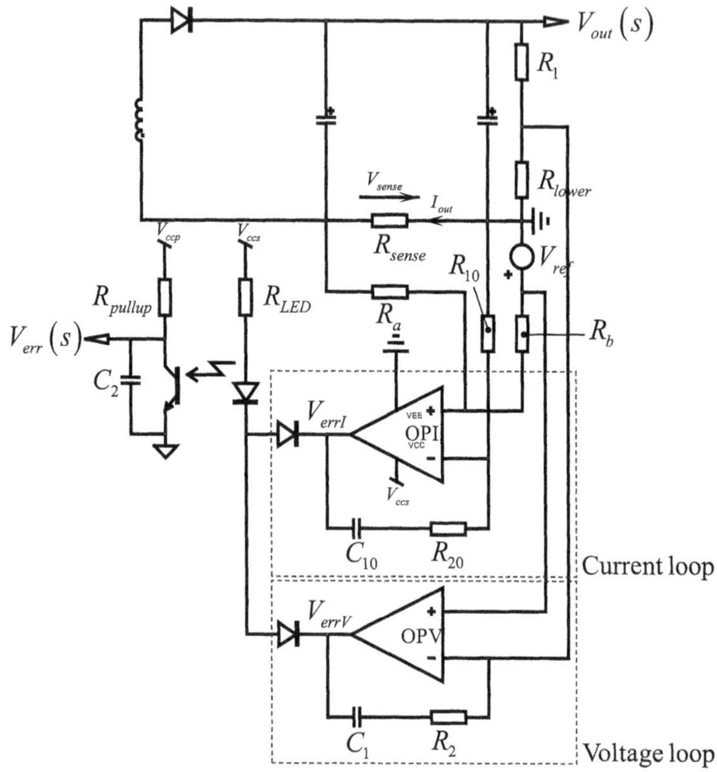

Figure 5.30 A typical charger secondary-side architecture showing both loops and their respective elements.

The trick is to sum the negative voltage across R_{sense} with a portion of the reference voltage so that the noninverting pin of the op amp reaches 0 when the current reaches the imposed limit. To analyze such a system, the best is to use the superposition theorem. The voltage on the noninverting pin is thus the sum of the voltage obtained when V_{ref} is shorted to ground and the voltage obtained once V_{sense}

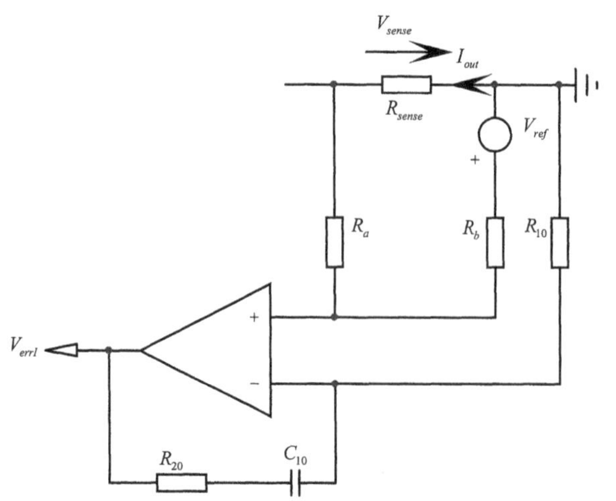

Figure 5.31 The zoomed current loop structure confirms a reference voltage connected to the output ground.

is shorted to ground. As the op amp will strive to maintain 0 V between both of its input pins, we have

$$V(+) = V_{ref}\frac{R_a}{R_a + R_b} + V_{sense}\frac{R_b}{R_a + R_b} = 0 \tag{5.134}$$

Solving for V_{sense}:

$$V_{sense} = -V_{ref}\frac{R_a}{R_b} \tag{5.135}$$

When referenced to the ground, $V_{sense} = -R_{sense}I_{out}$. We can thus extract the output current setpoint:

$$I_{out} = V_{ref}\frac{R_a}{R_{sense}R_b} \tag{5.136}$$

By selecting R_{sense} to minimize the losses but also to maintain a sufficient voltage drop for noise immunity (≈100 mV), R_a and R_b can be further calculated.

A CC-CV regulation circuit works by activating either the current loop if I_{out} is greater than the current setpoint, or the voltage loop, to maintain V_{out} within the target. When the voltage loop is active, the OPV op amp leads the regulation, while the op amp checking the current, OPI, keeps its output high, having its associated series diode blocked: the current regulation block is silent. When the current loop starts to detect that the current exceeds the limit imposed by (5.136), the OPI op amp pulls the optocoupler LED down and the output voltage goes down. The voltage-loop op amp OPV detects this event and increases its output, trying to force the converter to release more energy. As the two op amps outputs are wired in a logical OR configuration via series diodes, the output that pulls the most current out of the diodes leads: the current loop now regulates, V_{out} drops, op amp OPV naturally increases its output, and the voltage-loop diode blocks.

A typical example of such an application is a battery charger: when you connect a discharged battery to the converter, the output current is limited and kept constant as the battery voltage starts to increase. In this mode, the current loop leads. As the voltage increases and approaches the target, the voltage loop takes over and regulates the output level to a safe value.

The principle is depicted by the simplified sketch offered in Figure 5.32, where the switch pictures the action of the two diodes. As you can see, the ac signal coming out of either op amps crosses a common block $G_1(s)$ made of the optocoupler and its associated passive elements. Its transfer function has already been derived several times:

$$G_1(s) = \frac{V_{err}(s)}{V_{op}(s)} = \text{CTR}\frac{R_{pullup}}{R_{LED}}\frac{1}{1 + sR_{pullup}C_2} \tag{5.137}$$

Therefore, if you have decided to compensate the voltage loop according to certain criterion, $G_1(s)$ will be shaped to satisfy them. Unfortunately, this block will also affect the transfer function of the current loop when made active. Therefore, there are chances that the high-frequency pole purposely introduced by $R_{pullup}C_2$ for the voltage loop does not suit the compensation strategy you want for the current loop. However, the loops we want to compensate are almost identical in phase

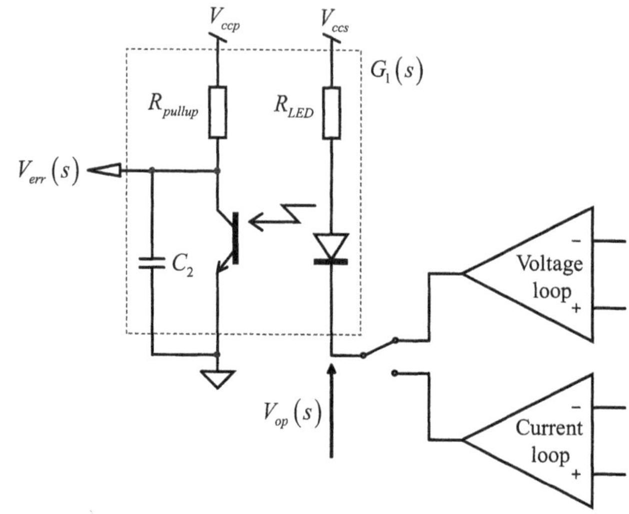

Figure 5.32 A CC-CV regulator features two loops, separately activated via a network of diodes, represented by a two-position switch. In ac, both loops cross a common block noted $G_1(s)$.

response at a given operating point. This is because the output current is the output voltage divided by the load resistor and transformed into a voltage by R_{sense}. We can show that the current loop ac response $H_i(s)$ is that of the voltage loop $H_v(s)$ scaled down by the sense resistor R_{sense} and the load resistor R_{load}:

$$H_i(s) = H_v(s)\frac{R_{sense}}{R_{load}} \tag{5.138}$$

A few iterations will thus be necessary to select a high-frequency pole that matches the voltage-loop requirements but also that of the current loop. We will see that in the design example. Capitalizing on the type 2a equation (5.39), the two separate transfer function expressions can be written the following ways:

$$G_v(s) = \frac{V_{err}(s)}{V_{out}(s)} = -G_1(s)\frac{R_2}{R_1}\left(1+\frac{R_2C_1}{s}\right) = -G_1G_{2v}\frac{1+\omega_{zv}/s}{1+s/\omega_p} \tag{5.139}$$

$$G_i(s) = \frac{V_{err}(s)}{V_{sense}(s)} = -G_1(s)\frac{R_{20}}{R_{10}}\left(1+\frac{R_{20}C_{10}}{s}\right) = -G_1G_{2i}\frac{1+\omega_{zi}/s}{1+s/\omega_p} \tag{5.140}$$

in which we have

$$G_1 = \text{CTR}\frac{R_{pullup}}{R_{LED}} \tag{5.141}$$

$$G_{2v} = \frac{R_2}{R_1} \tag{5.142}$$

$$G_{2i} = \frac{R_{20}}{R_{10}} \tag{5.143}$$

$$\omega_p = \frac{1}{R_{pullup}C_2} \tag{5.144}$$

$$\omega_{zv} = \frac{1}{R_2 C_1} \tag{5.145}$$

$$\omega_{zi} = \frac{1}{R_{20} C_{10}} \tag{5.146}$$

As expected, the high-frequency pole ω_p is common to both transfer functions. Once fixed for the first compensated loop, the designer will have to adjust the zero for the second loop in order to meet the required phase boost. Please note that we purposely adopted a control scheme where the fast lane has been suppressed: the LED anode is connected to an auxiliary voltage V_{ccs}. Furthermore, this auxiliary voltage, V_{ccs}, needs to be maintained when V_{out} collapses to zero; otherwise, the regulation would be lost with consequences depending on the control scheme. A few solutions helping to solve this problem will be further presented.

5.5.11 A Design Example

We have a discontinuous conduction mode (DCM) flyback converter operated in current-mode control, which delivers a 12 V output voltage. We want to set the constant current section to a maximum current of 1 A.

First, let us calculate the resistive network R_a, R_b, and R_{sense} of Figure 5.31. To arbitrarily limit the power dissipation to 100 mW on the sensing resistor, we can quickly derive its value:

$$R_{sense} = \frac{P_{R_{sense}}}{I_{out}^2} = \frac{0.1}{1} = 100 \text{ m}\Omega \tag{5.147}$$

Then, if the reference voltage in our integrated circuit is 2.5 V, we can fix one of the two resistors R_a or R_b and select the other one. If R_a is arbitrarily fixed to 2 kΩ and the sensing shunt is 100 mΩ, then according to (5.136), R_b equals

$$R_b = V_{ref} \frac{R_a}{R_{sense} I_{out}} = 2.5 \frac{2k}{0.1 \times 1} = 50 \text{ k}\Omega \tag{5.148}$$

To compensate the loops, we need open-loop plots of the converter we have designed, observing either the output voltage, $H_v(s)$, or the voltage image of the output current, $H_i(s)$. The results appear in Figure 5.33.

We have purposely selected two different crossover frequencies for the voltage and current loops, 1 kHz and 400 Hz, respectively, but equal values would also work. Let's tackle the voltage loop compensation first using the following data:

VOL = 0.2 V; the op amp minimum output voltage.
V_f = 1 V; the LED forward voltage.
$V_{CE,sat}$ = 0.3 V; the optocoupler saturation voltage.
V_{ccp} = 5 V; the primary-side pull-up V_{cc} level.
V_{ccs} = 8.2 V; the secondary-side V_{cc} level.
R_{pullup} = 20 kΩ; the optocoupler pull-up resistor given by the selected controller.

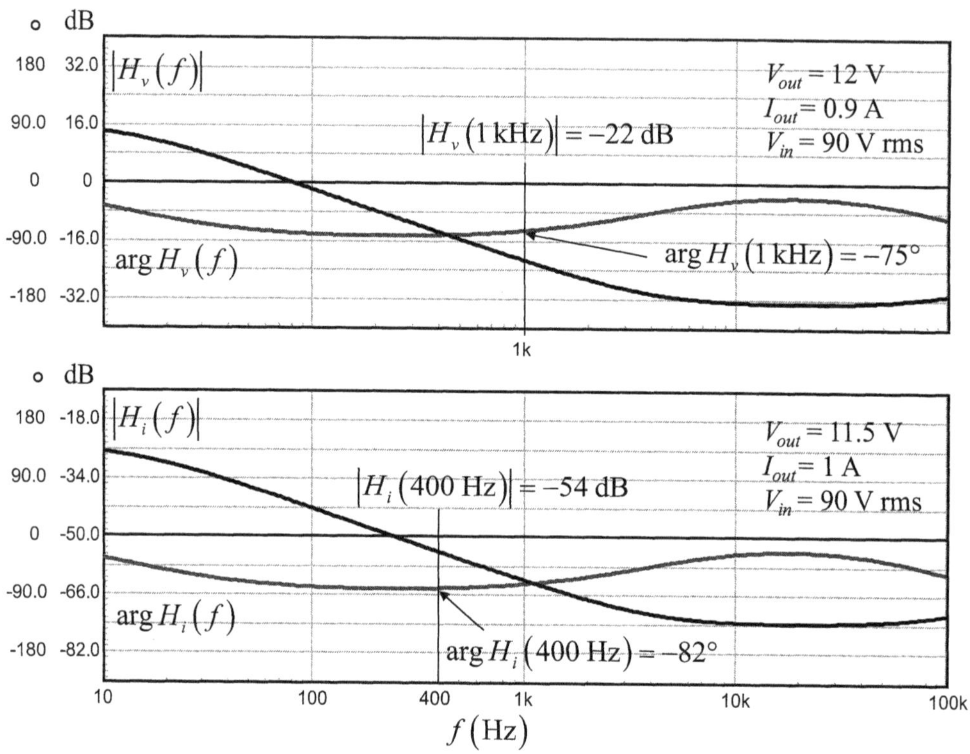

Figure 5.33 The output voltage and output current open-loop gains of a flyback converter operated in the DCM.

$\text{CTR}_{\text{min}} = 0.8$; the minimum optocoupler current transfer ratio.
$V_{out} = 12$ V; the converter output voltage.
$R_1 = 38$ kΩ; the upper resistor in the resistor bridge observing the output variable.
$R_{10} = 1$ kΩ; the resistor sensing the output current and connected to ground.
$f_{opto} = 6$ kHz; the optocoupler pole that has been characterized with R_{pullup}.

We first start by working the G_1 block, made of the optocoupler element. The maximum LED resistor value is computed using (5.121):

$$R_{LED,\max} = \frac{R_{pullup}(V_{ccs} - V_f - \text{VOL})\text{CTR}_{\min}}{(V_{ccp} - V_{CE,sat})} = \frac{20k \times (8.2 - 1 - 0.2) \times 0.8}{5 - 0.3} \approx 23.8 \text{ k}\Omega \tag{5.149}$$

Given the 20-kΩ pull-up resistor on the primary and its associated small pull-down current (235 μA), we have plenty of margin on the LED resistor selection. Let's adopt a 10-kΩ value. Following (5.141) the gain brought by G_1 is thus

$$G_1 = \text{CTR}\frac{R_{pullup}}{R_{LED}} = 0.8\frac{20k}{10k} = 1.6 \approx 4.1 \text{ dB} \tag{5.150}$$

For a 1-kHz crossover frequency, we need to lift up the gain curve by $G_v = 22$ dB (see Figure 5.33). As G_1 already equals 4.1 dB, G_{2v} will need to be

$$G_{2v} = G_v - G_1 = 22 - 4.1 = 17.9 \text{ dB} \tag{5.151}$$

Given a 38-kΩ resistor value for the upper resistor bridge divider, using (5.142) we have

$$R_2 = R_1 10^{\frac{G_{2v}}{20}} = R_1 10^{\frac{17.9}{20}} = 38k \times 7.85 = 298 \text{ k}\Omega \tag{5.152}$$

The phase lag of $H_v(s)$ is –75° at 1 kHz. For a 60° phase margin, we will need to extract the phase boost using the following formula, already seen in Chapter 4:

$$\arg H(f_c) - [270° - boost] = -360° + \varphi_m \tag{5.153}$$

The necessary phase boost thus amounts to

$$boost = -360° + \varphi_m - \arg H(f_c) + 270° = -90 + 60 + 75 = 45° \tag{5.154}$$

To satisfy this required boost at a 1-kHz crossover frequency, the pole will be placed at the following location:

$$f_p = \left[\tan(boost) + \sqrt{\tan^2(boost) + 1}\right] f_c = 2.4 \times 1k = 2.4 \text{ kHz} \tag{5.155}$$

Given a crossover frequency placed at the geometric mean between the zero and pole, the zero will be placed at

$$f_z = \frac{f_c^2}{f_p} = \frac{1k \times 1k}{2.4k} \approx 416 \text{ Hz} \tag{5.156}$$

We can now calculate capacitor C_1 to be placed in series with R_2 using (5.145):

$$C_1 = \frac{1}{2\pi R_2 f_z} = \frac{1}{6.28 \times 298k \times 416} \approx 1.3 \text{ nF} \tag{5.157}$$

Implementing (5.144), the capacitor taking place over the optocoupler collector is found to be

$$C_2 = \frac{1}{2\pi f_p R_{pullup}} = \frac{1}{6.28 \times 2.4k \times 20k} = 3.3 \text{ nF} \tag{5.158}$$

We know that the optocoupler parasitic capacitor C_{opto} contributes to C_2:

$$C_{opto} = \frac{1}{2\pi f_{opto} R_{pullup}} = \frac{1}{6.28 \times 6k \times 20k} = 1.3 \text{ nF} \tag{5.159}$$

We thus need to add a small extra capacitor C_{col} whose value is

$$C_{col} = C_2 - C_{opto} = 3.3n - 1.3n = 2 \text{ nF} \tag{5.160}$$

The voltage loop is now fully compensated. Let's take a look at the current loop now. Looking back to Figure 5.33, the gain deficiency is 54 dB at 400 Hz (G_i =

54 dB), together with a phase lag of 82°. According to (5.150), we need to provide an additional gain of

$$G_{2i} = G_i - G_1 = 54 - 4.1 = 49.9 \text{ dB} \tag{5.161}$$

Using (5.143), we can immediately calculate the series resistor R_{20} by arbitrarily affecting to R_{10} a value of 1 kΩ:

$$R_{20} = R_{10}10^{\frac{G_{2i}}{20}} = R_{10}10^{\frac{49.9}{20}} = 1k \times 312 = 312 \text{ k}\Omega \tag{5.162}$$

For a phase lag of 82° (see Figure 5.33), how much phase boost must we provide to reach a phase margin of 60°:

$$BOOST = -360° + \varphi_m - \arg H(f_c) + 270° = -90 + 60 + 82 = 52° \tag{5.163}$$

We know that a pole already exists, given by the presence of the pull-up resistor and capacitor C_2. For the voltage loop stability purposes, this pole is currently

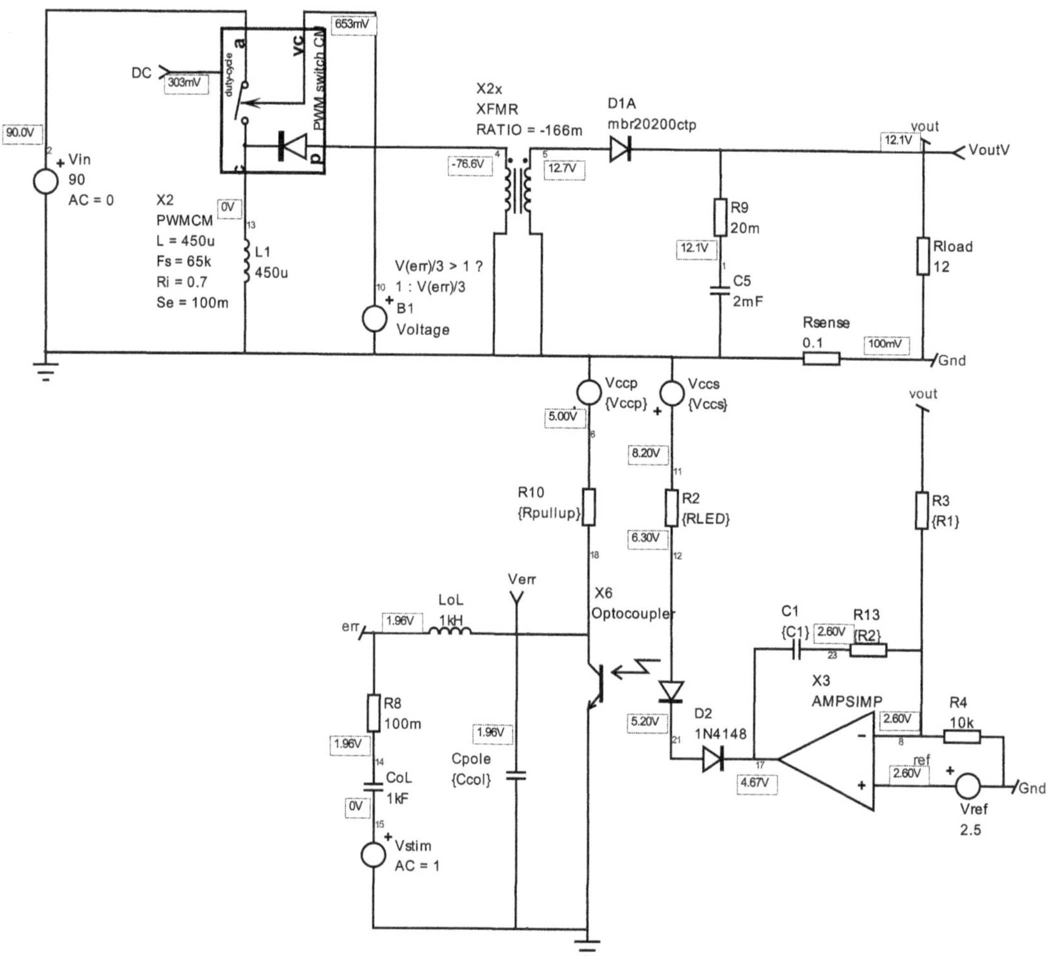

Figure 5.34 A simulation template helps us to check the final ac response once all components have been calculated. Here the voltage loop chain is not represented.

present at a 2.4-kHz frequency, as explained by (5.155). Where to place the zero in order to boost the phase by 52° at 400 Hz? As already explained in Chapter 4, we have

$$f_z = \frac{f_c}{\tan\left(boost + \tan^{-1}\left(\frac{f_c}{f_p}\right)\right)} = \frac{400}{\tan\left(52 + \tan^{-1}\left(\frac{400}{2.4k}\right)\right)} = 217 \text{ Hz} \tag{5.164}$$

Now using (5.146), we obtain

$$C_{20} = \frac{1}{2\pi f_z R_{20}} = \frac{1}{6.28 \times 217 \times 312k} = 2.35 \text{ nF} \tag{5.165}$$

That's it! We are all set now and can feed our simulation templates with the previously calculated values. This is what Figure 5.34 and Figure 5.35 show you.

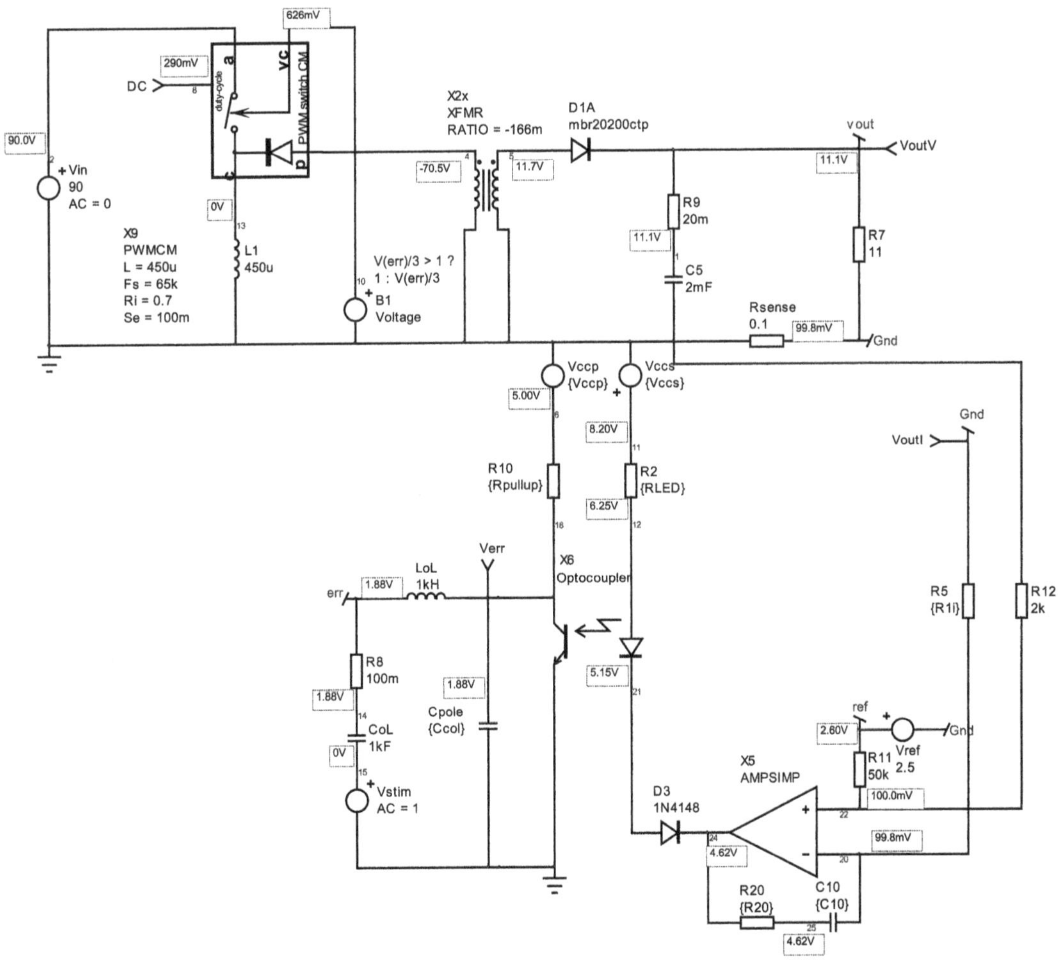

Figure 5.35 The current loop chain where the load is adjusted to consume 1 A, but V_{out} is slightly below the regulation point: this is the maximum output power.

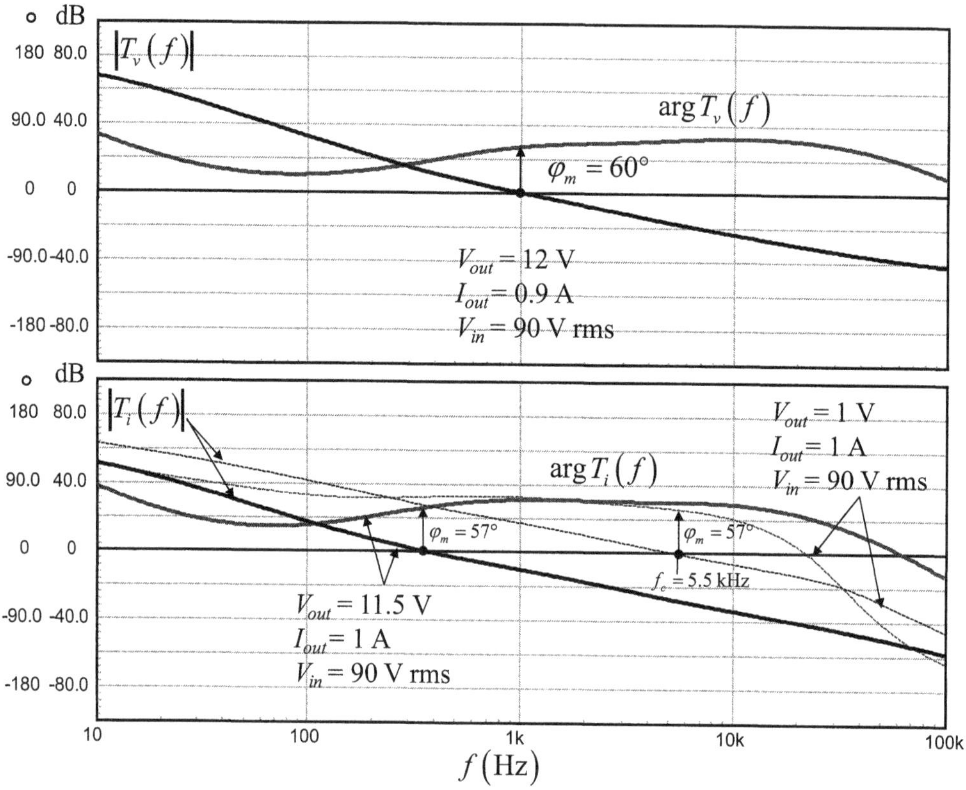

Figure 5.36 The loop gain response of both voltage and current lanes confirm the good choice of the compensation elements.

The complete open loop gain response for both chains appears in Figure 5.36 and confirms the pertinence of our compensation choices.

When the load resistance decreases in the constant-current section, the denominator of (5.138) goes down and the overall gain increases. As you can see in Figure 5.36, the loop gain of the current loop increases and extends the crossover to 5.5 kHz with a comfortable phase margin. The fact that we selected a lower bandwidth for the current loop compared to that of the voltage loop has naturally pushed the voltage-loop high-frequency pole away, leaving the current loop phase flat beyond 400 Hz. Thanks to this flat area and despite an increase in the crossover frequency, the phase margin in the current loop remains excellent.

We can now test the output voltage evolution versus the output current. This is what Figure 5.37 shows: we swept the output current while monitoring the output voltage. In the simulation tool, we simply replaced the load by a variable resistor and recorded both the output current and voltage. The graph testifies for an excellent transition between the two loops as I_{out} reaches 1 A.

There are several dedicated controllers for this type of constant-current/constant-voltage operation. ON Semiconductor offers the MC33341 or the NCP4300. One can also use a dual op amp like the LM358 to which a TL431 voltage reference is associated.

Whatever solution is used, you need to make sure the V_{cc} of these circuits always remains sufficient when V_{out} drops. Otherwise, troubles may occur when the

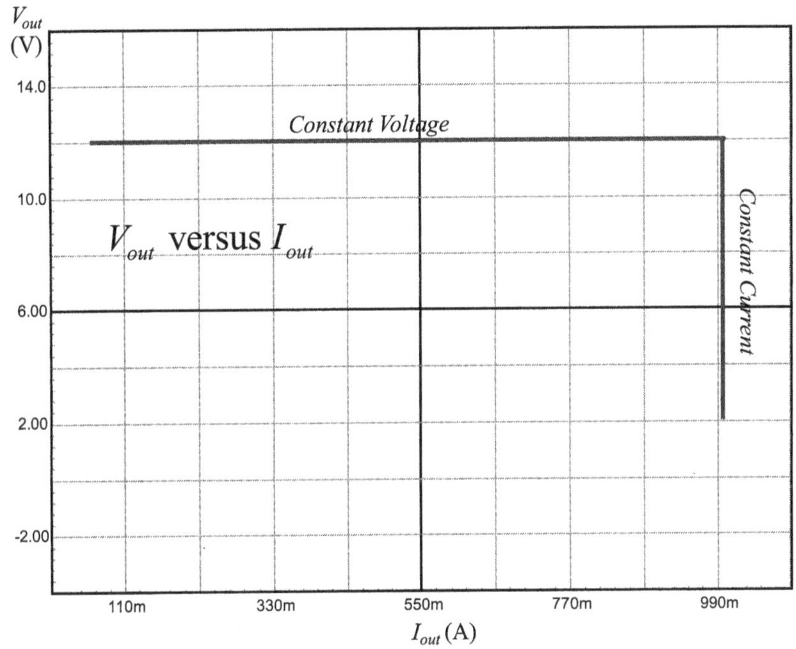

Figure 5.37 The output characteristics show a perfectly stable transition when I_{out} reaches 1A.

output voltage passes below the minimum operating voltage of these op amps. In a flyback converter application, one solution is to rotate the secondary-side rectifier as proposed in Figure 5.38 and observe that its cathode now swings to

$$V_k = V_{out} + NV_{in} \tag{5.166}$$

Therefore, when the output voltage disappears, NV_{in} is always there and supplies the op amp chain all the way down to a 0 V output. Unfortunately, the term NV_{in} can be significantly higher than the output voltage, in particular at the highest input line level. To cope with this large excursion, a simple regulator is implemented with a bipolar transistor Q_1 placed after a peak rectifying circuit (D_3C_1). To avoid an extra power dissipation when V_{out} is still present, let's say above 8 V, the Zener diode is selected such that Q_1 and D_1 are blocked when V_{out} is above 8 V. If we select an 8.2-V Zener diode, the voltage at D_1 cathode is roughly $8.2 - 2V_f \approx 6.9\ V$. When V_{out} plunges below this value, D_2 blocks and the regulator supplies the op amp chain. If the Zener voltage is sufficiently well decoupled from V_{out}, it can be used to supply the LED current chain and become the V_{ccs} point in Figure 5.30. R_z in this case will supply the bipolar base current plus the LED bias current.

5.6 The Type 2: Pole and Zero are Coincident to Create an Isolated Type 1

There are cases where no phase boost is necessary, just a high dc gain and a simple –1-slope to force a gain rolloff as the frequency increases. This is a simple integrator,

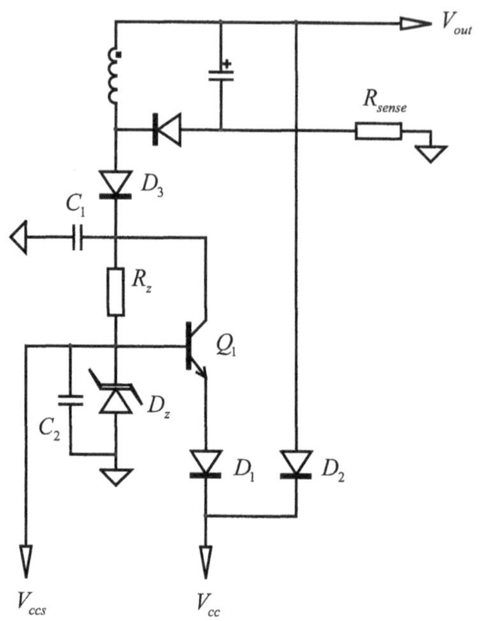

Figure 5.38 The rotation of the rectifier offers an easy means to self-supply the circuit despite the collapsing output voltage.

already identified as being a type 1 (see Section 5.1). An isolated type 2, in which both pole and zero coincide, can quickly be transformed into a type 1 compensator. However, we have seen before that the op amp pulling the LED cathode to ground (see Figure 5.21) is affected by a fast lane and sees its possible gain choices limited as shown with (5.104). The main offender in that case is the LED resistor, which combines a role in the bias point but also in the mid-band gain definition. However, in a type 1 configuration, there is no such thing as mid-bang gain: the 0-dB crossover pole f_{po} is selected to force a crossover at the desired frequency f_c. To unveil the 0-dB crossover pole frequency, we will rearrange the original isolated type 2 definition given by (5.93):

$$G(s) = -\frac{R_{pullup}}{R_{LED}}\text{CTR}\frac{1+sR_1C_1}{sR_1C_1(1+sR_{pullup}C_2)} = -\left(\frac{1}{s\dfrac{R_1R_{LED}}{R_{pullup}\text{CTR}}C_1}\right)\left(\frac{sR_1C_1+1}{1+sR_{pullup}C_2}\right) \tag{5.167}$$

This equation can be put under the following form where the mid-band gain definition disappears:

$$G(s) = -\frac{1}{s/\omega_{po}}\frac{1+s/\omega_z}{1+s/\omega_p} \tag{5.168}$$

In this expression, we have

$$\omega_{po} = \frac{1}{\dfrac{R_1R_{LED}}{R_{pullup}\text{CTR}}C_1} \tag{5.169}$$

$$\omega_p = \frac{1}{R_{pullup}C_2} \tag{5.170}$$

$$\omega_z = \frac{1}{R_1C_1} \tag{5.171}$$

As both the pole and the zero are placed at a similar location, (5.170) and (5.171) must be equal:

$$\frac{1}{R_{pullup}C_2} = \frac{1}{R_1C_1} \tag{5.172}$$

Solving for C_1:

$$C_1 = C_2\frac{R_{pullup}}{R_1} \tag{5.173}$$

Now, substituting (5.173) into (5.169) and solving for C_2, we find

$$C_2 = \frac{\mathrm{CTR}}{2\pi f_{po}R_{LED}} \tag{5.174}$$

In the previous expression, the selection of R_{LED} will be dictated by bias considerations only, still using (5.103) for this purpose:

$$R_{LED,\max} = \frac{R_{pullup}(V_{out} - V_f - \mathrm{VOL})\mathrm{CTR}_{\min}}{(V_{cc} - V_{CE,sat})} \tag{5.175}$$

The position of f_{po} depends on the wanted gain at crossover. Using (5.7) and (5.8), we can easily find its position:

$$f_{po} = f_cG \tag{5.176}$$

Once C_2 is found, (5.173) will lead to the value of C_1. That's it—we have our type 1 in a pull-down type. Time to see a design example.

5.6.1 A Design Example

Suppose we need to create an attenuation of –20 dB at a 100-Hz crossover frequency without any phase boost. The design parameters are the following ones:

VOL = 0.2 V; the op amp minimum output voltage.
V_f = 1 V; the LED forward voltage.
$V_{CE,sat}$ = 0.3 V; the optocoupler saturation voltage.
V_{cc} = 5 V; the primary-side pull-up V_{cc} level.
R_{pullup} = 20 kΩ; the optocoupler pull-up resistor.
$\mathrm{CTR}_{\min}$ = 0.3; the minimum optocoupler current transfer ratio.
V_{out} = 12 V; the converter output voltage.

$R_1 = 38$ kΩ; the upper resistor in the resistor bridge observing the output variable.

$f_{opto} = 6$ kHz; the optocoupler pole that has been characterized with R_{pullup}.

The maximum LED resistor value is computed using (5.175):

$$R_{LED,\max} = \frac{R_{pullup}(V_{out} - V_f - \text{VOL})\text{CTR}_{\min}}{(V_{cc} - V_{CE,sat})} = \frac{20k \times (12 - 1 - 0.2) \times 0.3}{5 - 0.3} \approx 13.8 \text{ k}\Omega \tag{5.177}$$

Applying a design margin, we chose a 10-kΩ value. Now, applying (5.7) and (5.176), we first determine the position of the 0-dB crossover pole f_{po}:

$$f_{po} = f_c G = 100 \times 10^{-\frac{20}{20}} = 10 \text{ Hz} \tag{5.178}$$

Placing a pole at 10 Hz implies a fairly large capacitor value for C_2 using (5.174):

$$C_2 = \frac{\text{CTR}}{2\pi f_{po} R_{LED}} = \frac{0.3}{6.28 \times 10 \times 10k} = 477 \text{ nF} \tag{5.179}$$

C_1 comes easily with (5.173):

$$C_1 = C_2 \frac{R_{pullup}}{R_1} = 477n \times \frac{20k}{38k} = 251 \text{ nF} \tag{5.180}$$

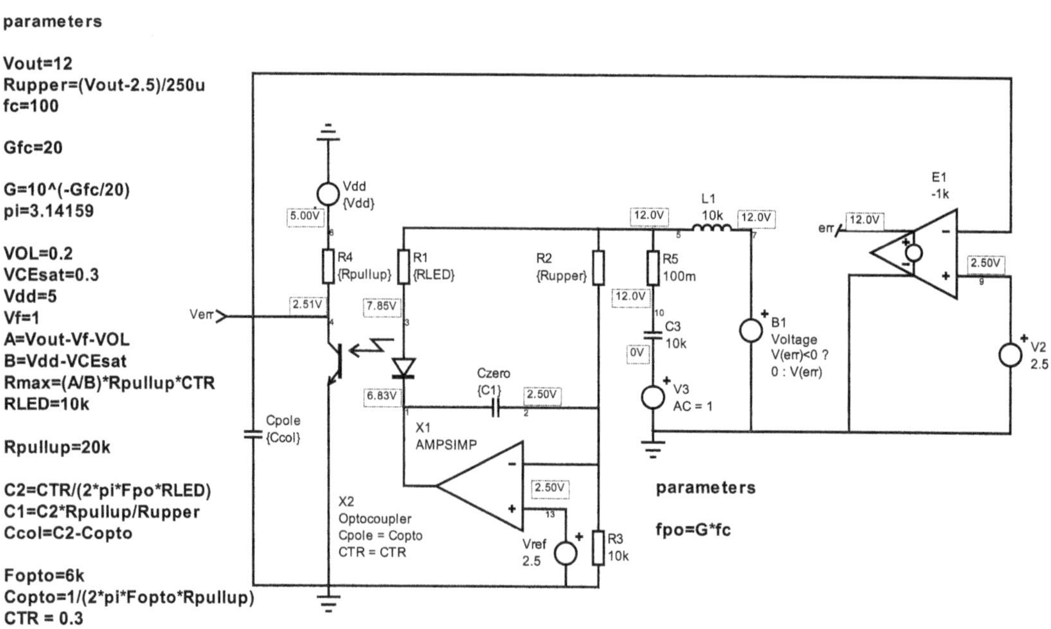

Figure 5.39 The isolated type 1 is formed by making the pole and the zero coincide. The LED resistor is calculated for bias purposes only.

The optocoupler parasitic capacitance is found to be

$$C_{opto} = \frac{1}{2\pi f_{opto} R_{pullup}} = \frac{1}{6.28 \times 6k \times 20k} = 1.3\,\text{nF} \tag{5.181}$$

Subtracted from C_2, the final capacitor C_{col} will be 475 nF or a normalized 0.47 μF. Now that we are all set, we can simulate the structure as demonstrated by Figure 5.39.

The results appear in Figure 5.40 and confirm the 20-dB attenuation at 10 Hz.

5.7 The Type 2: A Slightly Different Arrangement

In some cases, when the selected crossover frequency is low, with the help of their large values, the charging times of C_2 and C_1 create a natural soft-start effect: the error amplifier output slowly rises up and limits the power-on stress such as peak power dissipation in the switches. In a power factor corrector (PFC) circuit, it usually contributes to calm down the output overshoot. However, in a classical type 2 configuration, the charging capacitor is split between C_2 and C_1, with a resistor R_2 in series for C_1. In some high-power converters, the soft-start duration might not be long enough, despite large capacitor values. How could we keep the same crossover frequency and fully capitalize on the presence of these large capacitors? Thanks to a discussion initiated by my colleague Mr. Jim Young from ON Semiconductor, I found that a different arrangement was possible. Before unveiling it, let's understand the phenomenon in play at start-up.

At power up, as no output voltage exists on the converter output, the op amp pushes its output to the maximum. However, it quickly finds a limit as the voltage on its inverting pin immediately builds up to the reference voltage present on its noninverting input. This is because with both capacitors being discharged, the op amp output current crosses them and enters the divider bridge, imposing a "regulated" voltage on the inverting pin as any op amp should do. Therefore, rather than immediately delivering the full control to the control pin, the error voltage slowly rises, depending on the capacitor values and the current authorized by the resistors

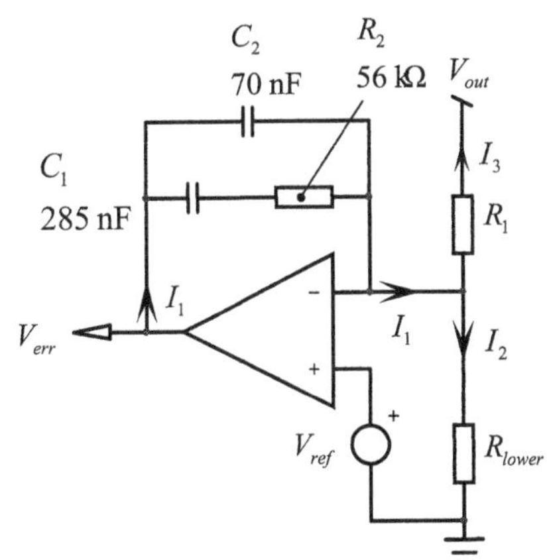

Figure 5.40 These ac results show the correct attenuation value at 100 Hz.

bridge. Figure 5.41 shows how the various currents split in the circuit during the startup sequence as V_{out} is null.

From this representation, we can write a few simple electrical laws:

$$I_2 = I_1 - I_3 = \frac{V_{ref}}{R_{lower}} \tag{5.182}$$

$$I_3 = \frac{V_{ref} - V_{out}}{R_1} \tag{5.183}$$

By substituting (5.183) into (5.182) and solving for the charging current $i_1(t)$, we have

$$i_1(t) = \frac{V_{ref}}{R_{lower}} + \frac{V_{ref} - v_{out}(t)}{R_1} \tag{5.184}$$

At startup, as $v_{out}(t)$ is zero, this equation can be arranged as

$$I_1 = \frac{V_{ref}}{R_{lower}} + \frac{V_{ref}}{R_1} = \frac{V_{ref}}{R_{lower} \parallel R_1} \tag{5.185}$$

In PFCs, R_1 is much larger than R_{lower} and the charging current simplifies to

$$I_1 \approx \frac{V_{ref}}{R_{lower}} \tag{5.186}$$

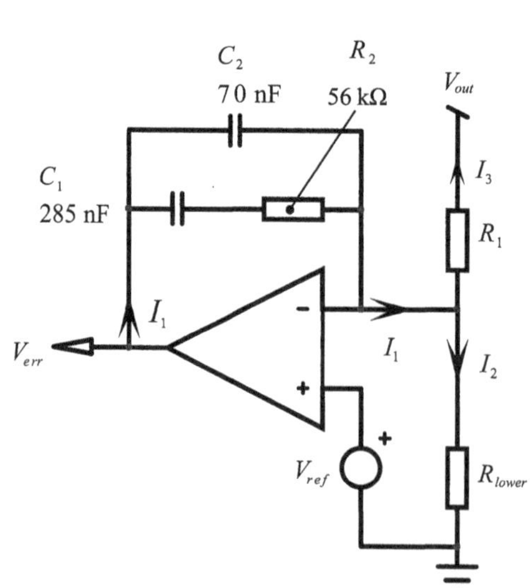

Figure 5.41 The op amp output voltage rise is not limited by the op amp output current capability but rather by the reference pin and the bottom resistor R_{lower}.

This current splits between C_2 and C_1, this latter being in series with the resistor R_2. The current has the choice between a short-circuit (C_2 is discharged) and a resistive path (C_1 discharged plus R_2), and most of the charging current flows within C_2, which is quickly charged by the current defined by (5.185). Powering Figure 5.41 configuration gives a linearly rising waveform, as Figure 5.42 confirms. It takes 1.2 ms before the converter can deliver its full power after the power-on sequence. For some PFC, this natural delay is not enough and cannot tame the output overshoot.

In the previous equations, the presence of R_2 prevents the current from immediately charging C_1. Why not change the configuration where C_2 would stay alone, and have the R_2C_1 network go over the upper resistor R_1? This is exactly what Figure 5.43 suggests.

The set of equations differs a little from what has been previously derived, but the spirit remains the same:

$$\frac{Z_f}{Z_i} = \frac{\frac{1}{sC_2}}{\left(\frac{1}{sC_1} + R_2\right) \| R_1} \tag{5.187}$$

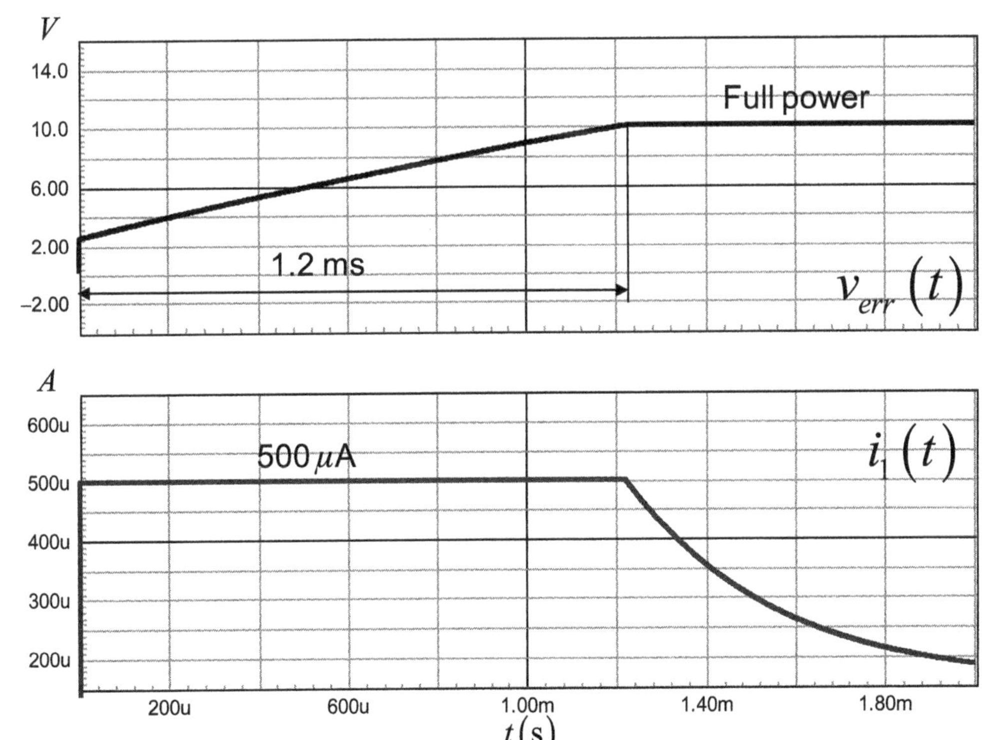

Figure 5.42 Despite a large capacitive value across the op amp (C_1 and C_2) value, the 500-μA charging current quickly brings the op amp output voltage up to its maximum level.

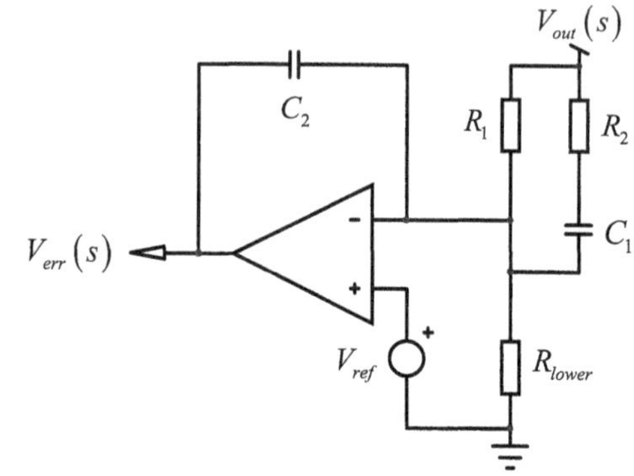

Figure 5.43 By moving the *RC* network across the upper resistor, the charging time is increased, providing additional free soft-start.

Developing and rearranging (5.187), we have

$$G(s) = -\frac{C_1(R_1+R_2)}{R_1C_2}\frac{\left(1+\frac{1}{s(R_1+R_2)C_1}\right)}{1+sR_2C_1} \tag{5.188}$$

In the previous equation, the pole, zero, and mid-band gain are immediately identified:

$$G_0 = \frac{R_1+R_2}{R_1}\frac{C_1}{C_2} \tag{5.189}$$

$$\omega_z = \frac{1}{(R_2+R_1)C_1} \tag{5.190}$$

$$\omega_p = \frac{1}{R_2C_1} \tag{5.191}$$

From the previous definition, we can extract the component values we are looking for:

$$C_1 = \frac{1}{2\pi f_z(R_1+R_2)} \tag{5.192}$$

$$C_2 = \frac{1}{2\pi f_z G_0 R_1} \tag{5.193}$$

$$R_2 = \frac{R_1 f_z}{f_p - f_z} \tag{5.194}$$

In (5.16), we have seen that the pole position was dependent upon R_2 but also the sum of C_1 and C_2. As in (5.191) the pole location now solely depends on C_2, for a similar pole position imposed in either Figure 5.5 or in Figure 5.43, C_2 in the second case will be much larger than in the first. Therefore, as the entire current defined by (5.185) will cross it at startup, we can expect a longer charging time than with the traditional type 2 configuration.

To verify this assumption, we have assembled a front-end power factor corrector averaged circuit whose return loop is made either by the classical type 2 approach (Figure 5.5) or by the rearranged configuration as in Figure 5.43. The schematic of the test fixture appears in Figure 5.44.

In this example, we have automated the calculations of the type 2 classical (R_2, C_2, and C_1) definitions and the proposed variation (R_{20}, C_{20}, and C_{10}). Then, we performed a startup sequence followed by a load step transition imposed by the output switch X_4. The pole and zero are, respectively, positioned at 61 Hz and 6.5 Hz. For the first case, Figure 5.5, C_1 equals 148 nF and C_2 is 16 nF. As expected, for a similar pole position, the second configuration (Figure 5.43) leads to a C_2 value of 154 nF, the sum of the previous C_1-C_2 values.

The waveforms collected in both configurations appear in Figure 5.45. The first configuration shows a rather fast output voltage rise $v_{out1}(t)$, affected by a 30-V overshoot. On the bottom of the simulation shot, the error voltage exhibits a sharp transition with almost no delay: the power is pushed to the maximum right after power-on. On the contrary, when the second configuration is tested, the output voltage settlement is much smoother and the overshoot is gone. This is testified by the observation of the error amplifier output showing an improved soft-start effect, due to the charging of the large capacitor C_2, now alone across the op amp output and inverting input. The transient response implemented shortly after the startup period shows similar responses with either options, confirming similar ac loop gain and phase.

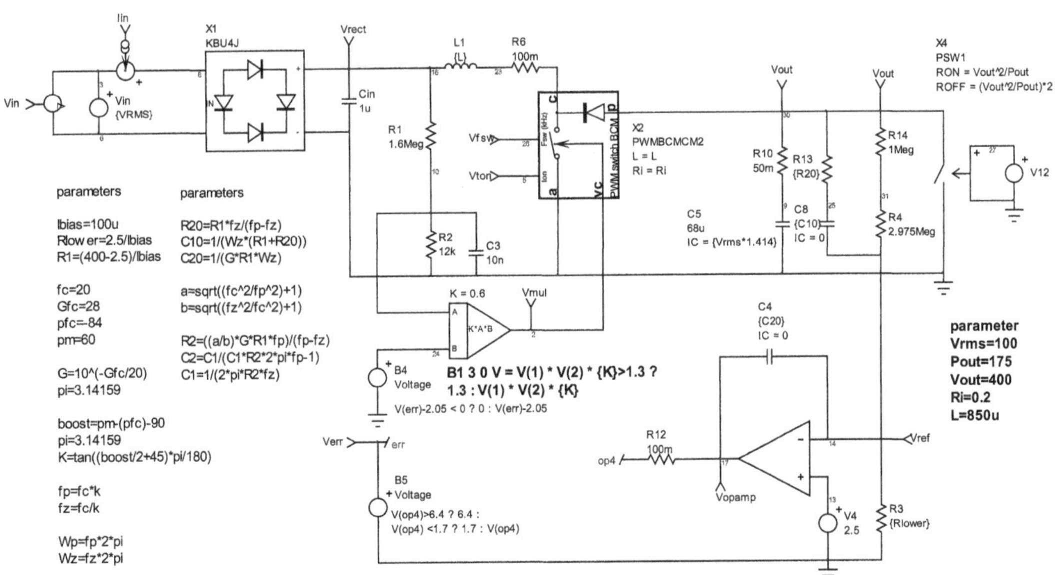

Figure 5.44 A classical PFC preconverter powered by a peak-current-mode controller, the MC33262.

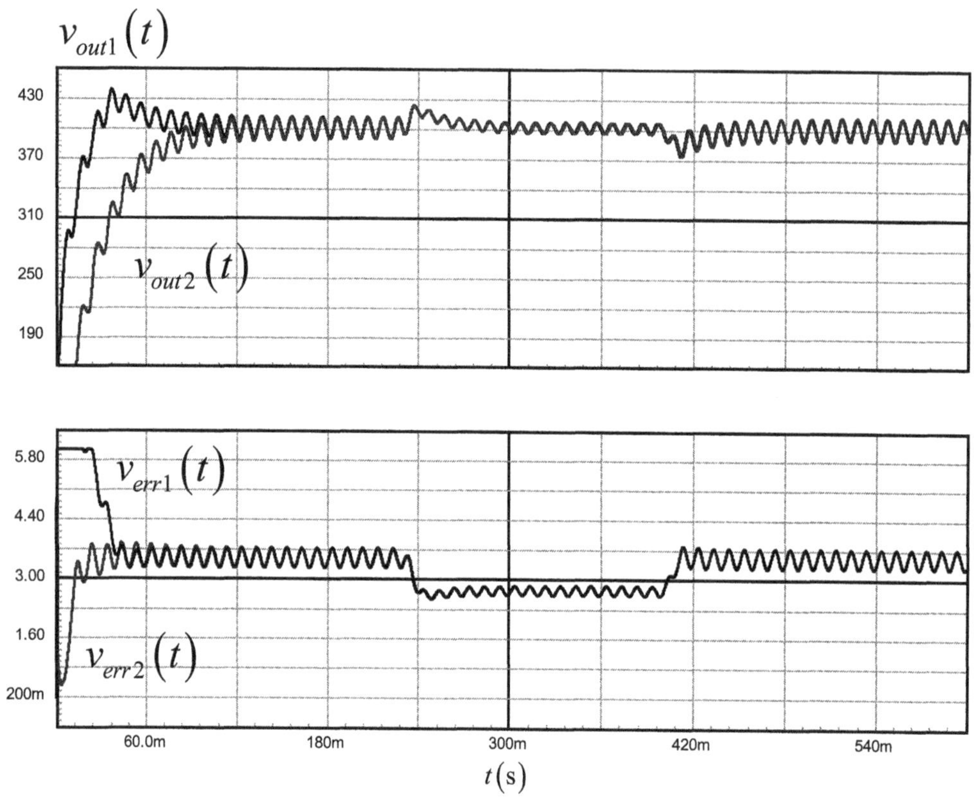

Figure 5.45 The results show the absence of overshoot in the adopted type 2 variation but a similar transient response, implying that both compensators position the poles and zeros at the exact same place.

5.8 The Type 3: An Origin Pole, a Pole/Zero Pair

The type 3 compensator is used when more than 90° of phase boost are necessary. By adding another zero/pole pair to the type 2 circuit, the type 3 can theoretically boost the phase up to 180°. Its architecture appears in Figure 5.46. The derivation of its transfer function does not really change; the principle remains the same: calculate the equivalent impedance Z_f placed across the op amp, and divide it by that of Z_i to further rearrange the equation:

$$Z_f = \left(\frac{1}{sC_1} + R_2\right)\frac{1}{sC_2} \bigg/ \left(\frac{1}{sC_1} + R_2\right) + \frac{1}{sC_2} \tag{5.195}$$

$$Z_i = \left(\frac{1}{sC_3} + R_3\right)R_1 \bigg/ \left(\frac{1}{sC_3} + R_3\right) + R_1 \tag{5.196}$$

Then, when dividing (5.195) by (5.196), we have

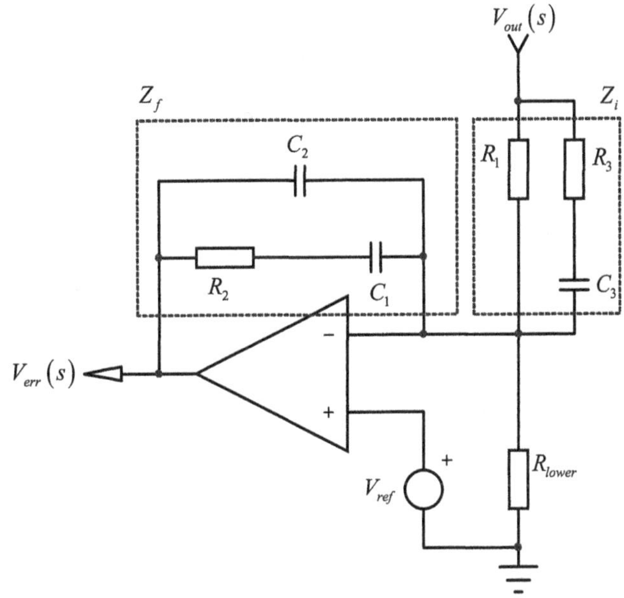

Figure 5.46 The type 3 compensator can boost the phase up to 180°.

$$\frac{Z_f}{Z_i} = G(s) = \frac{\left(\frac{1}{sC_1} + R_2\right)\frac{1}{sC_2} \bigg/ \left(\frac{1}{sC_1} + R_2\right) + \frac{1}{sC_2}}{\left(\frac{1}{sC_3} + R_3\right)R_1 \bigg/ \left(\frac{1}{sC_3} + R_3\right) + R_1} \tag{5.197}$$

By factoring sR_2C_1, we obtain a more familiar expression:

$$G(s) = -\frac{R_2C_1}{R_1(C_1 + C_2)} \frac{\frac{1}{sR_2C_1} + 1}{1 + sR_2\frac{C_1C_2}{C_1 + C_2}} \frac{sC_3(R_1 + R_3) + 1}{sR_3C_3 + 1} \tag{5.198}$$

This expression can be put under the normalized form:

$$G(s) = -G_0 \frac{\left(1 + \frac{\omega_{z1}}{s}\right)\left(1 + \frac{s}{\omega_{z2}}\right)}{\left(1 + \frac{s}{\omega_{p1}}\right)\left(1 + \frac{s}{\omega_{p2}}\right)} \tag{5.199}$$

where

$$G_0 = \frac{R_2}{R_1} \frac{C_1}{C_1 + C_2} \tag{5.200}$$

$$\omega_{z1} = \frac{1}{R_2C_1} \tag{5.201}$$

$$\omega_{z2} = \frac{1}{(R_1 + R_3)C_3} \tag{5.202}$$

$$\omega_{p1} = \frac{1}{R_2 \dfrac{C_1 C_2}{C_1 + C_2}} \tag{5.203}$$

$$\omega_{p2} = \frac{1}{R_3 C_3} \tag{5.204}$$

In the previous expressions, we have two zeros and two poles. They both can be coincident or spread apart of each other's. The design of such a compensator starts with the value of the series resistor R_2. To find it, we must derive the magnitude of $G(s)$ via its definition in (5.199):

$$|G(f_c)| = G_0 \frac{\sqrt{1+\left(\dfrac{f_{z1}}{f_c}\right)^2}\sqrt{1+\left(\dfrac{f_c}{f_{z2}}\right)^2}}{\sqrt{1+\left(\dfrac{f_c}{f_{p1}}\right)^2}\sqrt{1+\left(\dfrac{f_c}{f_{p2}}\right)^2}} = \frac{R_2 C_1}{R_1(C_1 + C_2)} \frac{\sqrt{1+\left(\dfrac{f_{z1}}{f_c}\right)^2}\sqrt{1+\left(\dfrac{f_c}{f_{z2}}\right)^2}}{\sqrt{1+\left(\dfrac{f_c}{f_{p1}}\right)^2}\sqrt{1+\left(\dfrac{f_c}{f_{p2}}\right)^2}} \tag{5.205}$$

By extracting C_1 and C_2 from (5.201) and (5.203), substituting them into (5.205) then solving for R_2, we have

$$R_2 = \frac{G R_1 f_{p1}}{f_{p1} - f_{z1}} \frac{\sqrt{1+\left(\dfrac{f_c}{f_{p1}}\right)^2}\sqrt{1+\left(\dfrac{f_c}{f_{p2}}\right)^2}}{\sqrt{1+\left(\dfrac{f_{z1}}{f_c}\right)^2}\sqrt{1+\left(\dfrac{f_c}{f_{z2}}\right)^2}} \tag{5.206}$$

The rest of the elements comes out easily:

$$C_1 = \frac{1}{2\pi f_{z_1} R_2} \tag{5.207}$$

$$C_2 = \frac{C_1}{2\pi f_{p_1} C_1 R_2 - 1} \tag{5.208}$$

$$C_3 = \frac{f_{p2} - f_{z2}}{2\pi R_{upper} f_{p2} f_{z2}} \tag{5.209}$$

$$R_3 = \frac{R_1 f_{z2}}{f_{p2} - f_{z2}} \tag{5.210}$$

In most of the applications, capacitor C_2 is much smaller than C_1, and resistor R_3 is also smaller than R_1. Therefore, the transfer equation (5.198) simplifies to

$$G(s) \approx -\frac{R_2}{R_1}\frac{\dfrac{1}{sR_2C_1}+1}{1+sR_2C_2}\frac{sC_3R_1+1}{sR_3C_3+1} = -G_0\frac{\left(1+\dfrac{\omega_{z_1}}{s}\right)\left(1+\dfrac{s}{\omega_{z2}}\right)}{\left(1+\dfrac{s}{\omega_{p_1}}\right)\left(1+\dfrac{s}{\omega_{p2}}\right)} \tag{5.211}$$

where

$$G_0 = \frac{R_2}{R_1} \tag{5.212}$$

$$\omega_{z_1} = \frac{1}{R_2C_1} \tag{5.213}$$

$$\omega_{z_2} = \frac{1}{R_1C_3} \tag{5.214}$$

$$\omega_{p_1} = \frac{1}{R_2C_2} \tag{5.215}$$

$$\omega_{p_2} = \frac{1}{R_3C_3} \tag{5.216}$$

The magnitude expression of the simplified type 3 now becomes

$$|G(f_c)| = \frac{R_2}{R_1}\frac{\sqrt{1+\left(\dfrac{f_{z1}}{f_c}\right)^2}\sqrt{1+\left(\dfrac{f_c}{f_{z2}}\right)^2}}{\sqrt{1+\left(\dfrac{f_c}{f_{p1}}\right)^2}\sqrt{1+\left(\dfrac{f_c}{f_{p2}}\right)^2}} \tag{5.217}$$

From which the value of R_2 can also be extracted:

$$R_2 = GR_1\frac{\sqrt{1+\left(\dfrac{f_c}{f_{p1}}\right)^2}\sqrt{1+\left(\dfrac{f_c}{f_{p2}}\right)^2}}{\sqrt{1+\left(\dfrac{f_{z1}}{f_c}\right)^2}\sqrt{1+\left(\dfrac{f_c}{f_{z2}}\right)^2}} \tag{5.218}$$

In (5.218) and (5.206), the term G represents the needed gain or attenuation at the selected crossover frequency and already defined by (5.7).

The expression of a type 3 configuration can be put under the following simple form:

$$G(s) = \frac{N(s)}{D(s)} \tag{5.219}$$

The argument of (5.219) is thus

$$\arg G(s) = \arg N(s) - \arg D(s) \tag{5.220}$$

In other words,

$$\arg N(j\omega) = \arg\left(-1 + j\frac{\omega_{z1}}{\omega}\right) + \arg\left(1 + j\frac{\omega}{\omega_{z2}}\right) \tag{5.221}$$

$$\arg D(j\omega) = \arg\left(1 + j\frac{\omega}{\omega_{p1}}\right) + \arg\left(1 + j\frac{\omega}{\omega_{p2}}\right) \tag{5.222}$$

Subtracting (5.222) from (5.221), we have

$$\arg G(j\omega) = \arg\left(-1 + j\frac{\omega_{z1}}{\omega}\right) + \arg\left(1 + j\frac{\omega}{\omega_{z2}}\right) - \arg\left(1 + j\frac{\omega}{\omega_{p1}}\right) - \arg\left(1 + j\frac{\omega}{\omega_{p2}}\right) \tag{5.223}$$

Solving the above leads to

$$\arg G(f) = \pi - \tan^{-1}\left(\frac{f_{z1}}{f}\right) + \tan^{-1}\left(\frac{f}{f_{z2}}\right) - \tan^{-1}\left(\frac{f}{f_{p1}}\right) - \tan^{-1}\left(\frac{f}{f_{p2}}\right) \tag{5.224}$$

In dc, because of the origin pole and the op amp inversion, the phase lags by

$$\lim_{f \to 0} \arg G(f) = \pi - \tan^{-1}(\infty) + \tan^{-1}(0) - 2\tan^{-1}(0) = \pi - \frac{\pi}{2} = \frac{\pi}{2} \text{ or } -270° \tag{5.225}$$

The phase boost at the crossover frequency f_c, is simply $\arg G(f_c) - \frac{\pi}{2}$. In other words, for our type 3 compensator, the phase boost alone is

$$boost = \frac{\pi}{2} - \tan^{-1}\left(\frac{f_{z1}}{f_c}\right) + \tan^{-1}\left(\frac{f_c}{f_{z2}}\right) - \tan^{-1}\left(\frac{f_c}{f_{p1}}\right) - \tan^{-1}\left(\frac{f_c}{f_{p2}}\right) \tag{5.226}$$

In the previous expression, the term $\frac{\pi}{2} - \tan^{-1}\left(\frac{f_{z1}}{f_c}\right)$ is simply $\tan^{-1}\left(\frac{f_c}{f_{z1}}\right)$ so

$$boost = \tan^{-1}\left(\frac{f_c}{f_{z1}}\right) + \tan^{-1}\left(\frac{f_c}{f_{z2}}\right) - \tan^{-1}\left(\frac{f_c}{f_{p1}}\right) - \tan^{-1}\left(\frac{f_c}{f_{p2}}\right) \tag{5.227}$$

It is important to note that in the previous formulas, the poles and zeros can be coincident or not; you can equally apply the given definitions by having $f_{p1} = f_{p2}/f_{z1} = f_{z2}$ or $f_{p1} \neq f_{p2}/f_{z1} \neq f_{z2}$.

5.8.1 A Design Example

In this example, we will design a type 3 compensator providing a 10-dB attenuation at 5 kHz together with a 145° phase boost. First, let us translate the 10-dB attenuation into a unit less value:

$$G = 10^{-\frac{10}{20}} = 316m \tag{5.228}$$

The next step is to place the poles and zeros to offer the desired 145° local boost at 5 kHz. From the previous chapters, we know how to position a double pole in relationship to the crossover frequency and the required phase boost:

$$f_{p_{1,2}} = \frac{f_c}{\tan\left(45 - \frac{boost}{4}\right)} = \frac{5k}{\tan\left(45 - \frac{145}{4}\right)} = \frac{5k}{154m} \approx 32.5 \text{ kHz} \tag{5.229}$$

The double zero is classically defined by remembering that the peak in phase boost occurs at the geometric mean between the double coincident pole/zero pair:

$$f_{z_{1,2}} = \frac{f_c^2}{f_{p_{1,2}}} = \frac{25k}{32.5k} \approx 769 \text{ Hz} \tag{5.230}$$

From (5.206), we can calculate the value of R_2, assuming an upper resistor R_1 of 10 kΩ.

$$R_2 = \frac{316m \times 10k \times 32.5k}{32.5k - 769} \frac{\sqrt{1+\left(\frac{5k}{32.5k}\right)^2}\sqrt{1+\left(\frac{5k}{32.5k}\right)^2}}{\sqrt{1+\left(\frac{769}{5k}\right)^2}\sqrt{1+\left(\frac{5k}{769}\right)^2}} = 498\ \Omega \tag{5.231}$$

The rest of the elements comes out easily:

$$C_1 = \frac{1}{2\pi f_{z_1} R_2} = \frac{1}{6.28 \times 769 \times 498} = 416 \text{ nF} \tag{5.232}$$

$$C_2 = \frac{C_1}{2\pi f_{p_1} C_1 R_2 - 1} = \frac{416n}{6.28 \times 32.5k \times 416n \times 498 - 1} = 10 \text{ nF} \tag{5.233}$$

$$C_3 = \frac{f_{p_2} - f_{z_2}}{2\pi R_{upper} f_{p_2} f_{z_2}} = \frac{32.5k - 769}{6.28 \times 10k \times 32.5k \times 769} = 20 \text{ nF} \tag{5.234}$$

$$R_3 = \frac{R_1 f_{z_2}}{f_{p_2} - f_{z_2}} = \frac{10k \times 769}{32.5k - 769} = 242\ \Omega \tag{5.235}$$

As usual, we have captured an automated type 3 simulation fixture as shown in Figure 5.47.

In the simulation examples, the poles and zeros are coincident, but nothing would prevent us from entering separate values for each of them. The simulation results appear in Figure 5.48 and confirm our calculations.

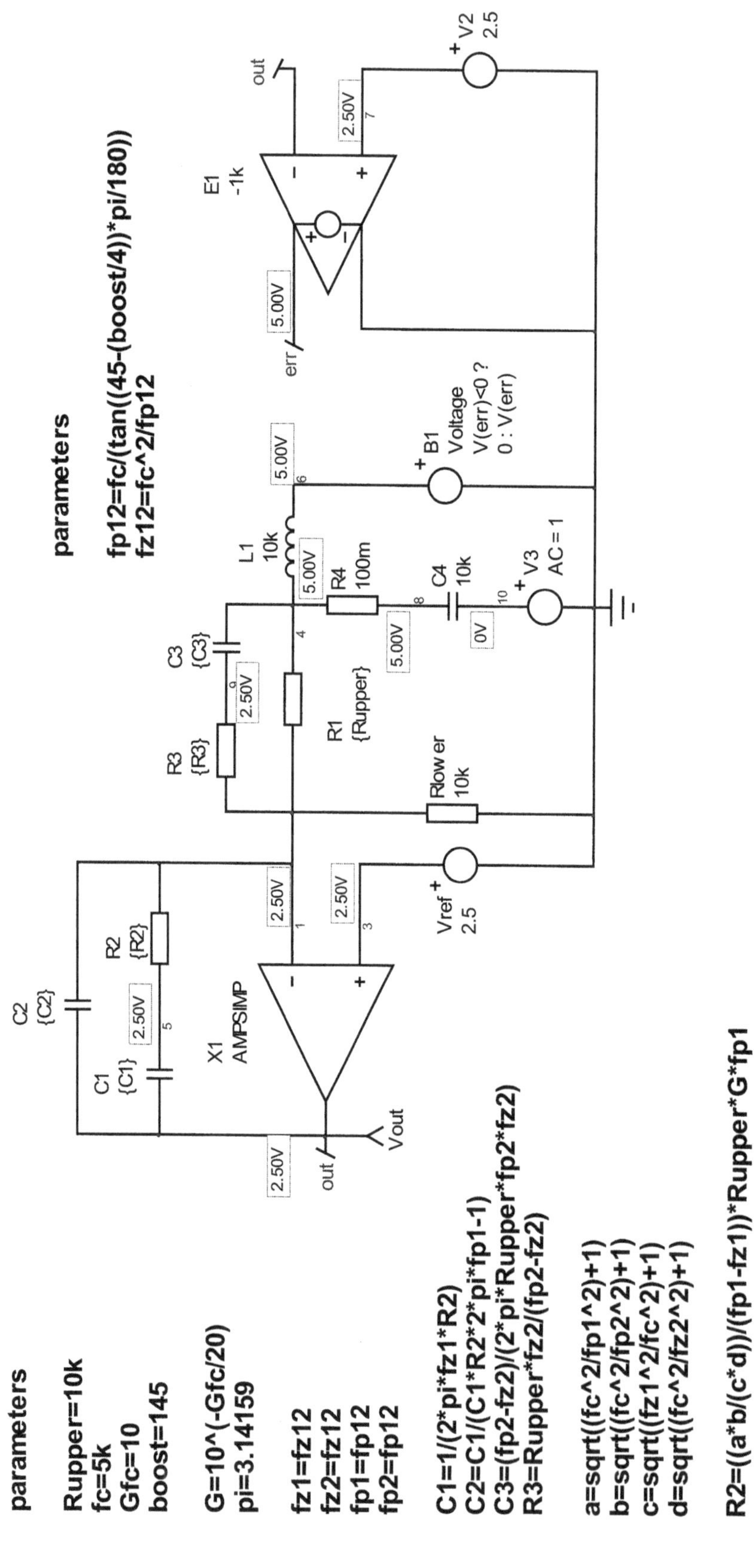

Figure 5.47 An automated type 3 simulation fixture calculates the component values for us.

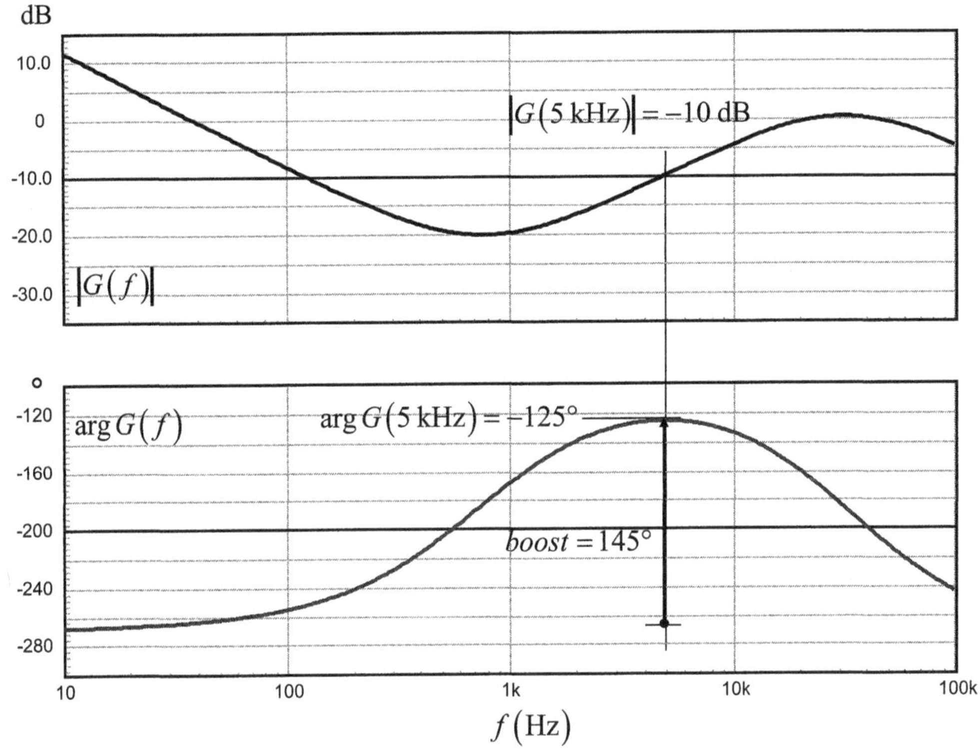

Figure 5.48 The ac simulation results of the type 3 compensator, confirming the 10-dB attenuation at 5 kHz.

5.9 The Type 3: Isolation with an Optocoupler

The isolated type 3 finds application in converters where a large phase boost is needed. This is the case for the voltage-mode buck-derived converters such as forward or push-pull types. The resonant poles must be damped, and the designer usually places a double zero right at the resonant peak to force a crossover with a –1-slope. Let's discover the various ways of wiring a type 3 with an optocoupler.

5.9.1 Optocoupler and Op Amp: The Direct Connection, Common Collector

The type 3 compensator is often used in voltage-mode forward converters, where the peaking of the *LC* network requires the placement of a double zero at the resonant frequency. Therefore, in isolated versions, it requires a connection to an opto-isolator, ending up again in various possible configurations. The first one is the direct link of the op amp output to the LED anode, as already discovered in the type 2 section. Such a configuration appears in Figure 5.49.

The type 3 compensator around the op amp has been transformed into a kind of type 3a, where the second pole brought by C_2 has been removed and moved to the optocoupler block. This change helps to include the optocoupler in the loop and

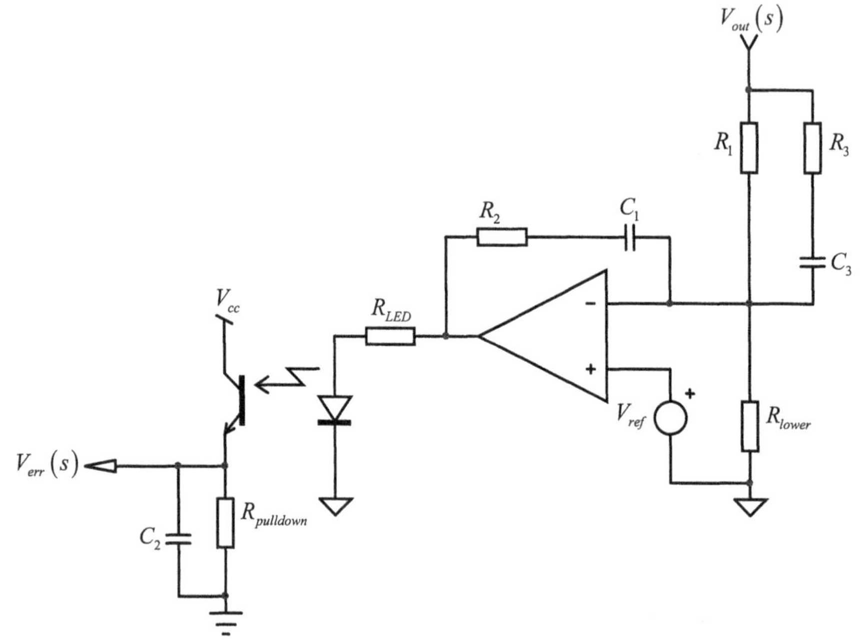

Figure 5.49 The op amp output of a type 3 configuration can directly drive the LED anode of the optocoupler.

also improves the noise immunity at the chain end. From the previous studies, it is possible to derive the compensator gain as follows:

$$G(s) = -\text{CTR}\frac{R_{pulldown}}{R_{LED}}\frac{R_2}{R_1}\frac{1+\dfrac{1}{sR_2C_1}}{1+sR_3C_3}\frac{1+sC_3\left[R_1+R_3\right]}{1+sR_{pulldown}C_2} = -G_1G_2\frac{\left(1+\dfrac{\omega_{z_1}}{s}\right)\left(1+\dfrac{s}{\omega_{z_2}}\right)}{\left(1+\dfrac{s}{\omega_{p_1}}\right)\left(1+\dfrac{s}{\omega_{p_2}}\right)} \tag{5.236}$$

where

$$G_1 = \text{CTR}\frac{R_{pulldown}}{R_{LED}} \tag{5.237}$$

$$G_2 = \frac{R_2}{R_1} \tag{5.238}$$

$$\omega_{z_1} = \frac{1}{R_2C_1} \tag{5.239}$$

$$\omega_{z_2} = \frac{1}{(R_1+R_3)C_3} \tag{5.240}$$

$$\omega_{p_1} = \frac{1}{R_3C_3} \tag{5.241}$$

$$\omega_{p2} = \frac{1}{R_{pulldown}C_2} \tag{5.242}$$

The definition for R_2 derived in (5.218) is still valid for this architecture:

$$R_2 = \frac{G}{G_1}R_1 \frac{\sqrt{1+\left(\frac{f_c}{f_{p1}}\right)^2}\sqrt{1+\left(\frac{f_c}{f_{p2}}\right)^2}}{\sqrt{1+\left(\frac{f_{z1}}{f_c}\right)^2}\sqrt{1+\left(\frac{f_c}{f_{z2}}\right)^2}} \tag{5.243}$$

In the previous expression, the term G/G_1 accounts for the optocoupler presence and its own network gain G_1 as described in (5.237). G is the total necessary gain or attenuation at crossover as described by (5.7). The rest of the element values is found using a system of equations involving (5.243), (5.239), (5.240), and (5.241):

$$C_1 = \frac{1}{2\pi f_{z1} R_2} \tag{5.244}$$

$$C_2 = \frac{1}{2\pi f_{p2} R_{pulldown}} \tag{5.245}$$

$$C_3 = \frac{f_{p1} - f_{z2}}{2\pi R_{upper} f_{p1} f_{z2}} \tag{5.246}$$

$$R_3 = \frac{R_1 f_{z2}}{f_{p1} - f_{z2}} \tag{5.247}$$

5.9.2 A Design Example

Suppose we want to create an isolated 10-dB amplification at 1 kHz, together with a 110° phase boost at this frequency. The design parameters are the following ones:

VOH = 10 V; the op amp maximum output voltage.
V_f = 1 V; the LED forward voltage.
$V_{CE,sat}$ = 0.3 V; the optocoupler saturation voltage.
V_{cc} = 5 V; the pull-up V_{cc} level.
$R_{pulldown}$ = 1 kΩ; the optocoupler pull-up resistor.
CTR_{min} = 0.8; the minimum optocoupler current transfer ratio.
R_1 = 10 kΩ; the upper resistor in the resistor bridge observing the output variable.
f_{opto} = 15 kHz; the optocoupler pole that has been characterized with $R_{pulldown}$.

Given the 110° phase boost at 1 kHz, we first assume to place coincident poles and zeros at a place to be determined. Capitalizing on what has already been derived, we will place a pole pair at the following position:

$$f_{p_{1,2}} = \frac{f_c}{\tan\left(45 - \frac{boost}{4}\right)} = \frac{1k}{315m} \approx 3.2\text{ kHz} \tag{5.248}$$

The double zero will be located at

$$f_{z_{1,2}} = \frac{f_c^2}{f_p} = \frac{1k}{3.2k} \approx 312\text{ Hz} \tag{5.249}$$

Then, with the pole and zero position being defined, we carry on by defining the upper limit for the LED resistor. This resistor ensures that the feedback voltage will always be able to swing up to a voltage close to V_{cc}-$V_{CE,sat}$, despite the minimum CTR of the optocoupler and the op amp output upper stop. This equation has already been derived in (5.73):

$$R_{LED,\max} = \frac{R_{pulldown}(\text{VOH} - V_f)\text{CTR}_{\min}}{(V_{cc} - V_{CE,sat})} = \frac{1k(10-1)}{5-0.3}0.8 = 1.5\text{ k}\Omega \tag{5.250}$$

Adopting a 20 percent safety margin and choosing a normalized value, we finally adopt R_{LED} = 1.2 kΩ. With this resistor known, we can evaluate the gain brought by the optocoupler chain alone and defined by

$$G_1 = \frac{R_{pulldown}}{R_{LED}}\text{CTR}_{\min} = \frac{1k}{1.2k}0.8 = 0.666 \tag{5.251}$$

Thanks to (5.243), the value for R_2 is obtained:

$$R_2 = \frac{G}{G_1}R_1\frac{\sqrt{1+\left(\frac{f_c}{f_{p_1}}\right)^2}\sqrt{1+\left(\frac{f_c}{f_{p_2}}\right)^2}}{\sqrt{1+\left(\frac{f_{z_1}}{f_c}\right)^2}\sqrt{1+\left(\frac{f_c}{f_{z_2}}\right)^2}} = \frac{10^{\frac{10}{20}}}{0.666}10k\frac{\sqrt{1+\left(\frac{1k}{3.2k}\right)^2}\sqrt{1+\left(\frac{1k}{3.2k}\right)^2}}{\sqrt{1+\left(\frac{312}{1k}\right)^2}\sqrt{1+\left(\frac{1k}{312}\right)^2}} \approx 15\text{ k}\Omega \tag{5.252}$$

We are now ready to evaluate the rest of the components using (5.244) through (5.247):

$$C_1 = \frac{1}{2\pi f_{z_1}R_2} = \frac{1}{6.28\times 312\times 15k} = 34\text{ nF} \tag{5.253}$$

$$C_2 = \frac{1}{2\pi f_{p_2}R_{pulldown}} = \frac{1}{6.28\times 3.2k\times 1k} \approx 50\text{ nF} \tag{5.254}$$

$$C_3 = \frac{f_{p_1} - f_{z_2}}{2\pi R_1 f_{p_1} f_{z_2}} = \frac{3.2k - 312}{6.28 \times 10k \times 3.2k \times 312} \approx 46 \text{ nF} \quad (5.255)$$

$$R_3 = \frac{R_1 f_{z_2}}{f_{p_1} - f_{z_2}} = \frac{10k \times 312}{3.2k - 312} = 1.08 \text{ k}\Omega \quad (5.256)$$

The capacitor put across the optocoupler, in total, must equal 50 nF (C_2). The optocoupler pole is located at 15 kHz. Given a pull-down resistor of 1 kΩ, we can calculate its parasitic contribution:

$$C_{opto} = \frac{1}{2\pi f_{opto} R_{pulldown}} = \frac{1}{6.28 \times 15k \times 1k} = 10.6 \text{ nF} \quad (5.257)$$

When subtracted from (5.254), you obtain the value for C_{col}, the final capacitor installed across the optocoupler:

$$C_{col} = C_2 - C_{opto} = 50n - 10.6n \approx 39 \text{ nF} \quad (5.258)$$

We are all set now. We can capture Figure 5.49, automate its bias point with a simple op amp, E_1, and then run a simulation. This what we propose in Figure 5.50. The ac results appear in Figure 5.51 and agree very well with the calculations.

5.9.3 Optocoupler and Op Amp: The Direct Connection, Common Emitter

There are some cases where the control law polarity does not suit the converter control circuitry, and the direct connection of the op amp to the optocoupler does not work. As we already introduced with the type 2 configuration, an option consists of placing the optocoupler loading resistor in the collector rather than in the emitter (see Figure 5.52).

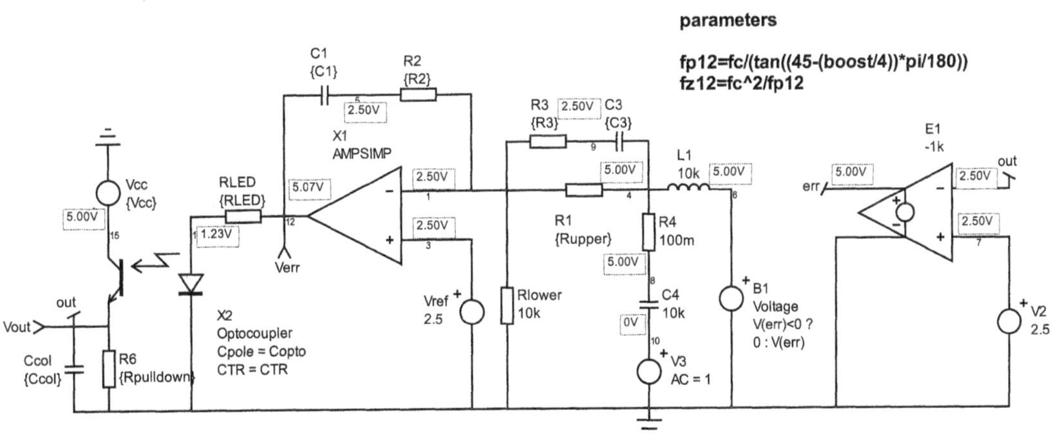

Figure 5.50 The simulation fixture includes an automated bias point adjustment. The elements are also automatically computed (not shown here).

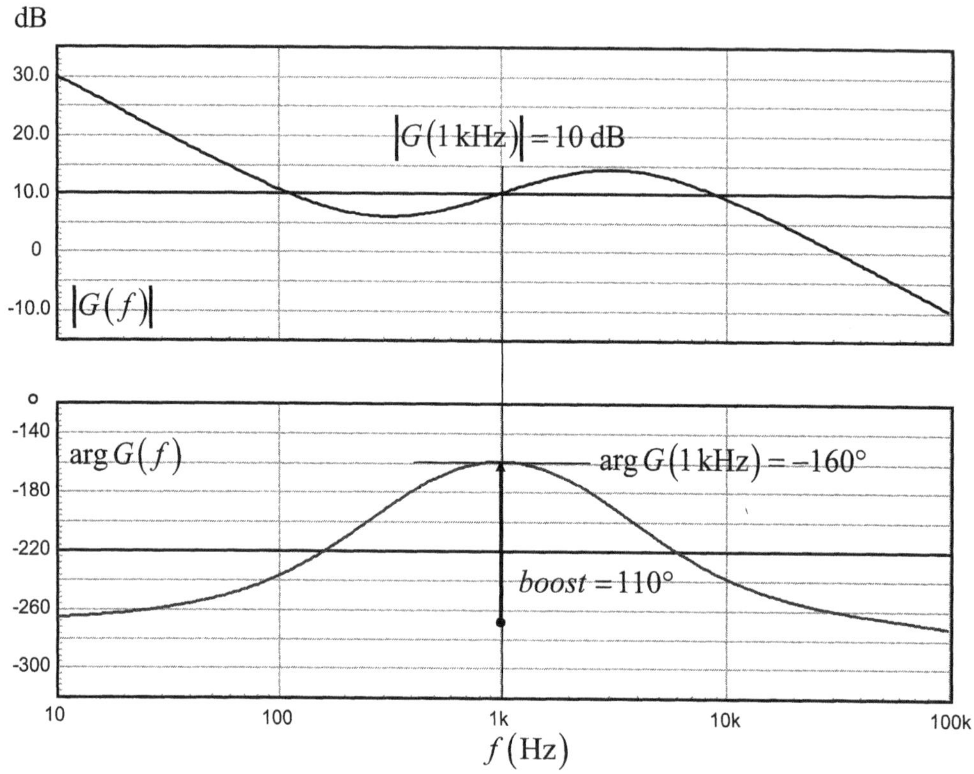

Figure 5.51 The result is excellent: the 10-dB line transition occurs right at 1 kHz, and the phase is boosted by 110°.

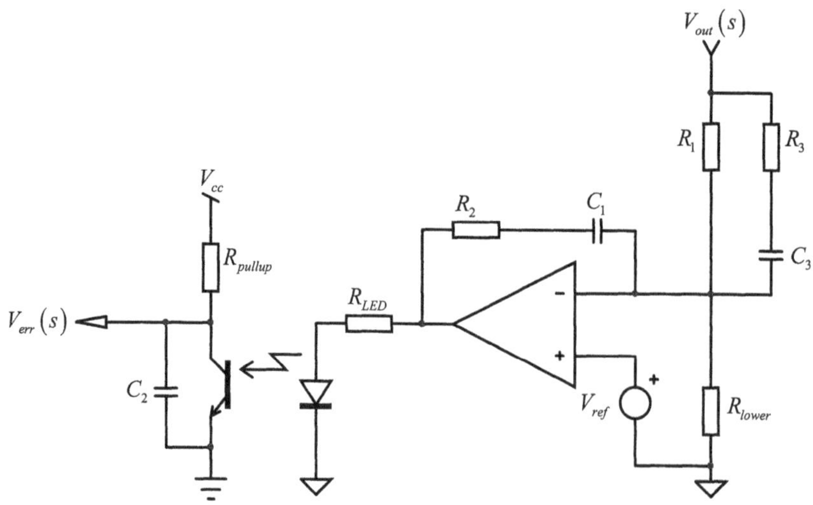

Figure 5.52 By placing the loading resistor in the collector rather than in the emitter, the compensation polarity is changed.

The transfer function is almost similar to that of (5.236), except that the negative sign is gone and the second pole depends on the pull-up resistor:

$$G(s) = \mathrm{CTR}\frac{R_{pullup}}{R_{LED}}\frac{R_2}{R_1}\frac{1+\dfrac{1}{sR_2C_1}}{1+sR_3C_3}\frac{1+sC_3[R_1+R_3]}{1+sR_{pulldown}C_2} = G_1G_2\frac{\left(1+\dfrac{\omega_{z_1}}{s}\right)\left(1+\dfrac{s}{\omega_{z_2}}\right)}{\left(1+\dfrac{s}{\omega_{p_1}}\right)\left(1+\dfrac{s}{\omega_{p_2}}\right)} \tag{5.259}$$

where

$$G_1 = \mathrm{CTR}\frac{R_{pullup}}{R_{LED}} \tag{5.260}$$

$$G_2 = \frac{R_2}{R_1} \tag{5.261}$$

$$\omega_{z_1} = \frac{1}{R_2C_1} \tag{5.262}$$

$$\omega_{z_2} = \frac{1}{(R_1+R_3)C_3} \tag{5.263}$$

$$\omega_{p_1} = \frac{1}{R_3C_3} \tag{5.264}$$

$$\omega_{p_2} = \frac{1}{R_{pullup}C_2} \tag{5.265}$$

The design methodology exposed in the previous design example remains valid for this type of arrangement.

5.9.4 Optocoupler and Op Amp: The Direct Connection, Common Emitter, and UC384X

The architecture already offered with the type 2 does not change for a type 3 configuration. The UC384X op amp is simply wired as an inverter. As its input impedance is that of the resistor connected to the optocoupler collector, it must be sufficiently high compared to the pull-up resistor. Otherwise, the resistor will change the pole location imposed by C_2 and R_{pullup}. Figure 5.53 depicts a type 3 driving an UC384X controller.

As we already pointed out, this type of control works if the op amp is powered from an auxiliary source already present at startup. Otherwise, the full bias of the LED at power-on (which pushes V_{err} to the maximum) will not occur, preventing the converter from starting up.

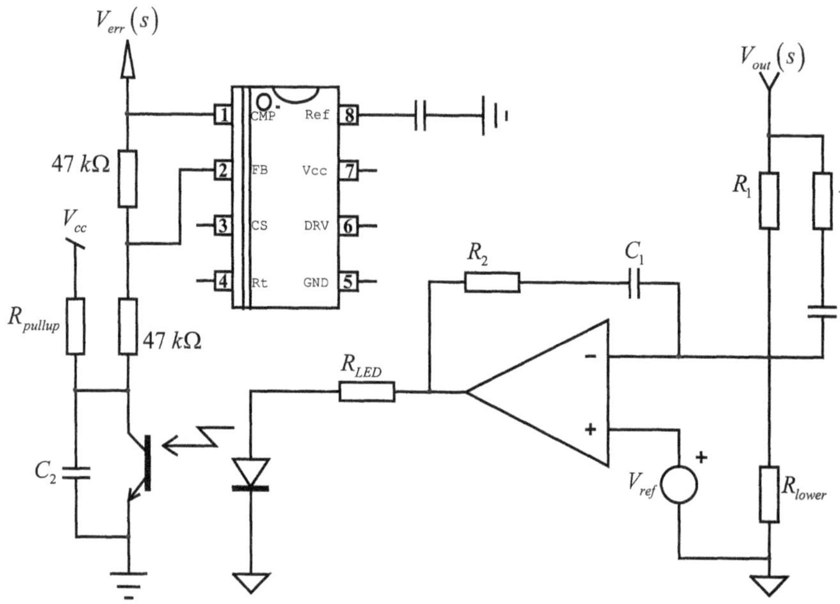

Figure 5.53 The UC384X can be used with a type 3 architecture. Wired as a follower, it does not perturb the chain.

5.9.5 Optocoupler and Op Amp: Pull-Down with Fast Lane

As introduced in Figure 5.20, some controllers do only accept the optocoupler wired in a common-collector configuration. In this case, the op amp must drive the LED in a pull-down association, pulling the LED cathode to ground. The application circuit appears in Figure 5.54. As we described in the type 2 section, the

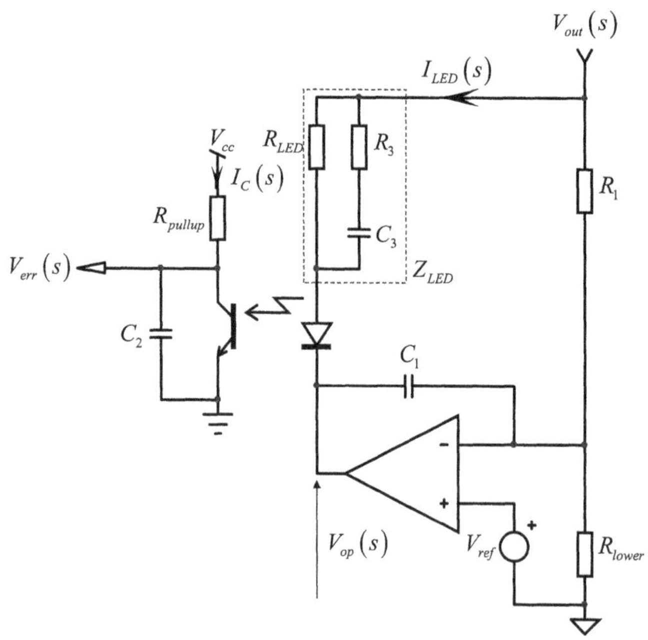

Figure 5.54 The application circuit of the op amp driving the optocoupler LED in a pull-down configuration.

pull-down configuration induces a fast-lane effect we have to deal with. This is the reason the network R_3C_3 must take place across the LED resistor and no longer across R_1. We will see in a moment how it affects the design flexibility.

Following the description of the type 2 configuration in Section 5.5.5, the LED ac current is defined by

$$I_{LED}(s) = \frac{V_{out}(s) - V_{op}(s)}{Z_{LED}} = (V_{out}(s) - V_{op}(s)) \frac{R_{LED} + \left(R_3 + \frac{1}{sC_3}\right)}{R_{LED}\left(R_3 + \frac{1}{sC_3}\right)} \tag{5.266}$$

The output voltage of the op amp is simply that of a type 1 compensator:

$$V_{op}(s) = -\frac{1}{sC_1R_1} V_{out}(s) \tag{5.267}$$

Substituting (5.267) into (5.266), we obtain

$$\begin{aligned} I_{LED}(s) &= \left(V_{out}(s) + V_{out}(s)\frac{1}{sR_1C_1}\right) \frac{R_{LED} + \left(R_3 + \frac{1}{sC_3}\right)}{R_{LED}\left(R_3 + \frac{1}{sC_3}\right)} \\ &= V_{out}(s)\left(\frac{1 + sR_1C_1}{sR_1C_1}\right) \frac{R_{LED} + \left(R_3 + \frac{1}{sC_3}\right)}{R_{LED}\left(R_3 + \frac{1}{sC_3}\right)} \end{aligned} \tag{5.268}$$

Rearranging the previous, we have

$$I_{LED}(s) = V_{out}(s)\left(\frac{1 + sR_1C_1}{sR_1C_1}\right) \frac{s(R_{LED} + R_3)C_3 + 1}{R_{LED}(sC_3R_3 + 1)} \tag{5.269}$$

Now realizing that the error voltage $V_{err}(s)$ is linked to the LED current by the optocoupler CTR, we have

$$V_{err}(s) = -I_{LED}(s) R_{pullup} \frac{1}{1 + sR_{pullup}C_2} \text{CTR} \tag{5.270}$$

Substituting (5.269) into (5.270) and rearranging, we obtain the transfer function we are looking for:

$$\frac{V_{err}(s)}{V_{out}(s)} = -\frac{R_{pullup}}{R_{LED}} \text{CTR} \frac{1 + 1/sR_1C_1}{1 + sR_{pullup}C_2} \frac{sC_3(R_{LED} + R_3) + 1}{1 + sC_3R_3} = -G_0 \frac{1 + \omega_{z_1}/s}{1 + s/\omega_{p_1}} \frac{1 + s/\omega_{z_2}}{1 + s/\omega_{p_2}} \tag{5.271}$$

In this equation, we can identify poles, zeros, and a static gain;

$$G_0 = \frac{R_{pullup}}{R_{LED}}\text{CTR} \tag{5.272}$$

$$\omega_{z_1} = \frac{1}{R_1C_1} \tag{5.273}$$

$$\omega_{z_2} = \frac{1}{(R_{LED} + R_3)C_3} \tag{5.274}$$

$$\omega_{p_1} = \frac{1}{R_{pullup}C_2} \tag{5.275}$$

$$\omega_{p_2} = \frac{1}{R_3C_3} \tag{5.276}$$

The magnitude of (5.271) is given by

$$|G(f_c)| = G_0\frac{\sqrt{1+\left(\frac{f_{z_1}}{f_c}\right)^2}\sqrt{1+\left(\frac{f_c}{f_{z_2}}\right)^2}}{\sqrt{1+\left(\frac{f_c}{f_{p_1}}\right)^2}\sqrt{1+\left(\frac{f_c}{f_{p_2}}\right)^2}} = \frac{R_{pullup}}{R_{LED}}\text{CTR}\frac{\sqrt{1+\left(\frac{f_{z_1}}{f_c}\right)^2}\sqrt{1+\left(\frac{f_c}{f_{z_2}}\right)^2}}{\sqrt{1+\left(\frac{f_c}{f_{p_1}}\right)^2}\sqrt{1+\left(\frac{f_c}{f_{p_2}}\right)^2}} \tag{5.277}$$

Then, combining (5.277), (5.273), (5.274), and (5.276) and solving for R_{LED}, R_3, C_1, C_2, and C_3, we can extract the definitions we need to design the compensator:

$$R_{LED} = \frac{R_{pullup}}{G}\text{CTR}\frac{\sqrt{1+\left(\frac{f_{z_1}}{f_c}\right)^2}\sqrt{1+\left(\frac{f_c}{f_{z_2}}\right)^2}}{\sqrt{1+\left(\frac{f_c}{f_{p_1}}\right)^2}\sqrt{1+\left(\frac{f_c}{f_{p_2}}\right)^2}} \tag{5.278}$$

$$C_3 = \frac{f_{p_2} - f_{z_2}}{2\pi R_{LED} f_{p_2} f_{z_2}} \tag{5.279}$$

$$R_3 = \frac{f_{z_2} R_{LED}}{f_{p_2} - f_{z_2}} \tag{5.280}$$

$$C_1 = \frac{1}{2\pi f_{z_1} R_1} \tag{5.281}$$

$$C_2 = \frac{1}{2\pi R_{pullup} f_{p_2}} \tag{5.282}$$

In the previous equations, G is the gain or the attenuation as defined by (5.7). The limiting factor here is, again, the LED resistor. This resistor must be sized to ensure that enough current can cross the LED to bring the error voltage down to the optocoupler $V_{CE,sat}$ level. This LED resistor still obeys (5.103) derived for the type 2 with the LED configured in a pull-down drive:

$$R_{LED,\max} = \frac{R_{pullup}(V_{out} - V_f - \mathrm{VOL})\mathrm{CTR}_{\min}}{(V_{cc} - V_{CE,sat})} \tag{5.283}$$

As this resistor also fixes the gain or the attenuation at the selected crossover frequency, it can be seen as a limiting factor for the possible gain selections. If we combine (5.277) and (5.283), we can solve for the minimum gain below which the system cannot go:

$$|G_{\min}| = \frac{V_{cc} - V_{CE,sat}}{V_{out} - V_f - \mathrm{VOL}} \frac{\sqrt{1+\left(\frac{f_{z1}}{f_c}\right)^2}\sqrt{1+\left(\frac{f_c}{f_{z2}}\right)^2}}{\sqrt{1+\left(\frac{f_c}{f_{p1}}\right)^2}\sqrt{1+\left(\frac{f_c}{f_{p2}}\right)^2}} \tag{5.284}$$

This $G_{\min}$ term illustrates the crossover gain below which the compensator cannot be designed. For instance, should you find a $G_{\min}$ of 5 dB with (5.284), you could simply not select a crossover frequency where you need to amplify by 3 dB. It would not work. However, 6 dB would. Changing the optocoupler CTR or R_{pullup} would not help to modify the result. Actually, (5.284) is made of two terms: the first one (involving VOL, V_{out}, and so on) corresponds to the minimum biasing conditions needed to operate the compensator. The second term involves the poles and zeros position—in other words, the needed phase boost. You can therefore see $G_{\min}$ as another limit for the maximum phase boost you can ask for a selected gain. We can derive an equation linking $G_{\min}$ and the phase boost you want. The result appears next:

$$G_{\min}(boost) = 20\log_{10}\left(\frac{V_{cc} - V_{CE,sat}}{V_{out} - V_f - \mathrm{VOL}}\sqrt{\cot^2\left(\frac{boost}{4} - 45\right)}\right) \tag{5.285}$$

If we consider the following standard values for the first term of the expression, we can plot (5.285), sweeping the phase boost between 0 and 180°:

VOL = 0.2 V; the op amp maximum output voltage.
V_{out} = 12 V; the regulated output voltage.
V_f = 1 V; the LED forward voltage.
$V_{CE,sat}$ = 0.3 V; the optocoupler saturation voltage.
V_{cc} = 5 V; the pull-up V_{cc} level.

The result appears in Figure 5.55 and a shows a gain starting from a minimum value (a null phase boost when both poles and zeros are coincident) and going up as the required phase boost increases. Therefore, when using this type of compensator,

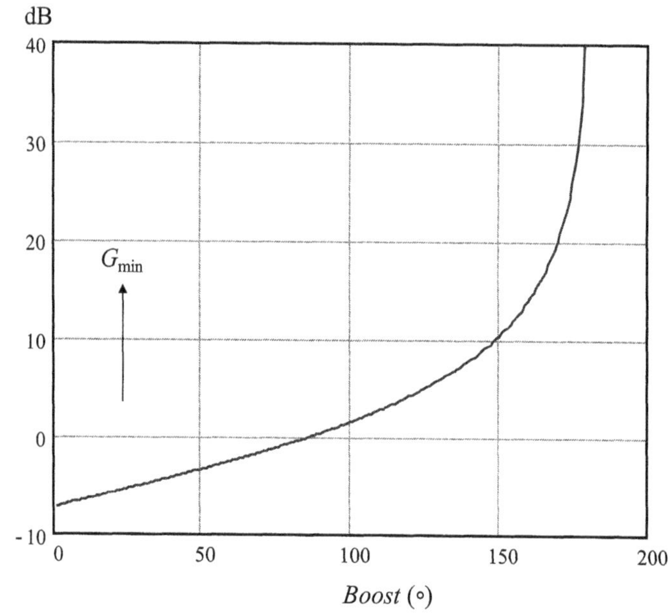

Figure 5.55 The minimum gain is dependent on the wanted phase boost.

it will be important to choose a crossover frequency where the needed gain and phase boost do not conflict with (5.284).

5.9.6 A Design Example

In this example, we want to cross over at a 1-kHz frequency with a 10-dB amplification. The needed phase boost is 120°. The component list appears next and does not change.

VOL = 0.2 V; the op amp minimum output voltage.
V_f = 1 V; the LED forward voltage.
$V_{CE,sat}$ = 0.3 V; the optocoupler saturation voltage.
V_{cc} = 5 V; the pull-up V_{cc} level.
V_{out} = 12 V; the converter output voltage.
R_{pullup} = 1 kΩ; the optocoupler pull-up resistor.
CTR_{min} = 0.8; the minimum optocoupler current transfer ratio.
R_1 = 38 kΩ; the upper resistor in the resistor bridge observing the output variable.
f_{opto} = 15 kHz; the optocoupler pole that has been characterized with R_{pullup}.

Given the needed 120° phase boost at 1 kHz, we first assume to place coincident poles and zeros at a place to be determined. Capitalizing on what has already been derived, we will place a pole pair at the following position:

$$f_{p_{1,2}} = \frac{f_c}{\tan\left(45 - \dfrac{boost}{4}\right)} = \frac{1k}{268m} \approx 3.7 \text{ kHz} \tag{5.286}$$

The double zero will be located at

$$f_{z_{1,2}} = \frac{f_c^2}{f_p} = \frac{1k}{3.7k} \approx 270\ \text{Hz} \tag{5.287}$$

Then, with the pole and zero position defined, we proceed by defining the upper limit for the LED resistor. This resistor ensures that the feedback voltage will always be able to deliver a voltage close to $V_{CE,sat}$, despite the minimum CTR of the optocoupler and the op amp output minimum swing. This equation has already been derived in (5.103):

$$R_{LED,\max} = \frac{R_{pullup}(V_{out} - V_f - \text{VOL})\text{CTR}_{\min}}{(V_{cc} - V_{CE,sat})} = \frac{1k \times (12 - 1 - 0.2)}{5 - 0.3} \times 0.8 = 1.8\ \text{k}\Omega \tag{5.288}$$

We now have to check whether the 120° phase boost at the selected 10-dB gain are compatible with the LED biasing conditions. Using (5.285), we have

$$G_{\min} = 20\log_{10}\left(\frac{5 - 0.3}{12 - 1 - 0.2}\sqrt{\cot^2\left(\frac{120}{4} - 45\right)}\right) = 20\log_{10}(435m \times 3.73) \approx 4.2\ \text{dB} \tag{5.289}$$

Since we want 10 dB, we are well above what (5.289) recommends, and we can thus continue with the design process. Applying the formulas we derived for R_2, C_2, C_3, C_1, and R_3, we have

$$R_{LED} = \frac{1k}{10^{\frac{10}{20}}} 0.8 \frac{\sqrt{1 + \left(\frac{270}{1k}\right)^2}\sqrt{1 + \left(\frac{1k}{270}\right)^2}}{\sqrt{1 + \left(\frac{1k}{3.7k}\right)^2}\sqrt{1 + \left(\frac{1k}{3.7k}\right)^2}} = 944\ \Omega \tag{5.290}$$

$$C_3 = \frac{3.7k - 270}{6.28 \times 3.7k \times 944 \times 270} \approx 580\ \text{nF} \tag{5.291}$$

$$R_3 = \frac{270 \times 944}{3.7k - 270} = 74\ \Omega \tag{5.292}$$

$$C_1 = \frac{1}{2\pi f_{z_1} R_1} = \frac{1}{6.28 \times 270 \times 38k} = 15.6\ \text{nF} \tag{5.293}$$

$$C_2 = \frac{1}{2\pi R_{pullup} f_{p_2}} = \frac{1}{6.28 \times 1k \times 3.7k} = 43\ \text{nF} \tag{5.294}$$

The capacitor across the optocoupler, in total, must equal 43 nF. The optocoupler pole is located at 15 kHz. Given a pull-up resistor of 1 kΩ, we can calculate its parasitic contribution:

$$C_{opto} = \frac{1}{2\pi f_{opto} R_{pullup}} = \frac{1}{6.28 \times 15k \times 1k} = 10.6\ \text{nF} \tag{5.295}$$

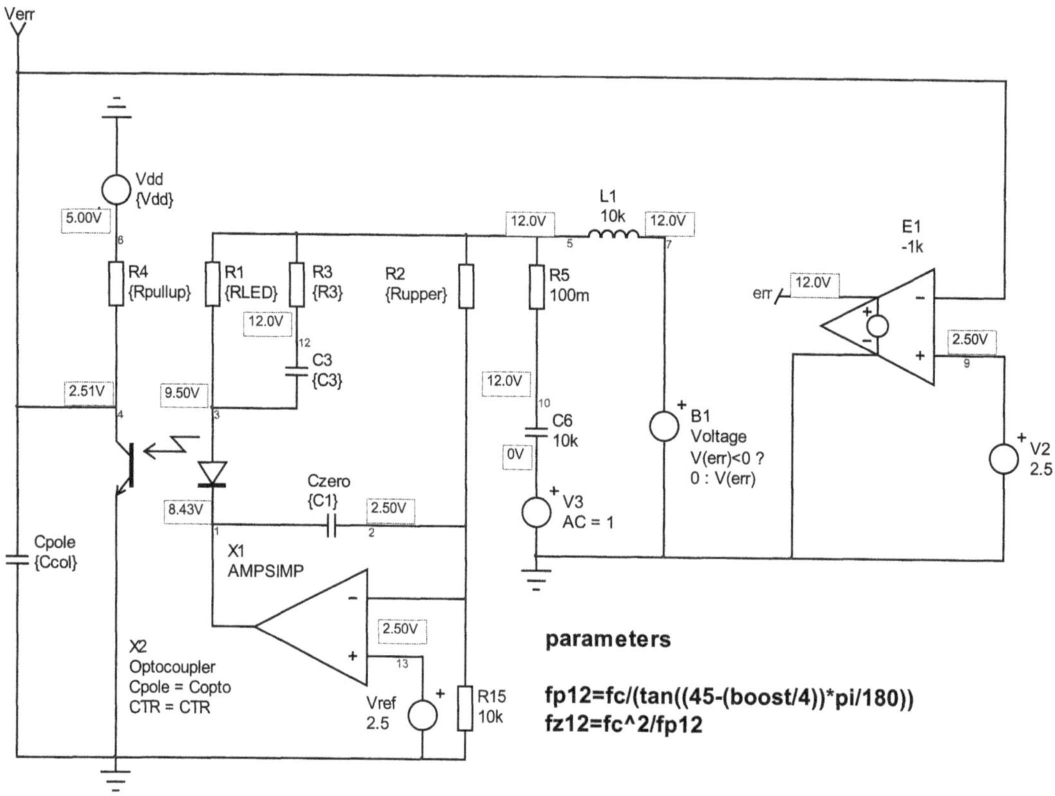

Figure 5.56 The bias-point-automated simulation template helps to test the calculated parameters for all our components.

When subtracted from (5.294), you obtain the value for C_{col}, the final capacitor installed across the optocoupler:

$$C_{col} = C_2 - C_{opto} = 43n - 10.6n \approx 32 \text{ nF} \tag{5.296}$$

The design is done now, and we can start the simulation as proposed by Figure 5.56. The parameter calculations are fully automated and we can change the needed boost or crossover gain on the fly and check if the results match our wishes. The simulation results appear in Figure 5.57 and confirm the design methodology we have adopted.

5.9.7 Optocoupler and Op Amp: Pull Down without Fast Lane

The fast lane presence really hampers the design, as it introduces another limiting variable. The problem comes from the access of $V_{out}(s)$ to the error path via the LED connection. To get rid of this problem, we must provide an ac isolation between the LED current and the regulated output variable. This is the solution already introduced in Figure 5.25 and updated in Figure 5.58.

Observing Figure 5.58 reveals the presence of a type 3a kind of compensator where the high-frequency pole has moved on the optocoupler. Let us write the

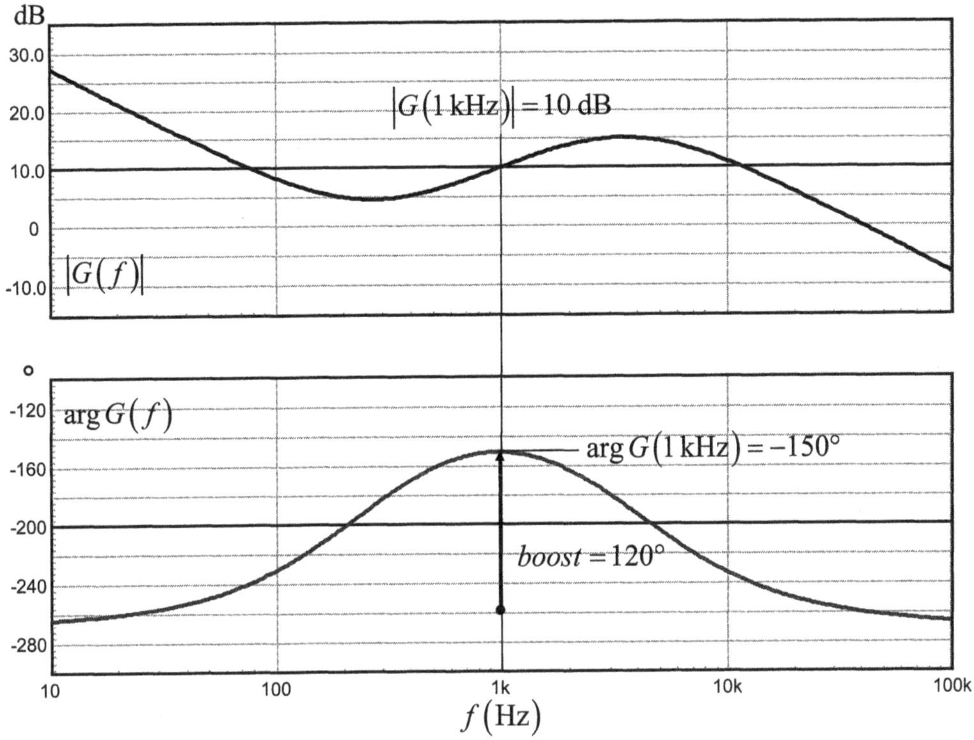

Figure 5.57 The simulation results confirm the correct phase boost as well as the perfect 10-dB crossover gain.

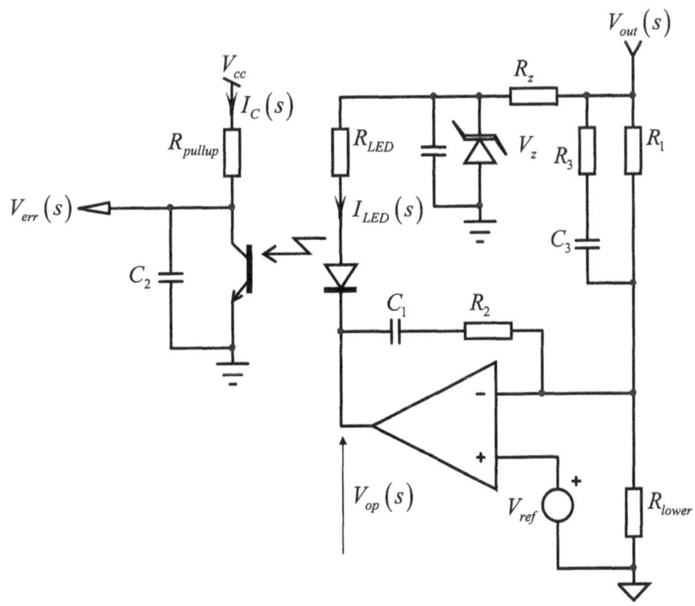

Figure 5.58 To get rid of this problem, we will apply the technique already seen in the type 2 section where the LED is ac-decoupled from V_{out} via a Zener diode.

output voltage of the op amp, $V_{op}(s)$, by deriving the impedance definitions for Z_i and Z_f:

$$Z_f = \left(\frac{1}{sC_1} + R_2\right) \tag{5.297}$$

$$Z_i = \left(\frac{1}{sC_3} + R_3\right)R_1 \Big/ \left(\frac{1}{sC_3} + R_3 + R_1\right) \tag{5.298}$$

By dividing (5.297) by (5.298) and factoring sR_2C_1, we obtain the transfer function of the op amp alone:

$$\frac{V_{op}(s)}{V_{out}(s)} = -\frac{R_2}{R_1}\frac{1+\dfrac{1}{sR_2C_1}}{1+sR_3C_3}(1+sC_3(R_1+R_3)) \tag{5.299}$$

With the op amp output driving the LED current, the error signal is thus that of the op amp translated by a gain brought by the optocoupler. We have already seen this in (5.114):

$$V_{err}(s) = -I_{LED}(s)R_{pullup}\frac{1}{1+sR_{pullup}C_2}\text{CTR} \tag{5.300}$$

The ac LED current being $-V_{op}(s)/R_{LED}$, the final transfer function definition pops up immediately:

$$\frac{V_{op}(s)}{V_{out}(s)} = -\frac{R_2}{R_1}\text{CTR}\frac{R_{pullup}}{R_{LED}}\frac{1+\dfrac{1}{sR_2C_1}}{1+sR_3C_3}\frac{1+sC_3(R_1+R_3)}{1+sR_{pullup}C_2} \tag{5.301}$$

This expression can be put under the normalized form:

$$G(s) = -G_0\frac{\left(1+\dfrac{\omega_{z_1}}{s}\right)\left(1+\dfrac{s}{\omega_{z_2}}\right)}{\left(1+\dfrac{s}{\omega_{p_1}}\right)\left(1+\dfrac{s}{\omega_{p_2}}\right)} \tag{5.302}$$

where

$$G_0 = \text{CTR}\frac{R_{pullup}}{R_{LED}}\frac{R_2}{R_1} = G_1G_2 \tag{5.303}$$

$$G_1 = \text{CTR}\frac{R_{pullup}}{R_{LED}} \tag{5.304}$$

$$G_2 = \frac{R_2}{R_1} \tag{5.305}$$

$$\omega_{z_1} = \frac{1}{R_2C_1} \tag{5.306}$$

$$\omega_{z_2} = \frac{1}{2\pi C_3(R_1 + R_3)} \tag{5.307}$$

$$\omega_{p_1} = \frac{1}{2\pi R_3 C_3} \tag{5.308}$$

$$\omega_{p_2} = \frac{1}{R_{pullup} C_2} \tag{5.309}$$

The magnitude of (5.302) is given by

$$|G(f_c)| = \text{CTR}\frac{R_{pullup}}{R_{LED}}\frac{R_2}{R_1}\frac{\sqrt{1+\left(\frac{f_{z_1}}{f_c}\right)^2}\sqrt{1+\left(\frac{f_c}{f_{z_2}}\right)^2}}{\sqrt{1+\left(\frac{f_c}{f_{p_1}}\right)^2}\sqrt{1+\left(\frac{f_c}{f_{p_2}}\right)^2}} \tag{5.310}$$

Then, combining and solving for R_{LED}, R_3, C_2 C_1, and C_3, we can extract the definitions we are looking for:

$$R_2 = \frac{GR_1R_{LED}}{R_{pullup}\text{CTR}}\frac{\sqrt{1+\left(\frac{f_c}{f_{p_1}}\right)^2}\sqrt{1+\left(\frac{f_c}{f_{p_2}}\right)^2}}{\sqrt{1+\left(\frac{f_{z_1}}{f_c}\right)^2}\sqrt{1+\left(\frac{f_c}{f_{z_2}}\right)^2}} \tag{5.311}$$

$$R_3 = \frac{R_1 f_{z_2}}{f_{p_2} - f_{z_2}} \tag{5.312}$$

$$C_1 = \frac{1}{2\pi f_{z_1} R_2} \tag{5.313}$$

$$C_2 = \frac{1}{2\pi f_{p_2} R_{pullup}} \tag{5.314}$$

$$C_3 = \frac{f_{p_2} - f_{z_2}}{2\pi R_1 f_{p_2} f_{z_2}} \tag{5.315}$$

With the fast lane being removed, the LED resistor no longer plays a role in the pole/zero positions. Its value now solely depends on acceptable bias conditions to give the error voltage the necessary swing, despite unavoidable dispersions on the optocoupler CTR. The formula derived in the type 2 case is still valid:

$$R_{LED,\max} = \frac{R_{pullup}(V_Z - V_f - \text{VOL})\text{CTR}_{\min}}{(V_{cc} - V_{CE,sat})} \tag{5.316}$$

In this expression, the V_Z term represents the selected Zener level obtained from the output. As discussed in Section 5.5.8, we recommend a Zener voltage equal to two-thirds of the output voltage. This tradeoff offers a limited voltage loss across the dropping resistor and ensures a good ac decoupling between the Zener voltage and the monitored output. Considering a Zener biasing current I_{Zbias}, the dropping resistor can be computed using an equation already derived:

$$R_Z = \frac{(V_{out} - V_Z)R_{pullup}\mathrm{CTR}_{\min}}{V_{cc} - V_{CE,sat} + I_{Zbias}R_{pullup}\mathrm{CTR}_{\min}} \tag{5.317}$$

All component values are now attached to a design formula; it is time for the design example!

5.9.8 A Design Example

In this example, we have to crossover at a 5-kHz frequency where a –10-dB attenuation is needed. At the crossover point, the phase must boosted by 150°. The parameter list appears next and does not change.

$\mathrm{VOL} = 0.2$ V; the op amp minimum output voltage
$V_f = 1$ V; the LED forward voltage.
$V_{CE,sat} = 0.3$ V; the optocoupler saturation voltage.
$V_{cc} = 5$ V; the pull-up V_{cc} level.
$V_{out} = 12$ V; the converter output voltage.
$R_{pullup} = 1$ kΩ; the optocoupler pull-up resistor.
$\mathrm{CTR}_{\min} = 0.8$; the minimum optocoupler current transfer ratio.
$R_1 = 38$ kΩ; the upper resistor in the resistor bridge observing the output variable.
$f_{opto} = 15$ kHz; the optocoupler pole that has been characterized with R_{pullup}.

Given the needed 150° phase boost at 5 kHz, we first assume we should place coincident poles and zeros at a location to be determined. Capitalizing on what has already been derived, we will place a pole pair at the following position:

$$f_{p1,2} = \frac{f_c}{\tan\left(45 - \frac{boost}{4}\right)} = \frac{5k}{131m} \approx 38 \text{ kHz} \tag{5.318}$$

The double zero will be located at

$$f_{z1,2} = \frac{f_c^2}{f_p} = \frac{25k}{38k} \approx 660 \text{ Hz} \tag{5.319}$$

As usual, let us pick a LED resistance that fits our bias requirements using (5.316):

$$R_{LED,\max} = \frac{R_{pullup}(V_Z - V_f - \mathrm{VOL})\mathrm{CTR}_{\min}}{(V_{cc} - V_{CE,sat})} = \frac{1k \times (8.2 - 1 - 0.2)}{5 - 0.3} \times 0.8 = 1.2 \text{ k}\Omega \tag{5.320}$$

Applying a 20 percent design margin, we select a 910 Ω resistor. For a 12 V output, we have selected an 8.2 V Zener diode. Considering a 1-mA bias current for this component, we can now calculate the associated dropping resistor using (5.317):

$$R_Z = \frac{(V_{out} - V_Z)R_{pullup}\text{CTR}_{\min}}{V_{cc} - V_{CE,sat} + I_{Zbias}R_{pullup}\text{CTR}_{\min}} = \frac{(12-8.2)\times 1k \times 0.8}{5-0.3+1m\times 1k\times 0.8} = 552\ \Omega \tag{5.321}$$

The type 3 passive elements are calculated using equations (5.311) through (5.315):

$$R_2 = \frac{10^{-\frac{10}{20}}\times 910}{1k\times 0.8}\frac{\sqrt{1+\left(\frac{5k}{38k}\right)^2}\sqrt{1+\left(\frac{5k}{38k}\right)^2}}{\sqrt{1+\left(\frac{660}{5k}\right)^2}\sqrt{1+\left(\frac{5k}{660}\right)^2}} = 1.8\ \text{k}\Omega \tag{5.322}$$

$$R_3 = \frac{R_1 f_{z2}}{f_{p2} - f_{z2}} = \frac{38k\times 660}{38k - 660} = 671\ \Omega \tag{5.323}$$

$$C_1 = \frac{1}{2\pi f_{z_1} R_2} = \frac{1}{6.28\times 660\times 1.8k} = 134\ \text{nF} \tag{5.324}$$

$$C_2 = \frac{1}{2\pi f_{p_2} R_{pullup}} = \frac{1}{6.28\times 38k\times 1k} = 4.2\ \text{nF} \tag{5.325}$$

$$C_3 = \frac{f_{p2} - f_{z2}}{2\pi R_1 f_{p2} f_{z2}} = \frac{38k - 660}{6.28\times 38k\times 38k\times 660} = 6.2\ \text{nF} \tag{5.326}$$

We know that the total capacitor connected across the optocoupler that forms C_2 is actually made of the optocoupler parasitic pole C_{opto} and the added capacitor C_{col}. The optocoupler pole is located at 15 kHz. Given a pull-up resistor of 1 kΩ, we can calculate its parasitic contribution:

$$C_{opto} = \frac{1}{2\pi f_{opto} R_{pullup}} = \frac{1}{6.28\times 15k\times 1k} = 10.6\ \text{nF} \tag{5.327}$$

When subtracted from (5.325), you obtain the value for C_{col}, the final capacitor installed across the optocoupler:

$$C_{col} = C_2 - C_{opto} = 4.2n - 10.6n \approx -6.4\ \text{nF} \tag{5.328}$$

This is a negative capacitor value, meaning that the second pole is entirely fixed by the optocoupler at 15 kHz rather than the expected one at 38 kHz. To ensure the presence of an extra capacitor across the optocoupler for an improved noise immunity, we can see that we do not have too many choices:

1. Reduce the crossover frequency until the added capacitor becomes positive again. Shoot for a value above 100 pF, placed very close to the controller. In

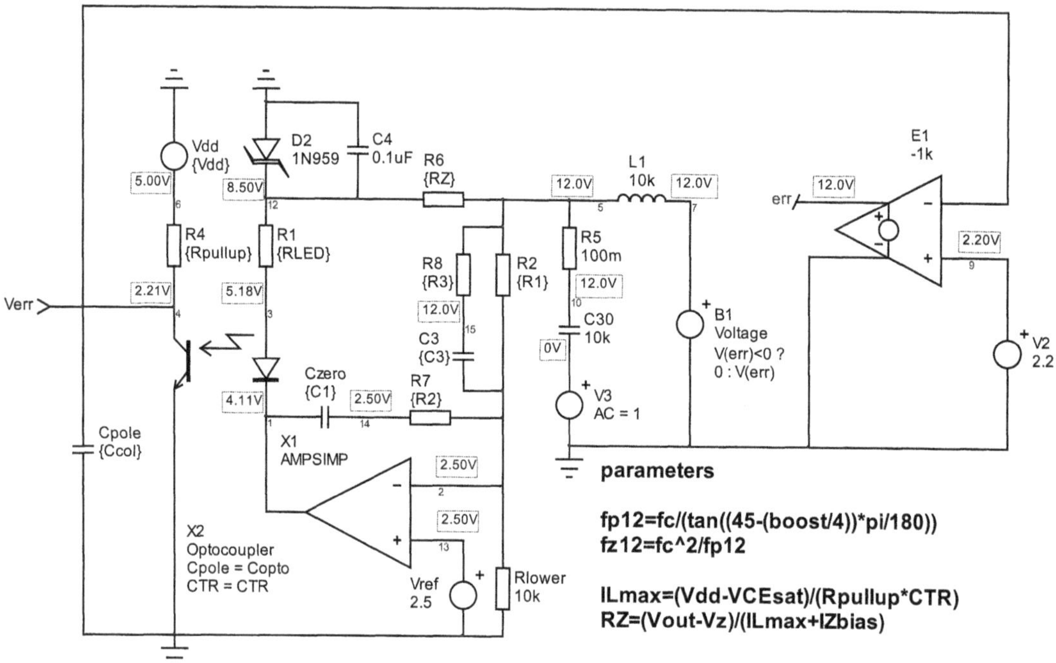

Figure 5.59 The type 3 featuring the Zener diode no longer shows the gain limit inherent to the fast lane.

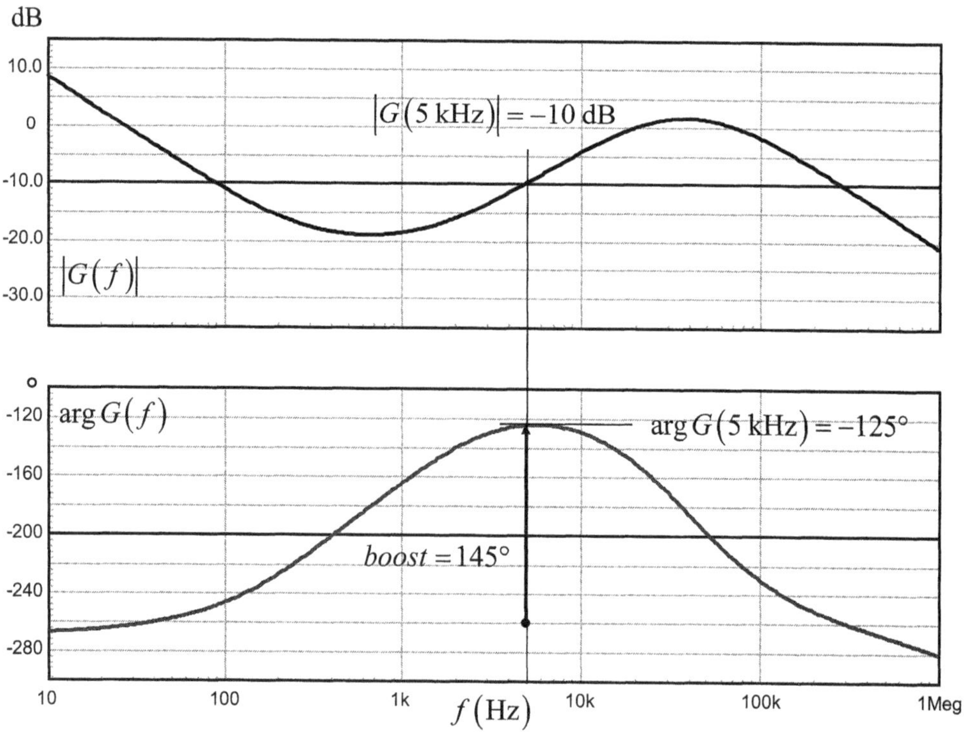

Figure 5.60 The simulation results of the type 3 compensator based on an op amp where the fast lane has been deactivated.

our example, reducing the crossover frequency to 1.8 kHz induces an extra capacitor value of 1 nF.

2. Adopt a faster optocoupler whose pole when associated with the selected pull-up resistor is beyond the high-frequency pole selected for the compensator. Another solution consists of using a cascode configuration.

Having these elements on hand, it is time for a simulation, as proposed by Figure 5.59.

The ac results show that the phase boost is slightly less than what was expected, mainly due to the mismatch in the second high-frequency pole that lead us to ignore C_{col} since the optocoupler pole fixes the second pole entirely (see Figure 5.60).

5.10 Conclusion

Op amp used in control loop can be wired in a lot of different ways. We believe this chapter covers most of the possible structures you will use during your design analysis. Whether you stabilize nonisolated dc-dc converters or ac-dc power supplies with an optocoupler, you should find the structure that suits your needs. If not, you have all the analysis examples that will help you derive the set of expressions you need. Finally, to let you strengthen your knowledge in the field of operational amplifiers and their usage in the power electronics field, [1–4] should put you on the right track.

References

[1] Mancini, R., "Op Amps for Everyone," Texas-Instruments Application note SLOD006b, August 2002.

[2] Carter, B., "Handbook of Operational Amplifier Functions," Texas-Instruments Application note SBOA092A, October 2001.

[3] Basso, C., *Switch Mode Power Supplies: SPICE Simulations and Practical Designs*, New York: McGraw-Hill, 2008.

[4] Mamano, B., "Isolating the Control Loop," Unitrode application note SLUP090, SEM700, 1990.

Appendix 5A: Summary Pictures

The following pictures summarize the component definitions associated with the structures described in the chapter.

Type 1
1 origin pole

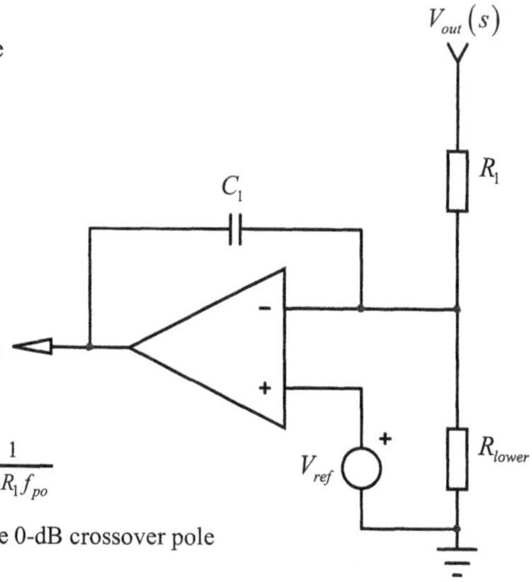

$$C_1 = \frac{1}{2\pi R_1 f_{po}}$$

f_{po} is the 0-dB crossover pole

Figure 5.61 The type 1.

Type 2
1 origin pole
1 pole f_p
1 zero f_z

$$R_2 = \frac{R_1 f_p G}{f_p - f_z} \frac{\sqrt{\left(\frac{f_c}{f_p}\right)^2 + 1}}{\sqrt{\left(\frac{f_z}{f_c}\right)^2 + 1}}$$

$$C_1 = \frac{1}{2\pi R_2 f_z}$$

$$C_2 = \frac{C_1}{2\pi f_p C_1 R_2 - 1}$$

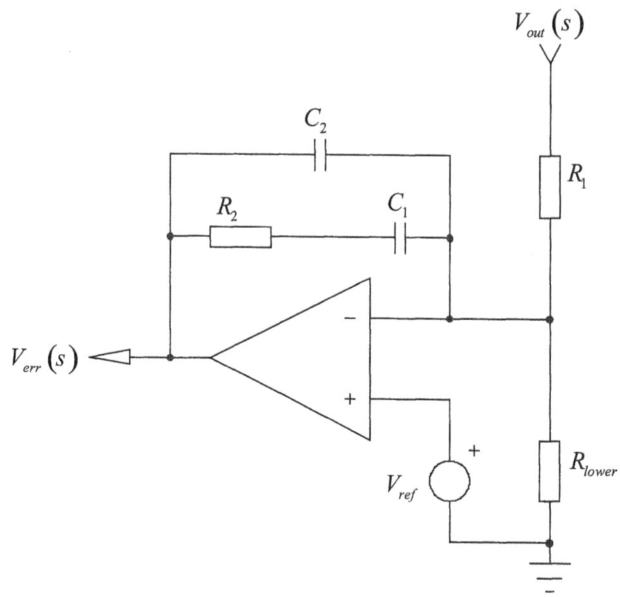

Figure 5.62 The type 2.

Isolated type 2 – direct drive
1 origin pole
1 pole f_p
1 zero f_z

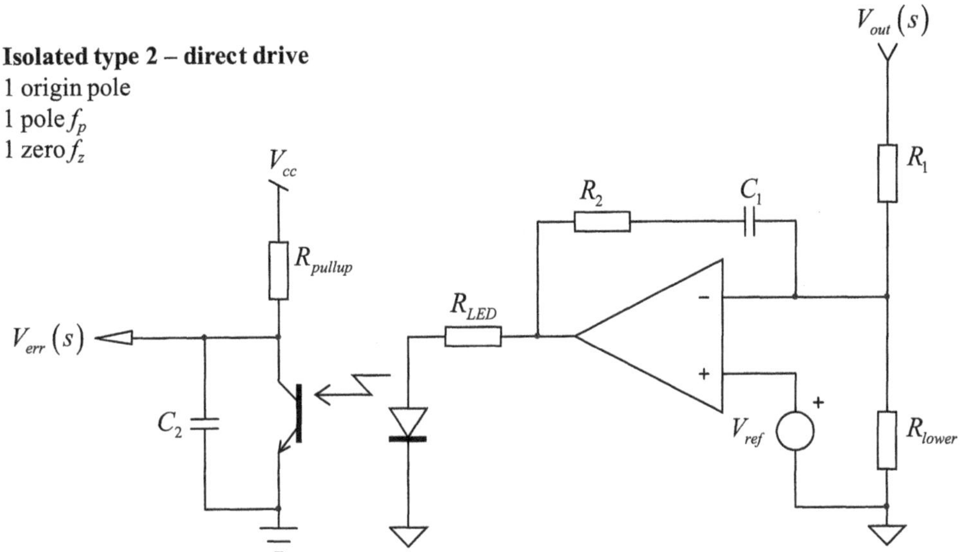

$$C_1 = \frac{1}{2\pi f_z R_2} \qquad C_2 = \frac{1}{2\pi R_{pullup} f_p} \qquad G_1 = \frac{R_{pullup}}{R_{LED}} \mathrm{CTR}_{\min} \qquad G_2 = \frac{G}{G_1} \qquad R_2 = R_1 G_2$$

$$C_{opto} = \frac{1}{2\pi f_{opto} R_{pullup}} \qquad C_{col} = C_2 - C_{opto}$$

Figure 5.63 An isolated type 2, direct optocoupler connection.

Isolated type 2 – fast lane
1 origin pole
1 pole f_p
1 zero f_z

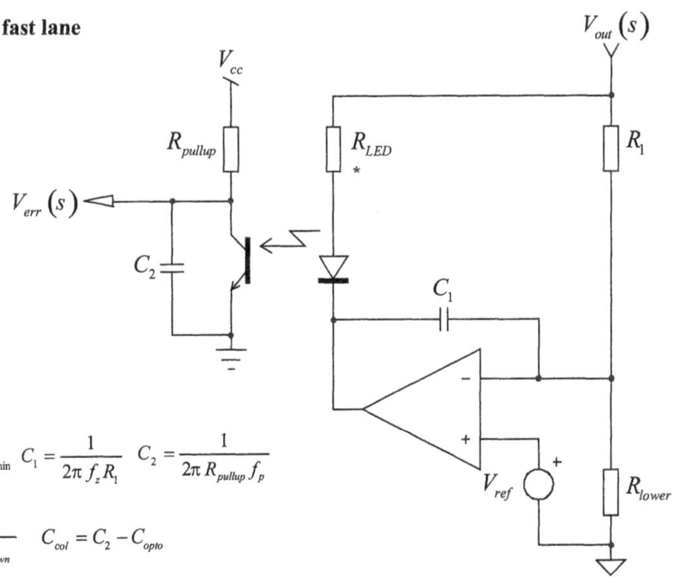

$$R_{LED} = \frac{R_{pullup}}{G} \mathrm{CTR}_{\min} \qquad C_1 = \frac{1}{2\pi f_z R_1} \qquad C_2 = \frac{1}{2\pi R_{pullup} f_p}$$

$$C_{opto} = \frac{1}{2\pi f_{opto} R_{pulldown}} \qquad C_{col} = C_2 - C_{opto}$$

* Check max value

Figure 5.64 An isolated type 2 with fast lane.

Isolated type 2 – no fast lane
1 origin pole
1 pole f_p
1 zero f_z

$$G_1 = \frac{R_{pullup}}{R_{LED}}\mathrm{CTR}_{\min} \quad G_2 = \frac{G}{G_1} \quad R_2 = R_1 G_2$$

$$C_1 = \frac{1}{2\pi f_z R_2} \qquad C_2 = \frac{1}{2\pi R_{pullup} f_p}$$

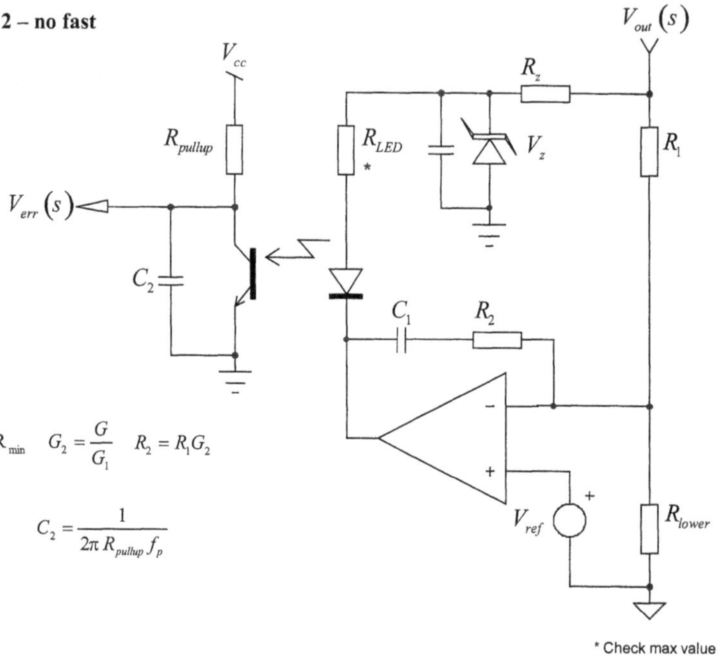

Figure 5.65 An isolated type 2 without fast lane.

Type 3
1 origin pole
2 poles f_{p1}, f_{p2}
2 zeros f_{z1}, f_{z2}

$$R_2 = \frac{GR_1 f_{p_1}}{f_{p_1} - f_{z_1}} \frac{\sqrt{1+\left(\frac{f_c}{f_{p_1}}\right)^2}\sqrt{1+\left(\frac{f_c}{f_{p_2}}\right)^2}}{\sqrt{1+\left(\frac{f_{z_1}}{f_c}\right)^2}\sqrt{1+\left(\frac{f_c}{f_{z_2}}\right)^2}}$$

$$C_1 = \frac{1}{2\pi f_{z_1} R_2} \quad C_2 = \frac{C_1}{2\pi f_{p_1} C_1 R_2 - 1} \qquad C_3 = \frac{f_{p_2} - f_{z_2}}{2\pi R_1 f_{p_2} f_{z_2}}$$

$$R_3 = \frac{R_1 f_{z_2}}{f_{p_2} - f_{z_2}}$$

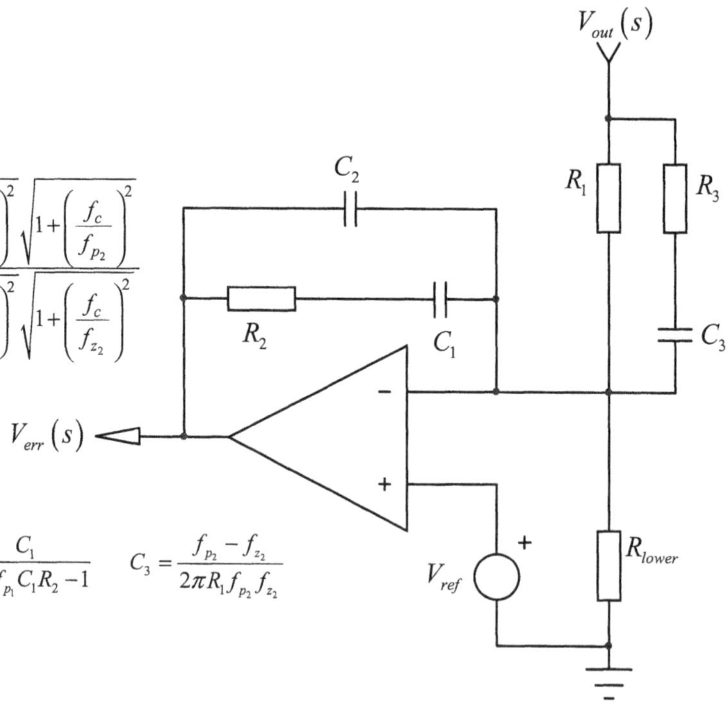

Figure 5.66 The type 3.

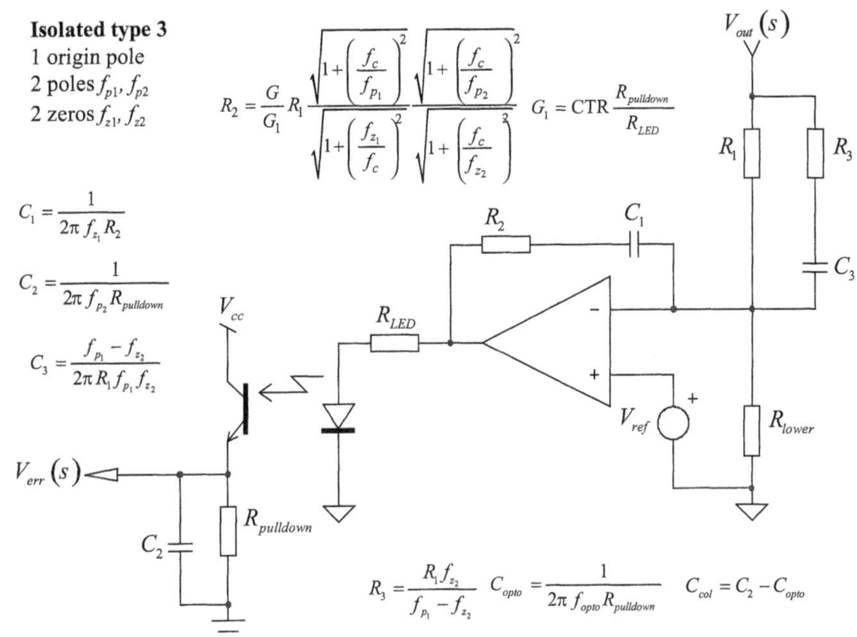

Figure 5.67 An isolated type 3, direct optocoupler connection.

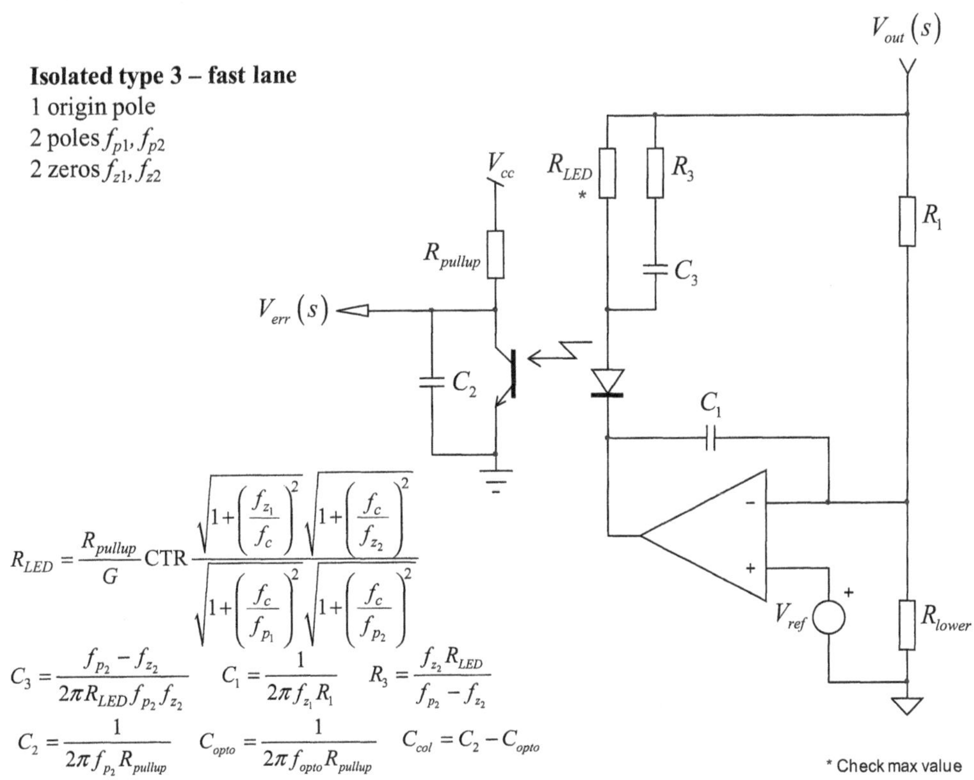

Figure 5.68 An isolated type 3 with fast lane.

Isolated type 3 – no fast lane
1 origin pole
2 poles f_{p1}, f_{p2}
2 zeros f_{z1}, f_{z2}

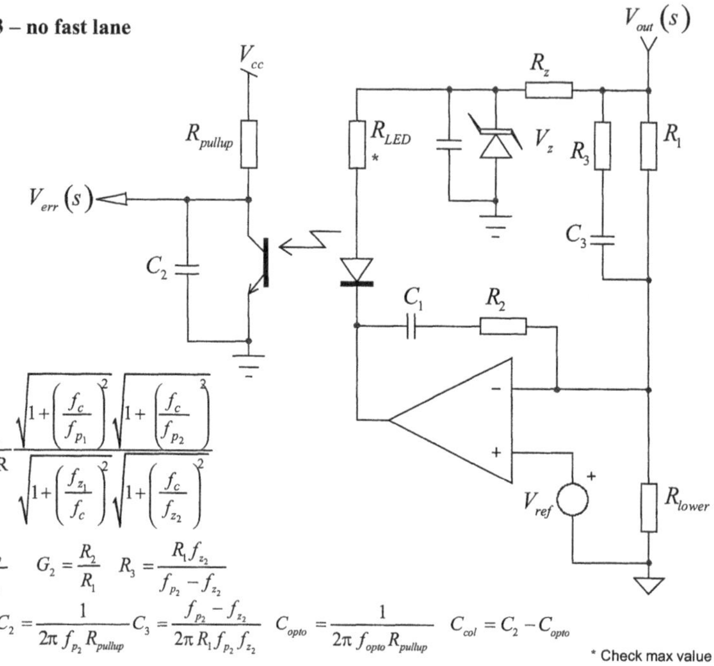

$$R_2 = \frac{GR_1R_{LED}}{R_{pullup}\text{CTR}} \frac{\sqrt{1+\left(\frac{f_c}{f_{p_1}}\right)^2}\sqrt{1+\left(\frac{f_c}{f_{p_2}}\right)^2}}{\sqrt{1+\left(\frac{f_{z_1}}{f_c}\right)^2}\sqrt{1+\left(\frac{f_c}{f_{z_2}}\right)^2}}$$

$$G_1 = \text{CTR}\frac{R_{pullup}}{R_{LED}} \quad G_2 = \frac{R_2}{R_1} \quad R_3 = \frac{R_1 f_{z_2}}{f_{p_2} - f_{z_2}}$$

$$C_1 = \frac{1}{2\pi f_{z_1} R_2} \quad C_2 = \frac{1}{2\pi f_{p_2} R_{pullup}} \quad C_3 = \frac{f_{p_2} - f_{z_2}}{2\pi R_1 f_{p_2} f_{z_2}} \quad C_{opto} = \frac{1}{2\pi f_{opto} R_{pullup}} \quad C_{col} = C_2 - C_{opto}$$

Figure 5.69 An isolated type 3, no fast lane.

Appendix 5B: Automating Components Calculations with *k* Factor

The *k* factor is a technique introduced by Dean Venable and described in [1]. Basically, it uses the formulas already derived in Chapter 4 that dictate the positions of the single or double pole/zero pair. Dean Venable's idea was then to include his *k* factor definition into the calculation of the *R* and *C* elements of any compensator. At that time, he only covered op amp-based circuits, and that is the limit of his analysis. The nice thing for beginners, however, is that you do not manipulate poles, zeros, or the 0-dB crossover pole position, as these values do not appear in the component definitions. The drawback of the method is that crossover is always placed at the geometric means of the poles and zeros positions: you cannot split them to cope with your particular design. This is the reason that in this book, we preferred manually placing individual poles and zeros for the best flexibility, whether or not you want coincident poles/zeros.

For the sake of explanation, but also because the method is popular, we will show how component values for op amp-based compensators were derived using the *k* factor method next.

Type 1

The type 1 does not need particular development, as we do not place a pole or a zero but rather a 0-dB origin pole. We can also think of the type 1 as a type 2 compensator where the poles and zeros would be made coincident. In that case, *k* as defined by

$$k = \tan\left(\frac{boost}{2} + \frac{\pi}{4}\right) \tag{5.329}$$

would simply be 1, as the phase boost in a type 1 is 0. The formulas given in the beginning of this chapter are those that Dean Venable derived in his paper: (5.7), (5.8), and (5.9).

Type 2

The type 2 from Figure 5.5 requires a little bit of algebraic efforts to unveil the value of k in the design formulas we have. The position of the pole, the zero, and the mid-band gain G for an op amp-based type 2 are the following:

$$G = \frac{R_2}{R_1} \frac{C_1}{C_1 + C_2} \tag{5.330}$$

$$\omega_z = \frac{1}{R_2 C_1} \tag{5.331}$$

$$\omega_p = \frac{1}{R_2 \dfrac{C_1 C_2}{C_1 + C_2}} \tag{5.332}$$

The relationship from Chapter 4 that links the pole and zero position with the crossover point is

$$\omega_p = k \cdot \omega_c \tag{5.333}$$

$$\omega_z = \frac{\omega_c}{k} \tag{5.334}$$

From (5.330), we can extract the value of R_2:

$$R_2 = \frac{G \cdot R_1 (C_1 + C_2)}{C_1} \tag{5.335}$$

A similar value can be obtained if we equate (5.332) and (5.333), and then solve for R_2. We should obtain

$$R_2 = \frac{C_1 + C_2}{C_1 C_2 \omega_c k} \tag{5.336}$$

Now, equating (5.335) and (5.336), we can extract the value of C_2:

$$\frac{G \cdot R_1 (C_1 + C_2)}{C_1} = \frac{C_1 + C_2}{C_1 C_2 \omega_c k} \tag{5.337}$$

Given the simplifications by $(C_1 + C_2)$ and C_1, it becomes

$$C_2 = \frac{1}{\omega_c G \cdot k \cdot R_1} = \frac{1}{2\pi f_c G \cdot k \cdot R_1} \tag{5.338}$$

Now, if we equate (5.331) and (5.334), we obtain another definition for R_2:

$$R_2 = \frac{k}{C_1 \omega_c} \tag{5.339}$$

This definition is similar to that given in (5.336); we can now solve for C_2:

$$\frac{k}{C_1 \omega_c} = \frac{C_1 + C_2}{C_1 C_2 \omega_c k} \tag{5.340}$$

We find

$$C_1 = C_2(k^2 - 1) \tag{5.341}$$

R_2 was already derived in (5.339) and equals

$$R_2 = \frac{k}{2\pi f_c C_1} \tag{5.342}$$

In these formulas, G is the gain/attenuation needed at the selected crossover frequency f_c.

Type 3

The type 3 was introduced in Figure 5.46. The positions of the poles and zeros pairs have been derived in this chapter. If we consider coincident poles and zeros, we have

$$\omega_{z_{1,2}} = \frac{1}{R_2 C_1} = \frac{1}{(R_1 + R_3) C_3} \tag{5.343}$$

$$\omega_{p_{1,2}} = \frac{1}{R_2 \dfrac{C_1 C_2}{C_1 + C_2}} = \frac{1}{R_3 C_3} \tag{5.344}$$

The crossover gain G involves the poles/zeros positions and is given by

$$G = \frac{R_2}{R_1} \frac{C_1}{C_1 + C_2} \frac{\sqrt{1 + \left(\dfrac{\omega_{z_{1,2}}}{\omega_c}\right)^2} \sqrt{1 + \left(\dfrac{\omega_c}{\omega_{z_{1,2}}}\right)^2}}{1 + \left(\dfrac{\omega_c}{\omega_{p_{1,2}}}\right)^2} \tag{5.345}$$

The relationship from Chapter 4 that links the pole and zero position is still

$$\omega_{p_{1,2}} = k \cdot \omega_c \tag{5.346}$$

$$\omega_{z1,2} = \frac{\omega_c}{k} \tag{5.347}$$

In (5.345), we can substitute the poles and zeros definitions in (5.346) and (5.347), and rewrite the expression as follows:

$$G = \frac{R_2 C_1 k^2 \sqrt{\frac{k^2+1}{k^2}}}{R_1(C_1+C_2)\sqrt{k^2+1}} = \frac{C_1 R_2 k}{R_1(C_1+C_2)} \tag{5.348}$$

As (5.343) and (5.347) are similar, we can extract the value of R_2 that we will substitute in (5.348):

$$R_2 = \frac{k}{\omega_c C_1} \tag{5.349}$$

$$G = \frac{k^2}{R_1 \omega_c (C_1+C_2)} \tag{5.350}$$

Now, if we equate (5.346) and the second term of (5.344), then substitute R_2 by its definition in (5.349), we have

$$\omega_c k = \frac{\omega_c(C_1+C_2)}{kC_2} \tag{5.351}$$

Solving for C_1, we obtain

$$C_1 = C_2(k^2-1) \tag{5.352}$$

Now that we have a link between C_1 and C_2, we can replace C_1 in (5.350) by its definition in (5.352). Then, solving for C_2, we obtain

$$C_2 = \frac{1}{2\pi f_c R_1 G} \tag{5.353}$$

Let's have a look at R_2 already given by (5.349). We can keep this value (equal to that of Dean Venable) or use (5.352) and (5.353) to develop it further:

$$R_2 = \frac{k}{\omega_c C_2(k^2-1)} = \frac{R_1 G \cdot k}{(k^2-1)} \tag{5.354}$$

R_3 and C_3 require a few more lines. From (5.344), we extract R_3:

$$R_3 = \frac{C_1 C_2 R_2}{C_3(C_1+C_2)} \tag{5.355}$$

From (5.343), we get R_2:

$$R_2 = \frac{C_3(R_1 + R_3)}{C_1} \tag{5.356}$$

and we substitute its definition in (5.355). We obtain

$$R_3 = \frac{C_2(R_1 + R_3)}{C_1 + C_2} \tag{5.357}$$

Solving for R_3, gives

$$R_3 = \frac{C_2 R_1}{C_1} \tag{5.358}$$

From (5.352), we can define C_2/C_1:

$$\frac{C_2}{C_1} = \frac{1}{k^2 - 1} \tag{5.359}$$

R_3 is now fully defined by

$$R_3 = \frac{R_1}{k^2 - 1} \tag{5.360}$$

C_3 is obtained by using (5.344) and (5.346):

$$\omega_c k = \frac{1}{R_3 C_3} \tag{5.361}$$

Thus,

$$C_3 = \frac{1}{2\pi f_c R_3 k} \tag{5.362}$$

We are all set! The difference in these equations with those derived by Dean Venable is the k value that we kept similar to that of the type 3 definition in (4.123) that Mr. Venable purposely squared:

$$k_{venable} = \left[\tan\left(\frac{boost}{4} + 45\right)\right]^2 \tag{5.363}$$

and thus uses $\sqrt{k}$ instead in all his equations. Final results are obviously similar. The following figures give a summary of the k factor components definitions in the three different compensator types.

Reference

[1] Venable, D., "The k Factor: A New Mathematical Tool for Stability Analysis and Synthesis," *Proceedings of Powercon 10*, 1983, pp. 1–12.

f_c is the crossover frequency
G is the gain/attenuation at crossover

$$C_1 = \frac{1}{2\pi f_c R_1 G}$$

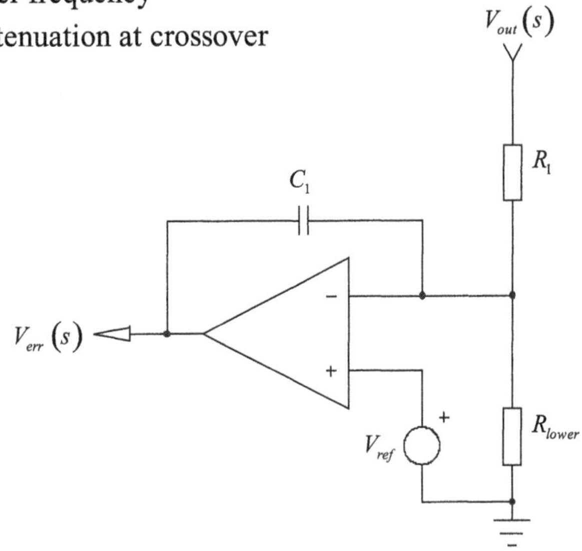

Figure 5.70 The type 1 is a simple compensator featuring an origin pole.

f_c is the crossover frequency
G is the gain/attenuation at crossover

$$k = \tan\left(\frac{boost}{2} + 45\right)$$

$$C_2 = \frac{1}{2\pi f_c G \cdot k \cdot R_1}$$

$$C_1 = C_2\left(k^2 - 1\right)$$

$$R_2 = \frac{k}{2\pi f_c C_1}$$

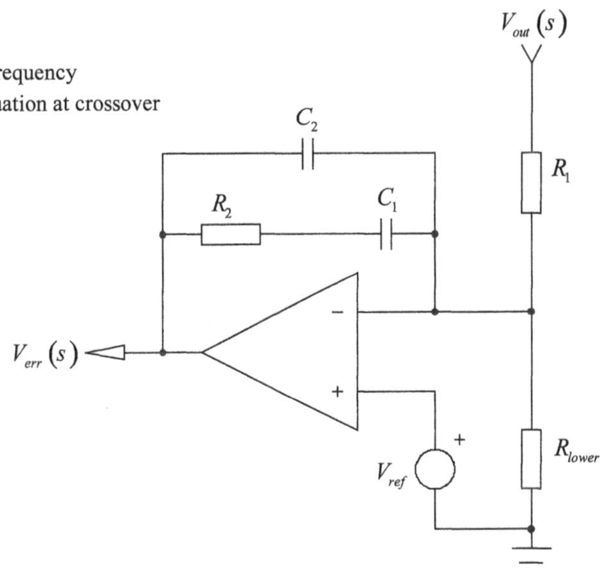

Figure 5.71 The type 2 compensator using components definitions involving the *k* factor.

f_c is the crossover frequency

G is the gain/attenuation at crossover

$$k = \tan\left(\frac{boost}{4} + \frac{\pi}{4}\right)$$

$$C_2 = \frac{1}{2\pi f_c R_1 G}$$

$$C_1 = C_2\left(k^2 - 1\right)$$

$$R_2 = \frac{k}{2\pi f_c C_1}$$

$$R_3 = \frac{R_1}{k^2 - 1}$$

$$C_3 = \frac{1}{2\pi f_c R_3 k}$$

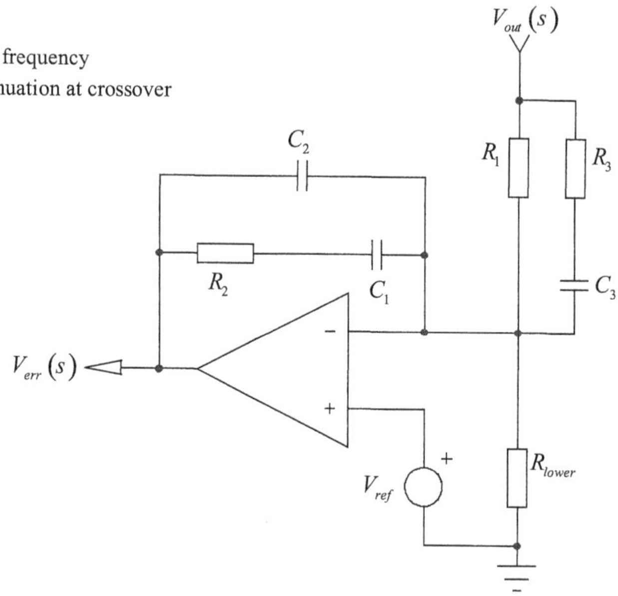

Figure 5.72 The type 3 compensator featuring the components definitions involving the *k* factor.

Appendix 5C: The Optocoupler

In isolated structures, the secondary-side information has to be conveyed across an isolation barrier to reach the nonisolated primary side. Several methods exist to cross that barrier, but the most popular uses an optical component called the optocoupler. Throughout the examples, we learned that the device affects the transfer function through several parameters, such as its current transfer ratio and transmission pole. Despite the care you will put in the selection of poles/zeros placement to build phase and gain margins, the insertion of the optocoupler can ruin your efforts if you do not realize its impact on the final transfer function. Knowing how to extract the optocoupler parameters and understand how they can change is vital to designing robust and reliable converters.

Transmitting Light

An optocoupler is made of a bipolar transistor and a gallium arsenide (GaAs) LED element. Encapsulated into a plastic package, it can provide galvanic isolation from 2.5 kV to 6 kV between a transformer-isolated secondary side and the primary side of a converter. The term galvanic comes after Luigi Galvani, an Italian scientist who studied the effects of electricity on muscles in the 18th century.

There are several manufacturing techniques related to the optocouplers. Reference [1] reviews the assembly techniques in great detail. Among them, the planar technique consists of laying the diode and transistor in the same plane and wire bonding them to a common leadframe. Figure 5.73 shows a simplified view of a planar optocoupler, where a silicone dome reflects the LED beam to further route it to the transistor collector-base junction. The photons emitted by the beam light are collected by the base of the transistor and give rise to a collector current. As no

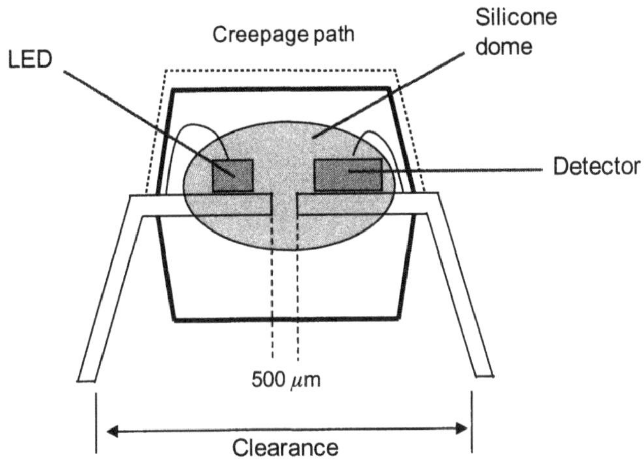

Figure 5.73 A simplified representation of an optocoupler construction.

electrical contact exists between the LED and transistor connections, the isolation is naturally created.

Current Transfer Ratio

The collector current I_c flowing in the transistor depends on the quantity of photons emitted by the LED. As the light intensity directly depends on the LED biasing current I_F, we have a relationship between both currents. This is the current transfer ratio (CTR), defined as follows:

$$\text{CTR}(\%) = \frac{I_c}{I_F}100 \tag{5.364}$$

The CTR is affected by a lot of external parameters: temperature, LED current, transistor gain dispersions, and so on. Figure 5.74 portrays the effect of the LED forward current on the optocoupler CTR. You can see the wide variations of these parameters, as the LED current changes. The characterized element is a CEL PS2913 optocoupler.

In modern consumer power supplies, where every milliwatt counts in the no-load standby power performance, the LED driving current is reduced down to a few hundreds of microamperes. As a result, the CTR collapses and can suffer from wide lots-to-lots dispersions. For a given optocoupler, a CTR range of 60–120 percent is not an uncommon value when the LED is biased in the vicinity of a few milliamperes. This number shrinks to less than 30 percent typically when operated at a 300-μA LED current, showing a division by 4 or a –12-dB attenuation when used in a gain chain!

In isolated ac-dc converters featuring an optocoupler in the return path and regardless of the implementation (op amp, TL431, and so on), the CTR plays a role in the gain expression. Usually, this mid-band gain compensates the gain deficiency of the output stage at a frequency where you want to cross over. If you ran your compensation calculations based on the highest CTR of 120 percent, you may experience a strong error in the crossover frequency if the CTR barely reaches

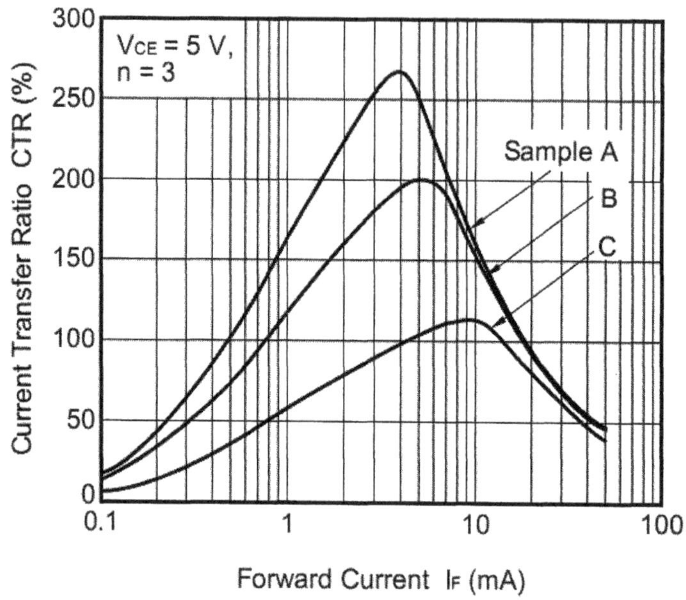

Figure 5.74 The CTR changes widely in relationship to the LED current.

30 percent at some point. In theory, the designer strives to ensure that the loop gain magnitude $|T(s)|$ passes the 0-dB axis with a –1-slope in order to keep the phase lag at this point under control. If the loop gain experiences a 12-dB reduction because the CTR jumps from 120 percent down to 30 percent, the crossover frequency reduces by a factor of four: you initially had 1 kHz, and you end up with 250 Hz! If the available phase margin is limited in this new crossover area, the converter can experience instability problems and will fail at final test. It is thus the designer's duty to understand the CTR variations of the device he or she selected and realize that unavoidable production dispersions can degrade the phase margin at crossover.

The Optocoupler Pole

The photons emitted by the LED are collected by the collector-base area of the bipolar transistor found in the optocoupler. To maximize the collected flux, the concerned area is purposely enlarged to the detriment of the parasitic capacitance between the collector and the base. Associated to the transistor gain β, the Miller equivalent capacitor can severely hamper the compensator phase margin when used in compensator circuits. To understand the impact of this element, a typical operating configuration—regardless of the compensator type, op amp, TL431, and so on—appears in Figure 5.75. In this approach, the compensation strategy recommends that you add a capacitor between the control pin of the considered controller (labeled FB in the picture) and ground. This is C_{col}, which introduces a pole equal to

$$f_p = \frac{1}{2\pi R_{pullup} C_{col}} \tag{5.365}$$

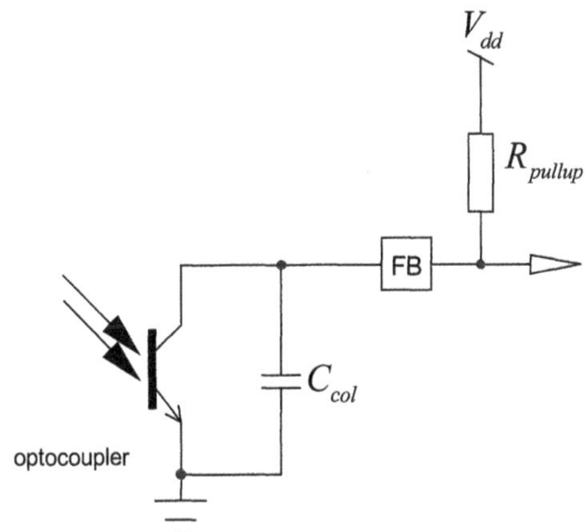

Figure 5.75 The optocoupler collector is often connected to a pull-up resistor where a stabilizing capacitor C_{col} brings a pole.

When you now connect the optocoupler to the control pin, you add its parasitic collector-emitter capacitor. This capacitor shows up in Figure 5.76, where a simplified, low-frequency, small-signal version of the optocoupler is proposed. As you can observe, this capacitor also couples with the pull-up resistor (or the pull down in a common-collector configuration) and introduces a pole at a frequency f_{opto}:

$$f_{opto} = \frac{1}{2\pi R_{pullup} C_{opto}} \tag{5.366}$$

Now, when the optocoupler capacitor couples to capacitor C_{col} you already have in place, the new pole is shifted to

$$f_p = \frac{1}{2\pi R_{pullup}(C_{col} + C_{opto})} \tag{5.367}$$

If this parasitic capacitor is large compared to C_{col}, you understand how it can distort the compensation strategy you shot for. Therefore, once the optocoupler capacitor is known, it must be subtracted from the needed capacitor you want (often

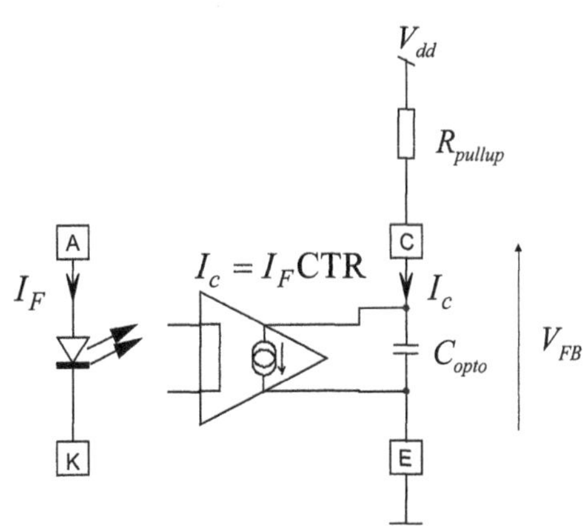

Figure 5.76 The simplified small-signal model of the optocoupler.

referred to as C_2 in the compensators) to make sure the sum of C_{opto} and C_{col} gives the right value:

$$C_{col} = C_2 - C_{opto} \tag{5.368}$$

Sometimes, it happens that C_{opto} is larger than the desired C_2 and (5.368) returns a negative value. The solution is to explore a new pole/zero combination, probably via the reduction of the selected crossover frequency. We recommend that C_{col} is at least 100 pF and placed close to the controller feedback pin. It greatly improves the immunity to spurious noises, as the feedback pin is usually high impedance and far from the optocoupler terminals.

Extracting the Optocoupler Pole

There are several ways to know the optocoupler pole position: you can read the data sheet and look for frequency response curves or timing diagrams, but the best way, in my opinion, is to set up a quick test fixture and ac sweep the optocoupler alone. That way, you know that the dc conditions and the component selection exactly match your converter implementation.

Figure 5.77 describes the way the optocoupler can be wired to unveil its pole position. The V_{bias} source fixes the dc operation point of this common-emitter configuration. You will tweak it to bring the optocoupler collector around $V_{dd}/2$ (for instance, 2 V if $V_{dd} = 5$ V), ensuring enough voltage dynamics when the ac sweep will begin. Please note that both R_{pullup} and R_{LED} can share a similar value, bringing the low frequency ac gain to the CTR value in this case. R_{bias} can be selected around a few kΩ.

The easiest way to ac sweep the circuit uses a network analyzer that will compute $20\log_{10}\frac{V(B)}{V(A)}$. The Bode plot will thus immediately appear in the computer screen. Looking for the –3-dB deviation from the low-frequency flat plateau will

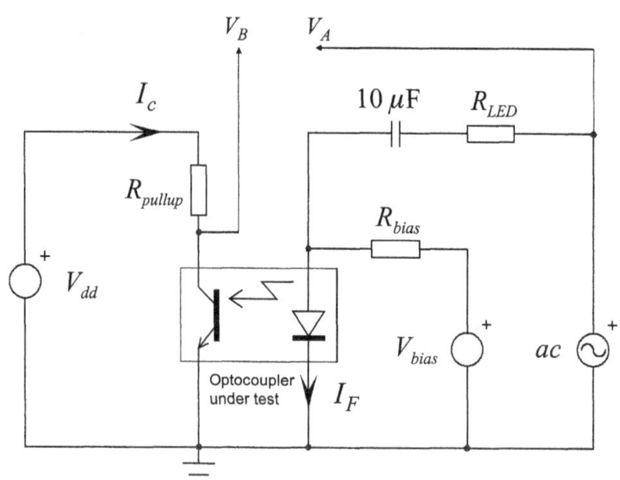

Figure 5.77 A simple test fixture biases the collector in the linear region and ac modulate the LED current.

indicate the pole position. This is what Figure 5.78 shows where the pole lies at 10 kHz. For this particular test, carried over a SFH615A-2, the pull-up resistor was set to 4.7 kΩ, imposing a maximum collector current around a milliampere from a 5-V V_{dd} bias source. Changing this resistor to 15 kΩ rolled the pole back to 3 kHz. With a 10-kHz pole and according to (5.366), the optocoupler capacitor is 3.4 nF. In Figure 5.78, as you can see, changing the dc operating point (different V_{ce} voltages) does not affect the pole position. Please note that the optocoupler features a second pole at higher frequency. We did not account for its presence in our simplified first-order model. It must be added if you target high crossover frequencies.

Without using a network analyzer, it is still possible to find the pole position. Use a sinusoidal function generator for the ac source and observe the collector voltage with an oscilloscope at, let's say, a 100-Hz frequency. Make sure the modulation is small enough to avoid distorting the observed signal. Tweak and offset the oscilloscope vertical channel to have the signal centered at the dc collector voltage, thus equally covering the five divisions up and down from the middle of the screen. Then, change the frequency and increase it until the modulation peak amplitude drops to around 3.5 divisions (total 7 divisions peak to peak). This point corresponds to a –3-dB drop from the reference point at 100 Hz: this is the pole frequency. Figure 5.79 shows an oscilloscope shot captured during a 1-kHz pole extraction following the previous procedure.

Watch for the LED Dynamic Resistance

In equations where the fast lane is involved, the overall gain expression depends only on the following terms: the optocoupler CTR, the pull-up resistor, and the LED series resistor—see (5.66), for instance. The LED series resistor is bounded by dc operating conditions imposed by the diode forward voltage (≈1 V) and the TL431 minimum operating voltage (2.5 V). As a result, in low output voltage applications

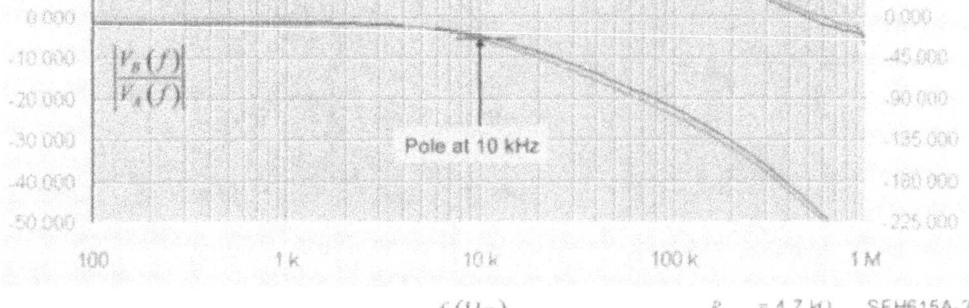

Figure 5.78 Looking at the point where the gain deviates from the low-frequency plateau by –3 dB indicates a 10-kHz pole presence.

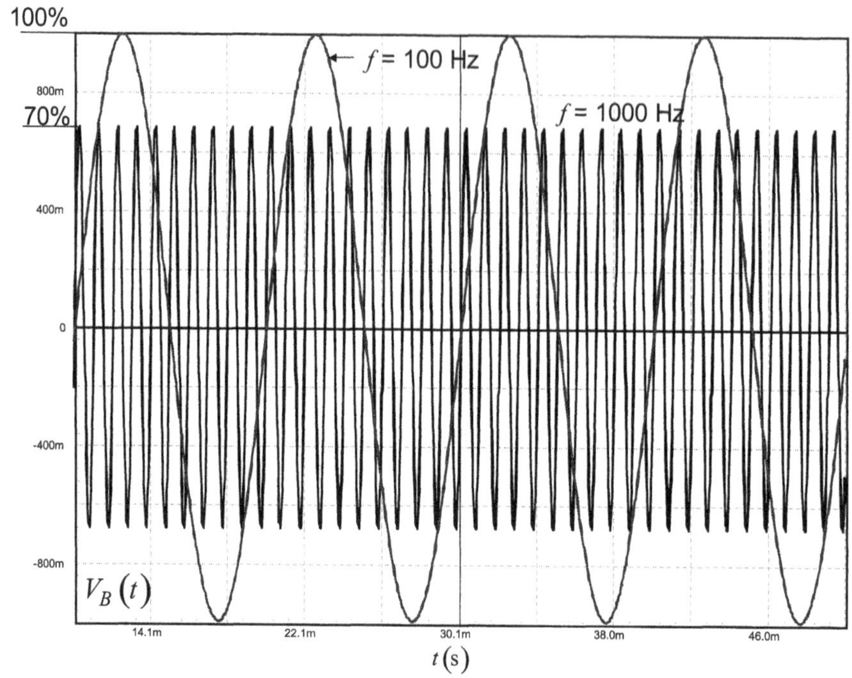

Figure 5.79 The pole extraction is straightforward with a simple oscilloscope—here, find a 1-kHz pole position.

(e.g., 5 V), this resistor can be of low value, in the vicinity of the hundred ohms. In that case, the LED dynamic resistance R_d can no longer be neglected. Furthermore, the bias resistor R_{bias} commonly installed over the LED derives a portion of the feedback current and also affects the total gain. Figure 5.80 portrays the simplified ac schematic, highlighting the elements around the LED. These small effects are often overlooked, but they can explain gain discrepancies observed in certain cases. A few equations can help us to formalize the role played by these elements and un-

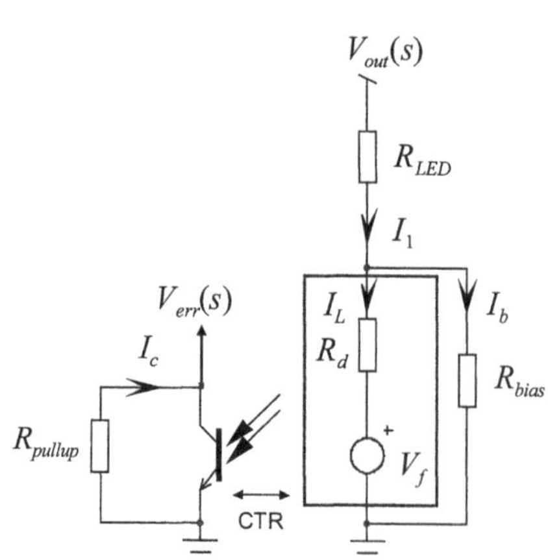

Figure 5.80 The full gain chain includes the LED series resistor and the bias generator. Both of them can affect the gain.

derstand how they interact with each other. The feedback voltage depends on the pull-up resistor and the current in it:

$$V_{err}(s) = -I_c(s)R_{pullup} = -I_L(s)R_{pullup}\mathrm{CTR} \tag{5.369}$$

The full ac current I_1 splits between the LED and the bias resistor R_{bias}. However, the LED current, alone, participates in the feedback chain. Therefore, the bias resistor "steals" away current from the loop. Ac wise (V_f is constant and equal to 0 in ac), the LED current is expressed by

$$I_L(s) = I_1(s)\frac{R_{bias}}{R_{bias}+R_d} = \frac{V_{out}(s)}{R_{LED}+R_{bias} \parallel R_d}\frac{R_{bias}}{R_{bias}+R_d} \tag{5.370}$$

Substituting (5.370) in (5.369), we can extract the transfer function of the optocoupler chain alone:

$$\left.\frac{V_{err}(s)}{V_{out}(s)}\right|_{s=0} = -\frac{R_{pullup}\mathrm{CTR}}{R_{LED}+R_{bias} \parallel R_d}\frac{R_{bias}}{R_{bias}+R_d} \tag{5.371}$$

In this expression, both R_d and R_{bias} play a role. R_{bias} is often set to 1 kΩ in order to provide the necessary milliampere current for the TL431 biasing current. If R_d is small compared to this value, less ac current will be diverted by R_{bias} and the gain chain will not suffer from its presence. On the contrary, if R_d becomes nonnegligible, the whole chain undergoes a gain reduction. What dynamic resistance value does an optocoupler LED exhibit?

Figure 5.81 shows the characterization of such a device at different bias currents and operating temperatures. As expected, the dynamic resistance varies depending on the operating current, like any diode would do. The dynamic resistance is extracted by looking at the curve in the vicinity of the operating point and computed as the voltage variation obtained by a small current change around the considered bias region:

$$R_d = \frac{\Delta V_f}{\Delta I_d} \tag{5.372}$$

From Figure 5.81, a dynamic resistance of ≈ 160 Ω is calculated with a 300-μA collector current. This is the case with modern PWM controllers that strive to reduce the consumed power in no-load conditions by hosting a high value internal pull-up resistor (usually between 10 and 20 kΩ). When the pull-up resistor is lowered to impose a 1-mA forward current (R_{pullup} = 1 kΩ), the dynamic resistance drops to ≈ 40 Ω. Applying (5.371) to a 5-V converter featuring the following element values—R_{LED} = 150 Ω, CTR = 0.3, R_{pullup} = 20 kΩ—we can compute the gain for various LED dynamic resistances:

$$\begin{aligned} G_1 \,|_{R_d=0\,\Omega} &\approx 32 \text{ dB} \\ G_2 \,|_{R_d=40\,\Omega} &\approx 30 \text{ dB} \\ G_3 \,|_{R_d=160\,\Omega} &\approx 25 \text{ dB} \end{aligned} \tag{5.373}$$

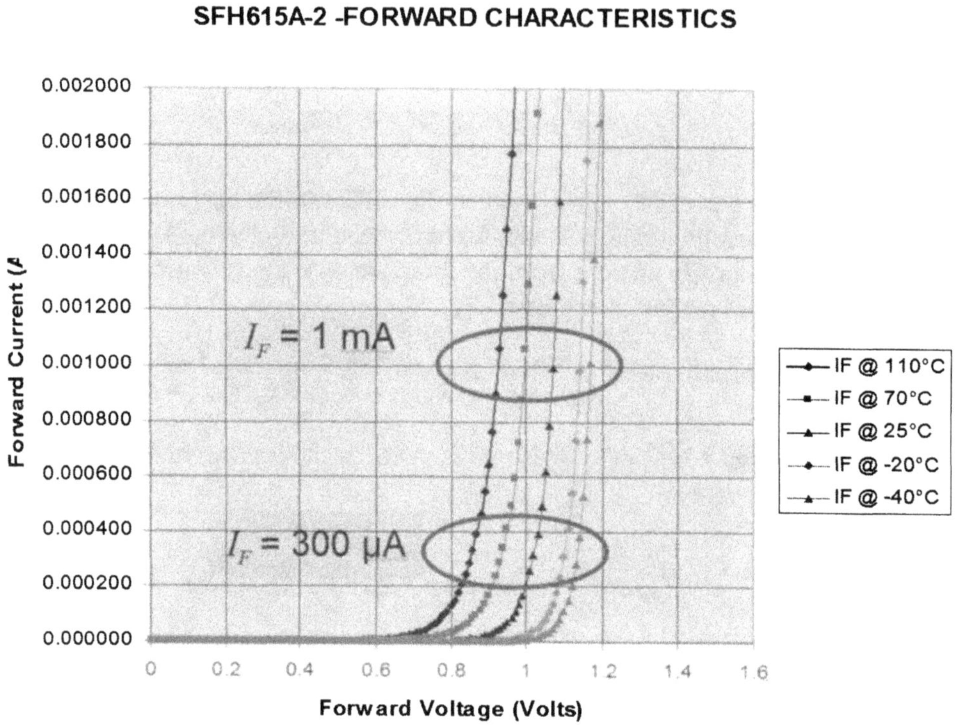

Figure 5.81 The LED dynamic resistance depends on its operating forward current.

As you can see, there is 7-dB gain difference from a calculation assuming a zero dynamic resistance and the reality of a LED operated at a low forward current. Again, a 7-dB difference in the mid-band gain can engender a crossover frequency mismatch of 2.2: you shoot for a 1-kHz crossover point and you end up below 500 Hz!

Good Design Practices

We have seen how the CTR, the LED dynamic resistance, and the parasitic pole can influence the optocoupler response. The key element to improving any of these offenders depends on the performance you are looking for. If an extremely low standby power is important while you are charging batteries (a notebook adapter, for instance), a wide bandwidth is probably not mandatory. Therefore, you can cope with a rather high pull-up resistor and a low collector current. The correspondingly low CTR value associated with a low frequency optocoupler pole will not hurt the performance in the end, as long as their natural variations are well accounted for in the design cycle. On the contrary, if time response and bandwidth are the key elements of your specifications, then make sure a low pull-up resistor value is selected (1 kΩ, for instance) to extend the optocoupler pole well beyond your crossover point and also reduce the LED dynamic resistance. You must, however, keep in mind that the optocoupler lifetime strongly depends on the LED forward current. As the optocoupler ages, the flow of emitted photons weakens and the transmission chain suffers. Operating the LED at a low current counteracts this aging process, at the expense of poor dynamic characteristics though. Some optocouplers compensate the LED deficiencies linked to its age, temperature, and so on. The IL300 from Vishay is one of them: a sensor monitors the LED light and modulates its operating current to keep the emitted flux constant. Please refer to the data sheet of this product for more information.

There are numerous industry-standard optocouplers manufactured by Sharp, Vishay, and CEL. Among these vendors, the PC817 and the SFH615 series are very popular. You find them in many notebook adapters and offline power supplies for TVs, DVD players, and so on. If you plan to design a high-bandwidth converter, optocouplers exhibiting high current transfer ratios (CTRs) are not recommended. To maximize the current transfer, the manufacturer will purposely grow the transistor area collecting the LED photons. By doing this, all the associated parasitic capacitances are increased, and switching times suffer. On the contrary, selecting low CTR devices will ensure a smaller internal transistor, naturally pushing away its inherent Miller capacitor.

Sometimes, when an optocoupler cannot be used, a magnetic link between the primary and secondary sides is the way to go. Reference [5] explores this solution in detail.

At the end, once the design is frozen, regardless of the isolation means, it is your duty to explore all the possible element variations with temperature and lots-to-lots dispersions in order to keep enough phase margin in all possible cases. Applying these rules is part of the recipe for a seamless mass production.

References

[1] Power 4-5-6, Ridley Engineering, www.ridleyengineering.com.

[2] Basso, C., *Switch Mode Power Supplies: SPICE Simulations and Practical Designs*, New York: McGraw-Hill, 2008.

[3] Basso, C., "Eliminate the Guesswork in Selecting Crossover Frequency," PET, August 1, 2008, http://powerelectronics.com/issue_20080801.

[4] Kek, T., and L. Tan, "Stacked LED Makes Compact Optocouplers," *EE Times Asia*, April 2005.

[5] Irving, B., and M. Jovanović, "Analysis and Design Optimization of Magnetic Feedback Control Using Amplitude Modulation," *IEEE Transactions on Power Electronics*, Vol. 24, No. 2, February 2009.

CHAPTER 6

Operational Transconductance Amplifier–Based Compensators

An operational transconductance amplifier (OTA) is a voltage-controlled current source circuit [1] affected by a transconductance factor, g_m, defined as

$$g_m = \frac{\Delta I_{out}}{\Delta V_{in}} \tag{6.1}$$

where ΔI_{out} is the amplifier output current change corresponding to a voltage change ΔV_{in} applied to its inputs.

In dc, the output current is obtained by multiplying the voltage difference ε between the noninverting pin and the inverting pin by the transconductance factor:

$$\varepsilon = V_{(+)} - V_{(-)} \tag{6.2}$$

$$I_{out} = \varepsilon g_m \tag{6.3}$$

The symbol for such an amplifier appears in Figure 6.1. You can see the transconductance value, g_m, expressed in amperes per volts or Siemens, in the international system unit. The symbols "mhos" and the reversed omega symbol ℧ are no longer in use. Suppose the g_m in the example is 200 μS; if you apply 1 V across the OTA inputs, it will deliver 200 μA, inducing 200 mV across a 1-kΩ resistor loading the output.

The output voltage is found using (6.3), noting that, as the inverting pin is grounded, $\varepsilon = V_{in}$:

$$V_{out} = I_{out} R_1 = V_{in} g_m R_1 \tag{6.4}$$

From this equation, the gain is simply

$$\frac{V_{out}}{V_{in}} = R_1 g_m \tag{6.5}$$

Due to this expression, we can see that by replacing R_1 with a complex network made of resistors and capacitors, we have a means to introduce poles and zeros, exactly as we did with the op amp approach.

OTAs are not very popular as standalone amplifiers. However, you often find them in power factor correction (PFC) controllers. The absence of virtual ground allows the inverting pin voltage to be used as a means to detect an overvoltage condition on the converter output. Also, designwise, they take much less die space than their op amp counterparts. Let's see the first simple OTA compensator circuit, the type 1.

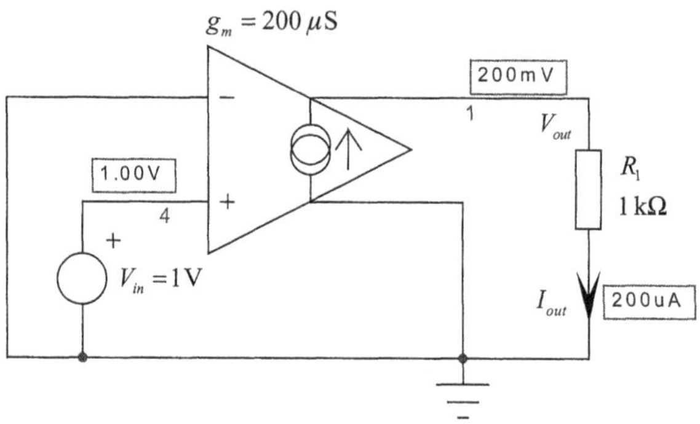

Figure 6.1 A voltage-controlled current source makes the simplest OTA. It converts the voltage measured between its inputs into an output current.

6.1 The Type 1: An Origin Pole

A type 1 circuit built on an OTA appears in Figure 6.2. The resistor of Figure 6.1 is replaced with a capacitor, C_1. The voltage on the inverting pin is now the converter output voltage V_{out} undergoing the voltage division brought by R_1 and R_{lower}:

$$V_{(+)}(s) - V_{(-)}(s) = V_{out}(s)\frac{R_{lower}}{R_{lower} + R_1} \tag{6.6}$$

The amplifier output voltage $V_{err}(s)$ is the output current multiplied by C_1 impedance:

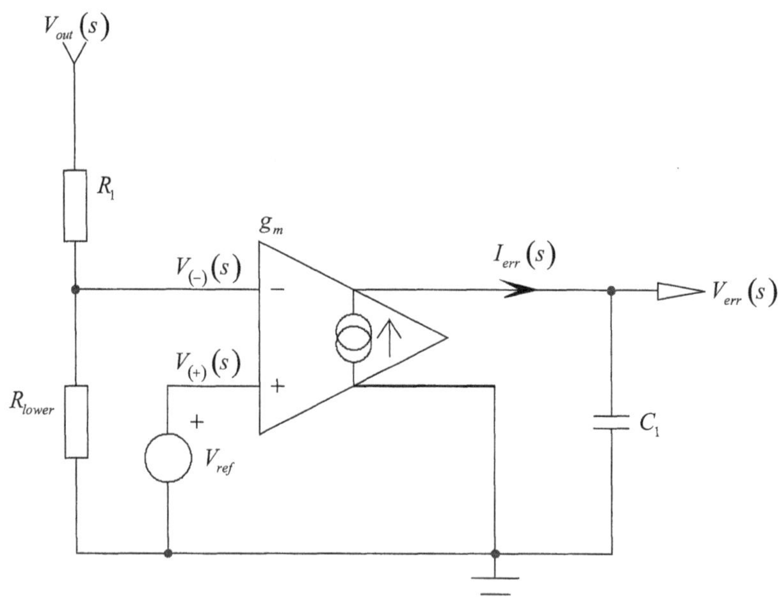

Figure 6.2 A type 1 with an OTA requires a simple capacitor connected from the output to ground.

$$V_{err}(s) = I_{err}(s)\frac{1}{sC_1} = g_m \frac{1}{sC_1}[V_{(+)}(s) - V_{(-)}(s)] \tag{6.7}$$

Substituting (6.6) into (6.7) and considering $V_{(+)}(s) = 0$ gives

$$V_{err}(s) = -g_m \frac{1}{sC_1}\frac{R_{lower}}{R_{lower}+R_1}V_{out}(s) \tag{6.8}$$

Rearranging this equation leads us to the transfer function we are looking for:

$$G(s) = \frac{V_{err}(s)}{V_{out}(s)} = -\frac{1}{sC_1\dfrac{R_{lower}+R_1}{g_m R_{lower}}} = -\frac{1}{sR_{eq}C_1} \tag{6.9}$$

where the equivalent resistor is defined as

$$R_{eq} = \frac{R_{lower}+R_1}{g_m R_{lower}} \tag{6.10}$$

Equation (6.9) can be put under the familiar form

$$G(s) = -\frac{1}{\dfrac{s}{\omega_{po}}} \tag{6.11}$$

In which the 0-dB crossover pole definition follows

$$\omega_{po} = \frac{1}{R_{eq}C_1} = \frac{g_m R_{lower}}{(R_{lower}+R_1)C_1} \tag{6.12}$$

The biggest change compared to the op amp approach can be seen in (6.12), where the divider network and the OTA g_m parameter now enter into the equations. Regarding the divider network, in the op amp case, thanks to the virtual ground effect, both pins were at a similar potential and the ac contribution of the lower resistor R_{lower} was nonexistent. In an OTA, we do not have a local feedback from V_{err} to the inverting pin, hence the absence of virtual ground. Therefore, we cannot ignore R_{lower} any longer, as confirmed by (6.6). The OTA g_m parameter can vary depending on the care put in the integrated circuit design. Its natural variation affects the position of the 0-dB crossover pole. These details are important, and you must check how the variation in g_m changes the 0-dB crossover pole for your application. It is time for a detailed design example.

6.1.1 A Design Example

Let's assume we want a compensator transfer function with an attenuation of 25 dB at 20 Hz. For a front-end PFC regulating a 400-V output with a 2.5-V reference voltage, the upper resistor R_1 could be 4 MΩ with a lower resistor of 25 kΩ. We have selected an OTA with a 100-μS transconductance. According to formulas derived in the op amp section type 1 example, the 0-dB crossover pole is chosen depending on the required attenuation/gain and the frequency at which it occurs:

$$f_{po} = f_c G = 20 \times 10^{-\frac{25}{20}} = 1.1\text{ Hz} \tag{6.13}$$

To place the crossover pole at this frequency, we calculate the capacitor value according to (6.12):

$$C_1 = \frac{g_m R_{lower}}{2\pi(R_{lower} + R_1)f_{po}} = \frac{100u \times 25k}{6.28 \times (25k + 4Meg) \times 1.1} \approx 90 \text{ nF} \tag{6.14}$$

With all the components known, we can run the simulation using the test circuit shown in Figure 6.3.

Please note that the OTA model is really the simplest possible model, a voltage-controlled current source. A more complex model would feature an upper and lower voltage clamp levels (given by the supply of the controller) plus a maximum output current limit. However, these extra components do not affect the ac response of the whole loop, and we can stay with the simplest representation.

The simulation results are delivered in Figure 6.4 and confirm our calculations. The phase lag is permanently set to 270°, which is normal for an integrator.

6.2 The Type 2: An Origin Pole plus a Pole/Zero Pair

The type 2 circuit has already been presented in the op amp section and also appears in [2]. It combines a pole zero/pair plus an origin pole for a high dc gain. The implementation using an OTA appears in Figure 6.5.

The voltage delivered by the OTA is its output current I_{err} multiplied by the loading impedance Z_L:

$$V_{err}(s) = I_{err}(s) Z_L(s) \tag{6.15}$$

The loading impedance Z_L is derived by associating two paralleled networks:

$$Z_L(s) = \left(R_2 + \frac{1}{sC_1}\right) || \frac{1}{sC_2} = \frac{\left(R_2 + \frac{1}{sC_1}\right)\frac{1}{sC_2}}{\left(R_2 + \frac{1}{sC_1}\right) + \frac{1}{sC_2}} \tag{6.16}$$

The current flowing in this network is the voltage present at the OTA inverting pin multiplied by the OTA transconductance g_m:

$$I_{err}(s) = -g_m V_{out}(s) \frac{R_{lower}}{R_{lower} + R_1} \tag{6.17}$$

Substituting (6.17) and (6.16) into (6.15), we have

$$G(s) = \frac{V_{err}(s)}{V_{out}(s)} = -g_m \frac{R_{lower}}{R_{lower} + R_1} \frac{\left(R_2 + \frac{1}{sC_1}\right)\frac{1}{sC_2}}{\left(R_2 + \frac{1}{sC_1}\right) + \frac{1}{sC_2}} \tag{6.18}$$

Rearranging (6.18), we obtain

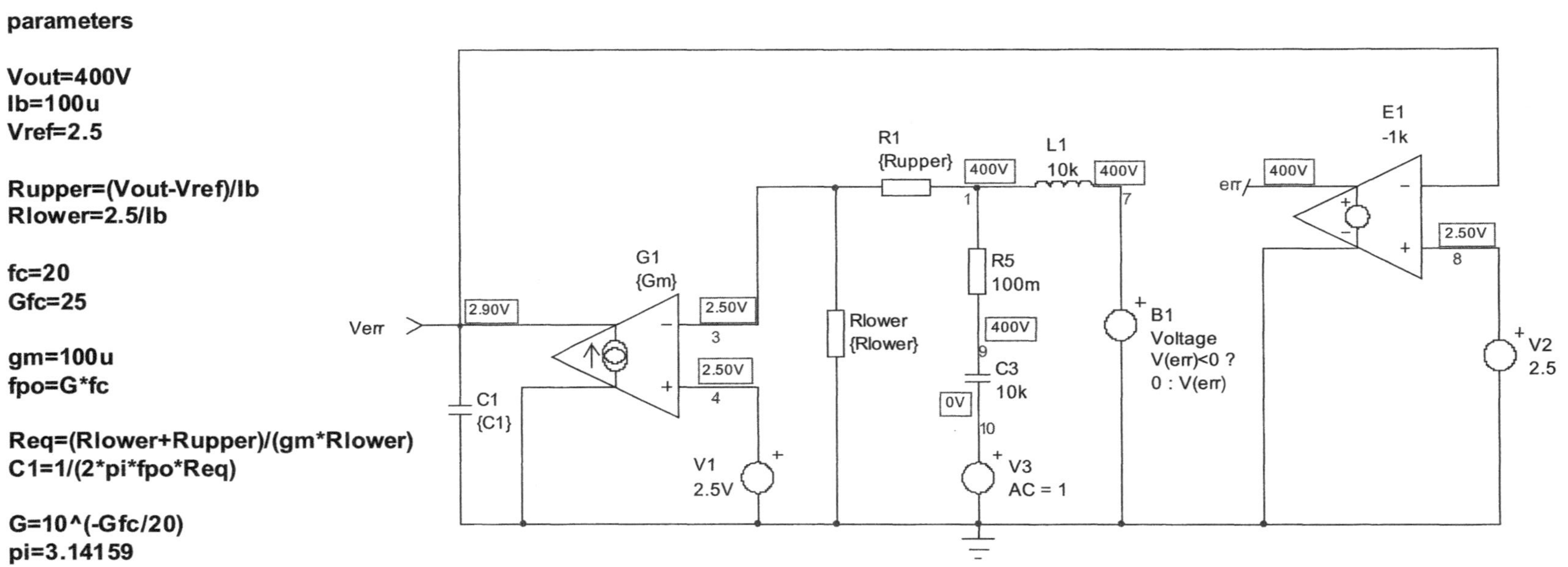

Figure 6.3 The OTA circuit is inserted within the automated bias point adjustment, confirming the 400-V output.

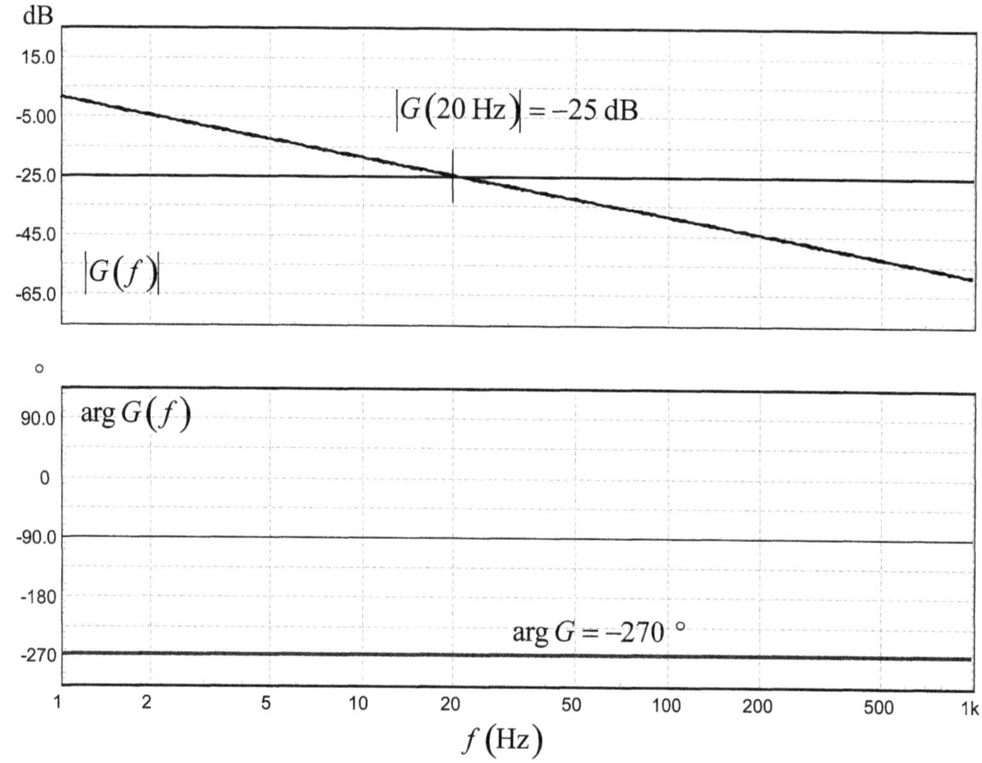

Figure 6.4 The ac results show the classical type 1 response, an integrator with a high dc-gain, crossing the −25-dB axis at 20 Hz.

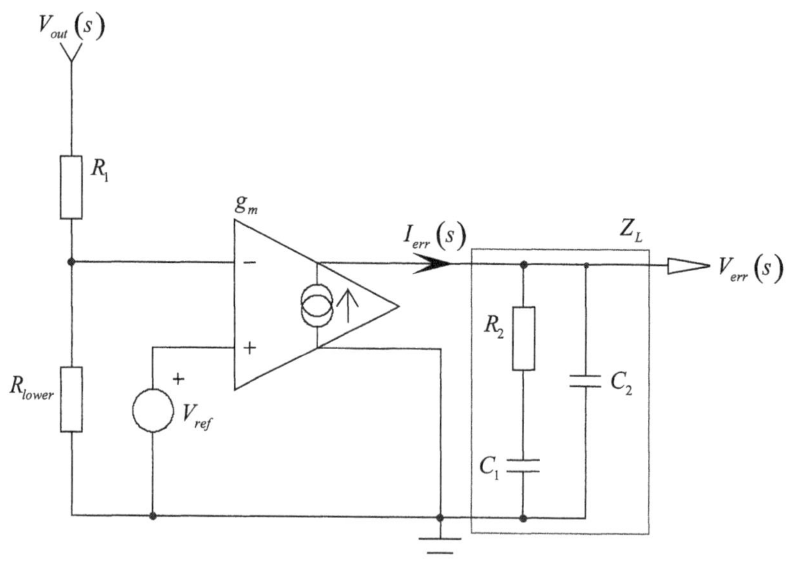

Figure 6.5 The type 2 with an OTA requires the addition of a *RC* network across the output capacitor C_2.

$$G(s) = -\frac{R_{lower} g_m}{R_{lower} + R_1} \frac{1 + sR_2C_1}{s(C_1 + C_2)\left(1 + sR_2 \frac{C_1C_2}{C_1 + C_2}\right)} \tag{6.19}$$

If we now factor sR_2C_1, we can easily unveil the poles and zeros definitions we are looking for:

$$G(s) = -\frac{R_{lower} g_m}{R_{lower} + R_1} \frac{R_2C_1}{C_1 + C_2} \frac{1 + \frac{1}{sR_2C_1}}{\left(1 + sR_2 \frac{C_1C_2}{C_1 + C_2}\right)} = -G_0 \frac{1 + \omega_z/s}{1 + s/\omega_p} \tag{6.20}$$

In this equation, we can identify the three pertinent terms:

$$G_0 = \frac{R_{lower}}{R_{lower} + R_1} \frac{g_m R_2 C_1}{C_1 + C_2} \tag{6.21}$$

$$\omega_z = \frac{1}{R_2C_1} \tag{6.22}$$

$$\omega_p = \frac{1}{R_2 \frac{C_1C_2}{C_1 + C_2}} \tag{6.23}$$

To extract each individual value, we have to derive the magnitude of (6.20) at the crossover frequency f_c:

$$|G(f_c)| = G_0 \frac{\sqrt{1 + \left(\frac{f_z}{f_c}\right)^2}}{\sqrt{1 + \left(\frac{f_c}{f_p}\right)^2}} \tag{6.24}$$

Now combining (6.21), (6.22), (6.23), and (6.24), we find the following parameters:

$$R_2 = \frac{f_p G}{f_p - f_z} \frac{(R_{lower} + R_1)}{R_{lower} g_m} \frac{\sqrt{1 + \left(\frac{f_c}{f_p}\right)^2}}{\sqrt{1 + \left(\frac{f_z}{f_c}\right)^2}} \tag{6.25}$$

$$C_1 = \frac{1}{2\pi f_z R_2} \tag{6.26}$$

$$C_2 = \frac{R_{lower} g_m}{2\pi f_p G (R_{lower} + R_1)} \frac{\sqrt{1 + \left(\frac{f_z}{f_c}\right)^2}}{\sqrt{1 + \left(\frac{f_c}{f_p}\right)^2}} \tag{6.27}$$

In these expressions, as usual, G represents the gain or attenuation you are looking for at the selected crossover frequency f_c.

In some cases, the capacitor C_2 is much smaller than C_1. Consequently, the gain equation greatly simplifies:

$$G(s) \approx -\frac{R_2 R_{lower} g_m}{R_{lower} + R_1} \frac{1+\dfrac{1}{sR_2C_1}}{(1+sR_2C_2)} = -G_0 \frac{1+\omega_z/s}{1+s/\omega_p} \tag{6.28}$$

Assuming a pole/zero pair and a crossover frequency placed at the geometric mean of the pole/zero position $\left(f_c = \sqrt{f_p f_z}\right)$, the component expressions are updated to

$$R_2 = \frac{G(R_{lower} + R_1)}{R_{lower} g_m} \tag{6.29}$$

$$C_1 = \frac{1}{2\pi f_z R_2} \tag{6.30}$$

$$C_2 = \frac{1}{2\pi f_p R_2} \tag{6.31}$$

Having all these elements on hand, we can proceed with a design example.

6.2.1 A Design Example

Let's re-use the type 1 OTA design example where we need to provide a 25-dB attenuation at 20 Hz. This time, we want a phase boost of 50°. Again, we assume we stabilize a PFC regulating at a 400 V output, featuring an upper resistor R_1 of 4 MΩ with a lower resistor R_{lower} of 25 kΩ. The selected OTA offers a 199-μS transconductance and the reference voltage is 2.5 V.

Where do we place our pole and zero pair? As we have detailed in the pole zero section, the pole can be placed at the following location:

$$f_p = \left[\tan(boost) + \sqrt{\tan^2(boost)+1}\right] f_c = 2.74 \times 20 = 54.8 \text{ Hz} \tag{6.32}$$

As the phase peaks at the geometric mean between the pole and the zero, this latter is placed at

$$f_z = \frac{f_c^{\,2}}{f_p} = \frac{400}{54.8} \approx 7.3 \text{ Hz} \tag{6.33}$$

Applying the design definitions available from (6.25) through (6.27), we find

$$G = 10^{\frac{G_{f_c}}{20}} = 10^{\frac{-25}{20}} = 0.056 \tag{6.34}$$

$$R_2 = \frac{54.8 \times 0.056}{54.8 - 7.3} \times \frac{25k + 4Meg}{25k \times 100u} \times \frac{\sqrt{\left(\frac{20}{54.8}\right)^2 + 1}}{\sqrt{\left(\frac{7.3}{20}\right)^2 + 1}} = 104 \text{ k}\Omega \tag{6.35}$$

$$C_1 = \frac{1}{2\pi f_z R_2} = \frac{1}{6.28 \times 7.3 \times 104k} = 211 \text{ nF} \tag{6.36}$$

$$C_2 = \frac{25k \times 100u}{6.28 \times 54.8 \times 56m \times (25k + 4Meg)} \times \frac{\sqrt{\left(\frac{7.3}{20}\right)^2 + 1}}{\sqrt{\left(\frac{20}{54.8}\right)^2 + 1}} = 32 \text{ nF} \tag{6.37}$$

The simulation schematic featuring automated calculations appears in Figure 6.6. The results of such a simulation appears in Figure 6.7 and show the -25-dB transition occurring at 20 Hz, as expected. The boost in phase is also 50°, as required.

6.3 Optocoupler and OTA: A Buffered Connection

An OTA coupled to an optocoupler is not a very common architecture. To the author's knowledge, applications where OTAs directly drive an optocoupler are mainly those found in cell phone chargers, where the secondary-side controller hosts a dual OTA whose outputs directly bias the optocoupler LED. This is the case for the MC33341 (ON Semiconductor) or the TLE-4305 (Infineon). With this latter, the designer couples the output of the OTA to the LED via an NPN transistor wired in a common-collector configuration. Should you try to drive the LED directly from the OTA, you would end up in low-gain type of compensator, without having the ability to benefit from an origin pole. The proposed configuration appears in Figure 6.8.

In this example, the output voltage divided by the resistive bridge biases the non-inverting pin of the OTA rather than its inverting pin. This is to maintain the correct polarity on the optocoupler collector (i.e., an error voltage going down as V_{out} increases). The ac current circulating in the LED is nothing other than the buffered OTA voltage divided by the LED series resistor, R_{LED}. However, R_{LED} is not the only ohmic path crossed by this current. We have a series combination of the transistor dynamic base resistor h_{11} or r_π with the LED dynamic resistor R_d. For the sake of simplicity we will ignore them, but they obviously will affect the precision of our calculations at the end. Let's start with the voltage seen on the OTA output, $V_{op}(s)$:

$$V_{op}(s) = V_{out}(s) \frac{R_{lower}}{R_{lower} + R_1} g_m \left(R_2 + \frac{1}{sC_1} \right) \tag{6.38}$$

In ac, this voltage is found on the transistor emitter and forces a LED current equal to

$$I_{LED}(s) = \frac{V_{op}(s)}{R_{LED}} = V_{out}(s) \frac{R_{lower}}{(R_{lower} + R_1)R_{LED}} g_m \left(R_2 + \frac{1}{sC_1} \right) \tag{6.39}$$

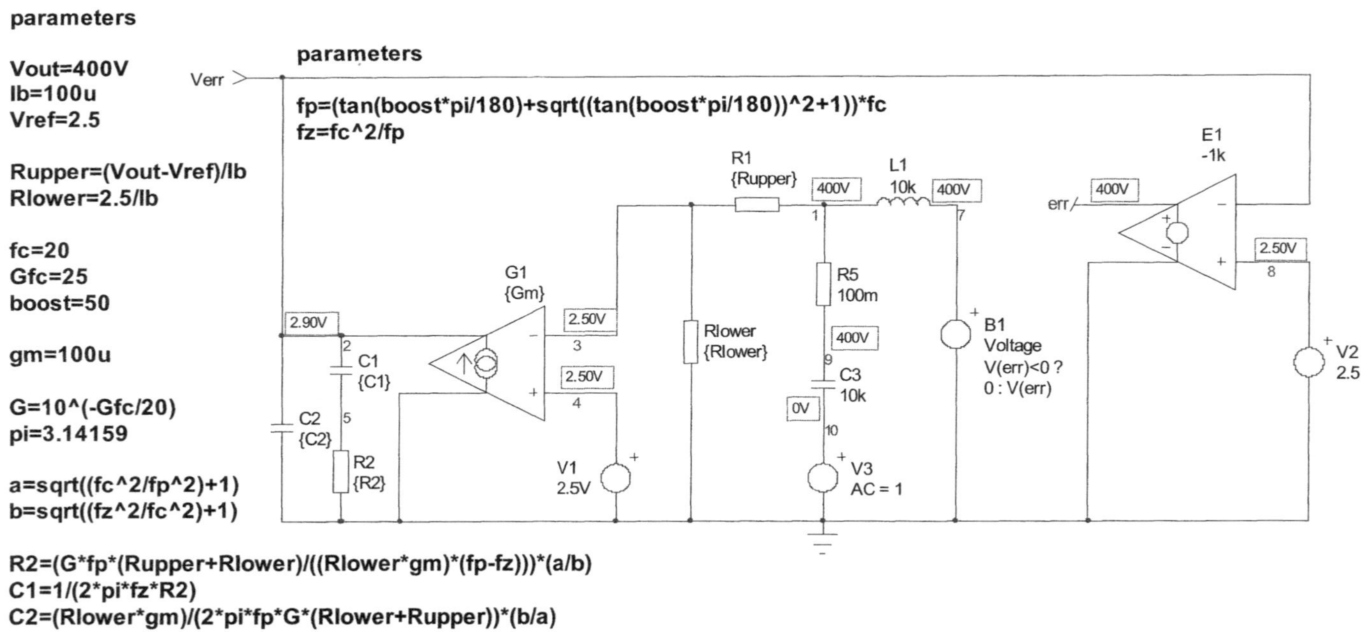

Figure 6.6 The type 2 OTA with automated calculations on the left side of the figure.

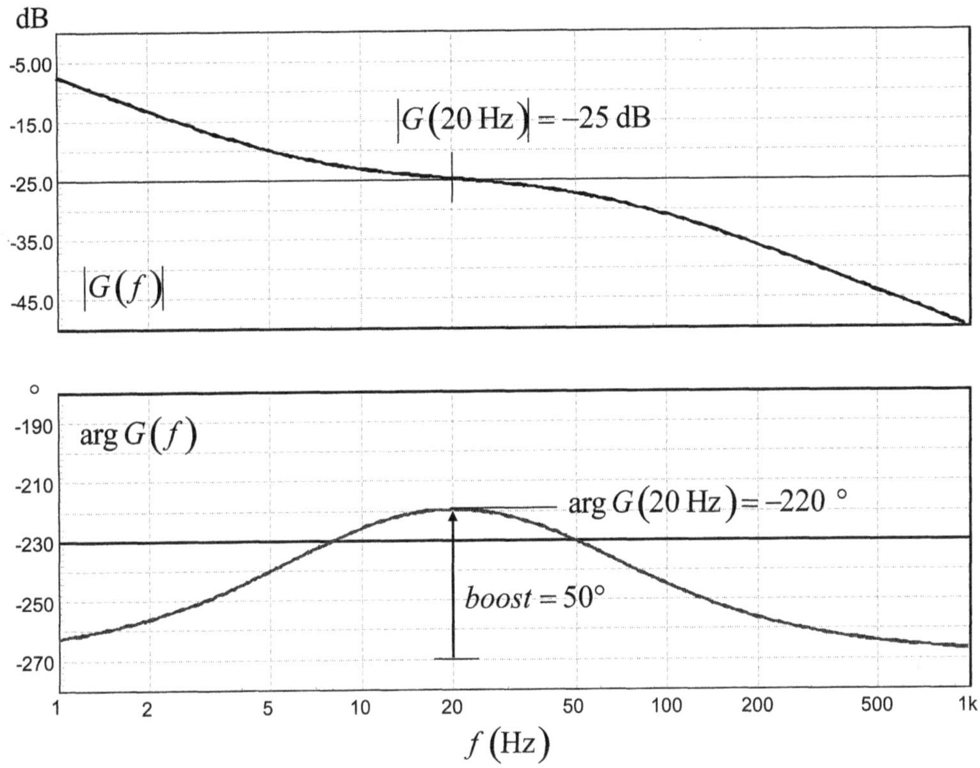

Figure 6.7 The ac simulation plots of the type 2 built with an OTA confirm the calculated results.

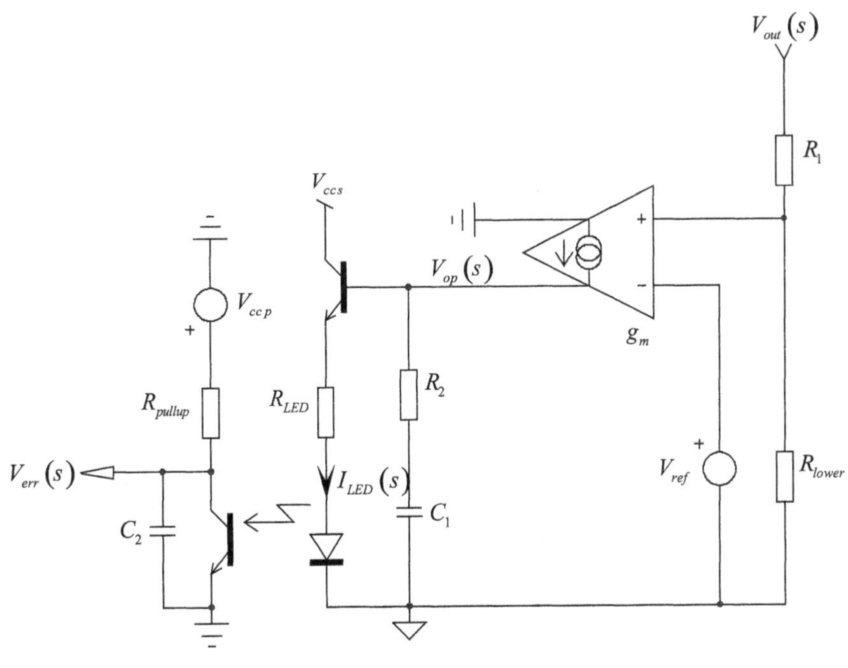

Figure 6.8 An OTA driving an optocoupler is not very common. It is sometimes found in CC-CV controllers such as the MC33341 or the TLE-4305. Please note the OTA inputs polarity as we connect the resistive divider to the noninverting pin.

When going through the optocoupler, this LED current undergoes its current transfer ratio action before crossing R_{pullup} in parallel with C_2, further generating the error voltage, $V_{err}(s)$:

$$V_{err}(s) = -V_{out}(s)\frac{R_{lower}R_{pullup}}{(R_{lower}+R_1)R_{LED}}g_m\text{CTR}\left(R_2+\frac{1}{sC_1}\right)\frac{1}{1+sR_{pullup}C_2} \tag{6.40}$$

By rearranging this equation, we can derive the complete compensator transfer function:

$$G(s) = \frac{V_{err}(s)}{V_{out}(s)} = -\frac{R_{lower}}{R_{lower}+R_1}\frac{g_m R_{pullup}R_2\text{CTR}}{R_{LED}}\frac{1+1/sR_2C_1}{1+sR_{pullup}C_2} \tag{6.41}$$

We can put it under a more familiar form as follows:

$$G(s) = -G_0\frac{1+\omega_z/s}{1+s/\omega_p} \tag{6.42}$$

Where we can identify

$$G_0 = \frac{R_{lower}}{R_{lower}+R_1}\frac{g_m R_{pullup}R_2\text{CTR}}{R_{LED}} \tag{6.43}$$

$$\omega_z = \frac{1}{R_2C_1} \tag{6.44}$$

$$\omega_p = \frac{1}{R_{pullup}C_2} \tag{6.45}$$

The OTA is a current generator but still, its upper output voltage capability, VOH, is limited by its dc supply. This imposes a design constraint on the LED resistor. The voltage across the OTA must be high enough to let the optocoupler collector pull the feedback pin to ground. Based on the exercise already done in the op amp section, the LED resistor must be smaller than:

$$R_{LED} \le \frac{R_{pullup}\text{CTR}_{\min}(\text{VOH}-V_{BE}-V_f)}{V_{ccp}-V_{CE,sat}} \tag{6.46}$$

In some circuits, the LED resistor is internally fixed and you will need to check if it complies with the recommended value in (6.46).

6.3.1 A Design Example

Let us try to stabilize the voltage loop of a converter operated in CC-CV. We assume a 10-dB gain at 1 kHz together with a 50° phase boost are required at this point. Suppose we have the following parameters:

VOH = 5 V; the OTA maximum output voltage.
V_f = 1 V; the LED forward voltage.

$V_{CE,sat}$ = 0.3 V; the optocoupler saturation voltage.
V_{BE} = 0.65 V; the OTA bipolar transistor buffer base-emitter voltage.
V_{ccp} = 5 V; the pull-up V_{cc} level on the primary side.
V_{ccs} = 5 V; the V_{cc} power supply level on the secondary side.
R_{pullup} = 20 kΩ; the optocoupler pull-up resistor.
g_m = 1 mS; the OTA transconductance value.
CTR_{min} = 0.3; the minimum optocoupler current transfer ratio.
R_1 = 38 kΩ; the upper resistor in the resistor bridge observing the output variable.
R_{lower} = 10 kΩ; the lower resistor in the resistor bridge observing the output variable.
f_{opto} = 6 kHz; the optocoupler pole that has been characterized with R_{pullup}.

With these elements on hand, let us calculate with (6.46) the maximum acceptable LED series resistor we can select:

$$R_{LED,\max} = \frac{R_{pullup}\mathrm{CTR}(\mathrm{VOH} - V_{BE} - V_f)}{V_{ccp} - V_{CE,sat}} = \frac{20k \times 0.3 \times (5 - 0.65 - 1)}{5 - 0.3} = 4.3\ \text{k}\Omega \tag{6.47}$$

The adopted integrated circuit, the TLE-4305, includes a 1-kΩ resistor in series with the transistor emitter, so we are safe. Now, let's see where to place the poles and zeros to obtain the 50° phase boost we need at the selected 1-kHz crossover frequency:

$$f_p = \left[\tan(boost) + \sqrt{\tan^2(boost) + 1}\right] f_c = 2.74 \times 1k = 2.74\ \text{kHz} \tag{6.48}$$

Since the phase peaks at the geometric mean between the pole and the zero, the latter is placed at

$$f_z = \frac{f_c^2}{f_p} = \frac{1k^2}{2.74k} \approx 365\ \text{Hz} \tag{6.49}$$

To properly adjust the mid-band gain, we need to evaluate the value of R_2. Starting from (6.42), we can extract the magnitude of the compensator and then solve for the resistor value. Following this option, we have

$$|G(f_c)| = G_0 \frac{\sqrt{1 + \left(\frac{f_z}{f_c}\right)^2}}{\sqrt{1 + \left(\frac{f_c}{f_p}\right)^2}} \tag{6.50}$$

Substituting G_0 from (6.43) in (6.50) and solving for R_2, we have

$$R_2 = \frac{G(R_{lower} + R_1)R_{LED}}{R_{lower}R_{pullup}g_m\mathrm{CTR}} \frac{\sqrt{1 + \left(\frac{f_c}{f_p}\right)^2}}{\sqrt{1 + \left(\frac{f_z}{f_c}\right)^2}} \tag{6.51}$$

In this equation, G corresponds to the gain/attenuation required at the crossover frequency. Here, we need a 10-dB gain:

$$G = 10^{\frac{Gf_c}{20}} = 10^{\frac{10}{20}} = 3.16 \tag{6.52}$$

If we introduce this value in (6.51), we find R_2 equals 2.5 kΩ. The capacitors values for C_1 and C_2 are straightforward, owing to (6.44) and (6.45):

$$C_1 = \frac{1}{2\pi R_2 f_z} = \frac{1}{6.28 \times 2.5k \times 365} = 174 \text{ nF} \tag{6.53}$$

The total capacitor needed to form the pole at 2.74 kHz with the pull-up resistor is found to be

$$C_2 = \frac{1}{2\pi f_p R_{pullup}} = \frac{1}{6.28 \times 2.74k \times 20k} = 2.9 \text{ nF} \tag{6.54}$$

We know that this value is actually made of an external capacitor placed in parallel with the parasitic capacitor of the optocoupler:

$$C_{opto} = \frac{1}{2\pi f_{opto} R_{pullup}} = \frac{1}{6.28 \times 6k \times 20k} = 1.3 \text{ nF} \tag{6.55}$$

As a result, C_{col} is evaluated to

$$C_{col} = C_2 - C_{opto} = 2.9n - 1.3n = 1.6 \text{ nF} \tag{6.56}$$

We now have everything on hand. The test fixture appears in Figure 6.9 and shows correct operating bias points for a 12-V output. Please note the presence of clamping diodes D_1 and D_2 that limits the upper and lower excursion of the voltage-controlled current source during the bias point calculation.

The ac results appear in Figure 6.10 and show a slight discrepancy between the target (10 dB of gain) and the obtained result. This is mainly due to the cumulative effects of the dynamic resistors from the base-emitter junction (h_{11} or r_π) and from the LED diode. The 1.5-dB error can easily be compensated by increasing the target by 1.5 dB, thus shooting for 11.5 dB. The boost of 50° is well respected, though.

6.4 The Type 3: An Origin Pole and a Pole/Zero Pair

By combining an RC network across the upper resistor R_1, it is possible to build a type 3 with an OTA. However, we will soon learn that the implication of the divider network plays a harmful role in the final transfer function. The proposed implementation appears in Figure 6.11. To obtain the transfer function, we must evaluate the input and loading impedances, noted Z_i and Z_L, respectively.

We know that the output current of the OTA depends on the voltage difference developed between its inputs. In ac, the voltage on the inverting pin depends on the divider network. Therefore,

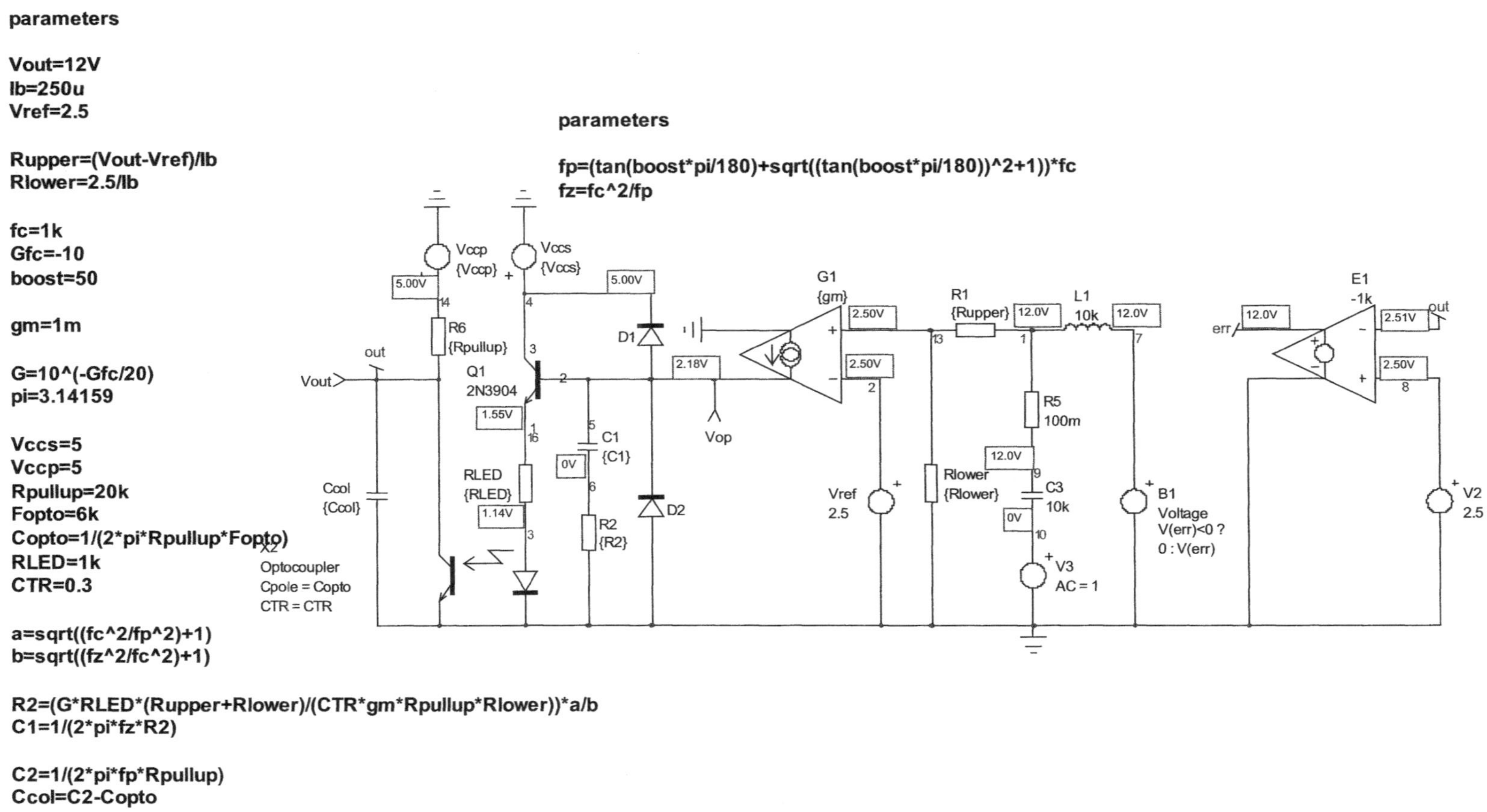

Figure 6.9 The test fixture includes the OTA driving the optocoupler LED via a NPN transistor.

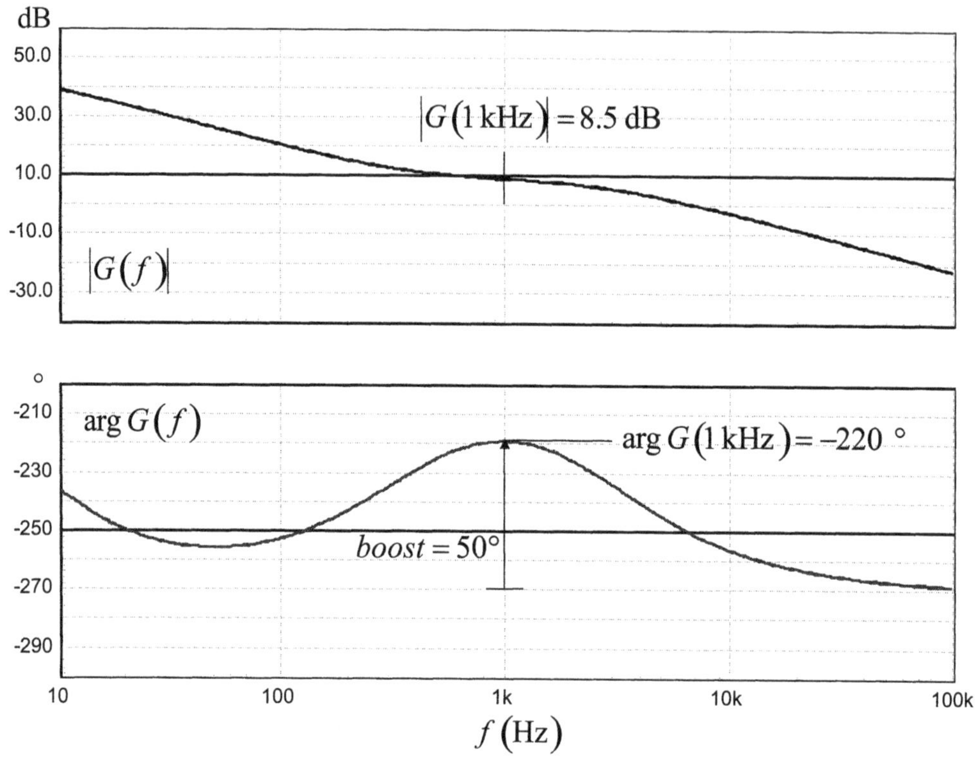

Figure 6.10 The ac sweep confirms the phase boost amplitude at 1 kHz, but we miss the target by 1.5 dB. This is because of the dynamic resistors offered by the bipolar transistors and the LED.

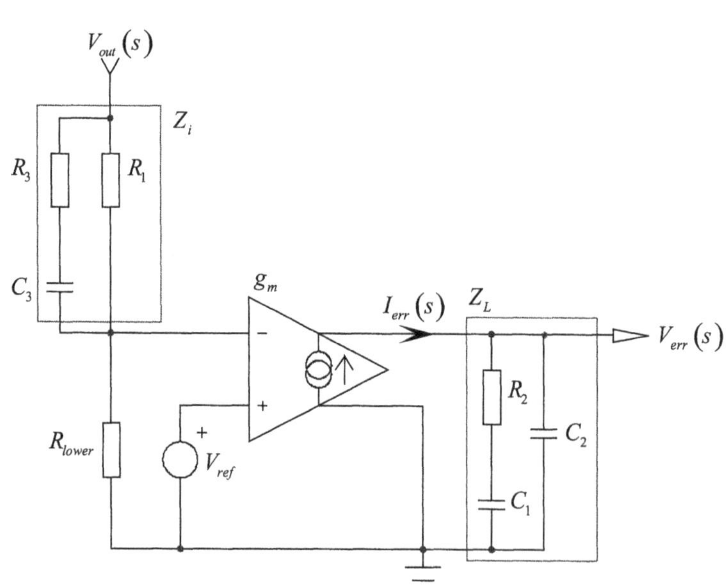

Figure 6.11 A type 3 can be built with an OTA; however, it is less flexible than with an op amp-based solution.

$$I_{err}(s) = -g_m V_{out}(s)\frac{R_{lower}}{R_{lower} + Z_i(s)} \tag{6.57}$$

The complex impedance $Z_i(s)$ can be calculated as follows:

$$Z_i(s) = \frac{\frac{sR_3C_3+1}{sC_3}R_1}{\frac{sR_3C_3+1}{sC_3}+R_1} = \frac{\frac{sR_3C_3+1}{sC_3}R_1}{\frac{sR_3C_3+1}{sC_3}+\frac{sR_1C_3}{sC_3}} = R_1\frac{sR_3C_3+1}{sC_3(R_3+R_1)+1} \tag{6.58}$$

We can now update (6.57):

$$I_{err}(s) = -g_m V_{out}(s)\frac{R_{lower}}{R_{lower}+\frac{R_1(sR_3C_3+1)}{sC_3(R_3+R_1)+1}} \tag{6.59}$$

Rearranging this equation gives

$$I_{err}(s) = -g_m\, V_{out}(s)\frac{R_{lower}}{R_{lower}+R_1}\frac{sC_3(R_3+R_1)+1}{sC_3\left(\frac{R_{lower}R_1}{R_{lower}+R_1}+R_3\right)+1} \tag{6.60}$$

As the OTA output current flows in a complex network Z_L, we can calculate its equivalent impedance and get the transfer function immediately:

$$Z_L(s) = (R_2 + C_1)\,||\,C_2 = \frac{\left(R_2+\frac{1}{sC_1}\right)\frac{1}{sC_2}}{\left(R_2+\frac{1}{sC_1}\right)+\frac{1}{sC_2}} \tag{6.61}$$

Since $V_{out}(s) = I_{err}(s)\, Z_L(s)$, we simply multiply (6.60) by (6.61) and obtain

$$G(s) = -g_m\frac{R_{lower}}{R_{lower}+R_1}\frac{sC_3(R_3+R_1)+1}{sC_3\left(\frac{R_{lower}R_1}{R_{lower}+R_1}+R_3\right)+1}\frac{\left(R_2+\frac{1}{sC_1}\right)\frac{1}{sC_2}}{\left(R_2+\frac{1}{sC_1}\right)+\frac{1}{sC_2}} \tag{6.62}$$

Developing and arranging (6.62) leads to

$$G(s) = -\frac{R_{lower}\,g_m}{R_{lower}+R_1}\frac{sC_3(R_3+R_1)+1}{\left[sC_3\left(\frac{R_{lower}R_1}{R_{lower}+R_1}+R_3\right)+1\right]}\frac{1+sR_2C_1}{\left[s(C_1+C_2)\left(1+sR_2\frac{C_1C_2}{C_1+C_2}\right)\right]} \tag{6.63}$$

Factoring $1 + sR_2C_1$, we have

$$G(s) = -\frac{R_{lower}}{R_{lower}+R_1}\frac{g_m R_2 C_1}{C_1+C_2}\frac{1+1/sR_2C_1}{\left[sC_3\left(\frac{R_{lower}R_1}{R_{lower}+R_1}+R_3\right)+1\right]}\frac{sC_3(R_3+R_1)+1}{\left[1+sR_2\frac{C_1C_2}{C_1+C_2}\right]} \quad (6.64)$$

This expression can be put in the normalized form:

$$G(s) = -G_0\frac{\left(1+\frac{\omega_{z_1}}{s}\right)\left(1+\frac{s}{\omega_{z_2}}\right)}{\left(1+\frac{s}{\omega_{p_1}}\right)\left(1+\frac{s}{\omega_{p_2}}\right)} \quad (6.65)$$

where

$$G_0 = \frac{R_{lower}}{R_{lower}+R_1}\frac{g_m R_2 C_1}{C_1+C_2} \quad (6.66)$$

$$\omega_{z_1} = \frac{1}{R_2C_1} \quad (6.67)$$

$$\omega_{z_2} = \frac{1}{(R_1+R_3)C_3} \quad (6.68)$$

$$\omega_{p_1} = \frac{1}{R_2\frac{C_1C_2}{C_1+C_2}} \quad (6.69)$$

$$\omega_{p_2} = \frac{1}{\left(\frac{R_{lower}R_1}{R_{lower}+R_1}+R_3\right)C_3} \quad (6.70)$$

From these values, we need to extract the design expressions that will let us calculate the passive component values. We first need the magnitude of G at the selected crossover frequency:

$$|G(f_c)| = G_0\frac{\sqrt{1+\left(\frac{f_{z_1}}{f_c}\right)^2}\sqrt{1+\left(\frac{f_c}{f_{z_2}}\right)^2}}{\sqrt{1+\left(\frac{f_c}{f_{p_1}}\right)^2}\sqrt{1+\left(\frac{f_c}{f_{p_2}}\right)^2}} = \frac{R_{lower}g_m}{R_{lower}+R_1}\frac{R_2C_1}{C_1+C_2}\frac{\sqrt{1+\left(\frac{f_{z_1}}{f_c}\right)^2}\sqrt{1+\left(\frac{f_c}{f_{z_2}}\right)^2}}{\sqrt{1+\left(\frac{f_c}{f_{p_1}}\right)^2}\sqrt{1+\left(\frac{f_c}{f_{p_2}}\right)^2}} \quad (6.71)$$

Then, by using (6.67), (6.68), (6.69), (6.70), and (6.71), we can solve for R_2, R_3, C_1, C_2 and C_3:

$$R_2 = \frac{G(R_1+R_{lower})f_{p_1}}{R_{lower}g_m(f_{p_1}-f_{z_1})}\frac{\sqrt{1+\left(\frac{f_c}{f_{p_1}}\right)^2}\sqrt{1+\left(\frac{f_c}{f_{p_2}}\right)^2}}{\sqrt{1+\left(\frac{f_{z_1}}{f_c}\right)^2}\sqrt{1+\left(\frac{f_c}{f_{z_2}}\right)^2}} \quad (6.72)$$

$$R_3 = \frac{R_1^2 f_{z2} - R_1 R_{lower}(f_{p2} - f_{z2})}{(f_{p2} - f_{z2})(R_1 + R_{lower})} \tag{6.73}$$

$$C_1 = \frac{1}{2\pi f_{z1} R_2} \tag{6.74}$$

$$C_2 = \frac{C_1}{2\pi C_1 R_2 f_{p1} - 1} \tag{6.75}$$

$$C_3 = \frac{1}{2\pi (R_1 + R_3) f_{z2}} \tag{6.76}$$

We can try to simplify these equations a little bit by recognizing that very often $C_2 \ll C_1$; however, the final equations will still remain quite complex. In fact, the equation defining R_3, (6.73), can become negative, simply meaning that the needed distance between the second pole and the second zero cannot be achieved. To determine what the limit is, let us observe the sign of (6.73) by checking its numerator. Since R_1 can be factorized and is always positive, we can exclude it from the sign study. Therefore, we have

$$R_1 f_{z2} - R_{lower}(f_{p2} - f_{z2}) > 0 \tag{6.77}$$

Solving this equation leads to

$$\frac{f_{p2}}{f_{z2}} < \frac{R_1}{R_{lower}} + 1 \tag{6.78}$$

The right term once multiplied by the reference voltage V_{ref} gives the targeted output voltage. Capitalizing on this remark, we have

$$\frac{f_{p2}}{f_{z2}} < \frac{V_{out}}{V_{ref}} \tag{6.79}$$

In other words, when the output voltage is large relative to the reference voltage, for instance, a 48-V output and a 2.5-V reference voltage, we have the ability to spread both the second pole and zero by a ratio of around 19, opening possibilities for the phase boost selection. Unfortunately, for a lower output voltage (e.g., a 5-V output with a 2.5-V reference), you cannot have more than a ratio of 2 between the second pole and the second zero, drastically limiting the total available phase boost. This limits the type 3 implementation with an OTA.

Let's try to derive the phase boost brought by a pole and a zero distant by a certain ratio r:

$$\frac{f_p}{f_z} = r \tag{6.80}$$

We know that the phase boost peaks at the geometric mean between the pole and the zero, and this is where the crossover frequency is usually placed. Otherwise stated, we have

$$f_c = \sqrt{f_p f_z} \tag{6.81}$$

Now, as the pole and the zero are linked by (6.80), a new relationship between the crossover frequency and the pole/zero pair can be derived:

$$f_c = f_z\sqrt{r} = \frac{f_p}{\sqrt{r}} \tag{6.82}$$

A transfer function featuring a pole and a zero obeys the following form:

$$G(s) = \frac{1+\dfrac{s}{\omega_z}}{1+\dfrac{s}{\omega_p}} \tag{6.83}$$

The phase boost at the crossover angular frequency ω_c or the frequency f_c brought by such pole/zero arrangement is found by subtracting the denominator argument to that of the numerator:

$$boost = \tan^{-1}\frac{\omega_c}{\omega_z} - \tan^{-1}\frac{\omega_c}{\omega_p} = \tan^{-1}\frac{f_c}{f_z} - \tan^{-1}\frac{f_c}{f_p} \tag{6.84}$$

If we now substitute the crossover frequency definition given by (6.82) into (6.84), we find that the pole and zero absolute locations disappear and only the ratio between them, r, remains

$$boost = \tan^{-1}\frac{f_z\sqrt{r}}{f_z} - \tan^{-1}\frac{f_p}{\sqrt{r}f_p} = \tan^{-1}\sqrt{r} - \tan^{-1}\frac{1}{\sqrt{r}} \tag{6.85}$$

To arrange this expression, we need to use a known trigonometric relationship:

$$\tan^{-1}\frac{1}{r} + \tan^{-1} r = \frac{\pi}{2} \tag{6.86}$$

From (6.86), we can extract the definition of $\tan^{-1}1/r$:

$$\tan^{-1}\frac{1}{r} = \frac{\pi}{2} - \tan^{-1} r \tag{6.87}$$

If we substitute this expression into (6.85), we have

$$boost = \tan^{-1}\sqrt{r} - \frac{\pi}{2} + \tan^{-1}\sqrt{r} = 2\tan^{-1}\sqrt{r} - \frac{\pi}{2} \tag{6.88}$$

Using the previous formula, we now have a means to predict the phase boost peak brought by a pole/zero pair depending on the distance between them. We consider that the peak occurs at the geometric mean. In a type 3 compensator, we have two pole/zero pairs. The first one involves the impedance loading of the OTA. This first pair is not affected by any limit implying that the pole and zero can be spread up to a maximum of 90° phase boost. On top of this phase boost, you add the second pole/zero pair; however, it is limited in spread by (6.79). The total phase boost will thus be 90° plus the one brought by the second pole/zero pair. Assume we have the two converters, one delivering 48 V and the second 5 V from a 2.5-V reference voltage. With the first one, given the ratio of 19 between V_{out} and V_{ref}, we could obtain an extra phase boost from the second pole/zero pair of

$$boost_{48V} = 2\tan^{-1}\sqrt{19} - 90° = 64° \tag{6.89}$$

bringing the total available phase boost to 90 + 64 = 154°. On the contrary, for the 5-V converter, given the ratio of 2.5 between V_{out} and V_{ref}, the extra phase boost would be limited to

$$boost_{5V} = 2\tan^{-1}\sqrt{2.5} - 90° = 25° \tag{6.90}$$

This would not allow a phase boost larger than 90° + 25°=115°.

6.4.1 A Design Example

We want to stabilize a 19-V power supply obtained from a 2.5-V reference voltage. The ratio between the output voltage and the reference is 19/2.5 = 7.6. To obtain this voltage from a 2.5-V reference voltage, we have R_1 equal to 66 kΩ and R_{lower} is 10 kΩ. Looking at (6.79), the pole and zero distance cannot be larger than

$$\frac{f_{p2}}{f_{z2}} < 7.6 \tag{6.91}$$

From (6.88), we can check what maximum phase boost we can expect from the second pole/zero pair position:

$$boost_{19V} = 2\tan^{-1}\sqrt{7.6} - 90° = 50° \tag{6.92}$$

In other words, the maximum phase boost we will get from the OTA wired in the type 3 configuration will have to stay smaller than

$$boost_{\max} = 90° + 50° = 140° \tag{6.93}$$

When selecting the crossover frequency, you will have to account from the previous value to make sure the phase boost at the selected crossover point will be realizable (e.g., less than 140°). Let's assume we want 130° at a 1-kHz crossover frequency, a point where the loop needs a 15-dB of amplification. G, in this case, is simply

$$G = 10^{\frac{Gf_c}{20}} = 10^{\frac{15}{20}} = 5.6 \tag{6.94}$$

Knowing that we can get 50° from the second pole/zero pair, we are going to spread the first pole/zero pair to reach 80° so that the sum of both phase boosts gives us the 130° we are looking for. To obtain the 80° phase boost from the pole f_{p1} and the zero f_{z1}, we will place them at a certain distance from the 1-kHz crossover frequency. This distance is calculated by extracting r from (6.88):

parameters

Vout=19V
Ib=250u
Vref=2.5

Rupper=(Vout-Vref)/Ib
Rlower=2.5/Ib

fc=1k
Gfc=-15
boost=130

fz1=87.5
fz2=378
fp1=11.4k
fp2=2.6k

gm=100u

G=10^(-Gfc/20)
pi=3.14159

a=sqrt((fc^2/fp1^2)+1)
b=sqrt((fc^2/fp2^2)+1)
c=sqrt((fz1^2/fc^2)+1)
d=sqrt((fc^2/fz2^2)+1)

aa=((a*b)/(c*d))
bb=(G*fp1*(Rupper+Rlower))/(Rlower*gm*(fp1-fz1))
R2=aa*bb

cc=((Rupper^2)*fz2)-Rupper*Rlower*(fp2-fz2)
dd=(fp2-fz2)*(Rupper+Rlower)
R3=cc/dd
C1=1/(2*pi*fz1*R2)
C2=C1/(2*pi*C1*R2*fp1-1)
C3=1/(2*pi*fz2*(Rupper+R3))

parameters

fp12=fc/(tan((45-(boost/4))*pi/180))
fz12=fc^2/fp12

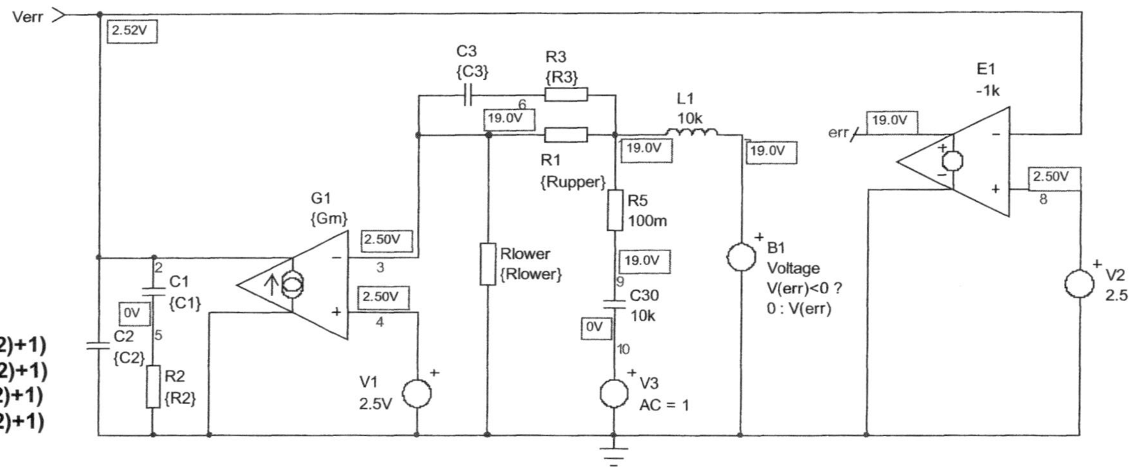

Figure 6.12 The OTA wired in a type 3 configuration is less flexible than its op amp counterpart.

$$r = \left[\tan\left(\frac{boost}{2} + \frac{\pi}{4}\right)\right]^2 \approx 130 \tag{6.95}$$

By rearranging (6.82), we can easily extract the first pole and the first zero positions in relationship to the crossover frequency f_c:

$$f_{z_1} = \frac{f_c}{\sqrt{r}} = \frac{1k}{\sqrt{130}} = 87.5 \text{ Hz} \tag{6.96}$$

$$f_{p_1} = f_c\sqrt{r} = 1k \times \sqrt{130} = 11.4 \text{ kHz} \tag{6.97}$$

From (6.91), we know that the distance between the second pole and zero have to be less than 7.6 (7 is used for safety margin). Positioned in relation to the 1-kHz crossover frequency, they have to be placed at the following positions:

$$f_{z_2} = \frac{f_c}{\sqrt{r}} = \frac{1k}{\sqrt{7}} = 378 \text{ Hz} \tag{6.98}$$

$$f_{p_2} = f_c\sqrt{r} = 1k \times \sqrt{7} = 2.6 \text{ kHz} \tag{6.99}$$

Applying (6.72) to (6.76), we find the following values for the passive elements placed around the OTA:

C_1 = 11.2 nF
C_2 = 86 pF
C_3 = 6.3 nF

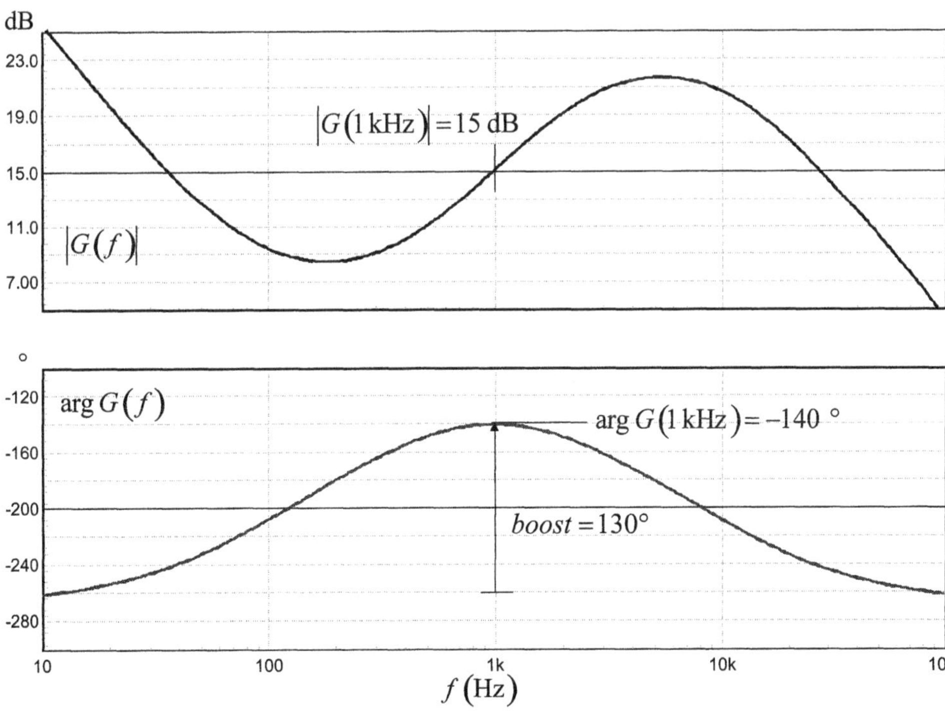

Figure 6.13 The ac response of the type 3 OTA confirms the realization of the desired phase boost.

$R_2 = 163\ \text{k}\Omega$
$R_3 = 1.05\ \text{k}\Omega$

The test fixture appears in Figure 6.12, where the automation of the component values resides on the left side of the schematic. The bias point confirms the 19-V input setpoint. Once the ac sweep has been performed, the results appear in Figure 6.13.

6.5 Conclusion

An OTA does not offer the same design flexibility as an op amp does. However, as explained, IC designers like the structure because of the small semiconductor die size it requires. The lack of virtual ground makes the lower side divider resistor enter the equations and requires attention when deriving poles and zeros placement. Depending on the ratio between the output and the reference voltages, it can hamper the second pole/zero pair placement in a type 3 configuration. For this reason, the OTA is rarely used in these configurations and must remain a designer choice for type 2 or 1 applications.

Appendix 6A: Summary Pictures

Figures 6.14 through 6.16 summarize the component definitions associated with the structures described in the chapter.

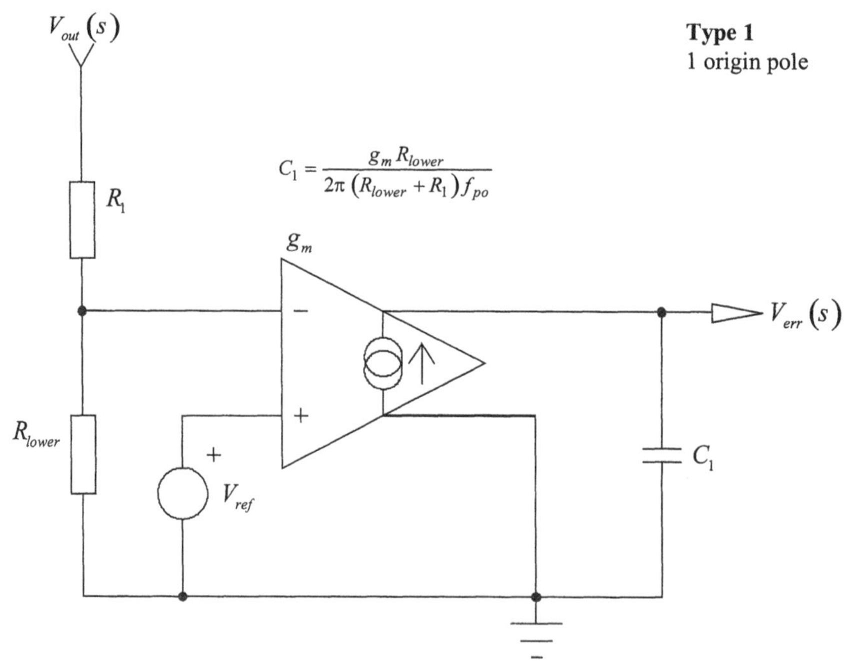

Figure 6.14 The type 1.

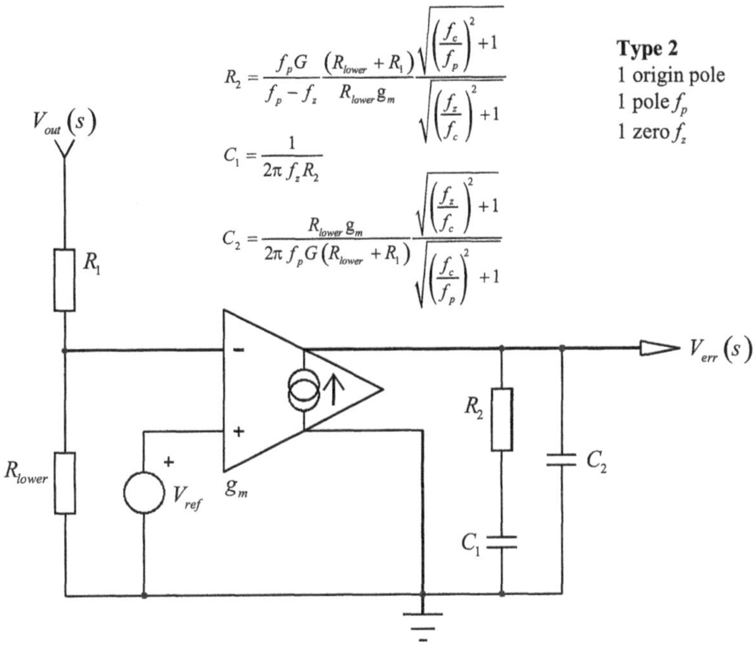

Figure 6.15 The type 2.

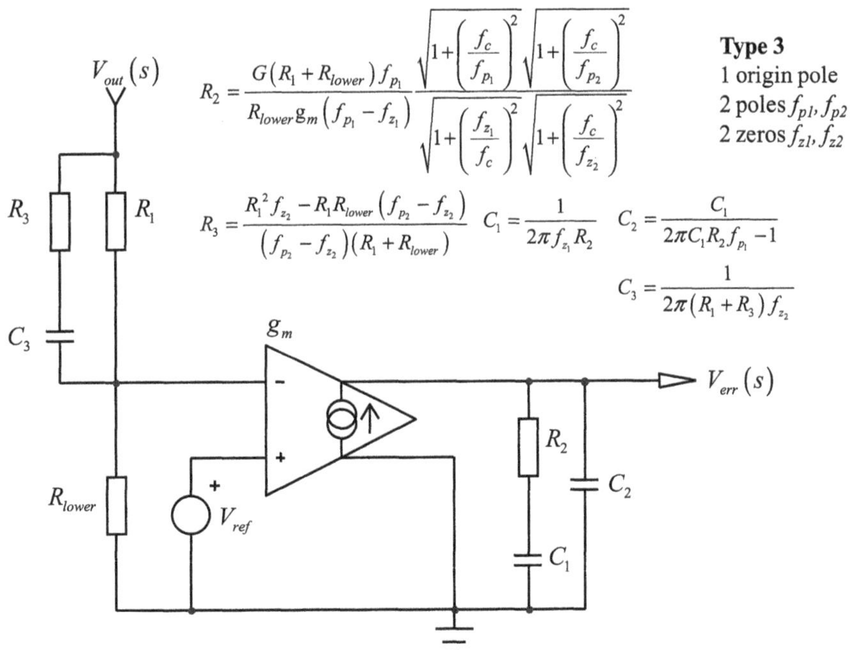

Figure 6.16 The type 3.

References

[1] Gratz, A., "Operational Transconductance Amplifiers," http://synth.stromeko.net/diy/OTA.pdf.

[2] Basso, C., Switch Mode Power Supplies: SPICE Simulations and Practical Designs, New York: McGraw-Hill, 2008.

CHAPTER 7

TL431-Based Compensators

The technical literature on loop control abounds with design examples of compensators implementing an operational amplifier (op amp). If the op amp certainly represents a possible way to generate an error signal, the industry choice for this function has been different for many years: almost all consumer power supplies involve a TL431 placed on the isolated secondary side to feed the error back to the primary side via an optocoupler. Despite its similarities with its op amp cousin, the design of a compensator around a TL431 requires a good understanding of the device operation. The introduction explores the internal structure of the device and details how its biasing conditions affect the loop performance. Then, the classical type 1, 2, and 3 structures associated with an optocoupler are explored in detail.

7.1 A Bandgap-Based Component

The TL431 equivalent architecture appears in Figure 7.1. It combines an open-collector op amp with a precise 2.5-V reference voltage. When the voltage present on the reference pin R exceeds the internal reference level, the bipolar transistor starts to conduct and a current is sunk between the cathode and anode pins. The TL431 is a self-contained op amp plus reference voltage device. As any active element, it needs a minimum voltage to operate as well as a certain amount of consumed current. For the TL431, this latter is called the bias current and must be set to at least 1 mA, as explained later. The same remark applies for the supply voltage, which cannot be lower than 2.5 V across the cathode and the anode of the component.

Figure 7.2 shows the internal schematic of a TL431 made in a bipolar technology. This circuitry was analyzed with the kind help of Mr. Kadanka, an integrated circuit designer at ON Semiconductor in the Czech Republic. The bias points are coming from a simple simulation setup where the device was used as a reference voltage [1]. This configuration implies a connection between it reference point *ref* and the cathode *k*, while the anode *a* is grounded, making the device behave as an active 2.5-V Zener diode.

In classical loop-control configuration, the TL431 observes a fraction of the output voltage seen by its *ref* pin and converts it into an output current sunk between the cathode and the anode. As such, the device can be considered as a transconductance amplifier. The TL431 internal reference circuit works around a popular structure called a *bandgap*. The description of a bandgap is outside the scope of this book but basically the principle consists of balancing the negative temperature coefficient of a junction (a transistor V_{BE}) by a thermal voltage, V_T, affected by a positive temperature coefficient. When summed together, these voltages nicely compensate to form a temperature-compensated reference voltage.

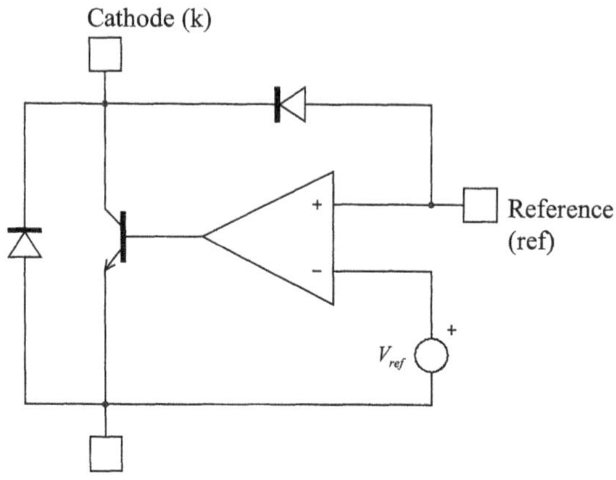

Figure 7.1 The equivalent circuit of a TL431 combines an op amp with a bipolar transistor.

To simplify the analysis, we will assume that the current gain β of all transistors used in the TL431 is very high, implying negligible base currents. The secret of operating the TL431 lies in the subtle equilibrium imposed by Q_9 and Q_1: when conditions are met (e.g., V_{out} reached its target and V_{ref} equals 2.5 V), both Q_9 and Q_1 share the same current I_1: V_{ka} remains constant. Any modification in this condition (for instance, brought by a setpoint variation or an increased output power demand

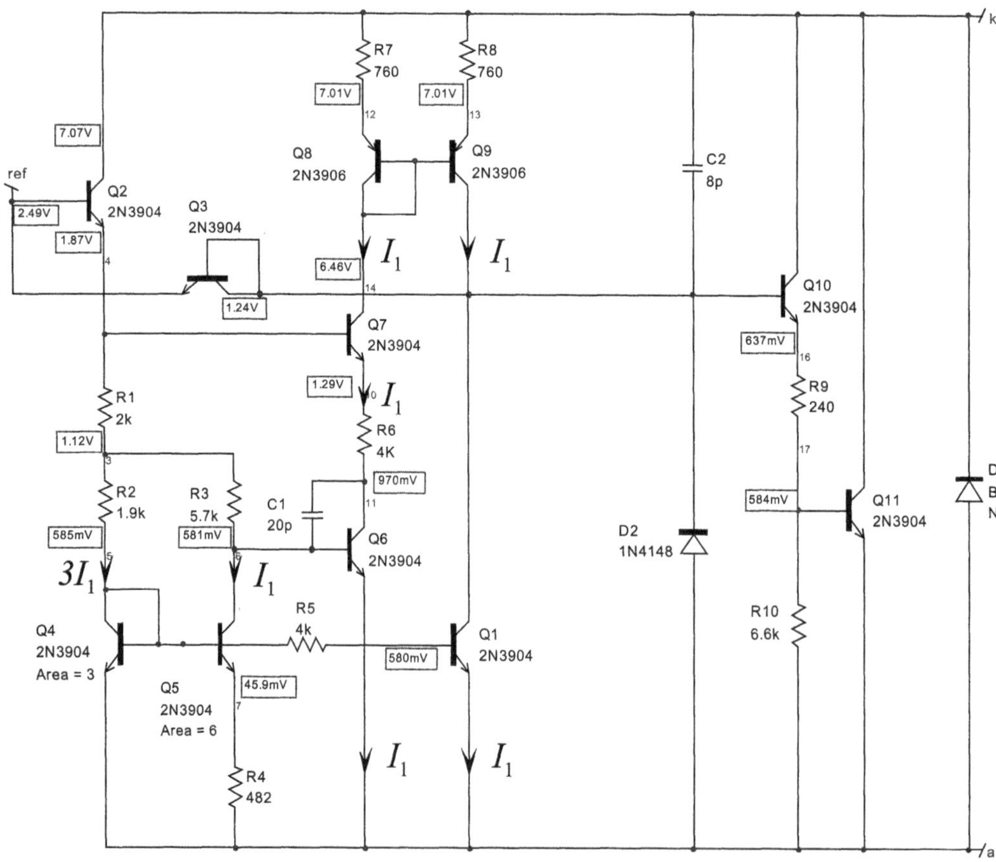

Figure 7.2 The internal schematic of a typical TL431 from ON Semiconductor where bias points in voltage and current have been captured at the equilibrium.

on the regulated converter) will either force either Q_9 to source more current or Q_1 to increase the current it sinks, changing the bias of the output darlington configuration made around Q_{10} and Q_{11}. This action brings V_{ka} down or up, respectively, and forces a current variation in the LED diode attached to the TL431 cathode (e.g., in an opto-isolated power supply application).

At the equilibrium, if we neglect the base currents, the current mirror brought by Q_8 and Q_9 duplicates I_1, which also circulates in Q_7 and Q_6. Because of the current mirror arrangement between Q_1 and Q_4, the same current but scaled up by a ratio of 1 to 3 flows in Q_4 as well. This is confirmed by identical voltage drops across R_2 and R_3 (≈530 mV), also featuring a 1 to 3 ratio (1.9 kΩ and 5.7 kΩ):

$$I_{C,Q_5} = \frac{I_{C,Q_4}}{3} = I_{C,Q_1} = I_1 \tag{7.1}$$

Q_7 is wired in a cascode configuration and helps to shield Q_6 voltage bias against the variations on the *k* terminal that would otherwise be duplicated minus a V_{BE} on Q_6 collector.

In Figure 7.1, the bandgap is made by associating transistors Q_4 and Q_5 together with the emitter resistor R_4. The *area* parameter on both devices indicates that Q_4 is "equivalent" to three transistors in parallel, whereas Q_5 is made of six paralleled transistors. Otherwise stated, the emitter size of the transistor Q_5 is twice that of transistor Q_4 given respective *area* parameters of 6 and 3. Therefore, not only the current densities J in their emitters are linked ($J_4 = 6J_5$), but their saturation currents I_S are also affected by this relationship:

$$I_{S,Q_5} = 2I_{S,Q_4} \tag{7.2}$$

7.1.1 The Reference Voltage

Now, it is interesting to calculate how the 2.5-V reference is actually established in the TL431. To do so, we need to start with current values flowing through Q_4 and Q_5. Capitalizing on the equilibrium, we know that I_1 circulates in Q_1 and $3I_1$ in Q_4 (7.1), but due to the ratio of resistors R_2 and R_3, I_1 also naturally flows in Q_5. We can write the following equation for voltage difference between base terminals of Q_4,Q_5 and the anode terminal:

$$V_{BE4} = V_{BE5} + I_1 R_4 \tag{7.3}$$

The base-emitter voltage of a bipolar transistor can be calculated from its collector current I_C according to

$$V_{BE} = V_T \ln\left(\frac{I_C}{I_S}\right) \tag{7.4}$$

where $V_T = \frac{kT}{q} \approx 25$ mV at a 27°C room temperature or 300 Kelvin. In this equation, I_S is the transistor saturation current (directly proportional to its emitter size), k is the Boltzmann constant $\left(1.38 \times 10^{-23}\right)$, and q is the electron charge equal to 1.601×10^{-19} C.

We know from (7.2) that the saturation current of Q_5 is twice the saturation current of Q_4. We can therefore update and substitute (7.4) in (7.3):

$$V_T \ln\left(\frac{3I_1}{I_S}\right) = V_T \ln\left(\frac{I_1}{2I_S}\right) + R_4 I_1 \tag{7.5}$$

Rearranging and factoring V_T gives

$$V_T\left[\ln\left(\frac{3I_1}{I_S}\right) - \ln\left(\frac{I_1}{2I_S}\right)\right] = R_4 I_1 \tag{7.6}$$

We know that $\ln a - \ln b = \ln\left(\frac{a}{b}\right)$; therefore,

$$V_T \ln\left(\frac{3I_1}{I_S}\frac{2I_S}{I_1}\right) = V_T \ln 6 = R_4 I_1 \tag{7.7}$$

From which we can extract the value of the current I_1:

$$I_1 = \frac{V_T \ln 6}{R_4} \tag{7.8}$$

Based on the value of R_4, which is 482 Ω, we can calculate the value of I_1:

$$I_1 = \frac{26m \times 1.79}{482} \approx 97\ \mu\text{A} \tag{7.9}$$

Knowing this current, the drop over the collector loads R_2 and R_3 can quickly be derived by ohm's law:

$$V_{R_2} = V_{R_3} = 3R_2 I_1 = R_3 I_1 = 1.9k \times 97u = 5.7k \times 97u = 553\ \text{mV} \tag{7.10}$$

which is not far away from the bias point calculated by SPICE and reflected in Figure 7.2. As R_1 is crossed by the sum of the currents flowing into Q_4 ($3I_1$) and Q_5 (I_1), its voltage drop is simply

$$V_{R_1} = 4R_1 I_1 = 2k \times 4 \times 97u = 776\ \text{mV} \tag{7.11}$$

Finally, if we stack up all the voltages we have calculated and assume transistors V_{BE} of 580 mV, we obtain the reference level we look for:

$$V_{ref} = V_{R_1} + V_{R_2} + V_{BE2} + V_{BE4} = 0.553 + 0.776 + 0.58 + 0.58 = 2.49\ \text{V} \tag{7.12}$$

This is the value displayed in Figure 7.1 over the *ref* node.

The frequency compensation of the TL431 is performed by the capacitors C_1, C_2, and the resistor R_6. For those interested, [2] details design information for bandgap-based circuits.

7.1.2 The Need for Bias Current

We explained that the condition for equilibrium is reached when V_{ref} equals 2.49 V. Now, if the voltage applied on the *ref* node changes (for instance, increases), the voltage change is propagated on the emitter of Q_2 and forces a current variation in R_1. This change will now be seen on Q_4 and Q_5 currents, conveying the information to Q_1 and Q_6, respectively. To that respect, you could see Q_1 and Q_6 working as a differential amplifier. However, as more current flows in Q_4, its path will naturally offer a larger gain to activate Q_1, sinking more current to ground that Q_6 does: V_{ka} is pulled down.

We could analytically calculate the transconductance gain of the TL431 but it is easier to deduce it from the characterization curve shown in Figure 7.3. From this curve, we can read a dc gain of 55 dB. This value, together with the 230-Ω pull-up resistor used for characterization implies a transconductance value of

$$g_m = \frac{10^{\frac{55}{20}}}{230} = 2.44 \text{ A/V} \tag{7.13}$$

As the caption details, the cathode current was set to 10 mA during the measurement. A traditional Zener diode needs some substantial bias current to make it operate far away from its knee. Otherwise, the dynamic impedance exhibited by the diode is affected, and the Zener voltage depends on the injected current. Despite being a kind of active Zener diode, the TL431 makes no exception to that particular rule: you need to inject current in its cathode to get the best from the device. We have run simulations on the transistor-level TL431 model to check where the knee sharpens. This is what appears in Figure 7.4.

In this drawing, we can identify a region below which the transconductance of the device is low, with values in the vicinity of 30 mA/V or 30 mS. As the cathode current increases, the transconductance improves and reaches up to 1 A/V. As more current is injected, values up to what is given by (7.13) are obtained.

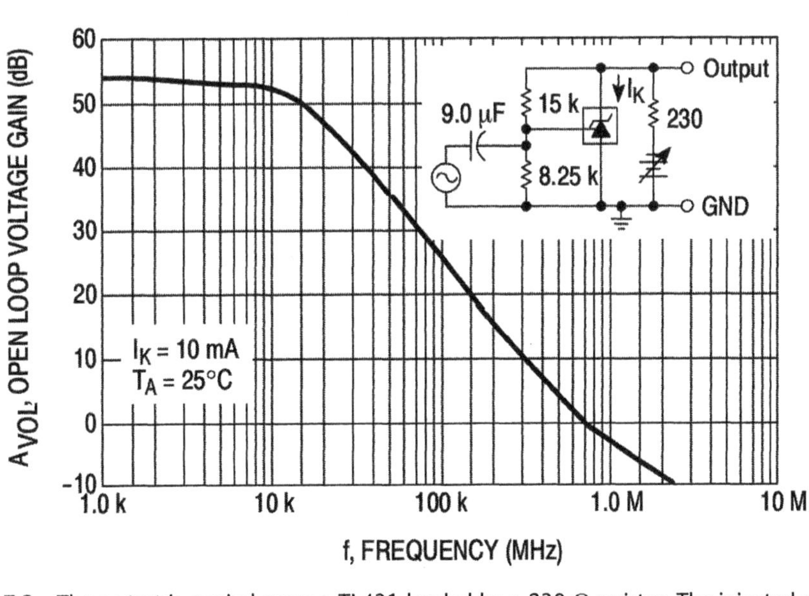

Figure 7.3 The ac test is carried upon a TL431 loaded by a 230-Ω resistor. The injected current is around 10 mA.

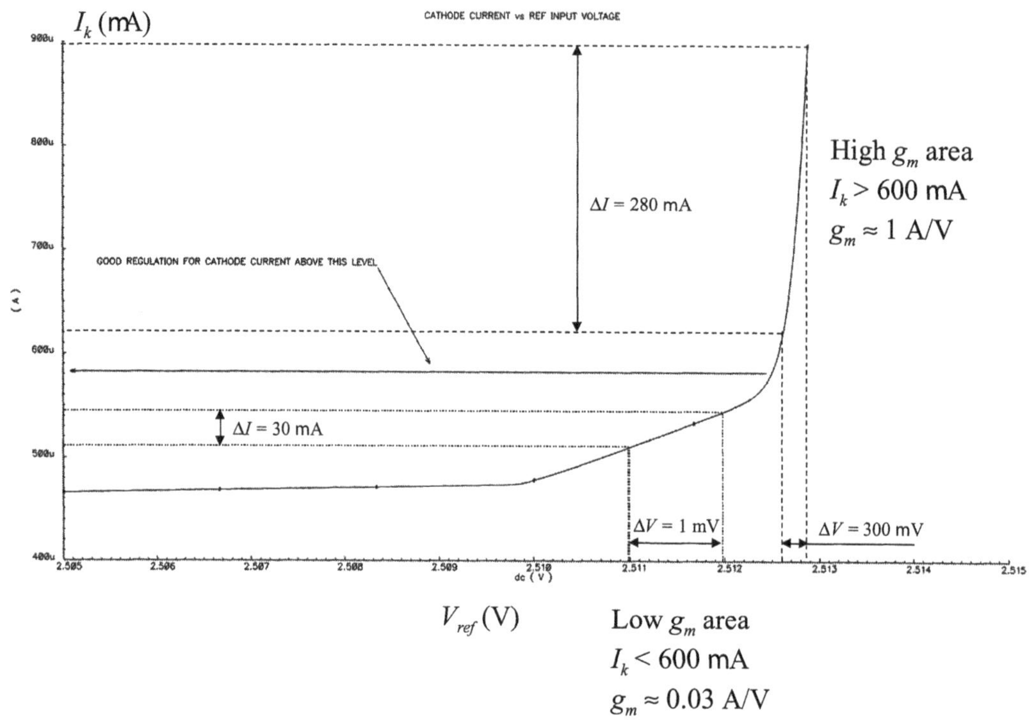

Figure 7.4 From this figure, we can clearly see the appearance of the knee below a 600-μA injected current. Below this value, the transconductance parameter g_m is rather poor.

One way to inject more current is described in Figure 7.5, where a simple resistor connected in parallel with the optocoupler LED increases the current injected in the TL431. As the LED forward drop is around 1 V, a paralleled 1-kΩ resistor forms a simple 1-mA generator, which sums up with the feedback current flowing through the LED. Please note that decreasing the LED series resistor R_{LED} does not change the TL431 current, as this current is imposed by the primary-side feedback current I_C, reflected in the LED by the optocoupler current transfer ratio (CTR). Changing R_{LED} value affects the mid-band gain but not the TL431 bias as the system operates in a closed-loop form. We will see that effect later.

Finally, what current shall we inject in the TL431? The specifications state a 1-mA minimum current, and one study claims a minimum of 5 mA to obtain decent performance [3]. What value must thus be selected? Well, besides the open-loop gain that is affected by the bias current, the consumption on the output voltage also plays a role in the selection. When you chase every tens of mW to stay below a 100-mW input power for an ac-dc adapter, you understand that you cannot afford to lose precious power in a bias current that is useless in no-load conditions. High-volume notebook adapters deal with a 1-mA extra bias current on top of the natural feedback current seen in the LED. The total bias in the TL431 is thus in the vicinity of 1.5 mA, and experience shows that it is often good enough to reach adequate performance without sacrificing the standby power.

To demonstrate this fact, we have built a simple test fixture where a compensator was assembled and a bias resistor wired as suggested by Figure 7.5. The

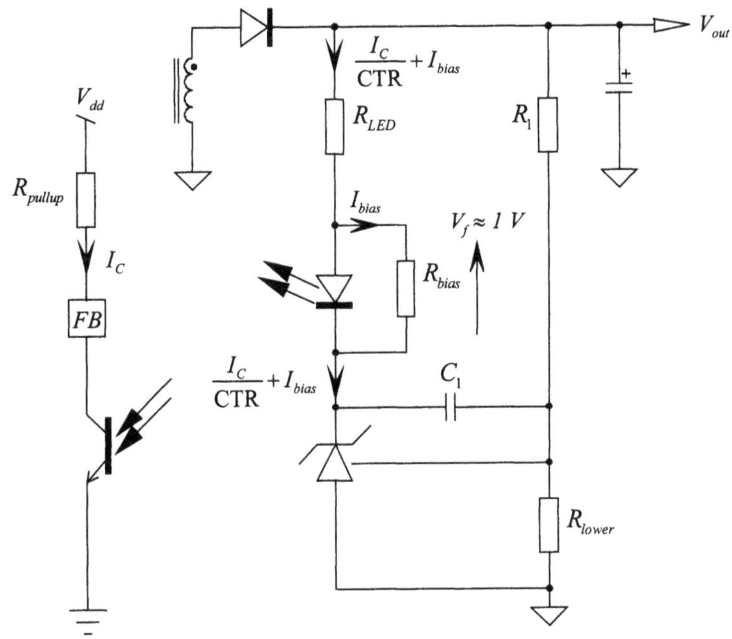

Figure 7.5 A simple resistor in parallel with the optocoupler LED creates a free current source generator.

collector current reflected through its CTR in the LED established to around 300 μA. We captured a plot without added bias current, letting these 300 μA only flow in the TL431. Then, a 1-kΩ resistor was put across the optocoupler LED to increase the bias current in the TL431 to roughly 1.3 mA. As one can immediately see in Figure 7.6, the open loop gain nicely benefited from this added bias current.

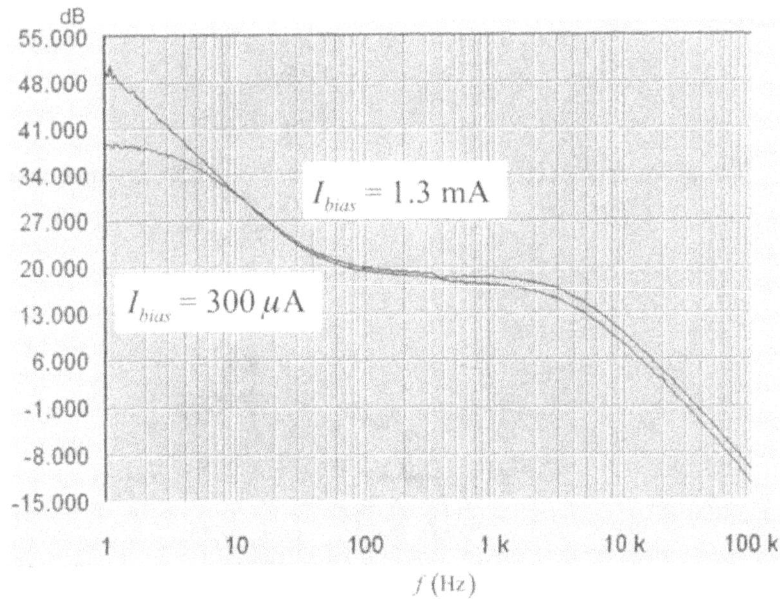

Figure 7.6 The dc gain of the TL431 clearly benefits from an added bias current.

7.2 Biasing the TL431: The Impact on the Gain

This is something we discussed in Chapter 5, but it makes sense to come back to it in this chapter because the TL431 needs an extra bias current to operate. As illustrated in Figure 7.5, the 1-mA bias is ensured by paralleling a resistor with the optocoupler LED. The compensator works by ac modulating the current circulating in the LED and transmitting it to the collector side via the optocoupler CTR. However, as the bias resistor is paralleled with the LED, it naturally steals away current, which returns to the TL431 but does not participate to the ac transfer since it is diverted from the LED. A simplified schematic appears in Figure 7.7. The V_f source illustrates the voltage drop of the LED but it does not play a role in ac as it is constant. This schematic involves the bias resistor and the LED dynamic resistor, R_d. From this representation, we can derive a few equations to show the impact of the bias current resistor.

The error voltage V_{err} is created from the circulation of the collector current in the pull-up resistor R_{pullup}:

$$V_{err}(s) = -I_c(s)R_{pullup} \tag{7.14}$$

The collector current is linked to that of the LED via the optocoupler current transfer ratio (CTR):

$$I_c(s) = I_{LED}(s)\mathrm{CTR} \tag{7.15}$$

By substituting (7.15) in (7.14), we have

$$V_{err}(s) = -I_{LED}(s)R_{pullup}\mathrm{CTR} \tag{7.16}$$

The LED current is actually the main current I_1, from which the bias current I_{bias} is removed. As we identify a resistive current divider, we have

$$I_{LED}(s) = I_1(s)\frac{R_{bias}}{R_{bias} + R_d} \tag{7.17}$$

The ac current I_1 (therefore ignoring the 1-V LED forward drop) is equal to

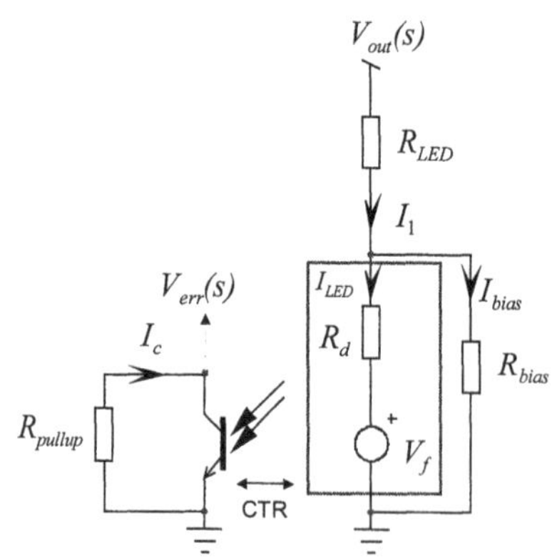

Figure 7.7 The bias resistor steals away current from the LED ac modulation.

$$I_1(s) = \frac{V_{out}(s)}{R_{LED} + R_{bias} \parallel R_d} \tag{7.18}$$

Using (7.17) and (7.18), we have

$$I_{LED}(s) = \frac{V_{out}(s)}{R_{LED} + R_{bias} \parallel R_d} \frac{R_{bias}}{R_{bias} + R_d} \tag{7.19}$$

Finally, by updating (7.16) and rearranging the result, we obtain the transfer function including all resistive paths:

$$\frac{V_{err}(s)}{V_{out}(s)} = -\frac{R_{pullup}\text{CTR}}{R_{LED} + R_{bias} \parallel R_d} \frac{R_{bias}}{R_{bias} + R_d} \tag{7.20}$$

In most of the cases, the LED dynamic resistor R_d is considered very small and (7.20) simplifies to

$$\frac{V_{err}(s)}{V_{out}(s)} \approx -\frac{R_{pullup}\text{CTR}}{R_{LED}} \tag{7.21}$$

The question now relates to the value of R_d. The dynamic resistor of a diode is linked to its operating current I_F. When operated far from the knee, the dynamic resistor is usually small. However, as the operating point shifts toward the knee, at low bias currents, R_d increases significantly. This is what happens when the pull-up resistor on the primary side is of high values in order to maintain a low-level of standby power. A small current in the LED is enough to vary the feedback level V_{err} and the dynamic resistor suffers.

To know what values we are talking about, we have performed the LED characterization of a popular optocoupler, the SFH615A-2. The results, collected at different temperatures, appear in Figure 7.8. With a 300-μA bias current, the dynamic resistor, $\Delta V_f/\Delta I_f$, is found to be 158 Ω. If the bias current increases to 1 mA, the resistor drops to 38 Ω. To check the impact of these values on the gain chain derived in (7.20), let's assume the following configuration:

R_{pullup} = 20 kΩ; the pull-up resistor loading the optocoupler collector.
R_d = 158 Ω; the LED dynamic resistor.
CTR = 0.3; the optocoupler CTR.
R_{LED} = 1 kΩ; the LED series resistor.
R_{bias} = 1 kΩ; the bias resistor placed in parallel with the LED.

The approximate gain expression where both the dynamic resistor and the bias action are neglected is

$$|G(s)| = \frac{20k \times 0.3}{1k} = 6 \text{ or } 15.6 \text{ dB} \tag{7.22}$$

Using the full expression, this gain value becomes

$$|G(s)| = \frac{20k \times 0.3}{1k + \frac{1k \times 158}{1k + 158}} \times \frac{1k}{1k + 158} = 5.28 \times 0.86 = 4.55 \text{ or } 13.2 \text{ dB} \tag{7.23}$$

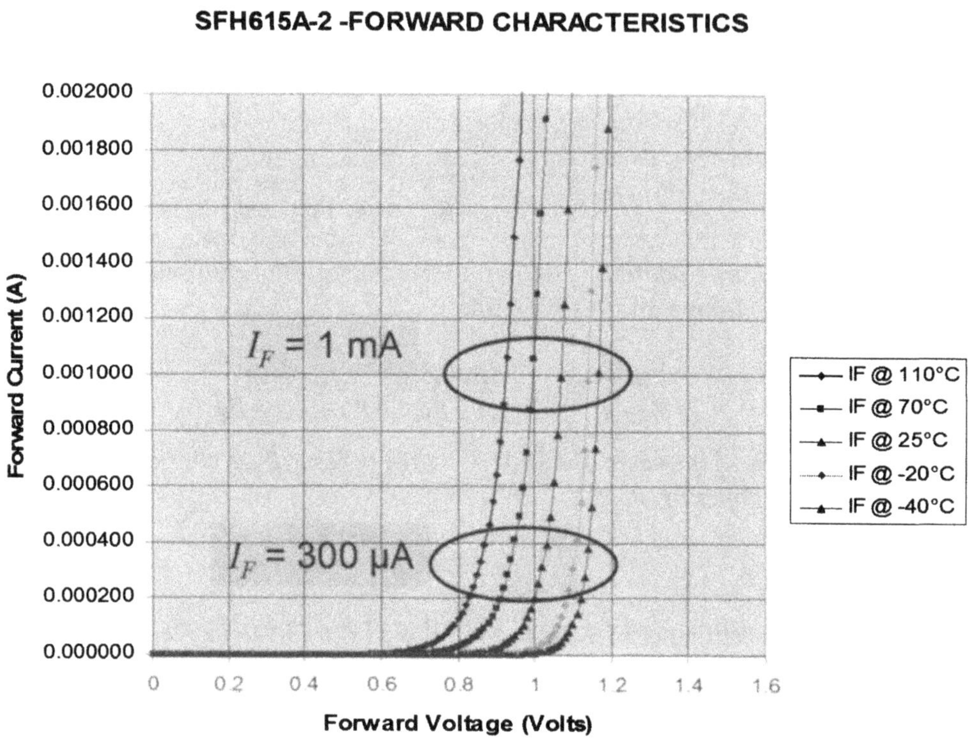

Figure 7.8 The dynamic resistor of the LED shows significant variations depending on where the operating point is.

Comparing both expressions, we can see a 2.4-dB difference. It's not a big number, but it can explain why the crossover frequency point is sometimes missed by a few decibels. To avoid this problem and extend the optocoupler bandwidth, you should force a sufficient static current in the LED by selecting low pull-up resistors on the primary side. In some cases, where the no-load standby power counts, you cannot use low R_{pullup} values, and the gain chain will be affected. When measuring the final open-loop response, it can explain the slight discrepancy you see between the computed crossover point and the final result.

7.3 Biasing the TL431: A Different Arrangement

If the parallel resistor eases the extra bias current generation for the TL431, we have just seen that it has an impact on the loop since it deviates current from the LED path. The nice thing, however, is that no calculations are required, and a simple 1-kΩ resistor placed across the LED terminals does the job perfectly. For those of you who do not like this technique because of its impact on the loop, another means exists via the direct connection of the bias resistor from the TL431 to V_{out}. This option, shown in Figure 7.9, does not affect the LED current.

The LED current depends on the current circulating in the optocoupler collector. This current is maximum when the collector is close to ground and is minimum when the collector is close to the V_{cc} point. The maximum feedback level, $V_{FB,max}$, occurs when the output power demand is high. In a peak-current mode controller,

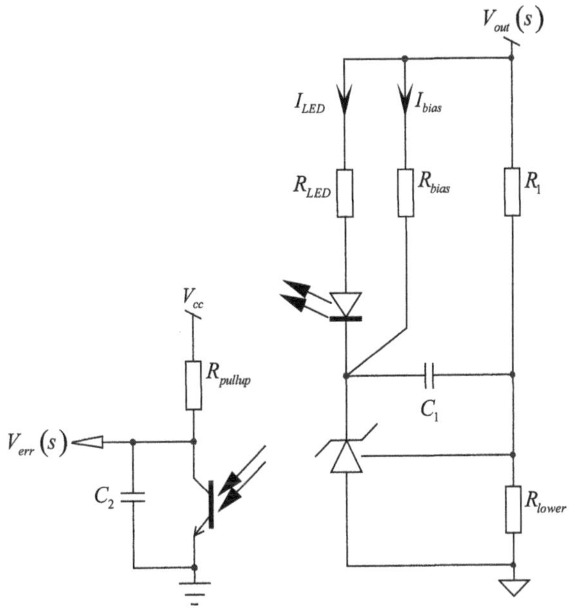

Figure 7.9 The extra resistor connected to the output voltage does not steal away current from the loop path.

the feedback voltage drives the peak current setpoint imposed to the storage inductor. This setpoint varies in relationship to the output power demand and the input voltage conditions. For safety reasons, the maximum current setpoint is internally limited to a certain value, in case the control loop would run away. In offline ac-dc controllers, this maximum setpoint is usually 1 V, safely limiting the current excursion in the sense resistor, and the inductor, to $1/R_{sense}$. To improve the voltage dynamics across the optocoupler collector and the immunity to noise, a divide-by-3 circuit is installed between the feedback pin and the peak current setpoint. Therefore, the feedback voltage will vary between 3 V (maximum power demand) and few hundred of millivolts at low power. Sometimes, one or two diodes are inserted in series with the feedback pin to offer a real 0 V setpoint when the optocoupler collector reaches its saturation level (≈300 mV). Such a configuration appears in Figure 7.10.

The collector current, and thus the error voltage, is controlled by the current circulating in the optocoupler LED. This current also flows in the TL431 and contributes to its bias point. This is I_{LED} in Figure 7.9. The minimum bias current occurs when the optocoupler CTR peaks at its maximum and the power demand is the highest. In that case, the feedback voltage is almost 3 V, assuming a common-emitter circuitry close to that in Figure 7.10 is implemented. Suppose we have the following parameters:

V_{cc} = 5 V; the pull-up resistor upper terminal voltage.
R_{pullup} = 20 kΩ; the collector pull-up resistance.
R_{LED} = 1 kΩ; the LED series resistance.
V_f = 1 V; the LED forward voltage.
I_{bias} = 1 mA; the wanted bias current in the TL431.
CTR = 30 percent to 120 percent; the optocoupler CTR variations.
$V_{FB,\max}$ = 3 V; the feedback voltage at maximum power.

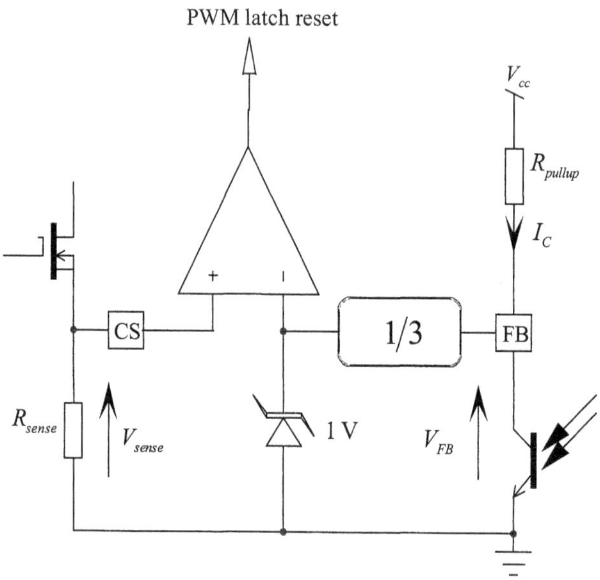

Figure 7.10 The internal circuitry of a peak current mode controller.

The minimum LED current in that case is simply

$$I_{LED,\min} = \frac{V_{cc} - V_{FB,\max}}{R_{pullup}\text{CTR}_{\max}} = \frac{5-3}{20k \times 1.2} \approx 83\ \mu\text{A} \tag{7.24}$$

As we can see, we are far away from the minimum bias current of 1 mA required in the TL431 data sheet. This is the purpose of the extra resistor R_{bias} shown in Figure 7.9—to provide this extra current. To derive a resistance value for R_{bias}, we first need to know the voltage across it. This voltage is that across the LED series resistance R_{LED} to which the LED forward drop V_f is added. If we divide this voltage by the needed bias current, we have our resistor value:

$$R_{bias} = \frac{R_{LED}I_{LED,\min} + V_f}{I_{bias}} = \frac{\frac{V_{cc} - V_{FB,\max}}{R_{pullup}\text{CTR}_{\max}} R_{LED} + V_f}{I_{bias}} \tag{7.25}$$

If we use the previous parameters, for a 1-mA bias current, the extra resistor value must be

$$R_{bias} = \frac{83u \times 1k + V_f}{1m} = 1083\ \Omega \tag{7.26}$$

This value is not far away from what a parallel resistor would give. However, this configuration has the following advantages:

1. As no ac current is stolen from the LED path, the presence of this resistor does not affect the loop gain, even in presence of large LED dynamic resistors.
2. To cope with the TL431 biasing requirements, the LED series resistor upper limit is bounded. If you add a bias resistor across the LED, to form the 1-mA bias generator, this upper limit is affected and it brings another

burden in the calculation process (see the following). On the contrary, when the bias resistor is directly connected to the output voltage, this drawback disappears.

3. Our experience shows that this configuration slightly reduces the output overshoot at startup. This is because the bias current is immediately present as V_{out} rises up. When using the 1-kΩ resistor in parallel with the LED, the bias current only occurs when the TL431 forces the LED conduction.

7.4 Biasing the TL431: Component Limits

The TL431 requires a minimum current to operate in favorable conditions. On the other side, it also needs a minimum voltage to deliver its performance. This minimum voltage is equal to the internal reference value and cannot be lower than 2.5 V. Figure 7.5 portrays a typical TL431 arrangement where the optocoupler collector pulls the feedback pin down as the LED current increases. The series resistor R_{LED} is there to limit the maximum LED current but also to fix the compensator mid-band gain as we will see in a few lines. Unfortunately, because of the biasing conditions imposed by the TL431 (minimum operating voltage of 2.5 V and the necessity to inject 1 mA at least), there is an upper limit on this resistor value. Let us derive this limit with a few equations. First, what is the maximum collector current on the optocoupler output necessary to bring the feedback voltage close to ground, actually the optocoupler saturation voltage $V_{CE,sat}$?

$$I_{C,\max} = \frac{V_{cc} - V_{CE,sat}}{R_{pullup}} \tag{7.27}$$

To circulate in the pull-up resistor, this current needs the photons emitted by the internal LED and collected by the transistor base area. The collector current I_C is linked to the LED current I_F by the current transfer ratio (CTR). In our case, to cover unavoidable dispersions, we need to use the minimum value of this CTR parameter. Looking at Figure 7.5 in more detail, we can see that the total current flowing through R_{LED} also includes the additional bias current brought by R_{bias}, provided we have adopted the bias generator featuring the 1-kΩ resistor across the LED. Therefore, (7.27) can be used to derive another equation:

$$I_{R_{LED},\max} = \frac{I_{C,\max}}{\mathrm{CTR}_{\min}} + I_{bias} = \frac{V_{cc} - V_{CE,sat} + I_{bias} R_{pullup} \mathrm{CTR}_{\min}}{R_{pullup} \mathrm{CTR}_{\min}} \tag{7.28}$$

The LED current depends not only on the output voltage, but also on the forward drop of the diode itself and the minimum operating voltage acceptable for the TL431:

$$I_{R_{LED},\max} = \frac{V_{out} - V_f - V_{\mathrm{TL431},\min}}{R_{LED}} \tag{7.29}$$

By equating (7.28) and (7.29) then solving for $R_{LED,\max}$, we have the maximum value this resistor cannot exceed:

$$R_{LED,\max} \leq \frac{V_{out} - V_f - V_{TL431,\min}}{V_{cc} - V_{CE,sat} + I_{bias}CTR_{\min}R_{pullup}} R_{pullup}CTR_{\min} \quad (7.30)$$

In the previous equations, we have

V_{out}, the output voltage
I_{bias}, the TL431 biasing current when the optocoupler LED is paralleled with a resistor (usually 1 kΩ for a 1-mA bias)
$V_{TL431,\min}$, the minimum voltage the TL431 can go down to (2.5 V)
V_f, the optocoupler LED forward drop (≈ 1 V)
$CTR_{\min}$, the minimum optocoupler current transfer ratio
$V_{CE,\,sat}$, the optocoupler saturation voltage (≈ 300 mV at a 1-mA collector current)
V_{cc}, the bias voltage where the pull-up resistor is connected to on the primary side

From the previous line, we can see that the dc operating conditions fix the upper excursion value of the LED series resistor. As we will later see, R_{LED} not only plays a role in dc, but also in the mid-band gain expression, hampering the TL431 application range. The fast lane is guilty . . .

7.5 The Fast Lane Is the Problem

Already discussed in the op amp section, it is interesting to come back on this typical characteristic of the TL431 configuration. In Figure 7.5, neglecting the bias current, the LED current is imposed by the voltage present on the output voltage, the diode forward drop, and the level on the TL431 cathode:

$$I_{LED} = \frac{V_{out} - V_f - V_{TL431}}{R_{LED}} \quad (7.31)$$

This current is actually made of two terms: a dc and ac current. The dc current fixes the operating point and corresponds to an output voltage equal to the assigned target. The ac current superimposes on the dc value and modulates the voltage on the FB node, $V_{err}(s)$. In this mode, the LED forward drop V_f no longer plays a role (it is constant and its derivative term is therefore null), and (7.31) can be updated as follows:

$$I_{LED}(s) = \frac{V_{out}(s) - V_{TL431}(s)}{R_{LED}} = \frac{V_{out}(s)}{R_{LED}} - \frac{V_{TL431}(s)}{R_{LED}} \quad (7.32)$$

In Figure 7.5, we can see a capacitor C_1 across the TL431 feedback path. At high frequencies, this capacitor becomes a short circuit and the ac voltage across the TL431 imputed to the output voltage variations sensed by R_1/R_{lower} goes to 0. However, as indicated by (7.31), the dc point is maintained and the converter keeps delivering the right output voltage V_{out}. In a classical configuration, because the feedback capacitor is a short circuit, we would suppose that the ac output of the whole system, $V_{err}(s)$, would also go down to zero.

Unfortunately, the LED ac current is made of two voltage terms, as indicated by (7.32): the one imposed by the TL431 (which is null at high frequencies because of C_1) but also that directly imposed by the output voltage. Mathematically, this fact can be expressed through the following equation:

$$\lim_{s\to\infty} I_{LED}(s) = \frac{V_{out}(s)}{R_{LED}} \tag{7.33}$$

As a result, despite a TL431 ac output being null, there is still a path from the output to the control via the optocoupler LED: this is the *fast lane* effect. Figure 7.11 shows a simplified representation of the system where the TL431 is replaced by a simple Zener diode, testifying for the disappearance of the ac link with V_{out} through the divider network. If V_{out} is modulated, then the LED current is also modulated and the perturbation propagates to the output, $V_{err}(s)$. In the light of (7.33), we can see that R_{LED} not only fixes the dc bias, but also interacts in the gain definition. This fact leads to situations where the selection of the LED series resistor based on gain/attenuation needs is not compatible with what (7.30) dictates. At this point, another solution will have to be found: the fast lane must be disabled.

7.6 Disabling the Fast Lane

If the fast lane represents the problem, we must find a way to get rid of it. How? By cutting the ac link between the observed variable (V_{out}) and the LED series resistor. This can be done by identifying in the board another dc source (e.g., 12 V), fully ac isolated from the observed variable. The output of a linear regulator will do perfectly, for instance, perhaps from another winding if necessary. Another solution consists of biasing a separated Zener-based network supplied by the output voltage we observe. If the Zener is properly biased and its voltage around two-thirds of V_{out} (to build enough ac isolation), then experience shows that it is a valid solution. This solution appears in Figure 7.12.

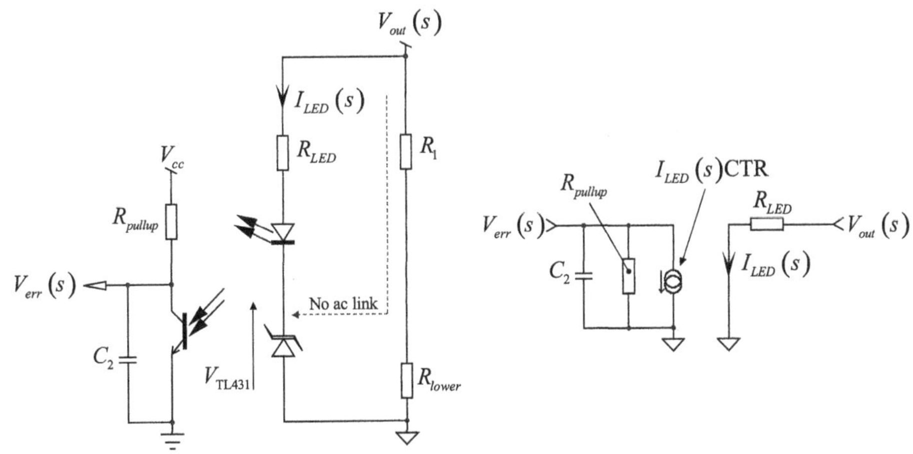

Figure 7.11 The fast lane is created because the LED current not only depends on the internal op amp ac output but also on the observed output voltage. The right side of the picture represents the equivalent ac small-signal model at high frequencies when the internal op amp ac output is zero.

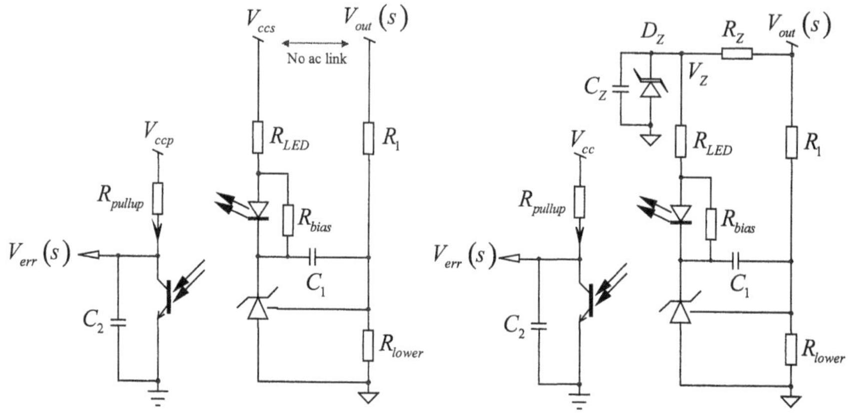

Figure 7.12 A solution is to hook the series resistor R_{LED} to a separate dc point, V_{ccs}, or build ac isolation from V_{out} via a Zener-based network.

The LED maximum value does not change from that derived in (7.30) except that V_{out} is replaced by the Zener voltage or V_{ccs} if you adopt Figure 7.12 left-side option:

$$R_{LED,\max} \leq \frac{V_Z - V_f - V_{\mathrm{TL431,min}}}{V_{ccp} - V_{CE,sat} + I_{bias}\mathrm{CTR}_{\min}R_{pullup}} R_{pullup}\mathrm{CTR}_{\min} \tag{7.34}$$

The bias resistor R_Z requires a design equation, since its role is to bias the Zener diode, but also to provide the TL431 operating current (feedback and bias currents). The Zener diode bias current I_{Zbias} has been selected so that the diode operates far enough from its knee where it will exhibit the lowest dynamic impedance and, thus, the best ac isolation from V_{out}. The LED maximum current is actually reached when the feedback voltage V_{FB} is pulled down to $V_{CE,sat}$. In this case, the LED current is simply

$$I_{LED,\max} = \frac{V_{cc} - V_{CE,sat}}{R_{pullup}\mathrm{CTR}_{\min}} \tag{7.35}$$

To this operating feedback current, we must add the 1-mA (or more) extra current I_{bias} brought by R_{bias} placed across the optocoupler LED. The Zener dropping resistor R_Z is therefore immediately derived through

$$R_Z = \frac{V_{out} - V_Z}{I_{LED,\max} + I_{bias} + I_{Zbias}} \tag{7.36}$$

Substituting (7.35) in (7.36), we obtain

$$R_Z = \frac{(V_{out} - V_Z)R_{pullup}\mathrm{CTR}_{\min}}{(V_{cc} - V_{CE,sat}) + (I_{Zbias} + I_{bias})R_{pullup}\mathrm{CTR}_{\min}} \tag{7.37}$$

In the previous equation, we have the following variables:

V_{out}, the output voltage

I_{bias}, the TL431 biasing current when the optocoupler LED is paralleled with a resistor (usually 1 kΩ for a 1-mA bias)

I_{Zbias}, the Zener diode biasing current
$\text{CTR}_{\min}$, the minimum optocoupler current transfer ratio
$V_{CE,sat}$, the optocoupler saturation voltage (≈ 300 mV at a 1-mA collector current)
V_{cc}, the bias where the pull-up resistor is connected to on the primary side

A 100-nF capacitor will be paralleled with the Zener diode to improve the filtering capabilities of the network. We are now all set to detail the design methods for the types 1, 2, and 3 using a TL431 regulator.

7.7 The Type 1: An Origin Pole, Common-Emitter Configuration

The type 1 is used to compensate power converters where no phase boost is necessary. The phase lag of a type 1 amplifier, whatever its construction (op amp, OTA, or TL431) is permanently flat to 270°. A type 1 can be built with a TL431 following the Figure 7.13 schematic:

In this circuit, the LED resistor is selected following (7.30) while the pole and zero are made coincident to cancel one another. This way, we create a simple integrator whose 0-dB crossover pole can be easily adjusted. If we neglect the LED dynamic resistor R_d, we can draw a simplified small-signal diagram from Figure 7.13. It appears in Figure 7.14.

The ac current circulating in the LED depends on V_{out} and the voltage on the internal op amp output terminal:

$$I_{LED}(s) = \frac{V_{out}(s) - V_{\text{TL431}}(s)}{R_{LED}} \tag{7.38}$$

The voltage output of the TL431 is a simple integrator and obeys the following transfer function:

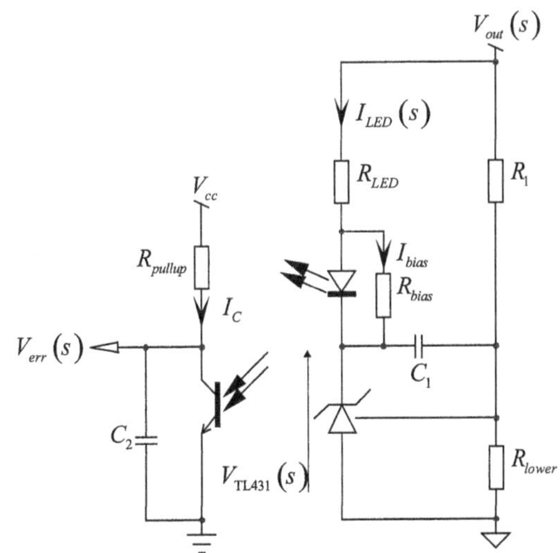

Figure 7.13 A type 1 built with a TL431 uses a type 2 configuration where the pole and zero have been made coincident.

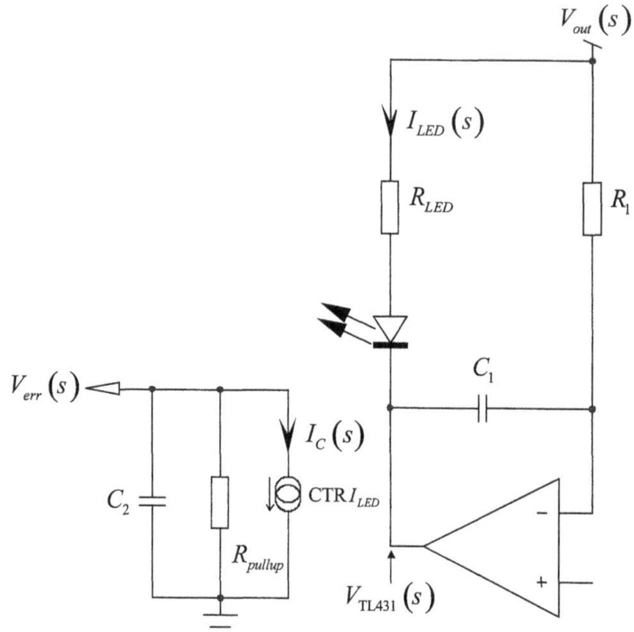

Figure 7.14 The small-signal version of the TL431 network reveals an integrator followed by a pole.

$$V_{TL431}(s) = -V_{out}(s)\frac{\frac{1}{sC_1}}{R_1} = -V_{out}(s)\frac{1}{sR_1C_1} \tag{7.39}$$

Substituting (7.39) into (7.38), we have

$$I_{LED}(s) = \frac{V_{out}(s) + V_{out}(s)\frac{1}{sR_1C_1}}{R_{LED}} = \frac{V_{out}(s)}{R_{LED}}\left(1 + \frac{1}{sR_1C_1}\right) = V_{out}(s)\left(\frac{1+sR_1C_1}{sR_{LED}R_1C_1}\right) \tag{7.40}$$

The previous expression shows the effect of the fast lane. By inspecting Figure 7.14, we would expect the presence of a simple integrator, given the position of C_1. However, as confirmed by (7.40), a zero clearly appears in the numerator of the expression. To build a type 1, we will then need to find a way to neutralize this zero by another pole. Let us carry on with the derivation. The output voltage, $V_{err}(s)$, is simply the collector current $I_C(s)$ circulating in a network made of the pull-up resistor and the capacitor C_2:

$$V_{err}(s) = -I_C(s)\frac{R_{pullup}\frac{1}{sC_1}}{R_{pullup} + \frac{1}{sC_1}} = -I_C(s)R_{pullup}\frac{1}{1+sR_{pullup}C_1} \tag{7.41}$$

We know that the collector current links to the LED current via the CTR:

$$I_C(s) = I_{LED}(s)\text{CTR} \tag{7.42}$$

By substituting (7.40) into (7.42), updating and rearranging (7.41), we have

$$\frac{V_{err}(s)}{V_{out}(s)} = -\frac{1}{s\dfrac{R_{LED}R_1}{R_{pullup}\text{CTR}}C_1}\left(\frac{1+sR_1C_1}{1+sR_{pullup}C_2}\right) \tag{7.43}$$

This equation can be put under a more familiar form such as

$$G(s) = -\frac{1}{s/\omega_{po}}\frac{1+s/\omega_z}{1+s/\omega_p} \tag{7.44}$$

where

$$\omega_{po} = \frac{R_{pullup}\text{CTR}}{R_{LED}R_1C_1} \tag{7.45}$$

$$\omega_z = \frac{1}{R_1C_1} \tag{7.46}$$

$$\omega_p = \frac{1}{R_{pullup}C_2} \tag{7.47}$$

To form a real type 1, we need to neutralize the zero and the pole to keep the 0-dB crossover pole alone. Therefore, (7.46) and (7.47) must be equal:

$$\frac{1}{R_1C_1} = \frac{1}{R_{pullup}C_2} \tag{7.48}$$

From the previous, we can extract the value of C_2:

$$C_2 = \frac{R_1C_1}{R_{pullup}} \tag{7.49}$$

From (7.45) we obtain a value for C_1:

$$C_1 = \frac{R_{pullup}\text{CTR}}{2\pi R_{LED}R_1 f_{po}} \tag{7.50}$$

Substituting C_1 definition into (7.49), we have an expression for C_2:

$$C_2 = \frac{\text{CTR}}{2\pi f_{po}R_{LED}} \tag{7.51}$$

The 0-dB crossover pole location must now be selected so that the magnitude G at the crossover frequency f_c exactly compensates the gain excess or deficiency read on the power stage Bode plot. The transfer function of an integrator obeys the following formula, in which ω_{po} represents the 0-dB crossover pole:

$$G(s) = \frac{1}{\dfrac{s}{\omega_{po}}} \tag{7.52}$$

From the previous equation, we can calculate the magnitude of $G(s)$ at the crossover frequency:

$$|G(f_c)| = \frac{f_{po}}{f_c} \tag{7.53}$$

We can now extract the 0-dB crossover pole location and feed (7.50)/(7.51) to get C_1 and C_2, respectively:

$$f_{po} = G \cdot f_c \tag{7.54}$$

7.7.1 A Design Example

Let us assume we want to compensate a 12-V power stage exhibiting a gain excess G_{f_c} of 25 dB at a 20-Hz frequency. This could be the case for a single-stage flyback converter, for instance. To reach this goal, we will define the position at which the 0-dB crossover pole must be located so that the magnitude of $G(s)$ is exactly –25 dB when the frequency reaches 20 Hz. These 25 dB of attenuation translate to

$$G = 10^{\frac{-G_{f_c}}{20}} = 10^{\frac{-25}{20}} = 56.2\text{m} \tag{7.55}$$

Applying (7.54), the 0-dB crossover pole must be placed at

$$f_{po} = G \cdot f_c = 0.0562 \times 20 = 1.12 \text{ Hz} \tag{7.56}$$

The single-stage flyback converter uses elements exhibiting the following values:

V_{out} = 12 V; the output voltage.
V_f = 1 V; the LED forward voltage.
I_{bias} = 1 mA; the TL431 biasing current when the optocoupler LED is paralleled with a resistor.
$V_{\text{TL431,min}}$ = 2.5 V; the minimum operating voltage of the TL431.
$V_{CE,sat}$ = 0.3 V; the optocoupler saturation voltage.
V_{cc} = 5 V; the pull-up V_{cc} level.
R_{pullup} = 10 kΩ; the optocoupler pull-up resistor.
CTR_{min} = 0.5; the minimum optocoupler current transfer ratio.
R_1 = 38 kΩ; the upper resistor in the resistor bridge observing the output variable.
f_{opto} = 10 kHz; the optocoupler pole that has been characterized with R_{pullup}.

Based on these values, we can immediately calculate the maximum LED series resistor via (7.30):

$$R_{LED,\max} \le \frac{12-1-2.5}{5-0.3+1m \times 0.5 \times 10k} 10k \times 0.5 \le 4.4 \text{ k}\Omega \tag{7.57}$$

Adopting a 20 percent derating factor, the final series resistor for the LED is 3.5 kΩ.

The capacitors calculations can now be undertaken:

$$C_1 = \frac{R_{pullup}\text{CTR}}{R_{LED}R_1 2\pi f_{po}} = \frac{10k \times 0.5}{3.5k \times 38k \times 1.12 \times 6.28} = 5.3\ \mu\text{F} \tag{7.58}$$

$$C_2 = \frac{R_1C_1}{R_{pullup}} = \frac{5.3u \times 38k}{10k} = 20.3\ \mu\text{F} \tag{7.59}$$

Given the optocoupler pole position value (10 kHz) with regard to the second pole (0.8 Hz), we can neglect its action on the compensator. The simulation schematic appears in Figure 7.15.

The ac response of the compensator appears in Figure 7.16 and confirms the target we set. The attenuation at 20 Hz is exactly 25 dB as expected. There is no phase boost; we knew it from the start.

7.8 The Type 1: Common-Collector Configuration

Wiring the optocoupler in a common-collector configuration does not change the operating bias point, but the transfer function loses its minus sign:

$$\frac{V_{err}(s)}{V_{out}(s)} = \frac{1}{s\dfrac{R_{LED}R_1}{R_{pulldown}\text{CTR}}C_1}\left(\frac{1+sR_1C_1}{1+sR_{pulldown}C_2}\right) \tag{7.60}$$

The calculations carried in the previous example are still valid, with R_{pullup} being replaced by $R_{pulldown}$.

7.9 The Type 2: An Origin Pole plus a Pole/Zero Pair

The type 2 configuration is the most popular configuration of the TL431 used in a compensator application. The application circuit does not change from that proposed in Figure 7.13. Therefore, the transfer function already derived for the type 1 remains valid. However, it is our interest to slightly rearrange it to make the mid-band gain appear:

$$\frac{V_{err}(s)}{V_{out}(s)} = -\frac{R_{pullup}\text{CTR}}{sR_{LED}R_1C_1}\left(\frac{1+sR_1C_1}{1+sR_{pullup}C_2}\right) \tag{7.61}$$

By factoring sR_1C_1 in the numerator, we recognize a classical type 2 transfer function:

$$\frac{V_{err}(s)}{V_{out}(s)} = -\frac{R_{pullup}\text{CTR}}{R_{LED}}\left(\frac{1+1/sR_1C_1}{1+sR_{pullup}C_2}\right) \tag{7.62}$$

This equation can be put under the form

$$G(s) = -G_0\frac{1+\omega_z/s}{1+s/\omega_p} \tag{7.63}$$

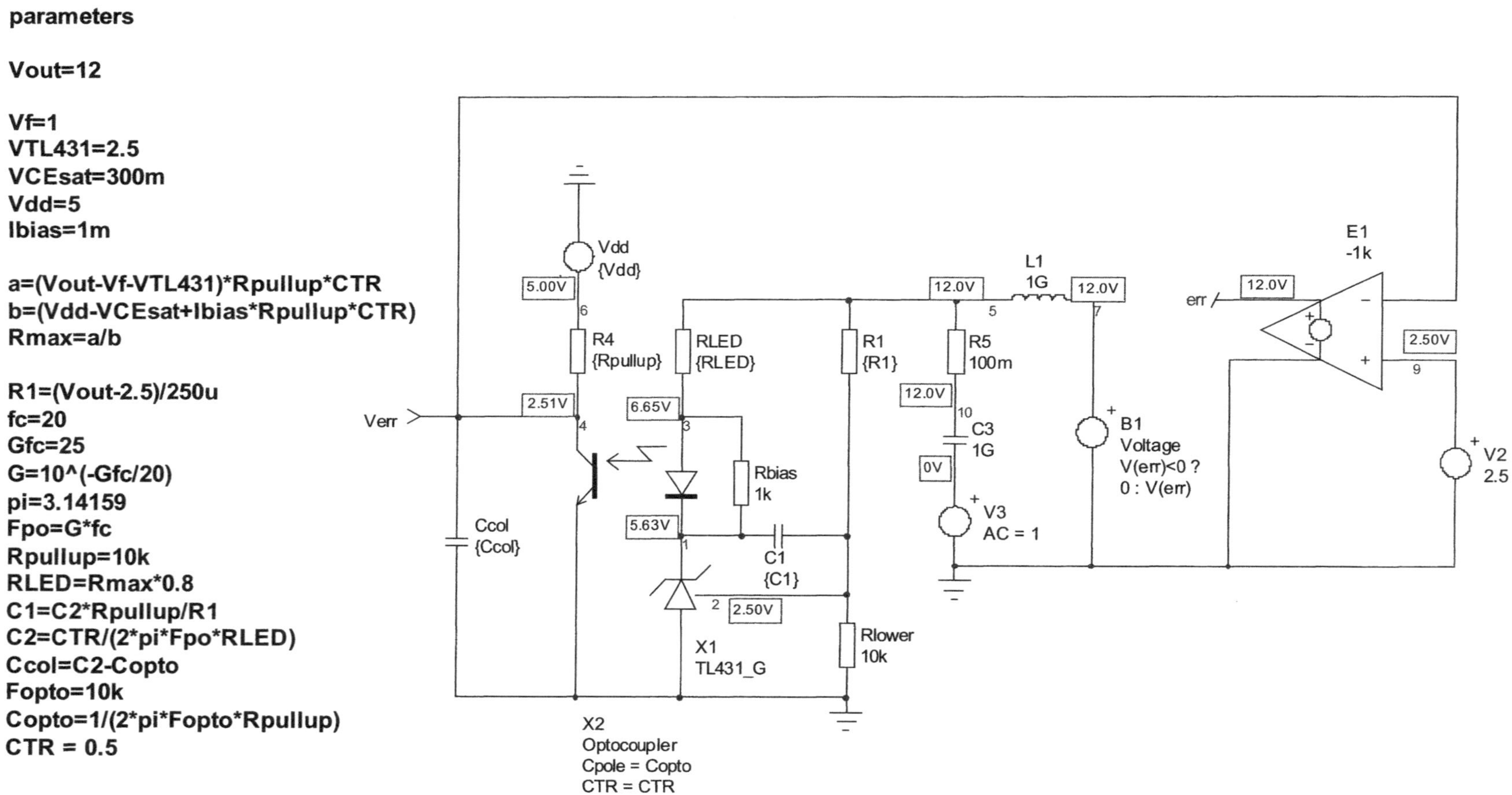

Figure 7.15 The type 1 configuration schematic implementing a TL431 does not cause any calculation difficulties.

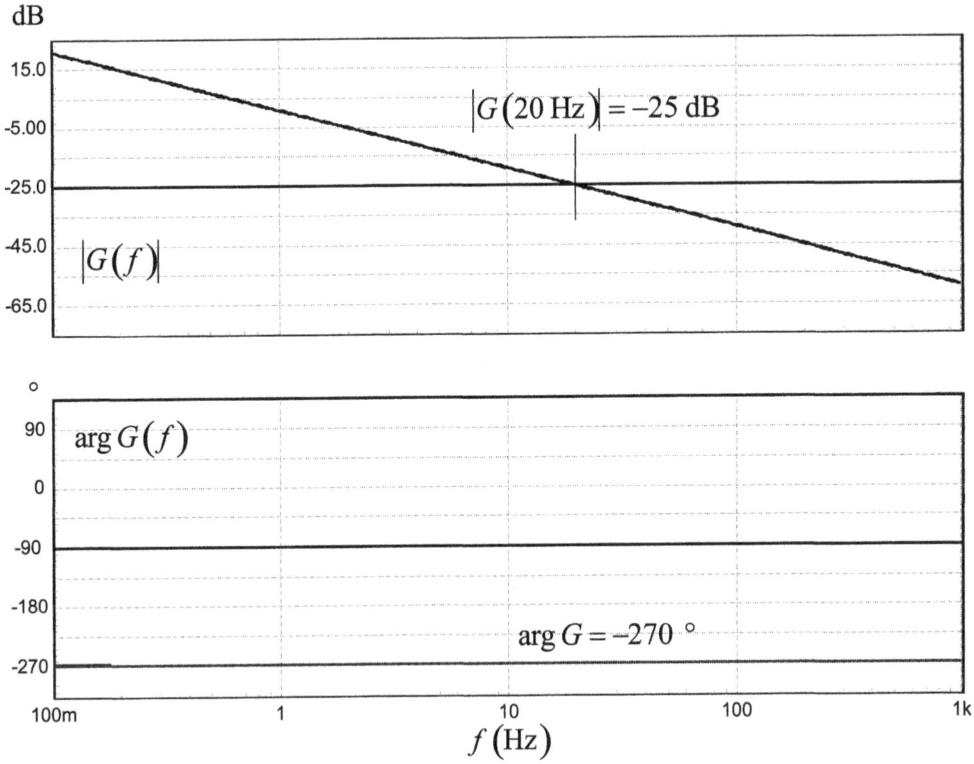

Figure 7.16 The ac response confirms the right attenuation at the selected 20-Hz crossover point.

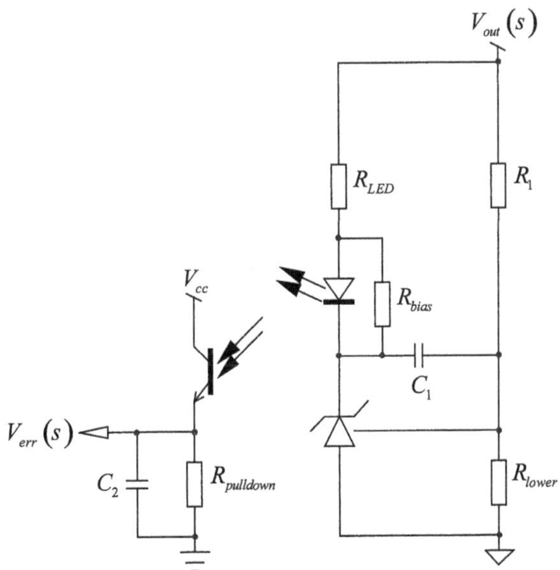

Figure 7.17 The common-collector configuration changes the signal polarity.

where

$$G_0 = \frac{R_{pullup}\text{CTR}}{R_{LED}} \tag{7.64}$$

$$\omega_z = \frac{1}{R_1 C_1} \tag{7.65}$$

$$\omega_p = \frac{1}{R_{pullup} C_2} \tag{7.66}$$

The expression for the mid-band gain, (7.64), requires some clarifications, though. We can see that the equation includes the LED resistor whose upper limit is bounded by (7.30). This means that the flexibility of choosing any gain or attenuation value for G_0 is hampered by the upper limit for R_{LED}. Again, this is an effect of the fast lane and there is nothing we can do to counteract this effect. Suppose (7.30) gives an upper limit of 860 Ω, together with a pull-up resistor of 20 kΩ and a CTR of 30 percent. In this case, the minimum gain imposed by G_0 is

$$G_0 > \text{CTR}\frac{R_{pullup}}{R_{LED,\max}} > 0.3\frac{20}{0.860} > 7 \text{ or } \approx 17 \text{ dB} \tag{7.67}$$

This means that you can only select a crossover frequency on the $H(s)$ Bode plot where the needed amplification is at least 17 dB. Should the selected point require a 10-dB gain, you could simply not use the TL431 with the selected pull-up and LED resistor values.

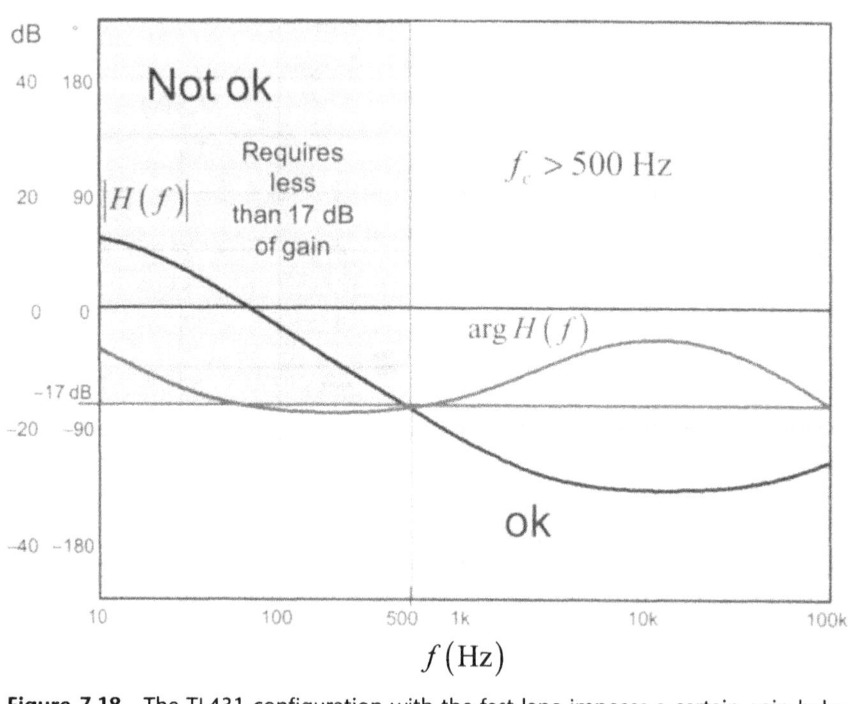

Figure 7.18 The TL431 configuration with the fast lane imposes a certain gain below which you simply cannot use the compensator. In the figure, the gain selection has to be greater than 17 dB.

Figure 7.18 shows the considered Bode plot and the possible working area. In this region, the type 2 built with the TL431 and operated with the resistors (860 Ω for R_{LED} and 20 kΩ for R_{pullup}) would work. However, as you can see, a crossover frequency below 500 Hz could not be accepted simply because you would need a gain below the minimum of 17 dB imposed by (7.67). What if you really need to cross over below 500 Hz? Well, if the phase lag is less than 45° at the considered point, you can use a type 1 and freely select the needed gain/attenuation, regardless of the LED series resistor. Otherwise, you need to get rid of the fast lane, as we will see in a few lines. First, let's study a design example of a type 2 configuration with a TL431.

7.9.1 A Design Example

Let's assume we need to compensate a 19-V output converter whose gain deficiency at 1 kHz is –15 dB. Given the phase margin we shoot for, we will boost the phase by 50° at 1 kHz. Where do we place our pole and zero pair? As we have detailed in the pole-zero section, the pole can be placed at the following location:

$$f_p = \left[\tan(boost) + \sqrt{\tan^2(boost) + 1}\right] f_c = 2.74 \times 1k = 2.74 \text{ kHz} \tag{7.68}$$

As the phase peaks at the geometric mean between the pole and the zero, this latter is placed at

$$f_z = \frac{f_c^2}{f_p} = \frac{1k^2}{2.74k} \approx 365 \text{ Hz} \tag{7.69}$$

The converter uses components that exhibit the following characteristics:

V_{out} = 19 V; the output voltage.
V_f = 1 V; the LED forward voltage.
I_{bias} = 1 mA; the TL431 biasing current when the optocoupler LED is paralleled with a resistor.
V_{TL431} = 2.5 V; the minimum operating voltage of the TL431.
$V_{CE,sat}$ = 0.3 V; the optocoupler saturation voltage.
V_{cc} = 5 V; the pull-up Vcc level.
R_{pullup} = 20 kΩ; the optocoupler pull-up resistor.
CTR_{min} = 0.3; the minimum optocoupler current transfer ratio.
R_1 = 66 kΩ; the upper element in the resistor bridge observing the output variable.
f_{opto} = 6 kHz; the optocoupler pole that has been characterized with R_{pullup}.

From these values, we can first check the maximum allowable LED series resistor:

$$R_{LED,\max} \le \frac{19 - 1 - 2.5}{5 - 0.3 + 1m \times 0.3 \times 20k} 20k \times 0.3 \le 8.7 \text{ k}\Omega \tag{7.70}$$

Then, let's translate the 15-dB gain into a decimal value and check if the corresponding LED resistor value complies with (7.70):

$$G = 10^{\frac{-G_{f_c}}{20}} = 10^{\frac{15}{20}} = 5.6 \tag{7.71}$$

From (7.64), we can extract the LED resistor:

$$R_{LED} = \frac{R_{pullup}\text{CTR}}{G} = \frac{20k \times 0.3}{5.6} = 1071\ \Omega \tag{7.72}$$

This resistor is well below the limit imposed by (7.70); we are safe. The first capacitor C_1 is found using (7.65) together with the 365-Hz zero we want to position:

$$C_1 = \frac{1}{2\pi f_z R_1} = \frac{1}{6.28 \times 365 \times 66k} = 6.6\ \text{nF} \tag{7.73}$$

We know that the pole capacitor C_2 is actually made of an extra capacitor C_{col} that we will place in parallel with the collector-emitter parasitic capacitor from the optocoupler:

$$C_2 = \frac{1}{2\pi R_{pullup} f_p} = \frac{1}{6.28 \times 20k \times 2.74k} = 2.9\ \text{nF} \tag{7.74}$$

Given the characterized optocoupler pole at 6 kHz, its parasitic capacitor is found to be

$$C_{opto} = \frac{1}{2\pi f_{opto} R_{pullup}} = \frac{1}{6.28 \times 6k \times 20k} = 1.3\ \text{nF} \tag{7.75}$$

As the total capacitance is 2.9 nF, the added capacitor C_{col} is simply the difference between (7.74) and (7.75):

$$C_{col} = C_2 - C_{opto} = 2.9n - 1.3n = 1.6\ \text{nF} \tag{7.76}$$

We are now all set and can proceed with the simulation schematic. It appears in Figure 7.19.

The automated calculation section on the left side helps to adjust the pole/zero pair on the fly in case some adjustments are further needed. The ac results appear in Figure 7.20.

We can see a slight gain mismatch that can be explained by the combined action of the LED dynamic resistor and the bias resistor. If necessary, it can be manually compensated by asking for a gain slightly higher than what is needed at 1 kHz. Practically, it will have no effect on the final loop performance.

7.10 The Type 2: Common-Emitter Configuration and UC384X

The popular UC384X lends itself very well to the implementation of a TL431 compensation network. Its internal op amp features a maximum output current capability of 1 mA: if you try to pull more current out of its pin, the op amp output collapses. Taking advantage of this, we can get rid of the op amp by wiring a pull-up resistor from its output to the 5-V reference voltage. With a maximum output current capability beyond 20 mA, we have enough headroom to select the pull-up resistor of our choice. Figure 7.21 shows the adopted configuration. The transfer function is that described by (7.62).

parameters

Vout=19
R1=(Vout-2.5)/250u
fc=1k
Gfc=-15

Vref=2.5
VCEsat=300m
Vf=1
Ibias=1m
Vdd=5
G=10^(-Gfc/20)
pi=3.14159
Fzero=365
Fpole=2.74k
A=Vout-Vf-Vref
B=Vdd-VCEsat+Ibias*CTR*Rpullup
Rmax=(A/B)*Rpullup*CTR
Rpullup=20k
RLED=CTR*Rpullup/G
C1=1/(2*pi*Fzero*R1)
C2=1/(2*pi*Fpole*Rpullup)
Ccol=C2-Copto
Fopto=6k
Copto=1/(2*pi*Fopto*Rpullup)
CTR = 0.3

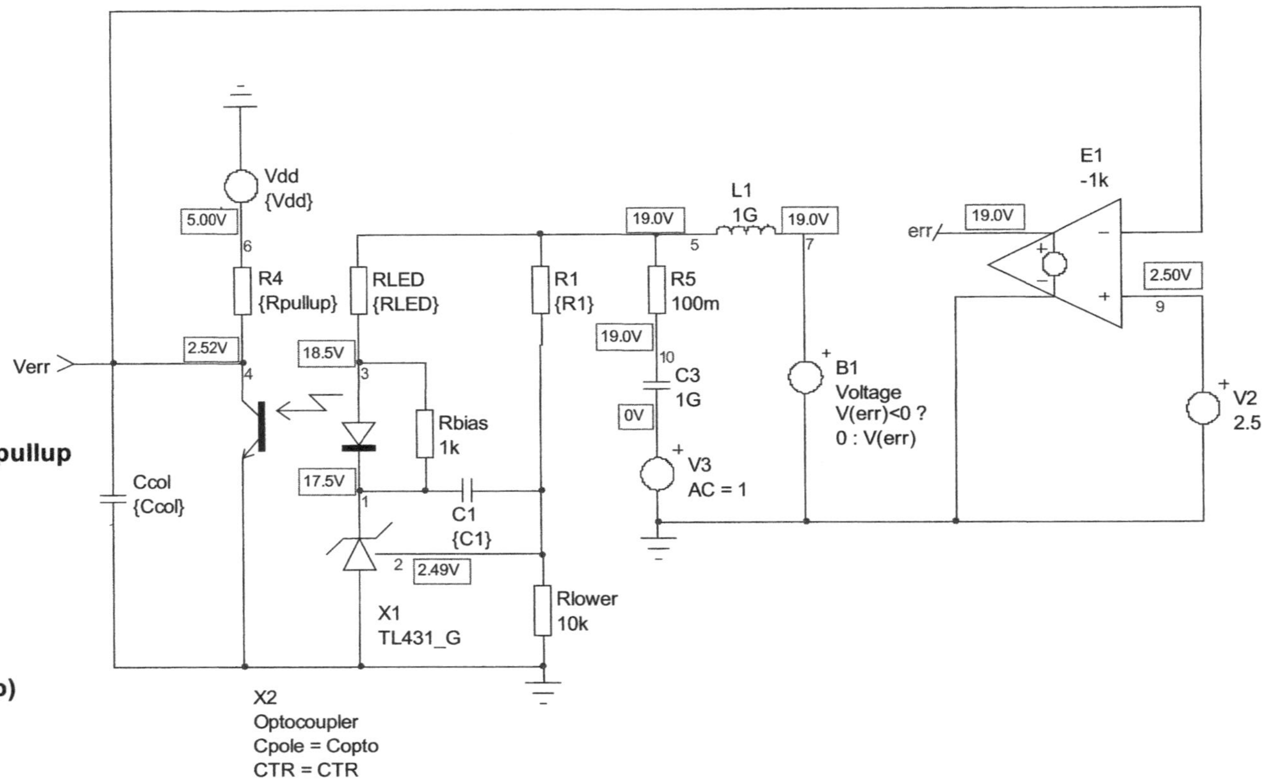

Figure 7.19 The simulation schematic of the TL431 wired in a type 2 configuration.

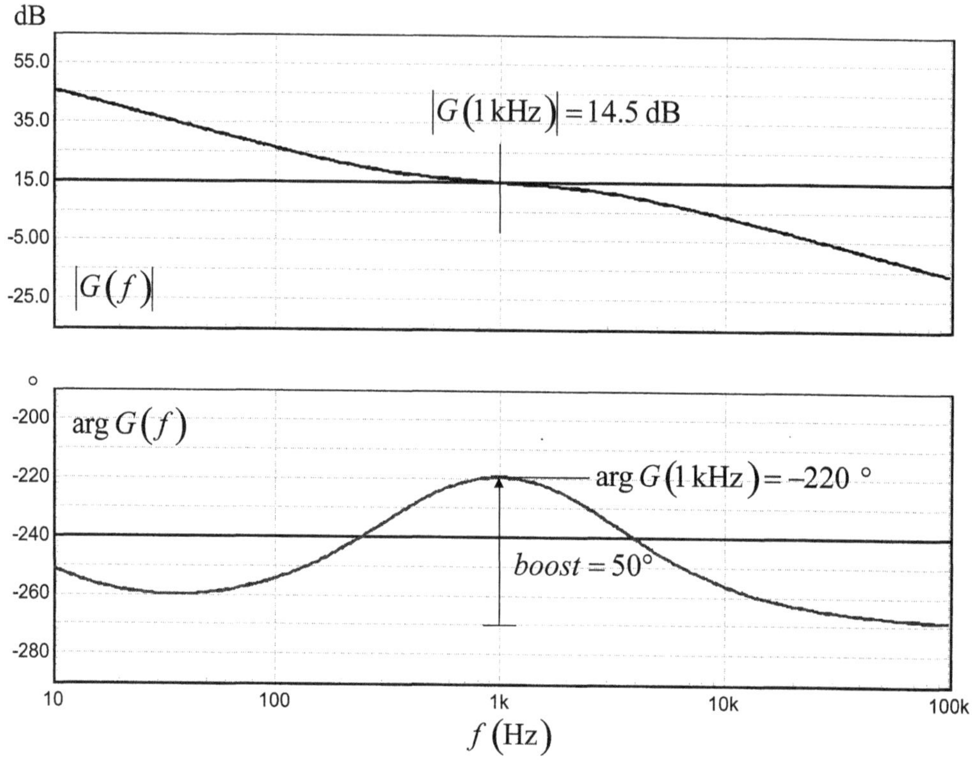

Figure 7.20 The ac results show a slight gain discrepancy, mainly imputable to the LED series resistor and gain deviation brought by the bias resistor.

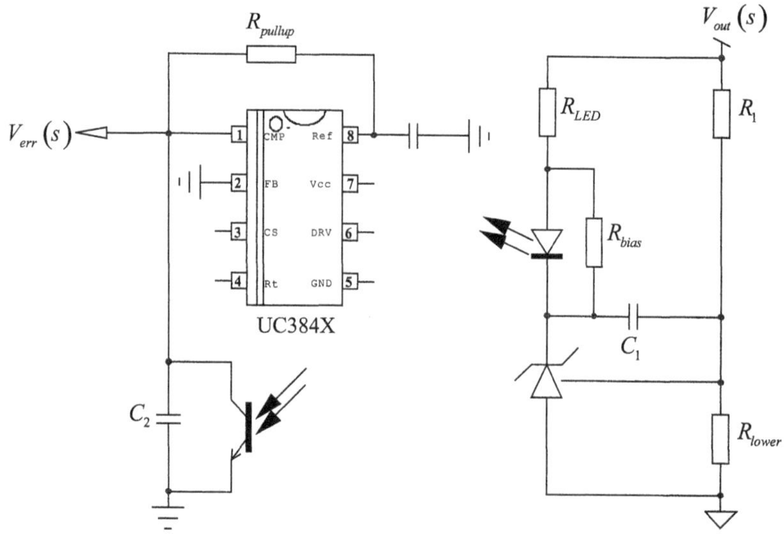

Figure 7.21 By disabling the internal 1-mA source via an extra pull-up resistor connected to the reference voltage, the TL431 can be used as a compensator with a UC384X.

7.11 The Type 2: Common-Collector Configuration and UC384X

The common-collector configuration reverses the control law polarity as indicated by (7.60): if V_{out} deviates from its target by going up, the error voltage increases. In most applications, the error voltage must go down in case of such a deviation. An inversion is thus needed to restore the proper polarity. The internal UC384X op amp can serve this purpose as shown in Figure 7.22. The op amp is wired as a unity gain inverter due to equal resistances values for R_i and R_f. The high-frequency pole is obtained by using C_2 placed across the pull-down resistor. We recommend physically placing this element as close as possible to the PWM controller to minimize noise pickup.

7.12 The Type 2: Disabling the Fast Lane

There are situations where you need flexibility in the selection of the crossover frequency. In other words, you need to select either a gain or an attenuation without being bounded by an upper R_{LED} limit. The way to avoid the limit is to disable the fast lane. As already explained, disabling the fast lane means cutting the parallel path that bypasses the op amp and drives the LED current directly from V_{out}. This is what is illustrated in Figure 7.12 and updated in Figure 7.23.

The transfer function derivation starts by calculating the current circulating in the LED, $I_{LED}(s)$. This current depends on R_{LED} and the TL431 cathode voltage. This is the voltage on the output of the internal op amp, as described by Figure 7.1, arranged in a type 2a configuration:

$$V_{TL431}(s) = -V_{out}(s)\frac{R_2 + \dfrac{1}{sC_1}}{R_1} = -V_{out}(s)\frac{\dfrac{sR_2C_1+1}{sC_1}}{R_1} = -V_{out}(s)\frac{sR_2C_1+1}{sR_1C_1} \tag{7.77}$$

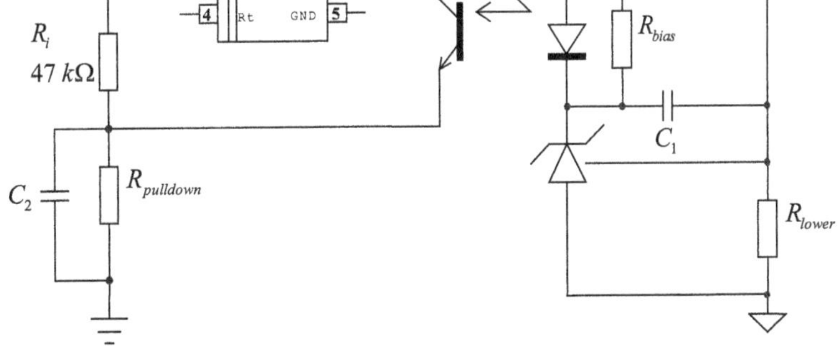

Figure 7.22 A TL431 type 2 wired in a common-collector configuration, driving a UC384X controller whose op amp is wired as an inverter.

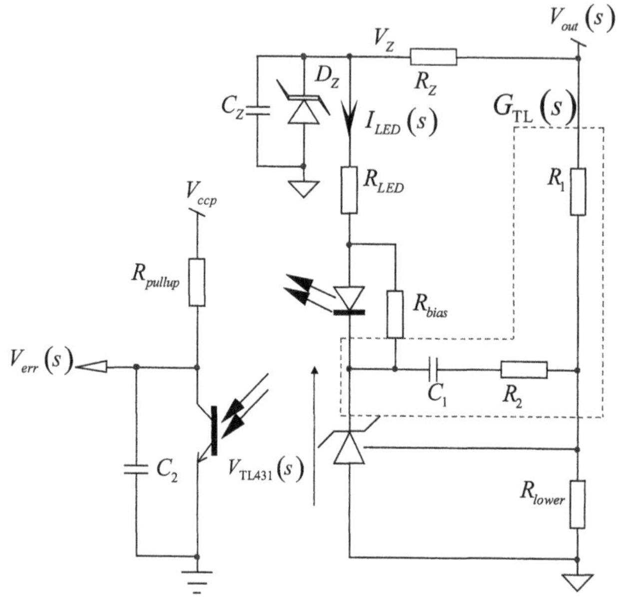

Figure 7.23 The fast lane can be disabled by using a simple Zener-based network or any other type of regulator if needed.

The link to the LED ac current comes easily:

$$I_{LED}(s) = -\frac{V_{TL431}(s)}{R_{LED}} = V_{out}(s)\frac{1}{R_{LED}}\frac{sR_2C_1+1}{sR_1C_1} \tag{7.78}$$

This current is conveyed on the optocoupler collector via its CTR. As the feedback voltage is linked to the circulation of the collector current in the network made of R_{pullup} and C_2 in parallel, we have

$$V_{err}(s) = -I_{LED}(s)\text{CTR}\frac{R_{pullup}\dfrac{1}{sC_2}}{R_{pullup}+\dfrac{1}{sC_2}} = -I_{LED}(s)\text{CTR}\,R_{pullup}\frac{1}{1+sR_{pullup}C_2} \tag{7.79}$$

The complete transfer function of our type 2 compensator in which the fast lane has been disabled is obtained by combining (7.78) and (7.79):

$$G(s) = -\text{CTR}\frac{R_{pullup}}{R_{LED}}\frac{1+sR_2C_1}{sR_1C_1(1+sR_{pullup}C_2)} \tag{7.80}$$

If we factor sR_2C_1, we can put the transfer function under the following normalized form:

$$G(s) = -\text{CTR}\frac{R_{pullup}}{R_{LED}}\frac{R_2}{R_1}\frac{1+1/sR_2C_1}{1+sR_{pullup}C_2} = -G_0\frac{1+\omega_z/s}{1+s/\omega_p} \tag{7.81}$$

where

$$G_0 = \text{CTR}\frac{R_{pullup}}{R_{LED}}\frac{R_2}{R_1} = G_1G_2 \tag{7.82}$$

$$G_1 = \text{CTR}\frac{R_{pullup}}{R_{LED}} \tag{7.83}$$

$$G_2 = \frac{R_2}{R_1} \tag{7.84}$$

$$\omega_z = \frac{1}{R_2 C_1} \tag{7.85}$$

$$\omega_p = \frac{1}{R_{pullup} C_2} \tag{7.86}$$

The difference with (7.62) is that the mid-band gain is no longer set by R_{LED} alone, but also by R_2 and R_1. As R_{LED} is fixed by bias conditions limits—see (7.30)—and R_1 depends on the output voltage setpoint, we have the freedom to adjust R_2 to match our gain requirements. The design thus consists of evaluating the gain block G_1 and checking where to place G_2 so that the product of both gives the mid-band gain we want. This is what the following design example will show.

7.12.1 A Design Example

We want to stabilize a 12-V converter whose transfer function $H(s)$ shows a gain excess of 22 dB at a crossover frequency of 20 Hz. Given the 50° phase boost we need, a type 2 configuration is a must for this design. The component characteristics are the following ones:

$V_{TL431,min}$ = 2.5 V; the TL431 minimum operating voltage.
I_{bias} = 1 mA; the additional TL431 bias current.
V_f = 1 V; the LED forward voltage.
$V_{CE,sat}$ = 0.3 V; the optocoupler saturation voltage.
V_{ccp} = 5 V; the pull-up Vcc level on the primary side.
R_{pullup} = 4.7 kΩ; the optocoupler pull-up resistor.
CTR_{min} = 0.8; the minimum optocoupler current transfer ratio.
V_{out} = 12 V; the converter output voltage.
R_1 = 38 kΩ; the upper resistor in the resistor bridge observing the output variable.
f_{opto} = 10 kHz; the optocoupler pole that has been characterized with R_{pullup}.
V_Z = 8.2 V; the Zener diode breakdown voltage.
I_{Zbias} = 2 mA; the Zener diode bias current.

First, we need to calculate the upper limit for the LED resistor. This is made by using (7.34):

$$R_{LED,\max} \le \frac{8.2-1-2.5}{5-0.3+1m\times 0.8\times 4.7k} 4.7k\times 0.8 \le 2.09\ \text{k}\Omega \tag{7.87}$$

Adopting a 20 percent safety margin, we select the normalized value of 1.5 kΩ. When associated with the CTR and the pull-up resistor, this network offers a gain G_1 equal to

$$G_1 = \text{CTR}\frac{R_{pullup}}{R_{LED}} = 0.8\frac{4.7k}{1.5k} = 2.5 \text{ or} \approx 8 \text{ dB} \tag{7.88}$$

The specification asks for an attenuation of –22 dB. Since we already have a 8-dB gain with G_1, G_2 must be calculated to provide an attenuation of –30 dB:

$$G_2 = 10^{\frac{-30}{20}} = 31.6m \tag{7.89}$$

G_2, as defined by (7.84), involves R_1 and R_2. As R_1 is fixed, we can immediately obtain the value of R_2:

$$R_2 = G_2 R_1 = 31.6m \times 38k = 1.2 \text{ k}\Omega \tag{7.90}$$

Where do we place the pole and the zero to obtain the 50° phase boost at 20 Hz? Applying the formulas already derived, we have

$$f_p = \left[\tan(boost) + \sqrt{\tan^2(boost) + 1}\right] f_c = 2.74 \times 20 = 54.8 \text{ Hz} \tag{7.91}$$

As the phase peaks at the geometric mean between the pole and the zero, this latter is placed at

$$f_z = \frac{f_c^{\,2}}{f_p} = \frac{400}{54.8} = 7.3 \text{ Hz} \tag{7.92}$$

Using (7.85) and (7.86), we can calculate the missing capacitor values:

$$C_1 = \frac{1}{2\pi f_z R_2} = \frac{1}{6.28 \times 7.3 \times 1.2k} = 18.2\ \mu\text{F} \tag{7.93}$$

We know that the pole capacitor C_2 is actually made of an extra capacitor C_{col} that we will place in parallel with the collector-emitter parasitic capacitor from the optocoupler:

$$C_2 = \frac{1}{2\pi R_{pullup} f_p} = \frac{1}{6.28 \times 4.7k \times 54.8} = 618.3 \text{ nF} \tag{7.94}$$

Given the characterized optocoupler pole at 10 kHz, the parasitic capacitor is found to be

$$C_{opto} = \frac{1}{2\pi f_{opto} R_{pullup}} = \frac{1}{6.28 \times 10k \times 4.7k} = 3.4 \text{ nF} \tag{7.95}$$

As the total capacitance is 2.9 nF, the added capacitor C_{col} is simply the difference between (7.74) and (7.75):

$$C_{col} = C_2 - C_{opto} = 618.3n - 3.4n = 615 \text{ nF} \tag{7.96}$$

It is however likely that a 0.68-μF capacitor will be used; therefore, the optocoupler contribution can be neglected.

Let us now have a look at the Zener network. Given the 12-V output, an 8.2-V Zener diode seems an appropriate pick. The choice of the Zener value is dictated by several arguments: the voltage needs to be sufficiently low compared to the output voltage in order to ensure a good ac isolation between the two dc lines. On the other hand, the Zener voltage must be high enough to offer sufficient bias headroom to the TL431 network. Experience shows that choosing V_Z as being two thirds of the output voltage represents a good tradeoff. By looking at the Zener diode specifications, a 2-mA bias current seems like the correct number to make the device operate far enough from its knee. With these parameters now known, we can calculate the value of the dropping resistor R_Z with the help of (7.37):

$$R_Z = \frac{(V_{out} - V_Z)R_{pullup}\mathrm{CTR}_{\min}}{(V_{ccp} - V_{CE,sat}) + (I_{Zbias} + I_{bias})R_{pullup}\mathrm{CTR}_{\min}} \tag{7.97}$$

$$= \frac{(12-8.2)\times 4.7k\times 0.8}{(5-0.3)+(2m+1m)\times 4.7k\times 0.8} = 894\ \Omega$$

To improve the decoupling action at low frequencies, a 10-μF capacitor is added in parallel with the Zener diode. We have everything now, and we can proceed with the application schematic shown in Figure 7.24. The dc values appearing on the sketch confirm the system is well biased. Let us have a look at the ac response, which is delivered in Figure 7.25.

As we can observe on the graph, the –22-dB target is missed by 3 dB. Despite the small error brought by the LED dynamic resistor R_d, which is neglected in the calculation (around 0.5 dB), the lack of perfect ac decoupling between the regulated voltage V_Z and the output level is the main offender. The mismatch can be compensated by increasing the target by 3-4 dB (e.g., asking for a 26-dB gain rather than the original 22 dB). Another option consists of using an active regulator based on a bipolar, as exemplified in Figure 7.26. Using this technique, the ac analysis shows an error mismatch reduced by 1 dB. This is what Figure 7.27 confirms, comparing both ac responses between Zener-based and a bipolar-based regulators.

As a conclusion, we can see that ac decoupling the fast lane affects the transfer function. Therefore, you must carefully design the ac decoupling circuitry if you plan to suppress the fast lane.

7.13 The Type 3: An Origin Pole plus a Double Pole/Zero Pair

Because of the fast lane presence, the TL431 does not lend itself very well to a type 3 architecture. As we will see in a few lines, the LED series resistor plays two roles, setting new limits on the design procedure. However, in some voltage-mode converters, there is a need for a phase boost higher than 90° where the type 3 with a TL431 still represents a possible candidate. Its architecture appears in Figure 7.28.

The derivation starts by the ac LED current definition, involving the impedance delimited as Z_{LED} in Figure 7.28. The ac current deviated through the bias resistor R_{bias} is purposely neglected:

```
parameters

Vout=12

R1=(Vout-2.5)/250u
fc=20
Gfc=22
boost=50

Vf=1
Ibias=1m
Vref=2.5
VCEsat=300m
Vccp=5
Vz=8.2
IZbias=1m
Rpullup=4.7k
Fopto=10k
Copto=1/(2*pi*Rpullup*Fopto)
CTR=0.8

G1=Rpullup*CTR/RLED
G2=10^(-Gfc/20)
G=G2/G1
pi=3.14159
C1=1/(2*pi*fz*R2)
C2=1/(2*pi*fp*Rpullup)
Ccol=C2-Copto
R2=G*R1
```

```
parameters

Rmax=((Vz-Vf-Vref)/(Vccp-VCEsat+Ibias*(Rpullup*CTR)))*Rpullup*CTR
RLED=Rmax*0.8
Za=(Vout-Vz)*Rpullup*CTR
Zb=(Vccp-VCEsat)+(Ibias+IZbias)*Rpullup*CTR
Rz=Za/Zb
fp=(tan(boost*pi/180)+sqrt((tan(boost*pi/180))^2+1))*fc
fz=fc^2/fp
```

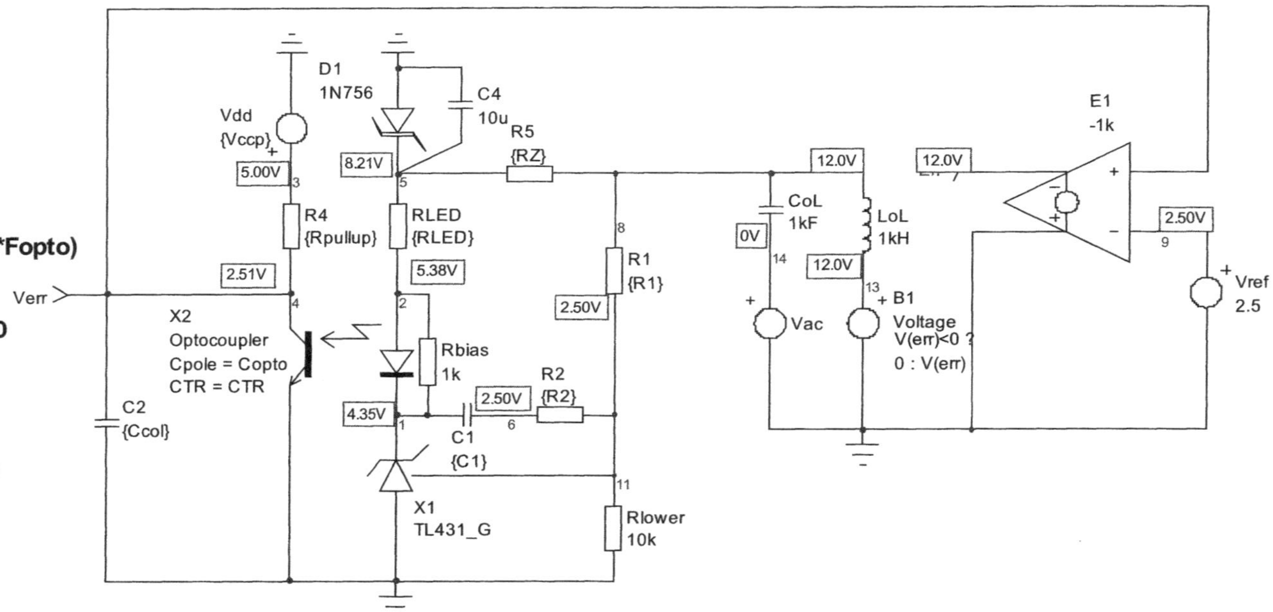

Figure 7.24 The application schematic of the TL431-based type 2 compensator where the fast lane has been disabled.

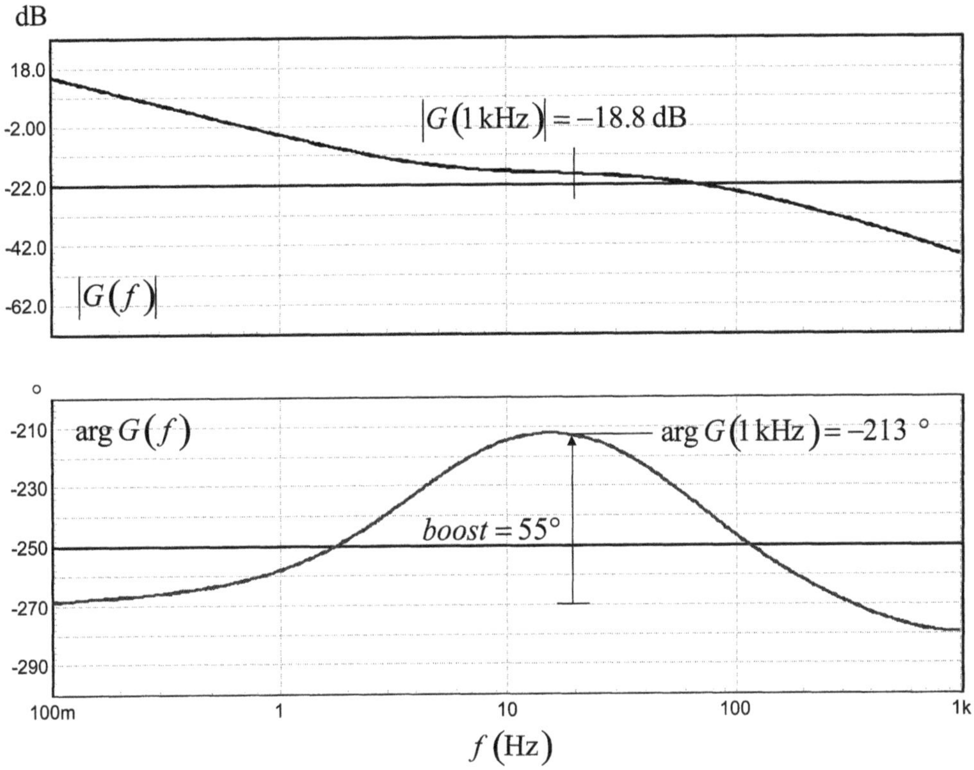

Figure 7.25 The dc response of the Zener-based schematic shows a target mismatch of 3 dB.

$$I_{LED}(s)=\frac{V_{out}(s)-V_{\mathrm{TL431}}(s)}{Z_{LED}}=(V_{out}(s)-V_{\mathrm{TL431}}(s))\frac{R_{LED}+\left(R_3+\frac{1}{sC_3}\right)}{R_{LED}\left(R_3+\frac{1}{sC_3}\right)} \tag{7.98}$$

The output voltage on the TL431 cathode, $V_{\mathrm{TL431}}(s)$, is that of a type 1 compensator:

$$V_{\mathrm{TL431}}(s)=-\frac{1}{sC_1R_1}V_{out}(s) \tag{7.99}$$

Substituting (7.99) into (7.98), we obtain

$$\begin{aligned}I_{LED}(s)&=\left(V_{out}(s)+V_{out}(s)\frac{1}{sR_1C_1}\right)\frac{R_{LED}+\left(R_3+\frac{1}{sC_3}\right)}{R_{LED}\left(R_3+\frac{1}{sC_3}\right)}\\&=V_{out}(s)\left(\frac{1+sR_1C_1}{sR_1C_1}\right)\frac{R_{LED}+\left(R_3+\frac{1}{sC_3}\right)}{R_{LED}\left(R_3+\frac{1}{sC_3}\right)}\end{aligned} \tag{7.100}$$

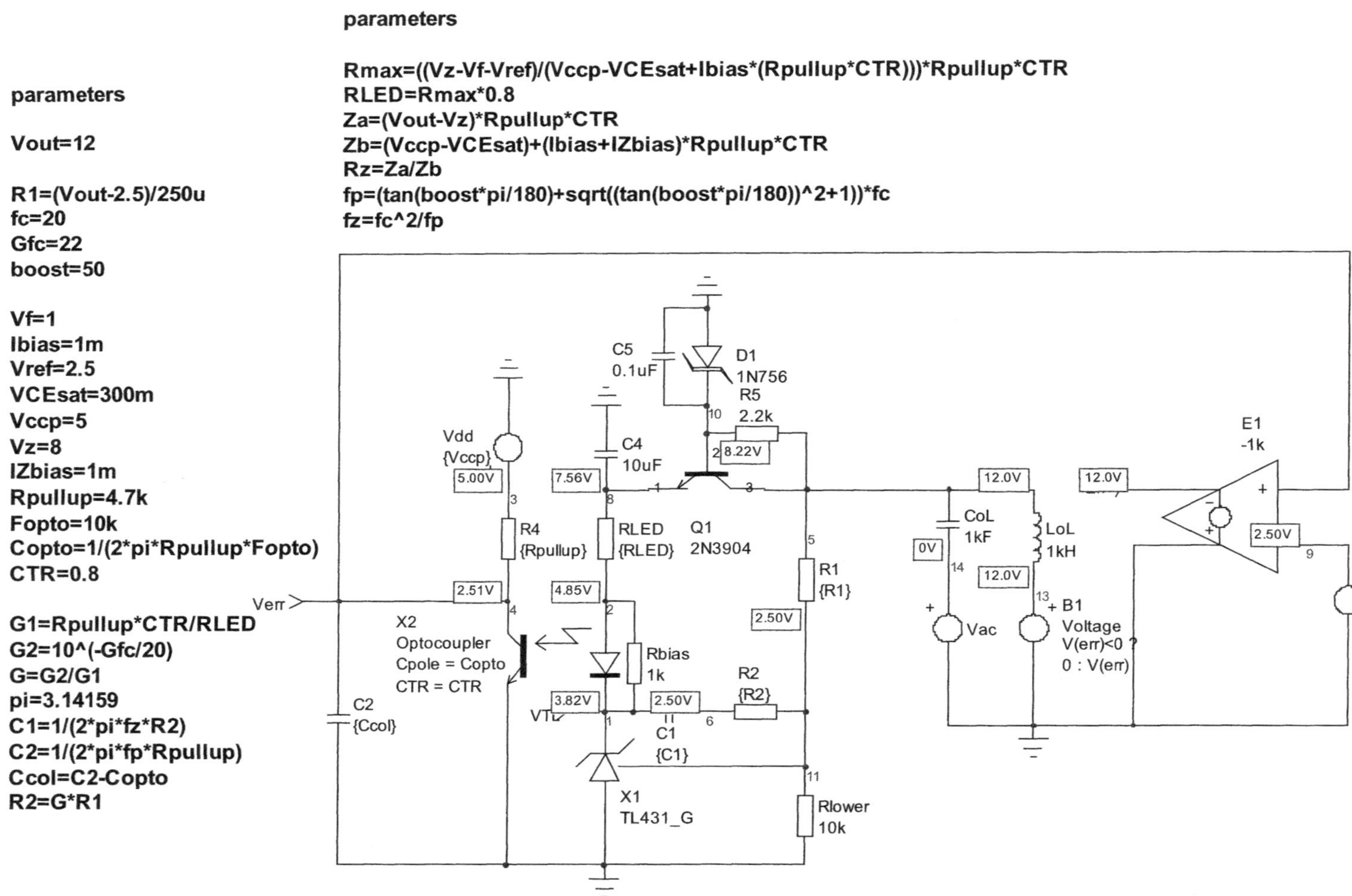

Figure 7.26 A bipolar transistor helps to better reject the ac modulation superimposed on the monitored output voltage.

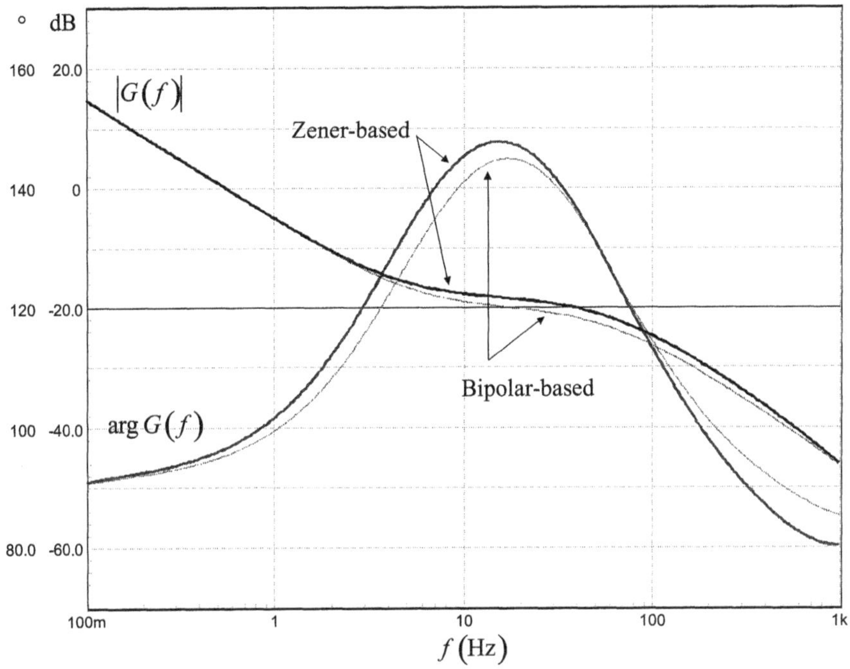

Figure 7.27 Ensuring a better ac decoupling between the TL431 bias and the observed variable clearly improves the overall shape.

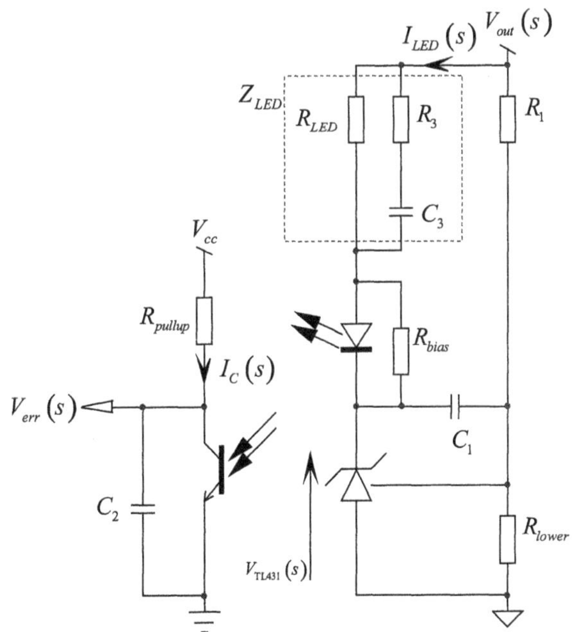

Figure 7.28 The TL431 can be wired in a type 3 configuration; however, the fast lane presence hampers its field of applications.

Rearranging the previous, we have

$$I_{LED}(s) = V_{out}(s)\left(\frac{1+sR_1C_1}{sR_1C_1}\right)\frac{s(R_{LED}+R_3)C_3+1}{R_{LED}(sC_3R_3+1)} \tag{7.101}$$

Now realizing that the error voltage $V_{err}(s)$ is linked to the LED current by the optocoupler CTR, we have

$$V_{err}(s) = -I_{LED}(s)R_{pullup}\frac{1}{1+sR_{pullup}C_2}\text{CTR} \tag{7.102}$$

Substituting (7.101) into (7.102) and rearranging, we obtain the transfer function we are looking for:

$$\frac{V_{err}(s)}{V_{out}(s)} = -\frac{R_{pullup}}{R_{LED}}\text{CTR}\frac{1+1/sR_1C_1}{1+sR_{pullup}C_2}\frac{sC_3(R_{LED}+R_3)+1}{1+sC_3R_3} = -G_0\frac{1+\omega_{z_1}/s}{1+s/\omega_{p_1}}\frac{1+s/\omega_{z_2}}{1+s/\omega_{p_2}} \tag{7.103}$$

In this equation, we can identify poles, zeros, and a static gain:

$$G_0 = \frac{R_{pullup}}{R_{LED}}\text{CTR} \tag{7.104}$$

$$\omega_{z_1} = \frac{1}{R_1C_1} \tag{7.105}$$

$$\omega_{z_2} = \frac{1}{(R_{LED}+R_3)C_3} \tag{7.106}$$

$$\omega_{p_1} = \frac{1}{R_{pullup}C_2} \tag{7.107}$$

$$\omega_{p_2} = \frac{1}{R_3C_3} \tag{7.108}$$

The magnitude of (7.103) is given by

$$|G(f_c)| = G_0\frac{\sqrt{1+\left(\frac{f_{z_1}}{f_c}\right)^2}\sqrt{1+\left(\frac{f_c}{f_{z_2}}\right)^2}}{\sqrt{1+\left(\frac{f_c}{f_{p_1}}\right)^2}\sqrt{1+\left(\frac{f_c}{f_{p_2}}\right)^2}} = \frac{R_{pullup}}{R_{LED}}\text{CTR}\frac{\sqrt{1+\left(\frac{f_{z_1}}{f_c}\right)^2}\sqrt{1+\left(\frac{f_c}{f_{z_2}}\right)^2}}{\sqrt{1+\left(\frac{f_c}{f_{p_1}}\right)^2}\sqrt{1+\left(\frac{f_c}{f_{p_2}}\right)^2}} \tag{7.109}$$

Then, combining (7.109), (7.105), (7.106), and (7.108) and solving for R_{LED}, R_3, C_1, C_2, and C_3, we can extract the definitions we need to design the compensator:

$$R_{LED} = \frac{R_{pullup}}{G}\text{CTR}\frac{\sqrt{1+\left(\frac{f_{z_1}}{f_c}\right)^2}\sqrt{1+\left(\frac{f_c}{f_{z_2}}\right)^2}}{\sqrt{1+\left(\frac{f_c}{f_{p_1}}\right)^2}\sqrt{1+\left(\frac{f_c}{f_{p_2}}\right)^2}} \tag{7.110}$$

$$C_3 = \frac{f_{p2} - f_{z2}}{2\pi f_{p2} f_{z2}} \frac{G}{R_{pullup}\mathrm{CTR}} \frac{\sqrt{1+\left(\frac{f_c}{f_{p1}}\right)^2}\sqrt{1+\left(\frac{f_c}{f_{p2}}\right)^2}}{\sqrt{1+\left(\frac{f_{z1}}{f_c}\right)^2}\sqrt{1+\left(\frac{f_c}{f_{z2}}\right)^2}} \tag{7.111}$$

$$R_3 = \frac{f_{z2}}{f_{p2} - f_{z2}} \frac{R_{pullup}}{G} \mathrm{CTR} \frac{\sqrt{1+\left(\frac{f_{z1}}{f_c}\right)^2}\sqrt{1+\left(\frac{f_c}{f_{z2}}\right)^2}}{\sqrt{1+\left(\frac{f_c}{f_{p1}}\right)^2}\sqrt{1+\left(\frac{f_c}{f_{p2}}\right)^2}} \tag{7.112}$$

$$C_1 = \frac{1}{2\pi f_{z_1} R_1} \tag{7.113}$$

$$C_2 = \frac{1}{2\pi R_{pullup} f_{p2}} \tag{7.114}$$

Unfortunately, as with any TL431-based designs, the limit for the LED series resistor values is set by the biasing conditions around the component. We have already derived the maximum value this resistor can take via (7.30). If we combine this definition with the gain expressed in (7.109), we obtain the minimum gain the compensator can be designed for:

$$|G_{\min}| = \frac{V_{cc} - V_{CE,sat} + I_{bias} R_{pullup} \mathrm{CTR}_{\min}}{V_{out} - V_f - V_{\mathrm{TL431,min}}} \frac{\sqrt{1+\left(\frac{f_{z1}}{f_c}\right)^2}\sqrt{1+\left(\frac{f_c}{f_{z2}}\right)^2}}{\sqrt{1+\left(\frac{f_c}{f_{p1}}\right)^2}\sqrt{1+\left(\frac{f_c}{f_{p2}}\right)^2}} \tag{7.115}$$

This $G_{\min}$ term illustrates the crossover gain below which the compensator cannot be designed. For instance, should you find a $G_{\min}$ of 5 dB with (7.115), you could simply not select a crossover frequency where a 3-dB gain is needed. It would not satisfy the biasing conditions for the TL431. However, 6 dB would. Changing the optocoupler CTR or R_{pullup} will not help to modify the result. Actually, (7.115) is made of two terms: the first one (involving V_{TL431}, V_{out}, and so on) corresponds to the minimum biasing conditions needed to operate the TL431-based compensator. The second term involves the poles and zeros position—in other words, the needed phase boost. You can therefore see $G_{\min}$ as another limit for the maximum phase boost you can ask for a selected gain. We have derived an equation linking $G_{\min}$ and the phase boost you want. The result appears next:

$$G_{\min}(boost) = 20\log_{10}\left(\frac{V_{cc} - V_{CE,sat} + I_{bias} R_{pullup} \mathrm{CTR}_{\min}}{V_{out} - V_f - V_{\mathrm{TL431,min}}} \sqrt{\cot^2\left(\frac{boost}{4} - 45\right)}\right) \tag{7.116}$$

If we consider the following standard values for the first term of the expression, we can plot (7.116), sweeping the phase boost between 0 and 180°:

V_{TL431} = 2.5 V; the TL431 minimum operating voltage.
V_{out} = 12 V; the regulated output voltage.
V_f = 1 V; the LED forward voltage.
$V_{CE,sat}$ = 0.3 V; the optocoupler saturation voltage.
V_{cc} = 5 V; the pull-up V_{cc} level.
I_{bias} = 1 mA; the TL431 extra biasing current.
R_{pullup} = 20 kΩ; the pull-up resistor loading the optocoupler collector.
CTR_{min} = 0.3; the CTR associated with the 20-kΩ pull-up resistor.

The result appears in Figure 7.29 and a shows a gain starting from a minimum value close to 0 dB (a null phase boost when both poles and zeros are coincident) and going up as the required phase boost increases. Therefore, when using this type of compensator, it will be important to choose a crossover frequency where the needed gain and phase boost do not conflict with (7.115). Should you need a phase boost of 150°, given the adopted component values, the graph instructs us that any crossover point requiring more than 18 dB of gain are possible candidates.

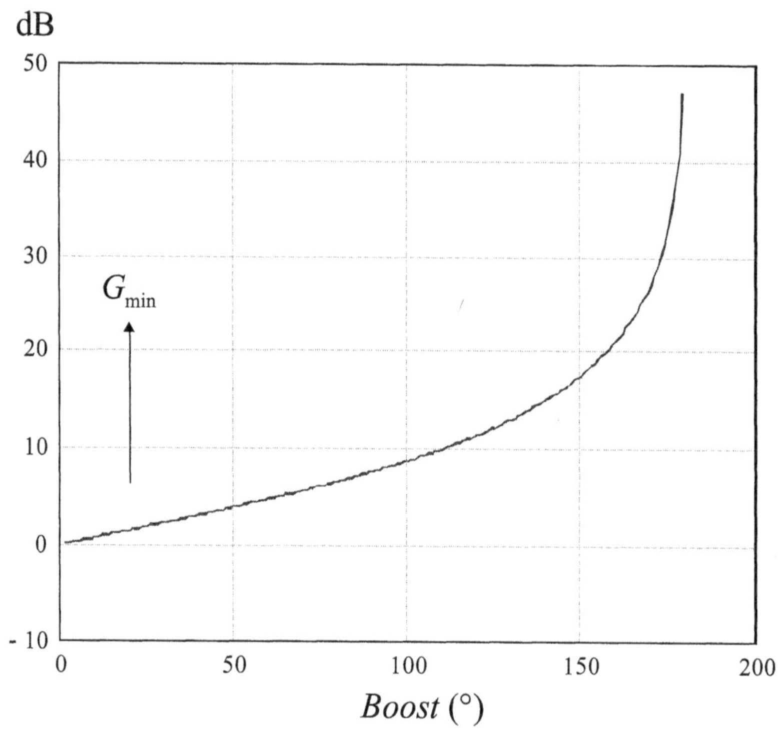

Figure 7.29 The burden of the LED series resistor seriously hampers the gain choices as the required phase boost increases.

7.13.1 A Design Example

In this example, we want to cross over at a 1-kHz frequency with a 15-dB amplification. The needed phase boost is 120°. The parameter list appears next; this is a 12-V converter:

$V_{TL431,min}$ = 2.5 V; the TL431 minimum operating voltage.
I_{bias} = 1 mA; the TL431 biasing current.
V_Z = 8.2 V; the Zener diode breakdown voltage.
I_{Zbias} = 2 mA; the Zener diode bias current.
V_f = 1 V; the LED forward voltage.
$V_{CE,sat}$ = 0.3 V; the optocoupler saturation voltage.
V_{cc} = 5 V; the pull-up V_{cc} level.
V_{out} = 12 V; the converter output voltage.
R_{pullup} = 20 kΩ; the optocoupler pull-up resistor.
CTR_{min} = 0.3; the minimum optocoupler current transfer ratio.
R_1 = 38 kΩ; the upper resistor in the resistor bridge observing the output variable.
f_{opto} = 6 kHz; the optocoupler pole that has been characterized with R_{pullup}.

Given the needed 120° phase boost at 1 kHz, we first assume to place coincident poles and zeros at a place to be determined. Capitalizing on what has already been derived, we will place a pole pair at the following position:

$$f_{p1,2} = \frac{f_c}{\tan\left(45 - \frac{boost}{4}\right)} = \frac{1k}{268m} \approx 3.7 \text{ kHz} \tag{7.117}$$

The double zero will be located at

$$f_{z1,2} = \frac{f_c^2}{f_p} = \frac{1k^2}{3.7k} \approx 270 \text{ Hz} \tag{7.118}$$

The maximum allowable series LED resistor is first evaluated to check the upper limit:

$$R_{LED,\max} \le \frac{12 - 1 - 2.5}{5 - 0.3 + 1m \times 0.3 \times 20k} 20k \times 0.3 \le 4.8 \text{ k}\Omega \tag{7.119}$$

The type 3 passive elements are calculated using (7.110) through (7.114):

$$R_{LED} = \frac{20k}{10^{\frac{15}{20}}} \times 0.3 \times \frac{\sqrt{1+\left(\frac{270}{1k}\right)^2}\sqrt{1+\left(\frac{1k}{270}\right)^2}}{\sqrt{1+\left(\frac{1k}{3.7k}\right)^2}\sqrt{1+\left(\frac{1k}{3.7k}\right)^2}} \approx 4 \text{ k}\Omega \tag{7.120}$$

Checking this value against that given by (7.119), we are below the limit, leaving us a little margin.

$$C_3 = \frac{3.7k - 270}{6.28 \times 3.7k \times 270} \times \frac{10^{\frac{15}{20}}}{20k \times 0.3} \times \frac{\sqrt{1+\left(\frac{1k}{3.7k}\right)^2}\sqrt{1+\left(\frac{1k}{3.7k}\right)^2}}{\sqrt{1+\left(\frac{270}{1k}\right)^2}\sqrt{1+\left(\frac{1k}{270}\right)^2}} = 138 \text{ nF} \quad (7.121)$$

$$R_3 = \frac{270}{3.7k - 270} \times \frac{20k}{10^{\frac{15}{20}}} \times 0.3 \times \frac{\sqrt{1+\left(\frac{270}{1k}\right)^2}\sqrt{1+\left(\frac{1k}{270}\right)^2}}{\sqrt{1+\left(\frac{1k}{3.7k}\right)^2}\sqrt{1+\left(\frac{1k}{3.7k}\right)^2}} = 308\ \Omega \quad (7.122)$$

$$C_1 = \frac{1}{2\pi f_{z_1} R_1} = \frac{1}{6.28 \times 270 \times 38k} = 15.5 \text{ nF} \quad (7.123)$$

$$C_2 = \frac{1}{2\pi R_{pullup} f_{p_2}} = \frac{1}{6.28 \times 20k \times 3.7k} = 2.15 \text{ nF} \quad (7.124)$$

We know that the total capacitor connected across the optocoupler that forms C_2 is actually made of the optocoupler parasitic pole C_{opto} and the added capacitor C_{col}. The optocoupler pole is located at 6 kHz. Given a pull-up resistor of 20 kΩ, we can calculate its parasitic contribution:

$$C_{opto} = \frac{1}{2\pi f_{opto} R_{pullup}} = \frac{1}{6.28 \times 6k \times 20k} = 1.3 \text{ nF} \quad (7.125)$$

When subtracted from (7.124), you obtain the value for C_{col}, the final capacitor installed across the optocoupler:

$$C_{col} = C_2 - C_{opto} = 2.15n - 1.3n \approx 850 \text{ pF} \quad (7.126)$$

We have everything on hand, we can now check the resulting ac response via our test fixture. It appears in Figure 7.30.

The simulation results are available in Figure 7.31 and show a slight gain discrepancy, explained, as usual, by the LED diode dynamic resistor added to the ac current deviated by the 1-kΩ bias resistor.

7.14 The Type 3: An Origin Pole plus a Double Pole/Zero Pair—No Fast Lane

In the previous example, we have seen that the LED resistor was slightly below its maximum value, leaving a weak design margin. If we had needed a lower gain or a

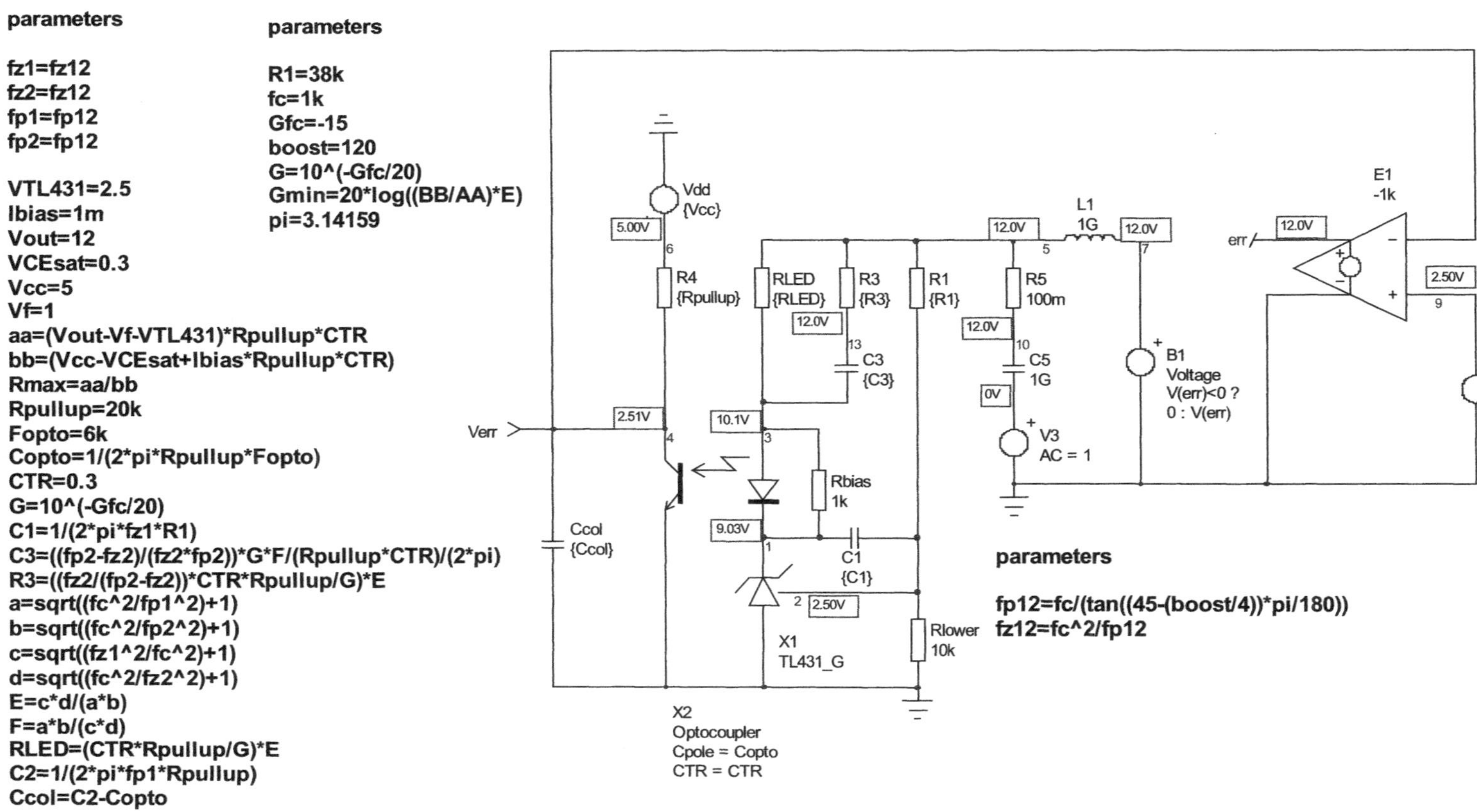

Figure 7.30 The type 3 made with a TL431 can work, but the phase boost and the allowable gain are intimately linked, hampering the flexibility of use.

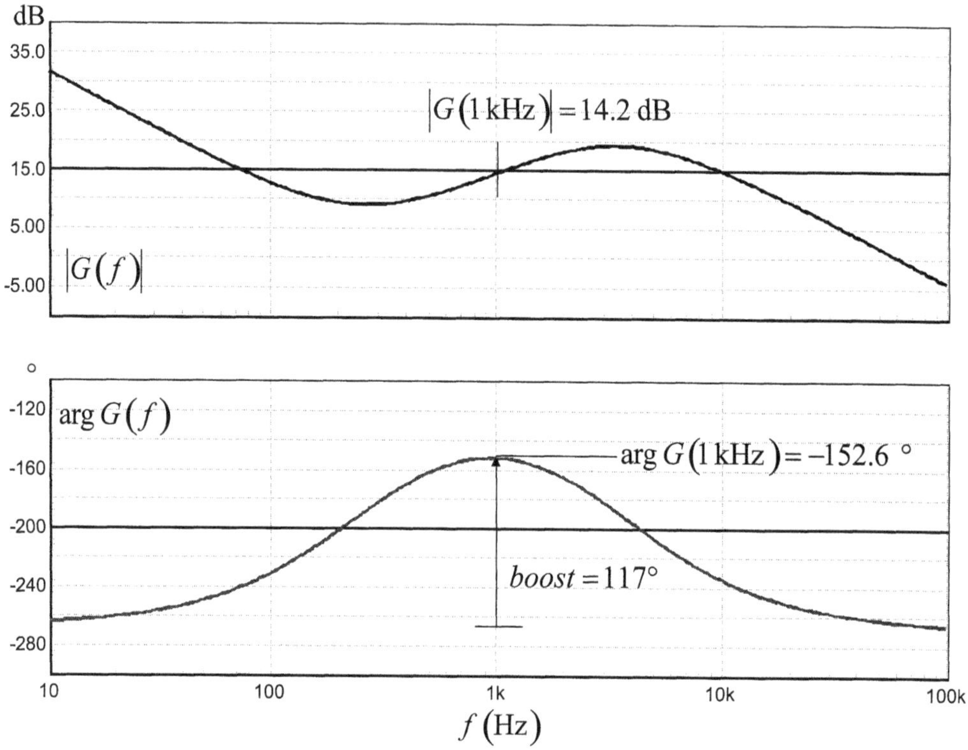

Figure 7.31 The ac response shows results that agree very well with the initial target.

larger phase boost at crossover, we would have hit the limit imposed by the biasing conditions around the TL431. Such problems can be easily overcome by disabling the fast lane. The type 3 with a disabled fast lane appears in Figure 7.32.

The schematic reveals a cascaded configuration where a kind of type 3a circuit appears in the $G_{TL}(s)$ transfer function (2 zeros, 1 origin pole and 1 pole), immediately followed by the optocoupler path, exhibiting a transfer function $O(s)$ affected by a gain and a pole. This latter is now well known and can be expressed as follows:

$$O(s) = \frac{V_{err}(s)}{V_{TL431}(s)} = \frac{R_{pullup}\text{CTR}}{R_{LED}} \frac{1}{1 + sR_{pullup}C_2} \tag{7.127}$$

The ac voltage across the TL431 is the ac output voltage multiplied by the ratio of impedances Z_f and Z_i, respectively, involving C_1R_2 and $R_1R_3C_3$:

$$Z_f = R_2 + \frac{1}{sC_1} = \frac{sR_2C_1 + 1}{sC_1} \tag{7.128}$$

$$Z_i = \left(\frac{1}{sC_3} + R_3\right) \| R_1 = \frac{R_1(1 + sR_3C_3)}{sC_3(R_1 + R_3) + 1} \tag{7.129}$$

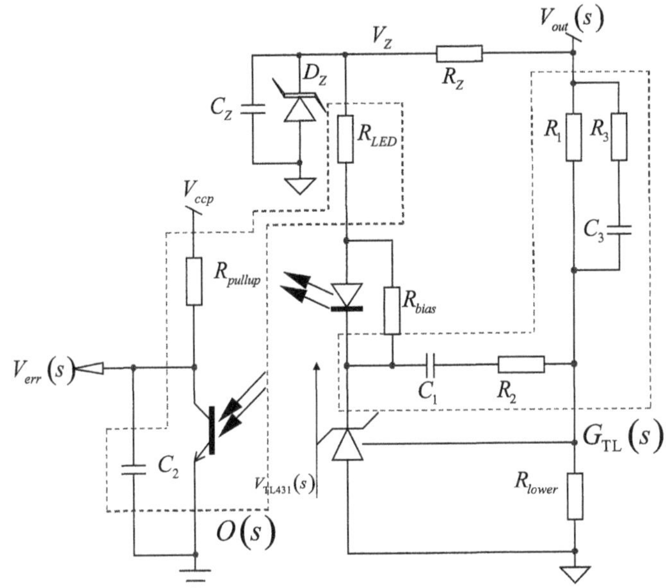

Figure 7.32 The fast lane must be disabled to obtain the best from the type 3 built with a TL431.

$$G_{TL}(s) = -\frac{Z_f}{Z_i} = -\frac{\dfrac{sR_2C_1+1}{sC_1}}{\dfrac{R_1(1+sR_3C_3)}{sC_3(R_1+R_3)+1}} \tag{7.130}$$

$$= -\frac{sR_2C_1+1}{sR_1C_1}\frac{sC_3(R_1+R_3)+1}{(1+sR_3C_3)} = -\frac{R_2}{R_1}\frac{\left(1+\dfrac{1}{sR_2C_1}\right)(1+sC_3(R_1+R_3))}{1+sR_3C_3}$$

Once (7.130) and (7.127) are multiplied together and the voltage to current gain from (7.78) is included, we obtain the complete transfer function we are looking for:

$$G(s) = -\frac{R_2R_{pullup}\text{CTR}}{R_{LED}R_1}\frac{\left(1+\dfrac{1}{sR_2C_1}\right)(1+sC_3(R_1+R_3))}{(1+sR_{pullup}C_2)(1+sR_3C_3)} \tag{7.131}$$

This expression can be put under the normalized form:

$$G(s) = -G_0\frac{\left(1+\dfrac{\omega_{z_1}}{s}\right)\left(1+\dfrac{s}{\omega_{z_2}}\right)}{\left(1+\dfrac{s}{\omega_{p_1}}\right)\left(1+\dfrac{s}{\omega_{p_2}}\right)} \tag{7.132}$$

where

$$G_0 = \text{CTR}\frac{R_{pullup}}{R_{LED}}\frac{R_2}{R_1} = G_1G_2 \tag{7.133}$$

$$G_1 = \text{CTR}\frac{R_{pullup}}{R_{LED}} \tag{7.134}$$

$$G_2 = \frac{R_2}{R_1} \tag{7.135}$$

$$\omega_{z_1} = \frac{1}{R_2 C_1} \tag{7.136}$$

$$\omega_{z_2} = \frac{1}{C_3(R_1 + R_3)} \tag{7.137}$$

$$\omega_{p_1} = \frac{1}{R_3 C_3} \tag{7.138}$$

$$\omega_{p_2} = \frac{1}{R_{pullup} C_2} \tag{7.139}$$

The magnitude of (7.132) is given by

$$|G(f_c)| = \text{CTR}\frac{R_{pullup}}{R_{LED}}\frac{R_2}{R_1}\frac{\sqrt{1+\left(\frac{f_{z1}}{f_c}\right)^2}\sqrt{1+\left(\frac{f_c}{f_{z2}}\right)^2}}{\sqrt{1+\left(\frac{f_c}{f_{p1}}\right)^2}\sqrt{1+\left(\frac{f_c}{f_{p2}}\right)^2}} \tag{7.140}$$

Then, combining and solving for R_{LED}, R_3, C_2, C_1, and C_3, we can extract the definitions we are looking for:

$$R_2 = \frac{GR_1R_{LED}}{R_{pullup}\text{CTR}}\frac{\sqrt{1+\left(\frac{f_c}{f_{p1}}\right)^2}\sqrt{1+\left(\frac{f_c}{f_{p2}}\right)^2}}{\sqrt{1+\left(\frac{f_{z1}}{f_c}\right)^2}\sqrt{1+\left(\frac{f_c}{f_{z2}}\right)^2}} \tag{7.141}$$

$$R_3 = \frac{R_1 f_{z2}}{f_{p2} - f_{z2}} \tag{7.142}$$

$$C_1 = \frac{1}{2\pi f_{z_1} R_2} \tag{7.143}$$

$$C_2 = \frac{1}{2\pi f_{p_2} R_{pullup}} \tag{7.144}$$

$$C_3 = \frac{f_{p2} - f_{z2}}{2\pi R_1 f_{p2} f_{z2}} \tag{7.145}$$

With the fast lane removed, the LED resistor no longer plays a role in the pole/zero positions. Its value now solely depends on acceptable bias conditions to give the TL431 the necessary voltage swing, despite unavoidable dispersions on the optocoupler CTR. The formula derived in (7.30) is valid:

$$R_{LED,\max} \leq \frac{V_Z - V_f - V_{\text{TL431,min}}}{V_{cc} - V_{CE,sat} + I_{bias}\text{CTR}_{\min}R_{pullup}} R_{pullup}\text{CTR}_{\min} \tag{7.146}$$

In this expression, the V_Z term represents the selected Zener level obtained from the output. We already derived its value and it appears next:

$$R_Z = \frac{(V_{out} - V_Z)R_{pullup}\text{CTR}_{\min}}{(V_{cc} - V_{CE,sat}) + (I_{Zbias} + I_{bias})R_{pullup}\text{CTR}_{\min}} \tag{7.147}$$

All component values are now attached to a design formula, so it is time for the design example!

7.14.1 A Design Example

In this example, we have to cross over at a 1-kHz frequency where a –10-dB attenuation is needed. At the crossover point, the phase must be boosted by 130°. The parameter list appears next and does not change.

$V_{\text{TL431,min}}$ = 2.5 V; the TL431 minimum operating voltage.
I_{bias} = 1 mA; the TL431 biasing current.
V_Z = 8.2 V; the Zener diode breakdown voltage.
I_{Zbias} = 2 mA; the Zener diode biasing current.
V_f = 1 V; the LED forward voltage.
$V_{CE,sat}$ = 0.3 V; the optocoupler saturation voltage.
V_{cc} = 5 V; the pull-up V_{cc} level.
V_{out} = 12 V; the converter output voltage.
R_{pullup} = 20 kΩ; the optocoupler pull-up resistor.
$\text{CTR}_{\min}$ = 0.3; the minimum optocoupler current transfer ratio.
R_1 = 38 kΩ; the upper resistor in the resistor bridge observing the output variable.
f_{opto} = 6 kHz; the optocoupler pole that has been characterized with R_{pullup}.

Given the needed 130° phase boost at 1 kHz, we first assume we must place coincident poles and zeros at a location to be determined. Capitalizing on what has already been derived, we will place a pole pair at the following position:

$$f_{p_{1,2}} = \frac{f_c}{\tan\left(45 - \dfrac{boost}{4}\right)} = \frac{1k}{221m} \approx 4.5 \text{ kHz} \tag{7.148}$$

The double zero will be located at

$$f_{z1,2} = \frac{f_c^{\,2}}{f_p} = \frac{1k^2}{4.5k} \approx 221 \text{ Hz} \tag{7.149}$$

First, let us select the LED series resistor using (7.146):

$$R_{LED,\max} \le \frac{8.2-1-2.5}{5-0.3+1m\times 0.3\times 20k} 20k\times 0.3 \le 2.6 \text{ k}\Omega \tag{7.150}$$

Including a 20 percent safety margin, we will choose a 1.8-kΩ resistor. Having this on hand, we can calculate the gain brought by the optocoupler path alone:

$$G_1 = \text{CTR}\frac{R_{pullup}}{R_{LED}} = 0.3\times\frac{20k}{2.1k} = 2.85 \text{ or } 9.1 \text{ dB} \tag{7.151}$$

Given the needed 10-dB attenuation, the attenuation we expect from the TL431 chain is simply

$$G_2 = G - G_1 = -10 - 9.1 = -19.1 \text{ dB or } 0.111 \tag{7.152}$$

The rest of the component values are derived using (7.141) through (7.145):

$$R_2 = \frac{0.11\times 38k\times 1.8k}{20k\times 0.3}\frac{\sqrt{1+\left(\frac{1k}{4.5k}\right)^2}\sqrt{1+\left(\frac{1k}{4.5k}\right)^2}}{\sqrt{1+\left(\frac{221}{1k}\right)^2}\sqrt{1+\left(\frac{1k}{221}\right)^2}} = 936\ \Omega \tag{7.153}$$

$$R_3 = \frac{R_1 f_{z2}}{f_{p2}-f_{z2}} = \frac{38k\times 221}{4.5k-221} \approx 2 \text{ k}\Omega \tag{7.154}$$

$$C_1 = \frac{1}{2\pi f_{z1}R_2} = \frac{1}{6.28\times 221\times 1.8k} \approx 777 \text{ nF} \tag{7.155}$$

$$C_2 = \frac{1}{2\pi f_{p2}R_{pullup}} = \frac{1}{6.28\times 4.5k\times 20k} \approx 1.8 \text{ nF} \tag{7.156}$$

$$C_3 = \frac{f_{p2}-f_{z2}}{2\pi R_1 f_{p2} f_{z2}} = \frac{4.5k-221}{6.28\times 38k\times 4.5k\times 221} \approx 18 \text{ nF} \tag{7.157}$$

We know that the total capacitor connected across the optocoupler that forms C_2 is actually made of the optocoupler parasitic pole C_{opto} and the added capacitor C_{col}. The optocoupler pole is located at 6 kHz. Given a pull-up resistor of 20 kΩ, we can calculate its parasitic contribution:

$$C_{opto} = \frac{1}{2\pi f_{opto} R_{pullup}} = \frac{1}{6.28 \times 6k \times 20k} \approx 1.3 \text{ nF} \tag{7.158}$$

Subtracted from C_2 value, we obtain the capacitor we need to install across the optocoupler:

$$C_{col} = C_2 - C_{opto} = 1.8n - 1.3n = 500 \text{ pF} \tag{7.159}$$

The Zener diode biasing resistor is found by using (7.147):

$$\begin{aligned} R_Z &= \frac{(V_{out} - V_Z) R_{pullup} \text{CTR}_{\text{min}}}{(V_{cc} - V_{\text{CE},sat}) + (I_{Zbias} + I_{bias}) R_{pullup} \text{CTR}_{\text{min}}} \\ &= \frac{(12 - 8.2) \times 20k \times 0.3}{5 - 0.3 + (2m + 1m) \times 20k \times 0.3} = 1 \text{ k}\Omega \end{aligned} \tag{7.160}$$

We now have everything on hand and can carry on with the simulation test fixture. It appears in Figure 7.33.

The operating points on the picture shows that the system is well stabilized. We can then start the ac simulation and its results appear in Figure 7.34. The transition point is well respected at 1 kHz, and we have around 10 dB of attenuation at that point. It would have been impossible to reach such an attenuation without disabling the fast lane. The boost is also very close to the target of 130°.

This last description ends our section on the TL431, where we have covered the majority of cases an engineer can face in his design work. Despite a rather low open-loop gain, compared to an op amp, the TL431 lends itself very well to the implementation of compensator once its bias limits are well understood and accounted for in the equations. This is what we have strived to do in the previous lines, showing how the part behaves in the classical type 1, 2, and 3 configurations.

7.15 Testing the Ac Responses on a Bench

All the ac results that we have presented so far are coming from simulations using a SPICE engine. We know by experience, that reality differs from simulation. To avoid unpleasant surprises, it is a necessity to check on the bench if the assumptions or the hypothesis adopted in a simulation context are validated by practical experiments. To do so, we must find a means to properly bias a TL431-based compensator and ac modulate its input to observe its output. The main difficulty with a high-gain compensator, however, is to maintain the right bias point and prevent the output from running in its upper or lower stops. In our case, we want a collector voltage around 2.5 V, the middle of its dynamic excursion when biased from a 5-V dc source. Assume a TL431 network designed to stabilize a 12-V converter. Even if you carefully adjust a dc power supply biasing the resistive divider network and delivering exactly 12.00 V, there will always be slow temperature drifts and noise that will inexorably push the circuit in its upper or lower stops. Therefore, rather than manually tweaking a dc source, why not automate the biasing process such as we did in our SPICE fixtures? This is exactly what we did in Figure 7.35. A

parameters

Vout=12

R1=(Vout-2.5)/250u
fc=1k
Gfc=10
boost=130
G=10^(-Gfc/20)
pi=3.14159

Vf=1
Ibias=1m
Vref=2.5
VCEsat=300m
Vcc=5
Vz=8.2
IZbias=2m
Rpullup=20k
Fopto=6k
Copto=1/(2*pi*Rpullup*Fopto)
CTR=0.3

G1=Rpullup*CTR/RLED
G2=G/G1

C1=1/(2*pi*fz1*R2)
C3=(fp2-fz2)/(2*pi*R1*fp2*fz2)
R3=R1*fz2/(fp2-fz2)

a=sqrt((fc^2/fp1^2)+1)
b=sqrt((fc^2/fp2^2)+1)
c=sqrt((fz1^2/fc^2)+1)
d=sqrt((fc^2/fz2^2)+1)

R2=(a*b/(c*d))*R1*G2
C2=1/(2*pi*fp1*Rpullup)
Ccol=C2-Copto

parameters

Rmax=((Vz-Vf-Vref)/(Vcc-VCEsat+Ibias*(Rpullup*CTR)))*Rpullup*CTR
RLED=Rmax*0.8
Za=(Vout-Vz)*Rpullup*CTR
Zb=(Vcc-VCEsat)+(Ibias+IZbias)*Rpullup*CTR
Rz=Za/Zb

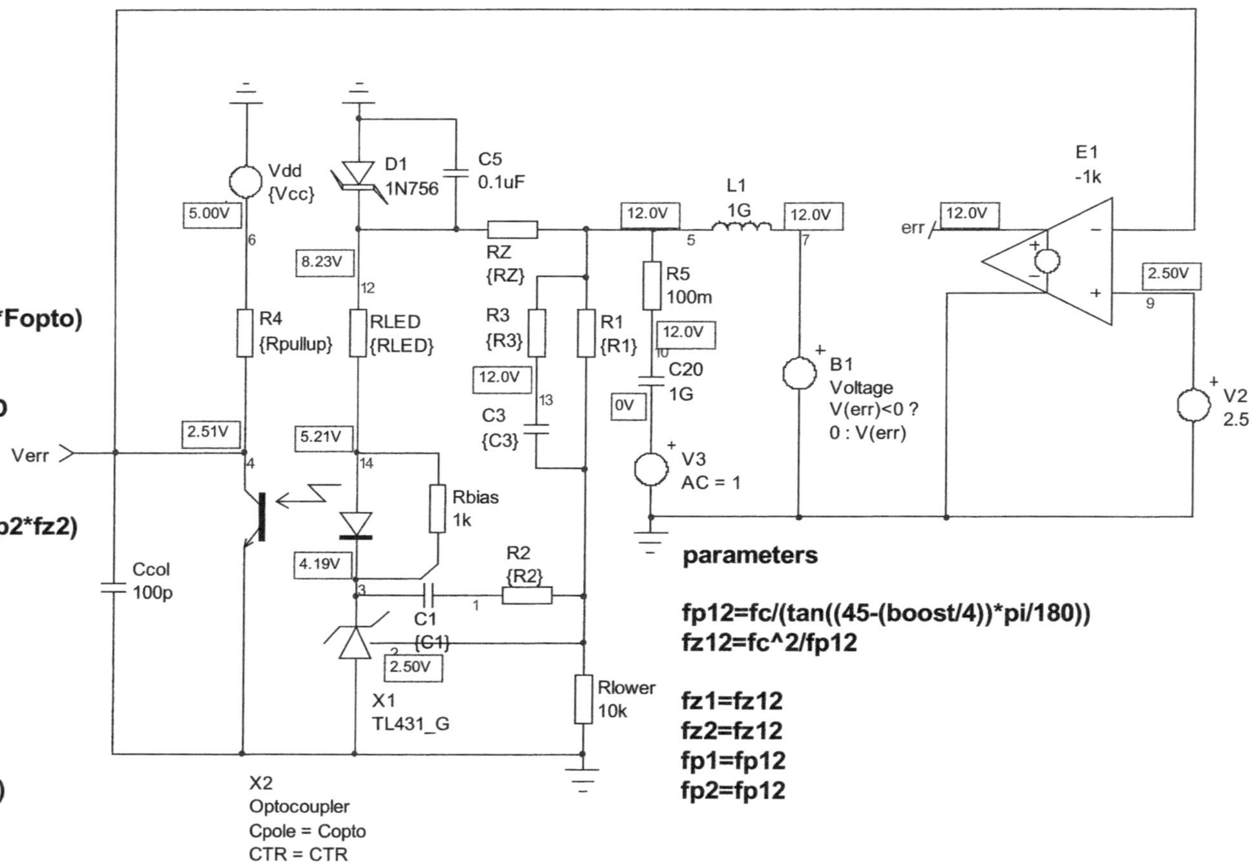

parameters

fp12=fc/(tan((45-(boost/4))*pi/180))
fz12=fc^2/fp12

fz1=fz12
fz2=fz12
fp1=fp12
fp2=fp12

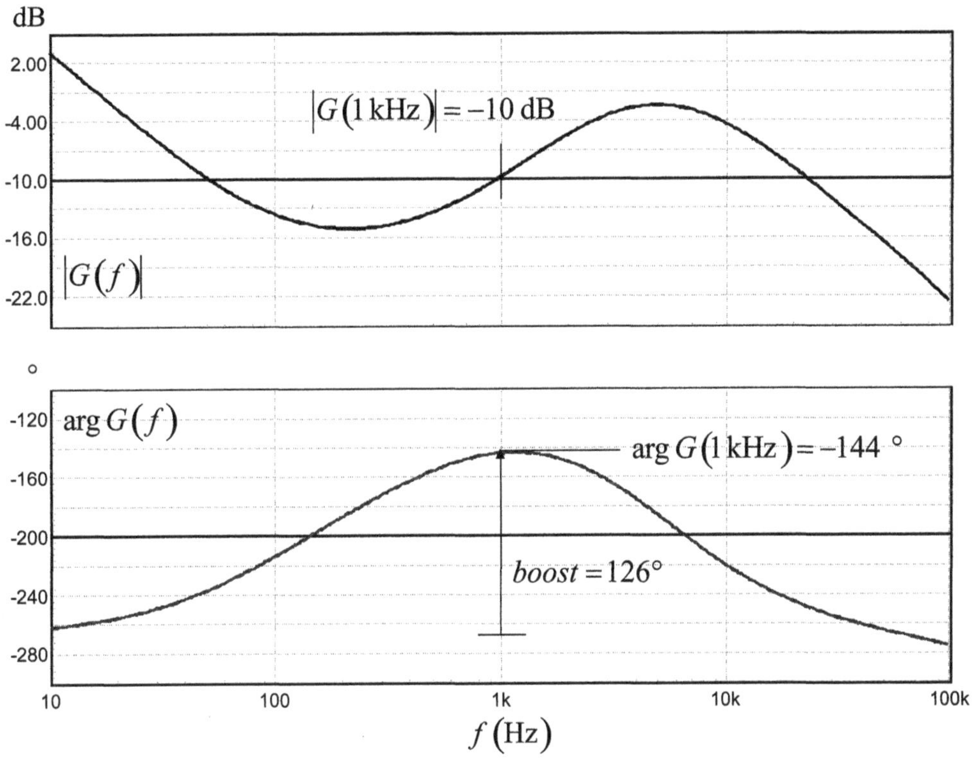

Figure 7.34 The ac results confirm the choices we made. The crossover point at 1 kHz exhibits the right attenuation, and the phase boost is within the specifications.

simple LM358 op amp monitors the optocoupler collector. It adjusts its output so that the collector voltage equals the 2.5-V setpoint present on the op amp inverting input. That way, if a drift occurs or a component is changed, the op amp will automatically adjusts its output to keep the collector at the right 2.5-V level. A 1000-μF

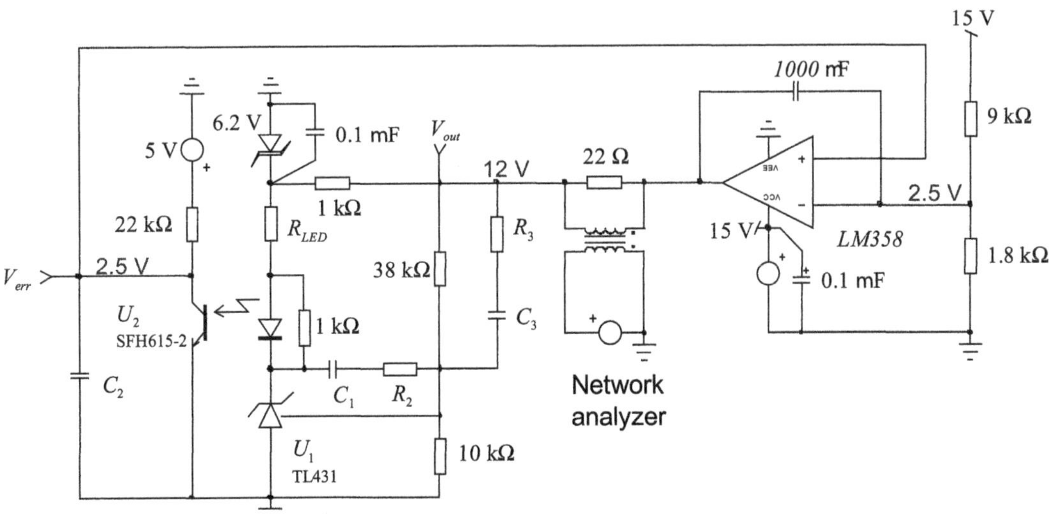

Figure 7.35 Rather than manually fixing the bias point, an op amp does the job automatically by maintaining the right dc voltage on the V_{out} node. R_3 and C_3 are only present for the type 3 study.

capacitor rolls off the loop gain and ensures the stability of the fixture. A network analyzer monitors V_{out} and V_{err} to produce the Bode plot we are looking for.

The first ac results are those of a type 2, and they appear in Figure 7.36. The compensator is designed to cross over at 1 kHz with a 0-dB gain, together with a 50° phase boost. Needless to say, this 0-dB transition would have been impossible to obtain without using the TL431 recommended configuration with the removed fast lane. The shape is in good agreement with the ac results we obtained by simulation.

The test has been carried on a type 3 compensator (see Figure 7.37), where the fast lane has also been disabled. Again, the results are very good, and the boost is up to nearly 120° in this example.

7.16 Isolated Zener-Based Compensator

Using the term compensator for a Zener diode–based design is not really adequate. The Zener diode is not an error amplifier and, besides ensuring a loose regulation, there is not much we can do with it. However, in cheap designs or when a raw dc voltage is good enough, it can be a possible choice. Figure 7.38 portrays the secondary side of a power supply featuring a Zener diode used as a regulator.

The transfer function of this simple arrangement is quite straightforward, capitalizing on what we have derived before:

$$\frac{V_{err}(s)}{V_{out}(s)} = -\text{CTR}\frac{R_{pullup}}{R_{LED}}\frac{1}{1+sR_{pullup}C_2} = -G_0\frac{1}{1+s/\omega_p} \tag{7.161}$$

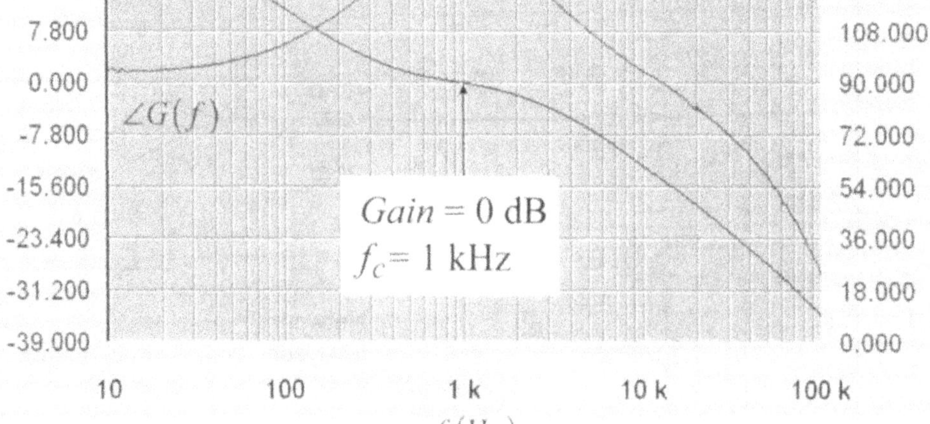

Figure 7.36 The Bode plot extracted with Figure 7.35 fixture confirms the type 2 nature of the tested configuration.

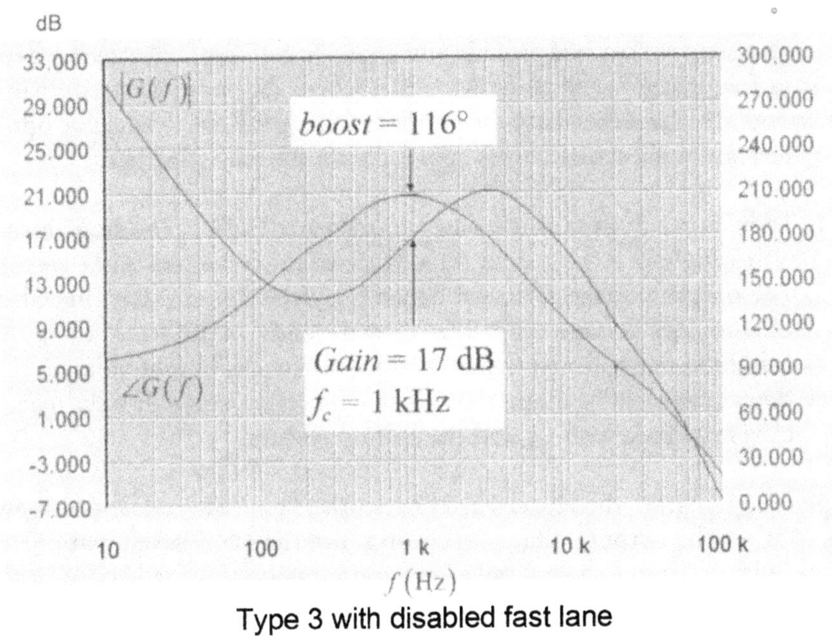

Figure 7.37 The ac sweep using the test fixture confirms the behavior of a type 3 compensator.

We have a bit of gain brought by G_0:

$$G_0 = \mathrm{CTR}\frac{R_{pullup}}{R_{LED}} \tag{7.162}$$

and a high-frequency pole:

$$\omega_p = \frac{1}{R_{pullup}C_2} \tag{7.163}$$

Of course, we could wire a capacitive network across the LED resistor, but with a lack of origin pole imposing a –1 slope, it would produce a bump in the gain curve

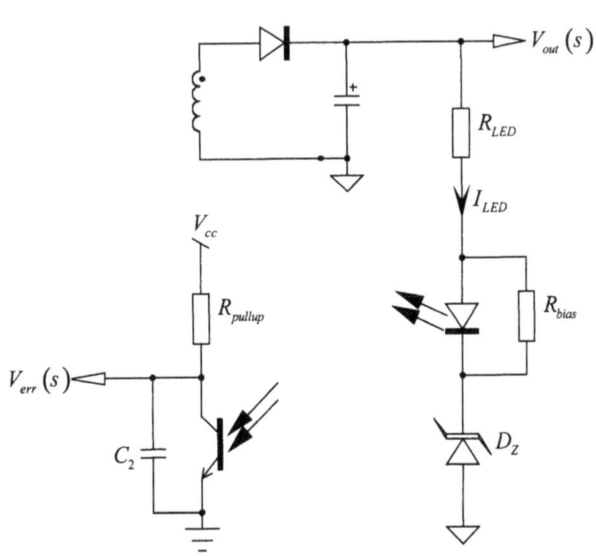

Figure 7.38 The Zener diode requires a sufficient bias current to properly operate.

and not a flat area as with the type 2 kind of curve. This Zener diode-based compensation circuitry does not provide any phase boost. You will thus have to select a crossover frequency on the plant Bode plot where the phase lag is compatible with that brought by the Zener-based network featuring the low-frequency optocoupler pole. The main weaknesses of this approach are the following ones:

1. There is no dc gain as with an op amp or a TL431. The static error can be rather high, and the output impedance of the converter is not very good.
2. The output voltage selection depends on the Zener diode, but also on the LED resistor voltage drop. This drop depends on the error voltage level (the current in the pull-up resistor) and influences the output voltage.
3. You cannot really shape the transfer function; you can only place a high-frequency pole with C_2 and the pull-up resistor.

Despite the remarks, the Zener-based network is popular in low-cost power supplies or when the output voltage precision is not the main design parameter. Actually, as indicated in the second bullet, the output voltage is made of several stacked contributors:

$$V_{out} = V_Z + R_{LED}I_{LED} + V_f \tag{7.164}$$

If V_Z and V_f are constant values, the main offender in the equation is the voltage drop across the LED resistor. This drop is a function of the current circulating in the LED and, thus, in direct relationship to the feedback level. The maximum current value in the LED occurs when the feedback is close to zero—actually the optocoupler saturation voltage. This level is defined by

$$I_{LED,\max} = \frac{V_{cc} - V_{CE,sat}}{R_{pullup}\text{CTR}_{\min}} + I_{bias} \tag{7.165}$$

On the contrary, at maximum power, the feedback voltage is close to V_{cc}, and that is where the LED current is minimum. The variable $V_{CE,\max}$ corresponds to the maximum value the feedback voltage can take when the maximum power is asked to the converter:

$$I_{LED,\min} = \frac{V_{cc} - V_{CE,\max}}{R_{pullup}\text{CTR}_{\max}} + I_{bias} \tag{7.166}$$

As we can see with these equations, the output voltage will slightly move in relationship to the feedback level. The important design parameter is to select an LED resistor so that its voltage drop remains reasonable compared to the Zener voltage. Otherwise, its contribution will affect V_{out}, and the final precision will seriously suffer. Once the voltage contribution of the LED series resistor is known, we can select the Zener voltage according to (7.164).

7.16.1 A Design Example

The transfer function of the converter we want to stabilize is given in Figure 7.39. On this graph, we can see that the maximum phase lag stays below 90°. Crossing over at 1 kHz requires a 20-dB gain. This is where we will focus our attention.

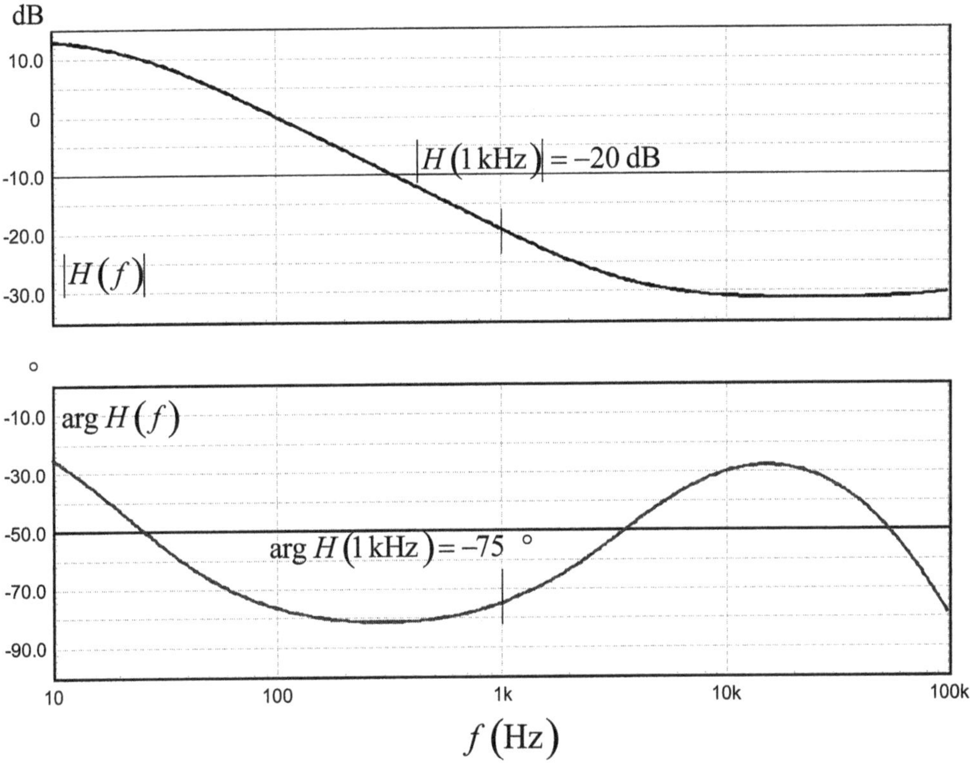

Figure 7.39 The transfer function of the converter *H(s)* shows a gain deficiency of −20 dB at 1 kHz.

Using (7.162), we can calculate the LED resistor we will install given the following parameters:

IZ_{bias} = 1 mA; the Zener diode biasing current.
V_f = 1 V; the LED forward voltage.
$V_{CE,sat}$ = 0.3 V; the optocoupler saturation voltage.
$V_{CE,\max}$ = 3 V; the optocoupler voltage at maximum power.
V_{cc} = 5 V; the pull-up V_{cc} level.
V_{out} = 12 V; the converter output voltage.
R_{pullup} = 20 kΩ; the optocoupler pull-up resistor.
$\mathrm{CTR}_{\min}$ = 0.3; the minimum optocoupler current transfer ratio.
$\mathrm{CTR}_{\max}$ = 1.2; the maximum optocoupler current transfer ratio.

$$R_{LED} = \mathrm{CTR}\frac{R_{pullup}}{G_{f_c}} = 0.3\frac{20k}{10^{\frac{20}{20}}} = 600\ \Omega \tag{7.167}$$

As explained, the voltage across this LED resistor will affect the output voltage given the current circulating in the optocoupler collector. The variations are calculated using (7.165) and (7.166):

$$V_{R_{LED,\max}} = R_{LED} I_{LED,\max} = R_{LED}\left(\frac{V_{cc} - V_{CE,sat}}{R_{pullup}\mathrm{CTR}_{\min}} + I_{bias}\right)$$
$$= 600 \times \left(\frac{5 - 0.3}{20k \times 0.3} + 1m\right) = 1.07 \text{ V} \quad (7.168)$$

$$V_{R_{LED,\min}} = R_{LED} I_{LED,\min} = R_{LED}\left(\frac{V_{cc} - V_{CE,\max}}{R_{pullup}\mathrm{CTR}_{\max}} + I_{bias}\right)$$
$$= 600 \times \left(\frac{5 - 3}{20k \times 1.2} + 1m\right) = 650 \text{ mV} \quad (7.169)$$

On average, the voltage variation across R_{LED} is around 860 mV. Considering a rather constant 1 V drop across the optocoupler LED, we can calculate the Zener voltage to get a 12 V output:

$$V_Z = V_{out} - V_f - V_{R_{LED},avg} = 12 - 1 - 0.86 \approx 10 \text{ V} \quad (7.170)$$

A 10-V Zener diode thus seems to be the right choice. In case the Zener voltage is not available, you still have the ability to slightly tweak the LED series resistance to match the wanted output voltage given the closest Zener voltage you have found. The test fixture appears in Figure 7.40 and confirms the right operating bias points. We are close to our 12-V target.

For noise reasons, we have placed a pole at 5 kHz so that a small capacitor can be placed across the optocoupler, close to the controller pins:

$$C_2 = \frac{1}{2\pi f_p R_{pullup}} = \frac{1}{6.28 \times 5k \times 20k} = 1.6 \text{ nF} \quad (7.171)$$

We know that the total capacitor connected across the optocoupler that forms C_2 is actually made of the optocoupler parasitic pole C_{opto} and the added capacitor C_{col}. The optocoupler pole is located at 6 kHz. Given a pull-up resistance of 20 kΩ, we can calculate its parasitic contribution:

$$C_{opto} = \frac{1}{2\pi f_{opto} R_{pullup}} = \frac{1}{6.28 \times 6k \times 20k} \approx 1.3 \text{ nF} \quad (7.172)$$

Subtracted from the C_2 value, we obtain the capacitor we need to install across the optocoupler:

$$C_{col} = C_2 - C_{opto} = 1.6n - 1.3n = 300 \text{ pF} \quad (7.173)$$

The ac response can be now tested by observing the V_{err} node. The answer appears in Figure 7.41. The slight 2-dB discrepancy in the gain can be explained by the LED dynamic resistor cumulated with that of the Zener diode itself. It can obviously be compensated, if necessary, by increasing the gain target in (7.167). If we combine both the Figure 7.39 and Figure 7.41 transfer functions, we can unveil the complete open-loop gain $T(s)$ in Figure 7.42. The open-loop dc gain is obviously

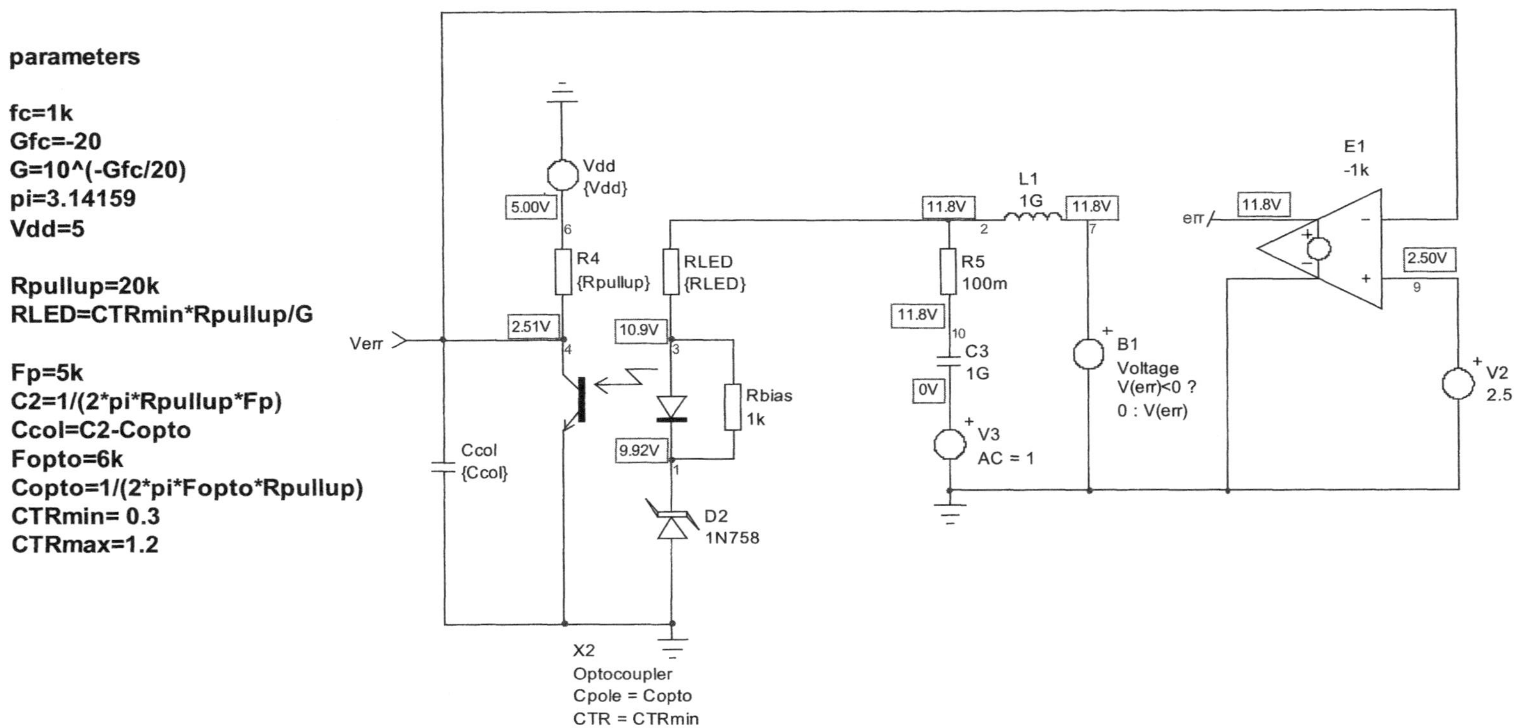

Figure 7.40 The Zener-based test fixture delivers the output voltage close to the target of 12 V.

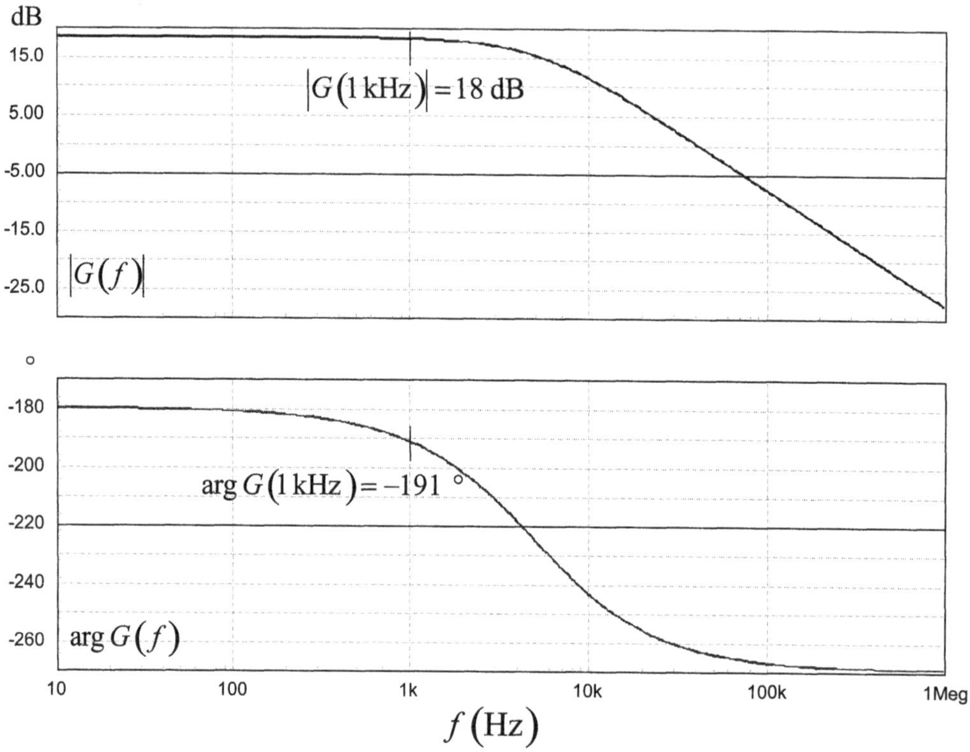

Figure 7.41 The ac response of the Zener-based network. No surprise, there is no origin pole . . .

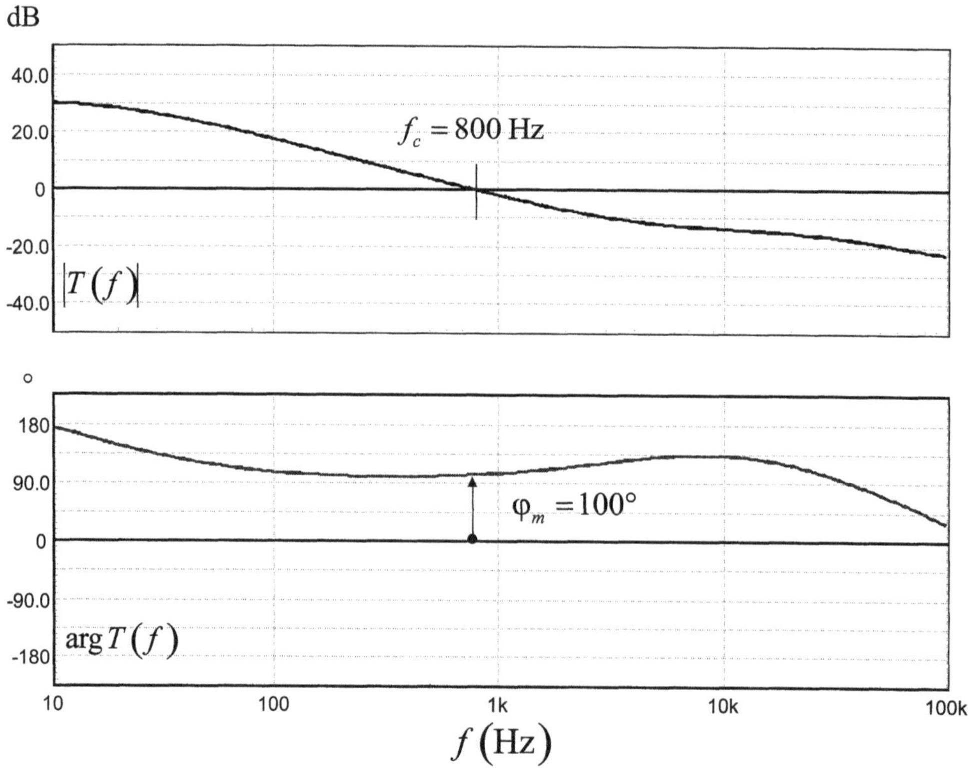

Figure 7.42 The loop gain shows a weak gain, but the phase margin is comfortable.

weak given the lack of origin pole. Owing to the missing 2 dB, we do not crossover at 1 kHz but at 800 Hz. The phase margin is comfortable with a 100° value.

Needless to say, the Zener diode dynamic impedance cannot be simulated with great accuracy, as inducing gain changes when going to the lab. However, experience shows that this first analytical debugging is helpful to figure out an LED series resistor value to wire on the prototype rather than pulling one out of thin air.

7.17 Nonisolated Zener-Based Compensator

In the case of nonisolated converter designs, the isolation barrier brought by the optocoupler is not necessary. Therefore, the implementation of the Zener-based circuitry slightly changes as indicated by Figure 7.43.

Rather than driving a single transistor, we used a current mirror made of two bipolar transistors. If the devices are well matched, the collector current circulating in the pull-up resistor is the same as the one circulating in the LED resistor. We are close to an optocoupler operation with a CTR of 1. The difference might lie in the h_{11} input parameter of transistor Q_1 base-emitter junction that appears in series with the Zener diode. If we neglect it and consider well-paired transistors, the gain is simply

$$\frac{V_{err}(s)}{V_{out}(s)} = -\frac{R_{pullup}}{R_{LED}} \tag{7.174}$$

The output voltage is then obtained by stacking up the various voltage drops:

$$V_{out} = V_Z + R_{LED}I_{LED} + V_{BE} \tag{7.175}$$

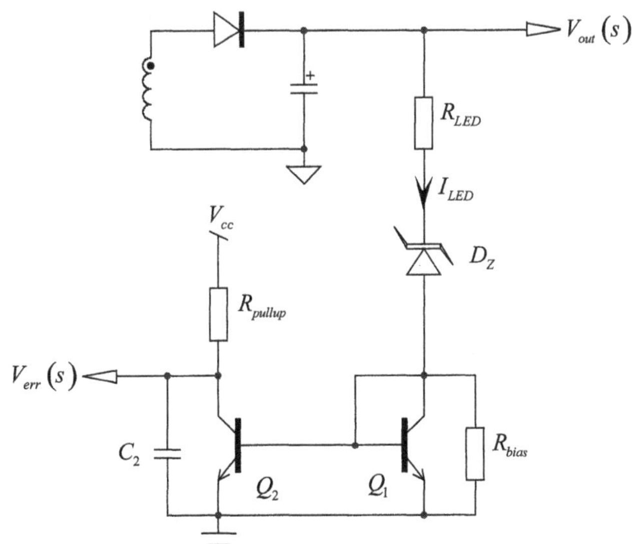

Figure 7.43 With the lack of an optocoupler, a current mirror is used to drive the feedback pin to ground.

The LED current is made of the feedback current plus the dc current imposed by the bias resistor, R_{bias}, placed across the base-emitter junction of Q_1. A V_{BE} of ≈650 mV can be considered for the design procedure at room temperature. The variation of this level is around –2 mV/°C.

$$I_{LED} = \frac{V_{cc} - V_{CE}}{R_{pullup}} + \frac{V_{BE}}{R_{bias}} \tag{7.176}$$

To limit the dc drifts, make sure the transistors are well paired and share the same junction temperature. To that respect, we recommend the usage of dual transistors such as the BC846BDW1T1G from ON Semiconductor. As these devices are manufactured from the same wafer and share a common leadframe, the thermal

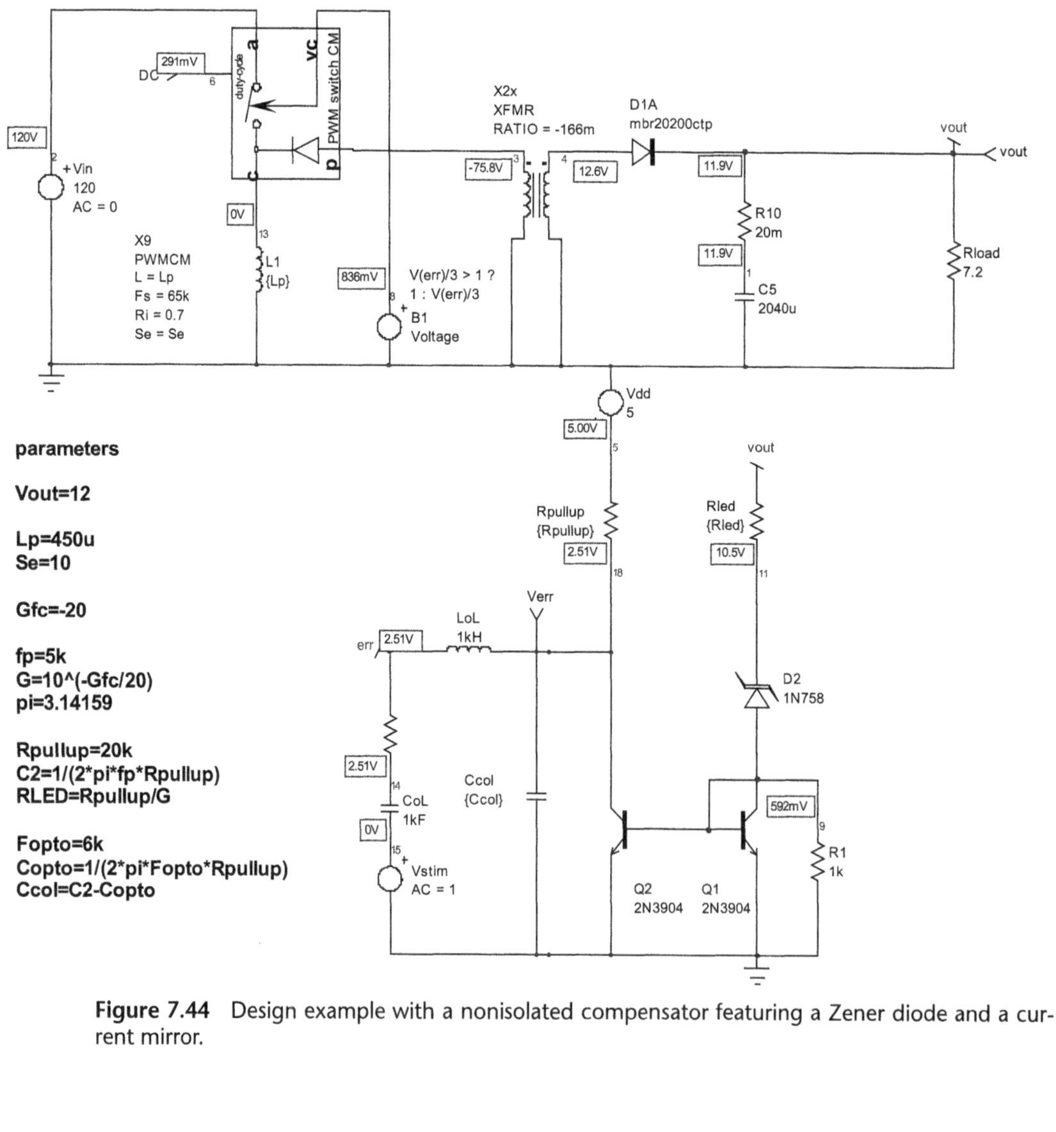

Figure 7.44 Design example with a nonisolated compensator featuring a Zener diode and a current mirror.

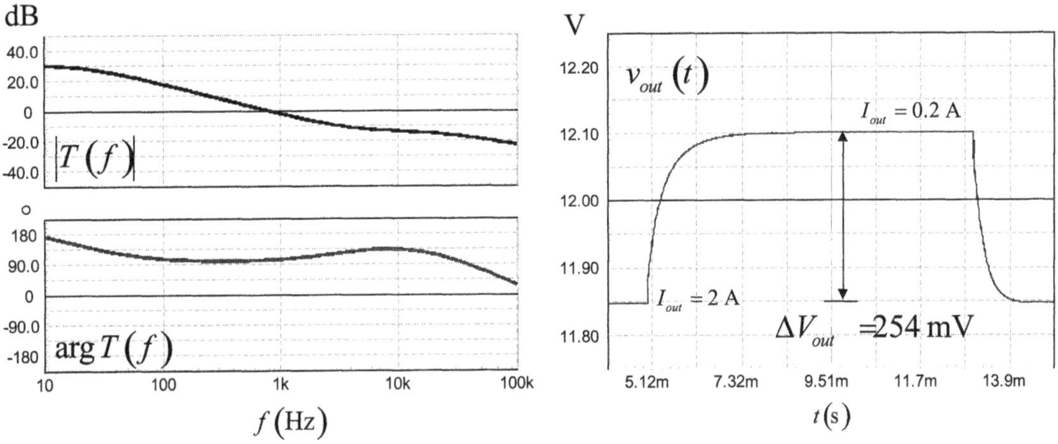

Figure 7.45 The ac and transient responses are not that ridiculous for such a simple compensator! The current is stepped from 2 A to 200 mA in 10 μs.

performance will be improved as their junction temperature (and thus their V_{BE} forward drops) will drift together. The dc output voltage precision depends on the V_{BE} contribution to the whole chain. Implementing this nonisolated technique where the V_{BE} contributes to a maximum of 10 percent of the dc output is a reasonable tradeoff. For instance, a 650-mV V_{BE} with a 12 V output is acceptable (5.4 percent). On the contrary, adopting this configuration for a 3.3 V output makes no sense at all. The design procedure is the same as the one described for the isolated version.

An application example appears in Figure 7.44 and shows good operating bias points. Despite its simplicity, the circuit works well and, as testified by Figure 7.45, the transient response is acceptable. Please note the square response to an output step, typical of a purely resistive output impedance brought by the lack of origin pole in the compensator transfer function.

7.18 Nonisolated Zener-Based Compensator: A Lower Cost Version

Before closing this section on Zener-based compensators, it is interesting to note the existence of an even cheaper compensator, based on a single bipolar transistor. Disclosed to me by Mr. Louvel, it appears in Figure 7.46 and combines a bipolar transistor with a Zener diode. Used in high-volume low-cost consumer applications, it offers a rather well temperature-compensated error amplifier. The Zener diode voltage is augmented by the series diode D1 forward drop, and it biases the emitter to $V_Z + V_f$. If we assume the V_{BE} of Q_1 to be around the same value of D_1 voltage drop V_f, then the voltage across R_{lower} is roughly that of D_Z. When the voltage on Q_1 base changes, it either forward biases Q_1 (it V_{out} decreases) and V_{err} increases, or it blocks it when V_{out} increases and V_{err} goes down. Based on this observation, the calculation of the elements is rather straightforward. R_Z must provide the

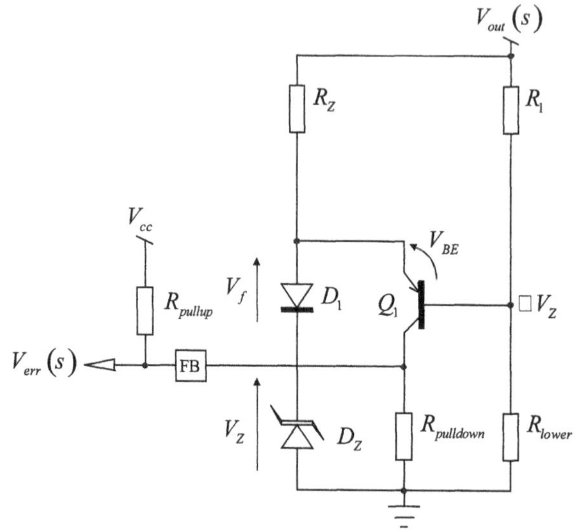

Figure 7.46 Based on a single bipolar transistor, this compensator allows cheap weighted compensation if necessary.

biasing current for the Zener diode but also the current to develop the error voltage across $R_{pulldown}$:

$$R_Z \leq \frac{V_{out} - V_f - V_z}{I_{Zbias} + \dfrac{V_{err,\max}}{R_{pulldown}}} \tag{7.177}$$

The divider network is calculated assuming V_Z appears across R_{lower}:

$$R_{lower} = \frac{V_Z}{I_{bridge}} \tag{7.178}$$

$$R_1 = \frac{V_{out} - V_Z}{I_{bridge}} \tag{7.179}$$

We recommend a bias current for the bridge to be at least 500 μA given the bipolar transistor base current. Regarding the Zener diode, a voltage of 6.2 V looks like a good choice, as it is the most stable value over temperature. A bias current for this device around 1–2 mA can be accepted.

Please note that this circuitry can only be coupled to a feedback input featuring a pull-up resistor, as in Figure 7.46. At startup, as V_{out} does not yet exist, there can be no voltage across $R_{pulldown}$. If no internal source biases the feedback pin to ask for power, the converter won't start. A controller like a NCP1200 can be a good candidate for this cheap compensator.

7.19 Conclusion

The TL431 is one of the most popular components used in the consumer field. Cheap and easy to bias, you must understand how it works before you can get the best of it. You must read the first section of this article to realize that the designer who thought of this self-contained op amp and reference voltage was a genius! The presence of the fast lane can be used as an advantage, making a type 2 compensator look quite simple at the end. Linked to an optocoupler, you realize that closing a loop with that device has never been that simple.

References

[1] Basso, C., *Switch Mode Power Supplies: SPICE Simulations and Practical Designs*, New York: McGraw-Hill, 2008.

[2] Circuit Sage, http://www.circuitsage.com/bandgap.html, last accessed June 11, 2012.

[3] Tepsa, T., and S. Suntio, "Adjustable Shunt Regulator Based Control Systems," *IEEE Power Electronics Letters*, Vol. 1, No. 4, December 2003.

Appendix 7A: Summary Pictures

Figures 7.47 through 7.51 summarize the component definitions associated with the structures described in the chapter.

Isolated type 1

1 origin pole

$$R_{LED,\max} \le \frac{\left(V_{out} - V_f - V_{\mathrm{TL431,min}}\right) R_{pullup} \mathrm{CTR}_{\min}}{V_{cc} - V_{CE,sat} + I_{bias} \mathrm{CTR}_{\min} R_{pullup}}$$

$$C_1 = \frac{R_{pullup}\mathrm{CTR}}{R_{LED} R_1 2\pi f_{po}}$$

$$C_2 = \frac{R_1 C_1}{R_{pullup}}$$

$$C_{opto} = \frac{1}{2\pi f_{opto} R_{pulldown}}$$

$$C_{col} = C_2 - C_{opto}$$

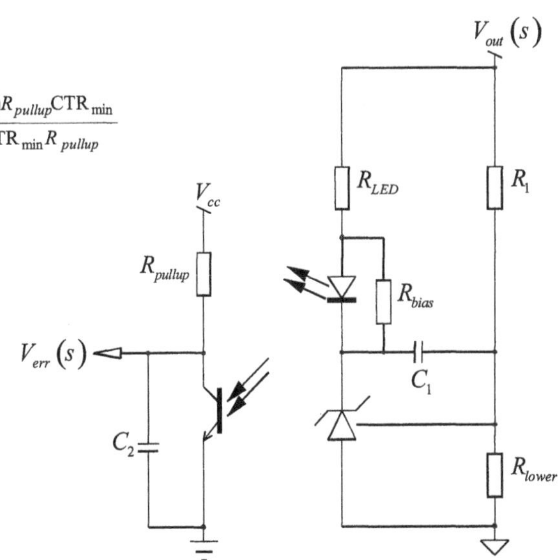

Figure 7.47 The isolated type 1.

Isolated type 2 – fast lane

1 origin pole

1 pole f_p

1 zero f_z

$$R_{LED} = \frac{R_{pullup}\text{CTR}}{G}$$

$$C_1 = \frac{1}{2\pi f_z R_1}$$

$$C_2 = \frac{1}{2\pi R_{pullup} f_p}$$

$$C_{opto} = \frac{1}{2\pi f_{opto} R_{pulldown}}$$

$$C_{col} = C_2 - C_{opto}$$

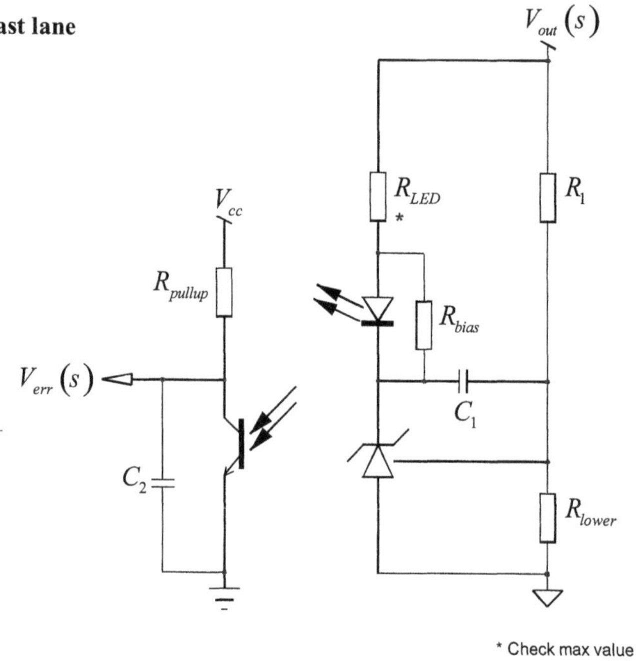

Figure 7.48 The isolated type 2 with fast lane.

Isolated type 2 – no fast lane

1 origin pole

1 pole f_p

1 zero f_z

$$R_{LED,\max} \square \frac{\left(V_Z - V_f + V_{\text{TL431,min}}\right) R_{pullup}\text{CTR}_{\min}}{V_{ccp} - V_{CE,sat} + I_{bias}\text{CTR}_{\min} R_{pullup}}$$

$$R_Z \square \frac{\left(V_{out} - V_Z\right) R_{pullup}\text{CTR}_{\min}}{\left(V_{cc} - V_{CE,sat}\right) + \left(I_{Zbias} + I_{bias}\right) R_{pullup}\text{CTR}_{\min}}$$

$$G_1 = \text{CTR}\frac{R_{pullup}}{R_{LED}} \quad G_2 = 10^{\frac{G-G_1}{20}}$$

$$R_2 = G_2 R_1$$

$$C_1 = \frac{1}{2\pi f_z R_2} \quad C_2 = \frac{1}{2\pi R_{pullup} f_p}$$

$$C_{opto} = \frac{1}{2\pi f_{opto} R_{pullup}} \quad C_{col} = C_2 - C_{opto}$$

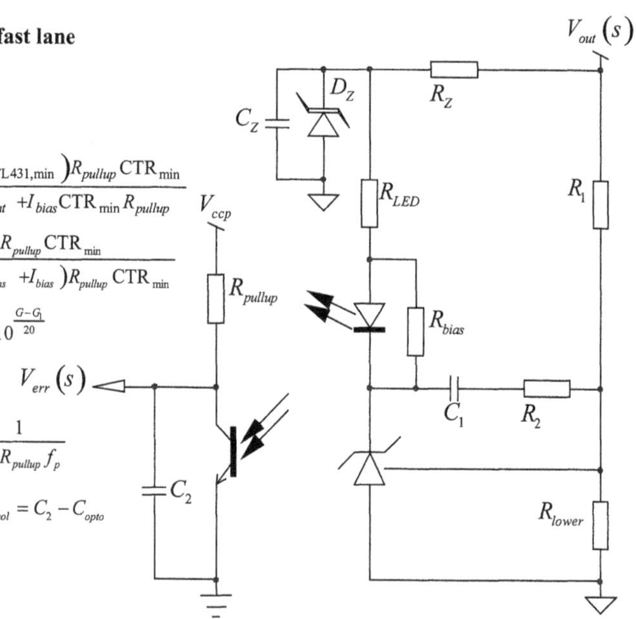

Figure 7.49 The isolated type 2 without fast lane.

Isolated type 3 – fast lane

1 origin pole

2 poles f_{p1}, f_{p2}

2 zeros f_{z1}, f_{z2}

$$R_{LED} = \frac{R_{pullup}}{G} \mathrm{CTR} \frac{\sqrt{1+\left(\frac{f_{z_1}}{f_c}\right)^2}\sqrt{1+\left(\frac{f_c}{f_{z_2}}\right)^2}}{\sqrt{1+\left(\frac{f_c}{f_{p_1}}\right)^2}\sqrt{1+\left(\frac{f_c}{f_{p_2}}\right)^2}}$$

$$C_3 = \frac{f_{p_2} - f_{z_2}}{2\pi R_{LED} f_{p_2} f_{z_2}} \qquad R_3 = \frac{f_{z_2} R_{LED}}{f_{p_2} - f_{z_2}}$$

$$C_1 = \frac{1}{2\pi f_{z_1} R_1} \qquad C_2 = \frac{1}{2\pi R_{pullup} f_{p_2}}$$

$$C_{opto} = \frac{1}{2\pi f_{opto} R_{pullup}} \qquad C_{col} = C_2 - C_{opto}$$

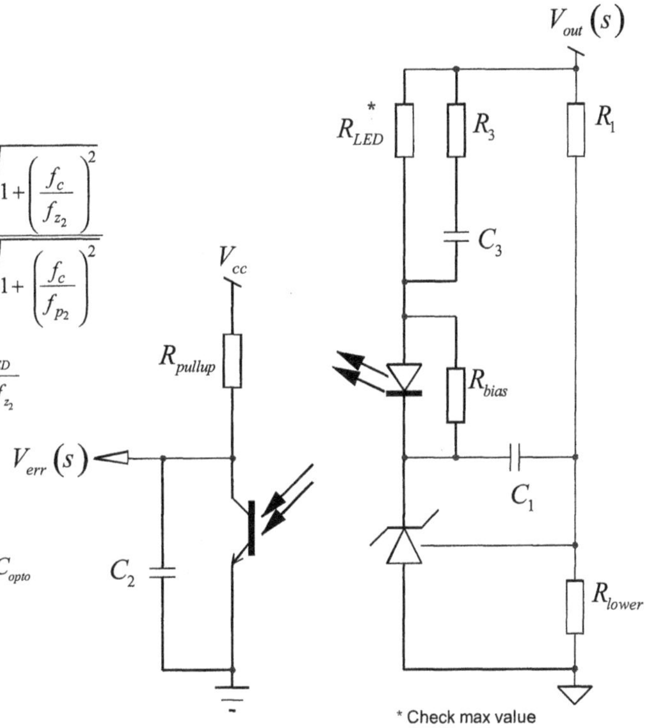

Figure 7.50 The isolated type 3 with fast lane.

Isolated type 3 – no fast lane

1 origin pole

2 poles f_{p1}, f_{p2}

2 zeros f_{z1}, f_{z2}

$$R_{LED,\max} \le \frac{\left(V_Z - V_f - V_{\mathrm{TL431,min}}\right) R_{pullup} \mathrm{CTR}_{\min}}{V_{cc} - V_{CE,sat} + I_{bias} \mathrm{CTR}_{\min} R_{pullup}}$$

$$R_2 = \frac{G R_1 R_{LED}}{R_{pullup} \mathrm{CTR}} \frac{\sqrt{1+\left(\frac{f_c}{f_{p_1}}\right)^2}\sqrt{1+\left(\frac{f_c}{f_{p_2}}\right)^2}}{\sqrt{1+\left(\frac{f_{z_1}}{f_c}\right)^2}\sqrt{1+\left(\frac{f_c}{f_{z_2}}\right)^2}}$$

$$R_3 = \frac{R_1 f_{z_2}}{f_{p_2} - f_{z_2}} \qquad C_1 = \frac{1}{2\pi f_{z_1} R_2}$$

$$C_2 = \frac{1}{2\pi f_{p_2} R_{pullup}} \qquad C_3 = \frac{f_{p_2} - f_{z_2}}{2\pi R_1 f_{p_2} f_{z_2}}$$

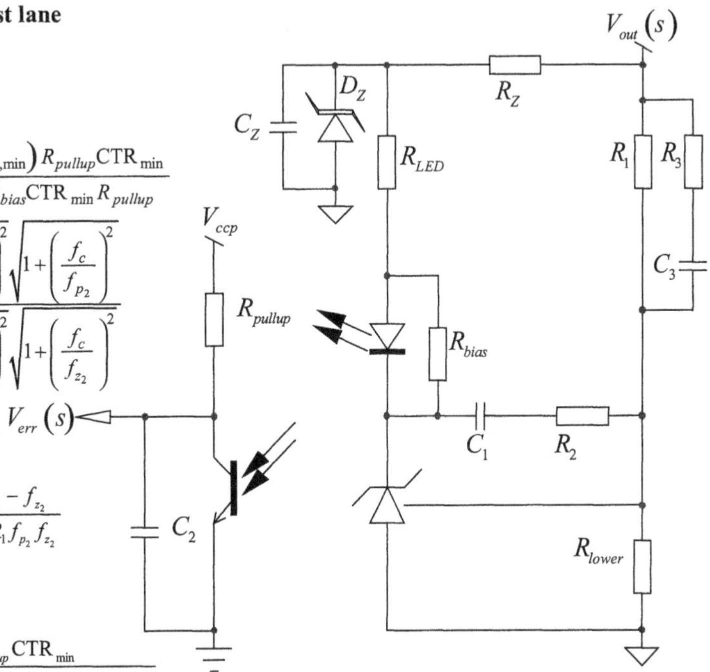

$$R_Z \le \frac{(V_{out} - V_Z) R_{pullup} \mathrm{CTR}_{\min}}{(V_{cc} - V_{CE,sat}) + (I_{Zbias} + I_{bias}) R_{pullup} \mathrm{CTR}_{\min}}$$

Figure 7.51 The isolated type 3 without fast lane.

Appendix 7B: Second Stage *LC* Filter

In all TL431-based compensation designs we have presented so far, there was no mention of a second-stage LC filter. As shown in Figure 7.52, this network is often included in flyback converters to filter out unwanted noise present on the output. As you can imagine, if the filter cutoff frequency is wrongly positioned, it can affect the loop response and degrade the stability margins.

In this chapter, we have shown how to isolate the TL431-based compensator and design it separately from the power stage: whether you use it in a flyback converter, a buck converter, or a boost converter, the equations set remains the same. Unfortunately, it is not possible to isolate a TL431-based compensator equipped with the post LC filter from the power stage it is attached to. The reason is that the insertion of the LC filter between the load and the power stage modifies the impedances you considered when extracting the converter small-signal model.

A simplified large-signal representation of a DCM fixed-frequency flyback converter appears in Figure 7.53. To calculate the open-loop transfer function, you first need to linearize the current source expression I_{out}. Then, the small-signal output voltage across C_1 is obtained by multiplying $\hat{i}_{out}$ with the equivalent impedance seen by the current source. As the LC filter is inserted, you can see how the circuit complicates since the inductor introduces a second-order network in the output path. However, if the series inductor impedance below the crossover frequency is much smaller than Z_1 and Z_2 impedances, then we can consider that the flyback power stage transfer function does not significantly change. This is the case when the added inductance is below 15 μH [1]. Choosing a low-value inductance makes sense from an efficiency point of view, as low-inductance magnetic components feature weak ESR values and remain small-size devices. Typical values used in flyback converters are up to a few micro-henries maximum.

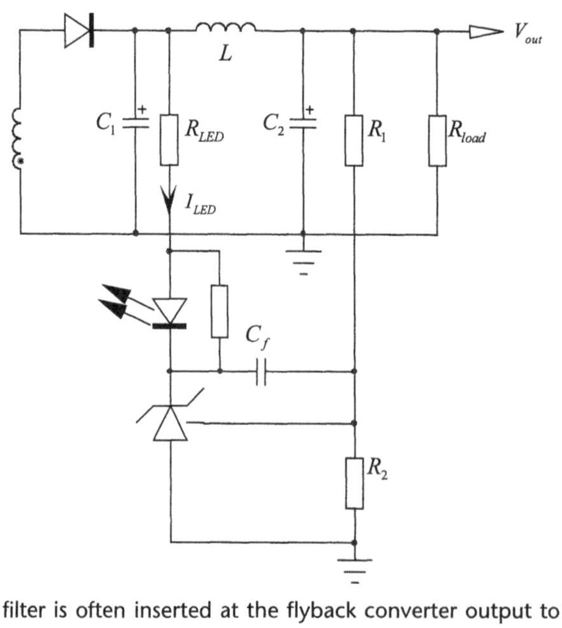

Figure 7.52 An *LC* filter is often inserted at the flyback converter output to filter out unwanted spikes.

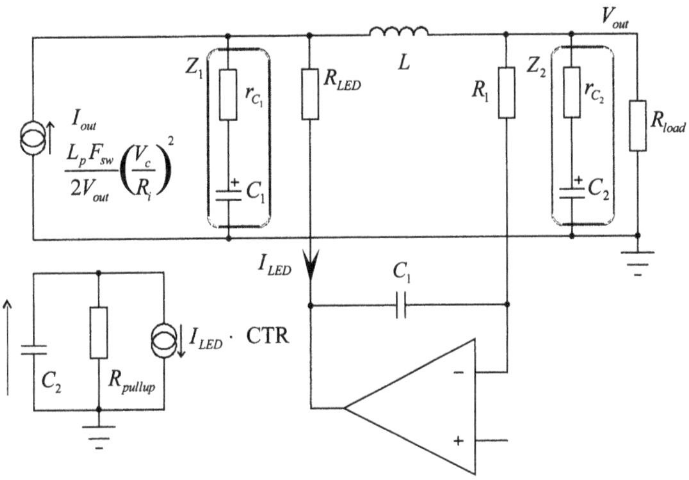

Figure 7.53 The inserted LC network modifies the impedance seen by the considered power stage. Here, we have a simplified large-signal DCM current-mode flyback converter.

A Simplified Approach

We can try, however, to gain insight by considering only the TL431-based transfer function, neglecting the converter loading changes brought by inserting the inductor L. The new equivalent schematic appears in Figure 7.54.

$H(s)$ lumps the LC filter effects loaded by the resistor R_{load}, damped by output capacitor ESR presence. We neglected the inductor dc losses for simplicity. The filter obeys the following equation:

$$H(s) = \frac{1}{1 + \left(\frac{s}{Q\omega_0}\right) + \left(\frac{s}{\omega_0}\right)^2} \tag{7.180}$$

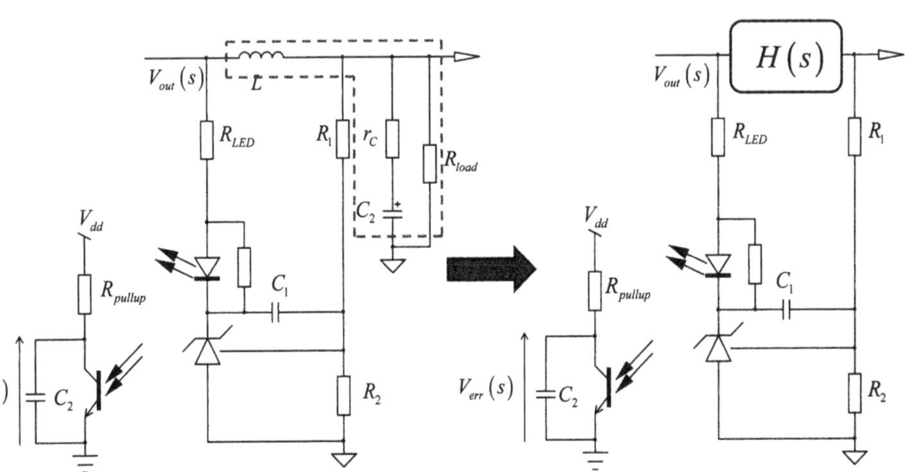

Figure 7.54 If we neglect the loading condition change on the power stage, we can extract the TL431 compensator featuring a post LC filter.

The denominator is that of the buck converter output impedance we have already derived in Chapter 4. Owing to its low entropy form, we can put r_L to zero and immediately get the definitions we need:

$$\omega_0 = \frac{1}{\sqrt{LC_2}}\sqrt{\frac{R_{load}}{r_C + R_{load}}} \tag{7.181}$$

$$Q = \frac{LC_2\omega_0(r_C + R_{load})}{L + C_2(r_c R_{load})} \tag{7.182}$$

If you follow the guidelines detailed in this chapter, you should be able to express the transfer function between V_{err} and V_{out} as follows:

$$\frac{V_{err}(s)}{V_{out}(s)} = -\frac{CTR \cdot R_{pullup}}{R_{LED}} \frac{\left(\frac{H(s)}{sR_1C_1} + 1\right)}{1 + sR_{pullup}C_2}$$

$$= -\frac{CTR \cdot R_{pullup}}{R_{LED}} \frac{1 + \left(\frac{s}{Q\omega_0}\right) + \left(\frac{s}{\omega_0}\right)^2 + \frac{1}{sR_1C_1}}{\left(1 + sR_{pullup}C_2\right)\left(1 + \left(\frac{s}{Q\omega_0}\right) + \left(\frac{s}{\omega_0}\right)^2\right)} \tag{7.183}$$

The transfer function is that of the traditional type 2 architecture to which a double pole and a double zero have been added. To check the effects of these extra terms, we have set up an automated simulation template of a type 2 compensator featuring the output filter. It is given in Figure 7.55. In this template, the inductor value is fixed at 2.2 μH, and C_2 is adjusted to modify the cutoff frequency. The ESR value is calculated so that the product r_C by C_2 is roughly equal to 70 μs, a rule of thumb that can help approximate the ESR value for a standard electrolytic capacitor.

In this compensator, the mid-band gain is set to 20 dB, and the phase boost peaks at 1 kHz. The integrator zero is located at 226 Hz, while the second pole is placed at 5.6 kHz. We have moved the resonant frequency from 10 kHz down to 100 Hz. The results are collected in Figure 7.56 ac plot. The *LC* network introduces a glitch in the magnitude at the resonant frequency. This glitch is acceptable as long as the resonant frequency occurs in a frequency region where the integrator path through the op amp and C_1 (Figure 7.54) has no gain, as pointed out in [2]: intuitively, the *LC* filter contribution has more difficulty altering the ac LED current if the integrator attenuation weakens the resonance effects. However, as the resonance approaches the region where the zero is located, the transfer function becomes distorted and the phase goes out of control (500 Hz and 100 Hz).

Simulation at Work

Figure 7.57 shows a 19-V/4-A current-mode flyback converter equipped with the *LC* filter. The crossover frequency is set to 5 kHz with 70° phase margin. The zero is posi-

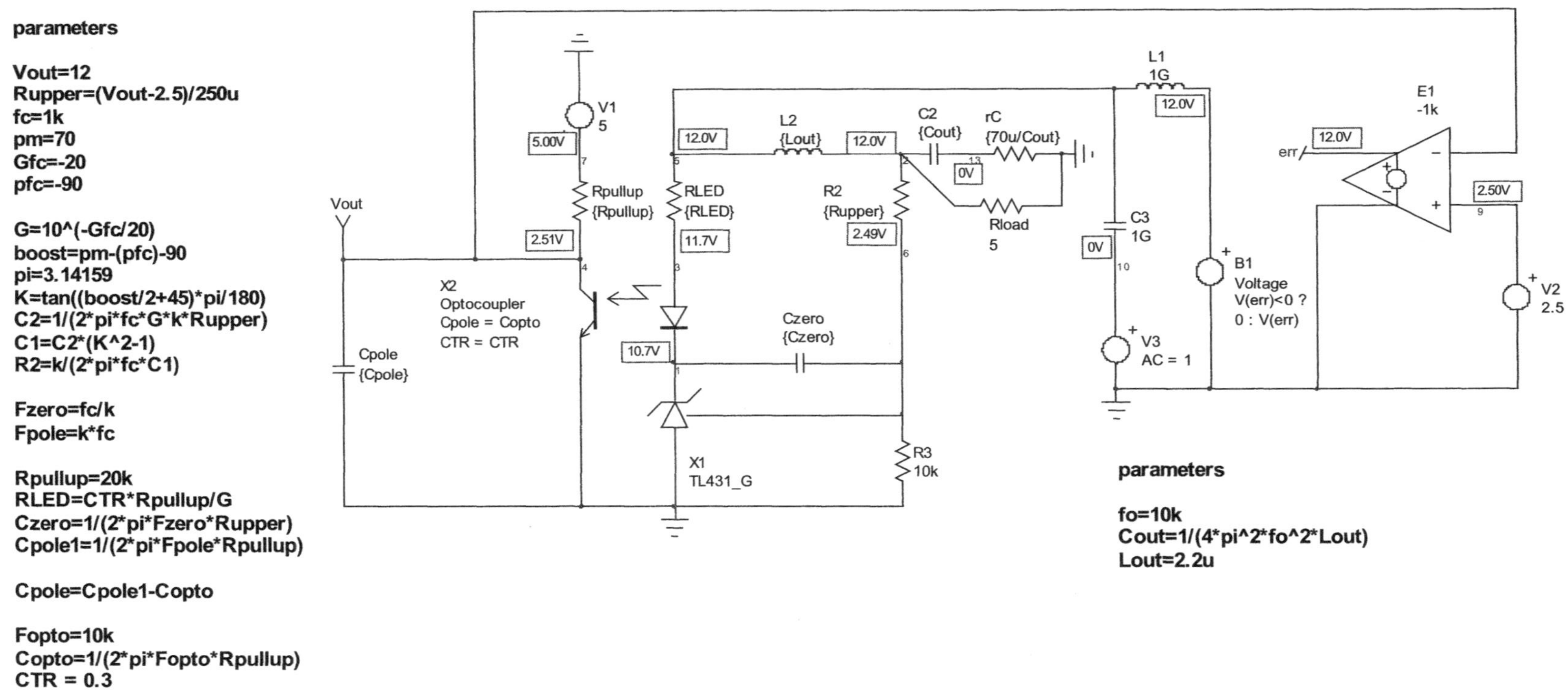

Figure 7.55 An automated type 2 compensator helps us to immediately see the effect of the *LC* network on the transfer function.

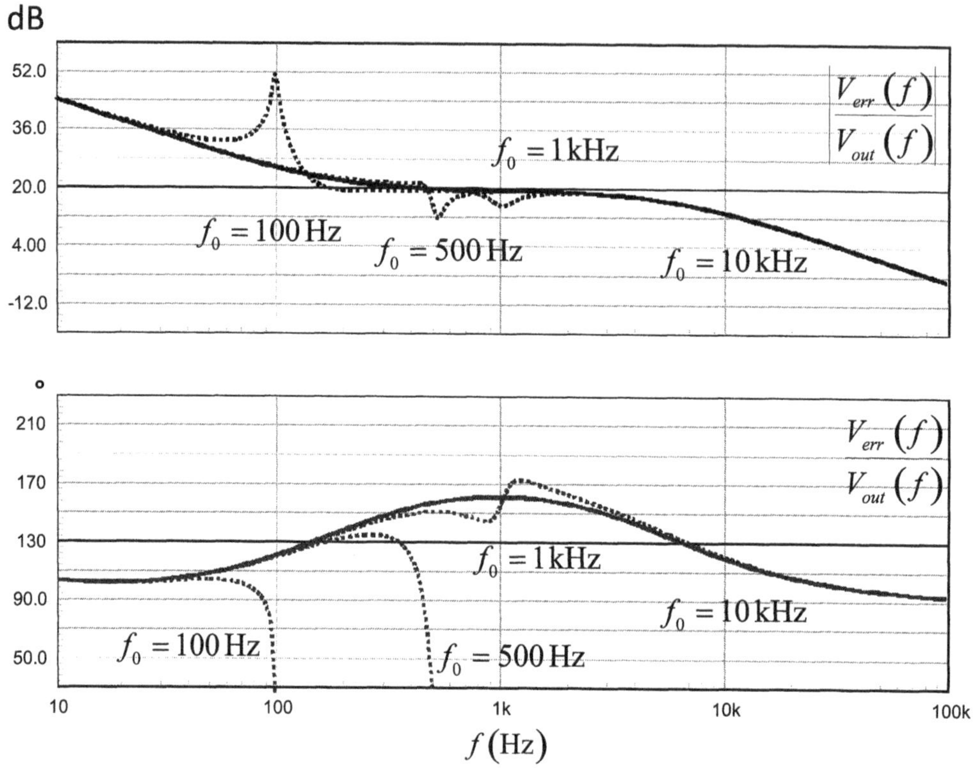

Figure 7.56 This ac plot shows the impact of the post *LC* filter resonant frequency on the TL431-based transfer function.

tioned at 800 Hz. With a fixed inductor value, the second capacitor C_2 is again automatically calculated to create a resonance at the selected frequency point. The ac results are given in Figure 7.58. The loop gain with a 2.2-μH/115-μF network (f_0 = 10 kHz) has no impact on the overall shape. When the resonant frequency is tuned to 5 kHz (L = 4.7 μH and C_2 = 215 μF), there is still no change of the loop response. Now, when the inductor is increased to 15 μH and C_2 to 1.7 mF, the cutoff frequency approaches the zero position and both gain and phase become out of control: the converter is unstable.

As a design recommendation, keep the *LC* network resonance frequency far away from the low-frequency zero location. Around a decade represents a good fit. As observed in the example where the zero seats at 800 Hz, the 10-kHz cutoff frequency does not change the overall response and will offer a good attenuation of the 65-kHz switching frequency ripple. Regarding the inductor value, keep in mind that its addition brings an extra annoying voltage undershoot in presence of sharp output current discontinuities. Values from 1 to 4.7 μH are usually observed in ac-dc adapters up to 100 W of output power.

As a final remark, when you install a *LC* filter, make sure the TL431 fast lane connection occurs before the network. You can see in Figure 7.55 that the upper terminal of R_{LED} connects before the inductor and not after. If you connect the resistor directly to the output, the LED current will have direct exposure to the phase and amplitude signals affected by the *LC* network and stability may suffer.

Loop measurements with the *LC* network require that you run multiple measurements in both lanes and vector-combine the data to reconstruct the overall loop

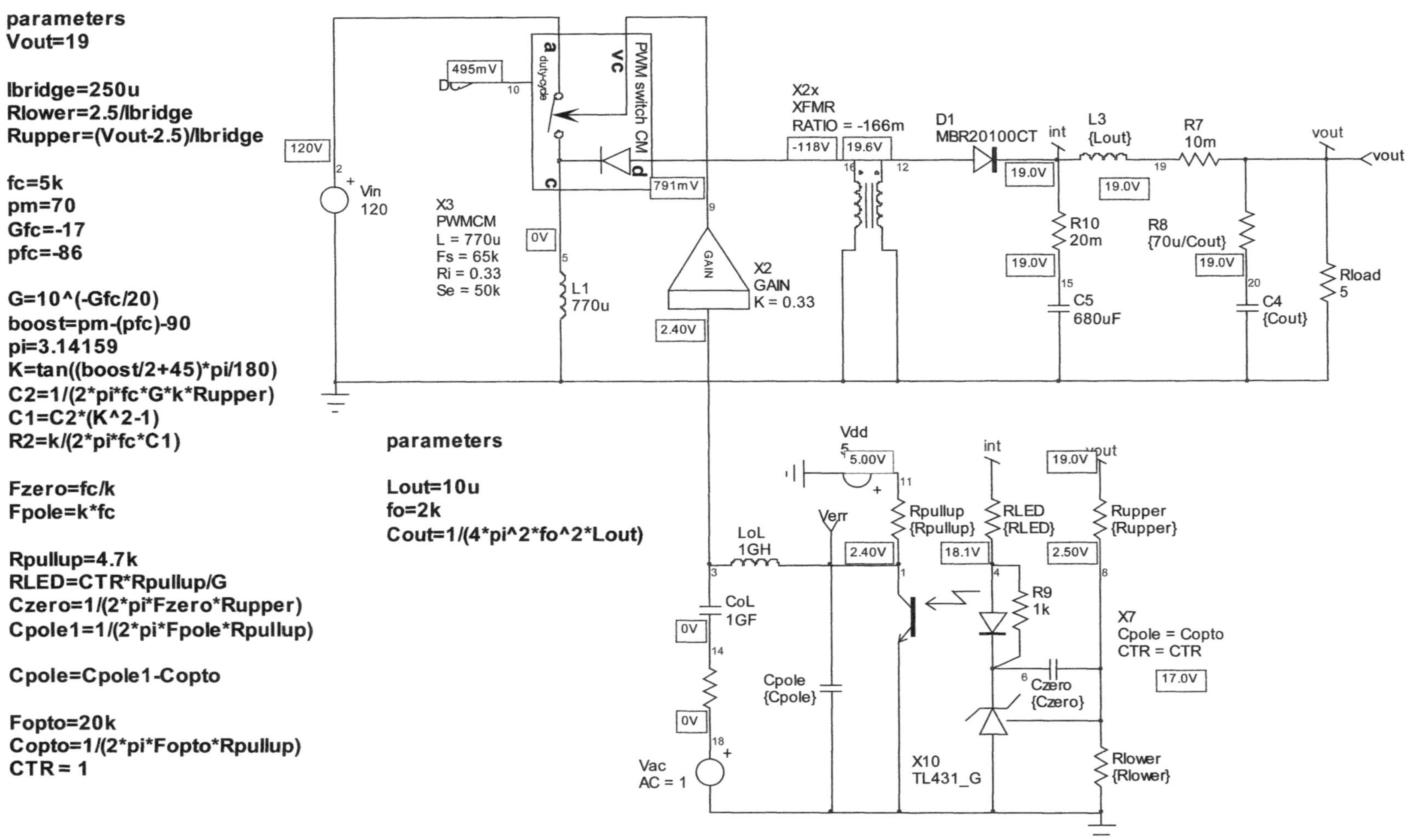

Figure 7.57 A SPICE simulation helps to understand the *LC* network impact at crossover.

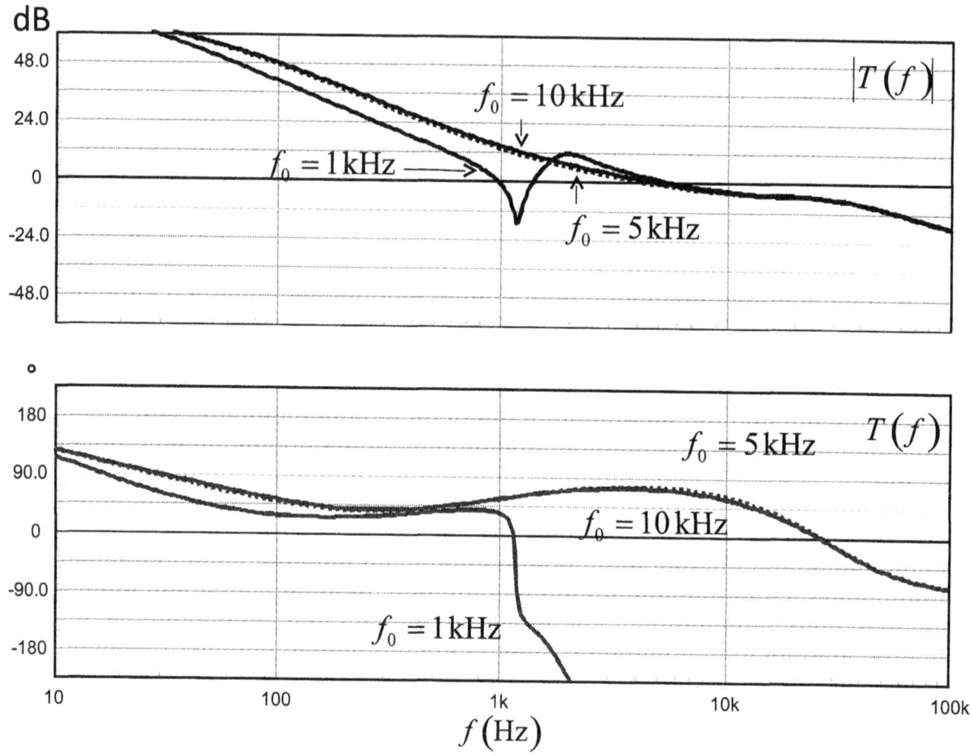

Figure 7.58 Ac results show no alteration at crossover when the *LC* network cutoff frequency stays away from the 1-kHz crossover point.

gain [3]. Another option is to open the loop at the optocoupler collector level as it naturally combines both signals. Please see Chapter 9 for more details.

References

[1] Irving, B., Y. Panov, and M. Jovanovic, "Small-Signal Model of Variable Frequency Flyback Converter," APEC 2003 Proceedings, Vol. 2, pp. 977–982.

[2] Ridley, R., "Designing with the TL431," Designer Series XV, www.switchingpowermagazine.com.

[3] Conseil, S., N. Cyr, and C. Basso, "Stability Analysis in Multiple Loop Systems," AND8327D, www.onsemi.com.

CHAPTER 8

Shunt Regulator–Based Compensators

In the previous chapters, we have studied compensators that deliver an error voltage, $V_{err}(s)$, which controls either the duty ratio D (voltage mode control) or the peak current setpoint (peak-current mode control). Popularized by Power Integrations over the past 15 years, the TOPSwitch® is a high-voltage switcher implementing the *shunt regulator*. Rather than controlling the duty ratio by changing a voltage level on a pin, the company combined the V_{cc} and the feedback pins together so that a single input could not only supply the control section but also drive the duty ratio excursion by monitoring the injected current: a three-pin integrated switcher was born. Even if the loop is controlled by a TL431 or a Zener diode, we dedicated a chapter to describe the compensator implementation given the peculiarity of the circuit. Figure 8.1 shows the representation of the shunt regulator as it can be found in a TOPSwitch device [1].

The circuit works like an active Zener diode featuring a dynamic resistor R_d of 15 Ω. When there is almost no injected current in the feedback pin (FB), the delivered duty ratio is maximum (67 percent). On the contrary, when more than 6 mA are injected, the duty ratio drops to its minimum value or 1.8 percent. Varying the injected current depending on the input/output conditions is the adopted means to adjust the duty ratio. To improve the noise immunity and also to shape the compensator response, a 7-kHz pole was included in the modulator path. We will need to account for its presence during the ac analysis. The curve depicting the modulator transfer function, D versus I_{FB}, appears in Figure 8.2.

Due to the information displayed on the curve, the small-signal gain of the whole modulator can be deduced. It is simply

$$S_{PWM} = \frac{\Delta D}{\Delta I_{FB}} = \frac{1.8-67}{6-2} \approx -16\%/\text{mA} \tag{8.1}$$

However, in the models we use, the duty ratio value is usually expressed in volts: a 1-V dc signal represents a 100 percent duty ratio. As current is primarily expressed in amperes rather than milliamperes, (8.1) needs to be reformulated to cope with our simulation models:

$$S_{PWM} = \frac{0.017-0.67}{6m-2m} = -163\text{ V/A} \tag{8.2}$$

Expressed in decibels, this slope becomes a gain of

$$G_{PWM} = 20\log_{10}(163) \approx 44\text{ dB} \tag{8.3}$$

This specific arrangement requires a little bit more of analysis compared to a traditional voltage-mode feedback. However, experience shows that it does not prevent

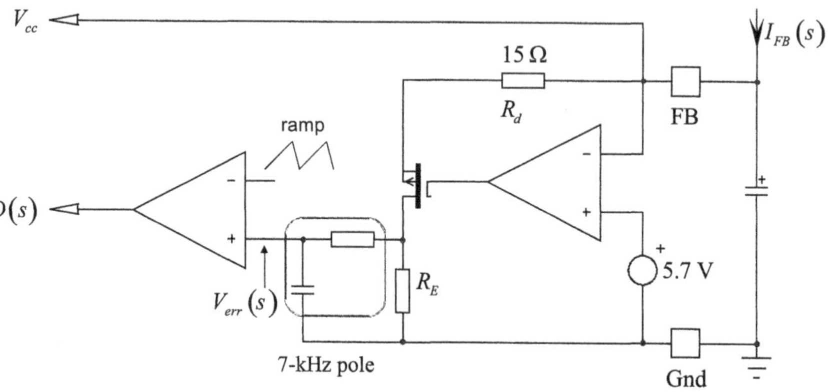

Figure 8.1 The internal structure of a TOPSwitch duty-cycle modulator. Unlike peak-current mode controllers whose setpoint is defined by a control voltage, the shunt regulator deals with a control current.

the implementations of the type 2 and 3 compensators as we described them in the previous chapters. The type 1 will not be described here, as a voltage-mode converter is usually compensated with a type 2 or even a type 3 when continuous conduction mode (CCM) is entered during an overload condition, for instance. If for a given reason a type 1 is needed, just use the type 2 formulas and set the phase boost to 0, and you have it!

8.1 The Type 2: An Origin Pole plus a Pole/Zero Pair

The duty ratio modulator input connects to the TL431-based compensator as suggested by Figure 8.3. The V_{CC} comes from an auxiliary winding, on the primary side of the converter. As the operating voltage of the switcher is around 5.7 V, this V_{CC} has to be above this value. 12 V is usually the adopted level.

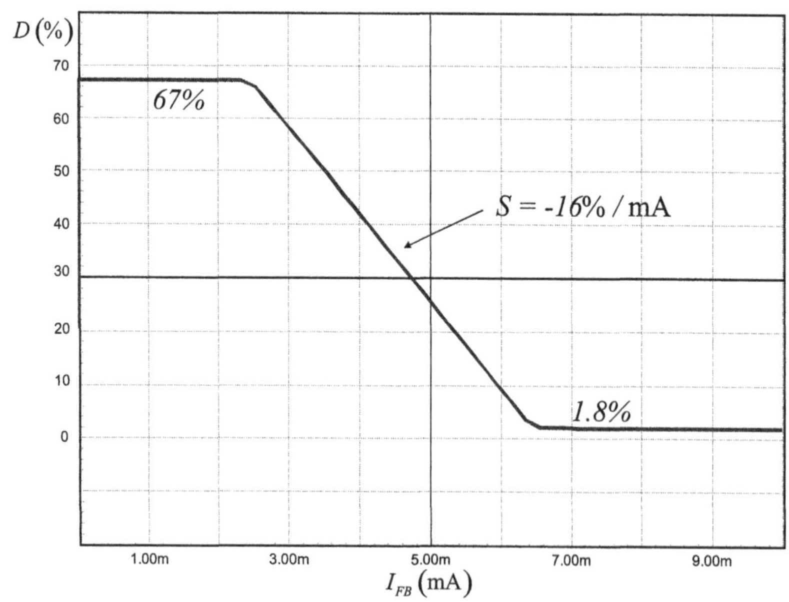

Figure 8.2 This curve shows the duty ratio variations in relationship to the injected current.

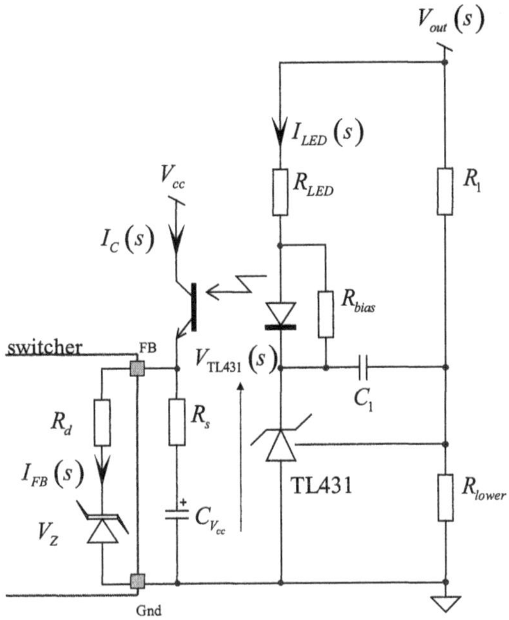

Figure 8.3 The TL431 regulates the output variable by injecting current into the feedback input of the power switcher. This current is derived from an auxiliary voltage.

When the output voltage deviates from its target, let's assume it increases, the current goes up in the optocoupler LED, and more current gets injected into the switcher feedback pin. According to Figure 8.2, the duty ratio diminishes. To ensure the self-supply of the switcher, a V_{cc} capacitor is connected from the feedback pin to the ground. It provides the necessary energy reservoir to help during the startup sequence and rectifies the high-frequency pulses when the converter operates. A 47-μF capacitor is a common choice for this part. Being wired between the feedback pin and ground, it also has an ac role as it introduces a pole. Since a zero might also be helpful for phase boost needs, it is of design practices to add a small resistor R_s in series with the capacitor. However, given the dc drop it introduces at startup, it is not recommended to go beyond 15 Ω unless an extra capacitor is added [2]. You are familiar with the rest of the arrangement around the TL431. Following these remarks, we can start the ac analysis. Let's start with the ac current circulating in the LED, neglecting the bias current contribution:

$$I_{LED}(s)=\frac{V_{out}(s)-V_{TL431}(s)}{R_{LED}}=\frac{V_{out}(s)+V_{out}(s)\left(\frac{1}{sR_1C_1}\right)}{R_{LED}}=\frac{V_{out}(s)}{R_{LED}}\frac{(1+sR_1C_1)}{sR_1C_1} \tag{8.4}$$

The LED current gives birth to the optocoupler collector current via the CTR of the device:

$$I_C(s)=\mathrm{CTR}\,I_{LED}(s) \tag{8.5}$$

This current then splits between the feedback pin and the capacitor network made of C_{Vcc} and R_s:

$$I_{FB}(s)=I_C(s)\frac{R_s+\frac{1}{sC_{V_{cc}}}}{R_d+R_s+\frac{1}{sC_{V_{cc}}}}=I_C(s)\frac{\frac{sR_sC_{V_{cc}}+1}{sC_{V_{cc}}}}{\frac{sC_{V_{cc}}(R_s+R_d)+1}{sC_{V_{cc}}}}=I_C(s)\frac{sR_sC_{V_{cc}}+1}{sC_{V_{cc}}(R_s+R_d)+1} \tag{8.6}$$

Now replacing $I_C(s)$ by its expression in (8.5) and (8.4), we have

$$\frac{I_{FB}(s)}{V_{out}(s)}=\frac{\text{CTR}}{R_{LED}}\frac{(1+sR_1C_1)}{sR_1C_1}\frac{sR_sC_{V_{cc}}+1}{sC_{V_{cc}}(R_s+R_d)+1} \tag{8.7}$$

Factoring $1+sR_1C_1$, we obtain

$$\frac{I_{FB}(s)}{V_{out}(s)}=\frac{\text{CTR}}{R_{LED}}\frac{1+1/sR_1C_1}{1+sC_{V_{cc}}(R_s+R_d)}(1+sR_sC_{V_{cc}}) \tag{8.8}$$

The current $I_{FB}(s)$ is injected into the duty ratio modulator whose transfer function includes a gain as described by (8.3), followed by a 7-kHz pole:

$$\frac{D(s)}{I_{FB}(s)}=-G_{PWM}\frac{1}{1+s/\omega_{p_2}} \tag{8.9}$$

The pole helps for the noise immunity but also offers the high-frequency pole f_{p2} we need for a type-2 configuration. Unfortunately, it is fixed and we will have to deal with it. If we now combine both expressions given by (8.8) and (8.9), we obtain the complete chain from the compensator input to the duty ratio generator:

$$\frac{D(s)}{V_{out}(s)}=-G_0\frac{1+\omega_{z_2}/s}{1+s/\omega_{p_1}}\frac{1+s/\omega_{z_1}}{1+s/\omega_{p_2}} \tag{8.10}$$

This is what Figure 8.4 depicts.

In equation (8.10), we have:

$$G_0=\frac{\text{CTR}}{R_{LED}}G_{PWM} \tag{8.11}$$

$$\omega_{z_1}=\frac{1}{R_sC_{V_{cc}}} \tag{8.12}$$

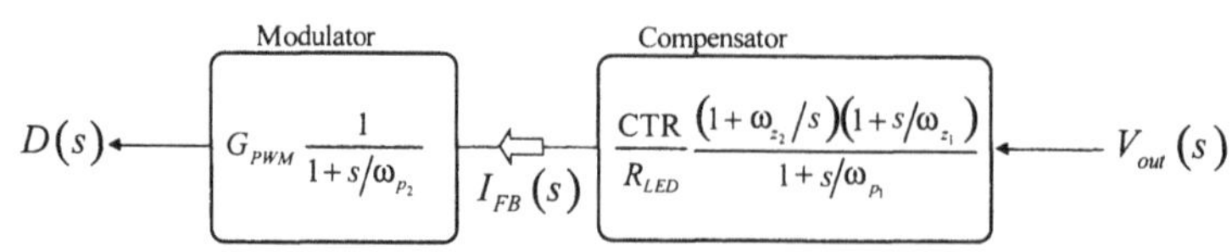

Figure 8.4 The chain from the observed variable to the duty ratio generator cascades two stages: the modulator and the compensator.

$$\omega_{z_2} = \frac{1}{R_1 C_1} \tag{8.13}$$

$$\omega_{p_1} = \frac{1}{(R_d + R_s)C_{V_{cc}}} \tag{8.14}$$

$$\omega_{p_2} = 44 \text{ krad/s} \tag{8.15}$$

This is two-pole and two-zero combination plus an origin pole. As we need a simple pole/zero pair and an origin pole, one pole and one zero will have to be canceled. We will see how in the design example. The resistor setting the mid-band gain is the LED series resistor. We need to extract its value for the magnitude definition of (8.8):

$$|G(f_c)| = \frac{\text{CTR}}{R_{LED}} G_{PWM} \frac{\sqrt{1+(f_{z_2}/f_c)^2}\sqrt{1+(f_c/f_{z_1})^2}}{\sqrt{1+(f_c/f_{p_1})^2}\sqrt{1+(f_c/f_{p_2})^2}} \tag{8.16}$$

From which we can extract the value of the LED resistor:

$$R_{LED} = \frac{\text{CTR}}{G} G_{PWM} \frac{\sqrt{1+\left(\frac{f_{z_2}}{f_c}\right)^2}\sqrt{1+\left(\frac{f_c}{f_{z_1}}\right)^2}}{\sqrt{1+\left(\frac{f_c}{f_{p_1}}\right)^2}\sqrt{1+\left(\frac{f_c}{f_{p_2}}\right)^2}} \tag{8.17}$$

Let's now have a look at the biasing limits. The maximum duty ratio is obtained when around 7 mA are injected in the feedback pin. As this current circulates in the optocoupler and the LED, its series resistor cannot be too high; otherwise, the compensator will fail to properly regulate in light load conditions where maximum current is needed. The maximum current the LED resistor will face occurs when I_C equals 7 mA:

$$I_{R_{LED},\max} = \frac{I_{C,\max}}{\text{CTR}_{\min}} + I_{bias} \tag{8.18}$$

The current circulating in the LED resistor is made of the reflected collector current plus the bias current I_{bias}. As the voltage on the TL431 cannot be lower than 2.5 V, we have

$$I_{R_{LED},\max} = \frac{V_{out} - V_f - V_{\text{TL431,min}}}{R_{LED}} \tag{8.19}$$

As (8.18) and (8.19) are equal, we can solve the maximum LED resistor value satisfying the proper operating conditions:

$$R_{LED,\max} \le \frac{\text{CTR}_{\min}\left(V_{out} - V_f - V_{\text{TL431,min}}\right)}{I_{C,\max} + I_{bias}\text{CTR}_{\min}} \tag{8.20}$$

Now that we have everything on hand, it is time for a design example. . .

8.1.1 A Design Example

The design example we will choose is a flyback converter operated in DCM. It appears in Figure 8.5 and shows how we combined the voltage-mode model and the shunt regulator sub circuit. The description of these two blocks is thoroughly documented in [3]. The ac sweep requires a little bit of care to choose the opening point. Remember that the loop regulates the output voltage by injecting a current in the feedback pin. In most of our SPICE analysis, we deal with voltage transfer loops. In our example, if we only have access to the feedback pin, we are stuck and need to observe the feedback current, probably through an extra resistor. This is not a big deal, but it requires extra manipulations to get the result. To circumvent this difficulty, we will open the loop after the modulator since it delivers a voltage between 18 mV and 680 mV for a duty ratio comprised between 1.8 percent and 67 percent. This is what the figure shows. This will help to observe the power stage trans-

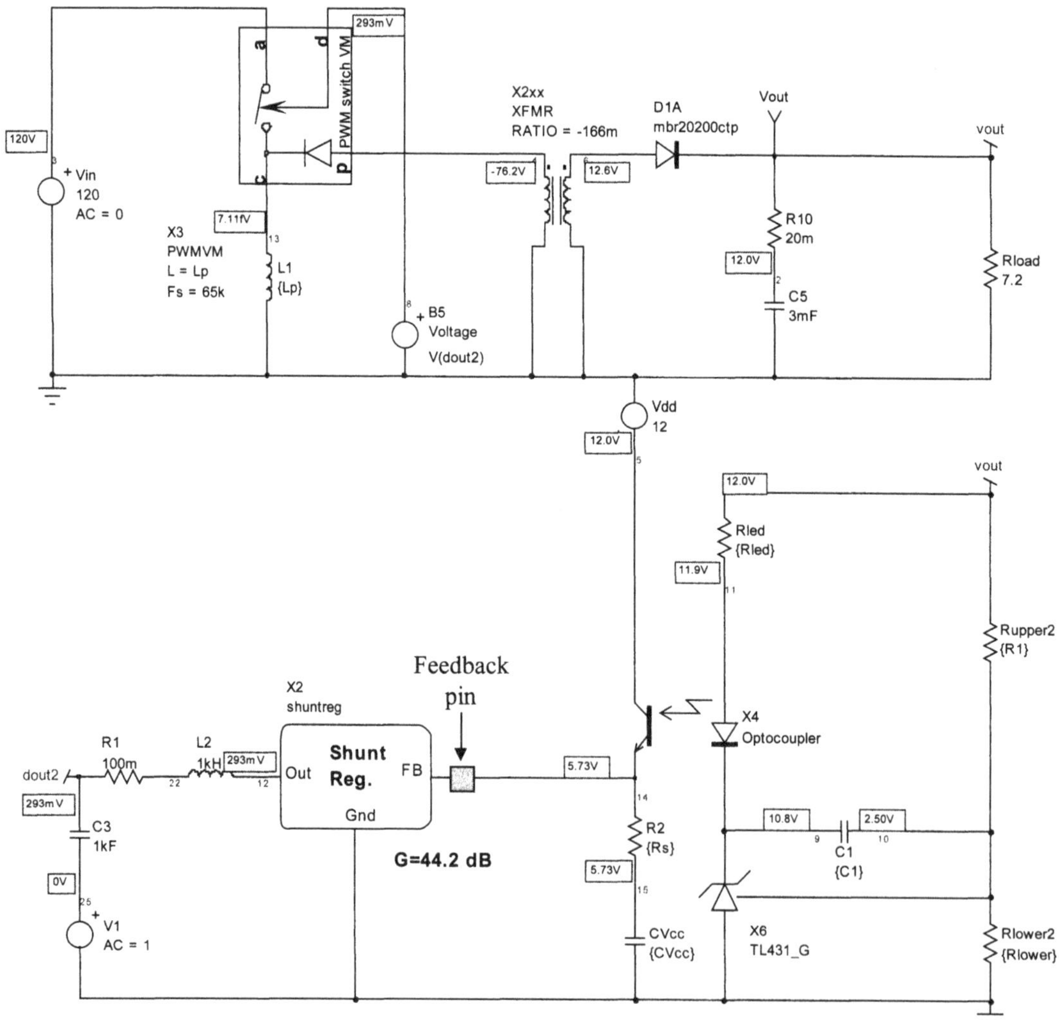

Figure 8.5 The flyback simulation of the TOPSwitch requires a particular loop opening, after the modulator.

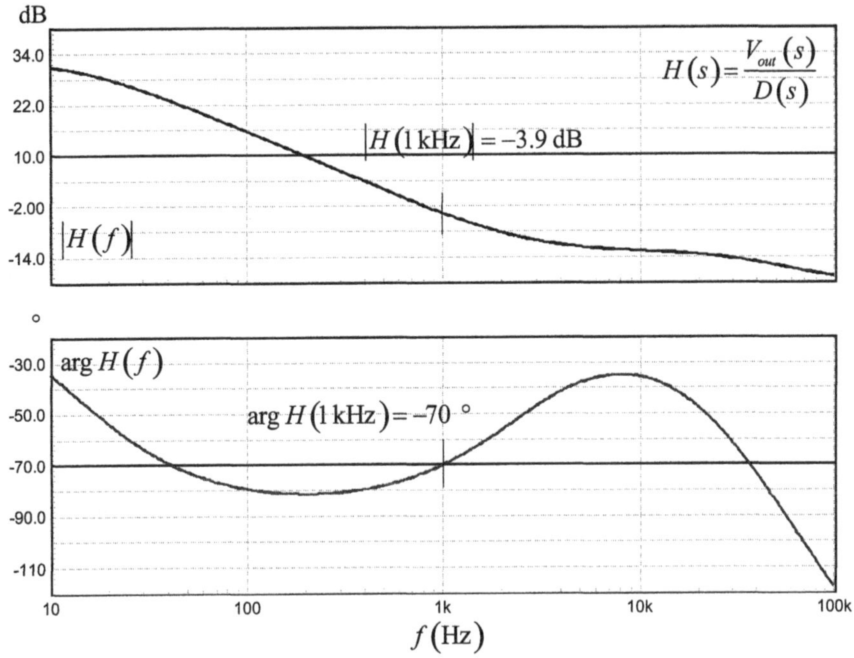

Figure 8.6 The power stage transfer function of the flyback operated in DCM. Please note that this is the output $V_{out}(s)$ to the duty ratio $D(s)$ input that is plotted.

fer function, excluding the modulator of course. The whole chain from V_{out} to D being available through (8.10), we have everything to carry on. One remark though. In real life, you will not be able to open the loop after the modulator since all is internal circuitry. Following guidelines in [4], it is likely that the injection occurs on the feedback pin directly, through a series resistor. In that case, the power stage transfer function will already be affected by the PWM modulator gain of 44 dB. Therefore, when calculating the LED series resistor value with the help of (8.17), the G_{PWM} term will have to disappear from the formula. The rest stays as it is.

The transfer function appears in Figure 8.6. If we select a 1-kHz crossover frequency with 60° phase margin, the gain deficiency at this point is –3.9 dB and the phase lag is –70°.

Let us debut the calculation by assessing the maximum LED resistor value, using (8.20) and assuming the following parameters:

V_{out} = 12 V; the output voltage.
V_f = 1 V; the LED forward voltage.
I_{bias} = 1 mA; the TL431 biasing current when the optocoupler LED is paralleled with a resistor.
$V_{\text{TL431,min}}$ = 2.5 V; the minimum operating voltage of the TL431.
CTR_{min} = 0.8; the minimum optocoupler current transfer ratio.
R_1 = 38 kΩ; the upper resistor in the resistor bridge observing the output variable.
C_{Vcc} = 47 μF; the selected V_{cc} capacitor according to the switcher data sheet.

$$R_{LED,\max} \le \frac{\text{CTR}_{\min}\left(V_{out} - V_f - V_{\text{TL431,min}}\right)}{I_{C,\max} + I_{bias}\text{CTR}_{\min}} \le \frac{0.8\times(12-1-2.5)}{7m+1m\times 0.8} \le 1.09\ \text{k}\Omega \tag{8.21}$$

The 3.9-dB of gain translate to

$$G = 10^{\frac{-Gf_c}{20}} = 10^{\frac{3.9}{20}} \approx 1.6 \tag{8.22}$$

Given the 70° phase lag at the selected crossover frequency and a desired phase margin of 60°, we need to boost the phase by 40°. Considering the 1-kHz frequency selection and the presence of a 7-kHz pole, we need to place a first zero at

$$f_{z1} = \frac{f_c}{\tan\left(boost + \tan^{-1}\left(\frac{f_c}{f_{p2}}\right)\right)} = \frac{1k}{\tan\left(40 + \tan^{-1}\left(\frac{1k}{7k}\right)\right)} = 896\ \text{Hz} \tag{8.23}$$

This first zero will serve to calculate the resistor R_s in series with the 47-μF V_{cc} capacitor:

$$R_s = \frac{1}{2\pi f_{z1} C_{V_{cc}}} = \frac{1}{6.28\times 896\times 47u} \approx 3.8\ \Omega \tag{8.24}$$

It is below the maximum of 15 Ω indicated by the switcher manufacturer, so we are safe. This resistor now combines with the shunt regulator input impedance to form a pole together with the V_{cc} capacitor as explained by (8.14):

$$f_{p1} = \frac{1}{2\pi(R_d + R_s)C_{V_{cc}}} = \frac{1}{6.28\times(15+3.8)\times 47u} = 180\ \text{Hz} \tag{8.25}$$

Equation (8.10) actually reveals a type 3 expression exhibiting a double pair of poles/zeros and an origin pole. For our type 2 compensation, one pair of pole/zero must be neutralized. This is the couple $f_{z2} f_{p1}$. We know that the first pole f_{p1} is positioned at 180 Hz; let's us calculate the value of C_1 to position the second zero f_{z2} right at this point:

$$C_1 = \frac{1}{2\pi f_{z2} R_1} = \frac{1}{6.28\times 180\times 38k} = 23.3\ \text{nF} \tag{8.26}$$

Given the neutralization of a pole/zero pair, we are now left with a type 2 system having a pole at 7 kHz, a zero at 896 Hz, and an origin pole. To finish our design, we just need to calculate the value of R_{LED} as expressed by (8.17):

$$R_{LED} = \frac{\text{CTR}}{G} G_{PWM} \frac{\sqrt{1+\left(\frac{f_{z2}}{f_c}\right)^2}\sqrt{1+\left(\frac{f_c}{f_{z1}}\right)^2}}{\sqrt{1+\left(\frac{f_c}{f_{p1}}\right)^2}\sqrt{1+\left(\frac{f_c}{f_{p2}}\right)^2}} = \frac{0.8}{1.6}\times 10^{\frac{44}{20}}$$
$$\times \frac{\sqrt{1+\left(\frac{180}{1k}\right)^2}\sqrt{1+\left(\frac{1k}{896}\right)^2}}{\sqrt{1+\left(\frac{1k}{180}\right)^2}\sqrt{1+\left(\frac{1k}{7k}\right)^2}} \approx 21\ \Omega \tag{8.27}$$

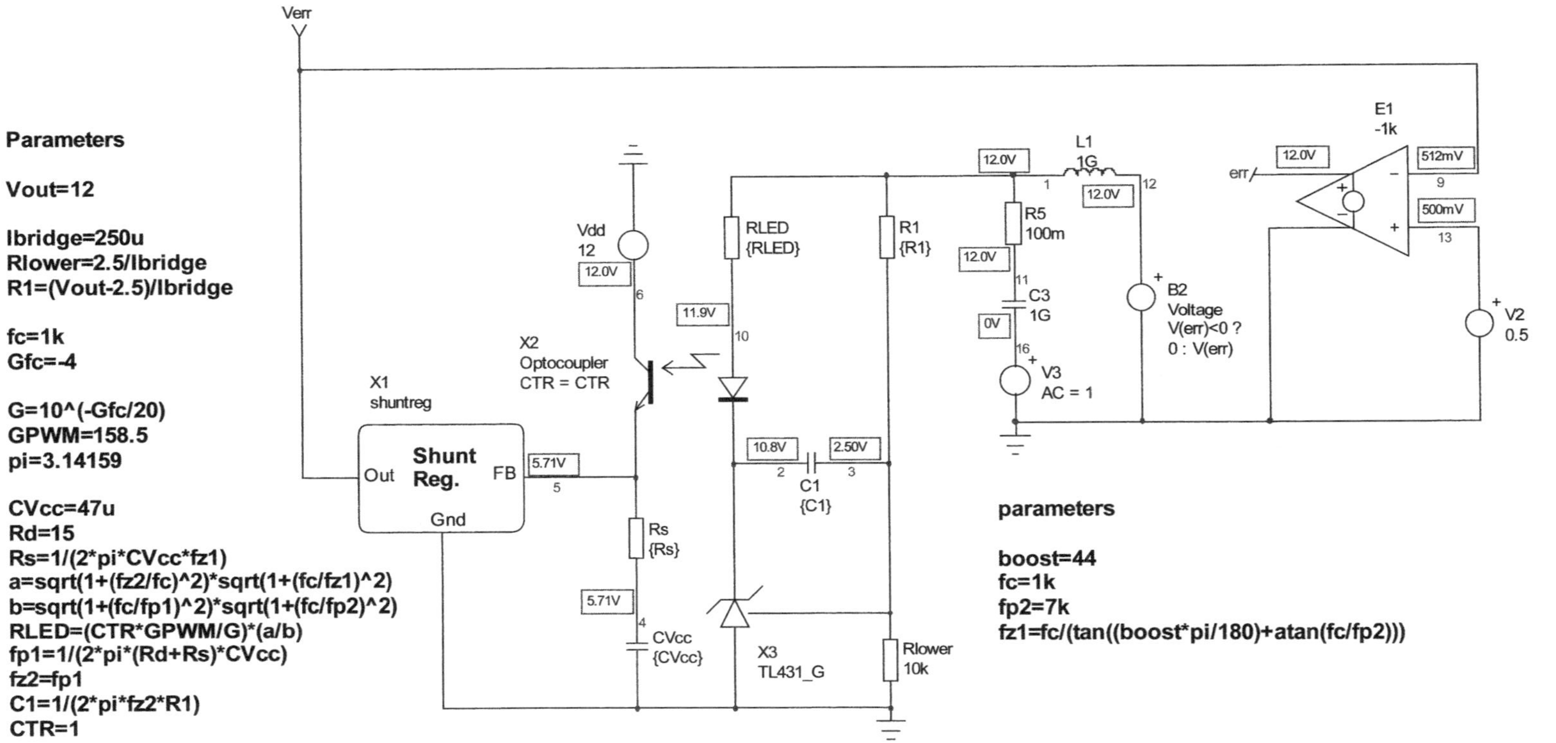

Figure 8.7 The compensator using the shunt regulator works well but requires care in selecting the surrounding elements.

Having all these elements calculated, we can now feed our test fixture and check the ac response of this compensator alone. This is what Figure 8.7 portrays. The bias points are good. The current in the LED resistor is around 3.5 mA, which according to Figure 8.2 must correspond to around 50 percent duty ratio. This is the value obtained on the output of the shunt regulator subcircuit: 512 mV corresponds to a 51.2 percent duty ratio.

The ac response of this compensator appears in Figure 8.8. The overall shape is ok, despite a small mismatch on the gain target. This is the effect of the LED dynamic resistor not accounted for in the calculations. As the series LED resistor is rather small, the contribution of the dynamic resistor is more significant. The phase boost is exhibiting the right value, but the peak is not occurring at the 1-kHz target. This is because of 7-kHz internal pole that we undergo cannot be adjusted at the right value. To benefit from the maximum phase boost, we should change the crossover frequency and put it to

$$f_{c,new} = \sqrt{f_{z1} f_{p2}} = \sqrt{896 \times 7k} = 2.5 \text{ kHz} \tag{8.28}$$

According to Figure 8.6, the gain target should then be changed to ≈ 9 dB in order to crossover at 2.5 kHz.

The compensated version of Figure 8.5 has been ac swept and also tested on transient response. The compound results are given in Figure 8.9 and show a good phase margin of nearly 60°. The output has been stepped from 0.8 A to

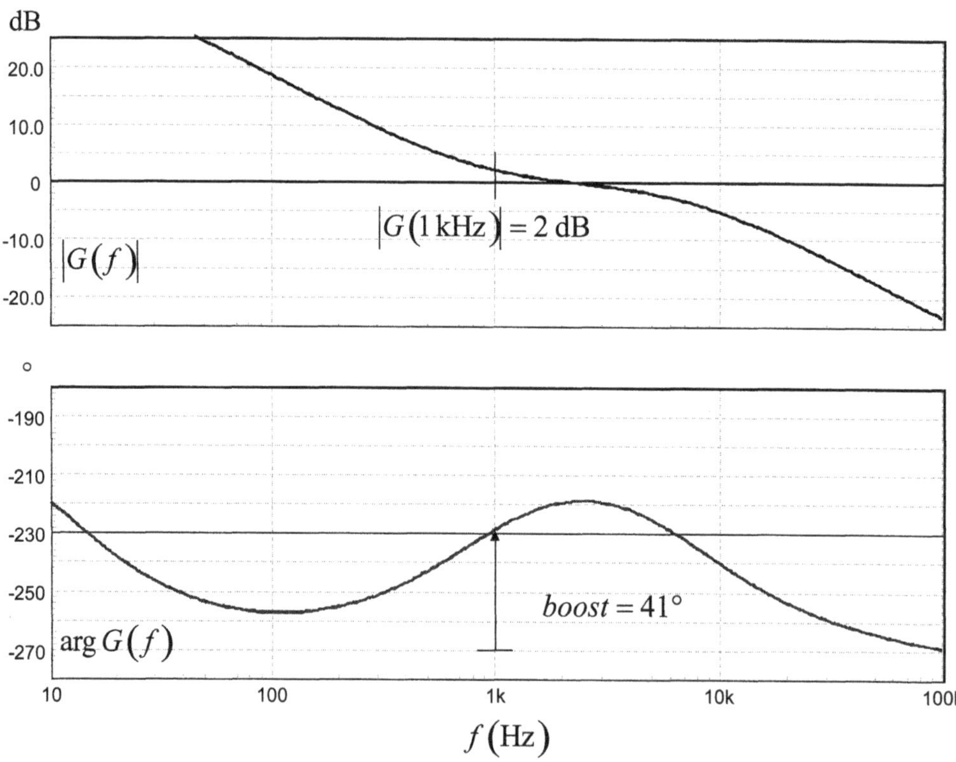

Figure 8.8 The ac response shows a slight discrepancy between the gain target and what is obtained. Again, the culprit is the LED dynamic resistor.

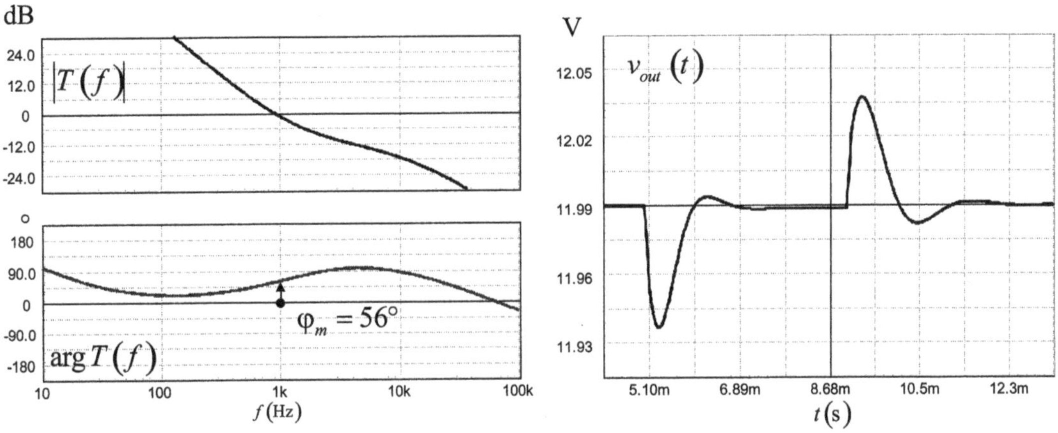

Figure 8.9 The open-loop gain shows a good phase margin, leading to an excellent transient response.

2 A with a slew-rate of 1 A/μs, and the deviation is very small: we have a stable power supply.

What limits affect this design? The first one relates to the LED series resistor value. In this design example, we have a 21-Ω resistor, so we have plenty of margin compared to the upper limit of 1 kΩ given by (8.21). The other limitation comes from the resistor inserted in series with the V_{CC} capacitor. This resistor introduces a zero, which is helpful to boost the phase in relationship to the 7-kHz internal pole. However, a resistor in series with the V_{CC} capacitor will affect the storage capabilities of the network during the startup sequence and the auto-recovery mode. This is because the voltage drop that appears across this series element can, in case it is too high, prematurely trip the undervoltage lockout circuitry. The data sheet advises a value less than 15 Ω in some TOPSwitch references. You can go beyond this value, but an extra capacitor has to be added (see [2] for more details). If we stick with a 15 Ω value and the recommended 47-μF capacitor, the zero you will add cannot be lower than

$$f_{z1,\max} = \frac{1}{2\pi R_{s,\max} C_{V_{cc}}} = \frac{1}{6.28 \times 15 \times 47u} \approx 226 \text{ Hz} \tag{8.29}$$

We know that the maximum phase boost peaks at the geometric mean between the considered zero and pole. For a 226-Hz zero and a 7-kHz pole, the peak will occur at

$$f_c = \sqrt{f_{z1} f_{p2}} = \sqrt{226 \times 7k} = 1.25 \text{ kHz} \tag{8.30}$$

At this frequency, the maximum phase boost you can obtain will be

$$Boost_{\max} = \tan^{-1}\left(\frac{f_c}{f_{z1}}\right) - \tan^{-1}\left(\frac{f_c}{f_{p2}}\right) = \tan^{-1}\left(\frac{1.25k}{226}\right) - \tan^{-1}\left(\frac{1.25k}{7k}\right) \approx 70° \tag{8.31}$$

8.2 The Type 3: An Origin Pole plus a Double Pole/Zero Pair

The type 3 compensator offers a larger phase boost than its type 2 counterpart. Unfortunately, because of the fixed internal 7-kHz pole present in the duty ratio modulator and the limit set for the resistor R_s, we cannot expect the full flexibility as with a traditional op amp approach. That being said, let us have a look at a type 3 implementation as it appears in Figure 8.10. The principle remains the same: how do we add another pole/zero pair given the fast lane presence? As we did for the TL431 with a voltage-mode feedback control, a RC network will be added in parallel with the LED series resistor. The network will add the zero/pole pair we are looking for in a type 3 architecture. As we already derived the transfer function of the type 2—see (8.8)—going to the type 3 simply requires us to replace R_{LED} with the equivalent impedance brought by the new network Z_{LED} highlighted in Figure 8.10:

$$Z_{LED} = \frac{R_{LED}\left(R_3 + \frac{1}{sC_3}\right)}{R_{LED} + \left(R_3 + \frac{1}{sC_3}\right)} = R_{LED}\frac{sR_3C_3 + 1}{sC_3\left(R_{LED} + R_3\right) + 1} \tag{8.32}$$

Now substituting this expression in (8.8), we have

$$\frac{I_{FB}(s)}{V_{out}(s)} = \frac{\text{CTR}}{R_{LED}}\frac{1 + 1/sR_1C_1}{1 + sC_{V_{cc}}\left(R_s + R_d\right)}\left(1 + sR_sC_{V_{cc}}\right)\frac{sC_3\left(R_{LED} + R_3\right) + 1}{sR_3C_3 + 1} \tag{8.33}$$

The current $I_{FB}(s)$ is injected into the duty ratio modulator whose transfer function includes a gain as described by (8.3), followed by a 7-kHz pole:

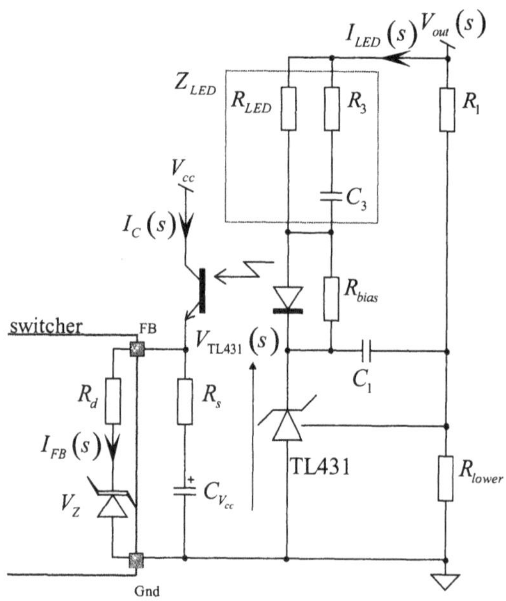

Figure 8.10 A type 3 implementation with a shunt regulator suffers from the limits imposed on R_s and the internal 7-kHz pole.

$$\frac{D(s)}{I_{FB}(s)} = -G_{PWM}\frac{1}{1+s/\omega_{p_2}} \tag{8.34}$$

If we now combine expressions given by (8.33) and (8.34), we obtain the complete chain from the compensator input to the duty ratio output:

$$\frac{D(s)}{V_{out}(s)} = -G_0\frac{(1+\omega_{z_2}/s)(1+s/\omega_{z_1})(1+s/\omega_{z_3})}{(1+s/\omega_{p_1})(1+s/\omega_{p_2})(1+s/\omega_{p_3})} \tag{8.35}$$

In (8.35), we have

$$G_0 = \frac{\text{CTR}}{R_{LED}}G_{PWM} \tag{8.36}$$

$$\omega_{z_1} = \frac{1}{R_s C_{V_{cc}}} \tag{8.37}$$

$$\omega_{z_2} = \frac{1}{R_1 C_1} \tag{8.38}$$

$$\omega_{z_3} = \frac{1}{sC_3(R_{LED}+R_3)} \tag{8.39}$$

$$\omega_{p_1} = \frac{1}{(R_d+R_s)C_{V_{cc}}} \tag{8.40}$$

$$\omega_{p_2} = 44 \text{ krad/s} \tag{8.41}$$

$$\omega_{p_3} = \frac{1}{sR_3C_3} \tag{8.42}$$

The resistor setting the mid-band gain is the LED series resistor. We first start by extracting the magnitude of the compensator gain, G(s), as derived in (8.35):

$$|G(f_c)| = \frac{\text{CTR}}{R_{LED}}G_{PWM}\frac{\sqrt{1+(f_{z_2}/f_c)^2}\sqrt{1+(f_c/f_{z_1})^2}\sqrt{1+(f_c/f_{z_3})^2}}{\sqrt{1+(f_c/f_{p_1})^2}\sqrt{1+(f_c/f_{p_2})^2}\sqrt{1+(f_c/f_{p_3})^2}} \tag{8.43}$$

From which we can extract the value of the LED resistor:

$$R_{LED} = \frac{\text{CTR}}{G}G_{PWM}\frac{\sqrt{1+\left(\frac{f_{z2}}{f_c}\right)^2}\sqrt{1+\left(\frac{f_c}{f_{z1}}\right)^2}\sqrt{1+\left(\frac{f_c}{f_{z3}}\right)^2}}{\sqrt{1+\left(\frac{f_c}{f_{p1}}\right)^2}\sqrt{1+\left(\frac{f_c}{f_{p2}}\right)^2}\sqrt{1+\left(\frac{f_c}{f_{p3}}\right)^2}} \tag{8.44}$$

The definitions to obtain C_3 and R_3 are rather simple to derive. Extract R_3 from (8.42):

$$R_3 = \frac{1}{2\pi f_{p_3} C_3} \tag{8.45}$$

Substitute (8.45) in (8.39), and solve for C_3:

$$C_3 = \frac{f_{p3} - f_{z3}}{2\pi R_{LED} f_{p3} f_{z3}} \tag{8.46}$$

The design methodology will remain similar to that of the type 2. As we have a system featuring three poles and three zeros now, a pole/zero pair has to be neutralized. These are f_{z_2} and f_{p_1}. Then the rest of the components follow easily, as we will see in the upcoming design example.

8.2.1 A Design Example

In this design example, let us assume the following component values:

V_{out} = 12 V; the output voltage.
V_f = 1 V; the LED forward voltage.
I_{bias} = 1 mA; the TL431 biasing current when the optocoupler LED is paralleled with a resistor
$V_{TL431,min}$ = 2.5 V; the minimum operating voltage of the TL431.
CTR_{min} = 0.8; the minimum optocoupler current transfer ratio.
R_1 = 38 kΩ; the upper resistor in the resistor bridge observing the output variable.
C_{Vcc} = 47 μF; the selected V_{cc} capacitor according to the switcher data sheet.

From the transfer function of the voltage-mode forward converter we want to stabilize, we need a gain of 10 dB at a 2-kHz frequency, while the needed phase boost is 120° at crossover. Is this feasible, given the limitations brought by the shunt regulator? We do not have too many choices. The easiest way is to position a second pole at 7 kHz and see where to position the double zero given the targeted phase boost. From our first chapters, we know that for a given phase boost and a known double pole position (7 kHz in our case), the double zero must be placed at the following place:

$$f_{z1,2} = \frac{f_c}{\tan\left(\frac{boost}{2} + \tan^{-1}\frac{f_c}{f_{p1,2}}\right)} \tag{8.47}$$

This formula is valid only as long as the denominator, $\tan(x)$, is different from 0. This would occur if x is either 0 or is equal to 90°. From (8.47), we can quickly solve the value of f_c for which x equals 90°:

$$f_c = \tan\left(90 - \frac{boost}{2}\right) f_{p1,2} \tag{8.48}$$

Given the 7-kHz double pole we have chosen and the 120° we are looking for, it gives us a maximum crossover frequency of

$$f_c = \tan\left(90 - \frac{120}{2}\right) \times 7k \approx 4 \text{ kHz} \tag{8.49}$$

Given the 2-kHz crossover value we want, we have some margin there. To reach the 120° phase boost at 2 kHz and assuming two coincident poles at 7 kHz, we will have to place the double zero at the following frequency:

$$f_{z1,2} = \frac{f_c}{\tan\left(\frac{boost}{2} + \tan^{-1}\frac{f_c}{f_{p1,2}}\right)} = \frac{2k}{\tan\left(\frac{120}{2} + \tan^{-1}\left(\frac{2k}{7k}\right)\right)} \approx 500\text{ Hz} \tag{8.50}$$

This value leads us to define the series resistor value, R_s:

$$R_s = \frac{1}{2\pi f_{z1} C_{V_{cc}}} = \frac{1}{6.28 \times 500 \times 47u} = 6.8\ \Omega \tag{8.51}$$

Together with the shunt regulator equivalent dynamic resistor, it creates a pole located at

$$f_{p1} = \frac{1}{2\pi (R_s + R_d) C_{V_{cc}}} = \frac{1}{6.28 \times (6.8 + 15) \times 47u} = 155\text{ Hz} \tag{8.52}$$

This pole has to be canceled by the second zero we are going to place with C_1:

$$C_1 = \frac{1}{2\pi f_{z2} R_1} = \frac{1}{6.28 \times 155 \times 38k} = 27\text{ nF} \tag{8.53}$$

The upcoming design steps require the knowledge of the LED series resistor. Applying (8.44) and using (8.3) for the duty ratio modulator gain, we have

$$R_{LED} = \frac{0.8}{10^{\frac{10}{20}}} \times 163 \times \frac{\sqrt{1+\left(\frac{155}{2k}\right)^2}\sqrt{1+\left(\frac{2k}{500}\right)^2}\sqrt{1+\left(\frac{2k}{500}\right)^2}}{\sqrt{1+\left(\frac{2k}{155}\right)^2}\sqrt{1+\left(\frac{2k}{7k}\right)^2}\sqrt{1+\left(\frac{2k}{7k}\right)^2}} = 50\ \Omega \tag{8.54}$$

As the values adopted in this example are the same as in the previous exercise for the type 2, the result of (8.21) still holds. We are safe; there is plenty of margin. The third pole has been chosen to be coincident with the internal one, f_{p2}, at 7 kHz, and the third zero will also be coincident with the first one at 500 Hz. Using (8.46), we can evaluate the value of C_3:

$$C_3 = \frac{f_{p3} - f_{z3}}{2\pi R_{LED} f_{p3} f_{z3}} = \frac{7k - 500}{6.28 \times 50 \times 7k \times 500} = 5.9\ \mu\text{F} \tag{8.55}$$

Having C_3, we can get the series resistor R_3:

$$R_3 = \frac{1}{2\pi f_{p3} C_3} = \frac{1}{6.28 \times 7k \times 5.9u} = 3.9\ \Omega \tag{8.56}$$

Parameters

Vout=12

Ibridge=250u
Rlower=2.5/Ibridge
R1=(Vout-2.5)/Ibridge

fz1=fz12
fz3=fz12
fp2=fp12
fp3=50k

Gfc=-10

G=10^(-Gfc/20)
GPWM=158.5
pi=3.14159

CVcc=47u
Rd=15
Rs=1/(2*pi*CVcc*fz1)

a=sqrt(1+(fz2/fc)^2)*sqrt(1+(fc/fz1)^2)
b=sqrt(1+(fc/fp1)^2)*sqrt(1+(fc/fp2)^2)
c=sqrt(1+(fc/fz3)^2)/sqrt(1+(fc/fp3)^2)

RLED=(CTR*GPWM/G)*(a/b)*c
fp1=1/(2*pi*(Rd+Rs)*CVcc)
fz2=fp1
C1=1/(2*pi*fz2*R1)
C3=(fp3-fz3)/(RLED*fz3*fp3*2*pi)
R3=1/(2*pi*C3*fp3)
CTR=1

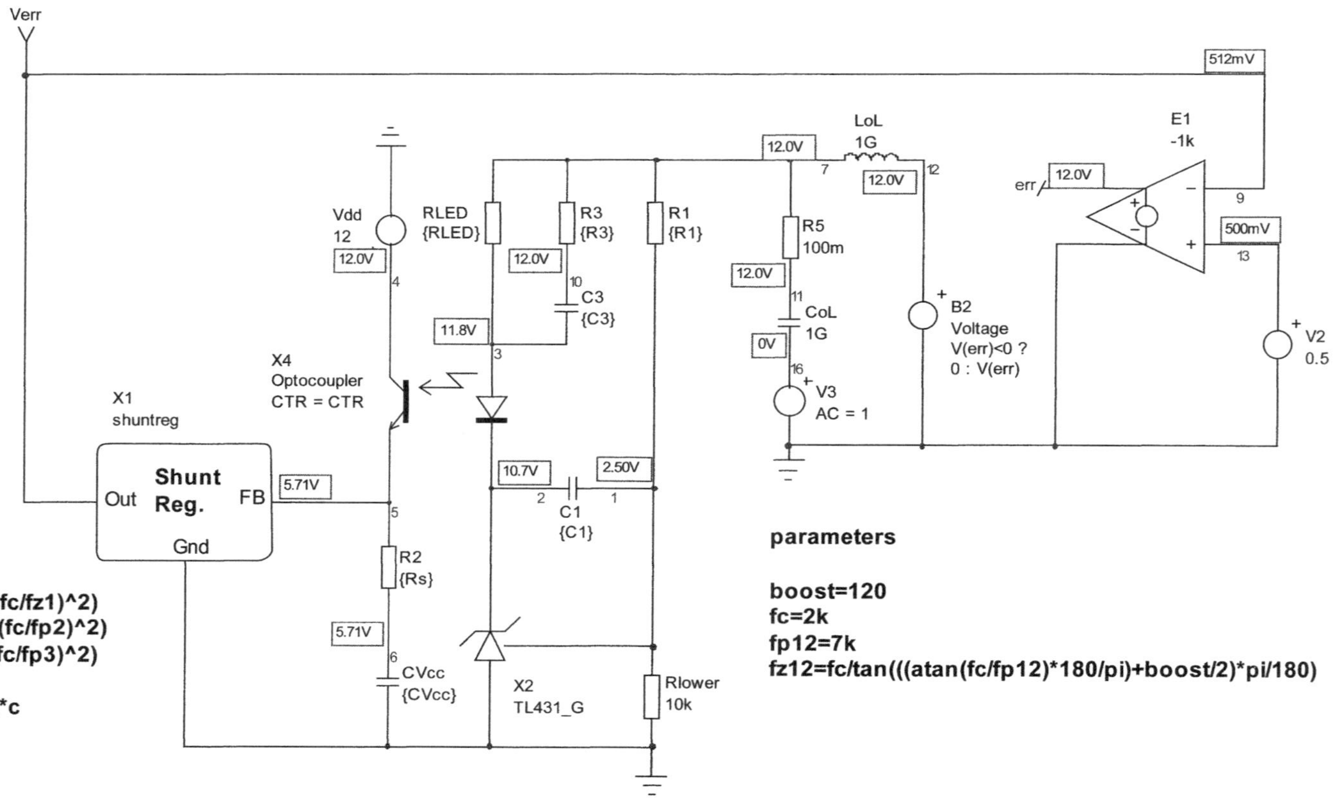

parameters

boost=120
fc=2k
fp12=7k
fz12=fc/tan(((atan(fc/fp12)*180/pi)+boost/2)*pi/180)

Figure 8.11 The shunt regulator can also be used in a type 3 compensator as it is shown here.

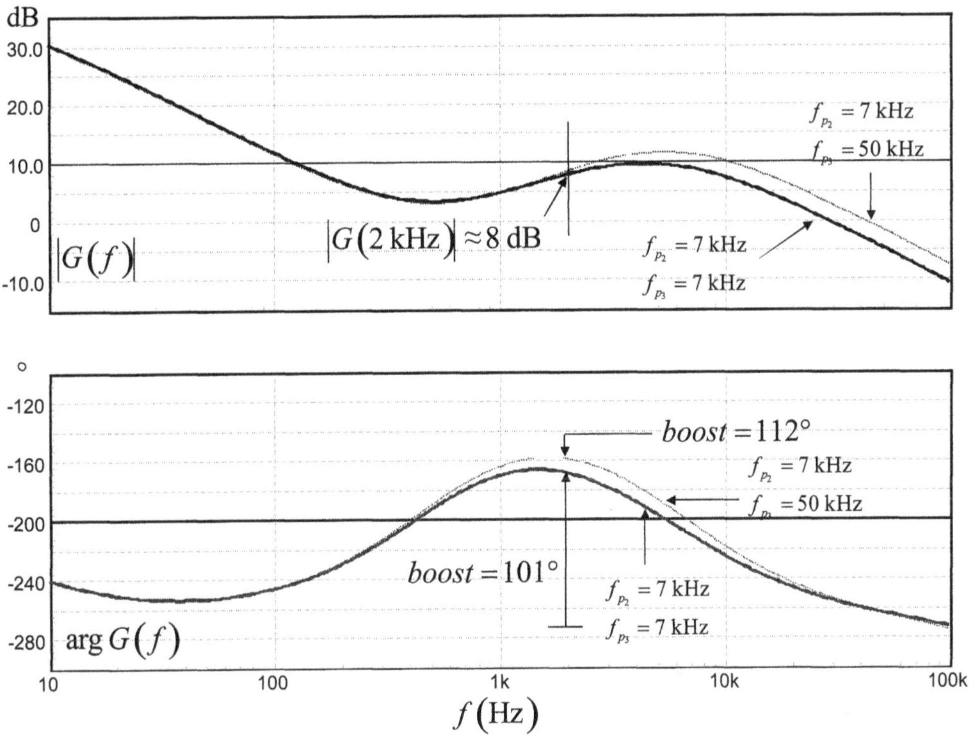

Figure 8.12 The ac response is a real type 3, despite a small mismatch on both the boost in phase and the crossover point.

We are all set now. The ac response of our compensator can now be explored thanks to the automated test fixture presented in Figure 8.11.

Once again, the automated calculations on the left panel help to change the compensator on the fly if needed. The ac response appears in Figure 8.12. There is a small mismatch on the crossover frequency, but this due to the LED dynamic resistance contribution, especially with LED series resistor of less than 100Ω. It also influences the phase boost as we can see on the plot. Since the equations are derived for the general case, we have the leisure to split the double pole and leave one to 7 kHz (we cannot move it; it is internal anyway) and push the second one to half the 100-kHz switching frequency, 50 kHz. As shown in the graph, it helps us to gain a little of phase boost, bringing the total to 112°.

8.3 The Type 3: An Origin Pole plus a Double Pole/Zero Pair—No Fast Lane

We know that the TL431 features a fast lane through the LED series resistor. This path offers to V_{out} a direct access to the duty ratio modulator, without passing through the TL431 op amp section. As we have detailed it in several other sections, the fast lane hampers the flexibility of the type 3 compensator by limiting the gain and the phase boost that can be generated. Figure 8.13 shows how the fast lane can be disabled via the implementation of a simple Zener-based regulator.

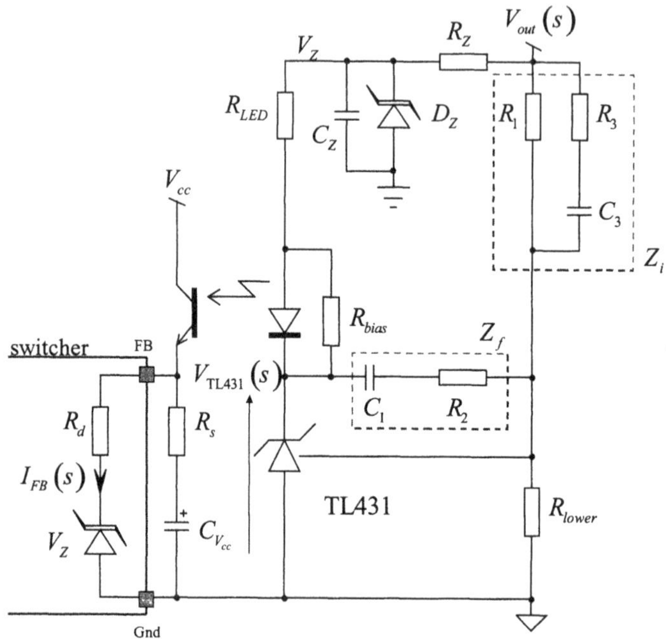

Figure 8.13 By using a simple Zener-based network, it is possible to get rid of the fast lane.

The principle behind the fast-lane deactivation is to remove the direct ac link between the LED ac current and the output voltage to be monitored. As shown in previous sections, and in particular in the TL431 chapter, a compensator based on a TL431 where the fast lane has been removed turns into a simple open-collector compensated op amp. The shunt regulator case does not change this rule. The output of the op amp under concern is the variable $V_{TL431}(s)$ in Figure 8.13. Its output voltage is the following:

$$V_{TL431}(s) = -V_{out}(s)\frac{Z_f}{Z_i} \tag{8.57}$$

Developing and rearranging this expression leads to

$$V_{TL431}(s) = -V_{out}(s)\frac{\dfrac{sR_2C_1+1}{sC_1}}{\dfrac{R_1(1+sR_3C_3)}{sC_3(R_1+R_3)+1}} = -V_{out}(s)\frac{sR_2C_1+1}{sR_1C_1}\frac{sC_3(R_1+R_3)+1}{(1+sR_3C_3)}$$

$$= -V_{out}(s)\frac{R_2}{R_1}\frac{\left(1+\dfrac{1}{sR_2C_1}\right)\left(1+sC_3(R_1+R_3)\right)}{1+sR_3C_3} \tag{8.58}$$

The ac LED current no longer depends on both $V_{out}(s)$ and the op amp voltage but solely on that latter. Therefore,

$$I_{LED}(s) = -\frac{V_{TL431}(s)}{R_{LED}} = \frac{V_{out}(s)}{R_{LED}}\frac{R_2}{R_1}\frac{\left(1+\frac{1}{sR_2C_1}\right)\left(1+sC_3\left(R_1+R_3\right)\right)}{1+sR_3C_3} \tag{8.59}$$

The expression of the current injected into the shunt regulator has been derived with (8.6):

$$I_{FB}(s) = I_C(s)\frac{sR_sC_{V_{cc}}+1}{sC_{V_{cc}}(R_s+R_d)+1} \tag{8.60}$$

As I_C and I_{LED} are linked by the optocoupler CTR, we have

$$\frac{I_{FB}(s)}{V_{out}(s)} = \frac{\text{CTR}}{R_{LED}}\frac{R_2}{R_1}\frac{sR_sC_{V_{cc}}+1}{sC_{V_{cc}}(R_s+R_d)+1}\frac{\left(1+\frac{1}{sR_2C_1}\right)\left(1+sC_3\left(R_1+R_3\right)\right)}{1+sR_3C_3} \tag{8.61}$$

Thanks to (8.9), we know the link between the injected feedback current and the obtained duty ratio. By linking this expression to (8.61), we obtain the complete transfer function of our compensator:

$$\frac{D(s)}{V_{out}(s)} = -G_{PWM}\frac{\text{CTR}}{R_{LED}}\frac{R_2}{R_1}\frac{\left(1+\frac{1}{sR_2C_1}\right)}{sC_{V_{cc}}(R_s+R_d)+1}\frac{sR_sC_{V_{cc}}+1}{1+s/\omega_{p_2}}\frac{\left(1+sC_3\left(R_1+R_3\right)\right)}{1+sR_3C_3} \tag{8.62}$$

We can put this expression under a more familiar form such as

$$\frac{D(s)}{V_{out}(s)} = -G_0\frac{(1+\omega_{z_2}/s)(1+s/\omega_{z_1})(1+s/\omega_{z_3})}{(1+s/\omega_{p_1})(1+s/\omega_{p_2})(1+s/\omega_{p_3})} \tag{8.63}$$

In this equation, we have

$$G_0 = G_{PWM}\frac{\text{CTR}}{R_{LED}}\frac{R_2}{R_1} \tag{8.64}$$

$$\omega_{z_1} = \frac{1}{R_sC_{V_{cc}}} \tag{8.65}$$

$$\omega_{z_2} = \frac{1}{R_2C_1} \tag{8.66}$$

$$\omega_{z_3} = \frac{1}{sC_3\left(R_1+R_3\right)} \tag{8.67}$$

$$\omega_{p_1} = \frac{1}{\left(R_d+R_s\right)C_{V_{cc}}} \tag{8.68}$$

$$\omega_{p_2} = 44 \text{ krad/s} \tag{8.69}$$

$$\omega_{p3} = \frac{1}{sR_3C_3} \tag{8.70}$$

The LED resistor can now be calculated based on the minimum biasing conditions only, and (8.20) is still valid. However, resistor R_2, which sets the mid-band gain, still needs to be derived according to the magnitude of (8.63):

$$|G(f_c)| = G_{PWM}\frac{\text{CTR}}{R_{LED}}\frac{R_2}{R_1}\frac{\sqrt{1+(f_{z2}/f_c)^2}}{\sqrt{1+(f_c/f_{p1})^2}}\frac{\sqrt{1+(f_c/f_{z1})^2}}{\sqrt{1+(f_c/f_{p2})^2}}\frac{\sqrt{1+(f_c/f_{z3})^2}}{\sqrt{1+(f_c/f_{p3})^2}} \tag{8.71}$$

Solving for R_2, we have

$$R_2 = \frac{GR_{LED}R_1}{G_{PWM}\text{CTR}}\frac{\sqrt{1+(f_c/f_{p1})^2}}{\sqrt{1+(f_{z2}/f_c)^2}}\frac{\sqrt{1+(f_c/f_{p2})^2}}{\sqrt{1+(f_c/f_{z1})^2}}\frac{\sqrt{1+(f_c/f_{p3})^2}}{\sqrt{1+(f_c/f_{z3})^2}} \tag{8.72}$$

The Zener resistor R_Z must authorize the operating current for the feedback loop to which the bias current for both the Zener diode itself and the TL431 must be added. The current circulating in the LED alone depends on the current sourced by the optocoupler. According to Figure 8.2, this current $I_{C,\max}$ amounts to 7 mA. It peaks to a maximum in the LED when the CTR is minimum. Using these data, we can now derive the Zener dropping resistor value:

$$R_Z = \frac{V_{out} - V_Z}{\frac{I_{C,\max}}{\text{CTR}_{\min}} + I_{Zbias} + I_{bias}} = \frac{(V_{out} - V_Z)\text{CTR}_{\min}}{I_{C,\max} + (I_{Zbias} + I_{bias})\text{CTR}_{\min}} \tag{8.73}$$

In the previous equation, we have the following variables:

V_{out}, the output voltage
I_{bias}, the TL431 biasing current when the optocoupler LED is paralleled with a resistor (usually 1 kΩ for a 1-mA bias)
I_{Zbias}, the Zener diode biasing current
$\text{CTR}_{\min}$, the minimum optocoupler current transfer ratio
$I_{C,\max}$, the maximum current injected into the switcher feedback pin (7 mA, depending on the switcher reference)

Everything needed has been derived, so let us continue with a design example.

8.3.1 A Design Example

Suppose we have the following component values:

V_{out} = 12 V; the output voltage.
V_f = 1 V; the LED forward voltage.
I_{bias} = 1 mA; the TL431 biasing current when the optocoupler LED is paralleled with a resistor.
$V_{\text{TL431,min}}$ = 2.5 V; the minimum operating voltage of the TL431.

I_{Zbias} = 3 mA; the Zener diode bias current.
V_Z = 8.2 V; the Zener diode breakdown voltage.
CTR_{min} = 0.8; the minimum optocoupler current transfer ratio.
R_1 = 38 kΩ; the upper resistor in the resistor bridge observing the output variable.
C_{Vcc} = 47 μF; the selected V_{cc} capacitor according to the switcher data sheet.

From the transfer function of the voltage-mode converter we want to stabilize, we need a gain of 10 dB at a 1-kHz frequency, while the needed phase boost is 130°. Given the internal 7-kHz pole brought by the duty ratio modulator and considering coincident poles, the second pole will also be placed at 7 kHz. Let us calculate the position of the double zeros, considering these coincident poles.

$$f_{z1,2} = \frac{f_c}{\tan\left(\frac{boost}{2} + \tan^{-1}\frac{f_c}{f_{p1,2}}\right)} = \frac{1k}{\tan\left(\frac{130}{2} + \tan^{-1}\left(\frac{1k}{7k}\right)\right)} = 303\text{ Hz} \tag{8.74}$$

This value leads us to define the series resistor value, R_s:

$$R_s = \frac{1}{2\pi f_{z1} C_{V_{cc}}} = \frac{1}{6.28 \times 303 \times 47u} = 11\ \Omega \tag{8.75}$$

Together with the shunt regulator equivalent dynamic resistor, it creates a pole located at

$$f_{p1} = \frac{1}{2\pi (R_s + R_d) C_{V_{cc}}} = \frac{1}{6.28 \times (11+15) \times 47u} = 130\text{ Hz} \tag{8.76}$$

At this point, we need to calculate the LED series resistor we will choose. Applying (8.20) and replacing the parameter V_{out} by the Zener voltage, we find a maximum value of

$$R_{LED,\max} \le \frac{CTR_{\min}(V_Z - V_f - V_{TL431,\min})}{I_{C,\max} + I_{bias}CTR_{\min}} \le \frac{0.8 \times (8.2 - 1 - 2.5)}{7m + 1m \times 0.8} \le 482\ \Omega \tag{8.77}$$

Assuming a 20 percent derating, we will pick a resistor of 385 Ω. Knowing this value, we can now use (8.72) and obtain a value for R_2:

$$R_2 = \frac{10^{\frac{10}{20}} \times 385 \times 38k}{163 \times 0.8} \frac{\sqrt{1+(1k/130)^2}}{\sqrt{1+(130/1k)^2}} \frac{\sqrt{1+(1k/7k)^2}}{\sqrt{1+(1k/303)^2}} \frac{\sqrt{1+(1k/7k)^2}}{\sqrt{1+(1k/303)^2}} = 234\text{ k}\Omega \tag{8.78}$$

Having this resistor on hand, we can evaluate the value of C_1 to position the zero canceling the pole f_{p1}:

$$C_1 = \frac{1}{2\pi f_{z2} R_1} = \frac{1}{6.28 \times 130 \times 234k} = 5.2\text{ nF} \tag{8.79}$$

The other elements, C_3 and R_3 are found using (8.55) and (8.56), respectively, where R_{LED} in the first equation is simply replaced by R_1:

$$C_3 = \frac{f_{p3} - f_{z3}}{2\pi R_1 f_{p3} f_{z3}} = \frac{7k - 303}{6.28 \times 38k \times 7k \times 303} \approx 13\ \text{nF} \tag{8.80}$$

Having C_3, we can get the series resistor R_3:

$$R_3 = \frac{1}{2\pi f_{p3} C_3} = \frac{1}{6.28 \times 7k \times 13n} = 1.7\ \text{k}\Omega \tag{8.81}$$

The compensator being ready now, let us check the Zener diode biasing resistor R_Z:

$$R_Z = \frac{(V_{out} - V_Z)\text{CTR}_{\min}}{I_{C,\max} + (I_{Zbias} + I_{bias})\text{CTR}_{\min}} = \frac{(12 - 8.2) \times 0.8}{7m + (3m + 1m) \times 0.8} = 298\ \Omega \tag{8.82}$$

We now have everything, and we can use our test fixture as portrayed in Figure 8.14 to test our ac results. The bias points are correct and the Zener voltage features the right value. We can display the response delivered by this compensator in Figure 8.15. A good crossover point can be observed, exactly at 10 dB, as expected. The phase boost is almost 130°, what we wanted from the original specifications.

8.4 Isolated Zener-Based Compensator

Presented in the TL431 chapter, a Zener diode can be implemented to regulate the output voltage of a switcher-based converter. However, this option is valid only for a raw dc output, as the cumulated precision of the stacked elements can barely be better than 10 percent. The application schematic featuring a Zener-based compensator appears in Figure 8.16.

Starting from the feedback current expression already derived in (8.6), we have

$$I_{FB}(s) = I_C(s)\frac{R_s + \frac{1}{sC_{V_{cc}}}}{R_d + R_s + \frac{1}{sC_{V_{cc}}}} = I_C(s)\frac{\frac{sR_sC_{V_{cc}} + 1}{sC_{V_{cc}}}}{\frac{sC_{V_{cc}}(R_s + R_d) + 1}{sC_{V_{cc}}}} = I_C(s)\frac{sR_sC_{V_{cc}} + 1}{sC_{V_{cc}}(R_s + R_d) + 1} \tag{8.83}$$

The collector current is linked to the LED ac current by the CTR. In this application, the dynamic resistor of the Zener diode R_{dZ} can no longer be neglected, as it can be as high as several tens of ohms, depending on the diode type and the biasing conditions. As a result, both the LED resistor R_{LED} and the Zener diode dynamic resistor R_{dZ} appear in series. Accounting for this fact, the LED ac current is

$$I_{LED}(s) = \frac{V_{out}(s)}{R_{LED} + R_{dZ}} = \frac{I_C(s)}{\text{CTR}} \tag{8.84}$$

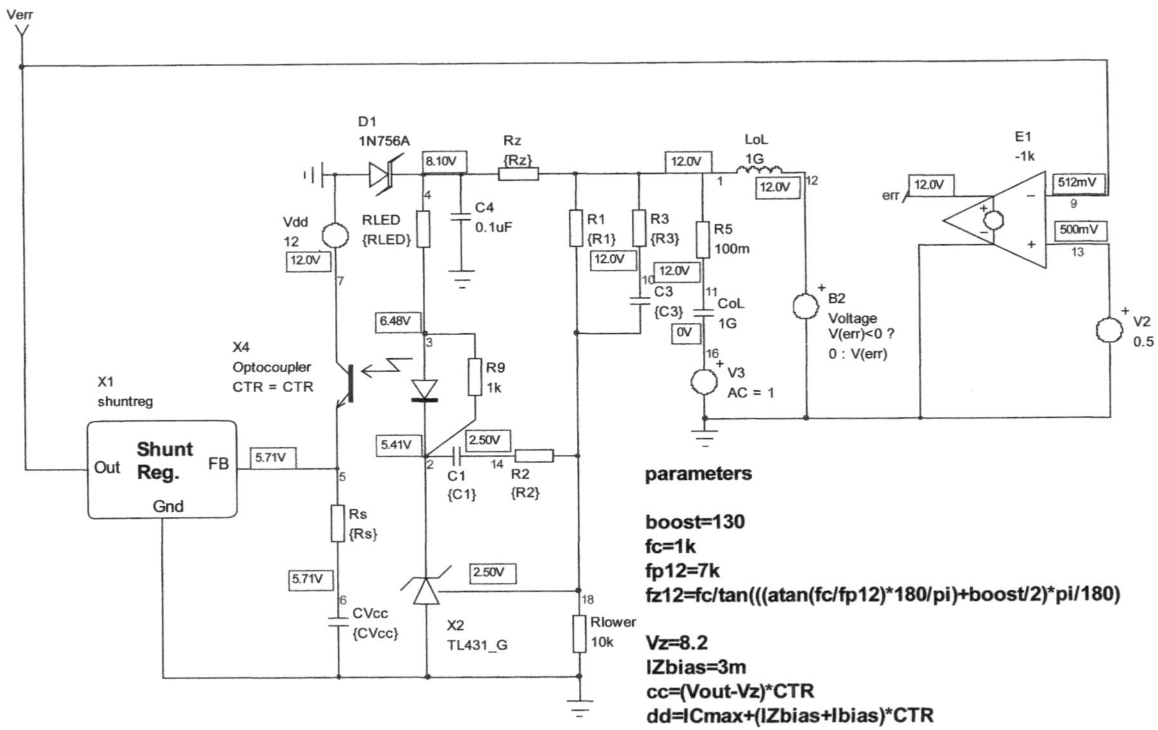

Parameters

Vout=12
Ibridge=250u
Rlower=2.5/Ibridge
R1=(Vout-2.5)/Ibridge
Vf=1
VTL431=2.5
ICmax=7m
Ibias=1m

aa=CTR*(Vz-Vf-VTL431)
bb=ICmax+Ibias*CTR
RLED=0.8*(aa/bb)

fz1=fz12
fz3=fz12
fp2=fp12
fp3=fp12
Gfc=-10

G=10^(-Gfc/20)
GPWM=158.5
pi=3.14159

CVcc=47u
Rd=15
Rs=1/(2*pi*CVcc*fz1)

a=sqrt(1+(fz2/fc)^2)*sqrt(1+(fc/fz1)^2)*sqrt(1+(fc/fz3)^2)
b=sqrt(1+(fc/fp1)^2)*sqrt(1+(fc/fp2)^2)*sqrt(1+(fc/fp3)^2)
R2=(G*RLED*R1/(GPWM*CTR))*b/a
fp1=1/(2*pi*(Rd+Rs)*CVcc)
fz2=fp1
C1=1/(2*pi*fz2*R2)
C3=(fp3-fz3)/(R1*fz3*fp3*2*pi)
R3=1/(2*pi*C3*fp3)
CTR=0.8

parameters

boost=130
fc=1k
fp12=7k
fz12=fc/tan(((atan(fc/fp12)*180/pi)+boost/2)*pi/180)

Vz=8.2
IZbias=3m
cc=(Vout-Vz)*CTR
dd=ICmax+(IZbias+Ibias)*CTR
Rz=cc/dd

Figure 8.14 The test fixture automates the various component values and lets the user change his or her mind on the fly to test other poles/zeros combinations.

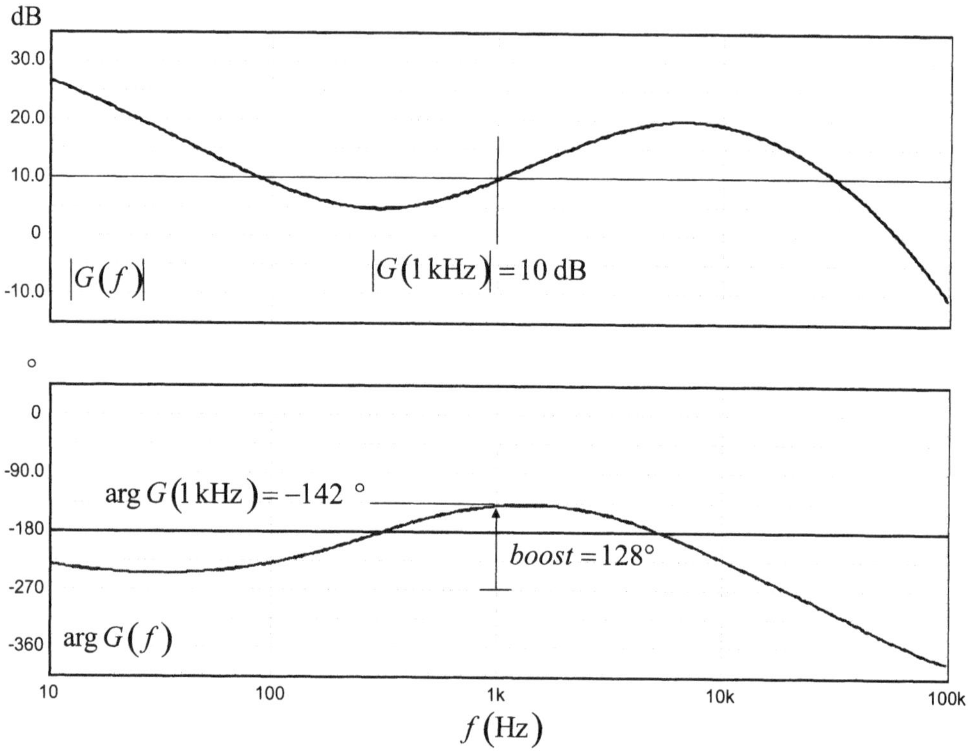

Figure 8.15 The ac response shows a good crossover point and a phase boost within specifications.

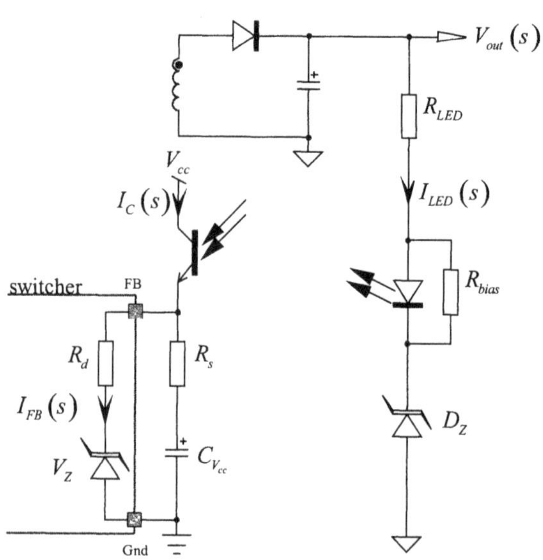

Figure 8.16 A Zener diode can be used in the secondary side to replace the TL431. This solution is only applicable when you expect a roughly regulated dc output level.

Extracting the collector current definition from (8.84) and substituting it in (8.83), we obtain a transfer function:

$$\frac{I_{FB}(s)}{V_{out}(s)} = \frac{\text{CTR}}{R_{LED} + R_{dZ}} \frac{sR_s C_{V_{cc}} + 1}{sC_{V_{cc}}(R_s + R_d) + 1} \tag{8.85}$$

Combined with the internal modulator function given by (8.9), we have

$$\frac{D(s)}{V_{out}(s)} = -G_{PWM} \frac{\text{CTR}}{R_{LED} + R_{dZ}} \frac{sR_s C_{V_{cc}} + 1}{sC_{V_{cc}}(R_s + R_d) + 1} \frac{1}{1 + s/\omega_{p_2}} = -G_0 \frac{1 + s/\omega_{z_1}}{1 + s/\omega_{p_1}} \frac{1}{1 + s/\omega_{p_2}} \tag{8.86}$$

In this expression, we can identify

$$G_0 = \frac{\text{CTR}}{R_{LED} + R_{dZ}} G_{PWM} \tag{8.87}$$

$$\omega_{z_1} = \frac{1}{R_s C_{V_{cc}}} \tag{8.88}$$

$$\omega_{p_1} = \frac{1}{(R_d + R_s) C_{V_{cc}}} \tag{8.89}$$

$$\omega_{p_2} = 44 \text{ krad/s} \tag{8.90}$$

We have double pole-single zero type of compensator. The pole and the zero defined by (8.88) and (8.89) are linked together since they share one common element, R_s:

$$\frac{f_{p_1}}{f_{z_1}} = \frac{1}{(R_s + R_d) C_{V_{cc}}} R_s C_{V_{cc}} = \frac{R_s}{R_d + R_s} \tag{8.91}$$

Unfortunately, they won't be of great help since they will cancel each other. It is, however, interesting to see how they influence the final phase lag. By definition, two poles and one zero bring the following argument:

$$\angle G(f_c) = \tan^{-1} \frac{f_c}{f_{z_1}} - \tan^{-1} \frac{f_c}{f_{p_1}} - \tan^{-1} \frac{f_c}{f_{p_2}} \tag{8.92}$$

The first pole and the first zero are linked as shown by (8.91), and it is interesting to sweep the position of f_{z1} and see how it affects the total phase lag. First, we must update f_{p1} capitalizing on (8.88) and (8.89). From (8.88), R_s is extracted and substituted into (8.89). After some changes, we obtain

$$f_{p_1} = \frac{f_{z_1}}{2\pi C_{V_{cc}} R_d f_{z_1} + 1} \tag{8.93}$$

When substituted into (8.92), we have an equation depending on f_{z1} only:

$$\angle G(f_c) = \tan^{-1} \frac{f_c}{f_{z_1}} - \tan^{-1} \frac{f_c(2\pi C_{V_{cc}} R_d f_{z_1} + 1)}{f_{z_1}} - \tan^{-1} \frac{f_c}{f_{p_2}} \tag{8.94}$$

It is now interesting to plot this formula for a given crossover frequency—let's say, 1 kHz—and see what recommendation we could extract on the zero placement. This is what Figure 8.17 portrays. As expected, the lower the zero, the less severe the phase lag

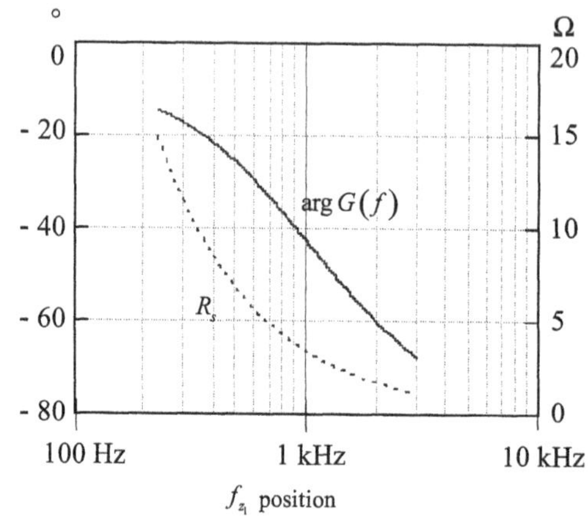

Figure 8.17 As the first zero slides in the low side of the spectrum, the total phase lag gets smaller.

is at 1 kHz. Given the recommendation on the series resistor R_s, whose value depends on the V_{cc} capacitor and the selected zero position, a value around 400 Hz looks like a good choice. The total phase lag at this point (–20°) cumulated with the phase reversal brought by the modulator chain will reach –200°. As indicated by Figure 8.6, at 1 kHz, the phase lag of the power stage is –70°, which when added to the –200° of the compensator should leave around 90° of phase margin at crossover (360° – 270°).

The output voltage of a converter stabilized by a Zener-based compensator will exhibit an output level equal to

$$V_{out} = V_Z + R_{LED} I_{LED} + V_f \tag{8.95}$$

The LED current will vary between two extremes, depending on the optocoupler CTR and the injected current in the feedback pin. At low output power, the injected current is maximum to reduce the duty ratio. At higher power conditions, the injected current is decreased to let the duty ratio increase:

$$I_{LED,\max} = \frac{I_{FB,\max}}{\mathrm{CTR}_{\min}} + I_{bias} = \frac{2.5m}{\mathrm{CTR}_{\min}} + I_{bias} \tag{8.96}$$

$$I_{LED,\min} = \frac{I_{FB,\min}}{\mathrm{CTR}_{\max}} + I_{bias} = \frac{7m}{\mathrm{CTR}_{\max}} + I_{bias} \tag{8.97}$$

If we consider the Zener voltage and the LED forward drop constant, the output voltage is likely to change depending on the loading conditions. Fortunately, as the LED series resistance is rather low by definition, the change will stay reasonable.

8.4.1 A Design Example

Suppose we have the following component values:

V_{out} = 12 V; the output voltage.
V_f = 1 V; the LED forward voltage.

I_{bias} = 1 mA; the Zener biasing current when the optocoupler LED is paralleled with a resistor.

CTR_{min} = 0.8; the minimum optocoupler current transfer ratio.

CTR_{max} = 1.2; the maximum optocoupler current transfer ratio.

C_{Vcc} = 47 μF; the selected V_{cc} capacitor according to the switcher data sheet.

The power stage transfer function is that of Figure 8.6. For a 1-kHz crossover point, we need to push the gain up by around 4 dB. Before that, we need to calculate the series resistor value R_s to position a zero at 400 Hz, as agreed before:

$$R_s = \frac{1}{2\pi f_{z_1} C_{V_{cc}}} = \frac{1}{6.28 \times 400 \times 47u} \approx 8.5\ \Omega \tag{8.98}$$

Associated with the shunt regulator input impedance of 15Ω, it leads to positioning the first pole at the following frequency:

$$f_{p_1} = \frac{1}{2\pi (R_s + R_d) C_{V_{cc}}} = \frac{1}{6.28 \times (15 + 8.5) \times 47u} = 144\ \text{Hz} \tag{8.99}$$

Having these values on hand, we can now derive the LED series resistor value from the magnitude of (8.86):

$$|G(f_c)| = \frac{CTR}{R_{LED} + R_{dZ}} G_{PWM} \frac{\sqrt{1 + (f_c/f_{z_1})^2}}{\sqrt{1 + (f_c/f_{p_1})^2}} \frac{1}{\sqrt{1 + (f_c/f_{p_2})^2}} \tag{8.100}$$

From this equation, we can extract the value of the LED resistor. The Zener dynamic resistor can be quickly characterized on the bench or its value can be extracted from the data-sheet parameters at a given bias current. The Zener diode we have selected exhibits a dynamic resistor of 10 Ω for a bias current of 10 mA.

$$R_{LED} = \frac{CTR\, G_{PWM}}{G} \frac{\sqrt{1 + (f_c/f_{z_1})^2}}{\sqrt{1 + (f_c/f_{p_1})^2}} \frac{1}{\sqrt{1 + (f_c/f_{p_2})^2}} - R_{dZ} = \frac{0.8 \times 163}{10^{\frac{4}{20}}}$$
$$\times \frac{\sqrt{1 + (1k/400)^2}}{\sqrt{1 + (1k/144)^2}} \frac{1}{\sqrt{1 + (1k/7k)^2}} - 10 \approx 21\ \Omega \tag{8.101}$$

With a 21-Ω resistor, we can calculate the corresponding voltage drops depending on the operating feedback conditions:

$$V_{R_{LED},\max} = R_{LED} I_{LED,\max} = R_{LED}\left(\frac{2.5m}{CTR_{min}} + I_{bias}\right) = 21 \times \left(\frac{2.5m}{0.8} + 1m\right) \approx 87\ \text{mV} \tag{8.102}$$

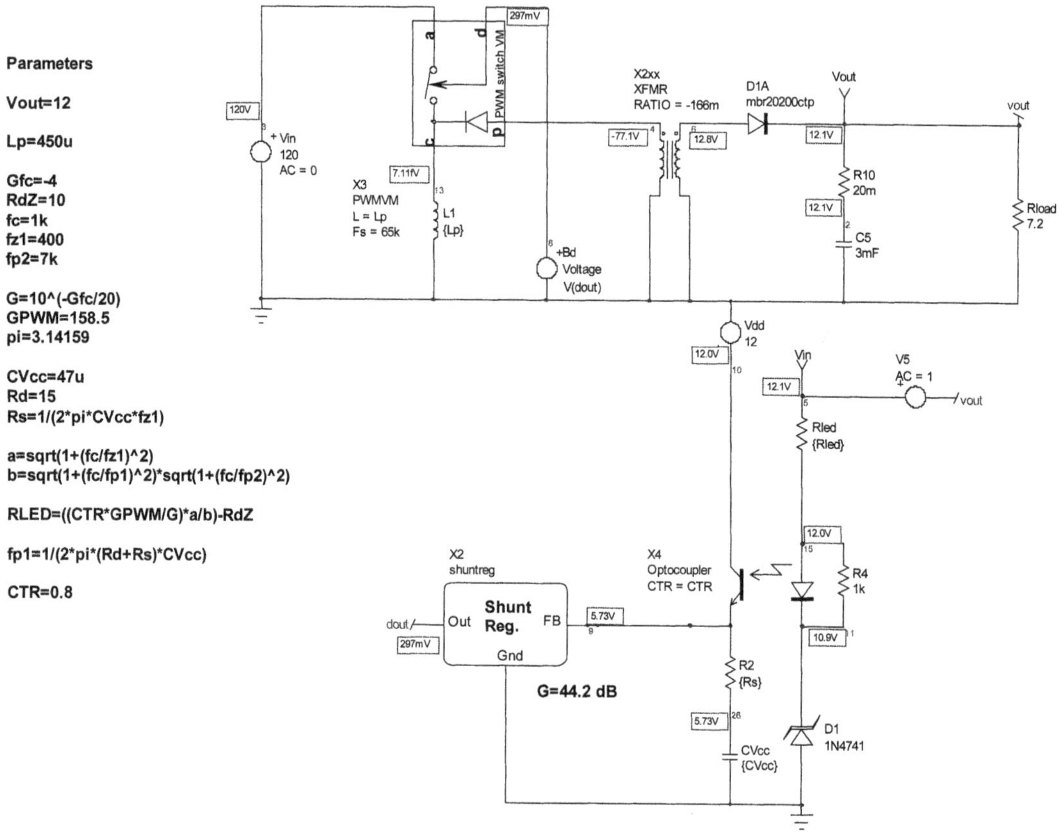

Figure 8.18 The flyback ac simulation fixture has been updated to implement a Zener-based compensator.

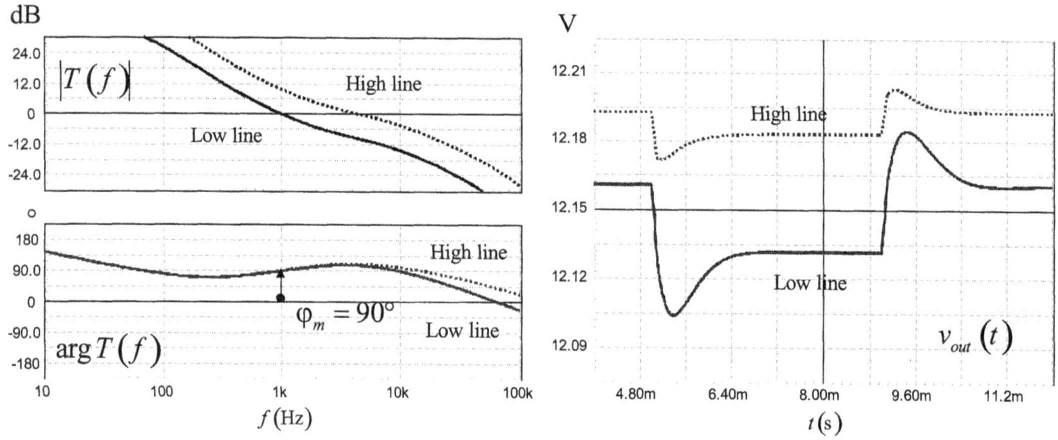

Figure 8.19 The open-loop gain shows a crossover frequency of 1 kHz with a comfortable phase margin of 90°, as expected.

$$V_{R_{LED},\min} = R_{LED} I_{LED,\min} = R_{LED}\left(\frac{7m}{\text{CTR}_{\max}} + I_{bias}\right) = 21\times\left(\frac{7m}{1.2} + 1m\right) = 144 \text{ mV} \quad (8.103)$$

The average value between these two extremes is around 116 mV. Considering a 12-V output and a 1-V voltage drop for the LED, the Zener voltage has to be selected to the following value:

$$V_Z = V_{out} - V_f - V_{R_{LED}} = 12 - 1 - 116m \approx 11\,\text{V} \quad (8.104)$$

With these value selected, we can now use the converter simulation test fixture proposed in Figure 8.5 and update it with the Zener-based compensator. This is what Figure 8.18 shows.

The bias points are good, the output voltage reaches 12 V, and the duty ratio establishes to around 30 percent (297 mV on the D input of the average model X_3). By inserting the ac source with the LED series resistor, we can sweep the whole chain. Keeping the ac source in place, it is also possible to run a transient test and assess the response. Both ac and transient results appear in Figure 8.19 and confirm the 1-kHz crossover point, together with a 90° phase margin. The transient response improves as the input voltage changes since the crossover frequency at high line increases to nearly 4 kHz. Please note the slight output voltage dc change at both line extremes. This phenomenon finds its root in the duty ratio setpoint change. As the injected current changes, it induces a different voltage drops across R_{LED} which, summed with the Zener voltage and LED voltage drop, bring a change in the output level.

8.5 Conclusion

This last design example ends our study on shunt regulator–based compensators. Despite the presence of an internal 7-kHz pole, reasonable crossover targets can be achieved as long as one understands the design methodology. In particular, some iterations are necessary to check whether the selected crossover frequency is compatible with the available phase boost.

References

[1] TOPSwitch data sheet, www.powerint.com.

[2] TOPSwitch Tips Techniques and Troubleshooting Guide, Power Integrations AN-14.

[3] Basso, C., Switch Mode Power Supplies: SPICE Simulations and Practical Designs, New York: McGraw-Hill, 2008.

[4] Ridley, R., "Loop Gain Measurement with Current Injection," *Switching Power*, 2005, http://www.ridleyengineering.com.

Appendix 8A: Summary Pictures

Figures 8.20 through 8.22 summarize the component definitions associated with the structures described in the chapter.

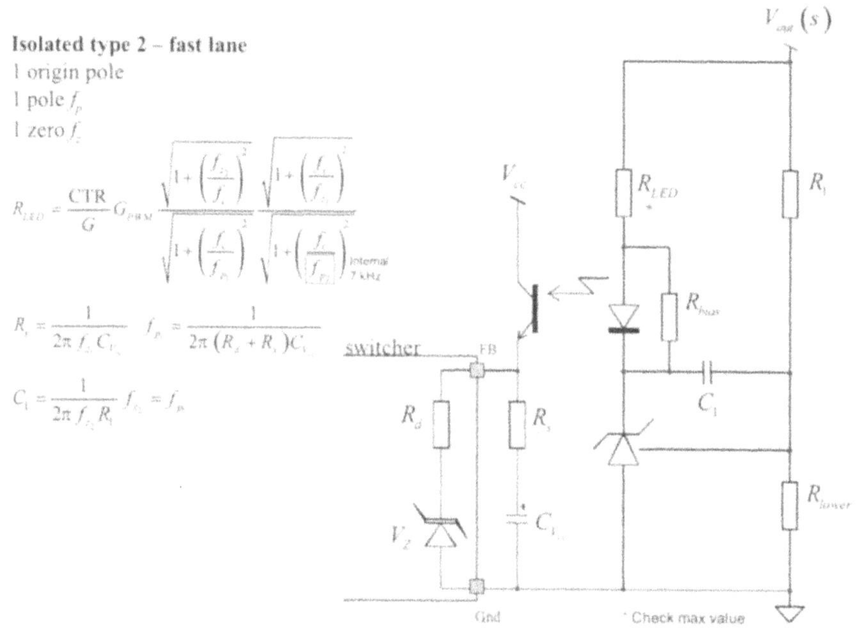

Figure 8.20 The isolated type 2 with fast lane.

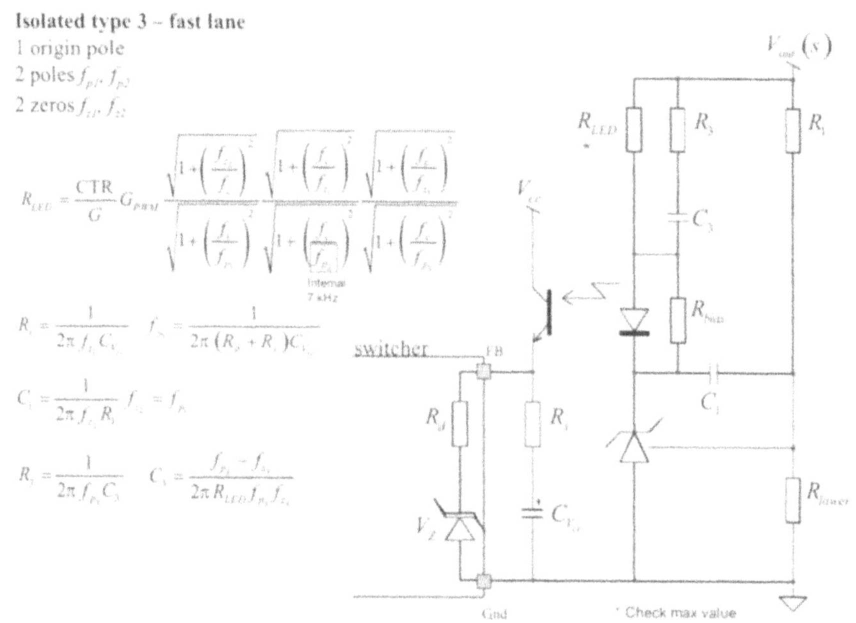

Figure 8.21 The isolated type 3 with fast lane.

Isolated type 3 – no fast lane

1 origin pole
2 poles f_{p1}, f_{p2}
2 zeros f_{z1}, f_{z2}

$$R_2 = \frac{GR_{LED}R_1}{G_{PWM}\text{CTR}} \frac{\sqrt{1+\left(f_c/f_{p_1}\right)^2}}{\sqrt{1+\left(f_{z_1}/f_c\right)^2}} \frac{\sqrt{1+\left(f_c/f_{p_2}\right)^2}}{\sqrt{1+\left(f_c/f_{z_2}\right)^2}} \frac{\sqrt{1+\left(f_c/f_{p_3}\right)^2}}{\sqrt{1+\left(f_c/f_{z_3}\right)^2}}$$

$$R_Z = \frac{(V_{out}-V_Z)\text{CTR}_{min}}{I_{C,max}+(I_{Zbias}+I_{bias})\text{CTR}_{min}}$$

$$R_s = \frac{1}{2\pi f_{z_1} C_{V_{cc}}} \quad f_{p_1} = \frac{1}{2\pi(R_d+R_s)C_{V_{cc}}}$$

$$C_1 = \frac{1}{2\pi f_{z_2} R_1} \quad C_3 = \frac{f_{p_3}-f_{z_3}}{2\pi R_1 f_{p_2} f_{z_3}}$$

$$R_{LED,max} \leq \frac{\text{CTR}_{min}\left(V_Z - V_f - V_{TL431}\right)}{I_{C,max} + I_{bias}\text{CTR}_{min}}$$

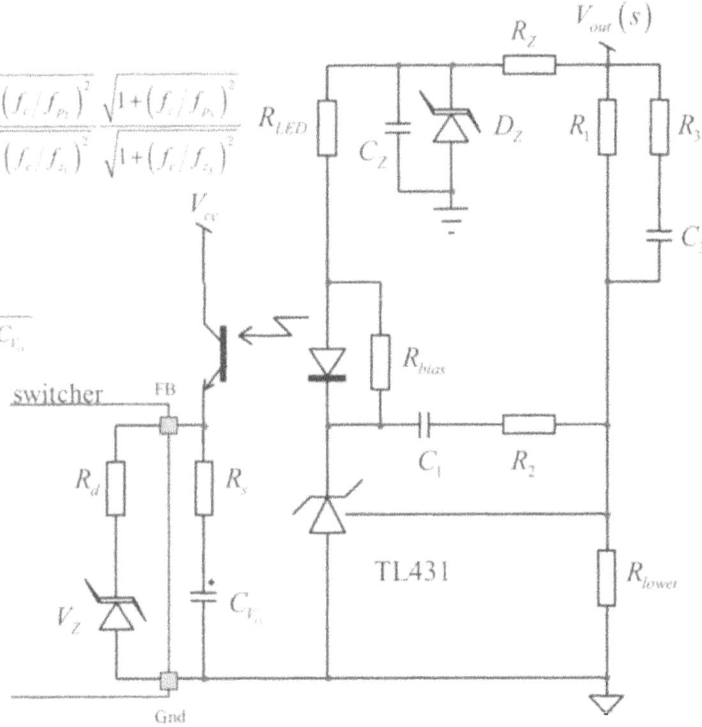

Figure 8.22 The isolated type 3 without fast lane.

CHAPTER 9

Measurements and Design Examples

Now that we understand why we need a compensator and how to design it, it is important to verify whether the prototype measurements on the bench comply with or at least approach our theoretical analysis. If not, then you need to understand where the discrepancy comes from so that you can update the analysis models with the new data. Let us start with the basics: how to open the loop on a prototype and measure the loop gain characteristics.

9.1 Measuring the Control System Transfer Function

In our quest to stabilize a control system, we have focused our energy on the open-loop gain transfer function analysis. However, once the control system operates, it is obviously in closed-loop conditions. To check if our theoretical approach was correct, we thus need a means to practically recreate the analysis conditions in which we purposely considered an open-loop path from the control input u to the output variable y. There are plenty of places in the control system where the return path can be broken. Two of them appear in Figure 9.1.

These are, however, block-based representations. We need a model closer to reality to understand what is at stake here. This updated model shows up in Figure 9.2. We see an op amp-based compensator whose error voltage V_{err} drives the control input of our power stage V_c. It can be a linear or switched converter—it does not change the picture. The loop can be opened either at the op amp output or in the resistive divider path. In both cases, there is a physical interruption of the return path and the control system no longer works as it should.

The loop can be broken for several reasons. For instance, you are at the beginning stage of your project and you need to extract the plant transfer function: there is no small-signal or analytical model for the converter under study. In this case, you want the plant Bode plot $H(s)$ to define the compensator structure given crossover and phase margin design goals. You can exclude the error amplifier and exclusively concentrate on the power stage frequency sweep first. This is the first method that we will explore.

The second option and the most popular one maintains the dc point by keeping the loop closed via a low-value resistor but offers the possibility to open the path in ac via the usage of a transformer. This is the best and most practical option you will use. Not only you will get the plant transfer function $H(s)$ but also the compensator $G(s)$ or the whole open-loop gain $T(s)$. Without this solution, exploring the loop gain, and thus combining H and G, it becomes extremely difficult, if not impossible, to maintain the control system in a linear and safe zone. It must be linear to avoid distortion when you ac modulate the circuit (possibly clipping the observed signal

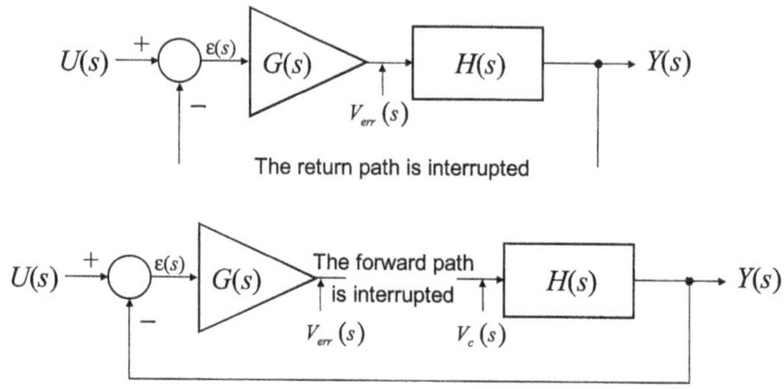

Figure 9.1 The analysis breaks the return path to observe the open-loop transfer function. Two possible examples where to break the loop.

because you are too close to the maximum or minimum output swing) and safe to avoid pushing the converter in one of its upper stops: a zero-volt output is probably okay but the output pushed to its maximum level is certainly dangerous. To avoid both scenarios, it is safer to apply the second method.

9.1.1 Opening the Loop with Bias Point Loss

When you physically open the loop, you obviously disturb the control system. How can it maintain its operating bias point to the targeted value if the return or direct path is broken? For instance, suppose you analyze a 12-V/1-A converter supplied from a 48-V input. If you open the loop, you will have to bias the power stage

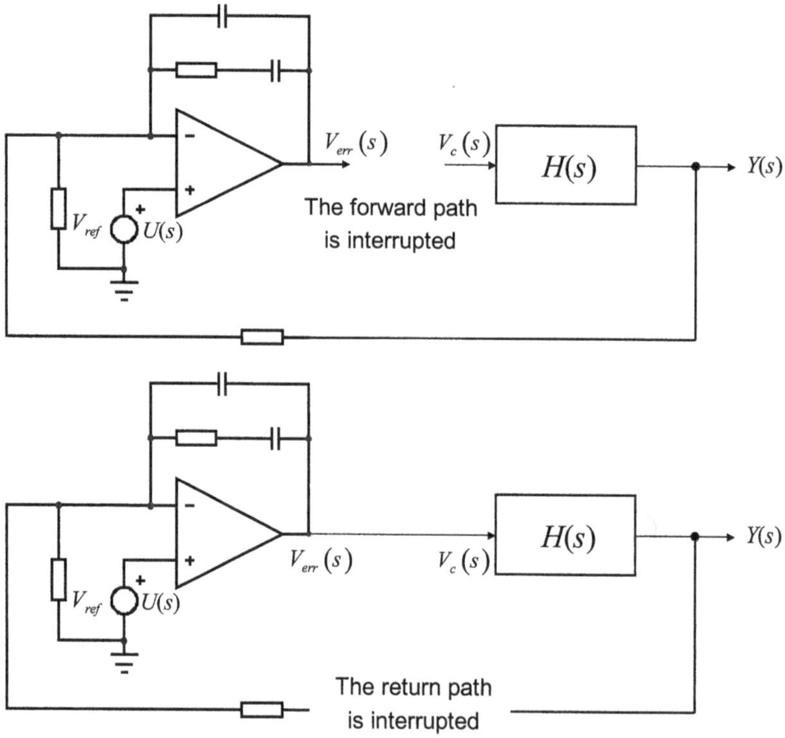

Figure 9.2 The op amp used in the compensator usually exhibits a large amount of gain.

control input so that it delivers the right output at the rated current. Then, once the dc bias is obtained, you will need to inject the ac modulation while observing its output. Ac injection can be done via an ac-coupling capacitor, for instance. This is the first method, described in Figure 9.3. The bias point is imposed by a resistive divider, which will give a finer control to bias the control input and provides impedance for the ac-coupling capacitor to modulate. In this example, we need 400 mV. Should you try to directly deliver this bias level via a dc source, it is very likely that the coarse output voltage adjustment makes the bias point setting a difficult exercise even more so since a switching power supply is a noisy place. It is best to drive the control input via a resistive divider. First, the 1-kΩ pull-down resistor wired close to the control input will ensure a low driving impedance, minimizing susceptibility to noise. Second, rather than tweaking the bias output voltage to 400 mV, with a 47-kΩ series resistor, you will adjust the dc source until you reach 12 V on the converter output, giving you a finer adjustment range than without the divider. The noise from the dc bias source will also be divided by the division ratio.

Once the converter output voltage has reached 12 V and delivers its 1 A current, you can hook an ac source through an ac-coupling capacitor and start the modulation. The ac source is part of a network analyzer. Such equipment is available from various manufacturers such as Ridley Engineering, Venable Instruments, or Omicron Lab. These devices also feature two inputs labeled A and B. Input A goes to the control input you sweep, while B observes the output signal. In Figure 9.3, as we want the plant transfer function, $H(s)$, A is connected to the converter control voltage and B collects the output signal data The network analyzer will automatically plot

$$|H(f)| = 20\log_{10}\left(\frac{|V_B(f)|}{|V_A(f)|}\right) \tag{9.1}$$

The modulating signal amplitude must be high enough so that the equipment can extract the useful signal from the noisy environment it is swamped in. However, it is important to keep the converter in a small-signal range. To verify this fact, hook

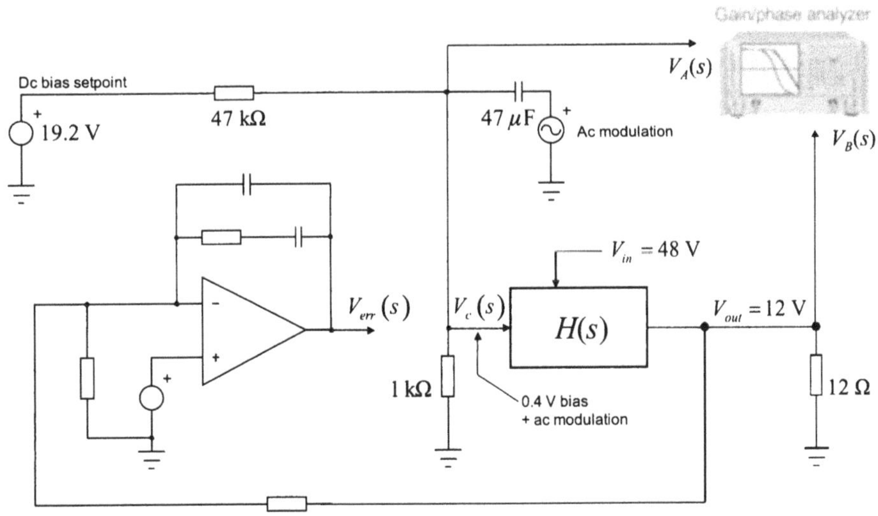

Figure 9.3 If you physically open the loop, you will have to externally force the converter at its operating bias point.

an oscilloscope probe to the output and tweak the modulating amplitude so that signal distortion is minimum (no clipping). No need to have the largest output signal; 1 to 5 percent of the output voltage is the kind of ac modulation amplitude you want. Let's see a practical example.

In Figure 9.4, we have shown an isolated flyback converter operated with a high-voltage controller, the NCP1200. The loop is physically open at the feedback pin level (the optocoupler is disconnected) and we apply an external dc bias via a recommended divider network, R_7 and R_8. The ac modulation is coupled through capacitor C_4. Pin 2 signal, actually the converter control voltage V_c, connects to input A of the network analyzer. The output signal is collected on the secondary side of the converter and goes to input B of the measurement instrument. As with any high-voltage converter, a few precautions must be taken before you apply power:

- This converter is powered from the ac grid and deals with lethal high-voltage levels. To prevent electric shocks and conflicts with the equipment grounds (oscilloscope, analyzer), you have to power the converter from an *isolated* source. Either ac or dc is fine, but it has to be isolated from the ac grid. For these experiments, we usually power our converters from a dc source. We have used the Xantrex XHR600-1.7 for years, and it is very well suited for these kind of measurements. You can safely clamp its output voltage excursion below 400 V, and the maximum output current can be programmed. In case of short circuit in the board, you naturally limit the damages.

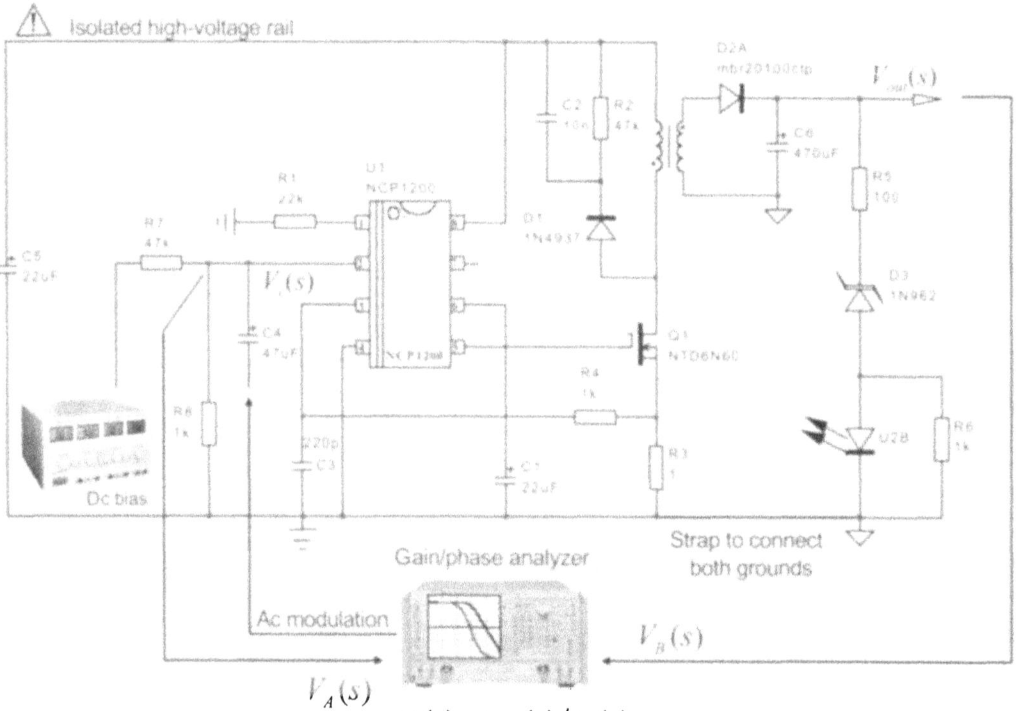

$$H(s)=V_{out}(s)/V_c(s)$$

Figure 9.4 In this flyback example, the converter uses a popular current-mode controller, the NCP1200. The loop is open at the feedback pin level (pin 2).

- You are dealing with a converter whose secondary side is isolated from the primary side through an optocoupler. To allow the power stage transfer function measurement, you must tie the secondary and primary grounds together. This is the low-side strap shown in Figure 9.4. You now better understand why an isolation from the grid is an absolute necessity.
- Because the ac-coupling capacitor C_4 is discharged, you should not connect the network analyzer source while the power supply is operating, as you will disturb the dc bias brought by R_7 and R_8 and potentially generate an output transient. To avoid such issue, make sure the analyzer and its ac source are connected and running when you power up the converter.
- It is recommended to use a true resistive load on the output whenever you can. If you do not have access to the required high-power resistor, turn the electronic load into resistance mode and not constant-current mode. We found the constant-current operation could lead to erroneous measurement results.

Now that everything is running, you should be able to obtain a graph such as that of Figure 9.5. The converter was loaded to operate in a continuous conduction mode (CCM). The sweep started at 1 Hz to obtain the dc gain and ended at 100 kHz while the analyzer filter was set to 1 Hz. We started with an ac modulation amplitude of 100 mV and increased it to 250 mV to reduce the spurious noise on the graph. Make sure that changing the ac modulation amplitude reduces the noise

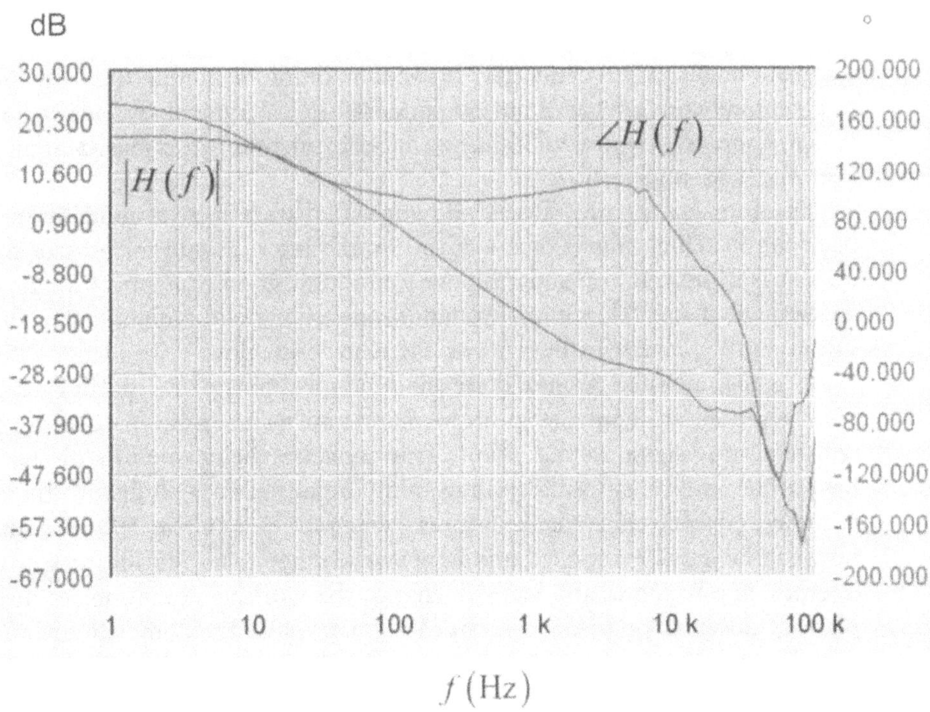

Figure 9.5 The typical ac response of a CCM flyback current-mode power stage.

but does not significantly change the overall response. Otherwise, it is indicative of a nonlinear operation at some point. As noted, if you visualize the output signal entering the analyzer and avoid clipping at all time, it is indicative of a linear operation, and you should be safe.

Measurement techniques require care and experience, especially in a noisy environment such as in a switching power supply. These techniques have been the objects of numerous publications, showing tips and tricks on how to get nice and reliable transfer function graphs. Please consult [1–6] at the end of this chapter for more information on the subject. All of these documents are available from the web.

The simple technique we presented lends itself well for low-gain power stages. Should you try to observe the error amplifier output (whether it is implemented with an op amp or a TL431), the technique becomes almost impossible to apply: any minor change in the dc control bias (noise, temperature drift, and so on), will push the error amplifier dc output in one of its stops (V_{cc} or ground) making an analysis impossible.

Loop interruption is also acceptable in converters where physically breaking the return path remains a safe operation, especially during the startup sequence. It is the case for most flyback-based converters of low to moderate output power levels. For more complex converters, especially high-power types, we do not recommend the physical opening of the loop.

9.1.2 Power Stage Transfer Function without Bias Point Loss

As explained, some converters do not accept the physical opening of their control loop. Sometimes, the converter is designed so a certain power-on sequence is fulfilled. In this case, a manual crank cannot easily be implemented. The case can also arise where the control input is not well accessible, and an opening represents a measurement risk. Fortunately, there is a simple technique that can be applied to a running converter without opening its loop. A typical secondary-side control system appears in Figure 9.6. It uses a TL431, but an op amp-based circuitry would keep the same approach.

The loop is normally closed, and you will disturb it by ac modulating the feedback pin in which impedance is given by the internal pull-up resistor R_4. In this case, it is 20 kΩ. If you superimpose a modulation on this pin through capacitor C_3, you will disturb the circuit around its operating point maintained by the TL431. Observing V_{out} and V_c will give you the plant Bode plot.

It is also possible to modulate the junction point of R_1 and R_2 with the network analyzer ac signal and plot the plant transfer function. A coupling capacitor will inject the signal at this point. This capacitor being initially discharged, you have to be careful, as brutally connecting the ac source will disturb the operating point and possibly engender an output overshoot. A 10-nF to 100-nF value should, however, be quickly charged and limit overvoltage risks. This technique keeps the operating point intact and lets you change the working conditions to measure the plant at different points. It does not let you explore the complete open-loop gain $T(s)$. However, if the cutoff frequency imposed by C_2 is too low, the high-frequency modulating signal cannot cross the transmission chain and the excitation at the control pin is drowned into noise. It then becomes difficult or impossible to extract

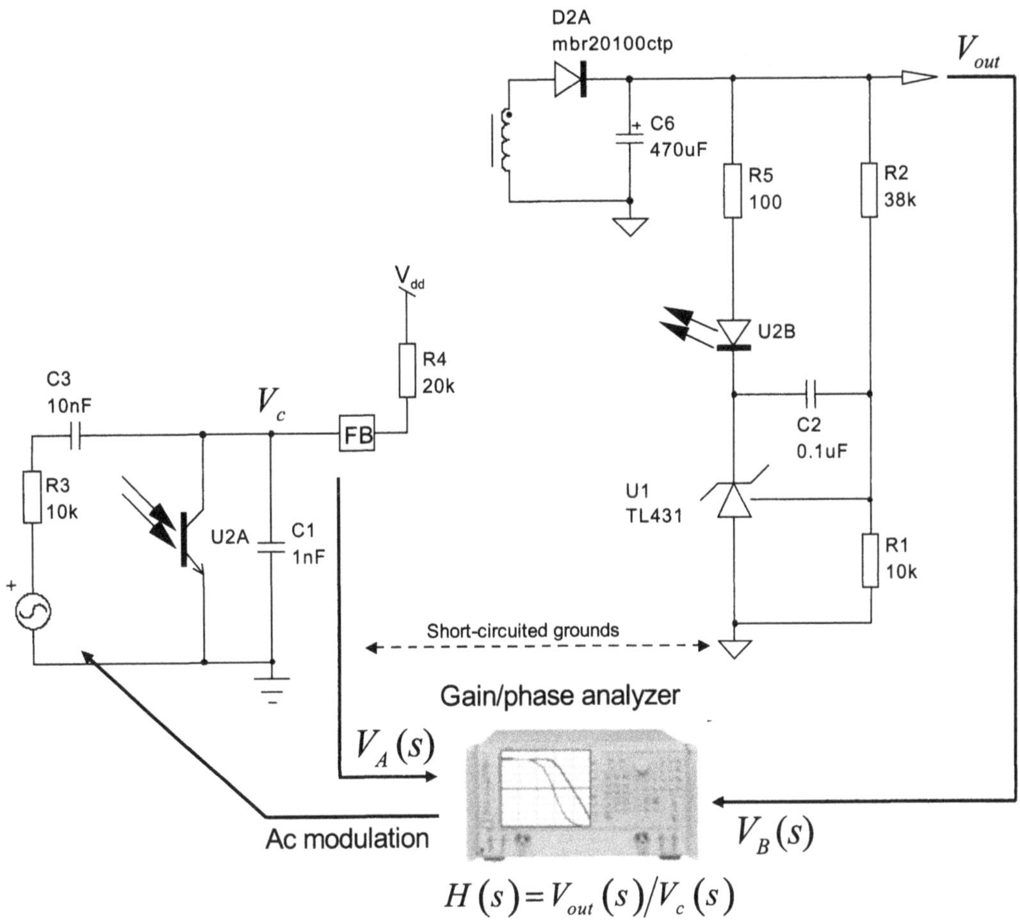

Figure 9.6 The loop is closed but can be perturbed by injecting a signal on the control pin FB.

the high-frequency gain, and you must either directly modulate the feedback pin or jump to the next paragraph.

9.1.3 Opening the Loop in ac Only

Keeping the operating point is important in an analysis, especially if you want to explore different working configurations: load changes, low or high input line, and so on. In any of these situations, the controlled variable (e.g., the output voltage) needs to be kept constant. If you apply the first method, where the loop is physically broken, any change in the input voltage or the loading conditions will require a duty ratio adjustment. You will do it manually via the external dc bias. When you want to cover numerous situations, it can quickly become a rather tedious exercise. By keeping the loop closed in dc, any change in the operating conditions will lead to a new operating point without any external tweak.

We have seen that we could keep the loop closed in dc as suggested in Figure 9.6 and unveil the plant transfer function. However, we could not graph the loop gain

$T(s)$. To maintain the dc point while probing whichever voltage transfer function we like, we need to apply an injection transformer-based method such as that described by Dr. Middlebrook in his founding paper. Please see [7] for details and download. The method preserves the dc point and does not require the physical interruption of the return path. Rather, it locally injects a series perturbation in the path across a resistor as documented in Figure 9.7.

The series resistor maintains the loop in a closed state. Its value typically ranges from 10 to 100 Ω and does not affect regulation. We have chosen 22 Ω in the example. Given the floating configuration of this resistor in both examples, we can only couple a ground-referenced ac source via a transformer. This is what the figure shows. This transformer applies the ac modulation signal across the resistor and is equivalent to the series insertion of a source in the loop. When brought back to a block-based drawing, we can easily picture this fact as illustrated in Figure 9.8. As the reference voltage (the control system setpoint) does not change during the ac sweep, its small-signal value is zero. The schematic simplifies to the lower part of Figure 9.8, which actually depicts a regulator. From this sketch, we can derive a few equations.

$$V_B(s) = -V_A(s)G(s)H(s) \tag{9.2}$$

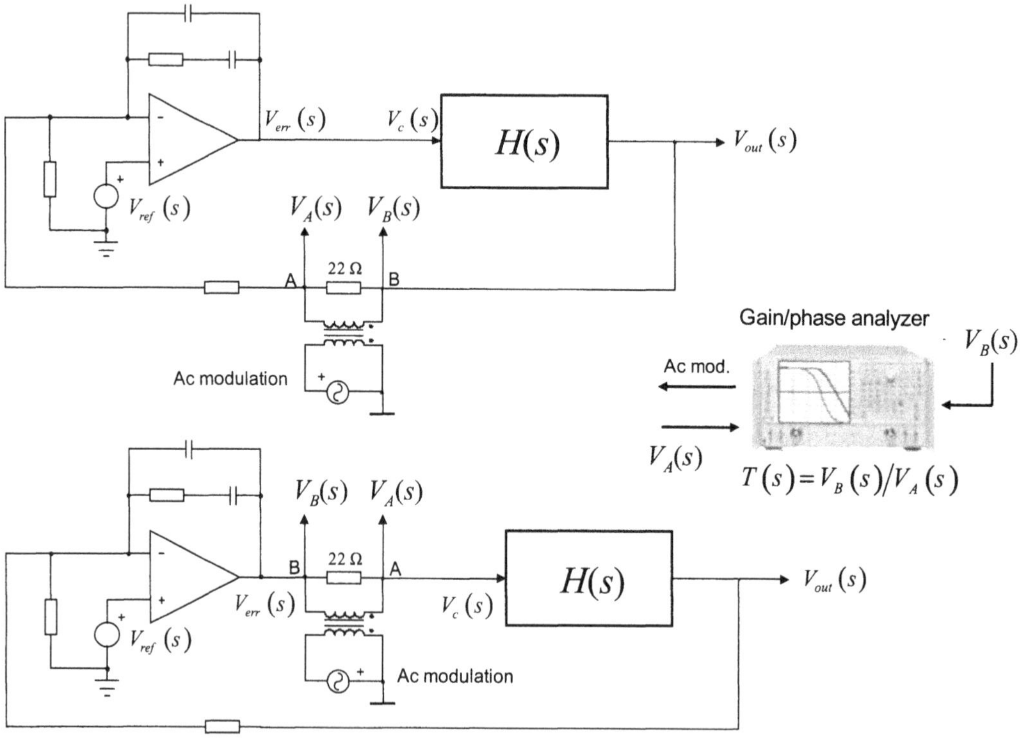

Figure 9.7 The perturbation is injected in series with the return or the direct path. The dc point is kept intact as no physical interruption occurs.

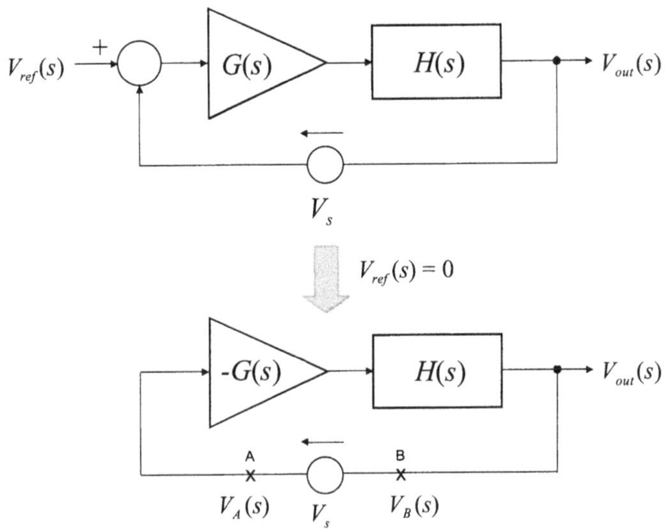

Figure 9.8 The inserted source appears like a perturbation in series with the returned information.

If we plot $V_B(s)/V_A(s)$, we have

$$\frac{V_B(s)}{V_A(s)} = -G(s)H(s) = T(s) \tag{9.3}$$

The minus sign illustrates what we already underlined in previous chapters. In the Figure 9.8 analysis, we naturally include the op amp phase reversal of 180° to which the loop gain argument (phase lag) is added. Therefore, the phase margin is no longer measured as the distance of the open-loop gain argument to the –180° limit but rather to –360° or 0°:

$$\varphi_m = \arg T(f_c) - 0° \tag{9.4}$$

In this configuration, we can consider the sinusoidal voltages as rotating vectors shown in Figure 9.9. The magnitude of the vector (also called its modulus) is the sine wave amplitude, while its argument is the counterclockwise angle formed by the vector with the horizontal axis. You will find more information on this so-called phasor notation in Wikipedia, for instance. In our case, the difference between the

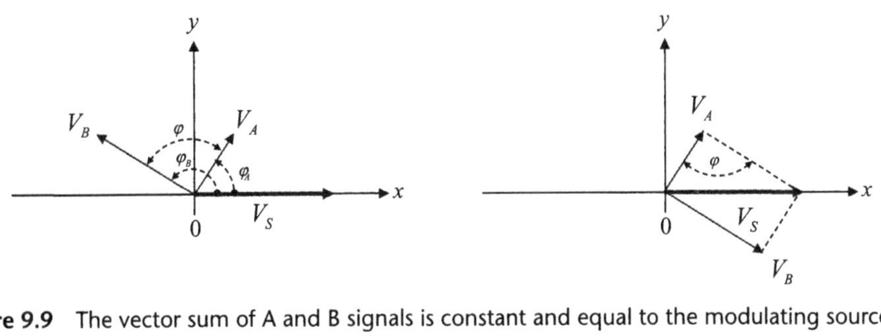

Figure 9.9 The vector sum of A and B signals is constant and equal to the modulating source amplitude VS.

injected signal in A and the returned signal in B is constant and always equates the source amplitude. We can therefore write

$$\vec{V_s} = \vec{V_A} - \vec{V_B} \tag{9.5}$$

The phase difference between these two signals is nothing other than the difference of the vectors arguments:

$$\angle\vec{B} - \angle\vec{A} = \varphi_B - \varphi_A = \varphi \tag{9.6}$$

On the left side of Figure 9.9, we have drawn the reference vector $\vec{V}_S$—the ac modulation—and the signals generated at points A and B. They are, respectively, affected by arguments φ_A and φ_B, the angle formed by the vector with the x-axis. From this picture, graphically drawing (9.5) is equivalent to summing:

$$\vec{V}_S = \vec{V}_A + \left(-\vec{V}_B\right) \tag{9.7}$$

This is what we did on the right side to show that the result equals the amplitude of vector V_S.

The vector sum of V_A and V_B always satisfies the right side of Figure 9.9. This is the key of operation. Therefore, if V_B changes in amplitude, as the loop gain magnitude and argument also vary, V_A automatically adjusts to keep the constant value imposed by V_S. To illustrate this operating principle, we have run a buck converter simulation template such as that of Figure 9.10. A ground-referenced VCO sinusoidal source sweeps the frequency between 10 Hz and 40 kHz. This source is transformed into a floating stimulus via the B-element B1 that scales down the 1-V source to 20 mV. The ac sweep results appear in Figure 9.11.

At low frequency, the dc gain is extremely high. Therefore, a small amplitude at point A is sufficient to generate a generous output signal at point B. When the crossover frequency approaches, both amplitudes tend to equalize. Exactly at crossover, they are equal, and the phase difference between both signals is our phase margin φ_m. As the frequency increases beyond crossover, the loop gain diminishes, asking for more signal at point A. Zooms of Figure 9.11 appear in Figure 9.12 and confirm the explanations. In the middle of the figure, the frequency is 4.97 kHz and the phase angle between the signals is:

$$\varphi_m = \frac{17}{201}360 = 30.4° \tag{9.8}$$

These numbers are in agreement with the calculated parameters in Figure 9.10 left-side template: 5 kHz crossover with a 30° phase margin.

9.1.4 Voltage Variations at the Injection Points

If you observe Figure 9.11, you can see a kind of peaking on signals A and B, at the left side of the graph. This peaking is direct a consequence of (9.5) that must always be satisfied. As such, the amplitude at points A and B permanently adjusts to satisfy this requirement (see [8] for an animated graph). However, as shown in Figure 9.12, the amplitude of one of the signals can shrink to an extremely low value. A few mV

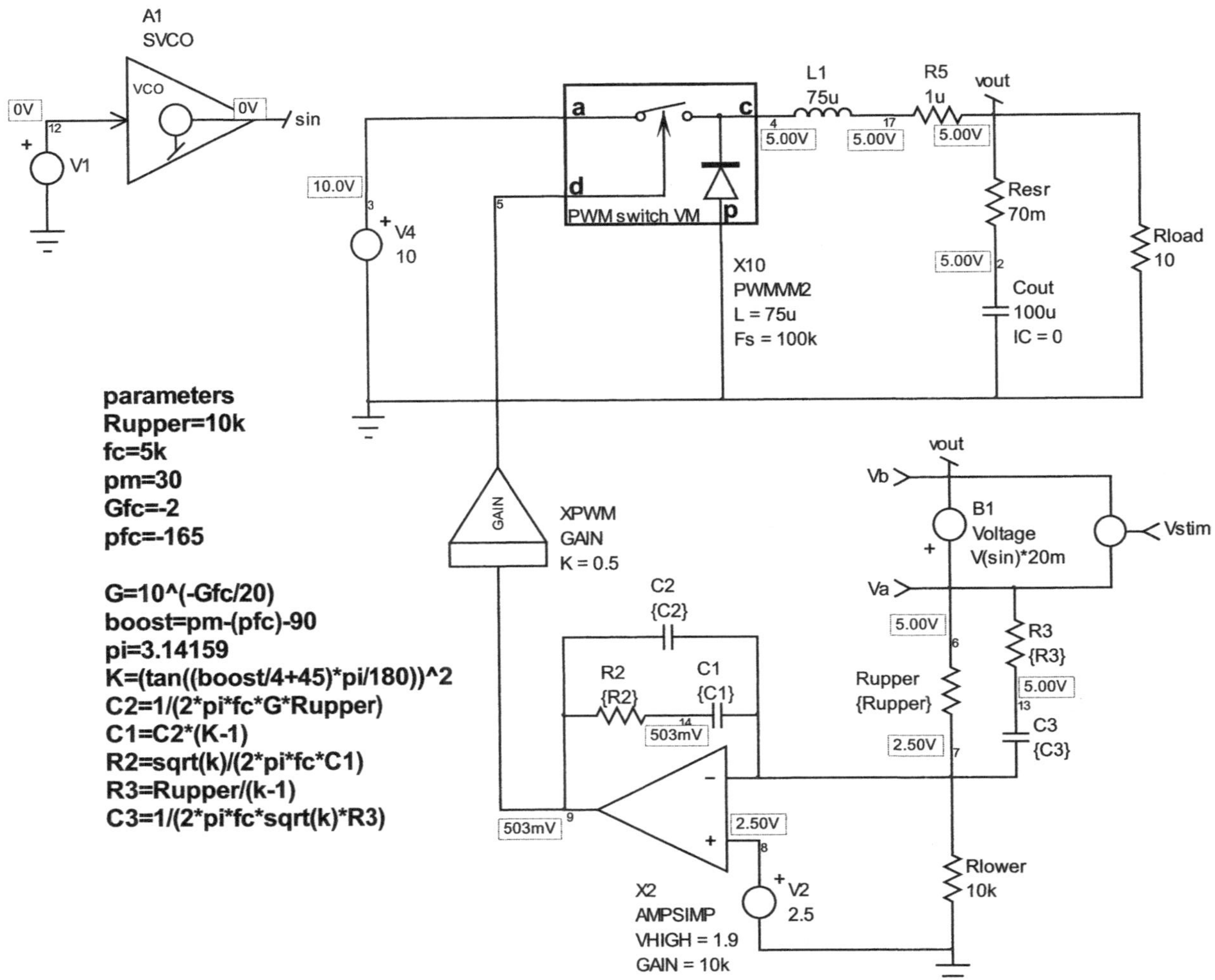

Figure 9.10 A simulation template helps to illustrate how points A and B move in relation to the loop gain variations along the frequency axis. The phase margin is set to 30° in this example.

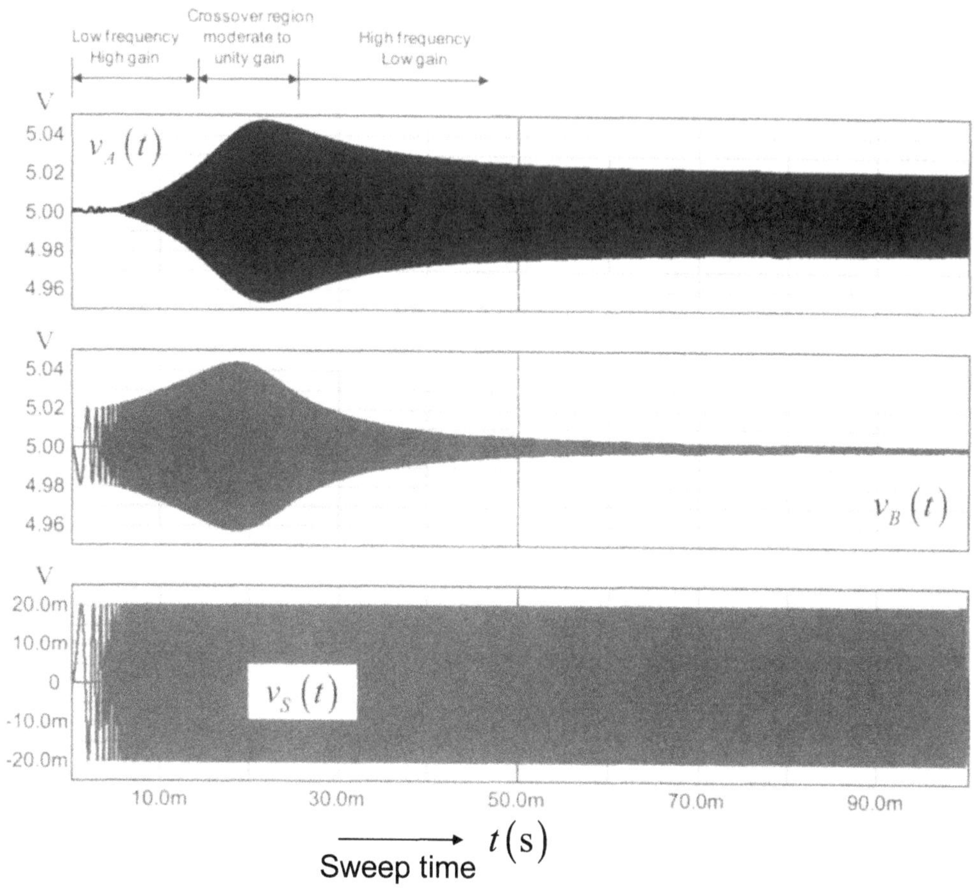

Figure 9.11 This graph shows how the control system permanently adjusts the A and B signals to maintain a constant ac source amplitude.

for signal A appear in the upper portion of the graph (low frequency, well below crossover) and around 1 mV for the low side of the plot (beyond the crossover frequency). How can we explain these changes?

There are two ways to look at the phenomenon. The first one deals with a simple block representation, actually rewriting equations around Figure 9.8. The basis is the simple Laplace equation linking points A and B amplitudes to the modulating signal:

$$V_S(s) = V_A(s) - V_B(s) \tag{9.9}$$

From this expression, we can easily extract the definitions of V_A and V_B:

$$V_B(s) = V_A(s) - V_S(s) \tag{9.10}$$

and

$$V_A(s) = V_S(s) + V_B(s) \tag{9.11}$$

To graphically represent these expressions, we have rearranged Figure 9.8 the way we propose in Figure 9.13.

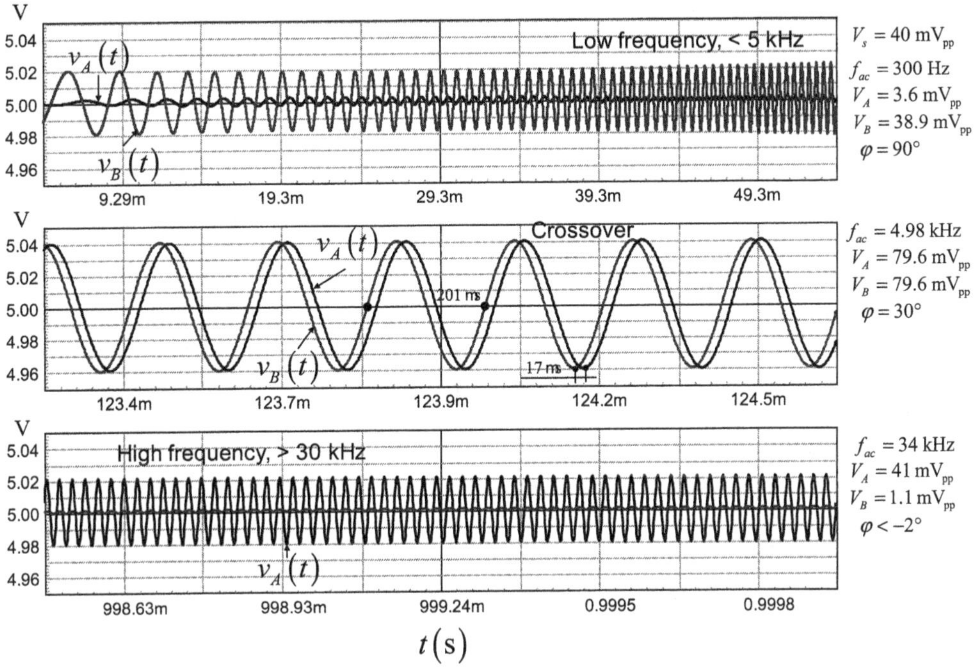

Figure 9.12 The signals amplitude are exactly equal at crossover and allow the phase margin measurement.

From this new picture, considering the inverting sign in the loop gain expression, we can write

$$V_B(s) = V_A(s)T(s) = T(s)[V_S(s) + V_B(s)] \tag{9.12}$$

$$V_B(s)[1 + T(s)] = T(s)V_S(s) \tag{9.13}$$

Finally, we have for V_B:

$$V_B(s) = V_S(s)\frac{T(s)}{1 - T(s)} \tag{9.14}$$

For V_A, the exercise remains similar:

$$V_A(s) = V_S(s) + V_B(s) = V_S(s) + V_A(s)T(s) \tag{9.15}$$

$$V_A(s)[1 - T(s)] = V_S(s) \tag{9.16}$$

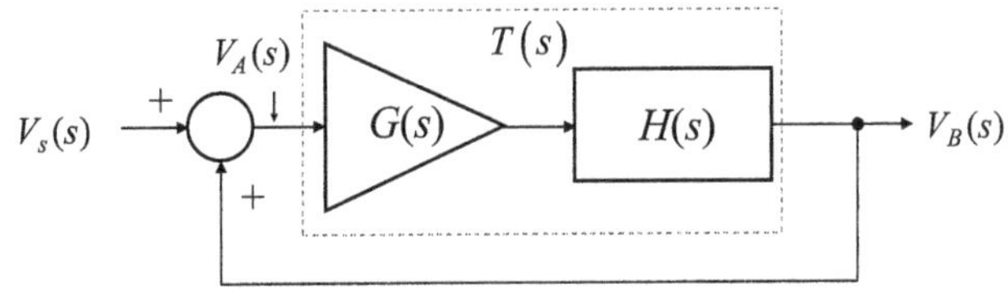

Figure 9.13 This simple representation lets us express V_A and V_B levels easily.

which leads to the following expression for V_A:

$$V_A(s) = V_S(s)\frac{1}{1-T(s)} \tag{9.17}$$

At very low frequency, the open-loop gain is important, several tens if not hundreds of decibels for an op amp-based circuit. In that case, (9.14) tells us that the output voltage V_B is almost that of the source level and phase reversed. On the contrary, the voltage at point A is extremely small as indicated by (9.17). Actually, the term multiplying V_S should ring a bell: the sensitivity function S tackled in Chapter 3. This function tells you how efficiently a system can reject a perturbation. Indeed, the insertion of the ac modulation is a perturbation that bothers the control system, which tries to reject it as much as it can. It succeeds at low frequency where the gain is strong. This is why the amplitude at point A is naturally small, almost drowned in the noise floor in high-gain systems. We will see that it is almost mandatory to increase the ac modulating source amplitude to make sure the analyzer can extract the signal. When T gets smaller as you slide the frequency axis, you will have to decrease this amplitude to avoid nonlinear operation since the rejection capability becomes weaker. Working with an analyzer that authorizes such source amplitude modulation is an obvious advantage.

The second explanation goes back to vector representation. Figure 9.14 graphs how the vectors sum up at three different frequencies: close to dc, at crossover, and beyond.

At dc, we know that the signal at point B in relation? to the modulating source is out of phase. It is represented as a horizontal vector pointing to the left if vector V_S is horizontal and points to the right. The signal at point A, in our compensation case, in delayed by 90°. Remember our ac plots with a type 3 compensator: we have the 180° of the inverting op amp (or –180° if you wish) and the –90° of the origin pole. The total argument at dc is therefore 90° or –270°. The vector sum of these two signals must equate the magnitude of the modulating signal V_S. At low frequencies, the loop gain is extremely high. Therefore, a small modulating signal at point A is sufficient to get a signal at point B. As both signals are affected by a 90° shift while vector B is horizontal, the only way to construct the vector sum is to have B of almost equal amplitude of the modulating signal while vector A magnitude is extremely small. The drawing in Figure 9.14(a) shows these signals where we purposely changed a little bit the argument of signal B; otherwise, vector A would

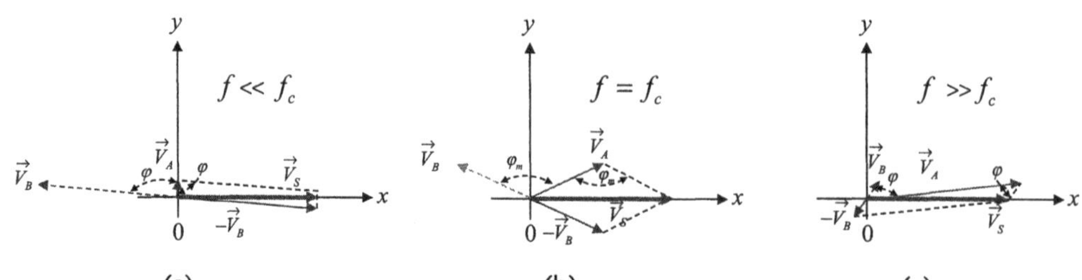

Figure 9.14 The amplitude of A and B adjusts depending on the phase difference between the signals. Their vector sum is constant and equal to the ac source magnitude V_S.

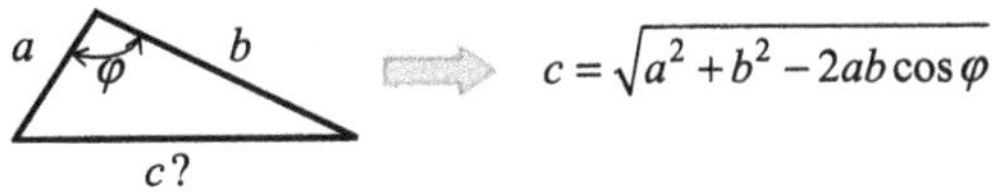

Figure 9.15 The Al-Kashi theorem works for all triangles, included right angled.

be invisible in the drawing. You have an idea of the signals amplitude on the right side of Figure 9.12.

At crossover, as expected, both signals amplitudes are similar since the magnitude of T at this frequency is 1. The phase between the two signals is the converter phase margin φ_m. The graphical representation appears in Figure 9.14(b). In this drawing, the phase margin is greater than 90°, and the amplitude of signals A and B is smaller than that of V_s. However, if you look back at comments in the middle of Figure 9.12, you see that the signals amplitudes exceed that of the source V_s by a ratio of two. How can it be?

To understand the phenomenon, we need to involve a theorem proposed by Persian scientist Al-Kashi (14th century). It is the generalization of the Pythagorean theorem and applies to any type of triangle.

Figure 9.15 represents a simple triangle where none of its angles is 90°. As indicated in the figure, the length c can be computed knowing the cosine of the angle φ. When this angle reaches 90°, the formula reduced to that of Pythagoras. This formula can be applied to our signal vectors, as drawn in Figure 9.16.

By applying Al-Kashi theorem, we immediately have:

$$\left|\overrightarrow{V_S}\right| = \sqrt{\left|\overrightarrow{V_A}\right|^2 + \left|\overrightarrow{V_B}\right|^2 - 2\cdot\left|\overrightarrow{V_A}\right|\cdot\left|\overrightarrow{V_B}\right|\cos\varphi} \tag{9.18}$$

V_A and V_B magnitudes are linked by the open-loop gain T:

$$\left|\overrightarrow{V_B}\right| = \left|\overrightarrow{V_A}\right|\cdot\left|T(s)\right| \tag{9.19}$$

And the angle φ is nothing more than the loop gain argument at the considered frequency. Combining the previous equations, we can extract the individual values of V_A and V_B as:

$$\left|V_A(s)\right| = \frac{\left|V_S(s)\right|}{\sqrt{1+\left|T(s)\right|^2 - 2\left|T(s)\right|\cos(\arg T(s))}} \tag{9.20}$$

$$\left|V_B(s)\right| = \frac{\left|V_S(s)\right|}{\sqrt{1+\dfrac{1}{\left|T(s)\right|^2} - \dfrac{2}{\left|T(s)\right|}\cos(\arg T(s))}} \tag{9.21}$$

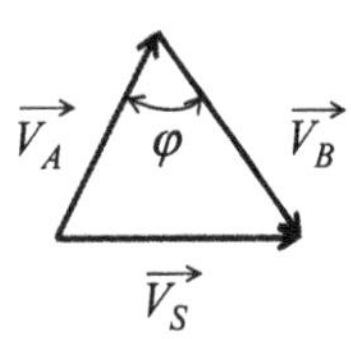

Figure 9.16 The signal amplitudes can be computed knowing the phase angle between the signals.

If we now consider the crossover frequency, it is a special case where both signal amplitudes are equal and the loop gain is 1. From (9.20), we can write:

$$|V_A(s)| = \frac{|V_S(s)|}{\sqrt{2-2\cos(\varphi_m)}} = \frac{|V_S(s)|}{\sqrt{2}\sqrt{1-\cos(\varphi_m)}} \tag{9.22}$$

From our school courses, we can remember that:

$$\sin\frac{\varphi}{2} = \sqrt{\frac{1-\cos\varphi}{2}} \tag{9.23}$$

From which we extract:

$$\sqrt{1-\cos\varphi} = \sqrt{2}\sin\frac{\varphi}{2} \tag{9.24}$$

If we substitute this definition in (9.22), we have:

$$|V_A(s)| = |V_B(s)| = \frac{|V_S(s)|}{2\sin\frac{\varphi_m}{2}} \tag{9.25}$$

Using this definition, we can now plot the amplitudes' evolution in relation to the phase margin at crossover. The graph appears in Figure 9.17. As we can see, a phase margin of 30°—as the one used in the Figure 9.10 template—gives a signal amplitude of 79 mV, very close to what we measured in Figure 9.12.

If we go back to our vector representations and combine again signal A and B of equal amplitudes but affected by three different phase margins, we obtain the drawing shown in Figure 9.18. In Figure 9.18(a), the margin is large, greater than 90°. The phase margin decreases in the middle and the vectors magnitudes have

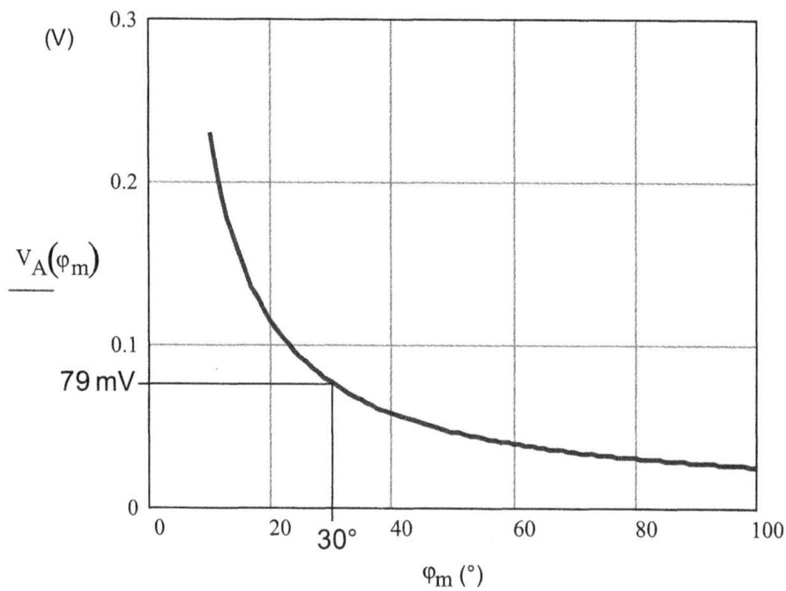

Figure 9.17 As the phase margin decreases at crossover, the signals' amplitudes tend to grow. At zero phase margin, their amplitude is theoretically infinite, indicative of an unstable system.

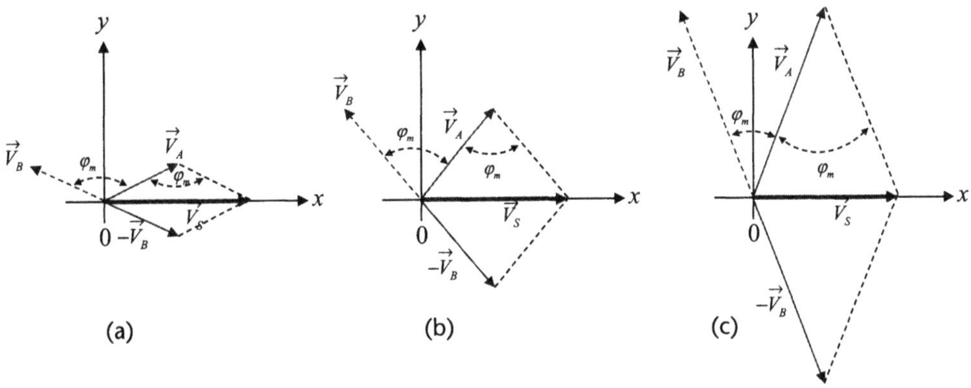

Figure 9.18 This simple drawing clearly illustrates how both V_A and V_B amplitudes run away as the phase margin reduces.

already grown. In Figure 9.18(c), the phase margin is even smaller and the vectors magnitudes exceed that of V_S.

With these two approaches, you now understand why the signals at points A and B change in amplitude across the frequency swept by the analyzer.

As explained, as our switching power supply is a noisy environment, a signal of low amplitude at point A can easily be drowned in the noise, leading to a poor signal-to-noise ratio. It is thus very common to increase the modulating signal in the low frequency portion of the analysis to improve the noise figure. This is particularly true for very high dc gain systems, where sometimes a modulating signal of several volts is necessary: remember, the higher the gain, the more difficult it is to shake the closed-loop system! Then, as crossover is approached, the amplitude must be seriously reduced. If it is not, overmodulation distortion can occur and the ac plot accuracy is affected. Most of the available network analyzers allow the user to adjust its ac modulation amplitude by segments. A typical amplitude pattern is given in Figure 9.19.

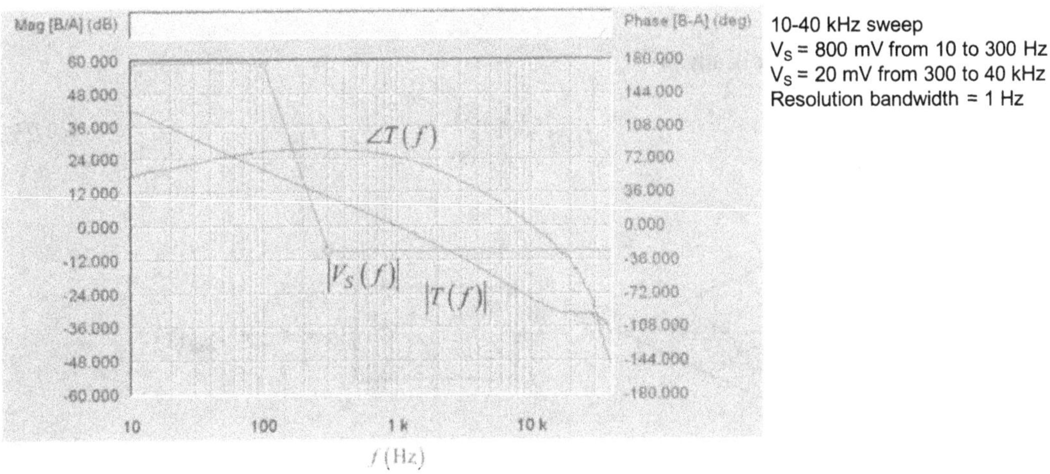

Figure 9.19 To ensure a good signal-to-noise ratio, the network analyzer lets you adjust the modulating signal ac amplitude along the frequency axis. Here is a screenshot obtained with an AP300 analyzer from Ridley Engineering.

It is not detailed here, but prior to any measurements you need to calibrate the probes across the frequency range of interest. It is an essential step to ensure correct results.

9.1.5 Impedances at the Injection Points

You can measure the loop gain by either physically interrupting the loop or by virtue of inserting a stimulus in series with the return path. In both cases, you need to consider the impedances at stakes when interrupting the path. They are the output impedance of the converter where you observe the controlled variable and the input impedance at the considered error amplifier. Let's look at Figure 9.20, where we interrupted the path at the error amplifier output. If we write the control to error transfer function, we will obtain the following expression:

$$T(s) = \frac{V_{err}(s)}{V_c(s)} = -G(s)H(s)\frac{Z_{in}(s)}{Z_{out}(s)+Z_{in}(s)} \tag{9.26}$$

This is the true loop gain affected by the impedances linking the output to the error amplifier input.

We learned in Figure 9.7 that the ac source could also be placed in series somewhere in the return path. To do that, we interrupt the wire between points A and B to insert the modulating source. Before the insertion, A and B were sharing a similar potential as they were connected together. With the source insertion, this relationship is no longer true: point B now drives point A with the ac source in series. What matters now is the impedance relationship between point A and B. Figure 9.21 updates Figure 9.20 by inserting the stimulus at points A and B.

To see the effects of these impedances, let's derive a few equations:

$$V_c(s) = -V_A(s)G(s) \tag{9.27}$$

$$V_B(s) = V_c(s)H(s) - Z_{out}(s)I_1(s) \tag{9.28}$$

The input current is simply:

$$I_1(s) = \frac{V_A(s)}{Z_{in}(s)} \tag{9.29}$$

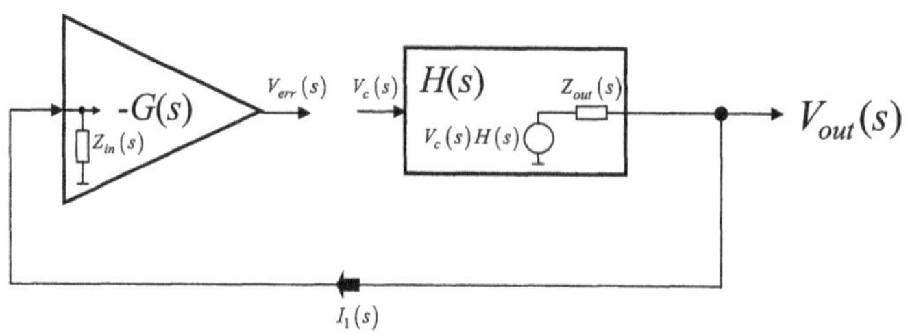

Figure 9.20 The loop is physically opened, but the resistive divider brought by the output/input impedances can alter the measurement.

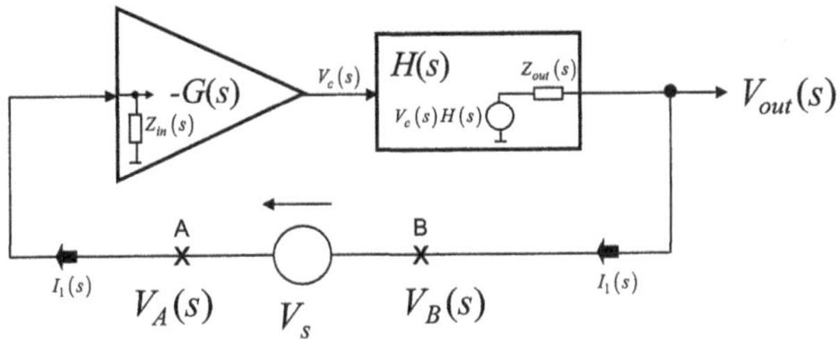

Figure 9.21 Point B must be a low-impedance point, whereas point A input current must be negligible.

Substituting (9.29) in (9.28), we have:

$$V_B(s) = -V_A(s)H(s)G(s) - \frac{Z_{out}(s)}{Z_{in}(s)}V_A(s) \tag{9.30}$$

Rearranging, we obtain the loop gain definition:

$$\frac{V_B(s)}{V_A(s)} = T(s) = -\left[H(s)G(s) + \frac{Z_{out}(s)}{Z_{in}(s)}\right] \tag{9.31}$$

This expression is different than the one derived in (9.26) and shows the effect of circuit impedances. To make the loop gain $T(s)$ independent from these impedances, you need to satisfy the following conditions:

$$Z_{out}(s) \ll Z_{in}(s) \tag{9.32}$$

$$\frac{Z_{out}(s)}{Z_{in}(s)} \ll T(s) \tag{9.33}$$

In other words, you must interrupt the loop at a position where point B output impedance is low and point A input impedance is much higher. Failure to do that will lead to a measurement distortion. References [7, 9] offer alternative methods to cope with a situation in which (9.32) and (9.33) cannot be respected. It naturally requires more manipulations to obtain the correct answer.

9.1.6 Buffering the Data

Most of modern ac-dc pulse width modulation (PWM) controllers no longer include the intelligence on the primary side. By intelligence, I mean the reference voltage and the error amplifier. These two elements are either combined in a TL431 placed on the secondary side or separately associated through an op amp and a discrete reference voltage. The typical architecture you will find is the one shown in Figure 9.22. It is that of a current-mode power supply.

The primary peak current is set by the voltage present on the feedback pin (labeled FB), further divided by R_1 and R_2 before reaching the peak current comparator. The FB pin level is adjusted by the current flowing in the optocoupler collector. This current depends on the TL431 action placed in the secondary side. Depending on the loading conditions, it automatically adjusts the LED current to program the

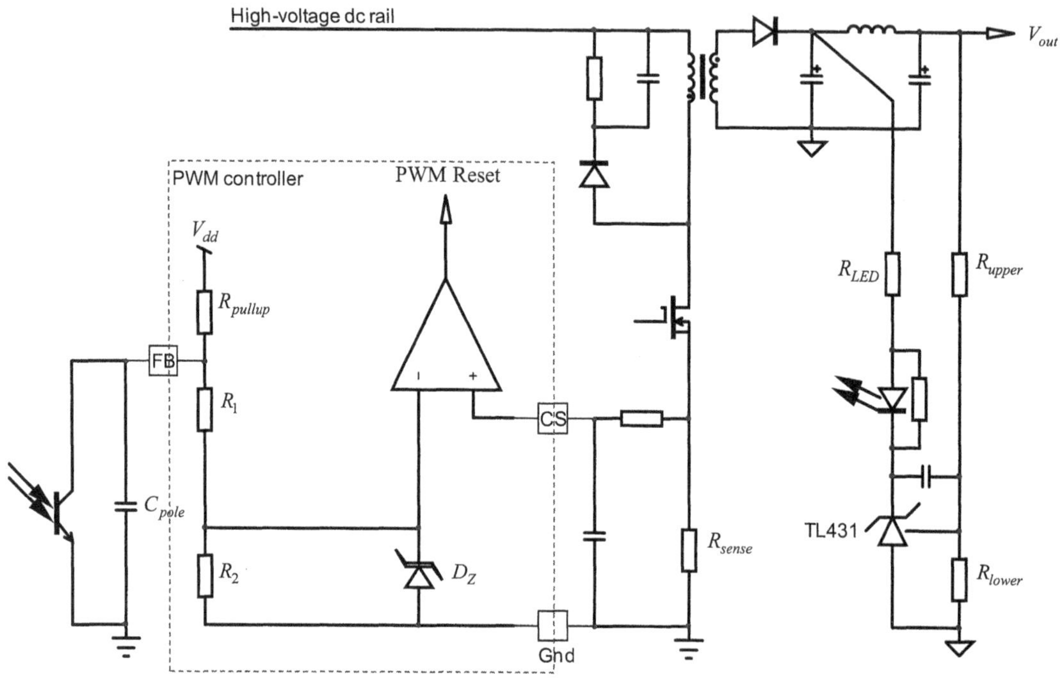

Figure 9.22 In modern power supplies, the intelligence is kept in the secondary side while the primary side features a simple feedback pin setting the inductor peak current or the duty ratio.

adequate peak current on the primary side. As explained in the TL431 chapter, we have two lanes, one fast and one slow. The fast one deals with R_{LED}, while the slow one deals with R_{upper}. One place to open the loop would be in series with R_{upper}, for instance, as it is usually a high-value resistance and the converter output is, by definition, a low-impedance point. Unfortunately, given the dual-lane architecture of the TL431 circuitry, ac modulating the system through R_{upper} would characterize only one path, the slow one. To form the complete chain, you would also need to characterize the other one (the slow lane) by inserting the ac source in series with R_{LED}. However, to apply the superposition theorem, you will need to disconnect one of the two paths and keep it dc biased while modulating the other one. Then, you will need to vector-combine both signals and reconstruct the complete loop gain. Nothing undoable, but it is a rather tedious and long process. Reference [10] shows you how to do that in detail.

If you want to simplify the measurement, the best way is exclude the output filter inductor. If you have selected it so that its cutoff frequency is high, the impact on the measurement is weak. Figure 9.23 illustrates how to connect the source. If you probe V_A and V_B as indicated, you will display the loop gain. Should you need the compensator transfer function including the optocoupler, move the V_B probe and place it at the feedback pin on the optocoupler collector. If you want the plant transfer function, keep V_B as in Figure 9.23 and move V_A to the feedback pin. Watch out: if you do that, you will have to connect both primary and secondary grounds as already suggested in Figure 9.4 and Figure 9.6. Make sure the high-voltage dc bias is fully isolated from the mains and apply safety rules in presence of these high voltages.

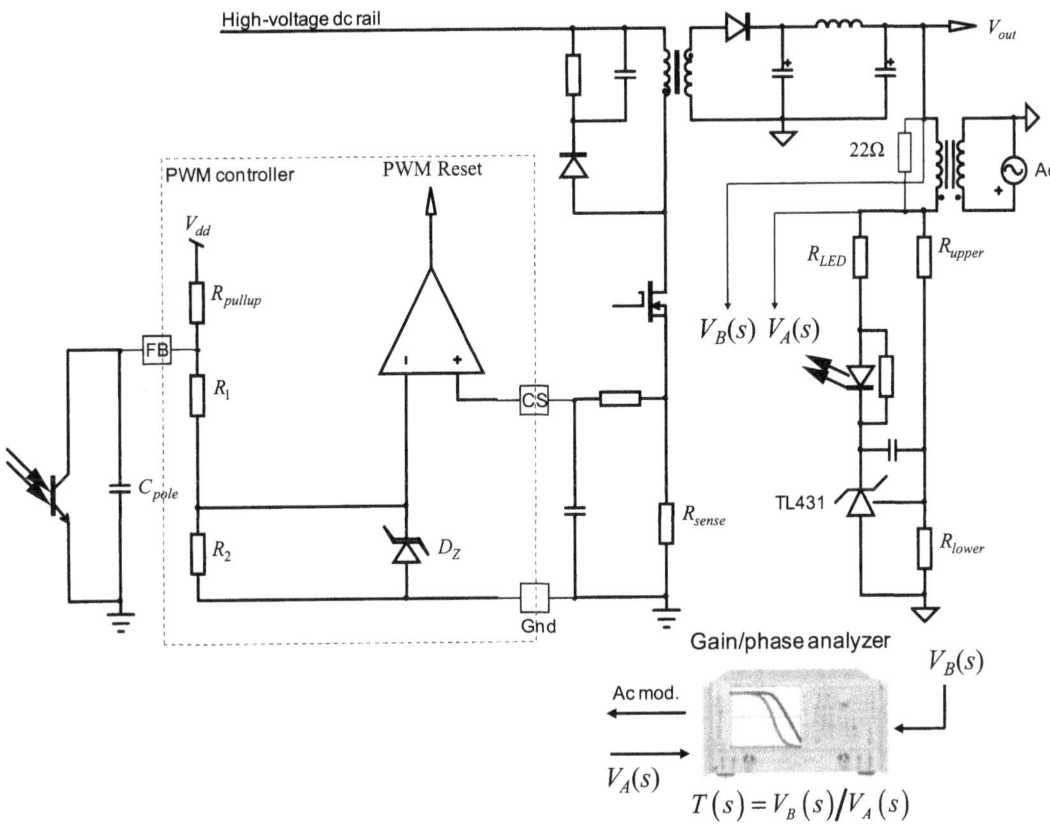

Figure 9.23 Connecting both lanes together helps to run a quick loop gain measurement.

Some designers may not like the fact that the fast lane connection before the output LC filter is ignored by this configuration. To obtain the complete loop gain without touching the secondary-side arrangement, one option is to open at the optocoupler collector. That way, we will capture the complete signal driven by the TL431 combining both lanes. Unfortunately, the optocoupler collector is of rather high output impedance, and if we disconnect it from the controller we lose all the internal bias conditions. The idea is to recreate these conditions externally and buffer the resulting signal before inserting the ac modulation. This is what Figure 9.24 presents. The optocoupler collector is loaded by the equivalent ac resistor R_{eq} present at the controller feedback pin. In our example, it is $R_{pullup} \,||\, (R_1 + R_2)$. The internal bias, V_{dd}, is usually 5 V and externally applied on this resistor to reproduce the original bias conditions.

The capacitor placed on the feedback pin creates a pole and has to be moved on the optocoupler collector side to realize the same pole. The NPN transistor buffers the whole and sources the collector signal under a low output impedance. We used a 1-kΩ resistor, but it can be lowered if necessary. Lowered sometimes to make sure the feedback pin can drop to a low voltage if asked by the loop. If R_{eq} is around 15–20 kΩ, a 1-kΩ pull-down works well. For high-bandwidth systems, forward converters (e.g., the pull-up resistor inside the controller) can be down to 4.7 kΩ or less. In that case, to offer the same regulation dynamics, the emitter resistor can be as low as 100 Ω if necessary. In a multi-output system, the insertion of this bipolar

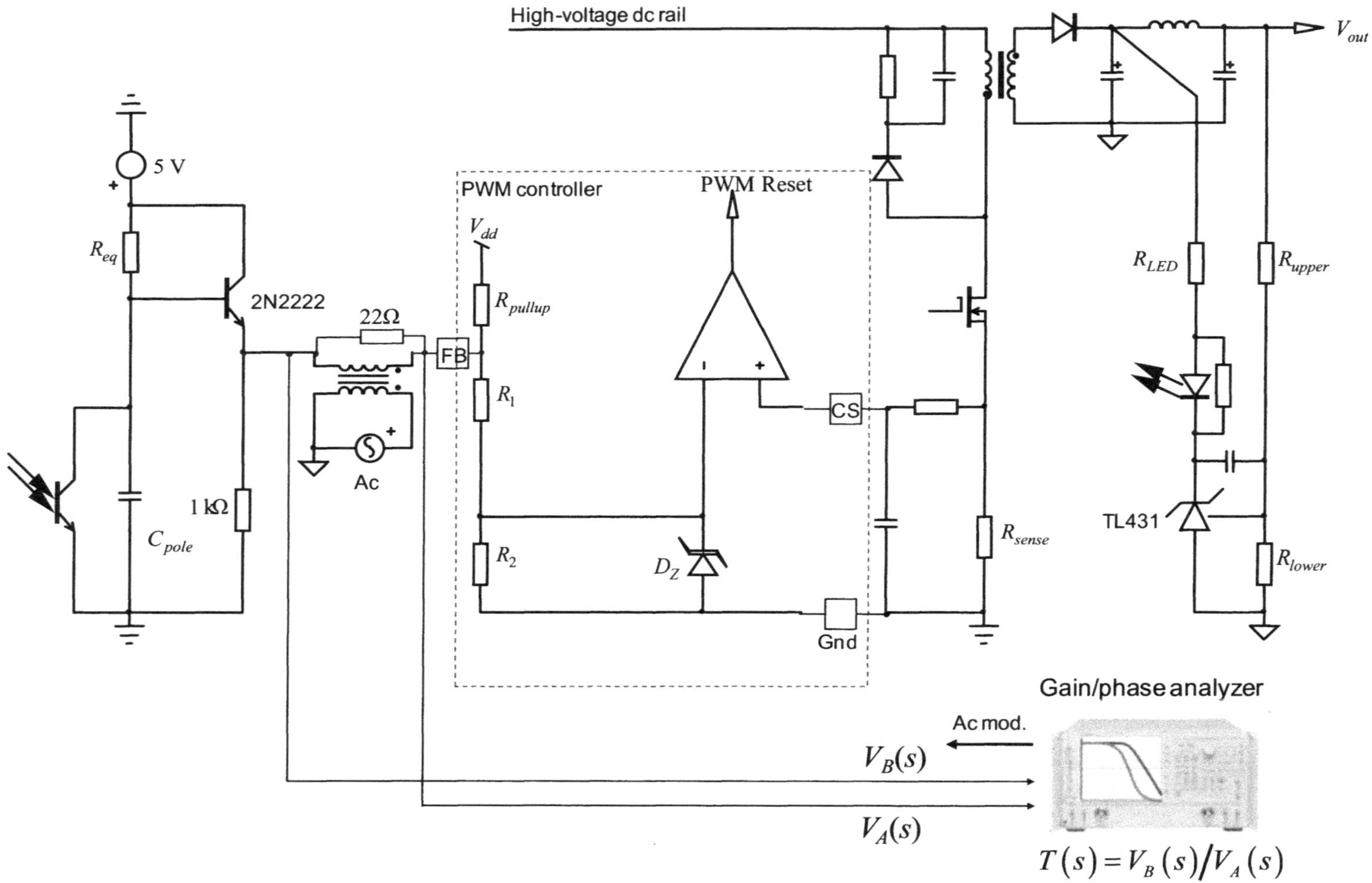

Figure 9.24 A simple bipolar buffers the optocoupler collector and drives the controller with a low output impedance.

transistor can bother the whole operation at startup. You will need some skill to slowly raise the voltages in order to ensure a proper regulation without tripping the protection controller. We have seen this issue in multi-output ATX power supplies, for instance.

Loop gain measurements require experience and patience. Fortunately, a lot of publications and manuals are available on the web to guide you in your experiments. Now that we know how to measure a loop, let's have a look at a few design examples. However, before closing this section, I want to express my gratitude to Dr. Jose Capilla from ON Semiconductor with whom I have had endless discussions on vectors and source amplitudes and whose internal seminar on the subject has been of great help [11].

9.2 Design Example 1: A Forward dc-dc Converter

This first design example covers a transformer-isolated dc-dc converter. Based on a single-switch forward topology, the converter delivers 5 V at a 20 A output current from a telecom network. Assume our converter specifications require a crossover frequency of 10 kHz, together with a 60° phase margin. This last requirement implies an opto-isolating device whose pole is naturally well beyond this crossover value. If we select a slow optocoupler, depending on the adopted feedback scheme, we may encounter difficulties to compensate the phase degradation as its pole kicks in. Adding another zero in the control loop looks like a possible option. However, if this zero provides phase help at first glance, its presence will hamper the recovery time, especially if it is located in the lower portion of the frequency spectrum. Using the inherent optocoupler pole as a companion rather than an enemy is the strategy we recommend.

9.2.1 Moving Parameters

Available optocoupler characterization curves is an element you might want to consider prior to the device selection. Collecting data points because of a weak data sheet content is something you want to avoid. To that respect, the PC817 specification sheet from Sharp offers a great deal of detail. Browsing it indicates that the A version represents a good pick: the CTR varies from 80 percent to 140 percent at a 25-°C junction temperature, and as Figure 9.25 confirms a 2-mA LED forward current brings it to 100 percent. Temperature variations shown in Figure 9.26 point out a total span of 60 percent, from a junction temperature ranging from –30 °C up to 100 °C.

As the parameters in the data sheet are guaranteed at 25°C only, the minimum CTR mixing lots-to-lots dispersions and temperature spreads must be updated to:

$$CTR_{min} = 80\% \times 60\% = 48\% \tag{9.34}$$

Therefore, we will have to shield our converter design against potential optocoupler CTR changes from 48 percent to around 140 percent, a corresponding 1:3.3 ratio. It is important to mention that these variations are just typical data and

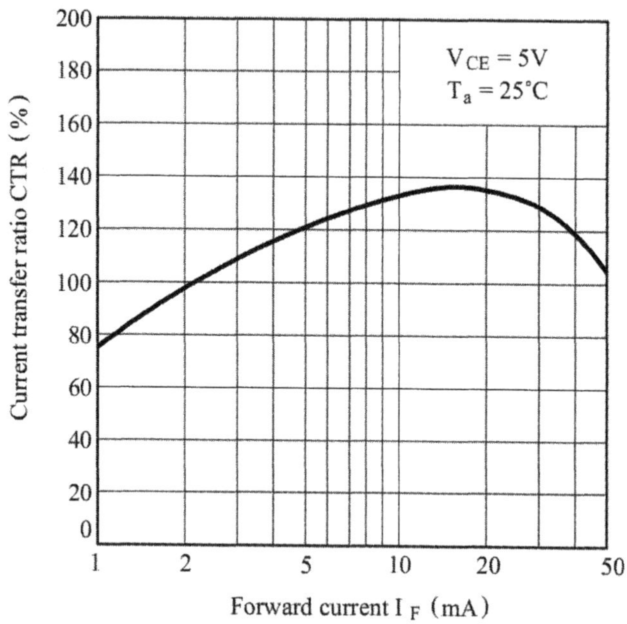

Figure 9.25 The CTR reaches a typical value of 100 percent for a 2-mA LED current.

often not guaranteed in the data sheet. Further discussions are thus necessary with the optocoupler manufacturer to obtain the real dispersion you should expect over the converter lifetime.

Bandwidth-wise, the data sheet offers a frequency characterization in relationship to the collector pull-up resistor. It appears in Figure 9.27 that a 1-kΩ pull-up resistor brings the pole to roughly 25 kHz. This is an excellent value for a 10-kHz

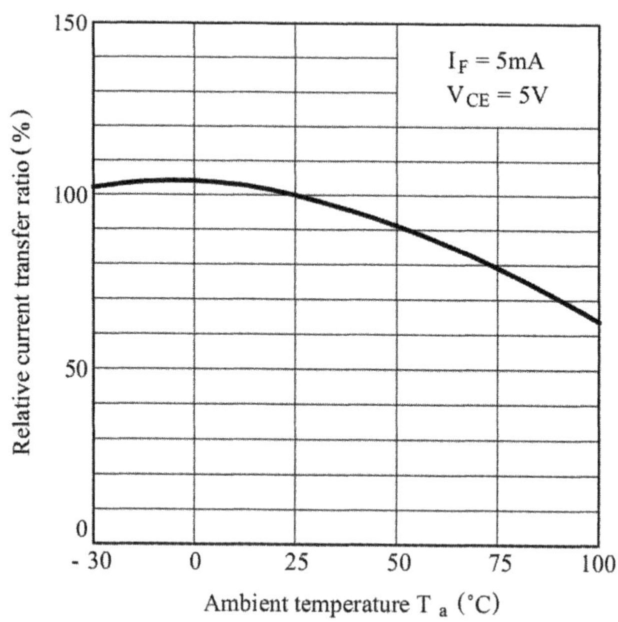

Figure 9.26 CTR versus temperature variations are also available and indicate a 60 percent reduction in the upper range.

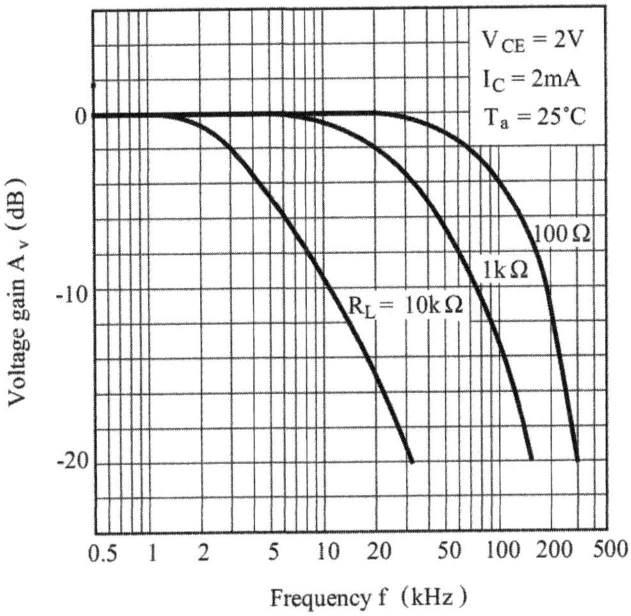

Figure 9.27 The frequency characterization of the PC817 reveals a 25-kHz pole when a pull-up resistor of 1 kΩ is adopted.

bandwidth project. Unfortunately, the document does not indicate the PC817 suffix this curve is attached to. For a serious study, once several optocoupler samples are received, it is the designer's task to set up a characterization fixture such as that described in Chapter 5's Appendix C to extract the pole positions in various operating conditions. In our case, as the selected controller featured a 3.3-kΩ pull-up resistor, we have recharacterized the chain and found a pole around 15 kHz.

Based on this information, we are able to determine the equivalent optocoupler parasitic capacitor value C_{opto} when its collector is loaded with a 3.3-kΩ resistor:

$$C_{opto} = \frac{1}{2\pi R_{pullup} f_{pole}} \approx 3.2\ nF \tag{9.35}$$

When lacking available bench data (e.g., you did not receive the optocoupler samples at the time of the study), information from Figure 9.27 can be used for a first-pass analysis that you will further refine by extensively testing and characterizing the prototype during laboratory experiments.

9.2.2 The Electrical Schematic

Our converter must deliver 5 V at 20 A from a telecom input power, implying a voltage variation from 36 V to 72 V dc. A forward converter belongs to the buck-derived topology. When operated in voltage-mode, this structure shows a resonating peak brought by the combination of the inductor and the output capacitor. When this resonant frequency is low, the compensation scheme must include a double zero located around the resonant peak. In terms of transient response, a zero involved in the compensation loop, $G(s)$, becomes a pole when you study the closed-loop transfer function. As we want a fast response with a good recovery time, we must avoid low frequency zeroes. For this reason, the current-mode topology looks like the best choice to go for as it turns the converter into a first-order system at frequencies below half of the switching frequency. Furthermore, the mode transition that occurs in light load conditions does not imply a particular care when designing the control loop. On the contrary, compensating a voltage-mode converter toggling between both modes is a difficult exercise, as the Bode plot dramatically changes when transitioning from the CCM to the discontinuous conduction mode (DCM). For budget reasons, a single-switch topology featuring a demagnetization winding was selected, but the example remains pertinent in case another choice would be adopted (two-switch forward, *RCD* clamp, and so on).

The NCP1252 from ON Semiconductor is a possible candidate for the PWM control section, as it packs a lot of good features. With its soft-start and frequency jittering capabilities, it can be considered a replacement of choice for future or existing UC384X-based designs. Furthermore, due to its integrated skip-cycle type of control, the circuit accepts no-load conditions without going into overvoltage as with other devices. The application schematic appears in Figure 9.28. Despite the involved power, there is not much around the controller. A simple pull-down resistor of 22 kΩ sets the switching frequency to 200 kHz, while the brown-out network made of R_2 and R_8 ensures the operating input voltage is always greater than 33 V.

The output inductor has been selected for a 10 percent ripple current when the controller switches at 200 kHz. The two output diodes accept up to 40 A, dropping 0.58 V when crossed by a 20-A current (Tj = 100°C). The control loop uses an operational amplifier driving the optocoupler LED cathode while its anode connects to the output voltage rail. This configuration is very close to that of a TL431 featuring two lanes: a slow lane implying the capacitor C_1 and a fast lane brought by R_3. The transfer function of this type 2 compensator has already been derived in Chapter 5 and equals:

$$\frac{V_{FB}(s)}{V_{out}(s)} = G(s) = -\left[\frac{sR_1C_1+1}{sR_1C_1(1+sR_{pullup}C_1)}\right]\frac{R_{pullup}\text{CTR}}{R_3} = -G_0\frac{1+s/\omega_z}{s/\omega_{po}(1+s/\omega_p)} \quad (9.36)$$

where

$$G_0 = \frac{R_{pullup}\text{CTR}}{R_3} \quad (9.37)$$

$$\omega_{po} = \frac{1}{R_1C_1} \quad (9.38)$$

$$\omega_z = \frac{1}{R_1C_1} \quad (9.39)$$

$$\omega_p = \frac{1}{R_{pullup}(C_{opto}+C_8)} \quad (9.40)$$

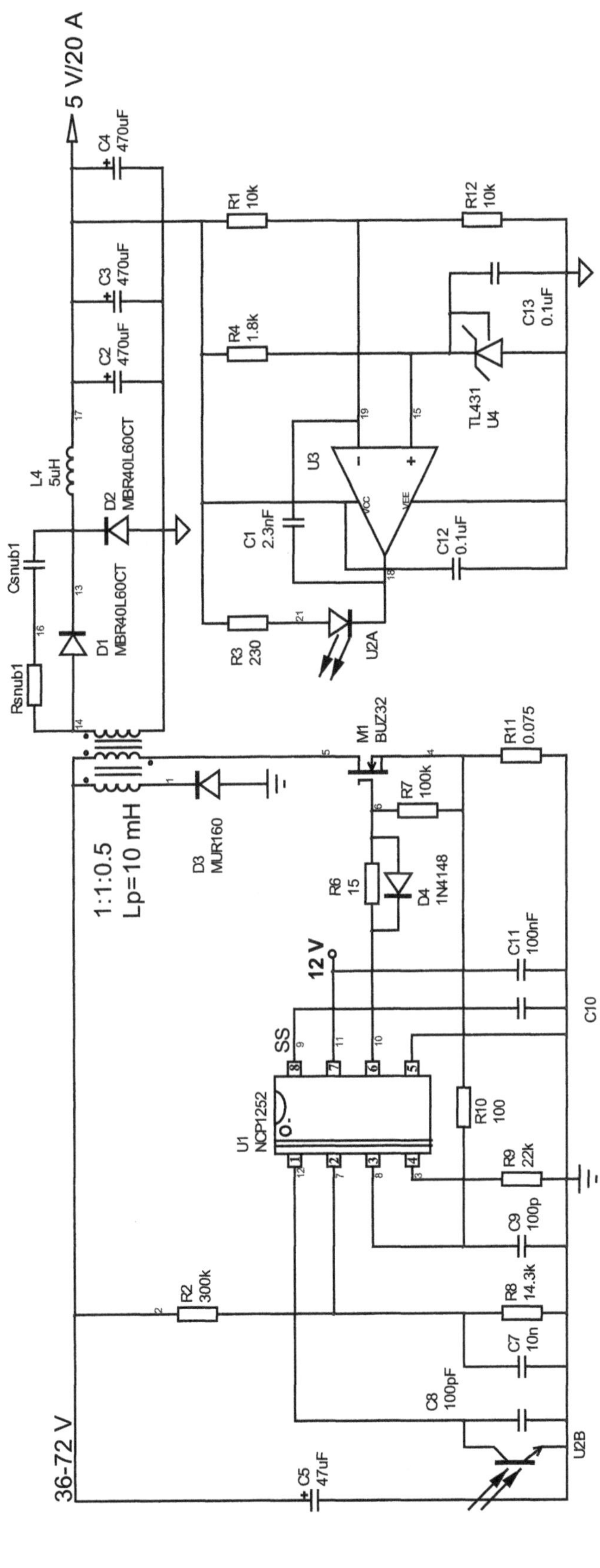

Figure 9.28 The application circuit of the single-switch forward converter using a NCP1252.

In this type of configuration, the dc conditions bounds R_3 values through the minimum output voltage the operational amplifier is capable of. We can show that the maximum value authorized for R_3 obeys the following formula:

$$R_{LED,\max} \leq \frac{V_{out} - V_f - V_{op\ amp,\min}}{V_{dd} - V_{CE,sat}} R_{pullup}\text{CTR}_{\min} \tag{9.41}$$

where:

$V_{opamp,min}$, the minimum output voltage the selected operational amplifier can go down to
V_{out}, the output voltage (5 V in our example)
V_f, the optocoupler LED forward drop (≈ 1 V)
$\text{CTR}_{\min}$, the minimum optocoupler current transfer ratio: roughly 50 percent for a PC817A as rounded from (9.34)
$V_{CE,sat}$, the optocoupler saturation voltage (≈ 300 mV at a 1-mA collector current) which imposes the minimum feedback voltage
V_{dd}, the internal bias of the pull-up resistor, usually 5 V
R_{pullup}, the internal pull-up resistor of 3.3 kΩ

If we consider a low-state voltage excursion for the op amp of 150 mV, the LED series resistor cannot exceed the following value:

$$R_{3,\max} \leq \frac{5-1-0.15}{5-0.3} \times 3.3k \times 0.5 \leq 1.35\ \text{k}\Omega \tag{9.42}$$

This condition, unfortunately, sets a minimum gain for the compensator featuring the op amp and the LED connected to the output voltage. According to (9.37), this minimum gain amounts to

$$G_0 = \frac{R_{pullup}\text{CTR}_{\max}}{R_3} = \frac{3.3k \times 1.4}{2.7k} = 1.7 \text{ or } 4.6\ \text{dB} \tag{9.43}$$

Some power stages require an attenuation at crossover. For instance, if your gain analysis of the power stage $H(s)$ reveals a +7-dB gain at 1 kHz, you will have to shape the compensator to make its magnitude equal –7 dB at 1 kHz. If you want to use the configuration described in Figure 9.28, you will be stuck given the result found in (9.43): 4.6 dB minimum gain. The limit is similar to that of the TL431 when its fast lane is used. To design such a compensator, you would have to (a) decouple the LED connection from V_{out} via a Zener or a regulator, or (b) drive the LED anode directly with R_3 connected to ground. This last option would certainly remove the output voltage influence on the compensator, but it would force you to wire the optocoupler in a common collector configuration to account for the phase reversal. Some controllers accept this; some do not.

9.2.3 Extracting the Power Stage Transfer Response

The output power stage ac transfer response can be obtained in different ways: bench measurement, analytical description, and SPICE simulation. Bench measurements require a rather complex setup plus an operational prototype. You must therefore wait for the various elements to be assembled before data can be exploited for the compensation purposes. On the other hand, the analytical derivation requires a simple sheet of paper and helps you to gain insight in the power stage transfer function. It shows you where the poles and zeros are and what external parameters can influence them. To that respect, going once through this kind of exercise is mandatory for someone serious about loop control. The main drawback is that you need two sets of equations depending on the operating mode (CCM or DCM). For this reason, the SPICE simulation of converters represents an excellent tradeoff. When the parasitic elements such as capacitor equivalent series resistors (ESR) are well extracted from the manufacturer data sheets (or, even better, you measure them through an impedance plot), experience proves that the predicted response matches very well with final measurements carried over the prototype. Furthermore, thanks to auto-toggling average models such as those described in [12], you will have the ability to explore both operating modes and check how your compensation strategies affect the transient response.

The first step is to extract the transfer function of the power stage, $H(s)$, at the lowest input voltage and maximum output current. The template in Figure 9.29 will give you this response. The optocoupler is a simplified model described in [12] but good enough to see the pole effect and its CTR impact on the gain. The B_1 source mimics the controller internals (i.e., the relationship between the feedback voltage and the peak current setpoint): a series diode drops 0.6 V and a divide-by-three bridge further affects the signals. This is what the equation describes, as well as clamping the maximum excursion to 1 V. It is very similar to what is found in a UC384x controller.

From the power stage transfer curve, we will obtain the information at the targeted crossover frequency (10 kHz, in our example) such as the magnitude and the phase lag. These data will feed our compensator calculation flowchart. Figure 9.30 portrays $H(s)$ Bode plot obtained in a few milliseconds of simulation time.

The accuracy of such a plot depends mainly on the parasitic elements. If you have properly extracted the capacitors ESR values, then there is a good chance that the ac simulation will accurately match the data collected in the laboratory. If needed, you can even refine the simple capacitor model by a more accurate one as described in the Kemet site (see [13]). The company offers various modeling software for different capacitor technologies such as Multi Layer Ceramic Capacitors. These comprehensive models reflects the ESR variations versus frequency and fit well simulations of high-quality dc-dc converters for space or military applications, for instance. In this domain, immunity to component variations is of extreme importance.

9.2.4 Compensating the Converter

From this picture, we can see that a 10-kHz crossover frequency requires a lift of 17.2 dB. As the phase lag is 51° at this frequency, we will need to tailor the compensator transfer function $G(s)$ to provide some phase boost to meet the needed phase margin target at crossover. To provide a good dc gain, we will use a compensator type featuring an origin pole in the transfer function. Its presence implies a permanent phase delay of 90° to which the op amp phase reversal adds another

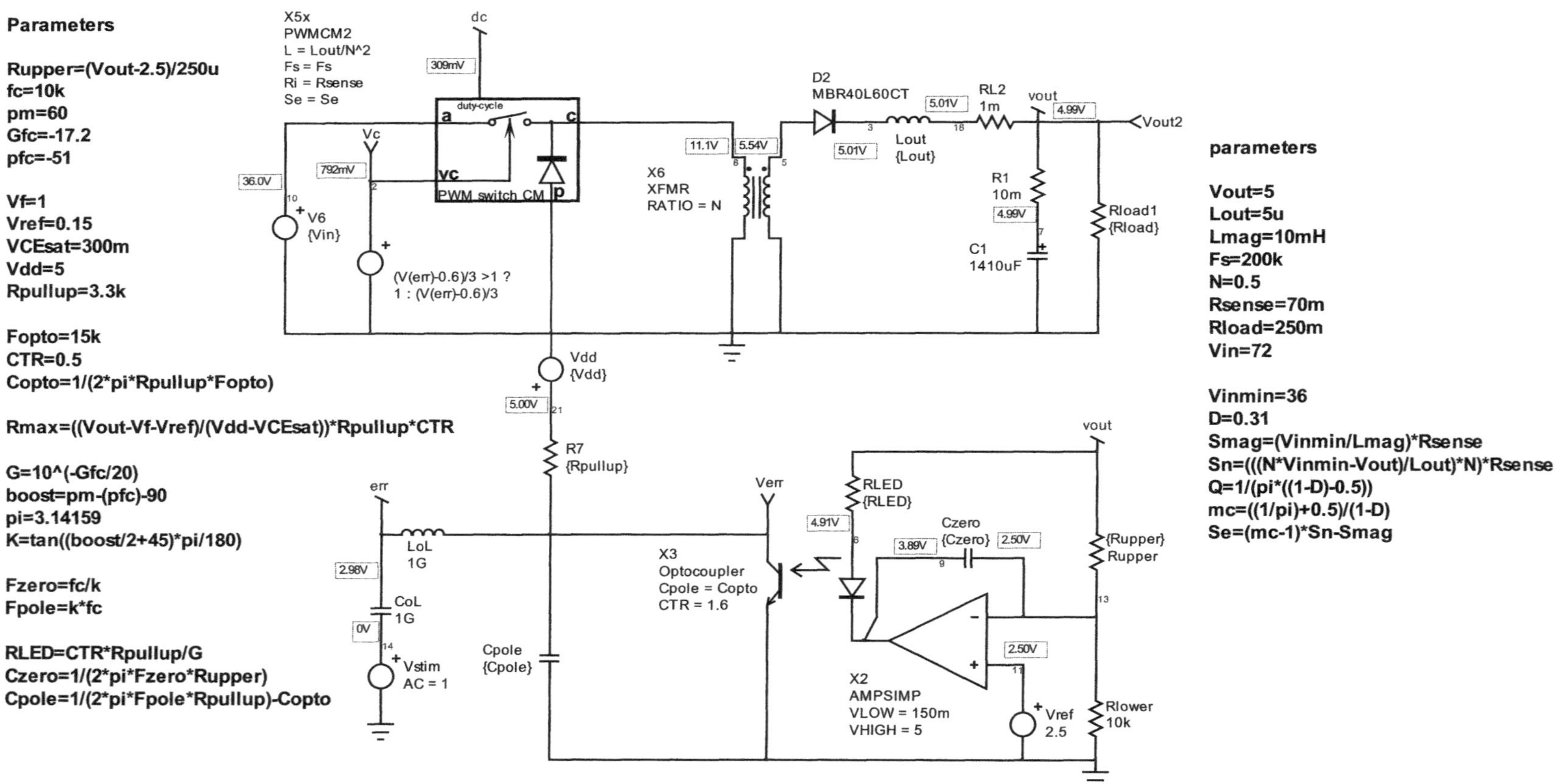

Figure 9.29 An average model operated in current-mode and associated with an op amp can quickly deliver the power stage transfer function.

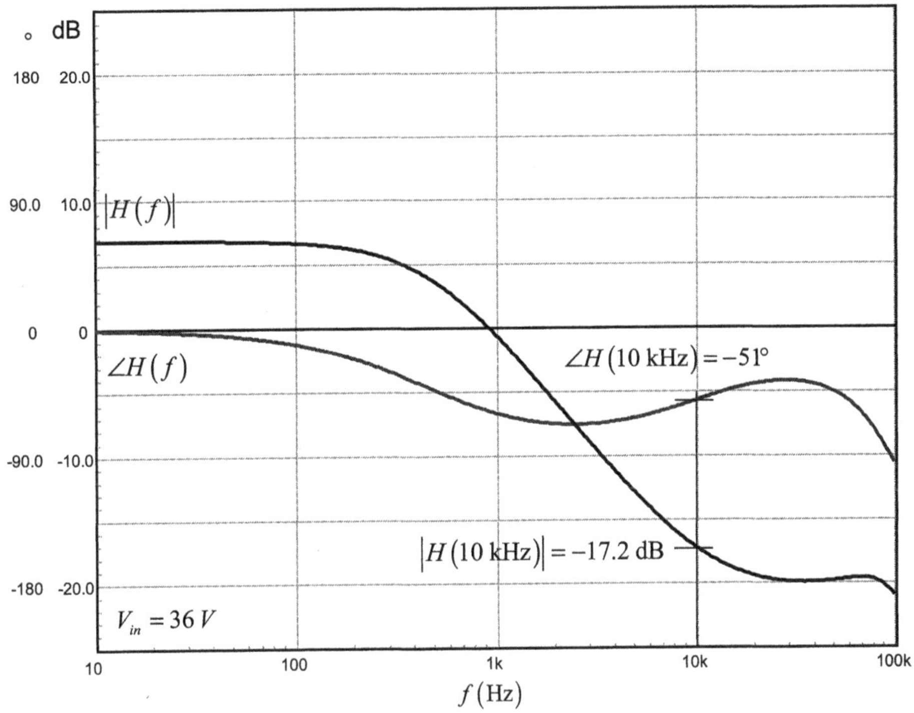

Figure 9.30 The transfer function $H(s)$ of the power stage at low input voltage and full load.

180°, making a total of –270°. If we want to obtain a 60° phase margin at 10 kHz, then the cumulated phase lag of arg $H(s)$ + arg $G(s)$ will need to be below $-360+60=-300°$. As the total argument of the power stage plus the origin pole of the compensator reaches $-270-51=-321°$, we need to provide a phase boost of 21° at 10 kHz. Otherwise stated:

$$BOOST = \varphi_m - \arg H(f_c) - 90° = 60+51-90 = 21° \tag{9.44}$$

Such a low phase boost can easily be obtained with a type 2 configuration. In the previous chapters, we have seen several methods to help you build the necessary phase boost. Using the k factor described in Chapters 4 and 5, we will place a zero at 6.8 kHz and a pole at 14.5 kHz. We can show that the phase boost brought by this configuration at 10 kHz is exactly:

$$\tan^{-1}\left(\frac{f_c}{f_z}\right) - \tan^{-1}\left(\frac{f_c}{f_p}\right) = 55.8 - 34.6 \approx 21° \tag{9.45}$$

Following equations (9.37), (9.39), and (9.40), we found a zero capacitor C_{zero} of 2.3 nF (C_1 in Figure 9.28) and a pole capacitor of 3.3 nF. According to (9.35), with an optocoupler parasitic capacitance of 3.2 nF, you just need to a add a small 100 pF for C_{pole} (C_8 in Figure 9.28) to reach the total value of 3.3 nF. Wire this capacitor as close as possible to the controller pins to locally improve the noise immunity. The gain is set by the LED series resistor (R_3 in Figure 9.28) to cope with the 17.2-dB requirement. Its value is 227 Ω and respects the limit set by (9.42).

Calculation on the power stage shows the necessity of external ramp compensation to damp the subharmonic poles. Once the natural ramp brought by the magnetizing current has been accounted for, a level of 16.6 kV/s can be adjusted in the NCP1252 via the current sense series resistor R_{10}. Figure 9.31 shows the loop gain when compensating elements have been placed over the operational amplifier. The upper and lower graph on the picture show the open-loop gain at low and high input line, respectively.

SPICE simulations of averaged models are extremely fast due to the lack of a switching component. This simulation speed helps to immediately assess the impact of the various parasitic elements on the converter stability. In Figure 9.31, the CTR has been changed from 50 to 160 percent to include some margin. It induces a crossover frequency variation from 9.3 kHz up to 23.7 kHz. If the end result keeps a good phase margin at both input voltage extremes, a 23-kHz crossover frequency represents a very aggressive target. The wider the bandwidth, the easier it becomes to collect parasitic noises and unwanted spurious signals. For this particular study, it would be wiser to reduce the crossover frequency around 7 kHz at low line/low CTR and manage a maximal excursion up to 15 kHz, a more reasonable value for a converter.

Once the compensation strategy is confirmed to deliver the right crossover frequency together with the needed phase margin, a transient step response can be performed. Again, thanks to the average modeling technique, the simulation time is flashing and the results appear in Figure 9.32 after a few milliseconds of computing time.

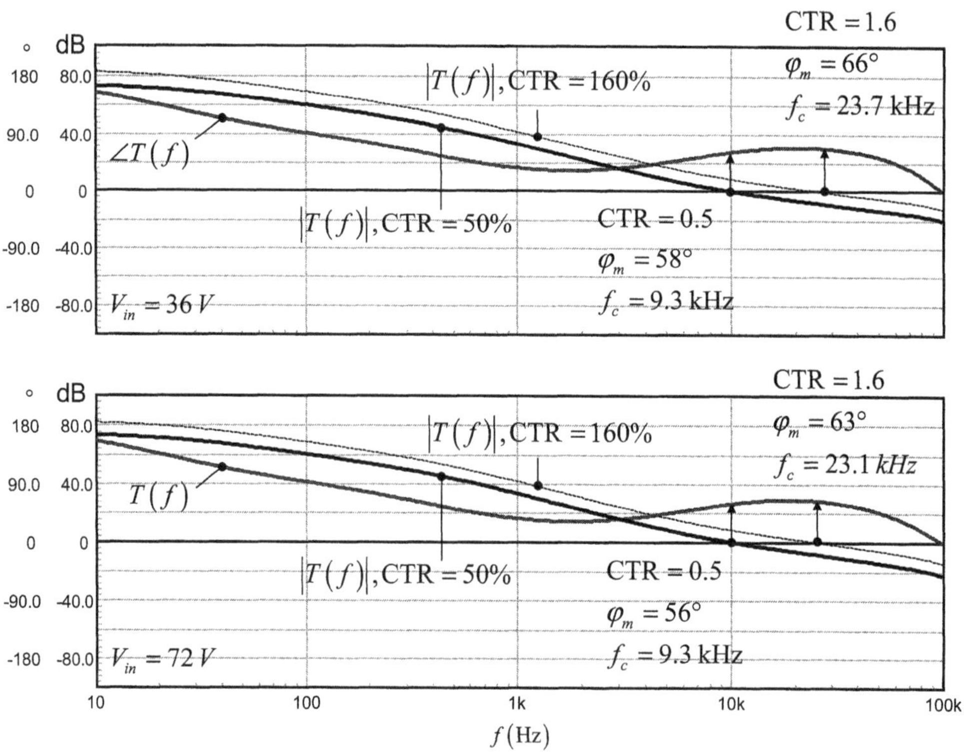

Figure 9.31 Thanks to simulation speed of average models, changing the parameters on the fly is an easy way to test the stability robustness. Here, the CTR is varied between 50 and 160 percent.

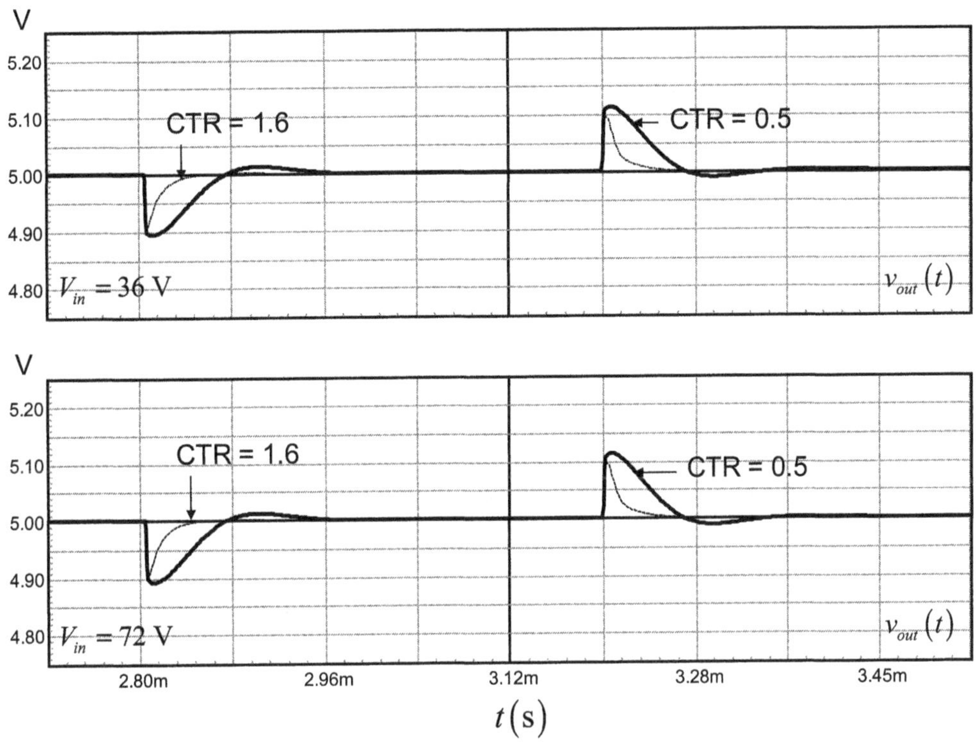

Figure 9.32 Transient response to a step from 10 A to 20 A with a slope of 1 A/μs.

As you can observe, the undershoot stays within a 150-mV variation. This variation slightly reduces as the bandwidth increases when the CTR peaks. This is because a direct relationship exists between the undershoot and the converter output impedance, as already explained in Chapter 3. The response is fast with almost zero overshoot in both input voltage conditions, and it satisfies our design requirements for this project. To refine the study, it is now possible to assign tolerances to the rest of the sensitive elements such as the output capacitors and their associated ESR and run a series of ac sweeps in a Monte Carlo style. Thanks to the fast simulation time, you will learn whether dangerous situations could exist when certain dispersions are combined together. Making sure the phase margin never gets too low in all situations will ensure a seamless mass-production period.

9.3 Design Example 2: A Linear Regulator

Power conversion implies not only switching converters but also linear regulators. In this case, the power element no longer operates in a switching mode but remains linear. This operating mode implies a decrease of efficiency but in lack of current and voltage discontinuities, the noise generated by such a converter remains extremely low compared to that of a switching converter.

Numerous application schematics exist for linear regulators. The first family was built around NPN series-pass bipolar transistors used as a dropping or a so-called ballast element. Still in service these days, this type of regulator requires several volts of input/output differential potential to operate in good conditions. The 78xx series (LM7805, MC7812, and so on) are good examples of these regulators. The typical schematic of such a device appears in Figure 9.33. Yes, I can hear linear regulator designers laughing from here: this is an oversimplified design for the sake of the example! You can see the ballast transistor, actually a darlington, whose base is driven by an op amp. Depending on the current sourced by the op amp, the transistor conducts more or less to feed the load at a constant output voltage. To maintain the transistor conduction, the base voltage needs to stay above the output voltage by two V_{BE}, around 1.3 V typically. Therefore, if you want to deliver 5 V, the minimum input voltage must be above 7 V if you include the various losses incurred to the driving stage (op amp output voltage saturation drop, for instance). This drawback is typical of non-low drop out (LDO) regulators whose conduction losses are rather high. On the contrary, LDO types use a P-channel or a PNP transistor and can accept input/output differential voltages down to a saturation voltage, a few hundred mV. In this mode, the gate or the base is easily brought to ground, and building a V_{GS} is no longer seen as a bias limit. The minimum drop in that case is given by the transistor $R_{DS(ON)}$ if a MOSFET is used. Such a simplified LDO appears in Figure 9.34 and shows how to wire the P-channel transistor so that its body diode does not bother us.

9.3.1 Extracting the Power Stage Transfer Function

The power stage transfer function of both circuits will be extracted using a simple SPICE template where the couple *LoL*/*CoL* helps to close the loop in dc but opens it in ac. As a reminder, the simulator shorts all inductors and opens all capacitors to compute the dc bias point before starting any type of simulation (ac or dc). This

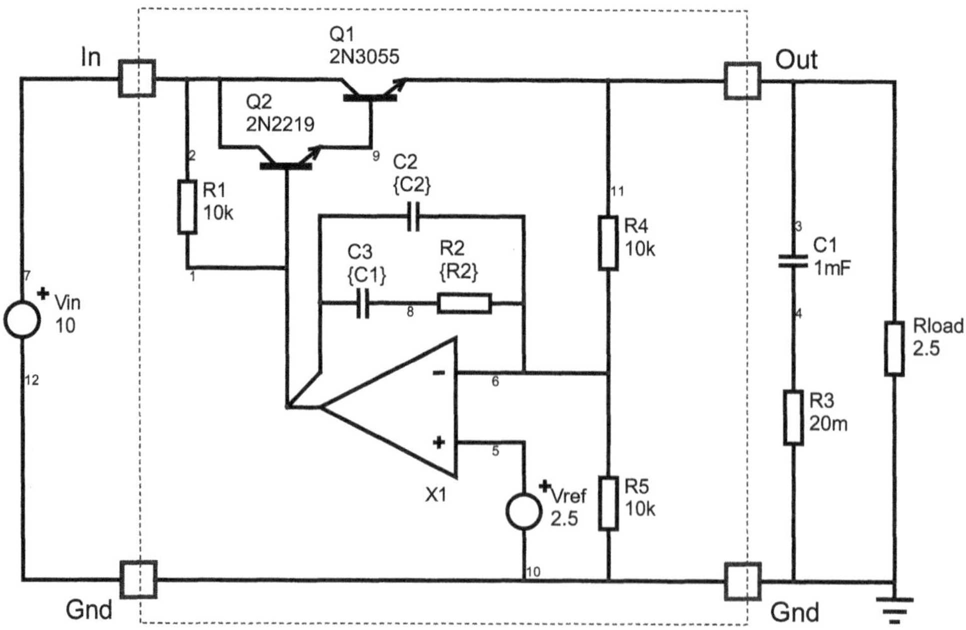

Figure 9.33 A simplified 5-V/2-A linear regulator using a series-pass transistor.

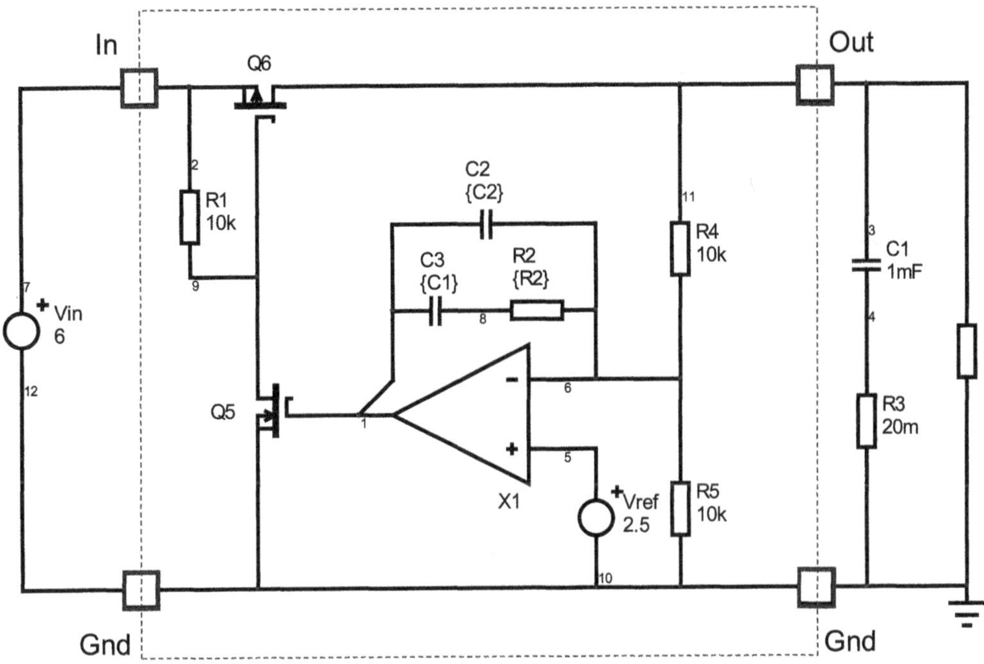

Figure 9.34 A simplified low-dropout regulator built around a P-channel MOSFET transistor.

bias point serves to further linearize the circuit as SPICE handles only linear equations. As a result, before the ac sweep starts, when SPICE computes the bias point, the loop is closed with *LoL* (*CoL* is an open circuit) and the bias point is automatically adjusted via the loop to meet the output target (5 V in our example). When the ac simulation starts, the *LoL*/*CoL* network forms an extremely low cutoff frequency filter: all ac modulation passes through *CoL* while the return path cannot cross *LoL*, whose high-value is equivalent to an open circuit. The loop is well closed in dc but open in ac. The power stage transfer function is probed in V_{out} while the loop gain, once compensated, is obtained by probing V_{err}. The traditional regulator simulation template is shown in Figure 9.35 while that of the LDO appears in Figure 9.36.

The bias points we have reflected on the schematic confirm the computed results are what we expect. The simulation results are given in Figures 9.37 and 9.38.

9.3.2 Crossover Frequency Selection and Compensation

In Chapter 3, we derived the relationship between the output response undershoot and the crossover frequency. When analyzing the step load output response, we found that the transient output deviation was made of the capacitive value, together with its parasitic elements. If reducing the di/dt will almost eliminate the ESL peak, the response deviation will still include the capacitor and its series ESR addition. Whatever the crossover frequency value, there is no possibility to avoid the ESR times the output current drop. However, rather than pushing the crossover frequency too far, we derived a simple formula at the end of Chapter 3. The formula

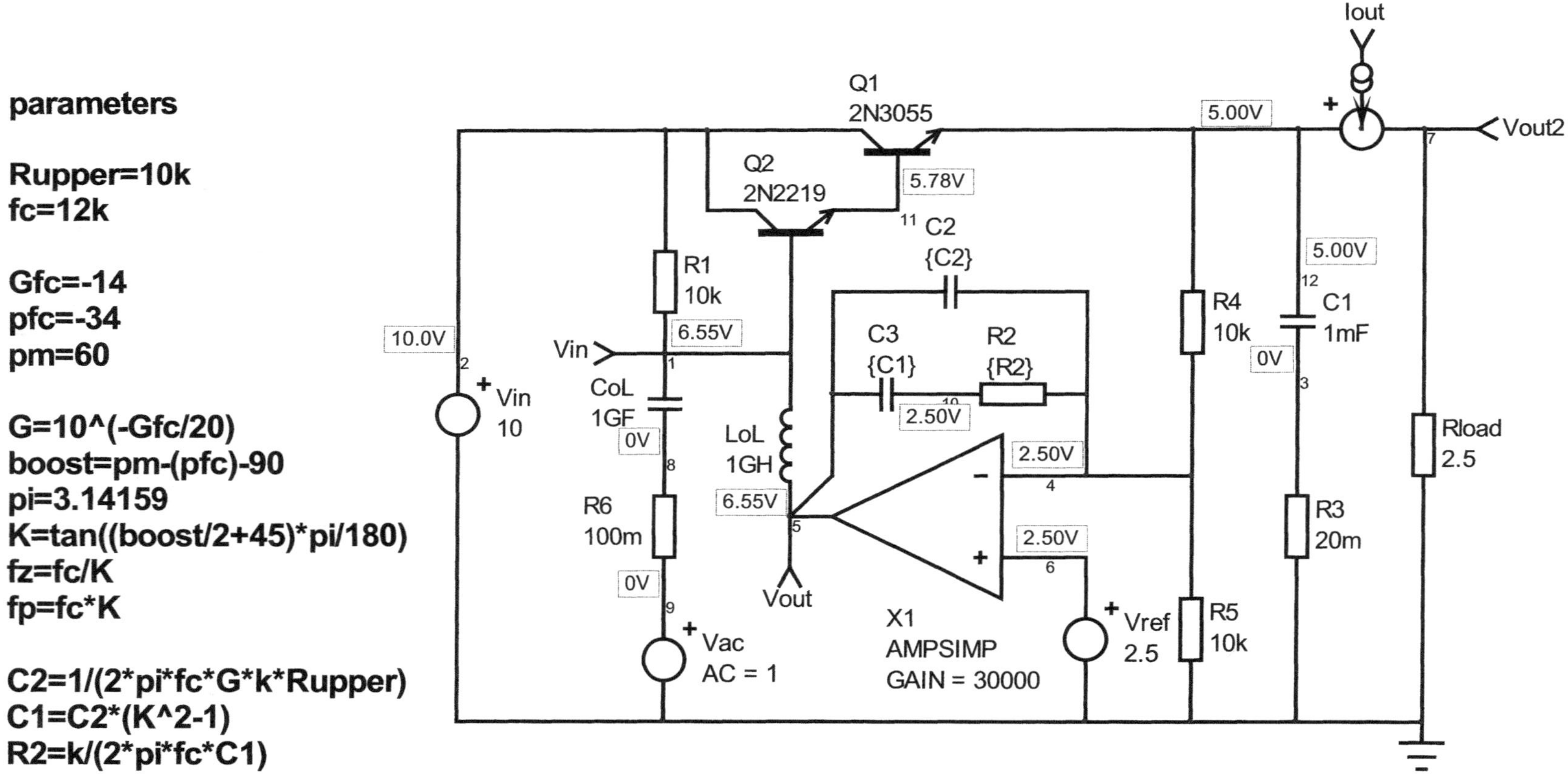

Figure 9.35 The loop is open behind the op amp, just before driving the darlington. We assume that the series-pass element driving circuit is embedded in the op amp.

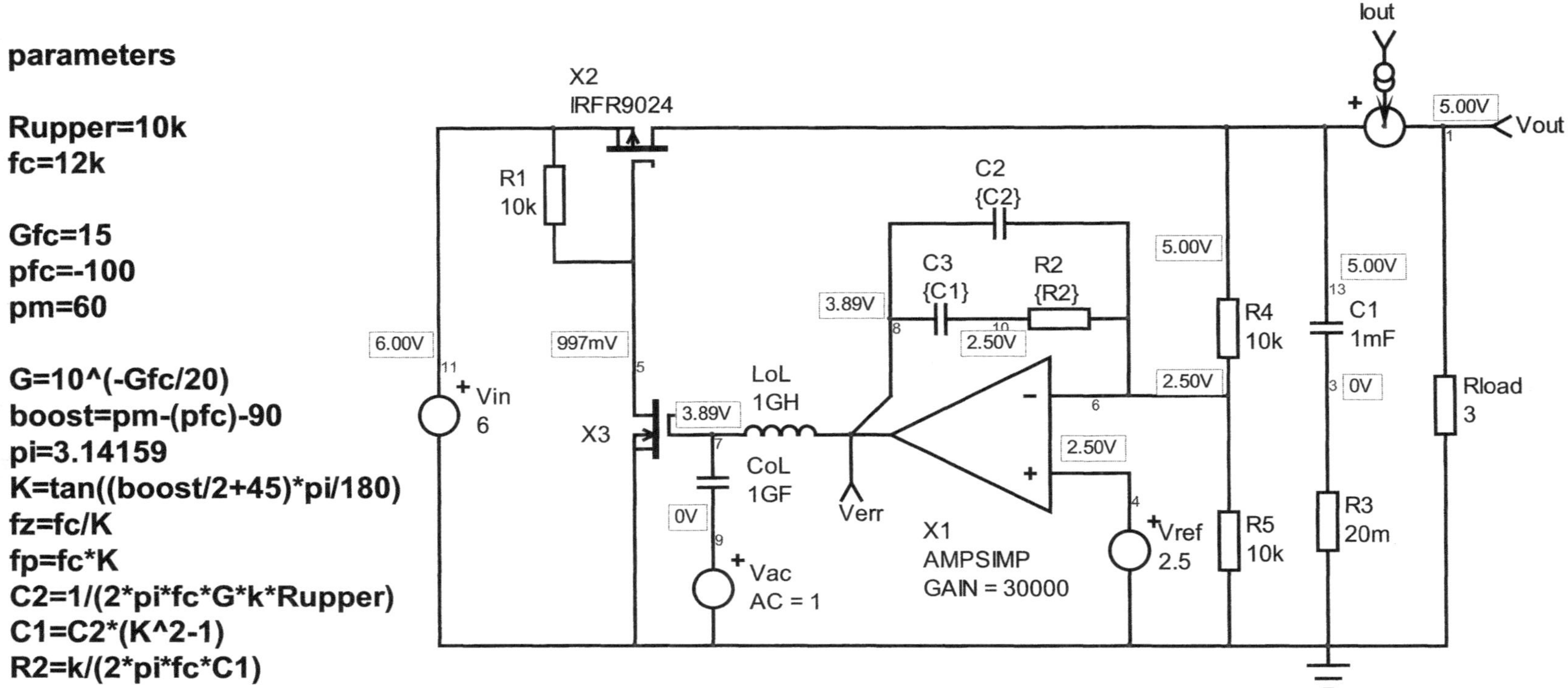

Figure 9.36 The simulation template for the P-channel-based LDO is very similar to that of the bipolar-based regulator.

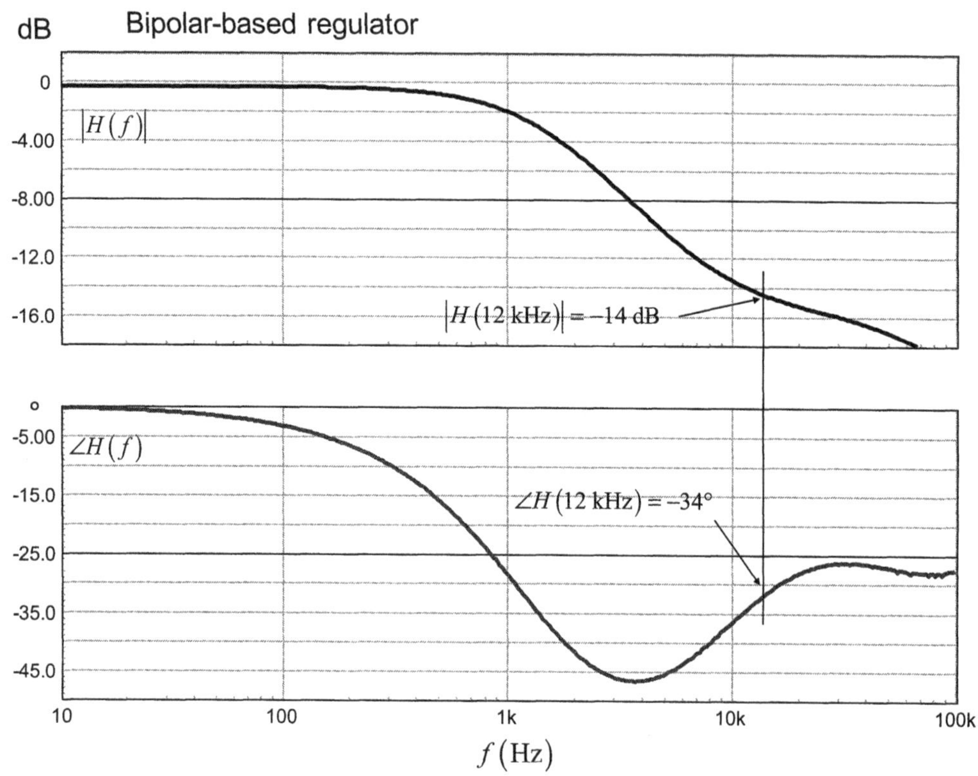

Figure 9.37 Ac results of the power stage show a gain deficit at the selected crossover frequency.

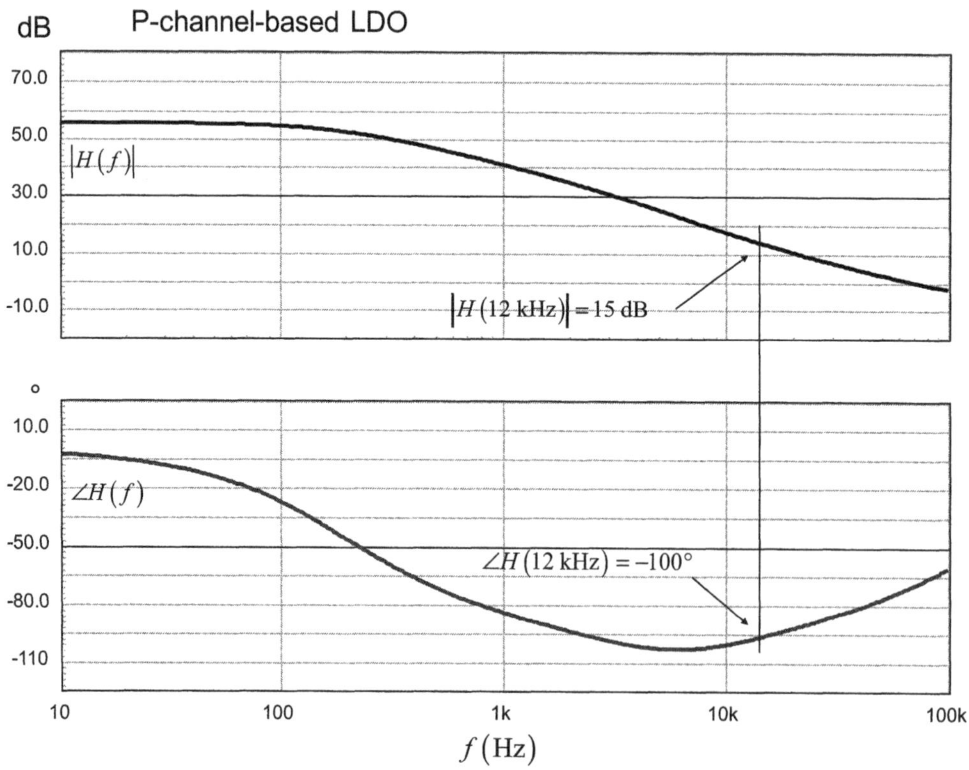

Figure 9.38 The power stage exhibits a large gain excess at the targeted crossover frequency.

expresses a crossover value high enough so that the capacitive contribution becomes negligible, leaving the unavoidable ESR undershoot. This approximate formula is:

$$f_c \approx \frac{0.24}{C_{out} r_C} \tag{9.46}$$

In both cases, we have an output capacitor of 1 mF, affected by a 20-mΩ ESR value. According to (9.46), we must place the crossover frequency at:

$$f_c \approx \frac{0.24}{1000u \times 20m} = 12 \text{ kHz} \tag{9.47}$$

Extracting the transfer function data at this frequency gives us for the bipolar-based regulator:

$$|H(12 \text{ kHz})| = -14 \text{ dB} \tag{9.48}$$

$$\angle H(12 \text{ kHz}) = -34° \tag{9.49}$$

For the LDO type, we have:

$$|H(12 \text{ kHz})| = 15 \text{ dB} \tag{9.50}$$

$$\angle H(12 \text{ kHz}) = -100° \tag{9.51}$$

The lack of voltage gain for the first power stage makes sense since we drive a common-collector architecture. For the second, the inverter behind the op amp provides gain and the series-pass element does as well.

To check what type of compensator we need, let's evaluate the necessary phase boost at crossover for a 60° phase margin objective. In the first case, we have:

$$boost = \varphi_m - \arg T(f_c) - 90° = 60 + 34 - 90 = 4° \tag{9.52}$$

There is almost no phase boost required; the compensation could be realized with a simple type 1 compensator. For the LDO type, the necessary phase boost amounts to:

$$boost = \varphi_m - \arg T(f_c) - 90° = 60 + 100 - 90 = 70° \tag{9.53}$$

In this second case, we need a type 2 compensator. For the sake of comparing similar structures, we have adopted two type 2 compensators.

For the first regulator, given the 4° phase boost, both the pole and zero will be almost coincident:

$$f_p = \tan\left(\frac{boost}{2} + \frac{\pi}{4}\right) f_c = \tan\left(\frac{4}{2} + 45\right) \times 12k = 12.9 \text{ kHz} \tag{9.54}$$

$$f_z = \frac{f_c}{\tan\left(\frac{boost}{2}+\frac{\pi}{4}\right)} = \frac{12k}{\tan\left(\frac{4}{2}+45\right)} = 11.2 \text{ kHz} \tag{9.55}$$

The mid-band gain is adjusted by the resistor R_2:

$$R_2 = \frac{R_4 f_p G}{f_p - f_z}\frac{\sqrt{\left(\frac{f_c}{f_p}\right)^2+1}}{\sqrt{\left(\frac{f_z}{f_c}\right)^2+1}} = \frac{10k\times 12.9k\times 10^{\frac{14}{20}}}{12.9k-11.2k}\frac{\sqrt{\left(\frac{12k}{12.9k}\right)^2+1}}{\sqrt{\left(\frac{11.2k}{12k}\right)^2+1}} = 384 \text{ k}\Omega \tag{9.56}$$

R_4 in the previous formula is the upper resistor in the divider bridge.

The capacitors values come easily:

$$C_1 = \frac{1}{2\pi R_2 f_z} = 37 \text{ pF} \tag{9.57}$$

$$C_2 = \frac{C_1}{2\pi f_p C_1 R_2 - 1} = 247 \text{ pF} \tag{9.58}$$

You could also apply the k factor technique as described in Appendix 5B, as it would lead to the exact same results. These capacitors values are rather low, and, if necessary, it would be easy to increase them by changing the divider network impedance. We currently have 10 kΩ for R_4 and R_5. Should we diminish these two values to 1 kΩ, C_1 and C_2 would, respectively, equal 370 pF and 2.47 nF while R_2 would drop to 38 kΩ.

The compensation for the LDO follows a similar path. To provide a 70° phase boost at a 12-kHz crossover frequency, we will place a pole and a zero at:

$$f_p = \tan\left(\frac{boost}{2}+\frac{\pi}{4}\right)f_c = \tan\left(\frac{70}{2}+45\right)\times 12k \approx 68 \text{ kHz} \tag{9.59}$$

$$f_z = \frac{f_c}{\tan\left(\frac{boost}{2}+\frac{\pi}{4}\right)} = \frac{12k}{\tan\left(\frac{70}{2}+45\right)} = 2.1 \text{ kHz} \tag{9.60}$$

The mid-band gain is adjusted by the resistor R_2:

$$R_2 = \frac{R_4 f_p G}{f_p - f_z}\frac{\sqrt{\left(\frac{f_c}{f_p}\right)^2+1}}{\sqrt{\left(\frac{f_z}{f_c}\right)^2+1}} = \frac{10k\times 68k\times 10^{-\frac{15}{20}}}{68k-2.1k}\frac{\sqrt{\left(\frac{12k}{68k}\right)^2+1}}{\sqrt{\left(\frac{2.1k}{12k}\right)^2+1}} = 1.8 \text{ k}\Omega \tag{9.61}$$

R_4 in the previous formula is the upper resistor in the divider bridge.

The capacitors values come easily:

$$C_1 = \frac{1}{2\pi R_2 f_z} = 49 \text{ nF} \tag{9.62}$$

$$C_2 = \frac{C_1}{2\pi f_p C_1 R_2 - 1} = 1.3 \text{ nF} \tag{9.63}$$

We can now run another ac simulation and check the loop gain response by probing the error amp voltage. The results for both configurations appear in Figure 9.39 and are very close to what we wanted. Some discrepancy can always happen, especially if we consider the op amp origin pole role.

9.3.3 Testing the Transient Response

We now have all of our component values. They are passed to the simulator as simple values or they can be automated as shown on the graph. Automation is interesting if you want to explore different phase margins on the fly: the corresponding transient response is immediate as all elements are recomputed before simulation start. To test our regulator step load response, we will sweep the output current from 1 A to 2 A in 10 μs. We purposely remove the inductive contribution to precisely see the resistive effect. The results appear in Figure 9.40.

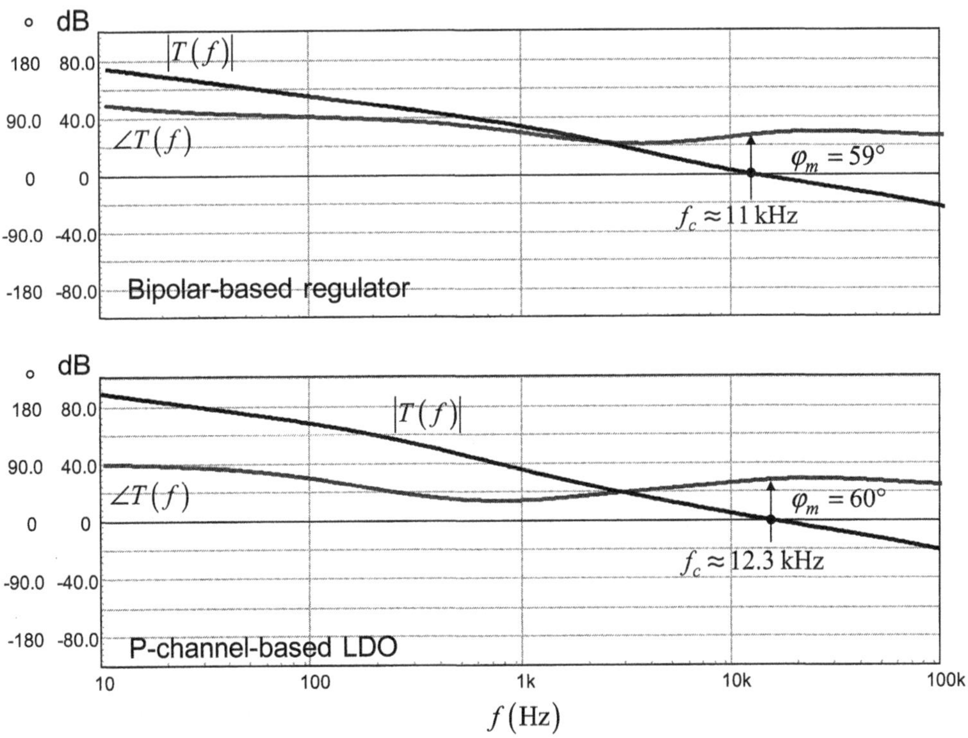

Figure 9.39 The ac responses show the right crossover frequencies and phase margins on both designs.

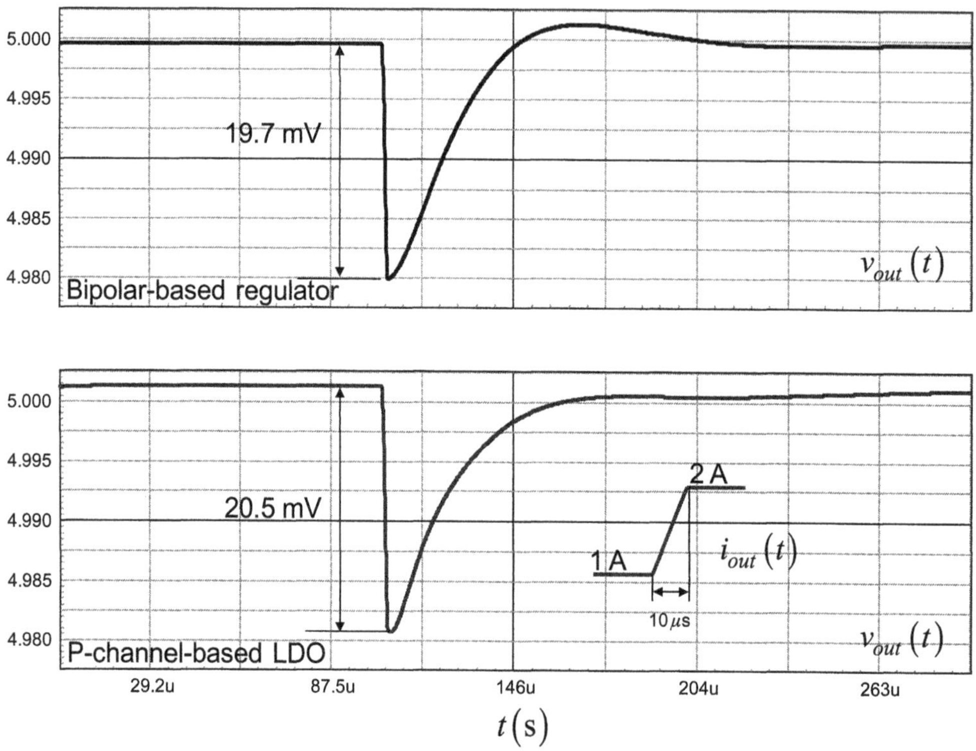

Figure 9.40 The transient response shows that the sole contributor to the voltage undershoot is the capacitor ESR.

The upper portion of the graph shows the bipolar-based regulator response. The compensation is performed by a pure integrator (pole and zero are almost coincident, turning the type 2 into a type 1). As expected, there is slight overshoot (0.03 percent), but the undershoot is exactly dictated by the ESR. The capacitive contribution does not exist. The following graph shows the response of the LDO for the same output stimulus. In this case, we have a zero located at 2 kHz. The recovery time is a bit longer (because of the zero presence in the low frequency portion) but the slight overshoot is gone. The undershoot is nicely given by the ESR alone, no capacitive contribution.

For the sake of the example, if we decrease the crossover frequency to 5 kHz we should see the capacitive contribution coming back in the picture. Figure 9.41 demonstrates this fact with the P-channel-based LDO. The whole response is no longer given by the ESR alone, but by the capacitor as well. As the crossover went down, the response time naturally expands.

9.4 Design Example 3: A CCM Voltage-Mode Boost Converter

Among switching converters, the boost topology is a popular design whose applications range from batteries-operated portable devices to high power ac-dc preconverters. Using two storage elements, this second-order converter can be designed to operate in the CCM at the lowest input voltage and the heaviest load. This design

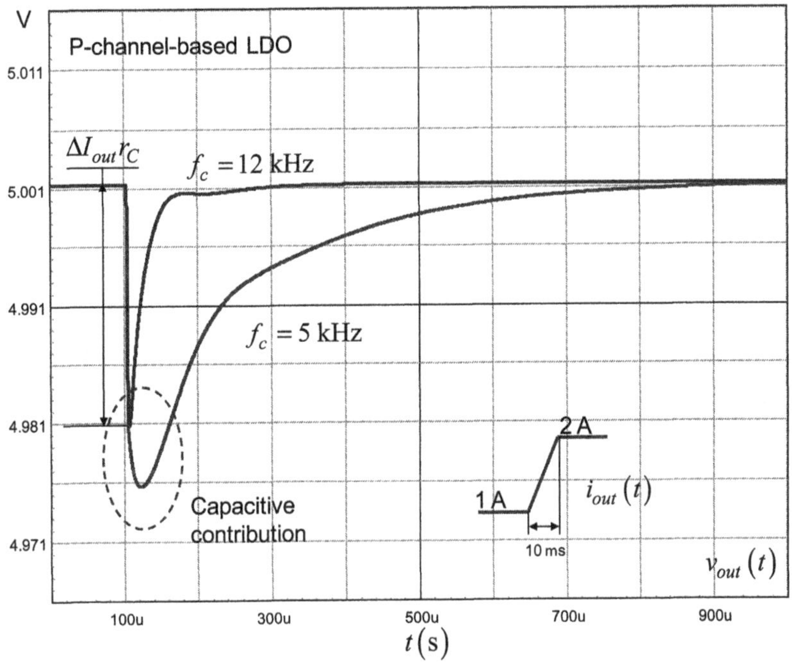

Figure 9.41 When the crossover frequency decreases to 5 kHz, the capacitive contribution shows up again.

example shows how to stabilize a CCM voltage-mode boost converter delivering 19 V from a car battery.

9.4.1 The Power Stage Transfer Function

You know the compensation of a given converter starts with its small-signal transfer function noted $H(s)$. Within this transfer function appear multiple poles, zeros, static gains, and so on. The dc operating point of the converter represents an important starting point which is, most of the time, required to compute the position of the aforementioned poles and zeros. With a boost converter operated in CCM and implementing voltage mode control, the dc transfer function M, when neglecting ohmic losses, is defined by:

$$M = \frac{V_{out}}{V_{in}} = \frac{1}{1-D} \tag{9.64}$$

where D represents the duty ratio that can be immediately extracted from (9.64):

$$D = \frac{V_{out} - V_{in}}{V_{out}} \tag{9.65}$$

The CCM boost transfer function appears in numerous textbooks and [12] has compiled and documented several other topologies in both voltage and current-mode controls. The control-to-output transfer function of the CCM boost converter operating in voltage mode appears in the following equation:

$$\frac{V_{out}(s)}{V_{err}(s)} = H_0 \frac{\left(1+\frac{s}{\omega_{z1}}\right)\left(1-\frac{s}{\omega_{z2}}\right)}{1+\frac{s}{Q\omega_0}+\left(\frac{s}{\omega_0}\right)^2} \tag{9.66}$$

By control-to-output, we mean how the ac voltage present on the error amplifier output propagates through the pulse width modulator (PWM) and drives the boost converter output voltage. Equation (9.66) indicates the presence of two zeros, one of them being our right half plane zero (RHPZ, ω_{z2}) plus a double pole located at ω_0, affected by a quality factor Q. The definitions of these elements are the following:

$$\omega_{z1} = \frac{1}{r_C C} \tag{9.67}$$

$$\omega_{z2} \approx \frac{R(1-D)^2}{L} \tag{9.68}$$

$$H_0 = \frac{V_{out}{}^2}{V_{in} V_{peak}} \tag{9.69}$$

$$\omega_0 \approx \frac{1-D}{\sqrt{LC}} \tag{9.70}$$

$$Q \approx \frac{\omega_0}{\frac{r_L}{L}+\frac{1}{C(r_C+R)}} \tag{9.71}$$

where r_C represents the output capacitor C ESR, V_{in} and V_{out} are, respectively, the input and output voltages; V_{peak} is the PWM modulator sawtooth amplitude; L is the boost converter inductor affected by r_L, its series resistance; D is the converter duty ratio; and Q the quality factor linked to the double pole.

To illustrate the boost converter ac response, we will take the example of a simple 60-W dc-dc converter elevating a 12-V battery voltage to a 19-V level, as in a notebook application powered from a car battery. Its specifications are as follows:

$V_{in,max} = 15$ V
$V_{in,min} = 11.5$ V
$V_{out} = 19$ V
$I_{out} = 3$ A
$R = 19/3 = 6.33\ \Omega$

Phase margin at cross over greater or equal to 60°, worse case.

The calculation steps using software from [14] give us the following values:

$F_{sw} = 100$ kHz
$V_{peak} = 2$ V
$L = 50\ \mu$H, $r_L = 10$ mΩ
$C = 1000\ \mu$F, $r_C = 20$ mΩ

From these values, we first evaluate the operating duty cycle at the lowest input voltage thanks to (9.65):

$$D = \frac{19 - 11.5}{19} = 0.395 \tag{9.72}$$

Then, we can proceed with the poles/zeros positions as defined by equations (9.67) through (9.71):

$$f_{z_1} = \frac{1}{2\pi \times 20m \times 1m} = 7.9 \text{ kHz} \tag{9.73}$$

$$f_{z_2} = \frac{6.33 \times (1 - 0.395)^2}{2\pi \times 50u} = 7.4 \text{ kHz} \tag{9.74}$$

$$20\text{Log}_{10}(H_0) = 20\text{Log}_{10}\left(\frac{19^2}{11.5 \times 2}\right) = 23.9 \text{ dB} \tag{9.75}$$

$$f_0 = \frac{1 - 0.395}{2\pi \times \sqrt{50u \times 1m}} = 430.8 \text{ Hz} \tag{9.76}$$

$$Q = \frac{2.7k}{\dfrac{10m}{50u} + \dfrac{1}{1m \times (20m + 6.33)}} = 7.57 \text{ or } 17.6 \text{ dB} \tag{9.77}$$

These results indicate the presence of a double pole placed at 430 Hz and affected by a 17.6-dB quality factor. This resonant frequency will move in relationship to the input voltage as it affects the duty ratio. The worst case occurs at the minimum input voltage and full load. However, once the compensation circuit is calculated, it is of the designer duty to make sure that the phase and gain margins do not degrade over the whole input voltage and load ranges. Also, as parasitic elements are involved, their impact on the final loop gain must also be carefully evaluated.

Once we have the numerical results, we have several choices to plot the CCM boost converter Bode response. One possibility consists of using scientific software such as Mathcad®, which can directly handle modules and imaginary notation via dedicated functions. This is the simplest and fastest approach. Unfortunately, for people who do not own such a software, we need to offer something simpler. For instance, directly extracting the module and argument of (9.66) via the poles and zeros positions is a possible solution that can be easily used with Excel® for instance. Capitalizing on this remark, the module and argument of $H(s)$ appear through (9.78) and (9.79).

$$|H(f)| = 20\text{Log}_{10}\left[H_0 \frac{\sqrt{1+\left(\frac{f}{f_{z1}}\right)^2}\sqrt{1+\left(\frac{f}{f_{z2}}\right)^2}}{\sqrt{\left(1-\left(\frac{f}{f_0}\right)\right)^2+\left(\frac{f}{f_0 Q}\right)^2}}\right] \tag{9.78}$$

$$\arg H(f) = \tan^{-1}\left(\frac{f}{f_{z1}}\right) - \tan^{-1}\left(\frac{f}{f_{z2}}\right) - \tan^{-1}\left[\frac{f}{f_0 Q}\frac{1}{1-\left(\frac{f}{f_0}\right)^2}\right] \tag{9.79}$$

Once these equations are entered into the calculation tool of your choice, the complete Bode diagram can be unveiled, as shown in Figure 9.42.

We are now to the point where you need to select the crossover frequency. The crossover frequency of a CCM boost converter is limited in the upper range by the lowest RHPZ position. As explained in Chapter 2, we need to limit the duty ratio slew-rate by selecting a crossover frequency well below the worse case RHPZ position. Experience shows that below 30 percent of this position gives adequate results. In our case, based on what (9.74) tells us, the maximum crossover frequency we can reasonably obtain is:

$$f_c < 0.3 \times 7.4k < 2.2 \text{ kHz} \tag{9.80}$$

Another consideration concerns the resonant frequency. You must always consider a crossover frequency beyond the region where the peak occurs. Otherwise, the loop will not offer enough gain to damp the LC network peaking in the output impedance expression, bringing instability to your converter. A common rule is

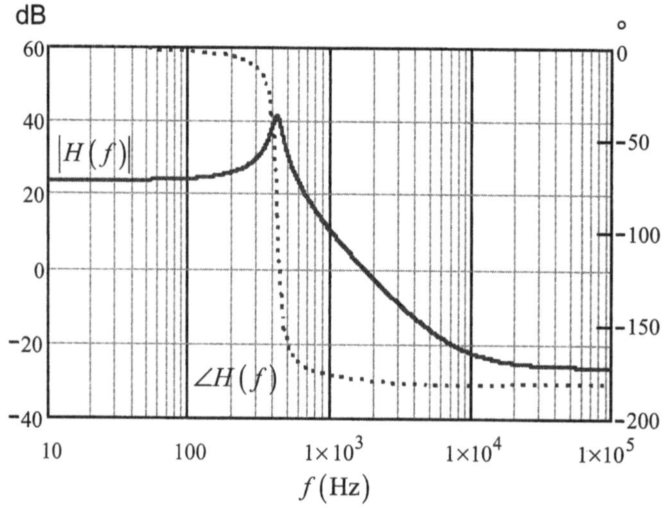

Figure 9.42 The boost converter CCM transfer function can quickly be plotted thanks to (9.78) and (9.79).

to recommend at least three times the maximum resonant frequency. In the CCM boost converter, this resonant frequency changes with the input conditions, reaching its maximum for a 15-V input level. In this case, (9.76) predicts a resonant peak placed at 562 Hz. Therefore, the crossover frequency must stay above:

$$f_c > 3 \times 562 > 1.68 \text{ kHz} \tag{9.81}$$

Combining (9.80) and (9.81), we will adopt a crossover point of 2 kHz. What if (9.81) would give an unrealistically high value? In that case, you would still have the solution to increase the output capacitor to push the resonance in a lower frequency region. Nevertheless, you must still make sure that the selected crossover frequency and the available capacitor lead to a theoretical undershoot compatible with your specification. Otherwise, another calculation iteration is necessary.

Now that we have selected a crossover frequency, you can either read the graph or use the magnitude/phase equations to extract the module and the phase rotation at 2 kHz for a maximum load and a 11-V input voltage. We obtain

$$|H(2k)| = -1.77 \text{ dB} \tag{9.82}$$

$$\arg H(2k) = -179° \tag{9.83}$$

9.4.2 Compensating the Converter

The loop stability exercise consists of shaping the compensator $G(s)$ to provide a gain at 2 kHz that compensates the gain deficiency (or excess) extracted at the crossover frequency. This way, we will satisfy $|H(f_c)G(f_c)| = 1$ implying a crossover at f_c. In our example, we must provide a gain of 1.77 dB when the frequency reaches 2 kHz. For the phase, it is a bit different. You need to provide a certain amount of excess phase at the crossover frequency to obtain the required phase margin *PM*. If we look for a 60° phase margin, this *boost* is calculated as follows:

$$boost = \varphi_m - \arg H(f_c) - 90 = 60 + 179 - 90 = 149° \tag{9.84}$$

The phase boost is obtained by placing poles and zeros at proper places. For a boost converter operating in CCM and a needed phase boost greater than 90°, you must use a type 3 compensator. Such an amplifier appears in Figure 9.43, together with the boost power stage configuration.

How do we adjust the elements to provide the necessary gain and where to place the poles and zeros? If the k factor works reasonably well for a first-order converter, it often leads to conditional stability in the case of a CCM converter featuring a resonant double pole. This is because the method solely focuses on the crossover region and not what happened before or beyond. For this reason, the manual placement is the preferred way of compensating when complex Bode plots

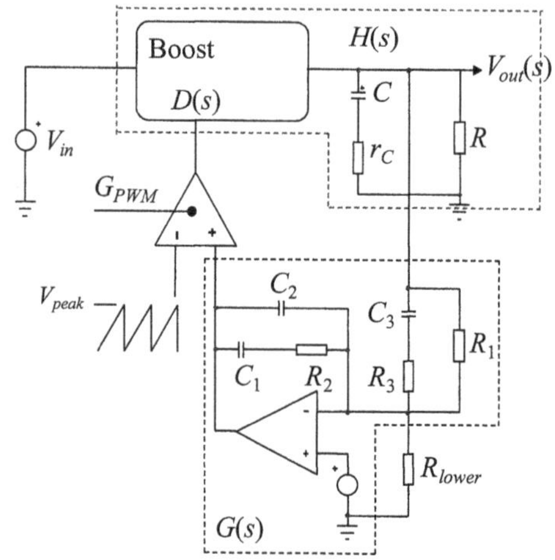

Figure 9.43 A type 3 compensator can be built with an operational amplifier. It can theoretically boost the phase up to 180°.

are in play. Before we present the method, we first need to introduce the transfer function of the type 3 amplifier. It can be shown that it looks as follows:

$$|G(f)| \approx 20 \cdot \mathrm{Log}_{10}\left[\frac{R_2}{R_3}\frac{\sqrt{1+\left(\frac{f_{z1}}{f}\right)^2}\sqrt{1+\left(\frac{f_{z2}}{f}\right)^2}}{\sqrt{1+\left(\frac{f}{f_{p1}}\right)^2}\sqrt{1+\left(\frac{f_{p2}}{f}\right)^2}}\right] \tag{9.85}$$

It brings a pole at the origin, a double zero, and a double pole. The pole placed at the origin is found on the vast majority of compensators. It is there to provide high gain in the dc region. This high dc gain reduces the output impedance and improves the static error and the input rejection. This origin pole naturally lags the phase by 90° (π/2) to which you must add the 180° inversion brought by the op amp (π). Capitalizing on these remarks, the *total* argument of the type 3 arrangement appears as described by (9.86).

$$\arg G(f) = \pi - \left(\tan^{-1}\frac{f}{f_{z1}} + \tan^{-1}\frac{f}{f_{z2}} - \tan^{-1}\frac{f}{f_{p1}} - \tan^{-1}\frac{f}{f_{p2}} - \frac{\pi}{2}\right) \tag{9.86}$$

The most troublesome contributor in the CCM boost converter transfer function is the double pole located at the resonant frequency. One possible solution is to place in G a double zero right at this location. In some converters, it is also interesting to split these zeros, one placed at the resonant frequency and the second slightly below it, to increase the phase margin in the DCM.

If the ESR zero occurs before the crossover frequency, we place a pole right at its location to force the gain decrease. The second pole will be placed at half the switching frequency to make sure enough gain margin exists as the phase margin vanishes to 0°. In our case, the ESR zero occurs after the crossover frequency, but not too far away from it. To show the impact of this first pole position in the compensator, we will cover two possible strategies.

Strategy 1

1. A double zero f_{z1} and f_{z2} is placed at the lowest resonant frequency (e.g., 430 Hz).
2. A first pole f_{p1} is placed right at the ESR zero occurrence, 7.9 kHz. The purpose of this pole is to force a –1 slope despite the presence of the boost converter resonant pole cancelled by the double zero.
3. A second pole f_{p2} will be placed at half of the switching frequency (50 kHz) to ensure a proper gain rolloff as the phase rotation further degrades. This pole gives the necessary gain margin we need (at least 10–15 dB).
4. Once all locations are known, the resistor R_2 from (9.85) is extracted to provide the proper gain compensation at f_c.
5. Given the selection of the poles/zeros affecting $G(s)$, we will evaluate the loop $T(s)$ resulting phase margin and see if it complies with our specification.

Strategy 2

1. A double zero f_{z1} and f_{z2} is placed below the lowest resonant frequency (e.g., 300 Hz).
2. A first pole f_{p1} is now computed to provide the required phase margin (e.g., 60°).
3. A second pole f_{p2} will be placed at half of the switching frequency (50 kHz) to ensure a proper gain rolloff as the phase rotation further degrades.
4. Once all locations are known, the resistor R_2 from (9.85) is calculated to provide the proper gain compensation at f_c.

When $H(s)$ and $G(s)$ are combined together, we observe the total loop noted $T(s)$. The loop phase margin PM after compensation is evaluated as the distance of the total phase rotation at f_c i.e. arg $T(f_c)$ and the 0° limit. Mathematically, it can be expressed as follows:

$$\varphi_m = \arg H(f_c) + \arg G(f_c) - 0° \tag{9.87}$$

Strategy 1 applied to (9.86) gives us a total phase lag for the compensator G at 2 kHz equal to:

$$\arg G(f) = 180 + 2\tan^{-1}\frac{2k}{430} - \tan^{-1}\frac{2k}{7.9k} - \tan^{-1}\frac{2k}{50k} - 90 \approx 229° \tag{9.88}$$

From (9.87), it leads to a final loop phase margin of:

$$\varphi_m = \arg H(f_c) + \arg G(f_c) = -179 + 229 = 50° \tag{9.89}$$

with 50° at an 11.5-V input voltage, increasing to 56° when the boost converter is supplied from 15 V.

For strategy number 2, we will fix the double zero to 300 Hz and adjust the position of the first pole to cope with our 60° phase margin requirement. We know from (9.84) that if the compensator two zeros and the second pole are fixed, the first pole can be placed to offer the required boost of 149°. In (9.86), the equation includes the origin pole phase lag plus the op amp phase reversal. However, the real boost brought by the compensator G is calculated as the phase bump amplitude above the –270° or –3π/2 limit (see Figure 9.44). Thus, from (9.86), we can extract the contribution of the first pole f_{p1} by removing the –3π/2 term. Therefore, we have:

$$\tan^{-1}\left(\frac{f_c}{f_{p1}}\right) = \tan^{-1}\left(\frac{f_c}{f_{z1}}\right) + \tan^{-1}\left(\frac{f_c}{f_{z2}}\right) - \tan^{-1}\left(\frac{f_c}{f_{p2}}\right) - boost \tag{9.90}$$

The pole position is now easily derived and follows (9.91):

$$f_{p1} = \frac{f_c}{\tan\left[\tan^{-1}\left(\frac{f_c}{f_{z1}}\right) + \tan^{-1}\left(\frac{f_c}{f_{z2}}\right) - \tan^{-1}\left(\frac{f_c}{f_{p2}}\right) - boost\right]} = 9.9 \text{ kHz} \tag{9.91}$$

Using this second strategy, we can recompute our phase margin by calculating the compensator argument at 2 kHz summed with the –179° of the power stage:

$$\arg G(f) = 180 + 2\tan^{-1}\frac{2k}{300} - \tan^{-1}\frac{2k}{9.9k} - \tan^{-1}\frac{2k}{50k} - 90 \approx 239° \tag{9.92}$$

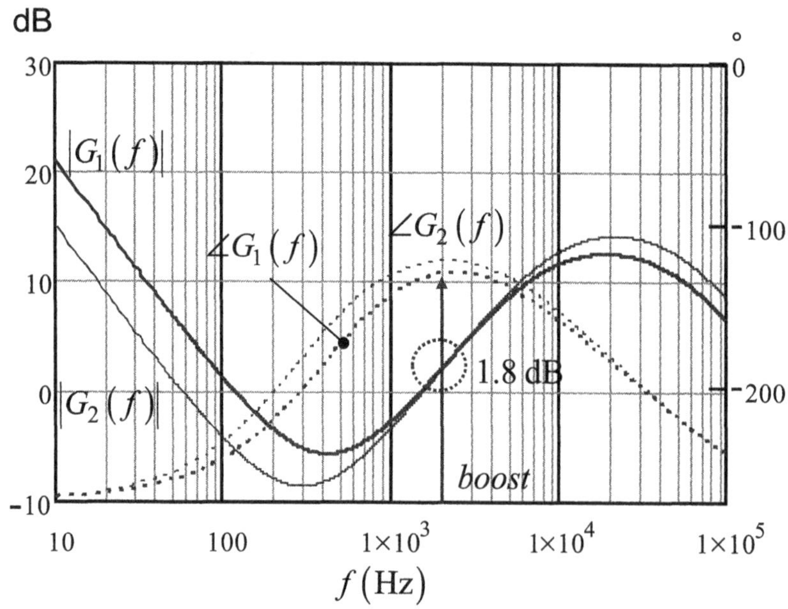

Figure 9.44 The type 3 compensator boosts the phase between the zeros and the poles. The first strategy offers a phase boost of 139°, whereas the second strategy increases it up to 149°, but to the detriment of the low frequency gain.

And check what the phase margin will be for a 11.5-V input voltage:

$$\varphi_m = \arg H(f_c) + \arg G(f_c) = 239 - 179 = 60° \tag{9.93}$$

If we compute it again at a 15-V input, we now have 66°, which is above the specification target. Both compensation options G_1 and G_2 appear in Figure 9.44.

If strategy 2 offers a more comfortable phase boost than strategy 1, please note the gain reduction in the low frequency region. This is the typical effect of moving the zeros farther down the frequency axis: you obtain a larger phase boost, but it will naturally affect the transient response by slowing down the system. Please note that both gain curves give exactly ≈1.8 dB at 2 kHz, compensating the gain deficiency of the power stage.

9.4.3 Plotting the Loop Gain

Once the compensator response is shaped according to your needs, you can now combine it with the boost converter frequency response and obtain the open-loop Bode plot we are looking for. It appears in Figure 9.45. Both strategies exhibit a crossover frequency exactly of 2 kHz, together with the phase margins we predicted.

Again, strategy 1 offers a larger low frequency gain but gives a phase margin below 60°. Strategy 2 improves the phase margin but slightly lowers the gain in the low frequency region.

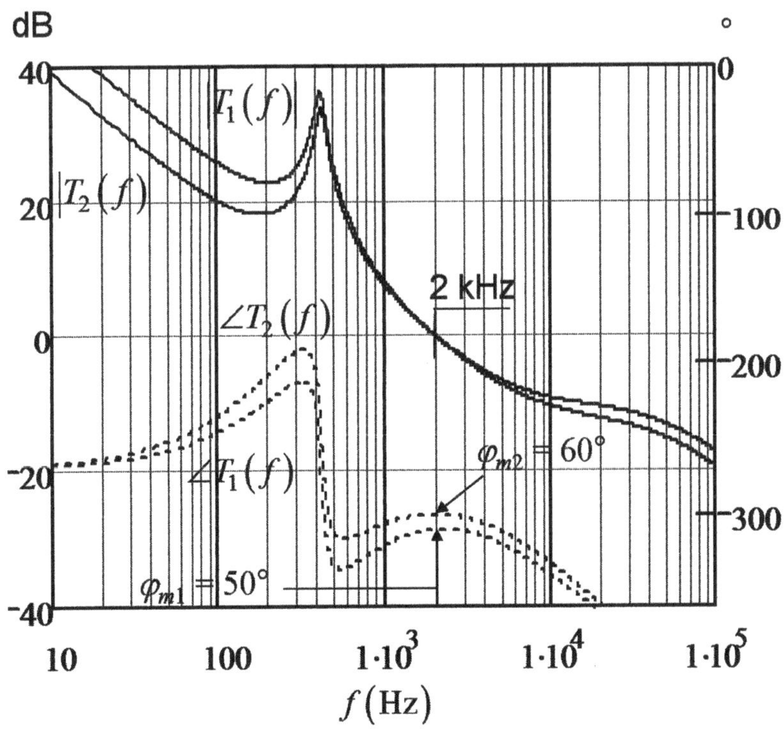

Figure 9.45 The loop gain combines *H*(*s*) and *G*(*s*) together. Strategy 1 and 2 give exactly a 2-kHz crossover frequency.

The calculations we have made are related to the typical ESR value of the output capacitor. As everyone knows, this ESR varies from lots to lots but also in temperature. The manufacturer gives you some indications on the minimum and the maximum of this ESR. It is important to perform the stability analysis with the ESR variations fully accounted for. In Table 9.1, we have assumed three different temperatures leading to three different ESR values. Then, thanks to the automated spreadsheet, we have entered these ESR values and collected the resulting phase margins depending on the two selected strategies.

As one can see, strategy 1 fails to offer the minimum acceptable phase margin at high temperature: we are below 45°. It does not say that the circuit will fail in production, but you start to have a design that can potentially be marginally stable. This kind of situation is not recommended for high-volume productions. On the contrary, strategy 2 offers enough margin, even at high temperature where we are still above 50°.

The complete study includes the operation of the boost converter when operated in light load conditions. Unfortunately, the converter will change its mode, transitioning from CCM to DCM. It can be shown that the transition occurs when the load reaches its critical point:

$$R_{\text{crit}} = \frac{2LF_{sw}}{D(1-D)^2} \tag{9.94}$$

It corresponds to a load of 69 Ω at a 11.5-V input voltage (5.2 W) or 76 Ω at a 15-V level (4.7 W). This is where the analytical analysis finds its limit: the transfer functions, both in ac and dc, change in DCM and all the equations must be updated. The new transfer function in DCM becomes:

$$\frac{V_{out}(s)}{V_{err}(s)} = H_0 \frac{\left(1+\frac{s}{\omega_{z_1}}\right)}{\left(1+\frac{s}{\omega_{p_1}}\right)} \tag{9.95}$$

where

$$H_0 = \frac{V_{in}}{V_{peak}} \frac{2}{2M-1} \sqrt{\frac{M(M-1)}{\tau_L}} \tag{9.96}$$

$$M = \frac{V_{out}}{V_{in}} \tag{9.97}$$

$$\tau_L = \frac{2L}{RT_{sw}} \tag{9.98}$$

Table 9.1 Phase Margin Variations Versus ESR Spreads

Temperature (°C)	*ESR* (mΩ)	*PM, strategy* 1 (°)	*PM, strategy* 2 (°)
0	40	62	72
25	20	50	60
70	10	43	53

$$\omega_{z_1} = \frac{1}{r_C C} \tag{9.99}$$

$$\omega_{p_1} = \frac{2M-1}{M-1}\frac{1}{RC} \tag{9.100}$$

You will have to take the compensator transfer function calculated to stabilized the CCM power stage and associate it with the transfer function of the power stage in DCM and check crossover and phase margin at both input voltage extremes. If the phase margin is acceptable, then you are done. If not, you will have to try different poles/zeros combinations so that stability is ensured regardless of the operating mode. This is not a Herculean labor, of course, but it clearly complicates and lengthens the stability analysis. This is where an auto-toggling SPICE model is of great help. Furthermore, it helps us predict the transient response in relationship to a set of operating conditions applied to the selected strategy.

9.5 Design Example 4: A Primary-Regulated Flyback Converter

There are ac-dc applications where using an optocoupler is not allowed. It can be for component cost reasons but also in cases where reliability is important, naturally banning optocoupler usage. For low-power applications, around 10W, the classical approach uses a primary-regulated flyback converter. Such a configuration appears in Figure 9.46. You can see two secondary windings: V_{out} is the main one, connected to the output load; it delivers the main power. V_{aux} is the auxiliary voltage supplying the controller and used for regulation purposes. As you can see, they do not share common ground for safety purposes. Isolation between the grounds is solely brought by the transformer. As indicated several times, modern PWM

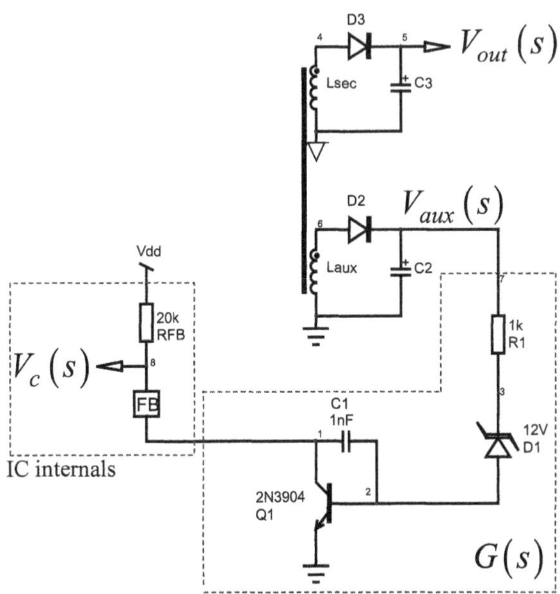

Figure 9.46 A primary-regulated flyback converter uses a simple transistor to regulate.

controllers do not embark op amp of any kind onboard. Capitalizing on the fact that all the regulation intelligence is kept on the secondary side, a simple feedback pin is enough to hook an optocoupler. Lacking this element, we will use a simple bipolar transistor to do pull-down job. Its base is driven by a Zener diode. It is a kind of low-gain shunt regulator, actually. The input of this error processing chain is the upper terminal of R_1, while the output is simply the control voltage V_c.

9.5.1 Deriving the Transfer Function

Most of the designers that I know have implemented this simple circuit by throwing component values based on their feelings! This is the wrong approach: you have to understand how the information is processed and see where weaknesses could hide. Otherwise, how can you stabilize a converter if you do not have a clue of the loop gain transfer function? Trial and error is not acceptable if you are a serious designer.

Based on this remark, we can draw a simplified small-signal schematic using the Ebers-Moll model for Q_1. The schematic is given in Figure 9.47.

At first glance, it looks like a piece of cake to solve such a transfer function. Well, we will see that fast analytical techniques will be of great help to carry the analysis. First, we count only one storage element, as this is a first-order system. Without knowing if it includes zeros or poles, it must obey the following transfer function:

$$G(s) = G_0 \frac{1 + s/\omega_z}{1 + s/\omega_p} = G_0 \frac{N(s)}{D(s)} \tag{9.101}$$

The first step in the process of deriving the transfer function is to get the dc gain G_0. This is done by opening all capacitors and shorting inductors if any. The new schematic becomes that of Figure 9.48.

The first equation to derive is the base current I_b:

$$I_b(s) = \frac{V_{aux}(s)}{r_d + R_1 + r_\pi} \tag{9.102}$$

Looking at the right side of the circuit, we can express the output voltage V_c:

$$V_c(s) = -\beta \cdot I_b(s) R_{FB} \tag{9.103}$$

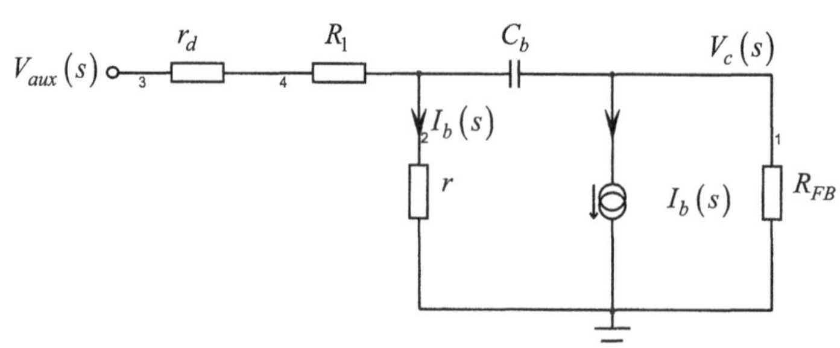

Figure 9.47 If we replace the transistor by its small-signal model, we obtain a simple sketch.

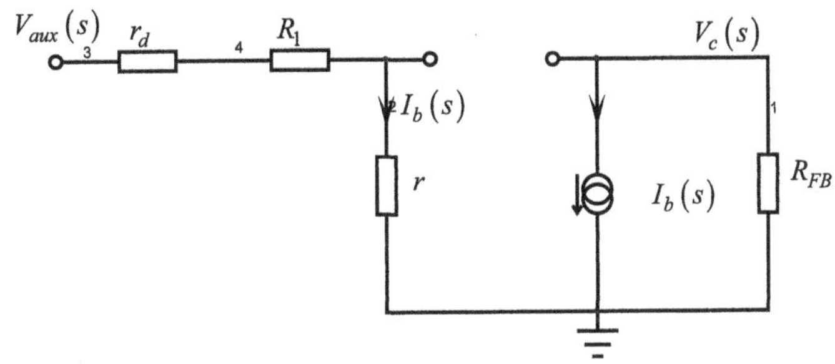

Figure 9.48 The dc transfer function is obtained by opening the capacitor.

Combining both equations, we now have for $s = 0$:

$$\frac{V_c(s)}{V_{aux}(s)} = G_0 = -\beta \cdot \frac{R_{FB}}{r_d + R_1 + r_\pi} \tag{9.104}$$

The capacitor is now put back in place and the schematic is slightly updated as shown in Figure 9.49. To check whether we have zero(s) in the transfer function, let's see if something in the signal path could prevent the excitation V_{aux} from reaching the output V_c.

If we have no response, it means that its upper terminal is 0 and no current flows through R_{FB}. The voltage across capacitor C_b is thus that across r_π:

$$V_\pi = r_\pi I_b(s) \tag{9.105}$$

$$V_{C_b} = \beta \cdot I_b(s)\frac{1}{sC_b} \tag{9.106}$$

As both voltages are equal, we have:

$$\beta \cdot I_b(s)\frac{1}{sC_b} = I_b(s) r_\pi \tag{9.107}$$

If we factor $I_b(s)$, we obtain the following expression:

$$I_b(s)\left[\beta \cdot \frac{1}{sC_b} - r_\pi\right] = 0 \tag{9.108}$$

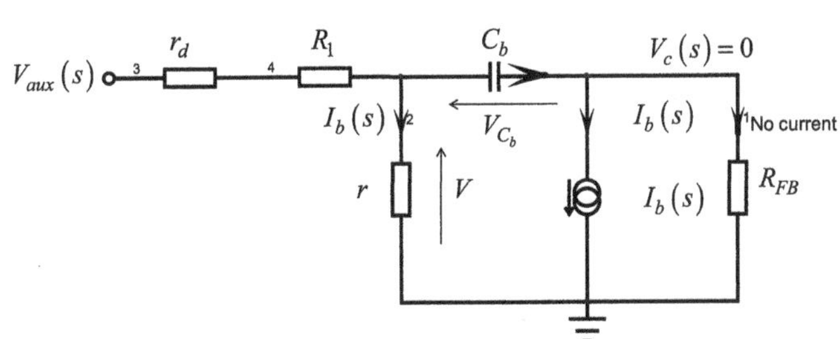

Figure 9.49 The exercise consists of finding what prevents the excitation from reaching the output.

which can only be realized if:

$$\beta \cdot \frac{1}{sC_b} - r_\pi = 0 \tag{9.109}$$

Solving for s gives us our zero location:

$$\omega_z = \frac{\beta}{C_b h_{11}} \tag{9.110}$$

This is a positive root; we have a nice RHPZ. The numerator can thus be put under the following form:

$$N(s) = 1 - s/\omega_z \tag{9.111}$$

Now that we have our numerator, we can look at the denominator $D(s)$. The denominator of a network depends only on its structure, not on the excitation signal. Regardless of whether you excite it at different inputs, let's say to unveil an output impedance, an input admittance, and so on, the denominator of the obtained transfer function will remain unchanged. The excitation signal can therefore be put to zero: open a current source and short circuit a voltage source. This is what is shown in Figure 9.50, further updated in Figure 9.51 where we lumped r_d and R_1 into a single resistor R:

$$R = r_d + R_1 \tag{9.112}$$

To get the denominator expression, we need to find the time constant in which C_b is associated with the equivalent resistor R_e driving it. In other words, let us derive the resistor expression seen at the capacitor terminals when we removed it from the picture. To derive an impedance, we excite its terminals by a current source (the "excitation") and we observe the voltage developed across the current source terminals: the "response." In Figure 9.51, we respectively called these signals $I_T(s)$ and $V_T(s)$.

Let's write some basic equations, starting with the voltage across the resistor R, on the left of the figure:

$$V_R(s) = (I_T(s) + I_b(s))R \tag{9.113}$$

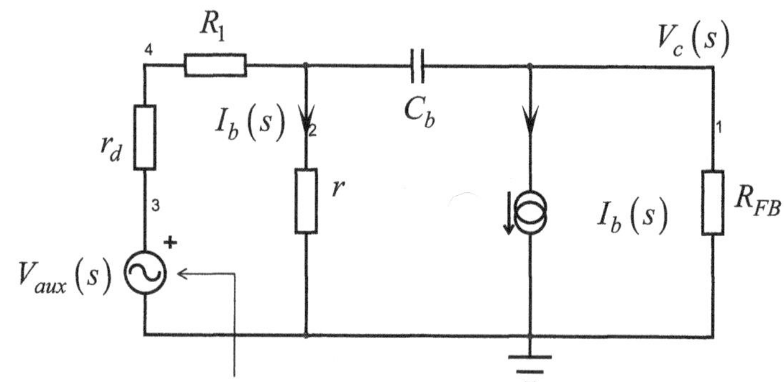

Figure 9.50 The denominator is obtained by setting the excitation signal to zero.

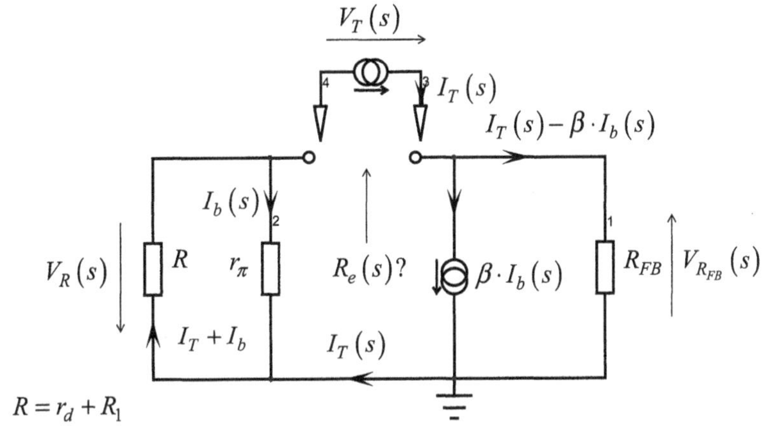

Figure 9.51 The excitation is now set to zero, and the lower terminal of r_d has been grounded.

This voltage is exactly the same that you will find across r_π but with an opposite sign:

$$V_R(s) = -r_\pi I_b(s) \tag{9.114}$$

Equating these two equations and solving for the have:

$$I_b(s) = -\frac{R \cdot I_T(s)}{R + r_\pi} \tag{9.115}$$

The voltage across the pull-up resistor *RFB* comes easily:

$$V_{R_{FB}}(s) = (I_T(s) - \beta \cdot I_b(s))R_{FB} \tag{9.116}$$

The voltage across the injection source V_T is not difficult either:

$$V_T(s) = V_{R_{FB}}(s) + V_R(s) \tag{9.117}$$

Now substituting (9.116), (9.114), and (9.115) into (9.117), we obtain:

$$V_T(s) = (I_T(s) - \beta \cdot I_b(s))R_{FB} - r_\pi I_b(s) \tag{9.118}$$

$$V_T(s) = \left(I_T(s) + \beta\frac{R \cdot I_T(s)}{R + r_\pi}\right)R_{FB} + r_\pi\frac{R \cdot I_T(s)}{R + r_\pi} = I_T(s)\left(\left(1 + \frac{\beta \cdot R}{R + r_\pi}\right)R_{FB} + \frac{R \cdot r_\pi}{R + r_\pi}\right) \tag{9.119}$$

Rearranging this equation, we have the port input impedance:

$$\frac{V_T(s)}{I_T(s)} = \left(1 + \beta\frac{R}{R + r_\pi}\right)R_{FB} + R \parallel r_\pi \tag{9.120}$$

The denominator can now be written under the following form:

$$D(s) = 1 + s/\omega_p \tag{9.121}$$

in which the pole is:

$$\omega_p = \frac{1}{C_b\left[\left(1+\beta\frac{r_d+R_1}{r_d+R_1+r_\pi}\right)R_{FB}+\left(r_d+R_1\right)\|\,r_\pi\right]} \tag{9.122}$$

The complete transfer function for Figure 9.47 is thus:

$$\frac{V_{FB}(s)}{V_{aux}(s)} = -\beta\cdot\frac{R_{FB}}{r_d+R_1+r_\pi}\frac{1-\dfrac{s}{\dfrac{\beta}{C_b h_{11}}}}{1+sC_b\left[\left(1+\beta\dfrac{r_d+R_1}{r_d+R_1+r_\pi}\right)R_{FB}+\left(r_d+R_1\right)\|\,r_\pi\right]} \tag{9.123}$$

This equation appears in a clear and ordered way, where you can see the pole and zero locations. This is the strength of the fast analytical techniques.

9.5.2 Verifying the Equations

When running this type of analysis, it is always possible to overlook a condition or a parameter, or simply make a mistake when developing equations. For all of these reasons, I will verify the ac response of (9.123) and confront it to a simple SPICE simulation. If all is correct, the curve must perfectly superimpose.

The Mathcad® sheet appears in Figure 9.52. With the values entered in the automated sheet, the RHPZ shows up at around 550 kHz. It is obviously of negligible influence if we plan to crossover below 10 kHz. We adopted arbitrary values for all components.

$\beta := 73 \quad h_{11} := 21.4\text{k}\Omega \quad R_1 := 1\text{k}\Omega \quad r_d := 12.6\text{k}\Omega \quad C_b := 1\text{nF} \quad R_{FB} := 20\text{k}\Omega$

$$G_0 := \frac{\beta\cdot R_{FB}}{h_{11}+r_d+R_1} = 41.714 \qquad 20\cdot\log(G_0) = 32.406$$

$$f_z := \frac{\beta}{C_b\cdot h_{11}\cdot 2\pi} = 542.912\text{kHz}$$

$$f_p := \frac{1}{2\cdot\pi\cdot C_b\cdot\left[\left(1+\beta\cdot\dfrac{r_d+R_1}{r_d+R_1+h_{11}}\right)\cdot R_{FB}+\dfrac{(r_d+R_1)\cdot h_{11}}{r_d+R_1+h_{11}}\right]} = 267.205\text{Hz}$$

$$G_1(s) := -G_0\cdot\frac{1-\dfrac{s}{2\cdot\pi\cdot f_z}}{1+\dfrac{s}{2\cdot\pi\cdot f_p}}$$

Figure 9.52 The Mathcad® sheet shows the presence of the pole and the RHPZ.

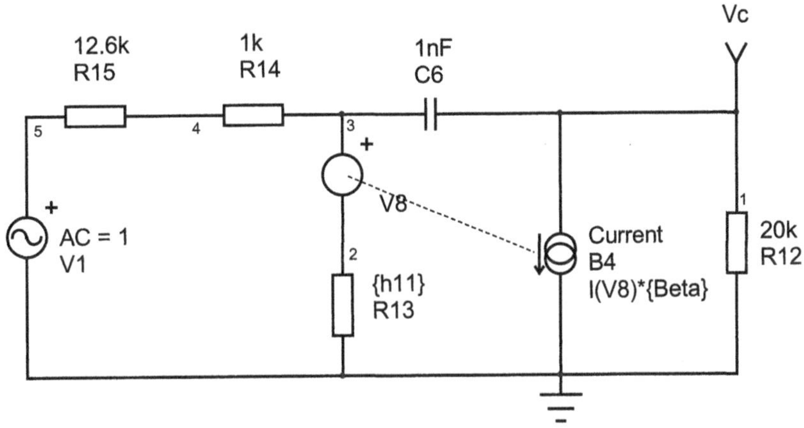

Figure 9.53 The SPICE simulation template is quickly captured.

The SPICE simulation of the Figure 9.47 circuit is not complicated, as shown in Figure 9.53. The simulation time is flashing; there is no switching component. We can now compare the Mathcad® results with those given by SPICE, and we can see in Figure 9.54 that the curves are similar leading to the conclusions that our equations are correct. It does not mean that the theory is correct! The real referee is the bench measurement that needs to agree with the theoretical results. What we have done here is just a simple analytic check that you need to do anyway.

9.5.3 Stabilizing the Converter

We have our compensator transfer function with (9.123). It is now time to tailor it in order to reach our crossover goal, together with an adequate phase margin. Needless to say, a simple bipolar transistor together with a Zener diode won't give us the same performance as with an op amp, but let's see what we can do anyway.

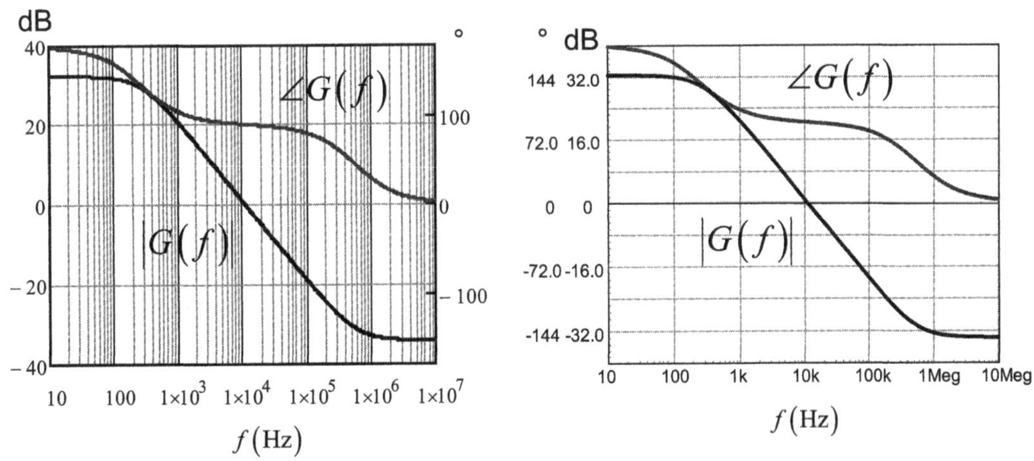

Figure 9.54 The results delivered by Mathcad® and SPICE are identical; the derived equations are correct.

First, in a primary-regulated converter, the winding on which you monitor the output voltage image is the auxiliary winding: you do not see directly the effects of the load on the output voltage, but an image translated by the turns ratio between the power and auxiliary windings. To perform small-signal analysis, you must account for this relationship and perform a full reflection of what is installed on the output to the auxiliary side. The implementation of such a reflection process appears in Figure 9.55. You have to individually reflect the power elements (the load and the filtering capacitor) to the winding where regulation is performed. As this auxiliary winding is already equipped with a capacitor and an equivalent resistor (the controller consumption for instance), the reflected elements combine with those already in place. Please note that the ESR and capacitors combinations shown in the schematic work only if the time constants $r_{C_1}C_1$ and $r_{C_2}C_2$ are equal. If they are not, it is an approximation. Furthermore, when reflecting the elements, we consider the diode dynamic resistance small enough to neglect it. This is not always the case, in particular in light-load operation.

Once you have reduced the current-mode flyback converter to a single winding type, you can use the small-signal CCM current-mode power stage transfer equation such as that given in [12]:

$$H(s) = H_0 \frac{\left(1+\frac{s}{\omega_{z1}}\right)\left(1-\frac{s}{\omega_{z2}}\right)}{1+\frac{s}{\omega_{p1}}} \tag{9.124}$$

$$H_0 = \frac{R_{eq}}{R_{sense}G_{FB}N} \frac{1}{\frac{(1-D)^2}{\tau_L}+2M+1} \tag{9.125}$$

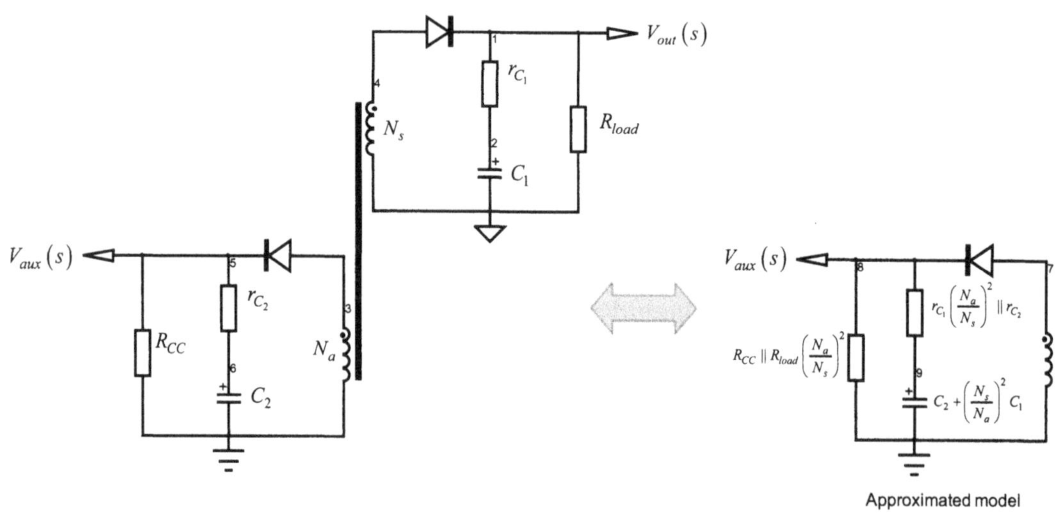

Figure 9.55 Reflection of the output elements across the auxiliary windings is necessary to perform the small-signal analysis.

$$f_{z_2} = \frac{(1-D)^2 R_{eq}}{2\pi D L_p \left(\dfrac{N_a}{N_p}\right)^2} \tag{9.126}$$

$$f_{p_1} = \frac{\dfrac{(1-D)^3}{\tau_L} + 1 + D}{2\pi R_{eq} C_{eq}} \tag{9.127}$$

$$f_{z_1} = \frac{1}{2\pi r_{C_{eq}} C_{eq}} \tag{9.128}$$

in which:

$$r_{C_{eq}} = r_{C_1}\left(\frac{N_a}{N_s}\right)^2 \parallel r_{C_2} \tag{9.129}$$

$$C_{eq} = C_2 + C_1\left(\frac{N_s}{N_a}\right)^2 \tag{9.130}$$

$$R_{eq} = R_{CC} \parallel R_{load}\left(\frac{N_a}{N_s}\right)^2 \tag{9.131}$$

$$\tau_L = \frac{L_p\left(\dfrac{N_a}{N_p}\right)^2}{R_{eq} T_{sw}} \tag{9.132}$$

You can clearly imagine how tedious it quickly becomes to calculate these contributions as you move some parameters to see their effects on stability. Even with an automated sheet, it is not that friendly. Furthermore, if the converter transitions to DCM, the equation set is no longer valid and must be updated.

The most convenient solution is to use SPICE with an auto-toggling DCM/CCM model derived for current mode (see [12]). The simulation template appears in Figure 9.56.

We have picked a standard NPN transistor to which a 12-V Zener diode has been added. The multiwinding transformer is built by adding a simple transformer model in parallel with the existing one. All the elements found on the power winding are automatically reflected on the auxiliary winding during simulation. Even better, the reflection includes the dynamic resistor of the rectifying diode that can easily change depending on the operating current: from an almost short circuit at high current, it can become a larger value in light load conditions, and our reflection formulas no longer work. On the contrary, SPICE will account for this change in the series resistance, leading to a correct power stage ac transfer function at all possible operating points. The power stage transfer function is unveiled in Figure 9.57. Please note that we probed node 10 (on the auxiliary winding) to graph the transfer function.

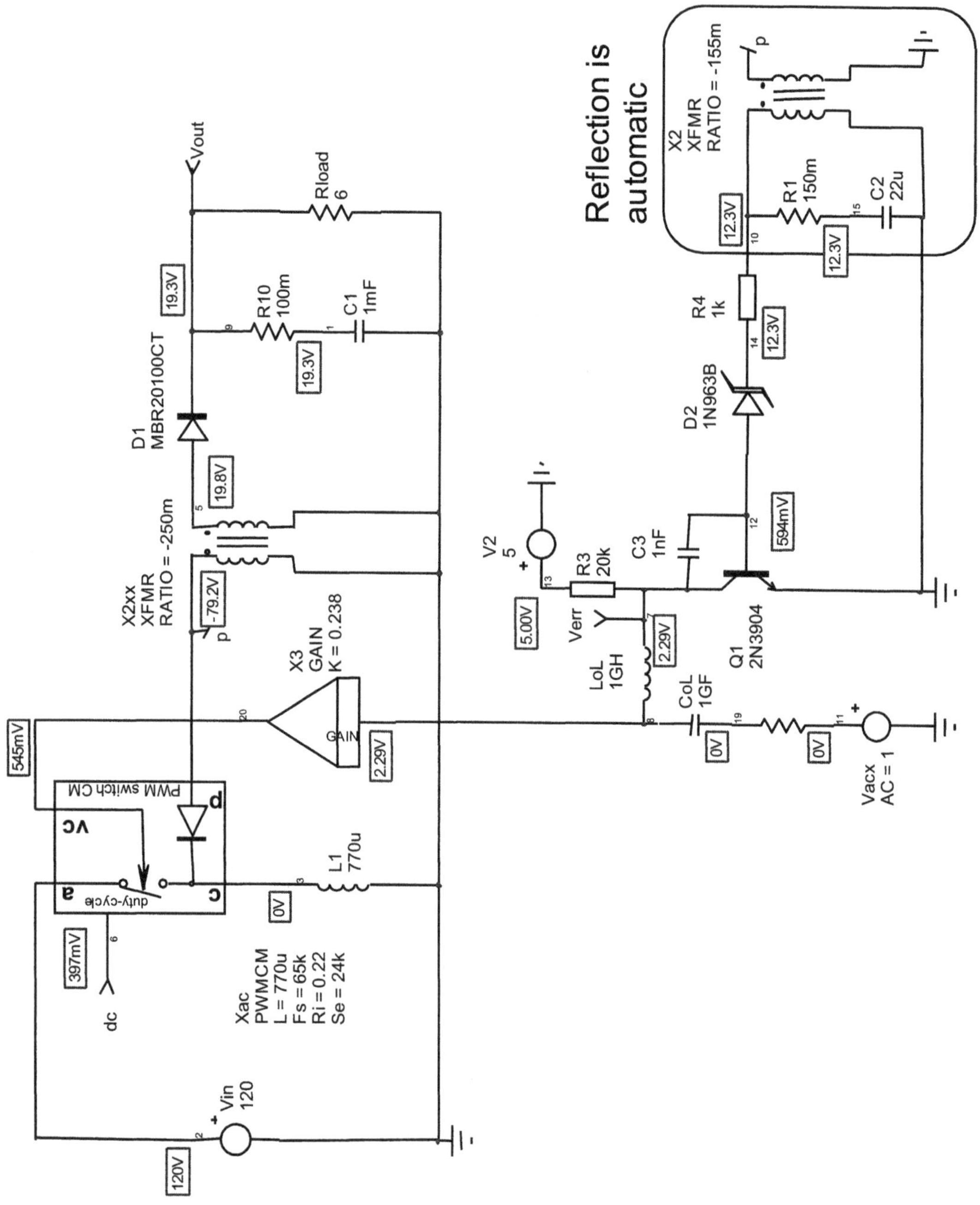

Figure 9.56 SPICE does all the reflections for you, and parameters sweep is a child's play.

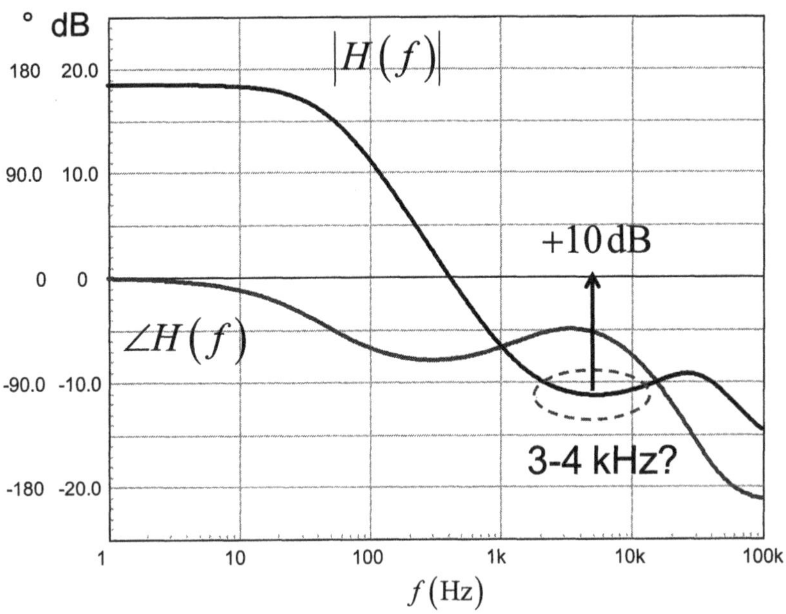

Figure 9.57 The power stage shows a first-order response with a minimum phase lag around 3–4 kHz.

As the adopted compensator cannot provide a phase boost of any kind, we will select a crossover area at a position where the phase lag is not too strong. Around 3–4 kHz seems to be the right region. If we neglect the high-frequency RHPZ contribution, we have a single-pole response as confirmed in Figure 9.54. We will therefore place the pole with C_b at a position where the gain provided by the compensator matches the 10-dB gain deficit indicated in Figure 9.57.

If we neglect the RHPZ contribution, the compensator transfer function described by (9.101) can be simplified to:

$$G(s) \approx G_0 \frac{1}{1+s/\omega_p} \tag{9.133}$$

The magnitude at crossover is found by replacing s by $j\omega_c = j2\pi f_c$:

$$|G(f_c)| \approx G_0 \frac{1}{\sqrt{1+\left(\frac{f_c}{f_p}\right)^2}} \tag{9.134}$$

From this value, we can extract the position of the pole based on the needed gain at f_c:

$$f_p = \frac{f_c \cdot |G(f_c)|}{\sqrt{G_0{}^2 - |G(f_c)|^2}} \tag{9.135}$$

Now, assume the following components values:

$R_1 = 1\ \text{k}\Omega$; Zener series resistor

$R_{FB} = 20\ \text{k}\Omega$; pull-up resistor

$r_d \approx 10\ \text{k}\Omega$; Zener diode dynamic resistor

$\beta \approx 70$; transistor current gain

$r_\pi \approx 21\ \text{k}\Omega$; transistor dynamic base resistance

We can calculate the compensator plateau gain G_0:

$$G_0 = \beta \cdot \frac{R_{FB}}{r_d + R_1 + r_\pi} = 73 \times \frac{20k}{10k + 1k + 21k} = 45.6 = 33\text{dB} \tag{9.136}$$

According to (9.135), the pole will have to be placed at

$$f_p = \frac{3.3k \times 3.16}{\sqrt{45.6^2 - 3.16^2}} \approx 230\text{Hz} \tag{9.137}$$

To calculate the capacitor value, we will use (9.122) from which we calculate the equivalent resistor value driving the capacitor C_b:

$$R_{eq} = \left(1 + \beta \frac{r_d + R_1}{r_d + R_1 + r_\pi}\right) R_{FB} + (r_d + R_1) \| r_\pi \approx 595\ \text{k}\Omega \tag{9.138}$$

The capacitor value is thus:

$$C_b = \frac{1}{2\pi f_p R_{eq}} = \frac{1}{6.28 \times 230 \times 595k} \approx 1.2n\text{F} \tag{9.139}$$

If we set this component values in our simulation template of Figure 9.56 and check the collector signal, we will have a view of the compensated loop gain. It appears in Figure 9.58.

It is not exactly the crossover frequency we wanted, but given the dispersion on the transistor gain and the various dynamic resistors, it is not that bad! A final step load simulation shows that the transient response is acceptable for such a simple system. The resulting waveform is in Figure 9.59. Please note that the average model does not include the various transformer leakage inductances that affect the transient response.

Some people will want to increase the Zener current, forcing it to work farther off its knee. In this case, simply install a resistor R_B across the transistor base-emitter. It will impose a Zener current at room temperature of:

$$I_Z \approx \frac{0.65}{R_B} \tag{9.140}$$

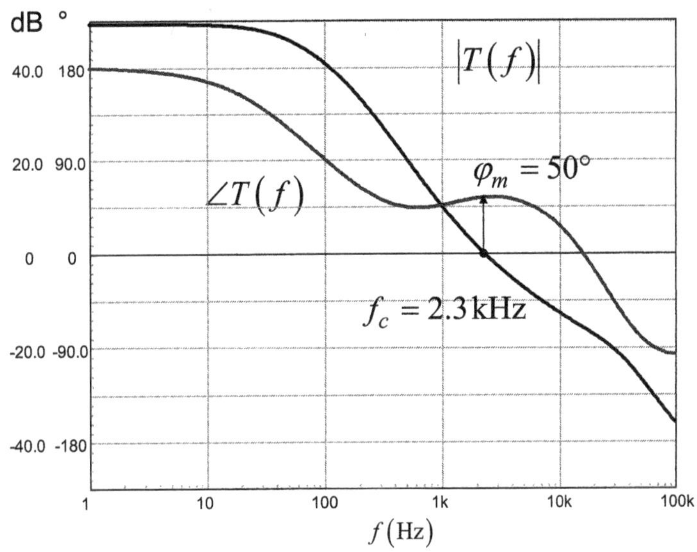

Figure 9.58 The compensated loop gain shows a good phase margin with a decent crossover frequency.

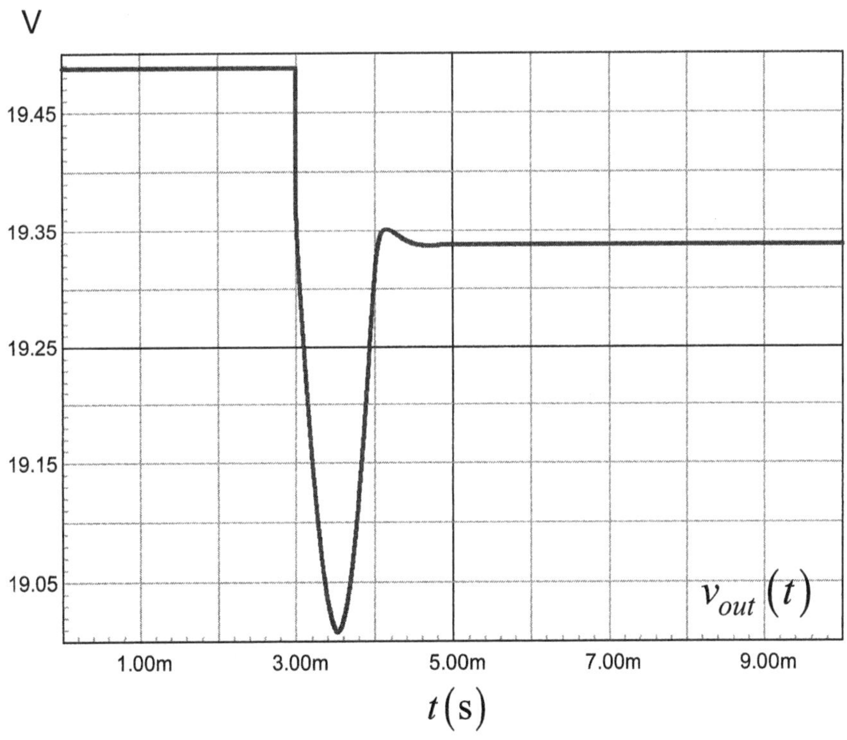

Figure 9.59 The transient response is acceptable for this primary-regulated converter. The step load is 1.5 A in 10 μs.

The transfer function is slightly modified by this resistor addition. If you go trough the equations again or look at page 53 of [15], you should find:

$$G_0 = \frac{1}{1+\frac{R}{R_B}} \frac{\beta R_{FB}}{r_\pi\left(1+\frac{R \parallel R_B}{r_\pi}\right)} \tag{9.141}$$

Its limit when R_B goes infinite is given by (9.104). The RHPZ position remains unchanged and (9.110) is still valid. The pole position is, however, slightly affected:

$$\omega_p = \frac{1}{C_b\left(R_{FB} + \left(1+\frac{\beta R_{FB}}{r_\pi}\right)\cdot r_\pi \parallel (R \parallel R_B)\right)} \tag{9.142}$$

In these expressions, $R = r_d + R_1$.

9.6 Design Example 5: Input Filter Compensation

Switching converters are noisy systems in essence. The sharp voltage or current interruptions generate differential or common mode noise that can couple to the supply wires in certain conditions. If low-noise systems share a common supply line with a switching converter, it is very likely that the switching converter pollutes the common supply rail and alters the sensitive equipments behaviors. To avoid high-frequency pulses to flow back to the dc source and pollute the supply rail, an input filter has to be installed. A typical filter appears in Figure 9.60. To minimize losses, it is usually made of an LC network. This network forms a second-order filter and prevents the noise from returning to the generator as ac currents are supposed to circulate in C while L opposes a high impedance to these currents.

Calculating the cutoff frequency of such a filter depends on the maximum input ripple specifications and is outside the scope of this book. Reference [12] presents several design examples based on a specification to fulfill. What we will see now is the problem brought by the filter insertion and how to cure it.

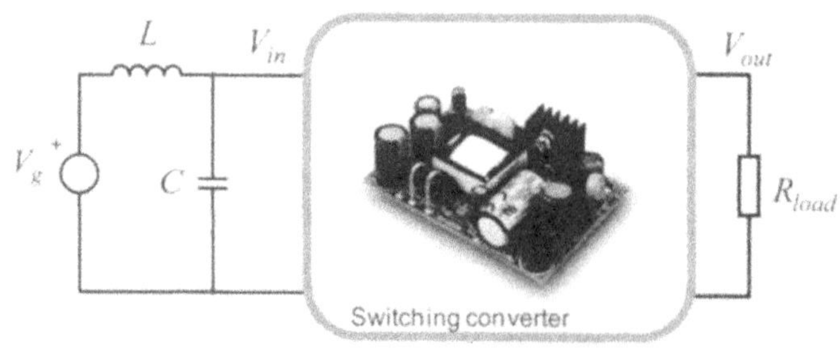

Figure 9.60 A typical filter is made by combining L and C elements.

9.6.1 A Negative Incremental Resistance

A switching converter is a closed-loop system. It makes sure that the output variable (e.g., V_{out}) stays in control regardless of incoming perturbations such as the input voltage V_{in} or the output current I_{out}: if you deliver a certain amount of power, whether it is low line or high line, the converter will feed the load at constant power. If we have 100 percent efficiency, we can write:

$$P_{in} = P_{out} = I_{in}V_{in} = I_{out}V_{out} \tag{9.143}$$

The input current can thus be derived as

$$I_{in}(V_{in}) = \frac{P_{out}}{V_{in}} \tag{9.144}$$

If we plot the input current versus the input voltage, we obtain the curve shown in Figure 9.61. If the input voltage increases as the output power is constant (10 W), then the input current must go down to satisfy (9.143). On the opposite, if the input voltage decreases, then the loop forces the current to increase. If we are interested in the slope of this curve, we must derive (9.144):

$$\frac{dI_{in}(V_{in})}{dV_{in}} = \frac{d\left(\frac{P_{out}}{V_{in}}\right)}{dV_{in}} = -\frac{P_{out}}{V_{in}^2} \tag{9.145}$$

This is the negative relationship that links the current to the voltage when the input line changes. The slope unit is amperes per volts, or a conductance. If we take the inverse of it, we have an expression that we will call the *incremental input resistance.*

Its expression is simply:

$$R_{in,inc} = -\frac{V_{in}^2}{P_{out}} \tag{9.146}$$

The concept requires a bit of abstraction. If the converter would offer a truly resistive input that absorbs power, then the current would exclusively depend on the input voltage. With a closed-loop converter, the input current depends on the input voltage, of course, but also on the constant output power controlled by the

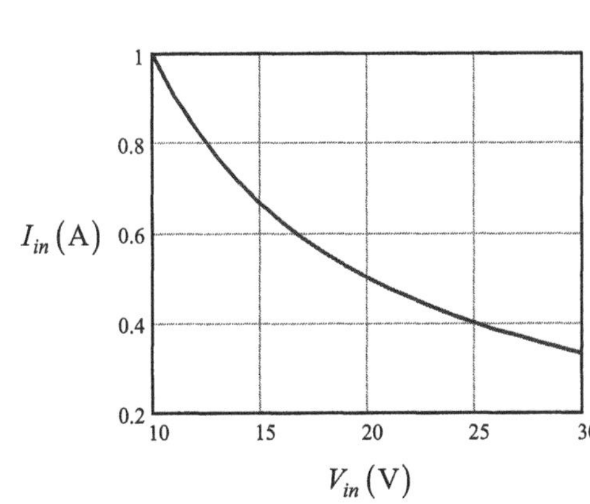

Figure 9.61 The relationship between the input current and the input voltage shows a negative slope.

system. The negative incremental resistance value expressed in (9.146) is the result of this process.

9.6.2 Building an Oscillator

An oscillator can be formed in different ways. One is to use a tuned LC network. If you bring an initial energy to the network with a fugitive stimulus, oscillations take place under the form of energy that swings back and forth between the storage elements L and C. Oscillations cease when the energy stored at the beginning of the oscillations have been totally dissipated in the ohmic paths such as capacitor and inductor ESRs. If you do not want the oscillations to cease, the energy transfer between L and C must be lossless. In other words, if you compensate the losses by canceling the circuit resistances, oscillations once started will be continuously maintained.

One way to cancel the ohmic path is to associate a negative resistor to the circuit. A negative resistor can be emulated using semiconductors (FETs, Gunn and tunnel-effect diodes, negative impedance converter op amp-based circuits, and so on). As an example, Figure 9.62 portrays an LC network where the losses are introduced by the loading resistor R. The quality factor in this example is adjusted to 5. The network is briefly excited by a 5-V step of a few μs duration. Figure 9.63 shows how the oscillations take place and die out after several cycles, indicative of energy dissipated in the resistor.

The graph shows you how the energy transfers back and forth between the capacitor and the inductor. The total energy stored in the circuit at any time (w_{tot}) is the sum of the energy individually stored in L (w_L) and C (w_C):

$$w_{tot}(t) = w_L(t) + w_C(t) = \frac{1}{2}Li_L^2(t) + \frac{1}{2}Cv_C^2(t) \tag{9.147}$$

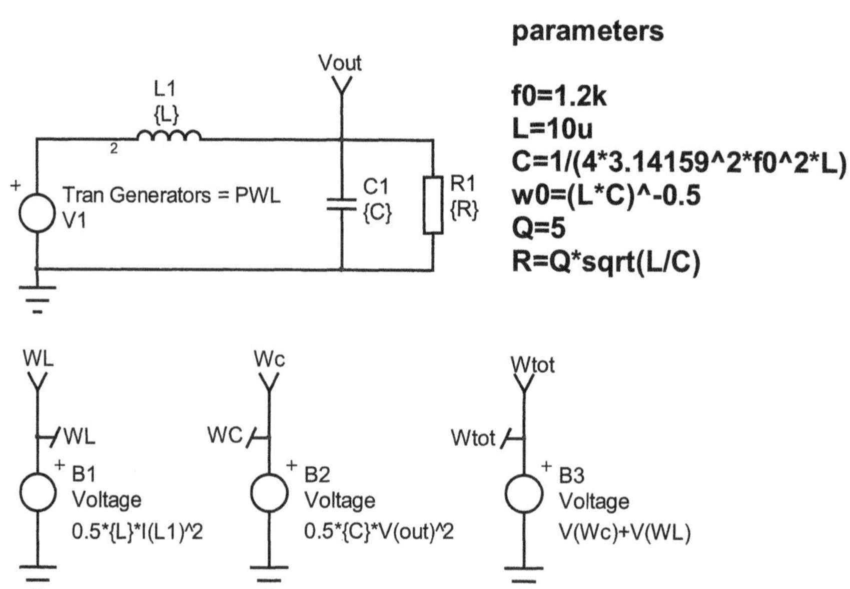

Figure 9.62 In this circuit, the losses damp the network and oscillations quickly die out.

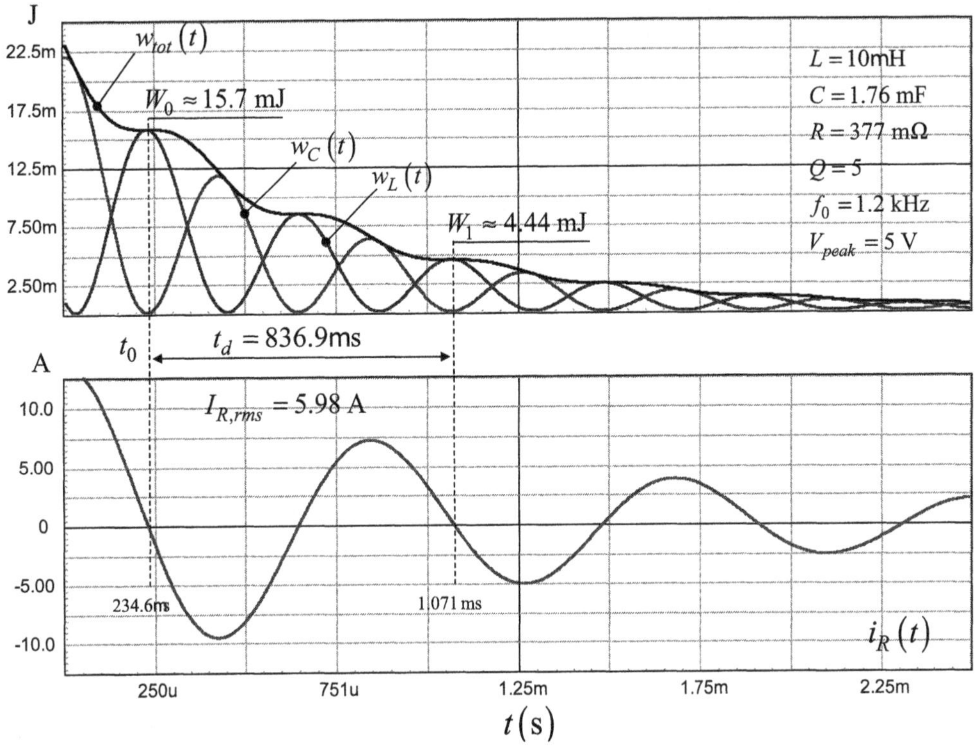

Figure 9.63 Because of the losses incurred to the resistor, the energy exchange between the inductor and the capacitor is affected and oscillations cease quickly.

As the series resistor R_1 dissipates the losses in heat, the amount of stored energy drops as time elapses. If we look at the figure, the energy lost between t_0 and $t_0 + t_d$ is:

$$\Delta W = w(t_0) - w(t_0 + t_d) = 15.7m - 4.44m = 11.26 \text{ mJ} \tag{9.148}$$

The power dissipated in the resistor during a complete pseudo period is simply:

$$P_{R_1} = R_1 I_{L,rms}{}^2 = 0.377 \times 5.98^2 = 13.467\text{W} \tag{9.149}$$

The corresponding energy lost in the resistor is given by:

$$W_{R_1} = P_{R_1} t_d = 13.48 \times 836.9u = 11.27 \text{ mJ} \tag{9.150}$$

This corresponds to what was found in (9.148).

From this graph, we can derive the quality factor. The quality factor Q is a parameter that quantifies the reactive component in the considered circuit. If high, reactive phenomena dominate over resistive losses. If low, the circuit is weakly reactive and losses are important. Q can be defined as follows:

$$Q = 2\pi \frac{\text{(stored energy)}}{\text{(energy dissipated per cycle)}} \tag{9.151}$$

In our circuit under a transient excitation, the considered energy is that stored in the capacitor and the inductor over the time t_d. By placing the cursors on the w_{tot} curve of Figure 9.63 and computing the average value over t_d, we read:

$$\langle w_{tot}(t)\rangle_{t_d} = 9.09 \text{ mJ} \tag{9.152}$$

If we apply (9.151), we have:

$$Q = 2\pi \frac{(\text{stored energy})}{(\text{energy dissipated per cycle})} = 6.28 \times \frac{9.09m}{11.27m} = 5.06 \tag{9.153}$$

The quality factor could also be obtained using the logarithmic decrement δ derived in Chapter 3. The added 2 in the numerator accounts for the energy terms rather than currents or voltages:

$$Q = \sqrt{\left(\frac{2\pi}{\ln\left(\frac{W_0}{W_1}\right)}\right)^2 + 0.25} = \sqrt{\left(\frac{2\,\pi}{\ln\left(\frac{15.715m}{4.444m}\right)}\right)^2 + 0.25} = 5 \tag{9.154}$$

Analytically, we can show that the quality factor of the *RLC* circuit from Figure 9.62 under harmonic excitation is:

$$Q = R\sqrt{\frac{L}{C}} \tag{9.155}$$

If we place in parallel with *R* a resistor value –*R*, the quality factor becomes

$$Q = \frac{-R^2}{R - R}\sqrt{\frac{L}{C}} = \infty \tag{9.156}$$

We have updated the simulation circuit from Figure 9.62 where a resistor of value –377 mΩ has been installed across R_1. The new simulation results appear in Figure 9.64 and confirm the presence of sustained oscillations. The current diverted by the paralleled combination of *R* and –*R* is zero. Should we derive the roots expression of the new *RLC* circuit featuring the added resistor of –*R* value, we would find imaginary poles, with nullified real parts. We have created a negative resistance oscillator.

9.6.3 Taming the Oscillations

In the previous example, a (positive) resistor R_1 was already in place across the output capacitor, efficiently damping the network (Q = 5). What if the negative resistor, alone, loads the *LC* filter? This is actually the Figure 9.60 illustration with a closed-loop converter loading an undamped filter. If you run the analysis of an *LC* filter loaded by a negative resistor, you will find that the poles are no longer pure imaginary values but have jumped on the right side of the vertical axis: they become right half plane poles or positive roots. If you remember the transient response of a system featuring RHP poles, the exponent of the exponential term(s) is positive,

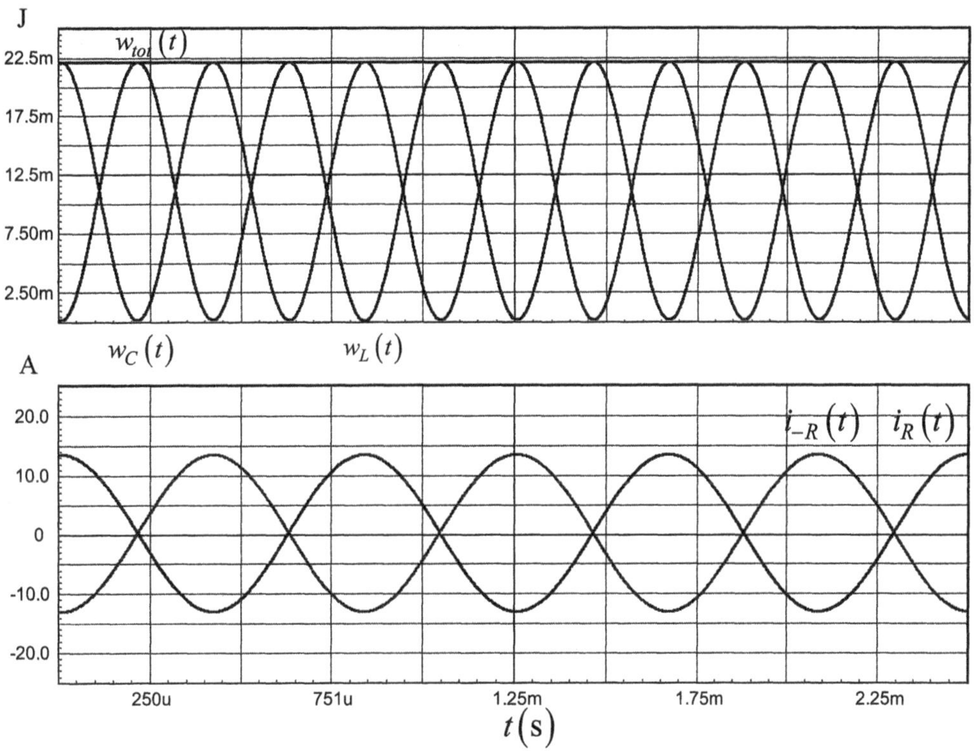

Figure 9.64 The negative resistance actually cancels the current diverted from the capacitor, bringing down the losses to 0 W.

meaning the system response diverges with time. This is what is confirmed by Figure 9.65. In a real system, the physical limits (i.e., power supply voltage) will clamp the excursion and the converter is likely to lock out. In high-power converters, the limit might be far away from the semiconductor maximum ratings: you should expect a loud noise followed by smoke.

To lower the quality factor, (9.151) tells us that we must create losses in the filter. These losses will dissipate some stored energy in heat, and oscillations will cease. Many solutions are possible, and [16] explores several circuits. The key is to maintain efficiency despite damping effects. One easy way to create a loss is to add a damping resistor across the inductor. As the inductor voltage at steady-state is zero, losses will only occur in presence of oscillations. Unfortunately, associating the damping resistor with the inductor introduces a zero in the transfer function counteracting one of the LC network double poles: from a second-order network, the filter becomes a first-order network, clearly affecting its filtering capabilities.

Another option consists of adding the damping resistor in parallel with the output capacitor. As ac damping is needed (when oscillations appear), you can insert this resistor in series with it a dc-block capacitor. Therefore, in dc, the resistor will be transparent to the circuit dissipating no power. Losses will occur only if ac oscillations arise. The updated filter sketch is given in Figure 9.66.

The calculation of the damping resistor is rather simple. Let's assume we have a 5-V/30-A ac-dc converter running from US mains. The lowest input voltage is 100 V

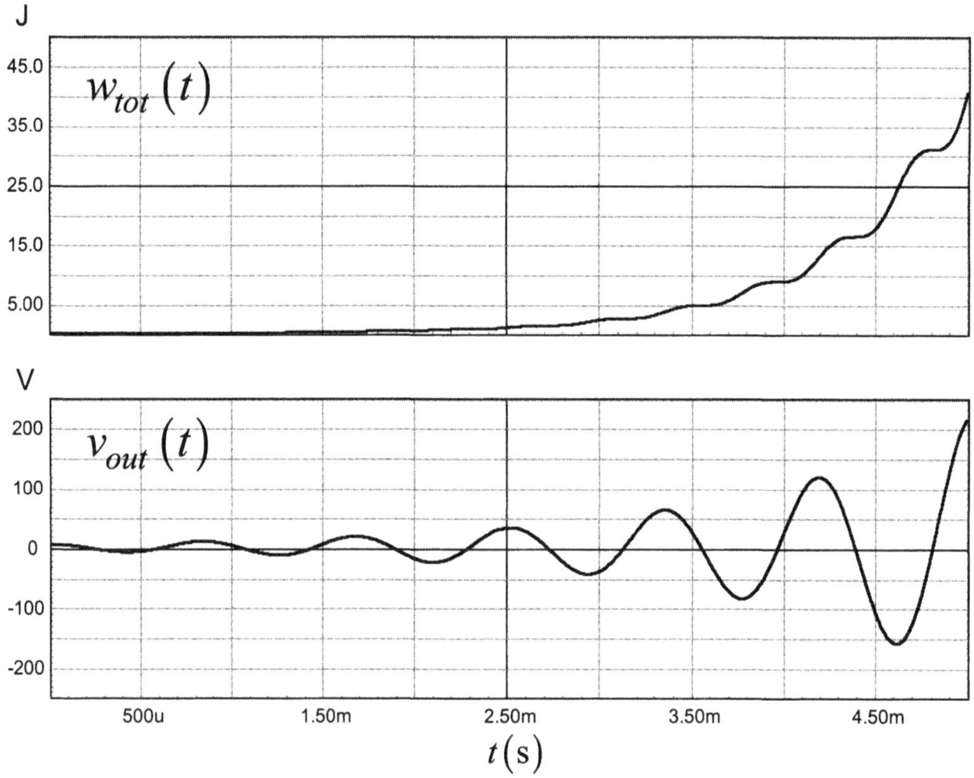

Figure 9.65 When the negative resistor alone loads the *LC* network, the output voltage diverges as *t* increases.

rms, or approximately a 100-V rectified dc rail, if we consider a 30 percent ripple on the bulk capacitor. Considering a 95 percent efficiency, the incremental input resistance of this converter is:

$$R_{in} = -\frac{V_{in}^2}{P_{out}}\eta = -\frac{100^2}{150}0.95 \approx -63\ \Omega \tag{9.157}$$

The *LC* network selected for this filter is a 150-μH inductor coupled to a 10-μF multilayer capacitor. We assume there is no ESR associated with these elements; this

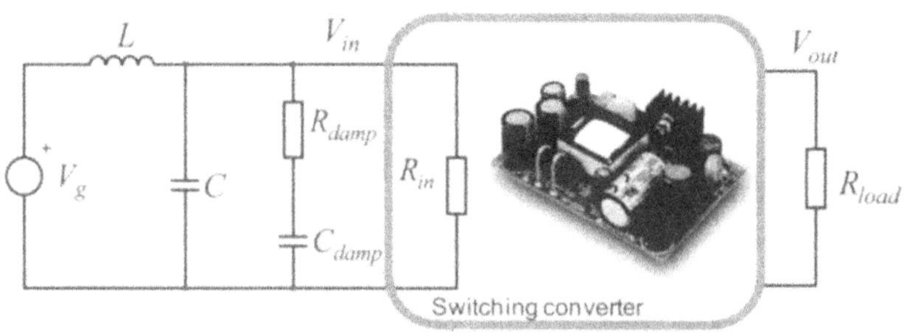

Figure 9.66 The resistor is dc-isolated from the supply rail and only couples in ac. Efficiency is less affected than with a series resistor.

is the worst case. If we use (9.155), we can update the quality factor expression by including the effects of the paralleled damping resistor R_{damp}:

$$Q \approx \frac{R_{in} \parallel R_{damp}}{\sqrt{\frac{L}{C}}} = \frac{R_{in}R_{damp}}{(R_{damp}+R)\sqrt{\frac{L}{C}}} \tag{9.158}$$

If we extract the damping resistor from this equation, we obtain

$$R_{damp} = \frac{QR_{in}\sqrt{\frac{L}{C}}}{R_{in} - Q\sqrt{\frac{L}{C}}} \tag{9.159}$$

To avoid oscillations, we can select a quality factor of 1 leading to a damping resistor of

$$R_{damp} = \frac{-63 \times \sqrt{\frac{150u}{10u}}}{-63 - \sqrt{\frac{150u}{10u}}} = 3.65\ \Omega \tag{9.160}$$

The dc-block capacitor is usually chosen to be 10 times the filter capacitor value. In our case, it corresponds to a 100-μF type. To test the damping effectiveness, we can first run an ac plot of the filter loaded by R_{in}. The simulation circuit appears in Figure 9.67, where all the elements have been installed. The negative resistance is emulated by a constant power source; see page 71 of [12]. The power is adjusted to 150 W while the dc rail is set to 100 V for the analysis.

The transfer function in Figure 9.68 shows how the added resistor effectively damps the filter. The capacitor can be selected to values different than 10 times the filter capacitor, but the damping efficiency then starts to weaken.

The various publications targeting on input filter problems have shown that having the filter output impedance well below the lowest locus of the converter

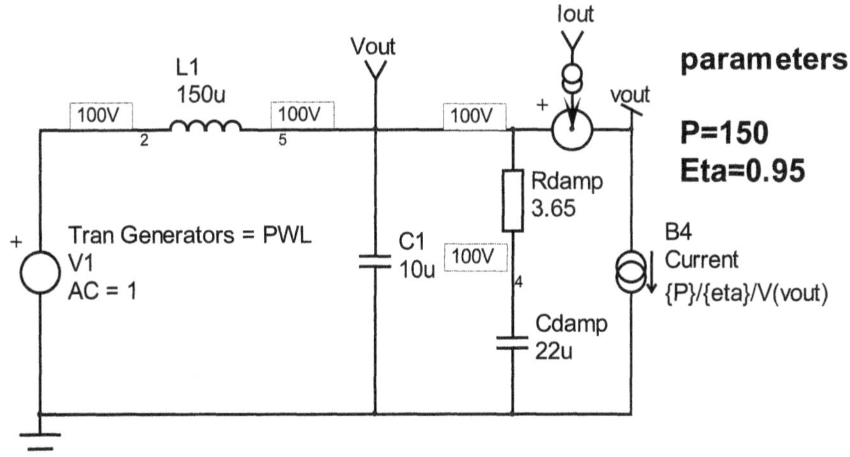

Figure 9.67 A constant-power source makes an ideal negative resistance load.

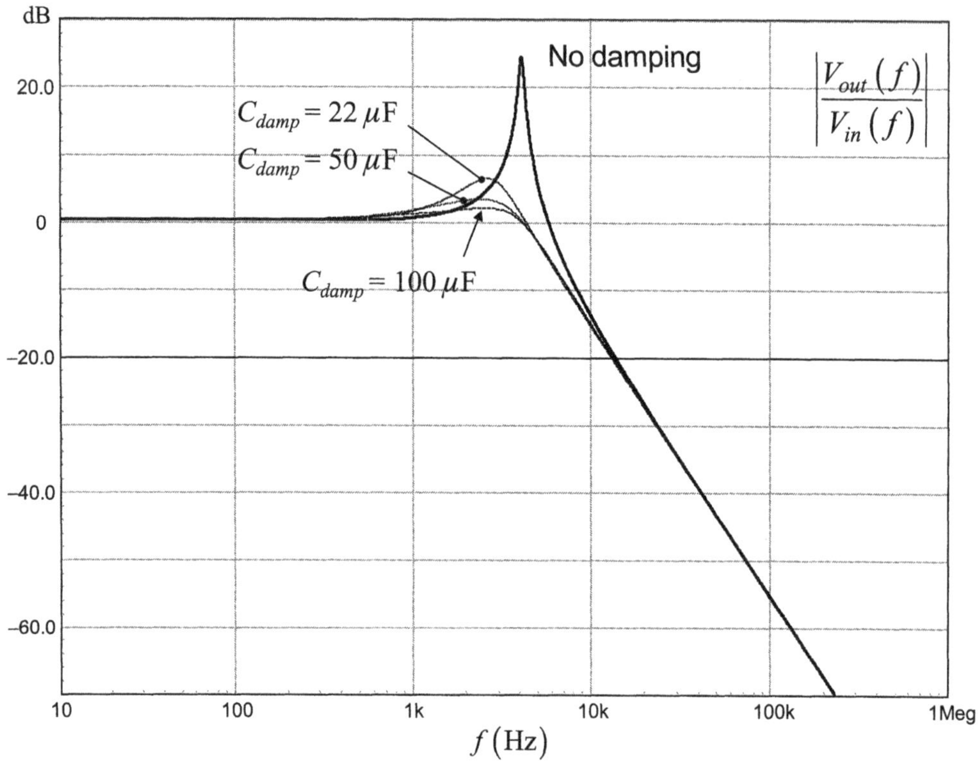

Figure 9.68 Without damping, you can clearly see the filter peaking. The resistor dissipates the ac energy and calms down any unwanted oscillations.

input impedance avoids potential instabilities. In other words, you have to damp your filter so that:

$$\left|Z_{out}(s)\right|_{\max} \ll \left|Z_{in}(s)\right|_{\min} \tag{9.161}$$

In this expression, the incremental resistance represents the dc value only. The complex input impedance depends on the closed-loop gain and must be carefully analyzed to check if (9.161) is satisfied along the frequency axis. Oscillations risks exist only up to the loop gain crossover frequency. Beyond that point, the negative incremental resistance reverts to a positive value as the converter is no longer able to correct the incoming high-frequency perturbations.

A simple check consists of plotting the damped filter output impedance and adding a line representing the negative incremental resistance defined by (9.157). This is the chart drawn in Figure 9.69. We have added the incremental resistance value, $20\log_{10}(63) = 36$ dBΩ as a horizontal reference. Without damping, the filter output impedance peaks so much that it exceeds the reference. Fortunately, once damping is added, a comfortable margin is created, making the design safe.

To further verify that the cure works as expected, we have stepped the input voltage in Figure 9.67 and checked the output voltage shape. This is what is shown in Figure 9.70 where the input voltage drops from 140 V to 100 V in 10 μs. When damping is adequate, the oscillations disappear immediately. If you reduce the

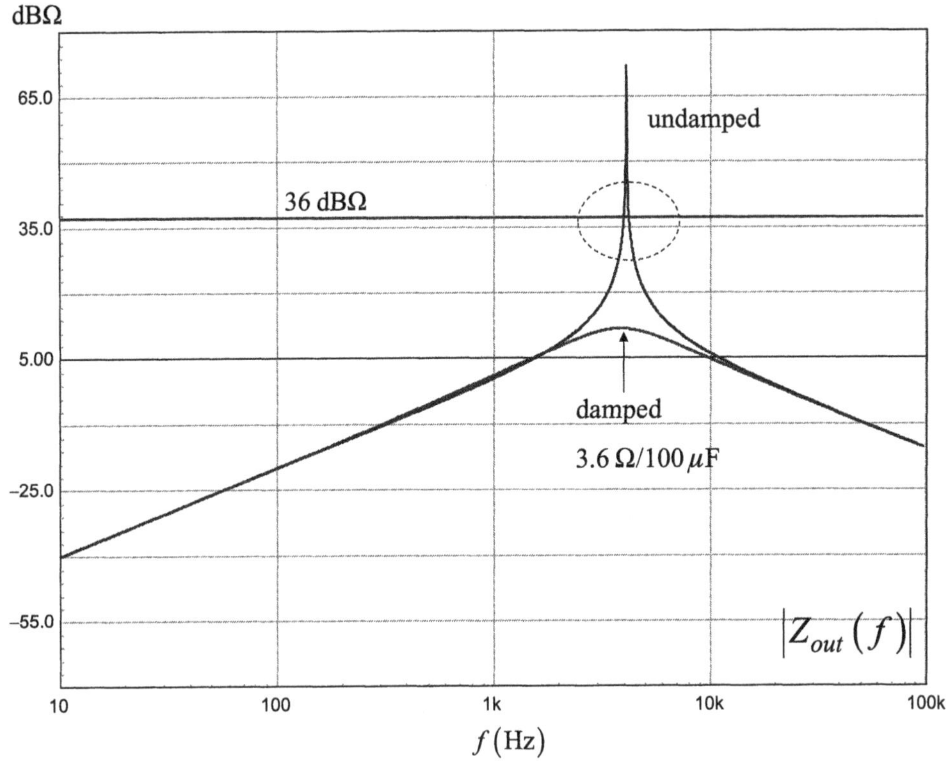

Figure 9.69 No overlap must exist between the incremental input resistance and filter output impedance.

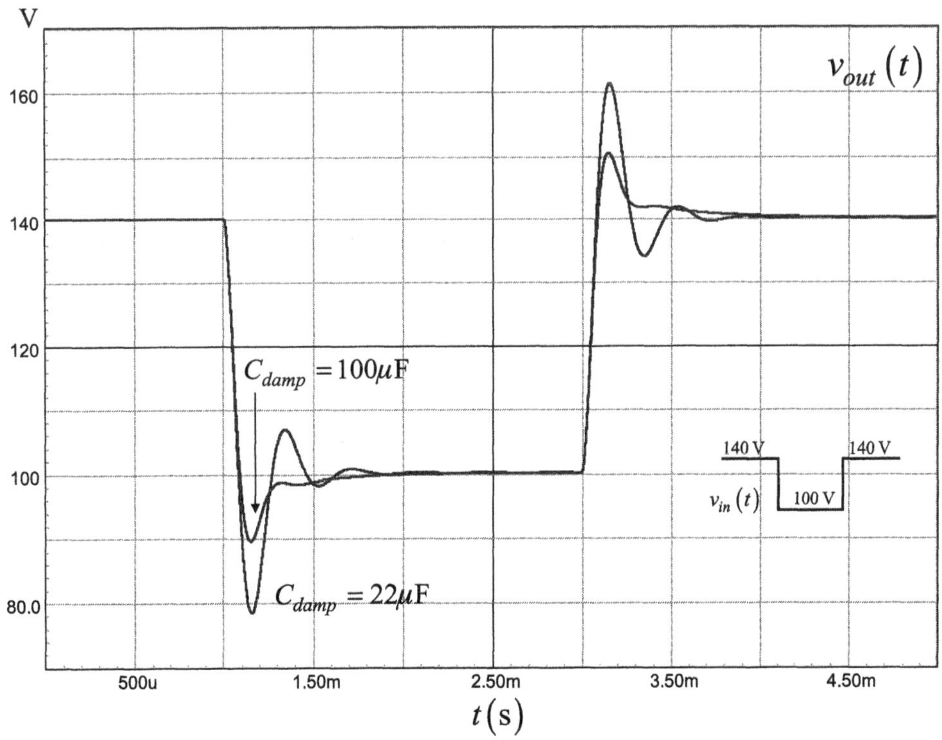

Figure 9.70 With a properly damped filter, oscillations are well damped and quickly die out.

100-μF capacitor to 22 μF, the damping becomes less efficient and a little ringing pops up.

The comprehensive analysis of such a converter will include the complete input impedance frequency sweep. More information on the subject will be found in [12, 16].

9.7 Conclusion

In this chapter, we learned how to measure a power converter loop gain. This stage is important, as it will let you know whether or not the assumptions you made during simulation or analytical analysis were confirmed by bench measurements. Some hidden parameters are difficult to predict, and these laboratories experiments are mandatory. At the beginning of the production, or during the pilot run, do not hesitate to sample some of the converters and recheck that stability is not jeopardized by components that were replaced at the last minute because of cheaper prices. You know that the ESR plays a role in the compensation scheme. Make sure high and low temperature transient responses are not too different; otherwise, this could indicate a parameter drift out of control. If you respect all these points and absolutely ban trial-and-error techniques, then you will be able to enjoy peaceful sleeps when mass production starts!

The design examples have shown that you can choose analytical analysis or simulation. Analytical analysis offers the advantage of unveiling the position of poles and zeros and how external parameters affect them. SPICE will give you the same response but without the insight brought by the analytical analysis. Combining both techniques, followed by real experiments, is the recipe for success.

References

[1] AP200 operating manual, AP Instruments, http://www.apinstruments.com.

[2] Ridley, R., "Measuring Frequency Response," *Switching Power*, 2006, http://www.switchingpowermagazine.com.

[3] Ridley, R., "Frequency Response Measurements for Switching Power Supplies," Texas-Instruments, Application note SLUP121, 2001.

[4] Venable Technical Reference Library, http://www.venable.biz/tr-papers2.php.

[5] Picotest website: https://www.picotest.com/blog.

[6] Omicron-lab website: http://www.omicron-lab.com/application-notes.html.

[7] Middlebrook, R. D., "Measurement of Loop Gain in Feedback Systems," *International Journal of Electronics*, Vol. 38, No. 4, 1975, http://www.ele.tut.fi/teaching/ele-3100/lk0809/tehol1/MiddleBrook75.pdf.

[8] http://en.wikipedia.org/wiki/Phasor.

[9] Panov, Y., and M. Jovanović, "Small-Signal Measurements in Switching Power Supplies," *Proceedings of Applied Power Electronics Conference*, Vol. 2, 2004, pp. 770–776.

[10] Basso, C., N. Cyr, and S. Conseil, "Stability Analysis in Multiple Loop Systems," Application note AND8327/D, ON Semiconductor.

[11] Capilla, J., "Opening an SMPS Regulation Loop with an ac Stimulus," ON Semiconductor internal seminar, September 2011, Toulouse, France.

[12] Basso, C., *Switch-Mode Power Supplies: SPICE Simulations and Practical Designs*, New York: McGraw-Hill, 2008.

[13] http://www.kemet.com/page/kemsoft.

[14] Power 4-5-6, Ridley Engineering, www.ridleyengineering.com.

[15] Vorpérian, V., *Fast Analytical Techniques for Electrical and Electronic Circuits*, Cambridge: Cambridge University Press, 2002.

[16] Erickson, B., and D. Maksimovic, *Fundamentals of Power Electronics*, New York: Springer 2001.

Conclusion

Writing a book on loop control is like a long, almost endless, journey: you know the destination but not all of the detours you are going to make before you reach it. The domain is so vast, and the exploration options so abundant, that to make one's way through all these meanders can quickly become an impossible exercise. By narrowing down the necessary knowledge to what is really important to our power conversion field, the work immediately becomes less complicated and no longer seemingly out of reach.

In this book, I have tried to give you the minimum set of necessary tools to either start designing a project or jump to a higher-level book. These tools, however, cannot be efficiently used if you do not understand the way they have been forged. Following the derivation lines not only helps you to progress in the field, but it also teaches the limits of the tools: in which cases you can use them and where it is necessary to select new ones. Therefore, as a recommendation, go through the lines I have derived and figure out how to reach the result yourself. It is time consuming, but this is the best way to learn.

Despite the length of the effort, I have really enjoyed writing this document. Not only because I, too, made progress in the area of control systems, but also because I hope I have shed a modestly different light on the field. If you, my reader friends, also feel this way after reading this book, then I will have accomplished my goal.

Appendix

Table of Variables and Acronyms Used in the Book

arg	The argument of a complex expression
BCM	Borderline conduction mode (same as CrM) or boundary conduction mode
BIBO	Bounded input bounded output
χ	The characteristic equation of a closed-loop transfer function, chi
CCM	Continuous conduction mode
CL	Closed loop, T_{CL} is the closed-loop gain for instance
CrM	Critical conduction mode
CTR	Current transfer ratio for an optocoupler
D	The converter averaged duty ratio
$d(t)$	The converter instantaneous duty ratio
$D(s)$	The Laplace expression of a transfer function denominator
δ	The logarithmic decrement, delta
ESR	Equivalent series resistance, also noted r_C or r_L respectively for a capacitor and an inductor
ESL	Equivalent series inductance
ε	The error voltage, epsilon
η	The converter efficiency, eta
f_c	The crossover frequency where $\lvert T(f_c)\rvert = 1$ or 0 dB
F_{sw}	The switching frequency
$G(s)$	The compensator Laplace expression
Gf_c	The gain deficit (or excess) at the selected crossover frequency
φ_m	The phase margin read at f_c
g_m	The transconductance of an operational transconductance amplifier (OTA)
GM	The gain margin in an open-loop Bode plot representation
$H(s)$	The plant Laplace expression
I_C	The capacitor dc current (0 at steady state)
$i_C(t)$	The instantaneous capacitor current
$\hat{i}_C$	The capacitor small-signal current
$i_d(t)$	The instantaneous diode current
I_d	The diode dc current
I_{in}	The input dc current of a given converter
I_L	The inductor dc current
$i_L(t)$	The instantaneous inductor current

$\hat{i}_L$	The inductor small-signal current
I_{out}	The dc output current
$i_{out}(t)$	The instantaneous output current
$\hat{i}_{out}$	The small-signal outputcurrent
k_d	The derivative term in a PID compensator
k_i	The integral term in a PID compensator
k_p	The proportional term in a PID compensator
L_p	The primary inductance of a transformer (usually in a flyback converter)
$\mathcal{L}\{f(t)\}$	The Laplace transform of a time-domain function f
LHP	Refers to the Left Half-Plane of the Argand chart
LHPP	A pole located in the LHP
LHPZ	A zero located in LHP
LTI	Linear time invariant
MIMO	Multi-input multi-output
$N(s)$	The Laplace expression of a transfer function numerator
OL	Open loop, T_{OL} is the open-loop gain for instance
ω	The angular frequency in radians per seconds, rad/s
ω_n or ω_0	The natural angular frequency in radians per seconds, rad/s
ω_d	The damped angular frequency in radians per seconds, rad/s
ω_r or ω_M	The resonant angular frequency in radians per seconds, rad/s
P_{in}	The converter input power
P_{out}	The converter output power
PID	Proportional integral derivative
PI	Proportional integral
Q	The quality factor of a filter
$r_n(t)$	The natural or free response (the excitation signal is set to zero)
$r_f(t)$	The forced response (the initial conditions are identically zero)
r_C	The series resistance of the capacitor
r_L	The series resistance of the inductor
RMS	Root mean square
R_{sense} or R_i	The sense resistor in a current-mode controlled converter. Sometimes called the burden resistor.
RHP	Refers to the right half-plane of the Argand chart
RHPP	A pole located in the RHP
RHPZ	A zero located in the RHP
s	$s = \sigma + j\omega$
SISO	Single output single input
SMPS	Switch mode power supply
SPICE	Simulation Program with Integrated Circuit Emphasis
τ	The time constant, tau, in second
$T(s)$	The loop gain Laplace expression
T_{sw}	The switching period
$v_C(t)$	The instantaneous voltage across a capacitor
V_C	The capacitor dc voltage
$v_c(t)$	The instantaneous control voltage
V_c	The dc control voltage

$v_L(t)$	The instantaneous voltage across an inductor
V_L	The inductor dc voltage (0 at steady state)
ζ	The damping ratio, zeta

Numbers and Prefixes in Operations

I have purposely used SPICE prefixes in all the operations to avoid scientific notations in equations. There is no space inserted between the number and the prefix. The prefixes are as follows:

$$1f = 10^{-15}$$

$$1p = 10^{-12}$$

$$1n = 10^{-9}$$

$$1u = 10^{-6}$$

$$1m = 10^{-3}$$

$$1k = 10^{3}$$

$$1Meg = 10^{6}$$

For example:

$$R_1 = 10\text{ k}\Omega \quad C_1 = 3\,\mu\text{F}$$

$$\tau = R_1C_1 = 10k \times 3u = 30m = 0.03\text{ s or }30\text{ ms}$$

About the Author

Christophe Basso is an application engineering director at ON Semiconductor in Toulouse, France. He has originated numerous integrated circuits, among which the NCP120X series has set new standards for low standby-power converters. SPICE simulation is one of his favorite subjects, and he authored two books on the subject. In his work, he promotes the combined usage of SPICE as a design companion, which, when associated with an equation-based approach, helps to understand how complex circuitries operate. This technique is appreciated and recognized by the numerous customers he visits worldwide. Developing new integrated circuits while helping and teaching design engineers is part of his professional activity in the field of ac-dc power conversion for the past 15 years.

Christophe holds a BSEE-equivalent from the Montpellier University (France) and a MSEE from the Institut National Polytechnique de Toulouse (France). He holds 22 patents on power conversion and often publishes papers in conferences and trade magazines. He also teaches professional seminars at international conferences such as APEC. He is an IEEE senior member. He maintains a web page where documents and models are available for download at http://cbasso.pagesperso-orange.fr/Spice.htm.

Index

www.ingramcontent.com/pod-product-compliance
Lightning Source LLC
LaVergne TN
LVHW062242200926
848484LV00006B/1595
9781608075577